# MODELING, ANALYSIS, AND CONTROL OF DYNAMIC SYSTEMS

# MODELING, ANALYSIS, AND CONTROL OF DYNAMIC SYSTEMS

**William J. Palm III**
*The University of Rhode Island*

*John Wiley & Sons*
*New York Chichester Brisbane Toronto Singapore*

*Library of Congress Cataloging in Publication Data:*

Palm, William J. (William John), 1944–
Modeling, analysis, and control of dynamic systems.

Includes bibliographical references and index.
1. Automatic control – Mathematical models. 2. Dynamics – Mathematical models. 3. System analysis. I. Title.
TJ213.7.P34 1983 629.8'312 82–8530
ISBN 0-471-05800-9 AACR2

Printed in the United States of America

10 9 8 7 6 5 4 3 2 1

To Mary Louise.
For Aileen, Billy, and Andy.

# Preface

This text is intended to be an introduction to dynamic systems and control for undergraduates in mechanical engineering as well as other engineering disciplines. It is assumed that the student has a background in calculus and college physics (mechanics, thermodynamics, and electrical circuits). Any other required material in physics and mathematics (e.g., differential equations, transforms, and matrices) is developed in the text and its appendixes.

Teaching and therefore writing a text in the area of dynamic systems and control present a great challenge for several reasons. The control engineer can be called on to develop mathematical models for a variety of applications involving hydraulic, pneumatic, thermal, mechanical, and electrical systems and must be familiar with the modeling techniques appropriate to each area. Here the challenge is to develop the student's ability to create a model that is sufficiently detailed to capture the dominant dynamic characteristics of the system, yet simple enough to be useful for design purposes. Technological developments in the control field have been rapid, as evidenced by the increasing use of digital computers as controllers. Treatment of digital control necessitates coverage of sampled-data systems, but without neglecting the fundamentals of continuous-time system analysis. In addition to hardware developments, the field has seen the rapid development of analytical techniques collectively known as "modern" control theory. Matrix formulations using the state-space approach provide a good basis for developing general algorithms that are well suited for computer-aided design purposes. On the other hand, the methods of "classical" control theory are still widely used and have significant advantages over the modern methods in many applications. Today's engineer must be familiar with the classical and modern approaches, and a balanced treatment of these topics is needed.

Here an attempt to satisfy these needs has been made by introducing the required concepts and methods in a balanced and gradual way. First, a unified energy-based approach to modeling is presented. Examples from different fields are given to show the student how to develop relatively low-order models that are adequate for many applications. Next is a presentation of analysis techniques for low-order models. In this way the student can be led to see the relation between the form of the model and the resulting analytical requirements. This approach also improves the students' understanding of the basic mathematics by allowing them to apply it immediately to previously developed models of physical systems. Digital simulation techniques using Euler, predictor-corrector, and Runge-Kutta integration are also introduced early in order to allow students to deal with the nonlinearities that occur in many applications.

An innovative feature of the text is the early introduction of sampled-data system analysis as a straightforward extension of Laplace transform methods for continuous-time models. Students thus can soon become more comfortable with discrete-time analysis by immediately building on their knowledge of continuous-time methods. This approach also allows a more effective integration of sampled-data methods into later chapters dealing with higher-order models and control-system design.

The pedagogic approach used in the text is a systematic progression from the simple to the more complex, with the model's order taken to be the measure of complexity. Most system-dynamics concepts can be explained with first-order models, and all the analytical techniques necessary for such models are treated early in the text. Models of second order and higher are then gradually introduced along with any required mathematical tools. In this way students can appreciate the need for such tools and can better understand their use. Because of this approach there are no separate chapters devoted to Laplace transform methods, frequency response, stability analysis, nonlinear systems, or state-space methods. Instead, students are led to apply all of these methods in a thorough analysis of each model form as it is introduced.

Chapter One establishes the viewpoint of dynamic systems analysis and control and its associated modeling requirements. Terminology and an overview of analytical techniques are presented. Zero-order system examples provide an easy introduction to system analysis because they require only algebraic models. Linearization, system diagrams, and the properties of feedback are introduced in this way.

The general structure of dynamic modeling based on energy principles and integral causality is developed in Chapter Two. Basic principles for modeling simple fluid, thermal, mechanical, and electrical systems are treated. Examples are given that lead to first-order models and that introduce commonly used hardware. The chapter also contains a discussion of models for natural resource management, biomedical, and socioeconomic applications.

The analysis of continuous-time models begins in Chapter Three. First-order models are used as a vehicle for introducing free and forced linear system response to arbitrary inputs in general – and to impulse, step, ramp, and sine functions in particular. Laplace transform solution methods, transfer functions, and system diagrams are extensively treated. Analysis of nonlinear models concludes the chapter. Some simple methods for obtaining closed-form solutions are given. When such solutions are not available, the linearization procedure presented often enables useful information to be obtained nevertheless. Finally, several easily programmed numerical integration schemes are derived for the purpose of investigating the effect of nonlinearities in applications where the assumptions required for linearization are too limiting.

The use of first-order models as a simple way of introducing general concepts continues in Chapter Four with the treatment of sampled-data systems and discrete-time models. Sampling, quantization, and coding effects in analog-to-digital conversion are discussed. The Laplace transform of a sampled time function provides a natural way of developing the $z$ transform. This transform is then used to obtain the forced response for common input functions. Sampled-data systems result when a digital-to-analog converter couples a digital device to a continuous-time system, such as in digital control of a mechanical load. The zero-order hold model of this coupling is used with the $z$ transform to develop methods for analyzing this important class of systems. Finally, the concept of digital filtering is introduced and analyzed with time- and frequency-domain techniques.

Second-order continuous-time models differ from first-order ones primarily in the possibility of oscillatory free response and resonance. These distinctions are emphasized in Chapter Five, and second-order models are used as a painless way of introducing matrix methods and stability analysis techniques useful for higher-order models. Response calculations for different input types are also used to reinforce the methods learned in Chapter Three. Analysis of continuous-time linear models is treated first, followed by nonlinear and discrete-time analysis in that order.

Chapter Six presents the basics of feedback control. The common two-position and proportional-integral-derivative types of control laws are analyzed for first- and second-order plants. Typical hardware is described, and the implementation of the control laws with electronic, pneumatic, and hydraulic elements is developed. The derivation of several digital control algorithms completes the chapter.

A determined effort has been made throughout the text to justify and to illustrate the analytical methods in terms of their practical applications. Book-length restrictions always prevent the treatment of design problems at the level of fine detail needed in practice, and some design considerations such as economics and reliability are difficult to treat within the scope of this study. Nevertheless, the discussion and examples should give students as much of a feel for the design process as is possible in an academic environment. To this end a number of topics that have not been given sufficient emphasis in the past are consolidated into Chapter Seven. Methods for selecting the controller gains, and for interpreting the effects of modeling approximations, are among the design issues discussed. All physical elements are power-limited, and design methods are given for taking the associated saturation nonlinearities into account. The classical design approaches of feed-forward, feedback, and cascade compensation are presented, followed by some newer methods such as state-variable feedback, decoupling control, and pseudo-derivative feedback. The latter topic shows that control system design is an exciting field with much to be discovered.

Control system design has always relied heavily on graphical methods. Chapter Eight develops the three most common of these: the root locus, Nyquist, and Bode plots. Computer programs are readily available for generating these plots, but designers must be able to sketch them first, at least roughly, so they can utilize the programs efficiently and interpret their results. The material in Chapter Eight is organized in this spirit. The graphical methods are then used to obtain a deeper understanding of control and compensation methods, and the important topics of lead and lag compensation are covered in detail. Extensions of the methods to digital control applications finish the chapter.

The full power of the state-variable matrix methods is demonstrated in Chapter Nine. These methods are well suited for computer applications and are especially useful for very high-order or multiinput systems. The eigenvalue problem and its relation to the free response are developed first, followed by computer-oriented techniques for determining the forced response. The important concepts of controllability and observability are then described with the eigenmode characterization of the system. These results are used in Chapter Ten for the design of advanced control algorithms such as pole placement, modal control, the linear-quadratic regulator, and state vector observers. An extension of both chapters' methods to discrete-time systems completes the coverage.

Six appendices are included. Appendix A deals with useful analytical techniques such as the Taylor and Fourier series, matrix analysis, and complex numbers. A self-contained development of the Laplace transform appears in Appendix B. Appendix C contains a listing of a useful FORTRAN program implementing the Runge-Kutta integration scheme. Reduction of complicated system diagrams is facilitated with Mason's rule (Appendix D). Analog simulation is the topic of Appendix E. Appendix F contains the Routh-Hurwitz stability criterion.

There is ample material in the book for a semester course in modeling and analysis of dynamic systems, followed by a semester course in control systems. At present there seems to be no generally accepted topic outline for a course sequence in these subjects. Most of the differences in existing sequences at various schools can be categorized by the relative emphasis given the following topics.

1. Modeling.
2. Sampled-data, discrete-time systems and digital control.
3. Nonlinear systems and digital simulation techniques.
4. Matrix methods.

The book has been designed to be flexible in terms of its use as a course text, and can accommodate all of the preceding differences, as follows.

For those wishing only a brief introduction to modeling, Sections 2.1 and 2.2 provide the necessary overview. It is suggested, however, that the material on mechanical and electrical systems (Sections 2.3 and 2.4) also be covered since most control systems are electromechanical.

For coverage of only continuous-time systems, Chapter Four can be omitted without affecting the student's ability to understand the later material. All of the material relevant to discrete-time systems has been placed at the end of each of the chapters following Chapter Four and can be skipped without loss of continuity. However, the differences between discrete and continuous-time analyses are often overshadowed by the similarities, as in the algebraic determination of system response via transform techniques. Thus the instructor is urged to consider an integrated treatment of these subjects, because of the importance of digital control and the increased student interest that this topic produces.

The material on nonlinear systems and digital simulation techniques mostly appears near the end of Chapters Three and Five. It is not a prerequisite for most other topics and can be skipped if desired. However, some understanding of linearization and nonlinear behavior is important, if only to provide a contrast to the linear theory. In addition, the use of digital simulation can generate a considerable amount of student enthusiasm.

Matrix methods and notation are now part of the standard vocabulary for the systems engineer, and all undergraduates should be exposed to them. They are introduced in Chapter Five and used at appropriate places thereafter. More intensive coverage is given in Chapters Nine and Ten, with Chapter Nine devoted to analysis and Chapter Ten to control system design. The analytical material does not rely on Chapters Six, Seven, and Eight. Thus the course coverage can jump from Chapter Five to Chapter 9 if an in-depth treatment of matrix methods is desired before control systems are introduced.

With the exception of Appendix C, no specific computer or calculator language has been used to present the concepts, and no assumptions have been made concerning the availability of such equipment to the student. However, if it is available, there are many opportunities to use it throughout the text. Programmable calculators can be put to good use in response calculations for evaluating complicated functions, for the simulation of nonlinear systems, and for root finding (a good example is to program the solution of a cubic polynomial since third-order systems are so common). Digital computers are also convenient for the applications mentioned earlier, especially with high-order systems. In addition, they can be used for matrix operations and for generating frequency-response and root-locus plots. Sources for such programs are given at the appropriate points in the text.

Successful creation of a textbook is a team effort, with the author only the most visible member. I would like to acknowledge the rest of the team. Three department chairmen at the University of Rhode Island, Dr. Charles D. Nash, Jr., Dr. Frank M. White, and Dr. Thomas J. Kim, provided consideration and encouragement throughout the project. Charlotte Woodhead, Terry Nelson, and Gail Colburn cheerfully and competently assisted in the preparation of the manuscript. I am indebted to several anonymous reviewers for their constructive criticism, suggestions, and enthusiasm. Perhaps someday they will reveal themselves, and I can thank them personally! I am also grateful to Carol Beasley, Ilene Zucker, Bill Stenquist, Rosamond Dana, and Cindy Stein of John Wiley & Sons for their guidance and patience when deadlines were missed. Finally, I want to thank my wife, Mary Louise, and our children, Aileen, Billy, and Andy, for their support and understanding during those numerous dislocations in both time and space.

**William J. Palm III**

# Contents

# MODELING, ANALYSIS, AND CONTROL OF DYNAMIC SYSTEMS

# CHAPTER ONE

# Introduction

Modeling, analysis, and control of dynamic systems have been of interest to engineers for a long time. Within recent years the subject has increased in importance for three reasons. Before the invention of the digital computer, the calculations required for meaningful applications of the subject were often too time consuming and error prone to be seriously considered. Thus gross simplifications were made, and only the simplest models of transient behavior were used, if at all. Now, of course, the widespread availability of computers, as well as pocket calculators, allows us to consider more detailed models and more complex algorithms for analysis and design.

Second, with this increased computational power, engineers have correspondingly increased the performance specifications required of their designs to make better use of limited materials and energy, for example, or to improve safety. This leads to the need for more detailed models, especially with regard to the prediction of transient behavior.

Finally, the use of computers as system elements for measurement and control now allows more complex algorithms to be employed for data analysis and decision making. For example, intelligent instruments with microprocessors can now calibrate themselves. This increased capability requires a better understanding of dynamic systems so that the full potential of these devices can be realized.

## 1.1 SYSTEMS

The term *system* has become widely used today, and as a result, its original meaning has been somewhat diluted. *A system is a combination of elements intended to act together to accomplish an objective.* For example, an electrical resistor is an element for impeding the flow of current, and it usually is not considered to be a system in the sense of our definition. However, when it is used in a network with other resistors, capacitors, inductors, etc., it becomes part of a system. Similarly, a car's engine is a system whose elements are the carburetor, the ignition, the crankshaft, and so forth. On a higher level the car itself can be thought of as a system with the engine as an element. Since nothing in nature can be completely isolated from everything else, we see that our selection of the "boundaries" of the system depends on the purpose and the limitations of our study. This in part accounts for the widespread use of the term *system*, since almost everything can be considered a system at some level.

### The Systems Approach

Automotive engineers interested in analyzing the car's overall performance would not have the need or the time to study in detail the design of the gear train. They most likely would need to know only its gear ratio. Given this information, they would then consider the gear train as a "black box." This term is used to convey the fact that the details of the gear train are not important to the study (or at least constitute a luxury they cannot afford). They would be satisfied as long as they could compute the torque and speed at the axle, given the torque and speed at the drive shaft.

The black-box concept is essential to what has been called the "systems approach" to problem solving. With this approach each element in the system is treated as a black box, and the analysis focuses on how the connections between the elements influence the overall behavior of the system. Its viewpoint implies a willingness to accept a less detailed description of the operation of the individual elements in order to achieve this overall understanding. This viewpoint can be applied to the study of either artificial or natural systems. It reflects the belief that the behavior of complex systems is made up of basic behavior patterns that are contributed by each element and that can be studied one at a time.

The behavior of a black-box element is specified by its *input-output relation.* An *input* is a *cause*; an *output* is an *effect* due to the input. Thus the input-output relation expresses the cause-and-effect behavior of the element. For example, a voltage $v$ applied to a resistor $R$ causes a current $i$ to flow. The input-output or causal relation is $i = v/R$. Its input is $v$, its output is $i$, and its input-output relation is the preceding equation.

## Block Diagrams

The black-box treatment of an element can be expressed graphically, as shown in Figure 1.1. The box represents the element, the arrow entering the box represents the input, and that leaving the box stands for the output. Inside the box we place the mathematical expression that relates the output to the input, if this expression is not too cumbersome, and is known. This graphical representation is a *block diagram.*

The diagram in Figure 1.1*a* represents the resistor with input $v$ and output $i$. They are related by the constant $1/R$. Figure 1.1*b* shows the representation of a spring whose resisting tensile force $f$ is proportional to its extension $x$ so that $f = kx$. Figure 1.1*c* shows how a force $f$ applied to a mass $m$ causes an acceleration $a$. The governing relation is Newton's law: $f = ma$. To obtain $a$ from $f$, we must multiply the input $f$ by the constant $1/m$. Thus the symbol in the box represents the operation that must be performed on the input to obtain the output.

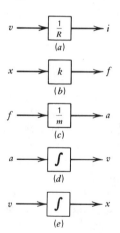

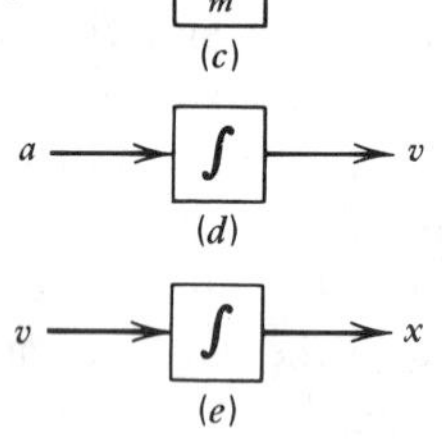


**Figure 1.1 Block diagrams of input-output relations. (*a*) Voltage-current relation for a resistor. (*b*) Displacement-force relation for a spring. (*c*) Force-acceleration relation for a mass. (*d*) Velocity as the time integral of acceleration. (*e*) Displacement as the time integral of velocity.**

Not all black-box representations must refer to actual physical elements. Since they express cause-and-effect relations they can be used to display processes as well as components. Two examples of this are shown in Figures 1.1*d* and 1.1*e*. If we integrate the acceleration $a$ over time, we obtain the velocity $v$, that is, $v = \int a\,dt$. Thus acceleration is the cause of velocity. Similarly, integration of velocity produces displacement $x$: $x = \int v\,dt$. The integration operator within each box in Figures 1.1*d* and 1.1*e* expresses these facts. Whenever an output is the time integral of the input, the element is said to exhibit *integral causality*. We will see that integral causality constitutes a basic form of causality for all physical systems.

The input-output relations for each element provide a means of specifying the connections between the elements. When connected together to form a system, the inputs to some elements will be the outputs from other elements. For example, the position of a speedometer needle is caused by the car's speed. Thus for the speedometer element, the car's speed is an input. However, the speed is the result of action of the drive-train element. The input-output relation can sometimes be reversed for an element, but not always. We can apply a current as input to a resistor and consider the voltage drop to be the output. On the other hand, the position of the speedometer needle can in no way physically influence the speed of the car.

The system itself can have inputs and outputs. These are determined by the selection of the system's boundary. Any causes acting on the system from the world external to this boundary are considered to be system inputs. Similarly, a system's outputs can be the outputs from any one or more of the elements, viewed in particular from outside the system's boundary. If we take the car engine to be the system, a system input would be the throttle position determined by the acceleration pedal, and a system output would be the torque delivered to the drive shaft. If the car is taken to be the system instead, the input would still be the pedal position, but the outputs might be taken to be the car's position, velocity, and acceleration. Usually our choices for system outputs are a subset of the possible outputs and are the variables in which we are interested. For example, a performance analysis of the car would normally focus on the acceleration or velocity, but not on the car's position.

A simple example of a system diagram is provided in Figure 1.2. Suppose that a mass $m$ is connected to one end of a spring. The other end of the spring is attached to a rigid support. In addition to the spring force $f_s$, another force $f_o$ acts on the mass. This force is considered to be due to the external world and acts across the system "boundary"; that is, it is not generated by any action within the system itself. It might be due to gravity, for example.

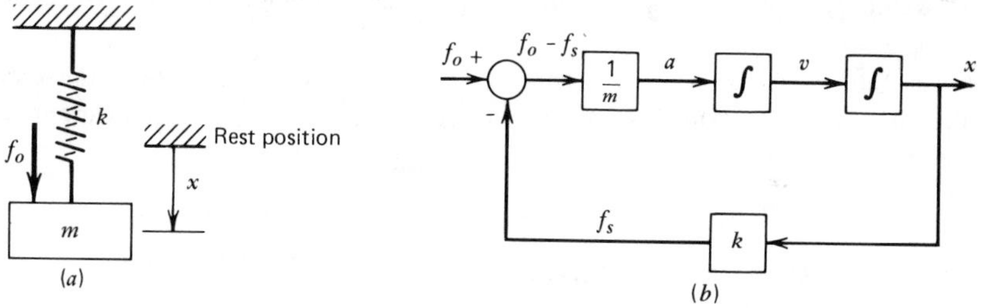


**Figure 1.2** (*a*) **Mass-spring system with an external force $f_o$. (*b*) Block diagram of the causal relations.**

The cause-and-effect relations can be summarized by the system diagram in Figure 1.2*b*. The net force on the mass in the direction of positive displacement $x$ is $f_o - f_s$, since the spring will pull up on the mass if the mass position is below the rest position ($x > 0$). The addition and subtraction of forces to produce the net force are represented by a new symbol, the *comparator* – a circle whose output is the signed sum of the inputs. The plus (+) sign indicates that $f_o$ is to be added; the minus (−) sign indicates that $f_s$ is to be subtracted to produce the net force.

The system input is $f_o$; its output could be any or all of the variables generated within the diagram. If we are interested only in the displacement $x$, then its arrow is shown leaving the system.

System diagrams such as this are a visually concise summary of the interplay between the causes and the effects. We will use them often.

### Static and Dynamic Systems

In general the present value of an element's output is the result of what has happened to the element in the past as well as what is currently affecting it. For example, the present position of a car depends on where it started and what its velocity has been from the start. We define a *dynamic* element to be one whose present output depends on past inputs. Conversely, a *static* element is one whose output at any given time depends only on the input at that time.

For the car considered as an element with an acceleration pedal position as input and car position as output, the preceding definition shows the element to be dynamic. On the other hand, we can consider a resistor to be a static element because its present current depends only on the voltage applied at present, not on past voltages. This is an approximation, of course, because the resistor cannot respond instantaneously to voltage changes. This is true of all physical elements, and we therefore conclude that a static element is an approximation. Nevertheless, it is widely used because it results in a simpler mathematical representation.

In popular usage the terms *static* and *dynamic* are used to distinguish situations in which no change occurs from those that are subject to changes over time. This usage conforms to the preceding definitions of these terms if the proper interpretation is made. A static element's output can change with time only if the input changes and will not change if the input is constant or absent. However, if the input is constant or is removed from a dynamic element, its output can still change. For example, if the car's engine is turned off, the car's position will continue to change because of the car's velocity (because of past inputs). A similar statement cannot be made for the electrical resistor.

In the same way we also speak of static and dynamic systems. A static system contains all static elements. Any system that contains at least one dynamic element is a dynamic system.

## 1.2 MODELING, ANALYSIS, AND CONTROL

We live in a universe that is undergoing continual change. This change is not always apparent if its time scale is long enough, such as with some geologic processes, but as engineers we often must deal with situations in which time-dependent effects are important. For example, design of high-speed production machinery for precise operation requires that the vibrational motions due to high accelerations be small in amplitude and die out quickly. Likewise, time-dependent behavior of fluid–flow and heat-transfer processes significantly affects the quality of the product of a chemical process. Even a relatively "stationary" object like a bridge must be designed to accommodate the motions and forces produced by a heavy rolling load.

## Modeling

In order to deal in a systematic and efficient way with problems involving time-dependent behavior, we must have a description of the objects or processes involved. We call such a description a *model.* A model for enhancing our understanding of the problem can take several forms. A physical model, like a scale model, helps us to visualize how the components of the design fit together and can provide insight not obtainable from a blueprint (which is another model form). Graphs or plots are still another type of model. They can often present time-dependent behavior in a concise way, and for that reason we will rely heavily on them throughout this study. The model type we will use most frequently is the *mathematical model*, which is a description in terms of mathematical relations. These relations will consist of differential or difference equations if the model is to describe a dynamic system.

The concept of a mathematical model is undoubtedly familiar from elementary physics. Common examples include the voltage-current relation for a resistor $v = iR$, and the force-deflection relation for a spring $f = kx$. One of our aims here is to introduce a framework that allows the development of mathematical models for describing the time-dependent behavior of many types of phenomena: fluid flow, thermal processes, mechanical elements, and electrical systems, as well as some nonphysical applications. In this regard, it is important to remember that the precise nature of a mathematical model depends on its purpose. For example, an electrical resistor can be subjected to mechanical deformations if its mounting board is subjected to vibration. In this case, the force-deflection spring model could be used to describe the resistor's mechanical behavior.

Thus we see that the nature of any object has many facets: thermal, mechanical, electrical, etc. No mathematical model can deal with all these facets. Even if it could, it would be useless because its very complexity would render it cumbersome. We can make an analogy with maps. A given region can be described by a road map, a terrain-elevation map, a mineral-resources map, a population-density map, and so on. A single map containing all this information would be cluttered and useless. Instead, we select the particular type of map required for the purpose at hand. In the same way, we select or construct a mathematical model to suit the requirements of the study.

The purpose of the model should guide the selection of the model's time scale, its length scale, and the particular facet of the object's nature to be described (thermal, mechanical, electrical, etc.). The time scale will in turn determine whether or not time-dependent effects should be included. (Tectonic plate motion constitutes time-dependent behavior on a geologic time scale but would not be considered by an engineer designing a bridge.) Similarly, the length scale partly dictates what details should or should not be included. The engineer analyzing the dynamics of high-speed machinery might treat a component as a point mass, whereas this approximation is useless to a metallurgist studying material properties at the molecular level.

Block diagrams are often used to display the mathematical model in a form that allows us to understand the interactions occurring between the system's elements. For example, the mathematical model of the mass-spring system shown in Figure 1.2*a* is

$$f_o - f_s = ma$$

$$f_s = kx$$

$$x = \int v\,dt$$

$$v = \int a\,dt$$

Each equation is a cause-and-effect relation for one part of the system.

The diagram is a valuable aid, but if we wish to solve for the displacement $x(t)$, we need the model in equation form. If we differentiate the last two relations we obtain

$$\frac{dx}{dt} = v \qquad \frac{dv}{dt} = a$$

or

$$a = \frac{d^2x}{dt^2}$$

Substitution of this and the second relation into the first gives

$$f_o - kx = m\frac{d^2x}{dt^2} \tag{1.2-1}$$

This differential equation is a quantitative description of the system. Given $f_o$, $k$, $m$, and the initial position $x(0)$ and velocity $v(0)$, we can use the methods of later chapters to solve the equation for $x(t)$.

## Analysis

A mathematical model represents a concise statement of our hypotheses concerning the behavior of the system under study. We can deal with the verification of the model in two ways. Verification by experiment or testing is ultimately required of all serious design projects. This is not always done at the outset of a study, however, especially if one is dealing with component types whose behavior is known to be well described by a specific model on the basis of past experience. For example, we are on firm ground in using the resistor model $v = iR$ without verification, as long as the operating conditions (voltage levels, temperatures, etc.) are not extreme. This is often the case for the types of problems we will be considering, since we will be concerned with the behavior of systems consisting of components whose individual behavior is often well understood.

Once we are satisfied with the validity of our chosen component models, they can be used to predict the performance of the system in question. Predicting the performance from a model is called *analysis*. For example, the current produced in a resistor by an applied voltage $v$ can be predicted to be $i = v/R$ by solving the resistor model for the unknown variable $i$ in terms of the given quantities $v$ and $R$. Most of our mathematical models will describe dynamic behavior and thus will consist of differential or difference equations. They will not be as easy to solve as the algebraic model just seen. Nevertheless, the techniques for analyzing such models are straightforward. These are introduced in Chapter Three.

Just as the model's purpose partly determines its form, so also does the purpose influence the types of analytical techniques used to predict the system's behavior. It is not possible to discuss these concepts with the simple resistor model, because its solution is so simple. However, we will be developing many types of analytical techniques whose applicability depends on the purpose of the analysis. Not all of these techniques will be brought to bear on any one problem, but the engineer should be familiar with all of them. They are the tools of the trade – a means to an end. We will not study them, as a mathematician would, for their inherent interest. Instead, we will focus on how they can help us predict the performance of a proposed design before it is built. Thus we can avoid a cut-and-try approach. This is especially important today since most modern engineering endeavors are too complex and expensive to allow them to be built without a thorough analysis beforehand.

## Control

The successful operation of a system under changing conditions often requires a control system. For example, a building's heating system requires a thermostat to turn the heating elements on or off as the room temperature rises and falls (Figure 1.3). Note that we have not shown the thermostat as one element because it has two functions: (1) to measure the room temperature and compare it with the desired temperature and (2) to decide whether to turn the furnace on or off. The variation in the outdoor environment is the primary reason for the unpredictable change in the room temperature. If the outside conditions (temperature, wind, solar insolation, etc.) were predictable, we could design a heater that would operate continuously to supply heat at a predetermined rate just large enough to replace the heat lost to the outside environment. No controller would be necessary. Of course, the real world does not behave so nicely, so we must adjust the heat-output rate of our system according to what the actual room temperature is.

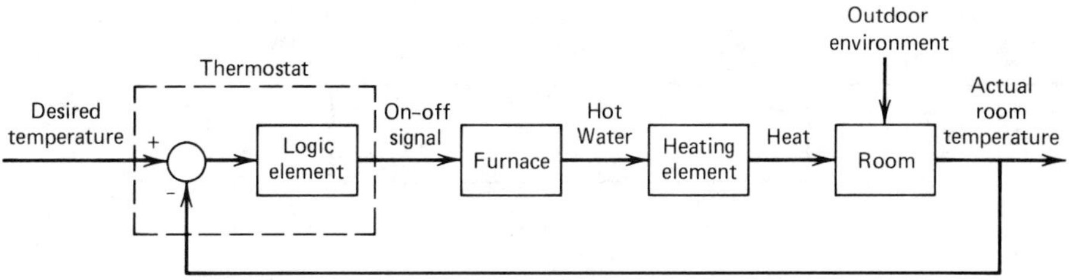


**Figure 1.3 Block diagram of the thermostat system for temperature control.**

We often wish to alter the desired operating conditions of a system. In our heating example the thermostat allows the user to specify the desired room temperature, say 68°F during the day and 60°F at night. When the thermostat setting is changed from 60°F to 68°F in the morning, the thermostat acts to bring the room temperature up to 68°F and to keep it near this value until the setting is changed again at night.

The term *control* refers to the process of deliberately influencing the behavior of an object so as to produce some desired result. The physical device inserted for this purpose is the *controller* or *control system.* Other common examples of controllers include:

1. An aircraft autopilot for maintaining desired altitude, orientation, and speed.
2. An automatic cruise-control system for a car.
3. A pressure regulator for keeping constant pressure in a water-supply system.

A cutaway view of a commonly used type of pressure regulator is shown in Figure 1.4 along with a block diagram of its operation. The desired pressure is set by turning a calibrated screw. This compresses the spring and sets up a force that opposes the upward motion of the diaphragm. The bottom side of the diaphragm is exposed to the water pressure that is to be controlled. Thus the motion of the diaphragm is an

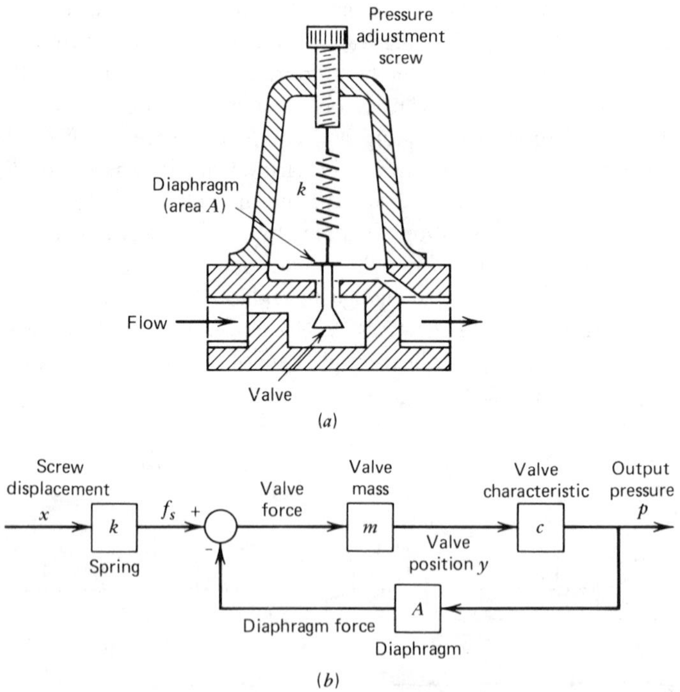


**Figure 1.4 Pressure regulator. (*a*) Cutaway view. (*b*) Block diagram.**

indication of the pressure difference between the desired and the actual pressures. It acts like a comparator. The valve is connected to the diaphragm and moves according to the pressure difference until it reaches a position in which the difference is zero.

From the preceding examples we see that the role of a controller is twofold.

1. It must bring the system's operating condition to the desired value.
2. It must maintain the desired condition in the presence of variations caused by the external environment.

In the terminology of the control engineer, we say that the controller must respond satisfactorily to changes in *commands* and maintain system performance in the presence of *disturbances.*

One or more controllers are often required in complex dynamic systems in order to make the system elements act together to achieve the intended goal. So the design of dynamic systems quite naturally involves the study of control systems. On the other hand, the variations produced by command changes and disturbances tend to upset the system. Thus control system design requires models that describe the dominant dynamic properties of the system to be controlled, and the analysis techniques must be capable of dealing with such a model. Modeling, analysis, and control of dynamic systems therefore constitute a unified area of study.

## 1.3 TYPES OF MODELS

As we have seen, a system model is a representation of the essential behavior of the system for the purposes at hand. In order to be useful it must contain the minimum amount of information necessary to achieve its purpose, and no more. This requirement is most immediately reflected in the choice of static- versus dynamic-element models. Those elements whose behavior is fast relative to other elements are often modeled as static elements in order to reduce the complexity of the model. For example, the switching time of a thermostat is fast compared to the time required for the room temperature to change appreciably. Thus the room temperature in Figure 1.2 would most likely be modeled as a dynamic element and the thermostat as a static one.

### Lumped and Distributed-Parameter Models

We have implicitly assumed in the heating example that the temperature in the room can be described by a single number, a temperature that is average in some sense. In reality the temperature varies according to location within the room, but if we did not choose to use a single representative temperature the required room model would be much more complicated.

Many variables in nature are functions of location as well as time. The process of ignoring the spatial dependence by choosing a single representative value is called *lumping* (room air is considered to be one "lump" with a single temperature).

Lumping an element is a technique usually requiring experience. It reflects the judgment of the engineer as to what is unimportant in terms of spatial variation.

It can be described as the spatial equivalent of the process of dividing a system into static and dynamic elements. The model of a lumped element or system is called a *lumped-parameter model.* If it is dynamic, the only independent variable in the model will be time; that is, the model will be an ordinary differential equation, like (1.2-1). Only time derivatives will appear, not spatial derivatives.

When spatial dependence is included, the independent variables are the spatial coordinates as well as time. The resulting model is said to be a *distributed-parameter model.* It consists of one or more partial differential equations containing partial derivatives with respect to the independent variables. The difference is illustrated in Figure 1.5$a$, which shows the temperature $T$ of a metal plate. If the plate is heated at one side, the temperature will be a function of location and time – $T = T(t, x, y, z)$ – and the model will be of the form

$$f\left(T, \frac{\partial T}{\partial t}, \frac{\partial^2 T}{\partial x^2}, \frac{\partial^2 T}{\partial y^2}, \frac{\partial^2 T}{\partial z^2}\right) = 0 \qquad (1.3\text{-}1)$$

But if the plate temperature is lumped with a single value, the model will be of the form

$$f\left(T, \frac{dT}{dt}\right) = 0 \qquad (1.3\text{-}2)$$

which is easier to handle mathematically.

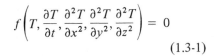

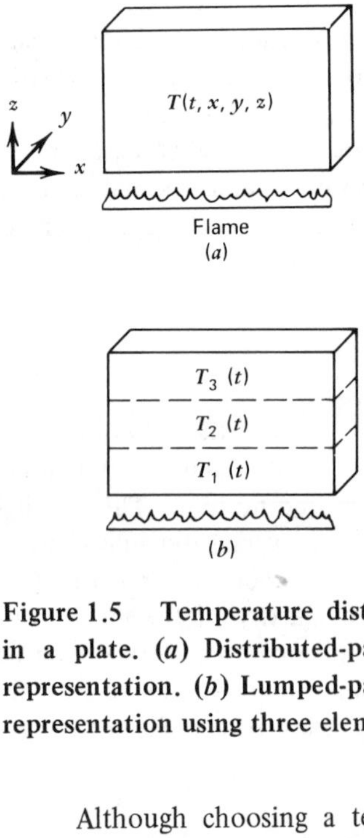


**Figure 1.5 Temperature distribution in a plate. (*a*) Distributed-parameter representation. (*b*) Lumped-parameter representation using three elements.**

Lumping may be done at several levels. For example, we may take a single temperature to represent the entire house. In this case a single differential equation would result. On the other hand, we may take a representative temperature for each room. In this case the total model would consist of a differential equation for each room temperature.

Although choosing a temperature for each room leads to several equations, the model is usually more manageable than if the lumping were not performed.

There are applications in engineering where a detailed model like (1.3-1) is required, and we do not dismiss such models as useless. However, we will see that in analyzing systems with many elements, distributed-parameter models of elements are a luxury we usually cannot afford because their complexity tends to prevent us from understanding the overall system behavior. We therefore will limit our treatment to lumped-parameter models. Note that if a more detailed model is needed, we can increase the number of lumped elements, such as is shown in Figure 1.5$b$, where we have used three temperature lumps in an attempt to model the dependence of the plate temperature as a function of distance from the flame. The resulting model would have three coupled ordinary differential equations, one for each $T_i(t)$.

## Linear and Nonlinear Models

We have seen that engineers should attempt to model elements as static rather than dynamic, and as lumped rather than distributed. The reason is that engineers eventually have to analyze the resulting system model, and its complexity can easily get out of hand if there is too much detail in each element model. In a similar vein we now discuss the distinction between model types based on *linearity*.

Let $y$ be the output and $x$ be the input of an element that can be either static or dynamic. Its model is written as

$$y = f(x) \tag{1.3-3}$$

where the function $f(x)$ may include operations like differentiation and integration. The model (or element) is said to be *linear* if, for an input $ax_1 + bx_2$, the output is

$$y = f(ax_1 + bx_2) = af(x_1) + bf(x_2) = ay_1 + by_2 \tag{1.3-4}$$

where $a$ and $b$ are arbitrary constants, $x_1$ and $x_2$ are arbitrary inputs, and

$$y_1 = f(x_1) \tag{1.3-5}$$

$$y_2 = f(x_2) \tag{1.3-6}$$

Thus linearity implies that multiplicative constants and additive operations in the input can be factored out when considering the effects on the output. The linearity property (1.3-4) is sometimes called the *superposition principle* because it states that a linear combination of inputs produces an output that is the superposition (linear combination) of the outputs that would be produced if each input term were applied separately. Any relation not satisfying (1.3-4) is nonlinear.

Let us consider some input-output relations to see if they are linear. The simple multiplicative relation $y = mx$ is linear because

$$y = m(ax_1 + bx_2) = amx_1 + bmx_2 = ay_1 + by_2$$

where $y_1 = mx_1$ and $y_2 = mx_2$. The operation of differentiation $y = dx/dt$ is linear because

$$y = \frac{d}{dt}(ax_1 + bx_2) = a\frac{dx_1}{dt} + b\frac{dx_2}{dt} = ay_1 + by_2$$

Similarly, integration is a linear operation. If $y = \int x\,dt$, then

$$y = \int (ax_1 + bx_2)dt = a\int x_1\,dt + b\int x_2\,dt = ay_1 + by_2$$

Any relation involving a transcendental function or a power other than unity is nonlinear. For example, if $y = x^2$,

$$y = (ax_1^2 + bx_2^2) = a^2x_1^2 + 2abx_1x_2 + b^2x_2^2 \neq ax_1^2 + bx_2^2$$

Similarly, if $y = \sin x$,

$$y = \sin(ax_1 + bx_2) \neq a \sin x_1 + b \sin x_2$$

The definition of linearity (1.3-4) can be extended to include functions of more than one variable, such as $f(x, z)$. This function is linear if and only if

$$f(ax_1 + bx_2, az_1 + bz_2) = af(x_1, z_1) + bf(x_2, z_2)$$

Differential equations represent input-output relations also, and can be classified as linear or nonlinear. The outputs (solutions) of the model depend on the outputs' initial values and on the inputs. We will see in Chapter Three that a superposition principle applies to linear differential equations. This is useful because it allows us to separate the effects of more than one input and thus to consider each input one at a time. It also allows us to separate the effects of the initial values of the outputs from the effects of the inputs. For these reasons we will always attempt to obtain a linear model for our systems provided that any approximations required to do so do not mask important features of the system's behavior.

A differential equation is easily recognized as nonlinear if it contains powers or transcendental functions of the dependent variable. For example, the following equation is nonlinear.

$$\frac{dy}{dt} = -\sqrt{y} + f$$

## Time-Variant Models

The presence of a time-varying coefficient does not make a model nonlinear. For example, the model

$$\frac{dy}{dt} = c(t)y + f$$

is linear. Models with constant coefficients are called *time-invariant* or *stationary* models, while those with variable coefficients are *time variant* or *nonstationary*. An example occurs if the mass $m$ in Figure 1.1 represents a bucket of water with a leak. Its mass would then change with time, and $m = m(t)$ in (1.2-1).

## Discrete and Continuous-Time Models

Sometimes it is inconvenient to view the system's dynamics in terms of a continuous-time variable. In such cases we use a discrete variable to measure time. Common examples of this usage include one's age (we usually express it in integer years, with no fractions) and interest computations on savings accounts (compounded quarterly, annually, etc.). For engineers the most important situation suggesting the use of discrete-time models occurs when a system contains a digital computer for measurement or control purposes. It is an inherently discrete-time device because it is driven by an internal clock that allows activity to take place only at fixed intervals. Thus a digital computer cannot take measurements continuously but must "sample" the measured variable at these instants.

If we choose to represent our system in terms of discrete time, the form of the model is a difference equation instead of a differential equation. For example, an amount of money $x$ in a savings account drawing 5% interest compounded annually will grow according to the relation

$$x(k+1) = 1.05x(k) \tag{1.3-7}$$

The index $k$ represents the number of years after the start of the investment. We will return to the analysis of discrete-time models in Chapter Four.

## Model Order

Equation (1.2-1) is called a *second-order* differential equation because its highest derivative is second order. It is equivalent to the relations

$$m \frac{dv}{dt} = f_o - kx \tag{1.3-8}$$

$$\frac{dx}{dt} = v \tag{1.3-9}$$

These two first-order equations are coupled to each other because of the $x$ term in the first equation and the $v$ term in the second. One cannot be solved without solving the other; they must be solved simultaneously. Taken together they thus form a second-order model.

We will organize our study of dynamic systems partly according to the order of the model. Chapters Three and Four treat first-order continuous- and discrete-time models, while Chapter Five covers higher-order models. In this way we can begin simply and gradually progress to more difficult topics.

## Stochastic Models

Sometimes there is uncertainty in the values of the model's coefficients or inputs. If this uncertainty is great enough, it might justify using a *stochastic* model. In such a model the coefficients and inputs would be described in terms of probability distributions involving, for example, their means and variances. Such a model would be useful for describing the effects of wind gusts on an aircraft autopilot. Although the wind is not random, presumably our knowledge of its behavior is poor enough to justify a probabilistic approach. However, the mathematics required to analyze such models is beyond the scope of this work, and we will not consider stochastic models further.

## A Model Classification Tree

Figure 1.6 is a diagram of the relationship between the various model types. We have extended only the branches that lead to linear time-invariant models since this is the type of most interest to us.

# 1.4 LINEARIZATION

Because of the usefulness of the superposition principle, we always attempt to obtain a linear model if possible. Sometimes this can be done from the outset of neglecting effects that would lead to a nonlinear model. A common example of this is the small angle approximation. If we assume that the angle of rotation $\theta$ of the lever in Figure 1.7 is small, the rectilinear displacement of its ends is roughly proportional to $\theta$ such that $x = L\theta$. The same is not true for a large enough value of $\theta$.

If such an approximation is not obvious, a systematic procedure based on the Taylor series expansion can be used (Appendix A). Let the input-output model for a

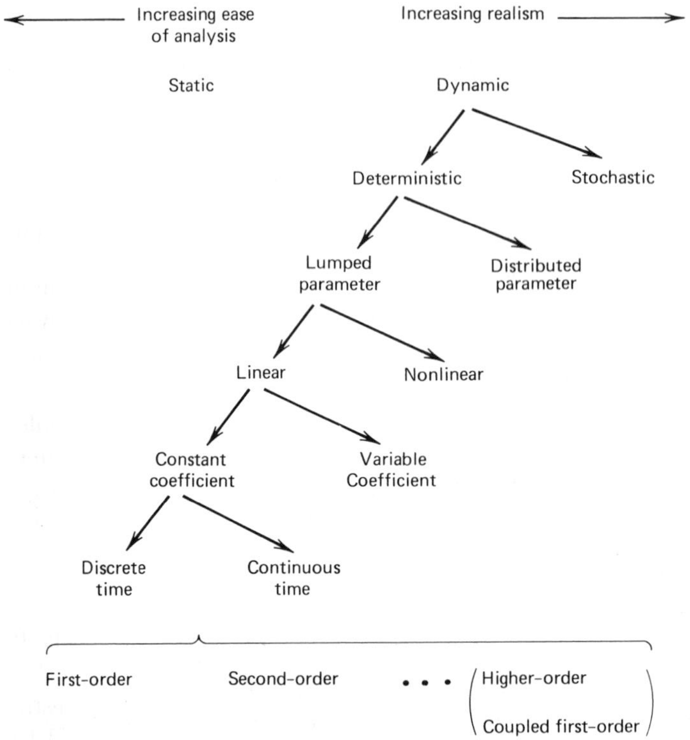


**Figure 1.6 Classification of mathematical models. We have not continued the tree on every branch for simplicity. For example, nonlinear models can also be classified as discrete or continuous time.**

static element be written as

$$w = f(y) \tag{1.4-1}$$

Its form is sketched in a general way in Figure 1.8. A model that is approximately linear near the reference point $(w_o, y_o)$ can be obtained by expanding $f(y)$ in a Taylor series near this point and truncating the series beyond the first-order term. The series is

$$w = f(y) = f(y_o) + \left(\frac{df}{dy}\right)_o (y - y_o) + \frac{1}{2!}\left(\frac{d^2f}{dy^2}\right)_o (y - y_o)^2 + \dots \tag{1.4-2}$$

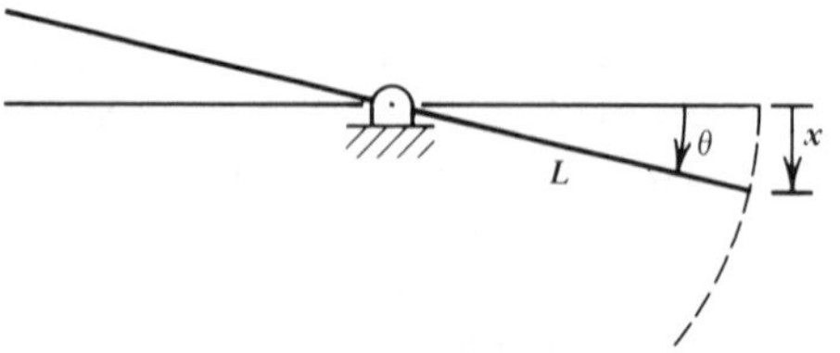


**Figure 1.7 Small-angle approximation for the displacement of a lever endpoint. For $\theta$ small, $x \cong L\theta$.**

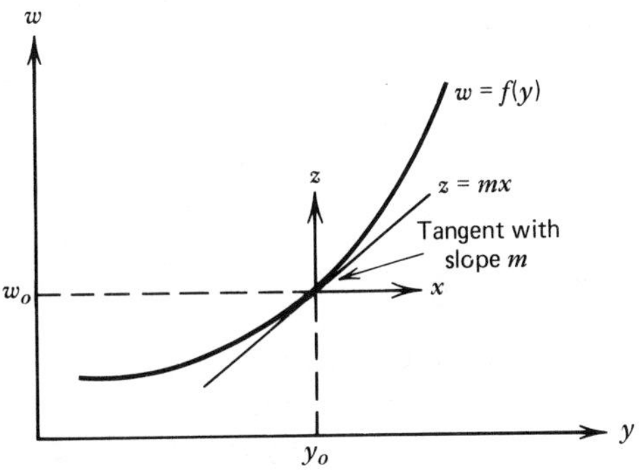


**Figure 1.8 Linearization of the function $w = f(y)$ about the point $(w_o, y_o)$.**

where the subscript "$o$" on the derivatives means that they are evaluated at the reference point $(w_o, y_o)$. If $y$ is "close enough" to $y_o$, the terms involving $(y - y_o)^i$ for $i \geqslant 2$ are small compared to the first two terms. Ignoring these higher-order terms gives

$$w = f(y) \cong f(y_o) + \left(\frac{df}{dy}\right)_o (y - y_o) \tag{1.4-3}$$

This is a linear relation. To put it into a simpler form let

$$m = \left(\frac{df}{dy}\right)_o \tag{1.4-4}$$

$$z = w - w_o = w - f(y_o) \tag{1.4-5}$$

$$x = y - y_o \tag{1.4-6}$$

Then (1.4-3) becomes

$$z \cong mx \tag{1.4-7}$$

The geometric interpretation of this result is shown in Figure 1.8. We have replaced the original function with a straight line passing through the point $(w_o, y_o)$ and having a slope equal to the slope of $f(y)$ at the reference point. With the $(z, x)$ coordinates, a zero intercept occurs and the relation is simplified.

## A Nonlinear Spring Example

No spring is linear over an arbitrary range of extensions. Instead, the force will increase nonlinearly with extension beyond some point, and the linear model used to obtain (1.2-1) will no longer be valid. Suppose the correct relation for a particular spring is $f = y^2$, where $y$ is the extension of the spring from its free length (Figure 1.9). Let it be attached to the mass as shown in Figure 1.2 and allow the mass to settle to its rest position $y_o$. At this position the weight of the mass will equal the spring force so that $mg = y_o^2$, or $y_o = \sqrt{mg}$. The Taylor series applied to the spring relation $f = y^2$ gives

$$\begin{aligned} f \cong y^2 &= y_o^2 + \left(\frac{dy^2}{dy}\right)_o (y - y_o) \\ &= y_o^2 + 2y_o(y - y_o) \end{aligned}$$

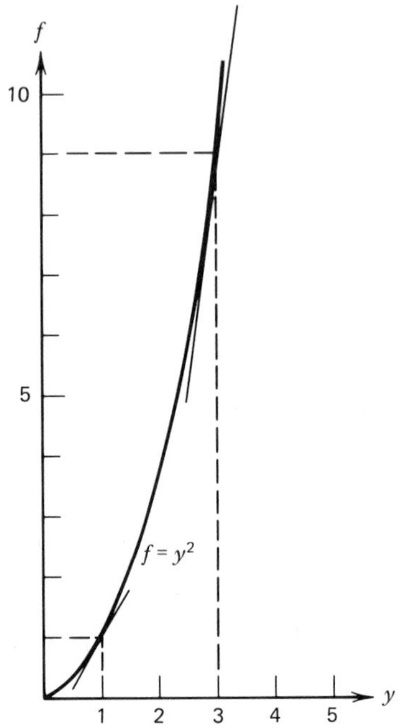


**Figure 1.9 Linearization of the function $f = y^2$ about the points (1, 1) and (9, 3). Note the different slopes for each linearization.**

Let $z = f - f_o = f - y_o^2$ and $x = y - y_o$. Thus

$$z \cong 2y_o x \qquad (1.4\text{-}8)$$

The variable $z$ is the change in spring force from its value at the rest position $y_o$. Equation (1.4-8) shows that the force change is approximately linear if $y$ is close to $y_o$; that is, if $x$ is small.

The free-body diagram of the mass is shown in Figure 1.10, with the gravity force and spring force being the only forces applied to the mass. From Newton's law,

$$m\frac{d^2x}{dt^2} = mg - (f_o + z)$$

Using (1.4-8) and the fact that $mg = y_o^2 = f_o$, we obtain

$$m\frac{d^2x}{dt^2} = -2y_o x \qquad (1.4\text{-}9)$$

which is a linear equation. If we had not approximated the spring force as a linear function, the system's differential equation would have been

$$m\frac{d^2y}{dt^2} = mg - y^2 \qquad (1.4\text{-}10)$$

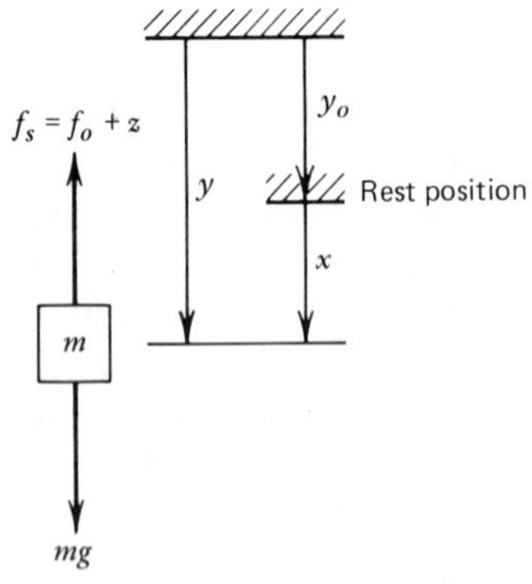


**Figure 1.10 Free-body diagram of the mass-spring system.**

which is nonlinear because of the $y^2$ term.

From (1.4-8) we see that the "spring constant" $k$ is $2y_o$. It depends not only on the spring's physical properties, which yield the constant 2, but also on the reference position $y_o$. The constant $2y_o$ is the slope of the force-extension curve of the spring at the position $y_o$. This position is determined by the spring's characteristics and the weight of the mass. If $mg = 1$, then $y_o = 1$ and $k = 2$. For a larger weight, say $mg = 9$, $y_o = 3$ and $k = 6$.

## The Multivariable Case

The Taylor series linearization technique can be extended to any number of variables. For two variables the function is

$$w = f(y_1, y_2) \tag{1.4-11}$$

and the truncated series is

$$w \cong f(y_{1o}, y_{2o}) + \left(\frac{\partial f}{\partial y_1}\right)_o (y_1 - y_{1o}) + \left(\frac{\partial f}{\partial y_2}\right)_o (y_2 - y_{2o}) \tag{1.4-12}$$

Define

$$z = w - w_o = w - f(y_{1o}, y_{2o}) \tag{1.4-13}$$

$$x_1 = y_1 - y_{1o} \tag{1.4-14}$$

$$x_2 = y_2 - y_{2o} \tag{1.4-15}$$

The linearized approximation is

$$z \cong \left(\frac{\partial f}{\partial y_1}\right)_o x_1 + \left(\frac{\partial f}{\partial y_2}\right)_o x_2 \tag{1.4-16}$$

The partial derivatives are the slopes of the function $f$ in the $y_1$ and $y_2$ directions at the reference point.

As an example, consider the perfect gas law

$$p = \frac{mRT}{V} \tag{1.4-17}$$

where $p$, $V$, $T$, and $m$ are the gas pressure, volume, temperature, and mass, respectively. The universal gas constant is $R$. If the gas is isolated in a flexible chamber, its mass is constant, but its volume, temperature, and pressure can change. For given reference values $T_o$ and $V_o$, a linearized expression for the pressure is

$$p \cong p_o + \left(\frac{\partial p}{\partial T}\right)_o (T - T_o) + \left(\frac{\partial p}{\partial V}\right)_o (V - V_o) \tag{1.4-18}$$

where $p_o = mRT_o/V_o$ and

$$\left(\frac{\partial p}{\partial T}\right)_o = \left(\frac{mR}{V}\right)_o = \frac{mR}{V_o} = a \tag{1.4-19}$$

$$\left(\frac{\partial p}{\partial V}\right)_o = \left(-\frac{mRT}{V^2}\right)_o = -\frac{mRT_o}{V_o^2} = -b \tag{1.4-20}$$

Sometimes the following notation is used to represent the variations from the reference values.

$$\delta p = p - p_o$$

$$\delta T = T - T_o$$

$$\delta V = V - V_o$$

In this case (1.4-18) through (1.4-20) give

$$\delta p \cong a\,\delta T - b\,\delta V \tag{1.4-21}$$

Since $a$ and $b$ are positive, (1.4-21) shows that the pressure increases if $T$ increases or if $V$ decreases.

Of course, we do not need the linearized form to calculate $p$ given $V, T, m$, and $R$. However, if $p$ were to appear in a differential equation with $T$ or $V$ as inputs or dependent variables, then the linearized form would be needed to obtain a linear differential equation.

## Linearization of Operating Curves

An element's input-output relation is not always given in analytical form, but might be available as an experimentally determined plot. For example, the operating curves of an electric motor might look something like those in Figure 1.11. For a fixed motor voltage $v_2$ and a fixed load torque $T_2$, the motor will eventually reach a fixed speed $\omega_2$ some time after the motor is started. This steady-state speed can be found from the curve marked $v_2$. For another load torque, say $T_a$, the resulting speed $\omega_a$ can be found from the same $v_2$ curve. However, if we fix the load torque but change the voltage, the resulting speed must be found by interpolating between the appropriate constant-voltage curves.

Suppose that the load torque $T$ and the voltage $v$ are to be inputs for a dynamic model that will have the speed $\omega$ as the output. Then we cannot use the curves as is but must convert them to an analytical expression. This expression must be linear if the dynamic model is to be linear.

A linearized expression can be obtained from the operating curves by using (1.4-16) and calculating the required slopes numerically from the plot. The partial derivative $(\partial f/\partial y_1)_o$ is computed with $y_2$ held constant at the value $y_{2o}$. It can be found approximately as follows (see Figure 1.12).

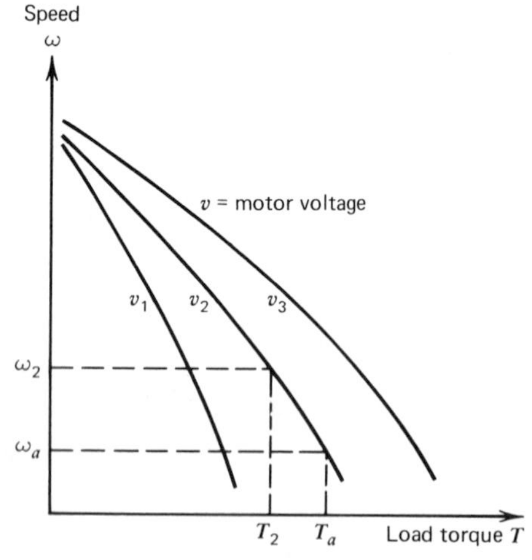


**Figure 1.11 Steady-state operating curves for an electric motor.**

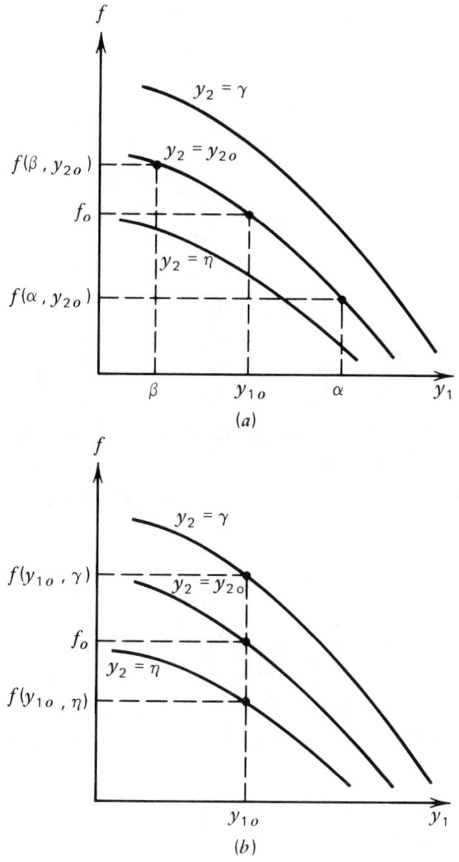


**Figure 1.12 Graphical computation of the linearization derivatives for the function $f(y_1, y_2)$ near the point $f_o = f(y_{1o}, y_{2o})$.**

$$\left(\frac{\partial f}{\partial y_1}\right)_o \cong \frac{f(\alpha, y_{2o}) - f(\beta, y_{2o})}{\alpha - \beta} \tag{1.4-22}$$

where the $y_1$ range $(\beta, \alpha)$ straddles $y_{1o}$. Similarly

$$\left(\frac{\partial f}{\partial y_2}\right)_o \cong \frac{f(y_{1o}, \gamma) - f(y_{1o}, \eta)}{\gamma - \eta} \tag{1.4-23}$$

The smaller $(\alpha - \beta)$ and $(\gamma - \eta)$ are made, the better the approximation, but this is limited by the ability to read the plot accurately for smaller increments.

## 1.5 FEEDBACK

A feature found in many static systems and in almost every dynamic system is one or more *feedback loops*, such as shown in Figures 1.2, 1.3, and 1.4. *Feedback* is the process by which an element's input is altered by its output. It occurs frequently in physiological and ecological systems and is deliberately employed in manufactured systems for several reasons, as we shall see. The human body utilizes many feedback loops for such purposes as body temperature control, blood pressure control, hand-eye coordination, and so forth. In nature, we can describe the alternating abundance and scarcity of prey and predators as being due to a feedback mechanism that prevents both species from becoming extinct.

Measurement of room temperature by a thermostat constitutes a feedback process because the measurement is used to influence the room temperature (Figure 1.3). Similarly the spring in Figure 1.2 acts as a feedback element. The greater the mass displacement, the greater the spring's restoring force attempting to return the mass to its rest position. The action of the diaphragm in the pressure regulator (Figure 1.4) combines the actions of a comparator and a sensor. As the pressure $p$ increases, the diaphragm motion acts to move the valve to decrease the pressure. Thus the output (the pressure) is made to influence itself.

Control systems rely heavily on the properties of feedback. We now explore these properties in more detail.

## Feedback Improves Linearity

As we indicated in Section 1.3, linear systems models will be the chief model form to be used in our study. One of the reasons for this is that the use of feedback often improves the linearity of the system. We can construct a system of elements whose individual behavior is nonlinear, but with proper use of feedback the resulting system's behavior will be approximately linear.

To illustrate this effect consider the nonlinear element shown in Figure 1.13*a*. Its input-output relation is

$$y = x^2 \tag{1.5-1}$$

If we introduce a feedback loop as in Figure 1.13*b*, we can write the following relations.

$$y = e^2$$

$$e = x - y$$

Thus

$$y = (x - y)^2 \tag{1.5-2}$$

The plot of $y$ versus $x$ for the original element and for the feedback system is shown in Figure 1.13*c*. The feedback system's input-output relation is closer to being a straight line and therefore is approximately linear over a wider range of $x$ than for the original nonlinear element. This wider range is what we mean by "improved linearity."

Feedback improves linearity in dynamic as well as static systems. Chapter Six provides several examples of this effect.

## Feedback Improves Robustness

The coefficient values and the form of a model are always approximations to reality and thus have some uncertainty associated with them. In addition, factors such as wear, heat, and pressure can cause the performance of a system to change with time.

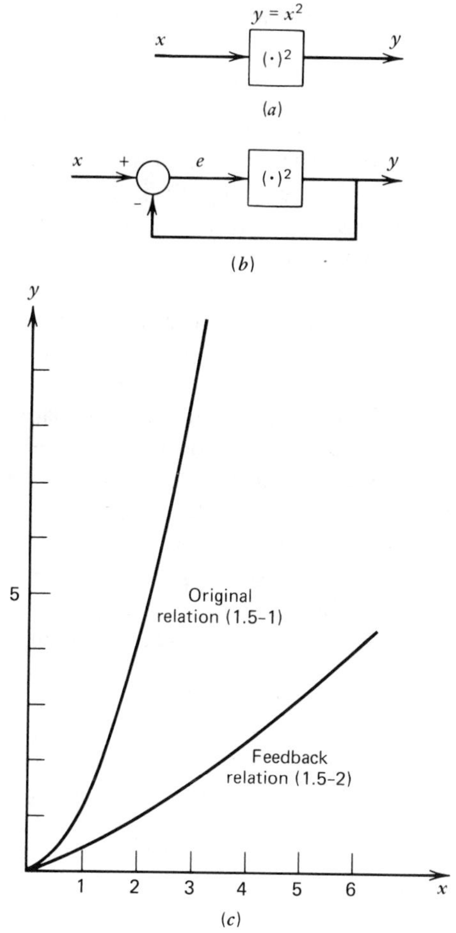


**Figure 1.13 Improvement of an element's linearity with feedback. (*a*) Original nonlinear element. (*b*) Original element with a feedback loop added. (*c*) Plots of the two input-output relations.**

Thus values of the design parameters that were optimal when the system was built might no longer give the desired performance. A vacuum-tube amplifier is an example of this effect. The heat generated causes the amplification factor to change in time. In light of this we should always investigate a prospective design to assess the sensitivity of its performance to uncertainties or variations in the system's parameters. Feedback can be used to improve the system's behavior in this respect.

We have seen that one purpose of the thermostat is to compensate for changes produced by variations in the outdoor environment – the system's "disturbances." This is another use for feedback, and systems that can maintain the output near its desired value in the presence of disturbances are said to have good *disturbance rejection.*

A system that has both good disturbance rejection and low sensitivity to parameter variations is said to be *robust*. We now illustrate how feedback can create or improve robustness.

## Parameter Sensitivity

First consider the reduction of parameter sensitivity. The element shown in Figure 1.14*a* has a proportional constant $G$ (called the *gain*). We wish the gain value to be $G = 10$ and select an element that has this nominal value. However, suppose that due to heat, wear, or poor construction, the actual value of $G$ can vary by ± 10%. In this case the input-output relation will be somewhere between $y = 9x$ and $y = 11x$, and this is considered unacceptable.

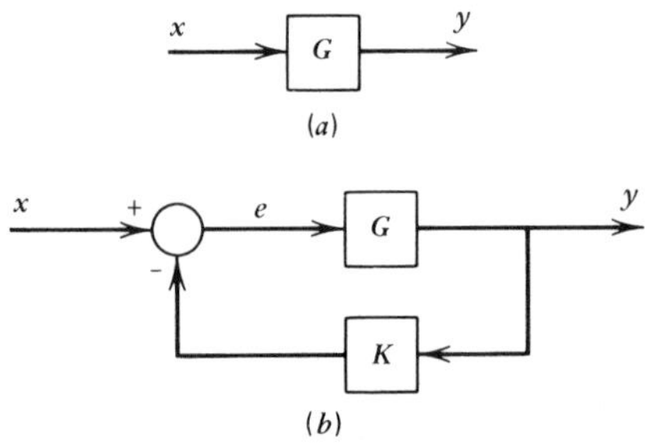


**Figure 1.14 Reduction of parameter sensitivity with feedback. (*a*) Original element with an uncertain gain $G$. (*b*) Insertion of a feedback element with a reliable gain $K$. If $KG \gg 1$, $y \cong \frac{1}{K}x$.**

To improve the situation we place a feedback loop with its own gain $K$ around the basic element $G$. Presumably the physical element we select to produce the gain $K$ will be relatively insensitive so that the value of $K$ will be constant and predictable. This arrangement is shown in Figure 1.14*b*. The governing relations are

$$y = Ge, \quad e = x - Ky$$

or

$$y = \frac{G}{1 + GK}x \tag{1.5-3}$$

Note that if we pick $G$ large enough so that $GK \gg 1$, then (1.5-3) becomes approximately

$$y \cong \frac{G}{GK}x = \frac{1}{K}x \tag{1.5-4}$$

The system's input-output relation becomes independent of $G$ as long as $GK$ is large! We now pick $K$ to obtain our desired input-output relation, here $y = 10x$. Thus $K$ must be $K = 0.1$, and we pick $G$ such that $0.1G \gg 1$ or $G \gg 10$.

Let us use $G = 1000$ and see what happens. From (1.5-3)

$$y = \frac{1000}{1 + 100}x = 9.901x$$

which is very close to the desired relation. Now if $G$ varies by ± 10% so that $G = 900$ and 1100, (1.5-3) gives

$$y = \frac{900}{1 + 90}x = 9.8901x$$

and

$$y = \frac{1100}{1 + 110}x = 9.91x$$

The sensitivity of the feedback system is much lower. This method of reducing the sensitivity is called *feedback compensation*.

A common engineering application of this approach is the stabilization of the gain of an electronic amplifier (Figure 1.15). Engineers working on this problem in the early part of the twentieth century discovered the principle of feedback compensation. The amplifier gain $A$ is large but is subject to some uncertainty. The feedback loop is created by a resistor. Part of the voltage drop across the resistor is used to raise the ground level at the amplifier input. From the resistor's voltage-current relation we obtain

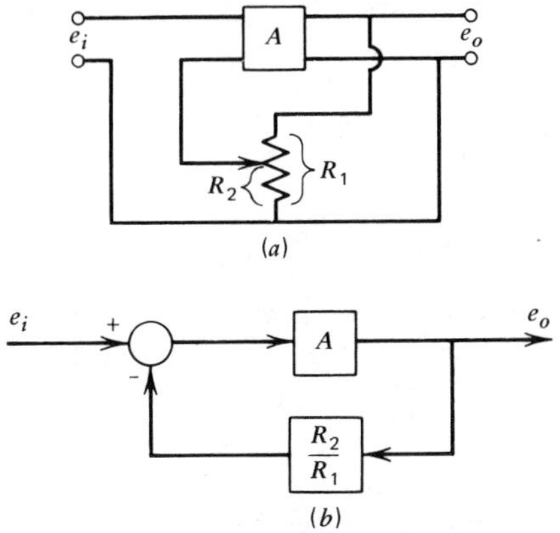


**Figure 1.15 Feedback compensation of an amplifier. (*a*) Circuit diagram. (*b*) Block diagram.**

$$e_o = A\left(e_i - e_o \frac{R_2}{R_1}\right) \qquad (1.5\text{-}5)$$

or

$$e_o = \frac{A}{1 + A\dfrac{R_2}{R_1}} e_i$$

If $AR_2/R_1 \gg 1$, then

$$e_o \cong \frac{R_1}{R_2} e_i \qquad (1.5\text{-}6)$$

Presumably the resistor values are sufficiently accurate and constant enough to allow the system gain $R_1/R_2$ to be reliable.

Another version of this application uses the *operational amplifier* (op amp). This is a voltage amplifier with a very large gain ($A = 10^5$ to $10^9$) that draws a negligible current. If resistors are placed in series and parallel around the op amp, as shown in Figure 1.16, the system's input-output relation will be

$$e_o \cong -\frac{R_2}{R_1} e_i \qquad (1.5\text{-}7)$$

This can be shown by writing the appropriate circuit equations. Since the current $i_3$ is negligible, the voltage $e_1$ is nearly zero, and $i_1 \cong i_2$. But

$$i_1 = \frac{e_i - e_1}{R_1}$$

$$i_2 = \frac{e_1 - e_o}{R_2}$$

Therefore

$$\frac{e_i - e_1}{R_1} = \frac{e_1 - e_o}{R_2}$$

Since $e_1 \cong 0$, it follows that

$$\frac{e_i}{R_1} \cong -\frac{e_o}{R_2}$$

which is equivalent to (1.5-7).

Op amps appear in many designs, and we will see more of them.

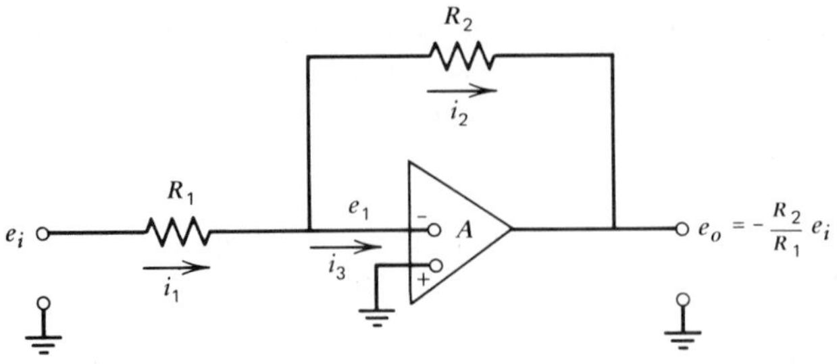


**Figure 1.16** Op amp multiplier. Note the sign inversion.

## Disturbance Rejection

Consider the system shown in Figure 1.17*a*. The disturbance is $u$, and the desired input-output relation between $y$ and $x$ is $y = 10x$. However, when $u \neq 0$, this relation is not obtained because

$$y = 5(2x - u) = 10x - 5u$$

This can be remedied by introducing a feedback loop and two gain elements, $B$ and $K$, as shown in Figure 1.17*b*.

We can use the superposition principle to find the output $y$ as a function of $x$ and $u$. First set $u = 0$ and solve for $y$ as a function of $x$.

$$y\Big|_{u=0} = \frac{10B}{1 + 10BK} x$$

Now replace $u$ and set $x = 0$. Solve for $y$ in terms of $u$ to obtain

$$y\Big|_{x=0} = \frac{-5}{1 + 10BK} u$$

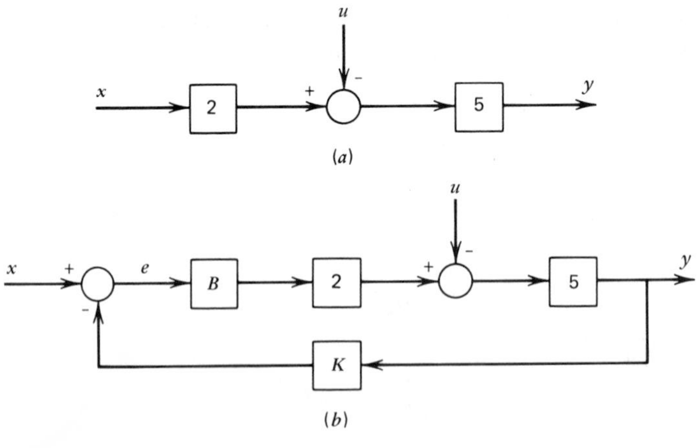


**Figure 1.17** Use of feedback for disturbance rejection. (*a*) Original open-loop system. (*b*) Feedback system.

Invoking superposition we obtain the general expression for $y$ by adding the above results.

$$y = \frac{10B}{1 + 10BK}x - \frac{5}{1 + 10BK}u \qquad (1.5\text{-}8)$$

We desire that $u$ have no effect on $y$, but this cannot be accomplished with finite values of $B$ and $K$. Therefore we back off somewhat and require only that no more than 10% of the value of $u$ show up in the output $y$; that is, we require that

$$\left| \frac{5}{1 + 10BK}u \right| \leqslant |0.1u|$$

This is satisfied if $5/(1 + 10BK) = 0.1$. From the desired relation $y = 10x$ we also have

$$\frac{10B}{1 + 10BK} = 10$$

The last two conditions give $B = 50$ and $K = 49/500$. This design meets the specifications and can be implemented if the values of $B$ and $K$ can be obtained physically and are reliable. The price paid for this improvement is that we have a more expensive system (due to the added elements $B$ and $K$) and a possibly less reliable system (since there are more elements to fail).

## The Importance of Dynamic Models for Feedback Systems

We have used static systems to illustrate the properties of feedback, but its most important applications occur in dynamic systems. Consider the motor operating curves shown in Figure 1.11. Suppose that the motor speed, voltage, and load torque are $\omega_2$, $v_2$, and $T_2$ initially, and that we wish to maintain the speed at $\omega_2$. If the load torque increases, the speed will decrease. However, the operating curves do not tell us *how long* it will take for the new speed to be established. The curves represent only the steady-state behavior of the system as a static element. The inertia of the motor obviously prevents the speed from changing instantaneously from the value $\omega_2$ to its new steady-state value.

In order to keep the speed near its desired value we would use a controller. A block diagram of the general situation is shown in Figure 1.18*a*. The controller would sense that the speed had decreased and would increase the motor voltage. Again, the curves give no information about the time it would take for the speed to return to its desired value $\omega_2$.

At this point it is not clear how we should design the controller to act. For example, should it change the voltage in proportion to the difference between the desired and actual speeds? Or should the voltage change be proportional to the *rate* of change of the speed difference? The fact that the motor and its load have inertia suggests that the time behavior of the speed depends to a great extent on the characteristics of the controller. For example, suppose we make the voltage change proportional to the *integral* of the speed difference such that

$$v = v_2 + K\int_0^t e\,dt \qquad (1.5\text{-}9)$$

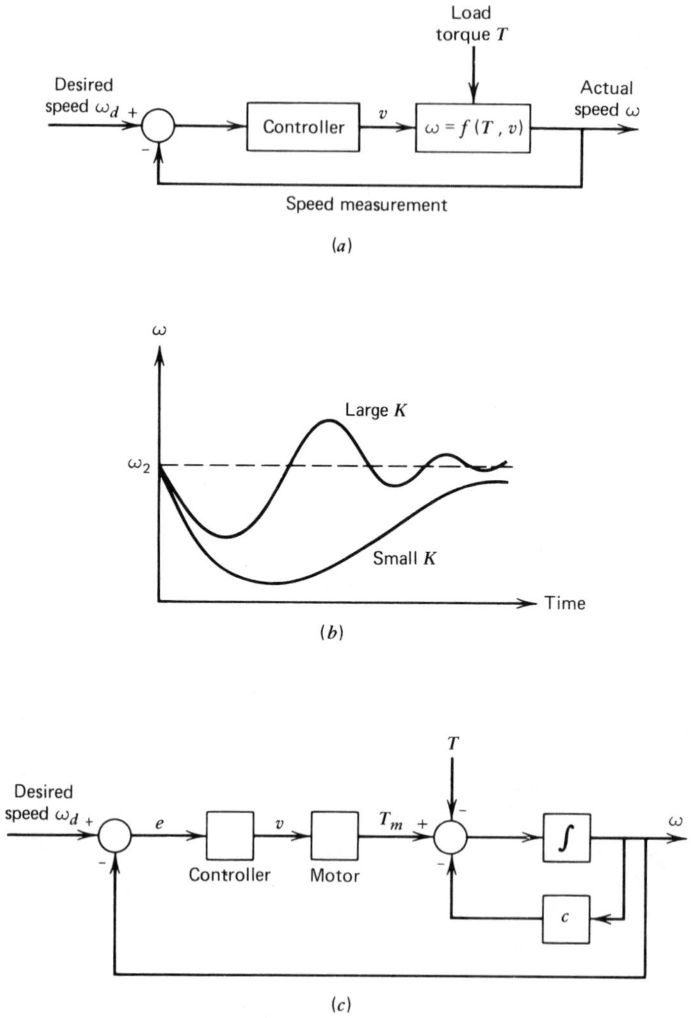


**Figure 1.18 Motor speed control. (*a*) General block diagram. (*b*) Transient behavior of speed for different values of the controller gain *K*. (*c*) Block diagram using a dynamic model of the load.**

where $e$ is the speed error $\omega_2 - \omega$. We will see in Chapter Six that no steady-state error due to a constant load-torque disturbance will exist if this control scheme is used. This is because the voltage will not stop increasing until the speed difference $e$ becomes zero. However, if the proportionality constant $K$ is made too large, the controller can "overcompensate." The result is an oscillation in speed about the desired value. This effect is shown in Figure 1.18*b*. The initial time in that plot is the time at which the load torque changes. On the other hand, if $K$ is made smaller, the oscillation does not occur, but the time to return $\omega$ to $\omega_2$ might be very long.

Obviously we will need a dynamic model that includes the effect of the inertia in order to design the controller properly. Such a model can be obtained by applying Newton's law. We assume that the motor and load inertias can be lumped into one inertia $I$. The motor torque is $T_m$, and this is resisted by a friction torque $T_f$, which can often be modeled as proportional to the speed; that is, $T_f = c\omega$. By summing the torques on the inertia we obtain the dynamic model for the load speed.

$$I\frac{d\omega}{dt} = T_m - T - c\omega \tag{1.5-10}$$

The block diagram of the system with the controller and motor is shown in Figure 1.18$c$. When used with a motor model that relates motor torque to motor voltage, the preceding equation is often sufficient to design the controller – for example, to pick the proper value of $K$ in (1.5-9).

The linearizing property of the feedback loop in the controller often allows us to model the system as a linear one. In addition, the reduction of the system's parameter sensitivity means that a lumped-parameter, low-order dynamic model is often satisfactory for the purpose of designing feedback systems. The model usually cannot be static, but must describe at least the dominant dynamic behavior of the system. We take up procedures for developing such models in the next chapter.

## PROBLEMS

1.1 What is the causal relation for the following elements with the given inputs and outputs?

(a) A capacitor (charge as input; voltage as output).

(b) An inertia (torque as input; angular acceleration as output).

(c) Angular acceleration as input; angular velocity as output.

(d) A water tank with vertical sides (water volume as input; water height as output).

(e) The heat energy stored in a body as input; the body temperature as output.

1.2 Draw a block diagram for the following models. The inputs are $u$ and $v$; the output is $y$. The variables $x$ and $e$ are internal variables. Show these on the diagram.

(a)
$$\begin{aligned} y &= 5x \\ x &= v + 3e - 4y \\ e &= u - 2y \end{aligned}$$

(b)
$$\begin{aligned} \dot{y} &= 6x \\ \dot{e} &= u - 2e \\ x &= 5v + 3y + e \end{aligned}$$

1.3 Obtain the input-output relations for each of the diagrams shown in Figure P1.3 The inputs are $u$ and $v$; the output is $y$.

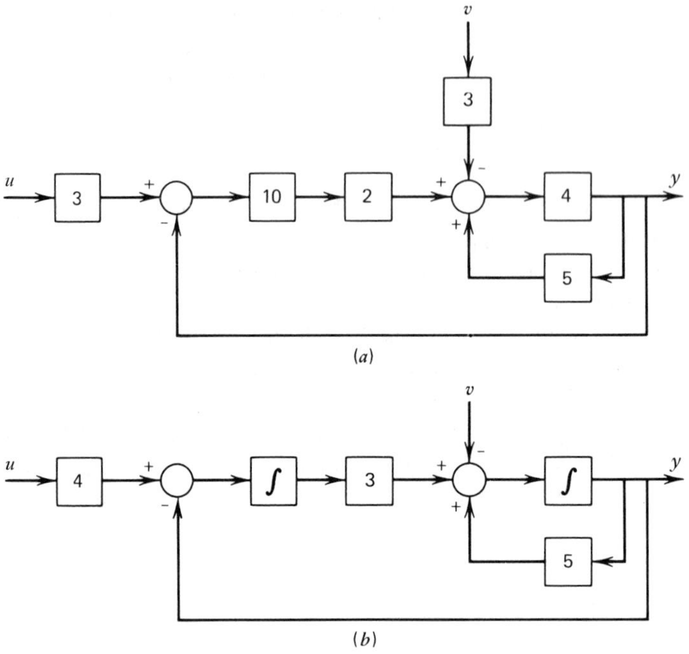


Figure P1.3

1.4 Water-level controllers represent the earliest examples of control systems. A simple version using a float and lever is shown in Figure P1.4.

(a) Discuss the system's operation. How can we adjust the water level that the system will maintain?

(b) Draw the block diagram with the desired level as the command input, the actual level as the output, and the change in water supply pressure as the disturbance.

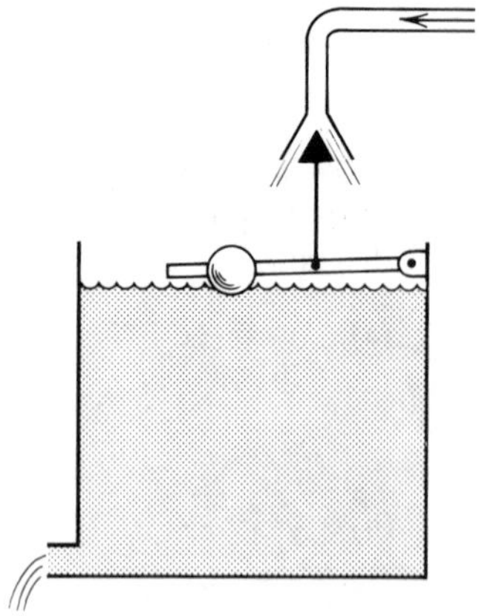

Figure P1.4

**1.5** A person attempting to balance a stick on end in the palm of a hand constitutes a control system. Draw the block diagram of the system. Is more than one measurement involved?

**1.6** A very long wire can have appreciable resistance. Should it be modeled as a distributed-parameter element?

**1.7** A certain cantilever beam has considerable mass. Should it be modeled as a distributed-parameter element? Discuss how an approximate lumped-parameter model might be developed.

**1.8** Is the following model nonlinear? Explain.

$$\frac{dx}{dt} = -3x + f, \qquad f = \begin{cases} +1 & \text{if } x \geqslant 0 \\ -1 & \text{if } x < 0 \end{cases}$$

**1.9** Consider the savings growth model (1.3-7).

**(a)** Suppose that $x(0) = \$1.00$. Find $x(5)$, the amount of money at the end of five years.

**(b)** Generalize the results of (a) to find an expression for $x(k)$ in terms of $x(0)$ for any integer $k$.

**1.10** Obtain a linearized expression for the following functions, valid near the given reference values.

**(a)** $w = \cos y, \quad y_o = 0$

**(b)** $w = \cos y, \quad y_o = \pi/4$

**(c)** $w = e^{3y}, \quad y_o = 1$

**(d)** $w = y_1^2 \sin y_2, \quad y_{1o} = 1, \quad y_{2o} = \pi/4$

**(e)** $w = y_1/y_2, \quad y_{1o} = 1, \quad y_{2o} = 3$

**1.11** The area $A$ of a rectangle is $A = y_1 y_2$, where $y_1$ and $y_2$ are the lengths of the sides.

**(a)** Obtain a linearized expression for $A$ if $y_{1o} = 2, y_{2o} = 5$.

**(b)** Give a geometric interpretation of the error in the linearization.

**1.12** Use the curves shown in Figure P1.12 to obtain a linearized expression for $w = f(y_1, y_2)$, valid near the point $y_{1o} = 1, y_{2o} = 10$.

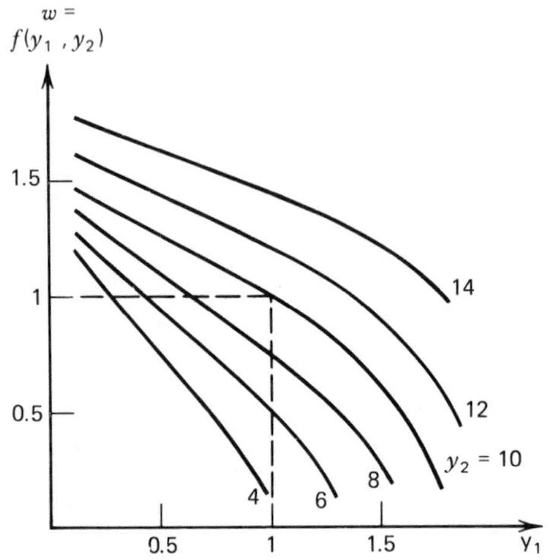


Figure P1.12

**1.13** (a) Show that the two systems in Figure P1.13 have the same input-output relation when the gain $K$ is 100.
(b) If the gain $K$ is subject to a ± 10% uncertainty, which system is the *least* sensitive?

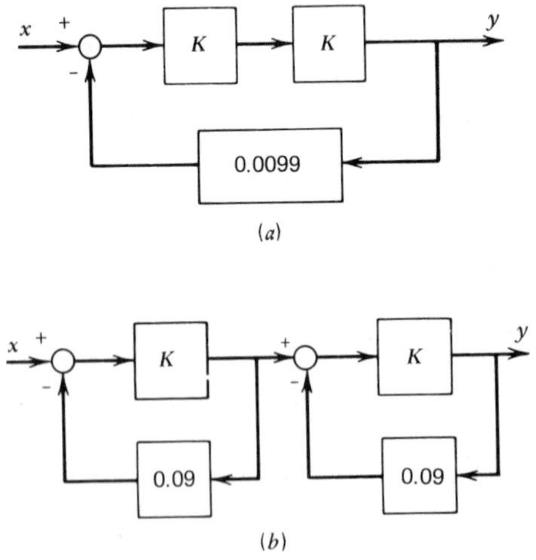


Figure P1.13

1.14 The following are common examples of control systems. Discuss their operation. Do they employ feedback?

(a) Toaster.
(b) Washing machine.
(c) Engine camshaft.
(d) Engine cooling system.
(e) Carburetor.
(f) Traffic light.

1.15 Explain the feedback action in the following systems.

(a) The law of supply and demand in economic systems.
(b) Temperature control in the human body.
(c) Predator-prey interactions.

1.16 This story circulates in several forms. A worker at a factory was in charge of activating the noon whistle at the plant. Being very conscientious and trusting, every day he phoned the research lab at the nearby university to set his clock according to their time. Eventually he became curious and asked the university researcher how he managed to keep his clock accurate. "Why, we set our clock by the noon whistle at the local factory" came the reply.

Is this a feedback process? If the university clock gains 2 minutes per day, what is the error in the timing of the "noon" whistle at the end of 5 days?

# CHAPTER TWO

# Modeling of Dynamic Systems

In Chapter One we discussed some approaches to modeling or describing the essential characteristics of an object or process. What might be an essential characteristic for your purposes might not be essential for an investigator whose goals differ from yours. Up to this point most of our models have not contained a description of the time-dependent behavior of the system being analyzed. For many applications, such a description offered by static or zero-order models is adequate. For the purposes of sizing the members of many structures, such as bridges or buildings, the methods of static force analysis are sufficient. However, if the structure is subjected to strong wind gusts or is located in an earthquake-prone area, it will be acted on by time-varying forces that can create stresses different from those predicted by the static model. The classic and widely used example of this effect is the Takoma Narrows Bridge, which collapsed due to the alternating aerodynamic forces produced by shedding vortices. In the past few years our knowledge of the behavior of soils subjected to the periodic forces generated by earthquakes has resulted in improved models that predict building motions under these conditions.

While these examples of structural collapse are vivid and should be kept in mind, the omission of time-varying aspects in a model rarely results in dramatic failures of engineering designs. Such omissions usually have more subtle consequences, and the engineer who chooses to ignore dynamic effects frequently misses the opportunity for an improved design. An example of this occurs in the design of large buildings and their heating, ventilating, and air-conditioning (HVAC) systems. The steady-state heat-transfer analysis traditionally used for this purpose is known to result in over-sized equipment with increased energy consumption (Reference 1). Criticism of such methods might come easy now, but we should view their use in the historical context of evolving design techniques. While engineers in the past might have been aware of the problems in neglecting dynamic models, they did not have the help of the computers and advanced mathematics frequently needed to analyze such models. Today we are much more fortunate.

Once the decision is made to incorporate time-dependent effects in a model, a hierarchy of time scales must be established. In HVAC design the modeler could be interested in the transient effects due to to the daily solar cycle, or might wish to analyze system performance on a seasonal basis. In the latter case the daily transients might be insignificant, and their inclusion would only result in a cluttered, difficult-to-analyze model. However, the design of a temperature-control system more likely requires a model of the daily cycle. In order to make these modeling decisions it sometimes helps to construct alternate models and to analyze their behavior under typical conditions. With the insight gained from this, a refined model is developed for the final design analysis. Modeling skill is enhanced by studying examples from many different fields, even including nontechnical areas such as biology and economics.

This chapter is an overview of the approaches available for developing a dynamic model. An understanding of its material will enable the reader to feel comfortable with most models encountered at the level of analysis of later chapters. The practicing engineer frequently is part of a team approach to model development for complex systems. In such cases deeper knowledge of the devices or processes is necessary, and the reader is directed to the references at the end of the chapter or to the appropriate specialist on the team. For example, on an aircraft design team, the propulsion, structural, and aerodynamics groups concentrate on their particular tasks, while the systems engineers are charged with team coordination and with the design of the automatic control systems to ensure that all subsystems operate compatibly. The type of modeling ability needed for this task is different from that needed by the other specialists. In this chapter an energy-based viewpoint is used to develop this ability. The concepts of effort and flow are employed with generalized resistance, capacitance, and inductance to develop a structure for models. The approach is then applied in turn to mechanical, electrical, fluid (compressible and incompressible), and thermal systems. The chapter concludes with a discussion of how these methods can be extended to nontechnical systems in biology, ecology, and socioeconomics.

## Systems of Units

None of the modeling techniques we will develop depend on a specific system of units. In order to make quantitative statements based on the resulting models, a set of units must be employed. Most engineering work in the United States has been based on the British Engineering system. However, the metric Systeme Internationale (the SI system) is rapidly becoming the worldwide standard. Until the changeover is complete, engineers in the United States will have to be familiar with both systems. In our examples we will use SI and British Engineering units in the hope that the student will become comfortable with both. Other systems are in use, such as the mks (meter-kilogram-second) and cgs (centimeter-gram-second) metric systems, and the British system, in which the mass unit is a pound. We will not use these because the British Engineering and the SI systems are the most common in engineering applications. We now briefly summarize these two systems.

The British Engineering system is an example of a "gravitational" system. The primary variable is force, and the unit of mass is derived from Newton's second law. The pound is selected as the unit of force and the foot and second as units of length and time, respectively. From Newton's law, force equals mass times acceleration, so the unit of mass must be

$$\text{mass} = \frac{\text{force}}{\text{acceleration}} = \frac{\text{pound}}{\text{foot}/(\text{second})^2}$$

This mass unit is named the *slug*. Energy has the dimensions of mechanical work; namely, force times displacement. Therefore, the unit of energy in this system is the foot-pound (ft-lb). Another energy unit in common use for historical reasons is the British thermal unit (Btu). The relation between the two is given in Table 2.1. Power is the rate of change of energy with time, and a common unit is *horsepower*. Finally, temperature in this system can be expressed in degrees Fahrenheit, or in absolute units, degrees Rankine.

TABLE 2.1 Systems of Units

| Quantity | Usual Symbol | British Engineering | SI Metric |
|---|---|---|---|
| Time | $t$ | second (sec) | second (sec or s) |
| Length | $x$ | foot (ft) | meter (m) |
| Force | $f$ | pound (lb) | newton (N) $\left(1\text{ N} = 1\,\dfrac{\text{kg-m}}{\text{sec}^2}\right)$ |
| Mass | $m$ | slug $\left(1\text{ slug} = 1\,\dfrac{\text{lb-sec}^2}{\text{ft}}\right)$ | kilogram (kg) |
| Energy[a] | $W$ or $Q_h$ | foot-pound (ft-lb) or British thermal unit (Btu) (1 Btu = 778 ft-lb) | newton-meter (N-m) or joule (J) (1 J = 1 N-m) |
| Power | $P$ | $\dfrac{\text{ft-lb}}{\text{sec}}$ or horsepower (hp) $\left(1\text{ hp} = 550\,\dfrac{\text{ft-lb}}{\text{sec}}\right)$ | watt (W) $\left(1\text{ W} = 1\,\dfrac{\text{N-m}}{\text{sec}}\right)$ |
| Temperature | $T$ | degrees Fahrenheit °F or degrees Rankine °R $T\text{°F} = (T + 460)\text{°R}$ | degrees Celsius °C degrees Kelvin K $T\text{°C} = (T + 273)\text{ K}$ |

[a] Energy in the form of mechanical work is usually denoted by $W$. Heat energy is typically represented by $Q_h$.

The SI metric system is an "absolute" system. This means that the mass is chosen as the primary variable, and the force unit is derived from Newton's law. Selecting the meter and the second as the length and time units, and the kilogram as the mass unit, the derived force unit is the newton. The common energy unit is the newton-meter, also called the joule, while the power unit is the joule/second, or watt. The difference between the boiling and freezing temperatures of water is 100 degrees Celsius, with zero degrees being the freezing point. The absolute temperature units are degrees Kelvin.

Table 2.2 lists the most commonly needed factors for converting between the

TABLE 2.2 Conversion Factors

| | | |
|---|---|---|
| Length | 1 m = 3.281 ft | 1 ft = 0.3048 m |
| Force | 1 N = 0.2248 lb | 1 lb = 4.4482 N |
| Mass | 1 kg = 0.06852 slug | 1 slug = 14.594 kg |
| Energy | 1 J = 0.7376 ft-lb | 1 ft-lb = 1.3557 J |
| Power | 1 W = $1.341 \times 10^{-3}$ hp | 1 hp = 745.7 W |
| Temperature | $T°\text{C} = \frac{5}{9}(T°\text{F} - 32)$<br>$T°\text{F} = \frac{9}{5}T°\text{C} + 32$ | |

British Engineering and the SI systems. For example, if the elastic constant of a certain spring is given as 50 lb/ft, its SI equivalent is found as

$$50\,\frac{\text{lb}}{\text{ft}} = 50\,\frac{4.4482\text{ N}}{0.3048\text{ m}} = 729.7\,\frac{\text{N}}{\text{m}}$$

## 2.1 MODEL DEVELOPMENT USING INTEGRAL CAUSALITY

The dynamics of a physical system result from the transfer, loss, and storage of mass or energy. Thus one way to develop a model of a system is to identify the flow paths and storage compartments of mass or energy and to describe quantitatively how these paths and compartments are connected.

### Rate, Quantity, and Integral Causality

*Rate* is the general term for one of the two variables on which we base our model. The other variable is termed *effort*, discussed shortly.

Rates are the natural choices to employ in a model intended to describe the time-dependent behavior of a system. For example, suppose that the tank shown in Figure 2.1 contains water and that we desire a model of the water mass as a function of time. We will employ the principle of conservation of mass to develop the model. This states that the mass flow rate into the tank, minus the outflow rate, equals the rate at which mass is stored in the tank; that is,

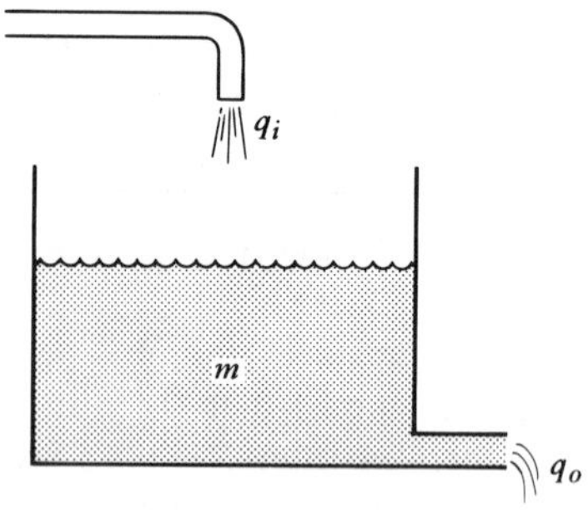


**Figure 2.1 Liquid level system.**

$$q_i - q_o = \frac{dm}{dt} \tag{2.1-1}$$

where $m$ is the water mass in the tank. The rate variable here is mass flow rate. In general, the rate variable is used to express the appropriate physical principle from which the model will evolve.

The choice of which physical principle to use, and therefore the appropriate rate variable, depends on the purpose for which the model is needed. Certainly the principles of thermodynamics also apply to the tank of water. Why not base the model instead on the principle of conservation of energy, and choose heat energy flow rate as the rate, since it describes the thermodynamics of the tank water? The answer is found in the purpose of the model. Our stated purpose is to develop a model of the water mass, and unless the water's heat energy affects the mass of water in the tank somehow, it has no role to play in our model. If, on the other hand, we needed a model of the water's temperature, then heat-energy flow rate would be an appropriate choice.

In addition to conservation of mass and energy, we frequently use other physical laws. One is conservation of charge in electrical circuits. Therefore, charge flow rate is a rate variable for models of such circuits. Another principle is Newton's second law of motion, which relates the impulse acting on a mass to the change in momentum, which is the product of mass and velocity. When using the second law, the appropriate rate is velocity.

We can classify a system as electrical, mechanical, fluid, or thermal, depending on what aspect of the system we are trying to model and what physical principles are used. Table 2.3 lists the rate variables for these system types. For example, a system is classified as "mechanical" if we are primarily interested in its impulse-momentum characteristics. This is a loose characterization intended to provide a structure for this chapter. Certainly fluids are subject to Newton's laws, but here when we describe a system as "fluid," we mean that we will be concerned not with the details of the fluid forces involved, but with the system dynamics due to the conservation of mass principle. (For an incompressible fluid such as a liquid, specification of its volume flow rate is equivalent to specification of its mass flow rate since the density is constant. Therefore, for incompressible fluids volume flow rate is sometimes used as rate.)

*Quantity* is the term used to describe the accumulation of rate over time. Rate and quantity are related as cause and effect through the principle of integral causality as discussed in Chapter One. In Figure 2.1 we can choose the mass flow rate as the rate variable. Then the quantity variable is the water mass, the time integral of the net

**TABLE 2.3 Primary System Variables and Their Usual Symbols**

| Type | Rate $r$ | Quantity $Q = \int r\,dt$ | Effort $e$ | Flux $\phi = \int e\,dt$ |
|---|---|---|---|---|
| Electrical | Current $i$ | Charge $Q$ | Voltage $v$ | Flux $\phi$ |
| Mechanical (translation) | Velocity $v$ | Displacement $x$ | Force $f$ | Impulse $M_x$ |
| Mechanical (rotation) | Angular velocity $\omega$ | Angular displacement $\theta$ | Torque $T$ | Angular impulse $M_\theta$ |
| Fluid (incompressible)[a] | Mass flow rate $q_m$ or volume flow rate $q$ | Mass $Q$ (or $m$) or volume $V$ | Pressure $p$ | None |
| Fluid (compressible) | Mass flow rate $q_m$ | Mass $Q$ (or $m$) | Pressure $p$ | None |
| Thermal | Heat flow rate $q_h$ | Heat energy $Q_h$ | Temperature $T$ | None |

[a] Mass and volume are readily interchangeable for incompressible fluids; frequently the symbol $Q$ denotes volume and $q$ mass flow rate.

mass flow rate. For mass conservation we see that at any time $t$, the mass in the tank is the mass initially present at $t = 0$ plus the mass added, minus the mass drained out. Thus

$$m(t) = m(0) + \int_0^t q_i \, dt - \int_0^t q_o \, dt$$

$$= m(0) + \int_0^t (q_i - q_o) \, dt \tag{2.1-2}$$

This could also be obtained by integrating both sides of (2.1-1). The present mass $m(t)$ depends therefore on the accumulation over time of the *net* rate (inflow minus outflow rate).

In the general case, quantity results from the time integration of the net rate. Quantity is the effect, rate is the cause, and the special way in which they are related is termed *integral causality*. It is a statement of the fact that the present condition of a physical system depends on the accumulation of what happened to it in the past. In general, if we denote quantity by $Q$ and net rate by $r$, integral causality states that

$$Q(t) = Q(0) + \int_0^t r \, dt \tag{2.1-3}$$

The quantity variable is easily determined once the rate has been selected. Table 2.3 summarizes the quantity variables and nomenclature for several system types. In thermal systems, quantity is heat energy. The increase in heat energy in a body is the time integral of the net heat flow rate. In electrical systems the quantity variable is charge, the time integral of current. Displacement is the quantity variable in mechanical systems, since it is the time integral of velocity. Note that mechanical systems in rotation are similar in principle to those in translation. Angular displacement and angular velocity play the roles of quantity and rate.

## Effort

Having noted that the outflow rate $q_o$ helps to cause the change in tank water mass in Figure 2.1, we now ask: what causes $q_o$? Obviously $q_o$ is caused by the hydrostatic pressure due to the weight of the water or, more precisely, by the pressure difference across the outlet. The hydrostatic pressure is an example of an effort variable. The term *effort* is used to describe the cause that is due to the storage of potential energy.

Force is the effort variable in mechanical systems. It is generated when a spring is compressed or when a mass is elevated against gravity. The hydrostatic pressure is an example of the latter effect; pressure is the effort variable in fluid systems. When allowed to act, force (or pressure) is capable of producing motion from the stored potential energy.

The effort variables are summarized in Table 2.3. Temperature difference is the effort variable that causes heat energy to flow from the higher to the lower temperature. Similarly, a voltage difference in an electrical system produces a current; voltage is an effort variable in such systems.

## Flux

We have seen that the relation between quantity and rate is one of integral causality. A similar causal relation is that between effort and *flux*, the time integral of effort (Table 2.3). The symbol for flux in general is $\phi$, and the relation is

$$\phi(t) = \phi(0) + \int_0^t e\,dt \tag{2.1-4}$$

Flux in mechanical systems is usually called *impulse* and is the time integral of force. The impulse $M_x$ is given in terms of the force $f$ by

$$M_x(t) = M_x(0) + \int_0^t f\,dt \tag{2.1-5}$$

But Newton's second law states that the change in momentum equals the applied impulse, or

$$m[v(t) - v(0)] = M_x(t) \tag{2.1-6}$$

where $m$ and $v$ are the mass and velocity. Therefore,

$$m[v(t) - v(0)] = M_x(0) + \int_0^t f\,dt \tag{2.1-7}$$

Differentiation with respect to time yields

$$m\frac{dv}{dt} = f \tag{2.1-8}$$

This corresponds to an alternative statement of Newton's law – that the product of mass and acceleration equals the applied force.

Flux in electrical circuits results from an accumulation of voltage over time. The change in current through an electrical inductor is related to the flux by

$$i(t) - i(0) = \frac{1}{L}\,\phi(t) \tag{2.1-9}$$

where $L$ is the inductance. The flux is related to the voltage across the inductor by

$$\phi(t) = \phi(0) + \int_0^t v\,dt \tag{2.1-10}$$

Combining these relations and differentiating gives

$$L\frac{di}{dt} = v \tag{2.1-11}$$

which is the commonly used voltage-current relation for an inductor.

In thermal and fluid systems no elements have a flux variable, unless we choose to consider fluid systems as a subcase of mechanical systems.

## Energy Storage and Dissipation

With the exception of thermal systems, the product of the effort and rate variables $e$ and $r$ is power $P$. For example, electrical power is the product of voltage and current; mechanical power is the product of force and velocity. So, except for thermal systems,*

$$P = er \tag{2.1-12}$$

Since energy is the time integral of power, we have

$$E = \int P\,dt = \int er\,dt \tag{2.1-13}$$

For example, the force-deflection relation of a linear mechanical spring is $f = kx$. Hence, with $e = f$ and $r = dx/dt$, (2.1-13) gives the expression for the potential energy $PE$ stored in the spring.

$$PE = \int f\frac{dx}{dt}\,dt = k\int x\,dx = \tfrac{1}{2}kx^2 \tag{2.1-14}$$

Similarly, the expression for the kinetic energy $KE$ stored in a mass $m$ moving with velocity $v$ can also be found from (2.1-13). With $e = f$, $r = v$, and $f = m(dv/dt)$ from Newton's law, we obtain

$$KE = \int fv\,dt = \int mv\frac{dv}{dt}\,dt = \tfrac{1}{2}mv^2 \tag{2.1-15}$$

The power expression (2.1-12) can also be used to compute the power dissipation in a system. For an electrical resistor, the effort $e$ is the applied voltage $v$, and the rate is the current $i$. So the power dissipated by a resistor is

$$P = iv = i^2R \tag{2.1-16}$$

where $v = iR$.

Expressions for potential and kinetic energy storage and power dissipation can be obtained for other system types in a similar way.

## State Variables

Because the dynamics of energetic systems are due to the dissipation and storage of energy, it is natural to construct a system model in terms of variables that describe these actions. The variables that describe a system's condition or *state* are its *state variables*. The quantity and flux variables are common choices for this role since they describe the potential and kinetic energy storage, but other choices are possible.

In general, the number of state variables required equals the number of energy storage compartments in the system. This number is the *order* of the system. When we speak of a first-order system, we mean that it has one way of storing energy; a second-order system has two energy compartments, and so on.

It will become apparent that the rate and effort variables of one compartment

---

*We could have chosen entropy flow rate as the rate variable in thermal systems, since its product with temperature does give power. However, this choice has rarely been used in practice.

must frequently be expressed as functions of the rate and effort variables of other compartments. For this reason the integrations in (2.1-3) and (2.1-4) cannot be computed directly, and thus we usually express the relation for each compartment as a differential equation. These equations take the following forms. Differentiating (2.1-3) gives

$$\frac{dQ}{dt} = r \tag{2.1-17}$$

Similarly (2.1-4) gives

$$\frac{d\phi}{dt} = e \tag{2.1-18}$$

The terms on the right-hand sides are the net rate and the net effort across the element. These terms are related to the other state variables of the system and to the system's input variables.

It follows that a first-order system is described by a single first-order differential equation (the highest derivative is first-order). A second-order system is described by a set of two coupled first-order equations (these can be combined to give a single equation that is second-order). The concept generalizes to higher-order systems. These coupled first-order equations are the *state equations.*

**Figure 2.2 Mass-spring system.**

For example, if a mass $m$ is connected to a support by a spring as in Figure 2.2, (2.1-17) states that

$$\frac{dx}{dt} = v \tag{2.1-19}$$

and (2.1-18) gives

$$\frac{dM_x}{dt} = f - f_s \tag{2.1-20}$$

where $f_s$ is the spring force. The force-deflection relation for the spring is

$$f_s = kx \tag{2.1-21}$$

and for the mass, from (2.1-6),

$$mv = M_x + mv(0) \tag{2.1-22}$$

Equations (2.1-20), (2.1-21), and (2.1-22) give

$$m\frac{dv}{dt} = f - kx \tag{2.1-23}$$

This equation and (2.1-19) are the state equations of the system in terms of the state variables $x$ and $v$. The system model is second-order because it contains two energy storage modes, one potential and one kinetic. The spring relation $f_s = kx$ serves to connect the describing equations of each mode. From (2.1-14) the state variable $x$ is seen to indicate the amount of potential energy stored. The state variable $v$ describes the kinetic energy storage [see (2.1-15)].

The choice of state variables for any given system is not unique and not always obvious. Before this choice can be made, we must first make simplifying assumptions based on the goals of our analysis and on the limitations of available information. Fortunately large classes of engineering problems can be solved with similar assumptions and approaches. Thus the engineer frequently can call upon previously verified models for many design applications. We will develop some of these in the sections to follow.

## 2.2 THE CONSTITUTIVE RELATIONS: RESISTANCE, CAPACITANCE, AND INDUCTANCE

The dynamic behavior of physical systems results from the interchange of energy between potential and kinetic forms and from the loss of energy through dissipation. For example, the motion in a mass-spring system is determined by the energy flow between the spring's potential energy and the mass's kinetic energy, and by the energy loss because of any friction present. Similarly, electrical capacitances and inductances form potential and kinetic energy storage compartments, while the resistances dissipate energy.

Electrical resistance, capacitance, and inductance are examples of *constitutive relations*, which are algebraic (static) relations usually developed at least partly by empirical methods. For example, the usual relation for an electrical resistor is

$$v = iR \tag{2.2-1}$$

where $v$ is the voltage difference across the resistor, $i$ is the current, and $R$ is the resistance. This equation arises not from the application of a basic physical law, but from the results of many observations (measurements) revealing that the voltage is directly proportional to current for typical resistor construction and materials.

The concepts of electrical resistance, capacitance, and inductance can be extended to develop general constitutive relations for other system types. These general relations provide links that couple the state equations to one another. We will see that the spring relation $f_s = kx$ is an example of a capacitance-type relation. In Section 2.1 it served to couple the differential equation (2.1-19) for the quantity variable $x$ to that for the rate variable $v$ (2.1-23).

### Resistance

Effort in physical systems is never completely free to act; it is always opposed in some manner. One way this occurs is with an element having the property of *resistance*. The friction of the fluid against the outlet pipe wall in Figure 2.3*a* resists the attempt of the pressure to force the fluid through the pipe, just as an electrical resistor opposes the current produced by a voltage difference. The magnitude of the frictional force depends on the fluid properties and on the flow conditions within the pipe. Thus for a given fluid, if a constant pressure difference is maintained across the pipe ends, the mass flow rate produced will depend on whether the flow condition is "smooth" (laminar) or "rough" (turbulent). If the pressure difference is changed, the flow rate will also change (Figure 2.3*b*). For the laminar case, the slope of the curve is $1/R$, where $R$ is the "laminar resistance." Its calculation is treated in Section 2.5.

Resistance in mechanical and fluid systems results from friction. Viscous friction

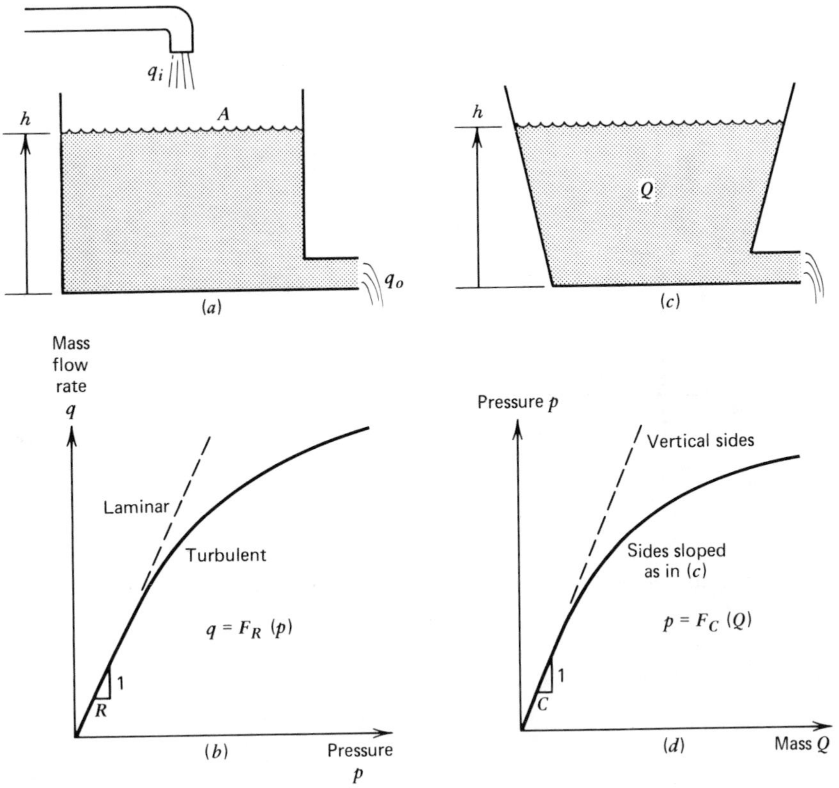


**Figure 2.3 Capacitance and resistance relations in a liquid system. (*a*) Tank with vertical sides (linear capacitance relation). (*b*) Resistance relations for laminar and turbulent flow. (*c*) Tank with sloped sides (nonlinear capacitance relation). (*d*) Capacitance relations for tanks shown in (*a*) and (*c*).**

in a damper or dashpot results from the fluid's viscosity (Figure 2.4). An example of such an element is the shock absorber on a car. The dashpot exerts a force $f$ proportional to the velocity difference $v$ across the element's endpoints, for relatively small velocities. Thus

$$f = cv \tag{2.2-2}$$

where $c$ is a constant, the *damping coefficient*.

The damper force varies as the square of the velocity difference for larger velocities. This phenomenon results from the difference between laminar and turbulent flow through the orifice as the piston moves. The square of the velocity does not change sign when the velocity difference is negative, but the resisting force does. Therefore the square law relation for the turbulent case must be written as

$$f = cv|v| \tag{2.2-3}$$

Note that the linear curve in Figure 2.4*b* is tangent to the nonlinear curve when the velocity difference is zero. That is, the linear relation is an approximation to the nonlinear one for small velocities.

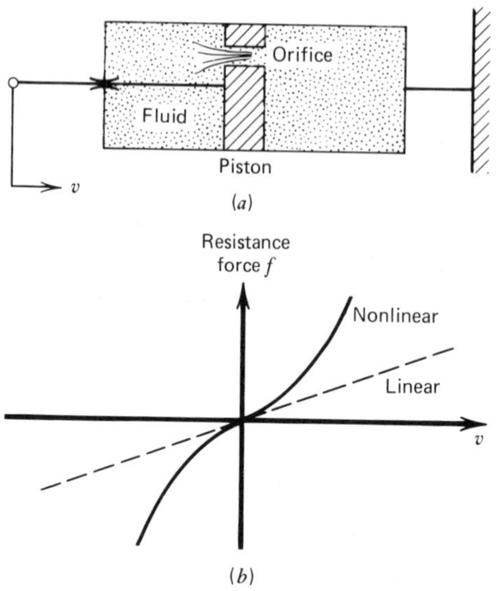


**Figure 2.4 Viscous damper. (*a*) Construction. (*b*) Constitutive relation.**

The direction of the cause-and-effect relationship is not fixed in a resistance element because the resistance relation is static; that is, it contains no time integrations or differentiations. The general resistance relation is of the form

$$r = F_R(e) \tag{2.2-4}$$

where $r$ is the rate variable, $e$ is the effort variable, and $F_R$ denotes the functional form of the relation. Figure 2.3*b* is a plot of this relation for pipe flow.

Because the cause-and-effect direction is not unique, we can also write (2.2-4) as the inverse relation

$$e = F_R^{-1}(r) \tag{2.5-5}$$

where $F_R^{-1}$ denotes the inverse function. Equation (2.2-3) is an example of this form, as is the plot shown in Figure 2.4*b*.

Resistance elements dissipate power. The resistance relation provides a way for the effort and rate variables to be coupled together in a system model, by means of (2.2-4).

### *Example 2.1*

Develop a dynamic model for the liquid mass $m$ in the tank system shown in Figure 2.3*a*. Assume that laminar flow exists in the outlet pipe.

The system consists of an element for storing potential energy (the liquid mass in the tank), a resistance element that opposes the flow from the tank (the outlet pipe), and an input flow rate $q_i$. The appropriate integral causality relation is the quantity-rate relation

$$\frac{dQ}{dt} = r \tag{2.2-6}$$

The quantity variable $Q$ is $m$, the liquid mass. The net flow rate into the tank is $r = q_i - q_o$. The effort variable is the hydrostatic pressure $p$, the force per unit area caused by the weight of the mass $m$. Thus $p = mg/A$.

The resistance element is described by the laminar flow relation given in Figure 2.3*b* as

$$q_o = \frac{1}{R}p \tag{2.2-7}$$

The model form (2.2-6) therefore reduces to

$$\frac{dm}{dt} = q_i - \frac{1}{R}p$$

or

$$\frac{dm}{dt} = q_i - \frac{g}{RA}m \tag{2.2-8}$$

Given $m(0)$ and $q_i$ as a function of time, we can solve (2.2-8) for $m(t)$.

## Capacitance

We have seen that the effort variable results from the storage of potential energy. The relation that describes how the effort depends on quantity is another constitutive relation known as *capacitance*. It is written in general as

$$e = F_C(Q) \tag{2.2-9}$$

For the linear case,

$$e = \frac{1}{C}Q \tag{2.2-10}$$

where $C$ is the capacitance of the element.

The constitutive relation for an electrical capacitor is the voltage-charge relation

$$v = \frac{1}{C}Q \tag{2.2-11}$$

Another example of a capacitance relation is given by the tanks shown in Figure 2.3. The hydrostatic pressure $p$ in a liquid is related to the liquid height $h$ by

$$p = \rho g h \tag{2.2-12}$$

where $\rho$ is the mass density of the liquid. If the bottom area of the tank is $A$, as in Figure 2.3*a*, the liquid mass in the tank is

$$m = \rho A h \tag{2.2-13}$$

thus

$$p = \frac{g}{A}m \tag{2.2-14}$$

Since $p$ and $m$ are the effort and quantity variables for this system, comparison of (2.2-14) with (2.2-10) shows the capacitance of the vertical-sided tank to be

$$C = \frac{A}{g} \tag{2.2-15}$$

If the tank's sides are sloped, as in Figure 2.3*c*, the relation (2.2-12) still holds but (2.2-13) does not. In this case the relation between $m$ and $h$ is nonlinear, and therefore so is the capacitance relation (Figure 2.3*d*).

The capacitance constitutive relation can be used to relate the effort and quantity variables in a model, as in Example 2.1, where the pressure $p$ was related to the variable of interest $m$ by the capacitance relation $p = gm/A$. This gave a model solely in terms of $m$ and the input variable $q_i$.

Quite often the variable of real interest in a problem is not one of the rate, quantity, effort, or flux variables, but is some other related variable. For example, we might be primarily interested in the liquid height $h$ in the tank of Figure 2.3*a*. If so, a relation analogous to a constitutive relation can be used to obtain the model in terms of the desired variable. In the present example this relation is $m = \rho A h$. Substituting this into (2.2-8) gives

$$\rho A \frac{dh}{dt} = q_i - \frac{\rho g}{R} h \qquad (2.2\text{-}16)$$

This can be solved for $h(t)$.

If in a mechanical system displacement acts against an elastic element such as a spring, potential energy is stored and a force is set up in the spring. The usual constitutive relation for an elastic element that undergoes small deflections is given by *Hooke's law*.

$$f = kx \qquad (2.2\text{-}17)$$

where $f$ is the elastic force, $x$ is the deflection, and $k$ is the elastic constant (the spring constant). In analogy with the previous relations we might say that the elastic capacitance is $1/k$, but $1/k$ is usually called the "compliance."

In mechanical systems, elastic elements such as beams, rods, and rubber mounts are usually represented pictorally as "springs." The linear relation between force and displacement given by (2.2-17) for such elements becomes less accurate as the deflection becomes larger. In such cases it is often replaced by

$$f = k_1 x + k_2 x^3 \qquad (2.2\text{-}18)$$

where $k_1 > 0$. Again, $x$ is the deflection of the spring from its free (unstressed) length and $f$ is the restoring fence set up in the spring by $x$. The spring is compressed if $x < 0$ and extended if $x > 0$. It is said to be *hard* if $k_2 > 0$ and *soft* if $k_2 < 0$. This relation is compared to the linear relation in Figure 2.5. Note that the hard and soft springs both have their own linear approximation tangent at the origin.

A similar process occurs when a rod is twisted. The torque so induced is proportional to the net angular deflection for small deflections. This type of element is called a "torsional spring."

Capacitance relations also exist for compressible fluid systems, in which the fluid mass is stored in a container under pressure, and in thermal systems in which energy is stored in an object in the form of heat. These phenomena will be discussed in Sections 2.7 and 2.8.

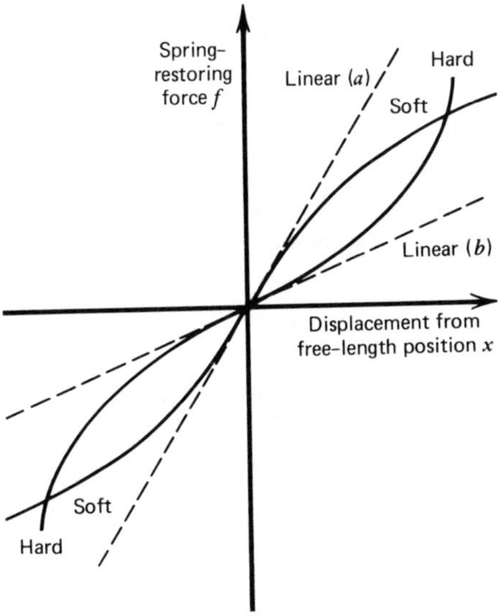


**Figure 2.5 Some constitutive relations for spring elements. Linear model $(a)$ is the small displacement model for the soft spring. Linear model $(b)$ approximates the relation for the hard spring.**

## Inductance

*Inductance* can be thought of as the property that allows the storage of kinetic energy. The general expression for this constitutive relation is

$$r = F_I(\phi) \tag{2.2-19}$$

In the linear case

$$r = \frac{1}{I}\phi \tag{2.2-20}$$

An example is the current-flux relation for an electrical inductor (2.1-9). Another is the impulse-momentum statement of Newton's second law (2.1-6), rewritten as

$$v(t) - v(0) = \frac{1}{m}M_x(t) \tag{2.2-21}$$

where the net rate is the velocity difference. Comparison of (2.2-20) with (2.2-21) shows that this mechanical inductance relation is linear and that the inductance of the system is the mass $m$.

The inductance constitutive relation can be used to relate flux and rate in a dynamic model. An example of this was seen with the use of (2.1-22), the inductance relation identical to (2.2-21), to eliminate the flux in (2.1-20). The result was a differential equation (2.1-23) in terms of rate and effort.

## Series and Parallel Elements

Because the constitutive relations of resistance, capacitance, and inductance elements are static, a system of several similar elements can be easily modeled as a single element by combining the constitutive relations of all the elements into a single expression.

This technique is perhaps familiar from electric circuit analysis. If two electrical resistors $R_1$ and $R_2$ are connected end to end as in Figure 2.6*a*, so that the same current $i$ flows through them, they can be modeled as one equivalent resistor as follows. Let the voltage drop across the resistors be $v_1$ and $v_2$, respectively. Then

$$i = \frac{1}{R_1} v_1 = \frac{1}{R_2} v_2$$

The total voltage drop $v$ across the pair is

$$v = v_1 + v_2 = (R_1 + R_2) i$$

Thus an equivalent resistor with a voltage drop $v$ and current $i$ has a resistance value of

$$R = R_1 + R_2 \qquad (2.2\text{-}22)$$

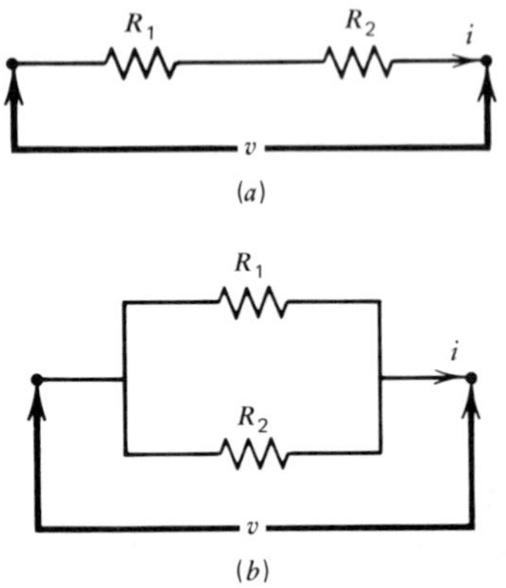


**Figure 2.6 Series and parallel resistances. (*a*) Series. (*b*) Parallel.**

If the resistors were connected side by side as in Figure 2.6*b*, so that the same voltage drop $v$ occurred across each, then

$$v = R_1 i_1 = R_2 i_2$$

where $i_1$ and $i_2$ are the currents in each resistor. The total current through the system is

$$i = i_1 + i_2 = \left( \frac{1}{R_1} + \frac{1}{R_2} \right) v$$

and the equivalent resistance $R$ is given by

$$\frac{1}{R} = \frac{1}{R_1} + \frac{1}{R_2} \qquad (2.2\text{-}23)$$

*Resistance* elements connected so that they experience the same *rate* variable are said to be *series* elements. If linear, the equivalent combined resistance is the sum of the individual resistances, as in (2.2-22). When connected to experience the same *effort* variable, the elements are in *parallel*, and the equivalent linear resistance is given by (2.2-23). These relations generalize to situations with more than two elements.

Note that elements connected end to end are not necessarily series elements. An example is given by two dashpots connected in such a manner (Figure 2.7*a*). They experience the same force, but not necessarily the same velocity; therefore, they are parallel elements.

*Capacitance* elements that experience the same *quantity* variable are in *series*; those with the same *effort* variable are in *parallel*. *Series inductance* elements experience the same *flux*; *parallel* elements have the same *rate* variable. If linear elements are involved, the constitutive relations (2.2-10) and (2.2-20) can be used to show that the equivalent values are given by

$$\frac{1}{C} = \frac{1}{C_1} + \frac{1}{C_2} \qquad (2.2\text{-}24)$$

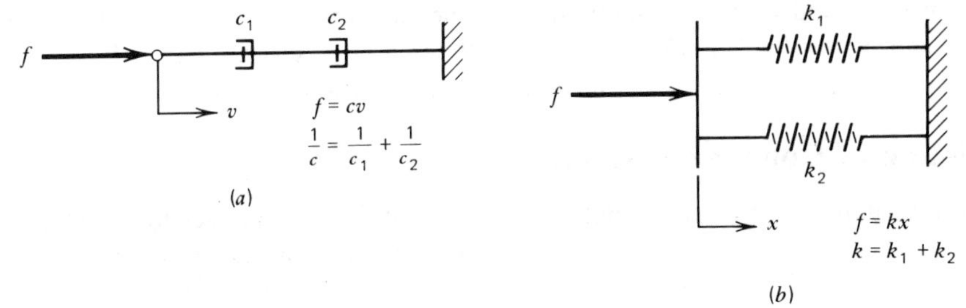


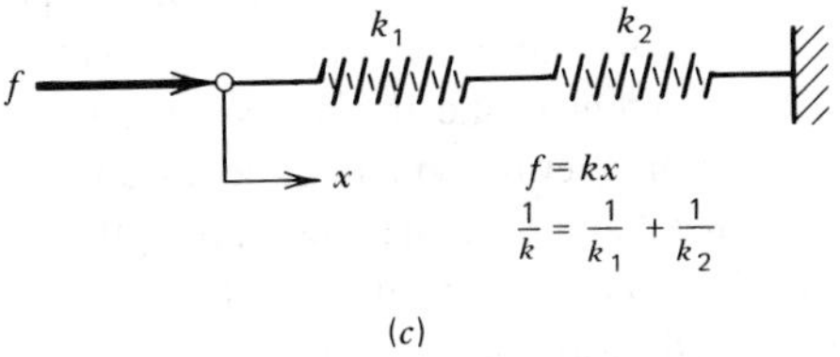


**Figure 2.7** Series and parallel mechanical elements. (*a*) Parallel dampers. (*b*) Series springs. (*c*) Parallel springs.

$$\frac{1}{L} = \frac{1}{L_1} + \frac{1}{L_2} \tag{2.2-25}$$

for *series* elements. For *parallel* elements,

$$C = C_1 + C_2 \tag{2.2-26}$$

$$L = L_1 + L_2 \tag{2.2-27}$$

Examples of series and parallel capacitance elements are shown in Figures 2.7*b* and 2.7*c*. In Figure 2.7*b* the two springs experience the same displacement (quantity) and are in series. Because spring capacitance is $C = 1/k$, their equivalent spring constant from (2.2-24) is

$$k = k_1 + k_2 \tag{2.2-28}$$

In Figure 2.7*c* the springs experience the same force (effort) and are in parallel. Thus from (2.2-26)

$$\frac{1}{k} = \frac{1}{k_1} + \frac{1}{k_2} \tag{2.2-29}$$

The relations (2.2-24) and (2.2-26) are used to determine the capacitance of an equivalent element with the same effort-quantity relation. They can be used, for example, to combine two electrical capacitances into one having the same voltage-*charge* relation. However, they cannot be used to compute an equivalent capacitance having the same voltage-*current* relation, because the latter involves a dynamic relation. Combination of elements involving dynamic relations requires the operator methods of Chapter 3.

## The Tetrahedron of State

The relations between the three energy elements – potential energy storage, kinetic energy storage, and energy dissipation – and the two integral causality relations can be summarized by the *tetrahedron of state* (Reference 2), shown in Figure 2.8. The arrows indicate the possible causal relations and the direction of causality.*

## Modeling of Complex Systems

We now treat more detailed examples of the use of these principles for developing models of mechanical, electrical, fluid, and thermal systems. Extremely detailed models are obviously beyond the scope of this work, and the reader is directed to the references and the many basic texts available in thermodynamics, heat transfer, fluid mechanics, dynamics, strength of materials, and electronics, for example. It should also be remembered that the viewpoint here is that of an engineer interested in the dominant time-dependent characteristics of a system, usually for the purposes of designing control schemes. Such designs usually require less detailed models than are commonly required for other purposes, because of the self-correcting nature of a properly designed feedback-control scheme.

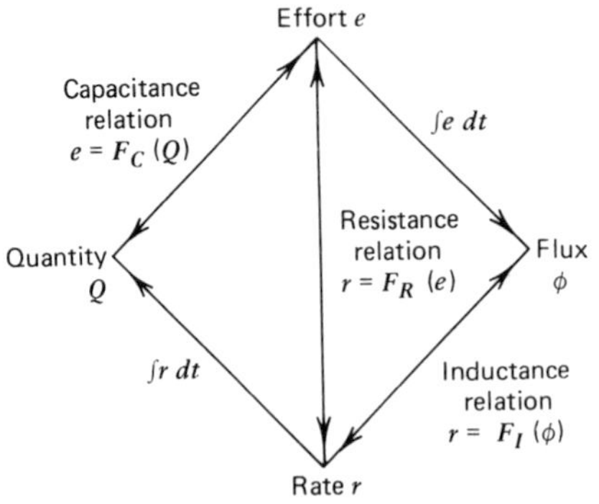


**Figure 2.8 Tetrahedron of state.**

# 2.3 MECHANICAL SYSTEMS

In the terminology of systems engineering, *mechanical* refers to the analysis of forces and motions in solid bodies. In many situations the internal forces in the body resulting from elastic deformation are small compared to the externally applied forces. In this case the mass or moment of inertia can be considered to be lumped at a point, and its motion is analyzed by "rigid-body dynamics."

The units and common symbols are given in Table 2.4. We assume the reader is familiar with the basic principles of rigid-body dynamics as found in an introductory course in physics or engineering mechanics. Our purpose here is to relate these concepts to the modeling framework provided by the energy and causality concepts of Section 2.1, and to show how mechanical components are modeled as resistance, capacitance, or inductance elements.

## The Laws of Motion

The Newtonian laws of motion for rigid bodies can be summarized as follows. The

*Some authors argue that causality is unidirectional in the capacitance and inductance relations (from quantity to effort and from flux to rate). This fine point seems to be irrelevant for most applications. For example, the spring deflection can be taken to be caused by an applied force, or the spring force can be treated as caused by the deflection. The point of view depends on the application at hand.

TABLE 2.4 Typical units in mechanical systems

| Quantity and Symbol | British Engineering | SI | Conversion[a] Factor | Other Common Units |
|---|---|---|---|---|
| Translational displacement $x$ | ft | m | 0.3048 | mile[b] |
| Translational velocity $v$ or $\dot{x}$ | ft/sec | m/sec | 0.3048 | mile/hr (mph)[c] |
| Angular displacement $\theta$ | radian (rad) | radian | – | degrees[d] |
| Angular velocity $\omega$ or $\dot{\theta}$ | rad/sec | rad/sec | – | revolutions/minute[e]<br>cycles/sec (cps) = hertz(Hz)[f] |
| Force $f$ | lb | N | 4.4482 | ounce (oz)[g] |
| Torque $T$ | lb-ft[h] | N-m[h] | 1.3557 | ounce-inch (oz-in.)[i] |
| Translational impulse $M$ or $M_x$ | lb-sec | N-sec | 4.4482 | – |
| Angular impulse $M$ or $M_\theta$ | lb-ft-sec | N-m-sec | 1.3557 | – |

[a] To convert from British engineering units to SI units, multiply the British units by this factor.
[b] 1 mile = 5280 ft.
[c] 1 mph = 1.46667 ft/sec.
[d] 1 degree = 0.017453 rad.
[e] 1 rpm = 0.1047198 rad/sec.
[f] 1 cps = 6.283185 rad/sec.
[g] 1 oz = 1/16 lb = 0.0625 lb.
[h] The dimensions of torque and energy are identical even though they are different physical quantities.
[i] 1 oz-in. = 0.005208 lb-ft.

vector sum of external forces acting on the body must equal the time rate of change of momentum. Thus

$$\frac{d\mathbf{P}}{dt} = \mathbf{f} \tag{2.3-1}$$

where **P** and **f** are the three-dimensional momentum and force vectors, respectively. If **v** is the vector velocity of the body's center of mass, and $m$ is the mass, the momentum is $\mathbf{P} = m\mathbf{v}$, and (2.3-1) becomes

$$m\frac{d\mathbf{v}}{dt} = \mathbf{f} \tag{2.3-2}$$

In addition, the vector sum of the external torques acting on the body must equal the time rate of change of the angular momentum. Thus

$$\frac{d\mathbf{\Omega}}{dt} = \mathbf{T} \tag{2.3-3}$$

where $\mathbf{\Omega}$ and **T** are the three-dimensional angular momentum and torque vectors.

*Chasle's theorem* states that the general motion of a rigid body can be considered as a translation plus a rotation about a suitable point. The center of mass is often taken to be this point. If the body's rotation and translation are not simply related to a convenient coordinate system, then (2.3-1) and (2.3-3) can result in complicated expressions. However, most of the systems of interest here involve rotation about a fixed axis. In this case the angular momentum can be expressed as

$$\mathbf{\Omega} = I\boldsymbol{\omega} \tag{2.3-4}$$

where $I$ is the body's mass moment of inertia about the axis, and $\boldsymbol{\omega}$ is the vector angular velocity. Thus (2.3-3) becomes

$$I\frac{d\boldsymbol{\omega}}{dt} = \mathbf{T} \tag{2.3-5}$$

If the coordinate system used to express $\boldsymbol{\omega}$ has one direction aligned with the axis of rotation, the vector relation (2.3-5) becomes a scalar one.

$$I\frac{d\omega}{dt} = T \tag{2.3-6}$$

The mass moment of inertia $I$ is given in Table 2.5 for several common objects.

For most of our problems involving rotation, (2.3-6) will suffice. In addition, the translational motion equation (2.3-2) can be written in scalar form as

$$m\frac{dv_i}{dt} = f_i \qquad i = 1, 2, 3 \tag{2.3-7}$$

where $v_i$ and $f_i$ are the velocity and force components along the rectilinear coordinates $x_1$, $x_2$, $x_3$. Further simplification results if the forces $f_i$ are not functions of the displacements or velocities in the other coordinate directions. In this case the set (2.3-7) becomes uncoupled. Each equation will be of the form

$$m\frac{dv_i}{dt} = f_i(v_i, x_i, t) \tag{2.3-8}$$

and can be solved independently of the other two.

Many problems involving both rotation and translation can be simplified by breaking them down into problems involving pure translation and pure rotation and by using (2.3-6) and (2.3-7).

## *Example 2.2*

A homogeneous cylinder of radius $R$ and mass $m$ is free to rotate about an axis that is connected to a support by a spring and damper (Figure 2.9*a*). Assume that the cylinder rolls without slipping on the surface of inclination $\alpha$. Find the equation of motion.

Since the problem statement does not specify the coordinates to be used, we are free to make a convenient choice. Choose $x_1$ to represent the tangential displacement of the cylinder's center of mass from its equilibrium position. In this position the spring force is balanced by the tangential component of the cylinder's weight. Let $x_2$ be the direction perpendicular to $x_1$, normal to the inclined surface. Choose $\theta$ to be the positive angle of rotation when the cylinder moves in the positive $x_1$ direction.

The free-body diagram is given in Figure 2.9*b*. The spring force is $kx_1$, the damper force is $cv_1$, the normal component of the reaction force due to gravity is

**TABLE 2.5 Mass Moment of Inertia**

**Definition**

$$I = \int r^2 \, dm \qquad r = \text{distance from reference axis to mass element } dm$$

| **Units for *I*** | *British* | | *SI* |
|---|---|---|---|
| | slug-ft$^2$ | = | 1.356 kg-m$^2$ |

**Values for common elements** ($m$ = element mass)

Cylinder

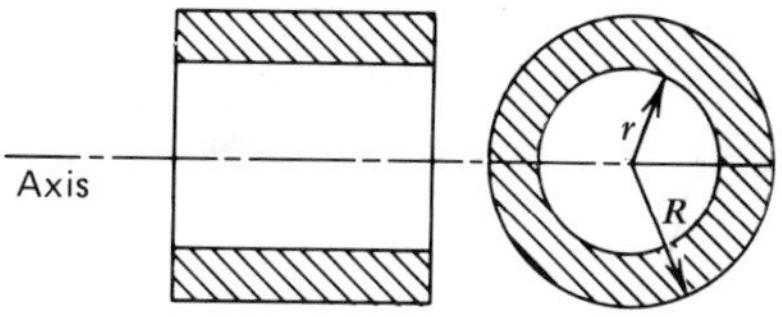


$$I = m \frac{R^2 + r^2}{2}$$

Point Mass

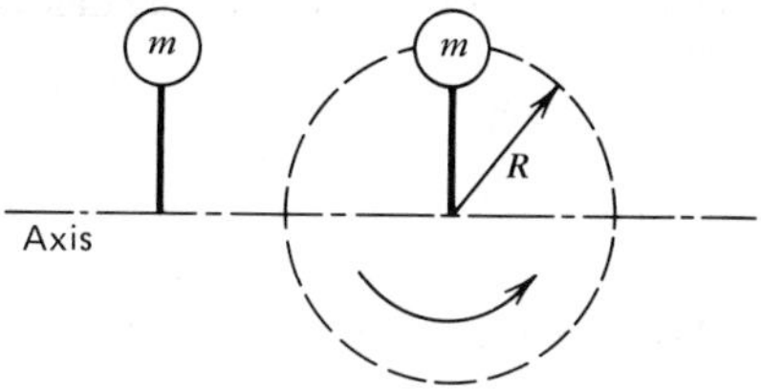


$$I = mR^2$$

Sphere

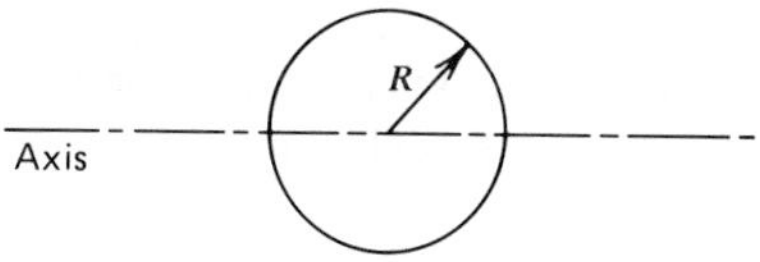


$$I = \frac{2}{5} mR^2$$

Lead screw

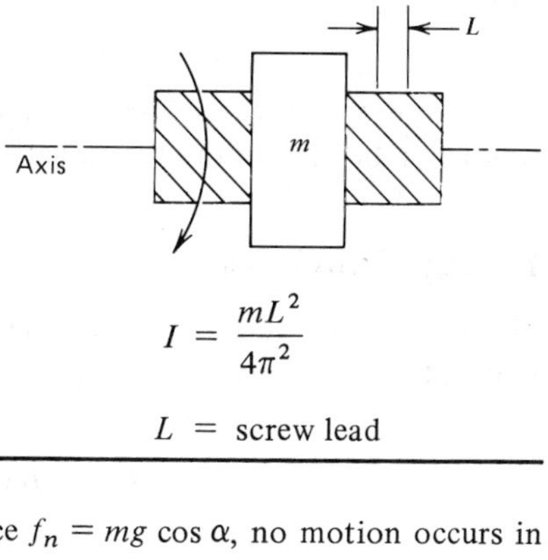


$$I = \frac{mL^2}{4\pi^2}$$

$L$ = screw lead

---

$f_n$, and the tangential reaction force is $f_t$. Since $f_n = mg \cos \alpha$, no motion occurs in the $x_2$ direction. For the $x_1$ direction, (2.3-8) gives

$$m \frac{dv_1}{dt} = -cv_1 - kx_1 + mg \sin \alpha - f_t \tag{2.3-9}$$

where $v_1 = \dot{x}_1$.* The rotational motion is given by (2.3-6) as

$$I \frac{d\omega}{dt} = Rf_t \tag{2.3-10}$$

where $\omega = \dot{\theta}$.

The condition of no slipping is $R\theta = x_1$. Thus $\dot{\omega} = \ddot{\theta} = \ddot{x}_1/R$, and (2.3-10) gives

*The dot notation is used to represent differentiation. Thus $\dot{x} = dx/dt$, $\ddot{x} = d^2x/dt^2$, etc.

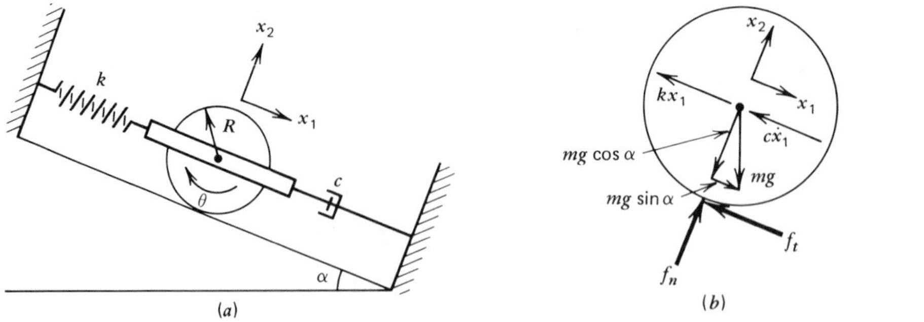


**Figure 2.9** (*a*) Cylinder-spring-damper system. (*b*) Free-body diagram of the cylinder.

$$f_t = \frac{I}{R^2} \ddot{x}_1$$

Substituting $f_t$ into (2.3-9) results in

$$\left(m + \frac{I}{R^2}\right) \ddot{x}_1 + c\dot{x}_1 + kx_1 = mg \sin \alpha$$

From Table 2.5, with $r = 0$, we have $I = \frac{1}{2}mR^2$, and the answer

$$1.5m\ddot{x}_1 + c\dot{x}_1 + kx_1 = mg \sin \alpha \tag{2.3-11}$$

Note that the effective mass for translation $(1.5m)$ is greater than for pure translation because of the additional effort required to rotate the cylinder.

## Energy Principles

The translational and rotational relations (2.3-6) and (2.3-8) can both be expressed in impulse-momentum form. For translation, (2.3-8) gives

$$mv(t) - mv(0) = M_x(t) = \int_0^t f(u)du \tag{2.3-12}$$

For rotation, (2.3-6) gives

$$I\omega(t) - I\omega(0) = M_\theta(t) = \int_0^t T(u)du \tag{2.3-13}$$

Since these relations have the same form, we will focus our discussion only on the translation case (2.3-12). But all of the following results also apply to rotational dynamics.

Alternate and more common forms of the second law are given by

$$m \frac{dv}{dt} = f \tag{2.3-14}$$

$$m\frac{d^2x}{dt^2} = f \tag{2.3-15}$$

where $x$ is the displacement and $v$ is the velocity of the mass. Equation (2.3-12) is the first integral of (2.3-14). Which of the three forms is most useful depends on how the force $f$ varies with $t$, $x$, and $v$. For example, if the force is given by

$$f(t) = at + b \tag{2.3-16}$$

the velocity from (2.3-12) is

$$\begin{aligned} v(t) &= v(0) + \frac{1}{m}\int_0^t (au + b)\,du \\ &= v(0) + \frac{a}{2m}t^2 + \frac{b}{m}t \end{aligned} \tag{2.3-17}$$

If the displacement is required, another integration must be made; that is,

$$\begin{aligned} x(t) &= x(0) + \int_0^t v(u)\,du \\ &= x(0) + \frac{a}{6m}t^3 + \frac{b}{2m}t^2 + v(0)t \end{aligned} \tag{2.3-18}$$

The wealth of information contained in Newton's second law (2.3-12) is remarkable. When the external force is derivable from a potential energy function that depends only on position, such as with gravity or a spring, the second law implies conservation of mechanical energy.

A *conservative force* is derivable from a potential energy function $V(x)$ such that

$$f = -\frac{dV}{dx} \tag{2.3-19}$$

In this case use this relation and the identity

$$\frac{dv}{dt} = \frac{dv}{dx}\frac{dx}{dt} = \frac{dv}{dx}v$$

in (2.3-14) to obtain

$$mv\frac{dv}{dx} = -\frac{dV}{dx}$$

This gives

$$mv\,dv = -dV$$

or

$$m\int v dv = -\int dV$$

The integration result is

$$m\frac{v^2}{2} + V(x) = C \tag{2.3-20}$$

This is a statement of conservation of mechanical energy; that is, the motion is such that the sum of the kinetic and potential energies is constant. The constant of integration $C$ is determined by the initial values of $v$ and $x$.

Equation (2.3-20) gives the velocity of the mass as a function of its position $x$ for conservative forces. If the velocity and position are desired as functions of time, (2.3-20) is not always the most convenient form to use. The following two examples illustrate this point.

## *Example 2.3*

Find the maximum height a ball of mass $m$ will reach if thrown vertically upward with an initial velocity $v(0)$. How long will it take to reach this height? Neglect air resistance.

Let the vertical displacement $x$ be measured from the launch point. The gravity force on the ball is $mg$ and is derivable from the potential energy function $V = mgx$. Thus (2.3-20) can be used to give

$$m\frac{v^2}{2} + mgx = C \tag{2.3-21}$$

The motion of the ball represents a flow of energy between the potential energy (capacitance) and the kinetic energy (inductance) compartments. From the given information, $v = v(0)$ when $x = 0$. Thus from (2.3-21)

$$C = \frac{m\,v^2(0)}{2}$$

The maximum height is achieved when $v = 0$. At that point (2.3-21) gives

$$x = \frac{v^2(0)}{2g}$$

which is independent of the mass.

In order to find the time to reach this height, we need a solution for $x$ or $v$ as a function of time. The gravity force is a special case of form (2.3-16) with $a = 0$ and $b = -mg$. So from (2.3-17)

$$v(t) = v(0) - gt$$

The required time is found by setting $v(t)$ equal to zero. The answer is

$$t = \frac{v(0)}{g}$$

The maximum height can be found by substituting this time into (2.3-18).

Equations (2.3-17) and (2.3-18) represent the complete description of the ball's dynamics. Two state variables, position and velocity, are required because of the two energy compartments involved. However, if only the velocity as a function of height is desired (not a complete description), the solution given by (2.3-20) is more convenient.

## *Example 2.4*

Consider the motion of the mass $m$ connected to a spring $k$ (Figure 2.2). Use an energy principle to determine how the velocity of the mass varies with position if it is initially released from position $x(0)$ with zero velocity. Assume the external force $f$ is zero.

Setting $f = 0$ in (2.1-23) and using (2.1-19), we obtain

$$m\frac{d^2x}{dt^2} = -kx \tag{2.3-22}$$

A complete description of the dynamics requires solving for the two state variables, $x$ and $v$, as functions of time. However, this solution is not needed to answer the question at hand. The spring force is derivable from the potential energy function

$$V(x) = k\frac{x^2}{2}$$

Thus (2.3-20) gives

$$m\frac{v^2}{2} + k\frac{x^2}{2} = C \tag{2.3-23}$$

With $v(0) = 0$, we get

$$C = k\frac{x^2(0)}{2}$$

The velocity as a function of position is found to be

$$v = \pm\sqrt{\frac{k}{m}[x^2(0) - x^2]} \tag{2.3-24}$$

where the $\pm$ sign refers to motion to the left or right. The maximum kinetic energy and thus the maximum velocity occurs when the potential energy is a minimum (at $x = 0$). The velocity is zero when $x = \pm x(0)$. Therefore the initial displacement is also the maximum displacement. Any time-related information, such as the period of the oscillation, must be obtained by solving the differential equation (2.3-22).

When the force is dependent only on velocity, we use the form (2.3-14).

## *Example 2.5*

Derive a model for the motion of a mass $m$ connected to a viscous damper with a damping coefficient $c$ (Figure 2.10). Assume that the mass is also subjected to a force $f$. The displacement $x$ is measured from an arbitrary fixed point.

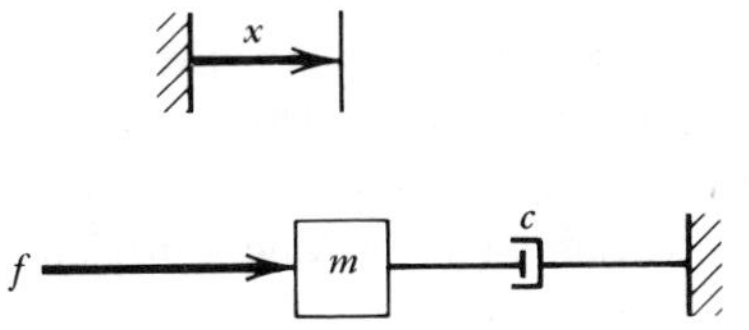


**Figure 2.10** **Mass-damper system.**

The net force in the positive direction of $x$ is $f - cv$. From (2.3-14), the required model is

$$m\frac{dv}{dt} = f - cv \tag{2.3-25}$$

If the force $f$ is identically zero or is a simple function of $v$, (2.3-25) can be integrated to obtain the velocity as a function of time. In this case the only energy storage is that due to kinetic energy, and only one state variable is required, the

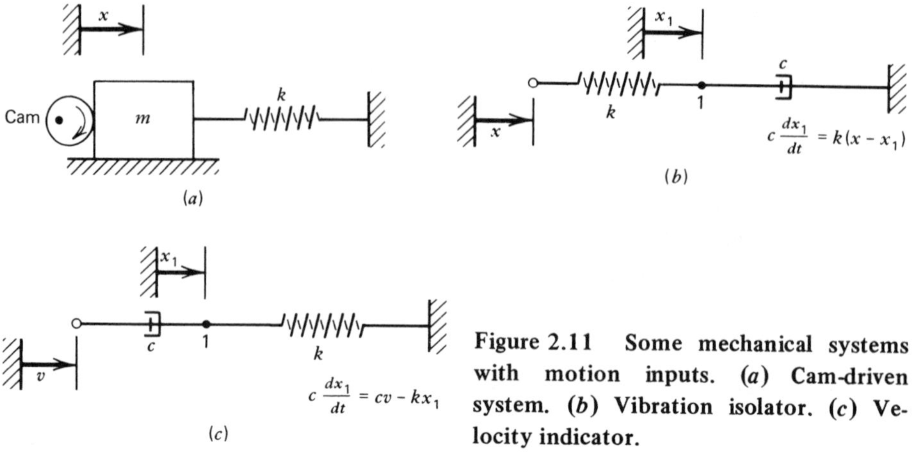


**Figure 2.11 Some mechanical systems with motion inputs. (*a*) Cam-driven system. (*b*) Vibration isolator. (*c*) Velocity indicator.**

velocity. The displacement $x$ does not affect the dynamics of the system, but simply serves as an auxiliary variable to relate the position of the mass to the location of an observer. The choice of a reference point for $x$ is irrelevant as long as the reference is an inertial one as required by Newton's laws.

## Prescribed Motions as Inputs

The system in Figure 2.10 has a force as the input. In many applications the input force is not known, but the device's motion is a specified function of time. In such cases we may take the given motion to be the input if proper care is used in deriving the model. In Newtonian mechanics, force acting on a mass is the cause, and motion is its effect. The specified motion is caused by some force that might be unknown and might in fact be of little interest to us. In other words, when a motion input is specified, we are saying that the driven system has no choice but to move as specified by the input. When postulating a displacement or velocity input to a mass, we implicitly assume the existence of a force capable of producing the specified motion – the case for the system shown in Figure 2.11*a*. A rotating cam drives the mass, and the displacement $x(t)$ is determined by the cam profile and its rotational speed. No differential equation model is needed here. The designer might simply be interested in computing the force exerted on the support as a result of the motion. This force is $kx(t)$.

Motions also appear as inputs to spring and damper elements. Figures 2.11*b* and 2.11*c* show some examples. The vibration isolator (Figure 2.11*b*) is used to prevent the effects of the input displacement $x$ from being transmitted to point 1. The arrangement in Figure 2.11*c* can be used in an instrument to provide a displacement at point 1 that is indicative of the input velocity magnitude.

## Spring Elements

Table 2.6 summarizes the series and parallel laws for linear spring elements, both translational and rotational. In each case the equivalent $k$ is that value required of a single element in order for it to be equivalent to the multielement system shown.

**TABLE 2.6 The Spring Constant**

| | Translational | Torsional |
|---|---|---|
| Symbol | 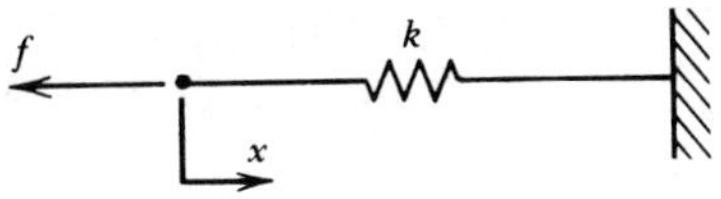  | 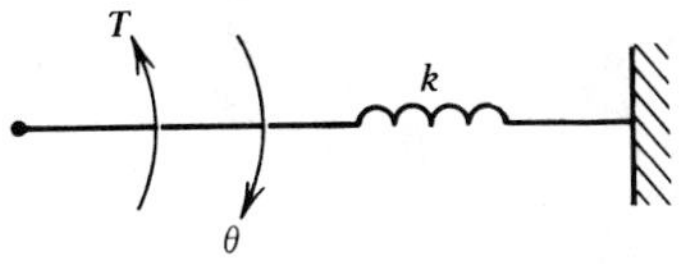  |
| Linear law | $f = kx$ | $T = k\theta$ |
| Parallel elements | 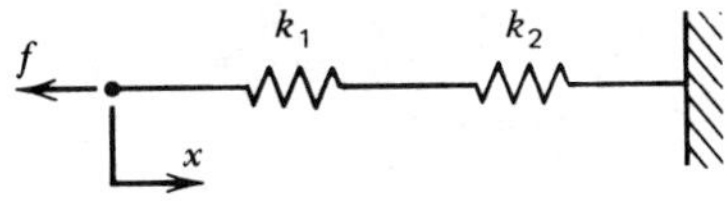  | 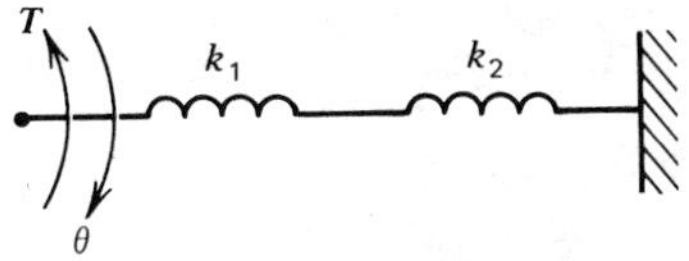  |
| | $k = \dfrac{1}{1/k_1 + 1/k_2}$ | |
| Series elements | 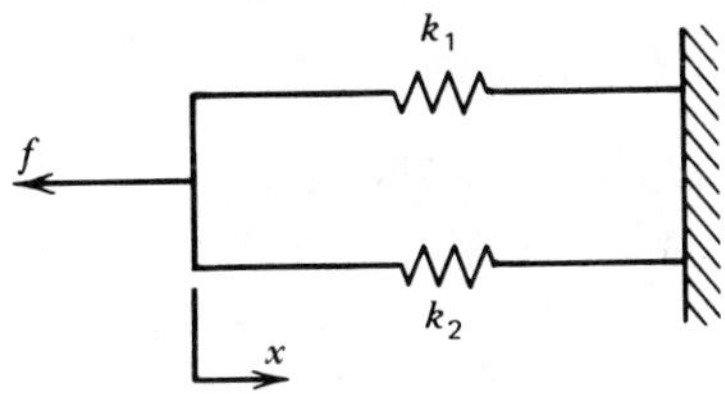  | 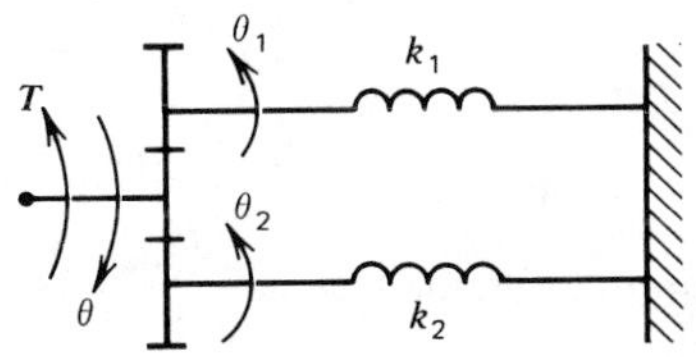  |
| | $k = k_1 + k_2$ | |

| **Units** | **British** | | **SI** |
|---|---|---|---|
| $k$ (translational) | lb/ft | = | 14.594 N/m |
| $k$ (torsional) | ft-lb/rad | = | 1.3557 N-m/rad |
| $E$ (modulus of elasticity) | lb/ft$^2$ | = | 47.88 N/m$^2$ |
| $G$ (shear modulus of elasticity) | lb/ft$^2$ | = | 47.88 N/m$^2$ |

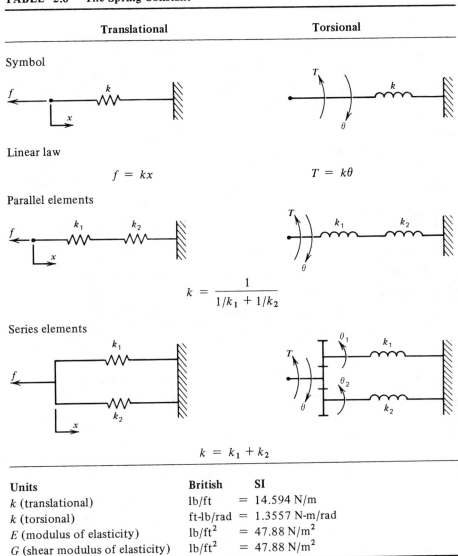

Table 2.7 lists the expressions for the spring constant for several common elements. These formulas are derived from the force-displacement relations given in elementary texts on mechanics of materials. The moduli of elasticity $E$ and $G$ are given for steel and aluminum in Table 2.8.

## Damping Elements

Table 2.9 summarizes the series and parallel laws for combining linear damping elements. Damping constants for three common elements are given in Table 2.10. The

**TABLE 2.7 Spring Constants for Common Elements**

Rod in compression/tension

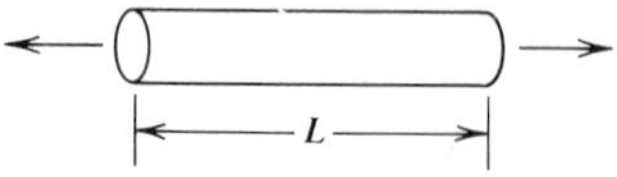


$$k = \frac{EA}{L}$$

$A$ = cross-sectional area

Rod in torsion

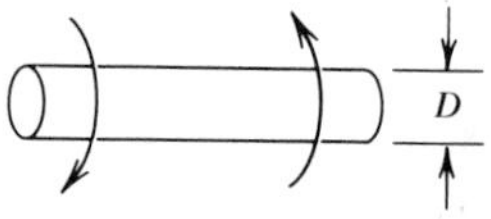


$$k = \frac{G\pi D^4}{32L}$$

$L$ = length

Helical wire coil

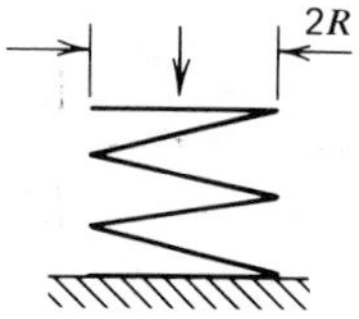


$$k = \frac{Gd^4}{64nR^3}$$

$d$ = wire diameter

$n$ = number of coils

Cantilever beam

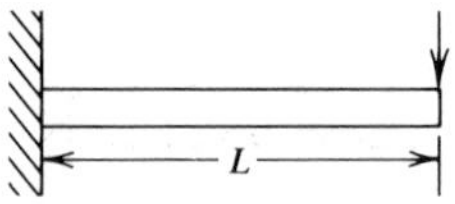


$$k = \frac{Ewh^3}{4L^3}$$

$w$ = beam width

$h$ = beam thickness

Doubly clamped beam

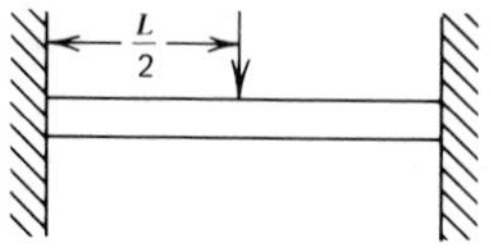


$$k = \frac{16Ewh^3}{L^3}$$

Air spring

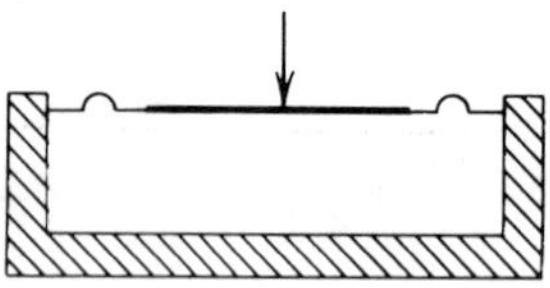

$$k = \frac{\gamma PA^2}{V}$$

$A$ = diaphragm area

$P, V$ = nominal pressure and volume

$\gamma$ = ratio of specific heats

first will be derived in Section 2.6. The dashpot relation for sliding viscous friction is a consequence of *Newton's law of viscosity*, which states that

$$\tau = \mu \frac{du}{dy} \tag{2.3-26}$$

where $\tau$ is the shear stress in the lubricating film, $\mu$ is the lubricant's viscosity coef-

**TABLE 2.8 Mechanical Properties of Steel and Aluminum**

| | Steel[a] | Aluminum |
|---|---|---|
| Mass density $\rho$[b] | 15.2 slug/ft$^3$ | 5.25 slug/ft$^3$ |
| Modulus of elasticity $E$[b] | $4.32 \times 10^9$ lb/ft$^2$ | $1.44 \times 10^9$ lb/ft$^2$ |
| Shear modulus of elasticity $G$[b] | $1.73 \times 10^9$ lb/ft$^2$ | $5.76 \times 10^8$ lb/ft$^2$ |

[a] Steel is the collective name for a family of alloys containing iron, carbon, and frequently other substances. These properties vary with composition but the values given here are representative.

[b] Factors for converting to SI units are given in Table 2.4.

**TABLE 2.9 The Damping Constant**

**Translational** | **Torsional**

Symbol

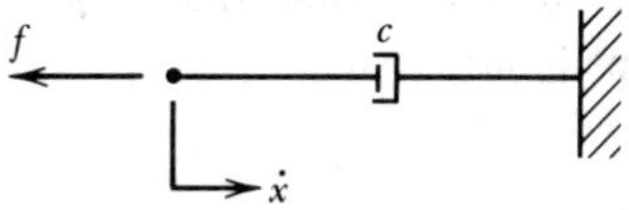


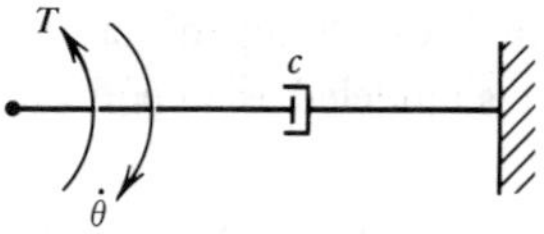


Linear law

$$f = c\dot{x} \qquad\qquad T = c\dot{\theta}$$

Parallel elements

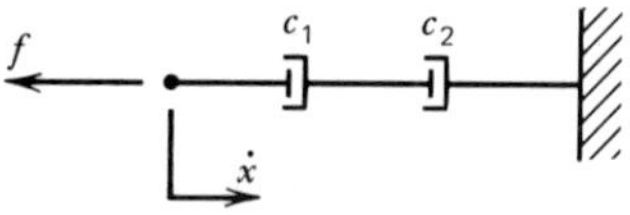


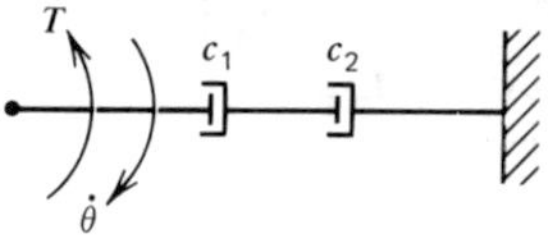


$$c = \frac{1}{1/c_1 + 1/c_2}$$

Series elements

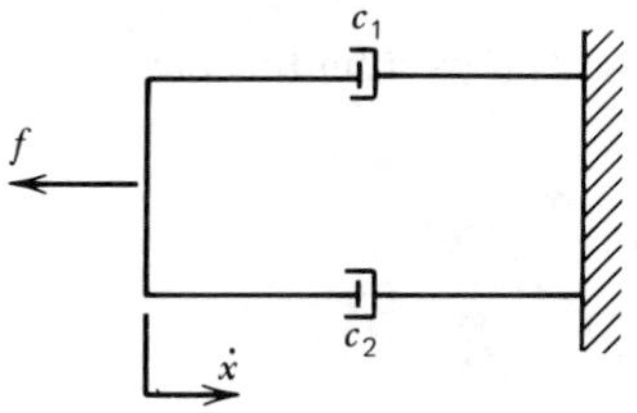


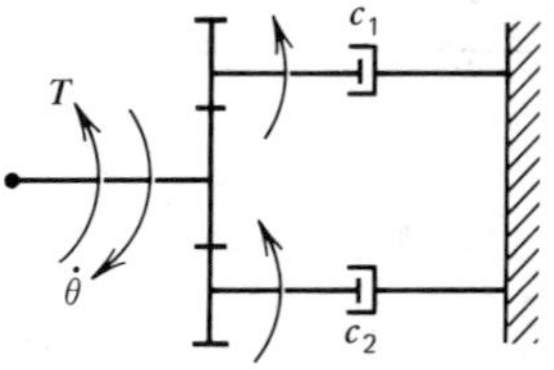


$$c = c_1 + c_2$$

| Units | British | | SI |
|---|---|---|---|
| $c$ (translational) | lb-sec/ft | = | 14.594 N-sec/m |
| $c$ (torsional) | ft-lb-sec/rad | = | 1.3557 N-m-sec/rad |

ficient, and $du/dy$ is the velocity gradient in the film normal to the surface. If we assume that the gradient is constant and that the fluid velocity near the surface is zero relative to the surface, it can be shown that the friction force is

$$f = \frac{A\mu}{d} v = cv \tag{2.3-27}$$

where $A$ is the lubricated area under the mass and $d$ is the film thickness. For high velocities these assumptions are not valid, and (2.3-27) must be replaced by a nonlinear relation similar to that shown in Figure 2.4*b*.

The journal-bearing relation shown in Table 2.10 is known as Petrov's law and is a consequence of Newton's viscosity law applied to a rotating shaft.

## Lumped-Parameter Equivalent Systems

In any real system the mass will be distributed throughout and not concentrated at a point with infinitesimal volume. If the forces of interest are external to the system, the principles of rigid-body dynamics can be applied by considering the system's mass to be concentrated or "lumped" at the center of mass. Thus the original distributed-mass system is modeled as an equivalent lumped-mass system.

**TABLE 2.10 Damping Constants of Common Elements**

Dashpot (laminar flow)

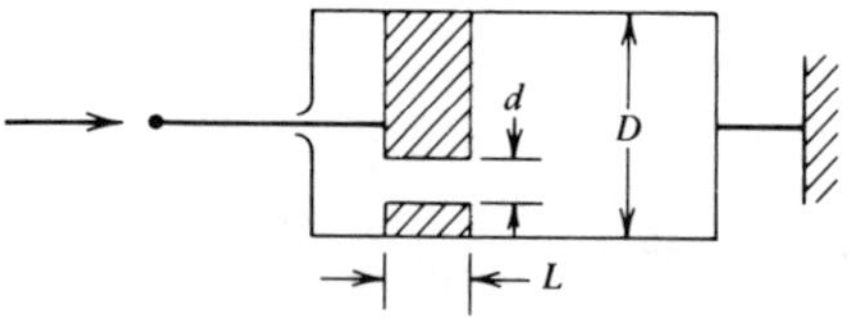


$$c = 8\pi\mu L\left[\left(\frac{D}{d}\right)^2 - 1\right]^2$$

Sliding viscous friction

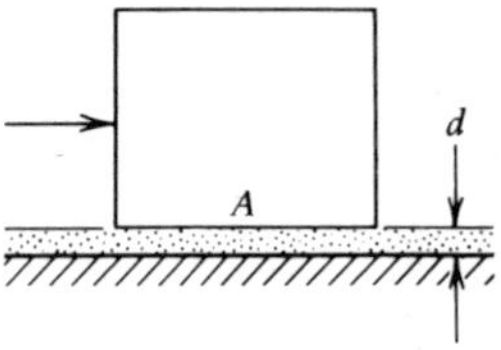


$$c = \frac{A\mu}{d}$$

$A$ = lubricated area

$d$ = film thickness

Journal bearing

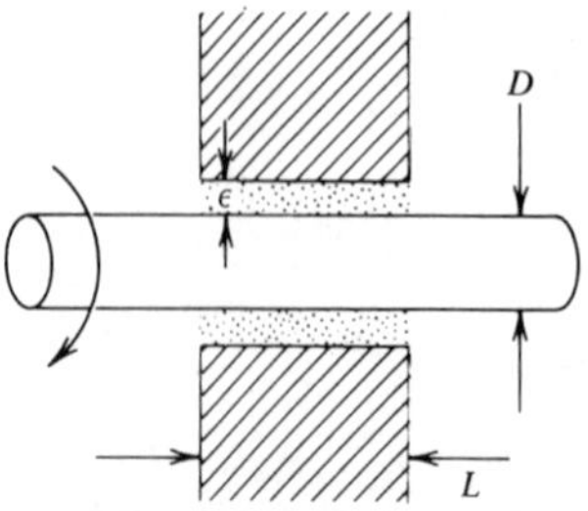


$$c = \frac{\pi D^3 L\mu}{4\epsilon}$$

$\epsilon$ = radial clearance

If the forces of importance are internal to the system, such as bending stresses in a beam, the situation is more complicated. The application of Newton's second law requires that a displacement variable be assigned to each mass, so that a distributed-mass model requires an infinite number of displacement variables, one for each infinitesimal mass element. This leads to a description in terms of partial differential equations. On the other hand, a lumped-mass model results in a set of ordinary differential equations, one second-order equation for each dimension in which motion occurs (rotational and translational). Because ordinary differential equations are easier to solve, we usually prefer a lumped model.

Mass is not the only property we seek to represent in a lumped fashion. For example, the stress resulting from bending in a cantilever beam is distributed throughout the beam length. The beam's equivalent spring constant $k$ is that value required of an ideal spring if it is to represent the force-deflection characteristics of the end-point of the beam. Similarily, viscous friction forces are distributed throughout a system. The equivalent damping constant $c$ is that value required for an ideal damper to represent the force-velocity characteristics at a chosen point in the system.

There are several methods for modeling equivalent systems. None are exact because no two systems can be identical in every respect. The equivalent masses in Table 2.11 were found by requiring that the kinetic energy of the equivalent system equal that of the original. The equivalent spring constants in Table 2.7 were obtained from the force-deflection characteristics of the original system at the point where the force was applied. An alternate means of computing the spring constant is to require equivalence of potential energy in the original and equivalent systems. Note that a spring constant found in this manner might not be the same as that found from the force-deflection characteristics. We will not go into the details of these methods, but will use the results for common elements as listed in the accompanying tables. Detailed discussion is given in Reference 3.

In the examples in Table 2.11, the origin of the coordinate system for the lumped model is taken to be at the location of the concentrated mass, and the expressions for the equivalent mass are valid only for that coordinate choice. If, for example, we wish to use a coordinate system located at the midpoint of the cantilever beam, we will need another expression for the equivalent mass $m_e$. Thus, in general, the values of the parameters in a lumped model depend on the location of the model's coordinate system relative to the original distributed-parameter system.

## Geared Systems

Important examples of equivalent systems are given by devices called *mechanical transformers*. Devices such as levers, gears, cams, and chains transform the motion at the input into a related motion at the output. Their analysis is simplified by describing the effects of their mass, elasticity, and damping relative to a single location, such as the input location. The result is a lumped-parameter model whose coordinate system is centered at that location.

We can characterize a transformer by a single parameter $b$ such that in terms of the general displacement variable $Q$,

$$Q_1 = bQ_2 \tag{2.3-28}$$

**TABLE 2.11 Lumped-Mass Equivalent Systems**

$m$ = concentrated mass
$m_d$ = distributed mass
$m_c$ = equivalent lumped mass

| | Actual System | Equivalent System |
|---|---|---|
| Helical spring or rod | 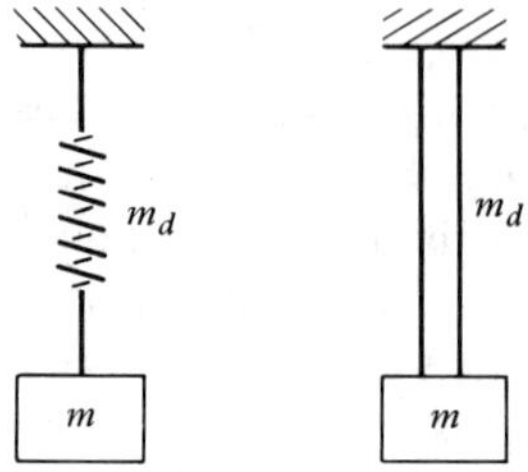 |  $m_e = m + \frac{1}{3} m_d$ |
| Cantilever beam | 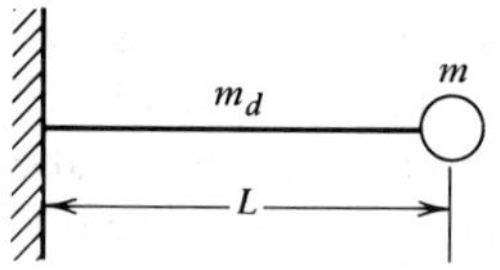 | 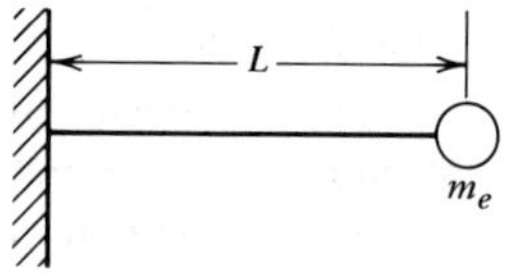 $m_e = m + 0.23 m_d$ |
| Fixed-end beam | 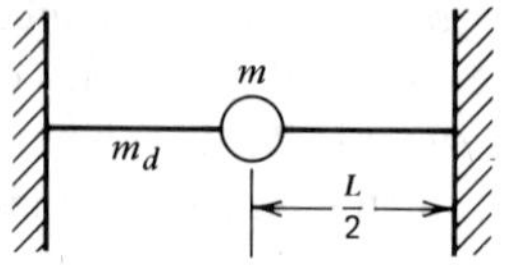 | 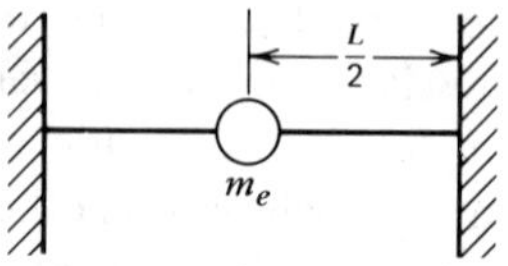 $m_e = m + 0.375 m_d$ |
| Simply supported beam | 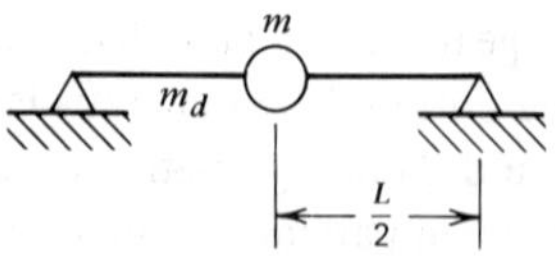 | 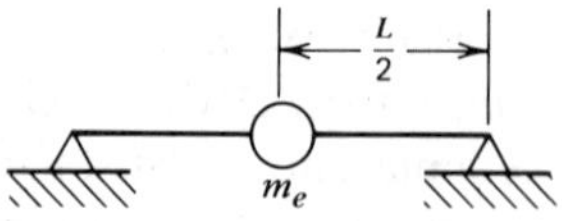 $m_e = m + 0.50 m_d$ |

where the subscripts 1 and 2 refer to the input and output states, respectively. Since the rate variable $r$ is related to $Q$ by $\dot{Q} = r$, we also have

$$r_1 = br_2 \tag{2.3-29}$$

if $b$ is constant. An ideal transformer neither stores nor dissipates energy. Therefore, the output power equals the input power, or $e_1 r_1 = e_2 r_2$. This shows the effort variables to be related as

$$e_1 = \frac{1}{b} e_2 \tag{2.3-30}$$

A specific example is the gear pair shown in Table 2.12, with $Q = \theta$, $e = T$ and $r = \omega$. The parameter $b$ is the gear ratio $N$. If $N > 1$ the pair is a speed reducer whose output torque $T_2$ is greater than the input torque $T_1$.

Suppose an element possessing inertia, elasticity, or damping properties is connected to the output shaft of a gear pair, as shown in Table 2.12. How can we represent the geared system by an equivalent gearless system? If we wish to reference all motion to the input shaft, we can proceed as follows. We assume that the inertia, twisting, and damping of the input shaft are negligible. If not, they can be included in a manner similar to the following. We derive an equivalent system by requiring that its kinetic energy, potential energy, and power dissipation terms be identical to those of the original system. The expressions for these quantities in the equivalent system are

$$\text{kinetic energy} = \tfrac{1}{2} I_e \omega_1^2$$

$$\text{potential energy} = \tfrac{1}{2} k_e \theta_1^2$$

$$\text{power dissipation} = T_1 \omega_1 = (c_e \omega_1)\omega_1 = c_e \omega_1^2$$

If we neglect the gear inertias, gear-tooth elasticity, and gear friction, the corresponding expressions for the original system are

$$\text{kinetic energy} = \tfrac{1}{2} I \omega_2^2$$

$$\text{potential energy} = \tfrac{1}{2} k \theta_2^2$$

$$\text{power dissipation} = T_2 \omega_2 = c \omega_2^2$$

Equating these expressions and using the gear ratio formulas, we get the parameters for the equivalent system.

$$I_e = I \left(\frac{\omega_2}{\omega_1}\right)^2 = \frac{1}{N^2} I \tag{2.3-31}$$

$$k_e = k \left(\frac{\theta_2}{\theta_1}\right)^2 = \frac{1}{N^2} k \tag{2.3-32}$$

$$c_e = c \left(\frac{\omega_2}{\omega_1}\right)^2 = \frac{1}{N^2} c \tag{2.3-33}$$

The next two examples illustrate the lumping process.

**TABLE 2.12 Gear Train Relations**

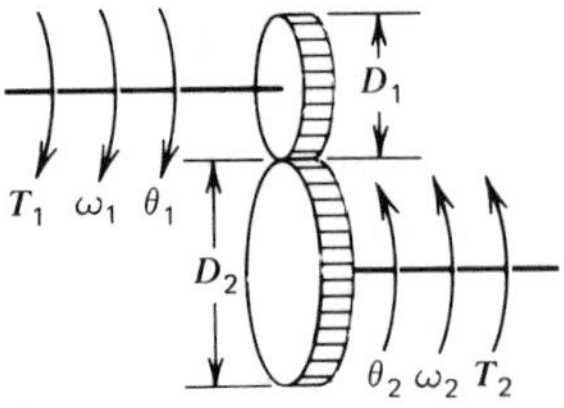


$n$ = number of gear teeth

Gear ratio $\dfrac{D_2}{D_1} = \dfrac{n_2}{n_1} = N$

Velocity ratio $\dfrac{\omega_2}{\omega_1} = \dfrac{n_1}{n_2} = \dfrac{1}{N}$

Displacement ratio $\dfrac{\theta_2}{\theta_1} = \dfrac{n_1}{n_2}$

$= 1/N$

Torque ratio $\dfrac{T_2}{T_1} = \dfrac{n_2}{n_1} = N$

(neglects friction loss)

| **Original Geared System** | **Equivalent Gearless System** |
|---|---|

Inertia

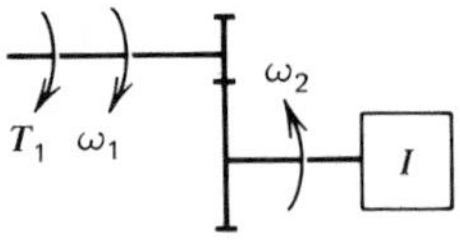


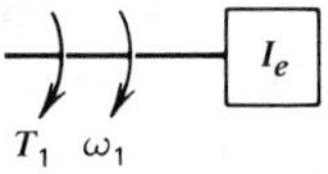


$$I_e = I\left(\frac{\omega_2}{\omega_1}\right)^2 = \frac{1}{N^2} I$$

(neglects gear inertias)

Spring constant

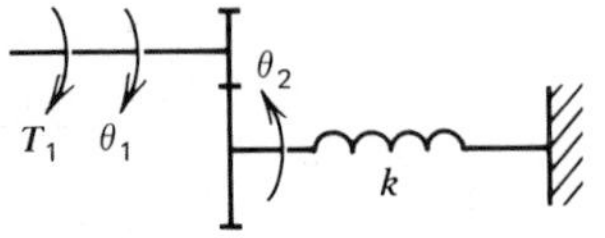


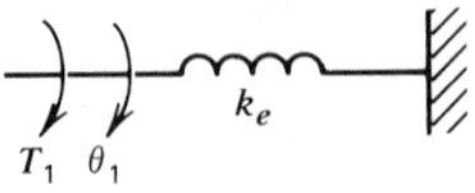


$$k_e = k\left(\frac{\theta_2}{\theta_1}\right)^2 = \frac{1}{N^2} k$$

(neglects gear-tooth elasticity)

Damping constant

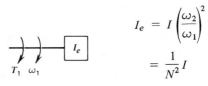

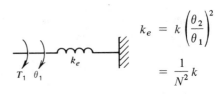

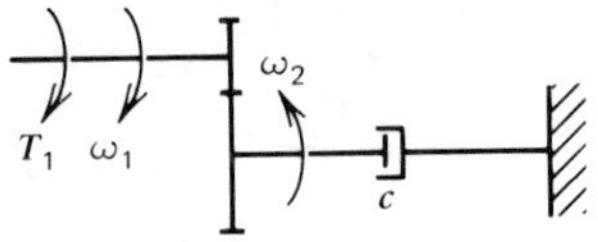


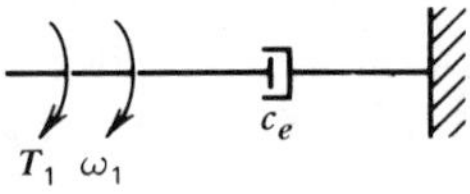


$$c_e = c\left(\frac{\omega_2}{\omega_1}\right)^2 = \frac{1}{N^2} c$$

(neglects gear friction)

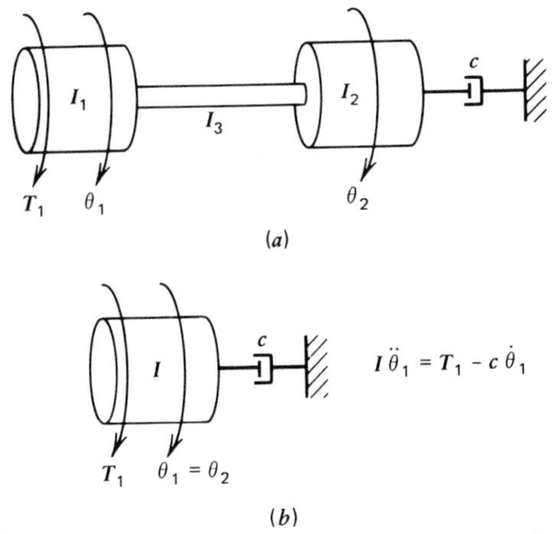


**Figure 2.12 System with three inertias. (*a*) Original system. (*b*) Lumped-parameter equivalent system.**

*Example 2.6*

For systems with more than one inertia on a single shaft, it is frequently possible to add the inertias to produce an equivalent system. A pump is represented in Figure 2.12. The pump impeller $I_2$ is driven by the shaft $I_3$ connected to an electric motor producing a torque $T_1$ on its rotor $I_1$. The viscous friction of the fluid acting on the impeller is represented by the damper. A model is desired for the behavior of the angular displacement $\theta_2$ due to the applied torque $T_1$. Discuss how a lumped-parameter model in the form of (2.3-6) can be developed (Figure 2.12*b*).

A good initial assumption might be to neglect the twist $\theta_1 - \theta_2$ between the inertias $I_1$ and $I_2$. This corresponds to neglecting the potential energy in the shaft and implies that $\theta_1 = \theta_2$ and $\omega_1 = \omega_2$. Equating the kinetic energy of the original and equivalent systems gives

$$\tfrac{1}{2} I\omega^2 = \tfrac{1}{2} I_1\omega_1^2 + \tfrac{1}{2} I_2\omega_1^2 + \tfrac{1}{2} I_3\omega_1^2$$

With a coordinate reference on the motor shaft $\omega = \omega_1$, and thus

$$I = I_1 + I_2 + I_3 \qquad (2.3\text{-}34)$$

The equivalent system model is given by (2.3-6) with $I$ as shown and $T = T_1 - c\omega$. If the damping force is large or if the torque $T_1$ is applied suddenly, the shaft twist will become important, and a model incorporating shaft elasticity will be needed.

*Example 2.7*

Suppose that the pump motor drives the impeller through a gear pair (Figure 2.13). Assume also that the gear and shaft inertias are significant. Neglect the elasticity of the gear teeth and shafts and assume the friction and backlash in the gears are small. Develop a lumped-inertia model of the impeller velocity as a function of the motor torque.

The angular displacements of the gears are related through the gear ratio $N$ such that

$$\theta_B = N\theta_C$$

$$N = \frac{D_C}{D_B} \qquad (2.3\text{-}35)$$

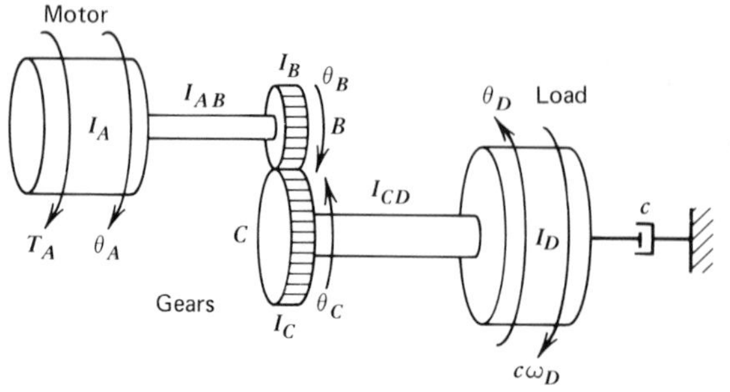


Figure 2.13 Geared system.

Note that a positive displacement $\theta_B$ produces a positive displacement $\theta_C$. Thus a negative sign is not needed in (2.3-35) even though the directions of rotation are opposite. From this we also see that $\omega_B = N\omega_C$.

With negligible shaft elasticity, the twist is zero, and thus

$$\theta_A = \theta_B$$

$$\omega_A = \omega_B$$

$$\theta_C = \theta_D$$

$$\omega_C = \omega_D$$

and

$$\omega_A = N\omega_D$$

Also, we can use the results of (2.3-34) to obtain $I_1$, the lumped inertia on shaft $AB$, as follows.

$$I_1 = I_A + I_{AB} + I_B$$

Similarly, the lumped inertia on shaft $CD$ is

$$I_2 = I_C + I_{CD} + I_D$$

For a single-inertia model we must reference all quantities to one shaft. Let us choose the motor shaft $AB$ as the reference. Then from Table 2.12, the inertia equivalent to $I_2$ referenced to shaft $AB$ is

$$I_{e2} = \frac{1}{N^2} I_2$$

The total equivalent inertia $I$ is the sum of all the inertias referenced to shaft $AB$, or

$$I = I_1 + I_{e2} \tag{2.3-36}$$

From Table 2.12, the viscous damping acting on shaft $CD$ has an equivalent value on shaft $AB$ of

$$c_e = \frac{1}{N^2} c \tag{2.3-37}$$

Therefore, the desired model is

$$I \frac{d\omega_A}{dt} = T_A - c_e \omega_A \tag{2.3-38}$$

with $I$ and $c_e$ given earlier.

This model can be used to compute the motor torque required to maintain the load at some desired constant speed. For constant speed, (2.3-38) shows the required torque to be

$$T_A = c_e\omega_A = c_e N\omega_D \tag{2.3-39}$$

The time to achieve this speed from rest can also be estimated with this model using the techniques of Chapter Three. The lumping of the two gears into one inertia prevents the model from giving information on the gear-tooth forces, the effects of shaft elasticity, gear-bearing friction, and backlash. Such information is obtained from the multimass, higher-order models considered in later chapters. However, in many applications, the simple model (2.3-38) is adequate.

## Levered Systems

Another example of a mechanical transformer is the lever shown in Table 2.13. The transformer parameter is the lever ratio $a/b$. For small angular displacement $\theta$, geometry shows that $x = a\theta$ and $y = b\theta$. The lever is an ideal transformer (no power loss) if its mass, elasticity, and pivot friction are negligible. In this case the input work $f_1x$ equals the output work $f_2y$. Thus $f_2 = (a/b)f_1$.

We can choose to lump a levered system in terms of any one of the coordinates $x$, $y$, or $\theta$. If the system's mass, elasticity, damping, and input force are concentrated at the left-hand end, the lever has no effect if we choose the coordinate $x$ (Table 2.13). An equivalent model in terms of either $y$ or $\theta$ can be found as for the gear pair, by equating corresponding energy terms. The results are given in Table 2.13. They can also be used for systems whose mechanical properties are located at more than one point, as in the following example.

*Example 2.8*

Derive an equivalent model in terms of the coordinate $x$, and another in terms of $\theta$, for the system shown in Figure 2.14.

Relative to the left-hand end, the equivalent values of $k_2$, $m_b$, and $f_2$ are

$$k_{2e} = k_2(b^2/a^2)$$

$$m_{be} = m_b(b^2/a^2)$$

$$f_{2e} = f_2(b/a)$$

Since $c$ and $m_a$ are located at the point of reference, they are unchanged. The parameters of the complete equivalent system are found by adding the values referenced to the left-hand end. They are

$$k_e = k_{2e}$$

$$c_e = c$$

$$f_e = f_{2e}$$

$$m_e = m_a + m_{be}$$

**TABLE 2.13 Lever Relations**

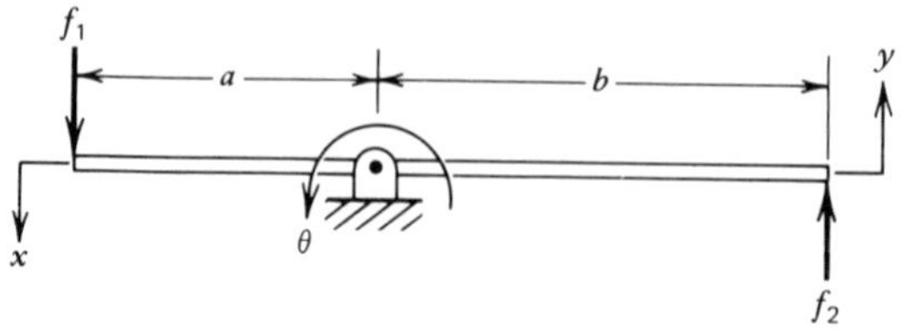


Lever ratio $\dfrac{x}{y} = \dfrac{a}{b}$ (for small displacements)

Force advantage $f_2 = \dfrac{a}{b} f_1$

**Original levered system**

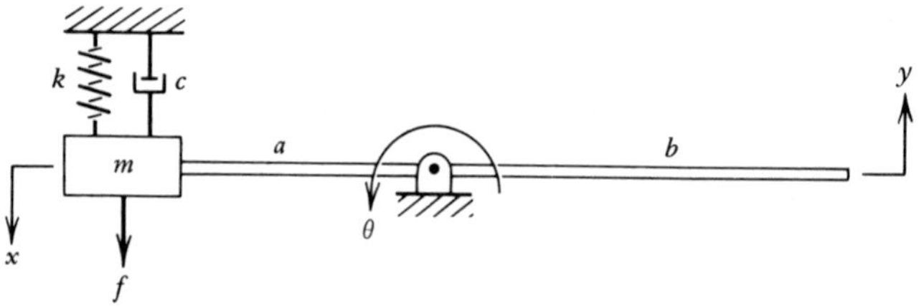


**Equivalent Systems**

(for small displacements and negligible lever mass, lever elasticity, and pivot friction)

Lumped for coordinate $x$

$$m_e = m$$
$$k_e = k$$
$$c_e = c$$
$$f_e = f$$

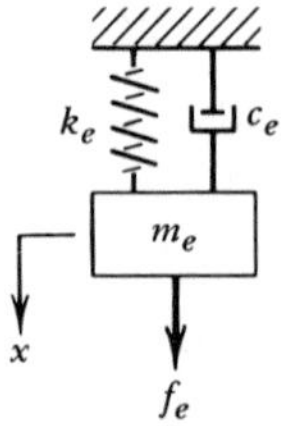


Lumped for coordinate $y$

$$m_e = m(a^2/b^2)$$
$$k_e = k(a^2/b^2)$$
$$c_e = c(a^2/b^2)$$
$$f_e = f(a/b)$$

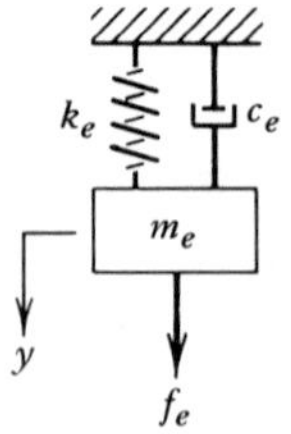


Lumped for coordinate $\theta$

$$I_e = ma^2 \qquad k_e = ka^2$$
$$c_e = ca^2 \qquad T_e = fa$$

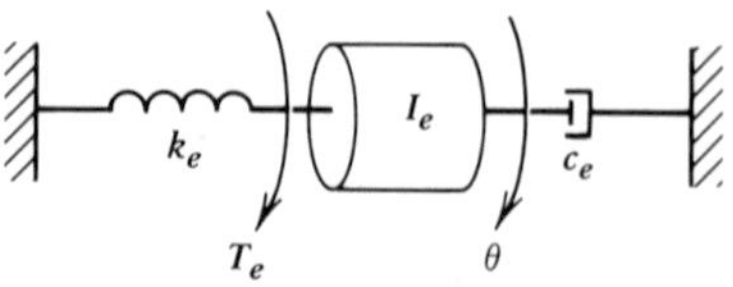

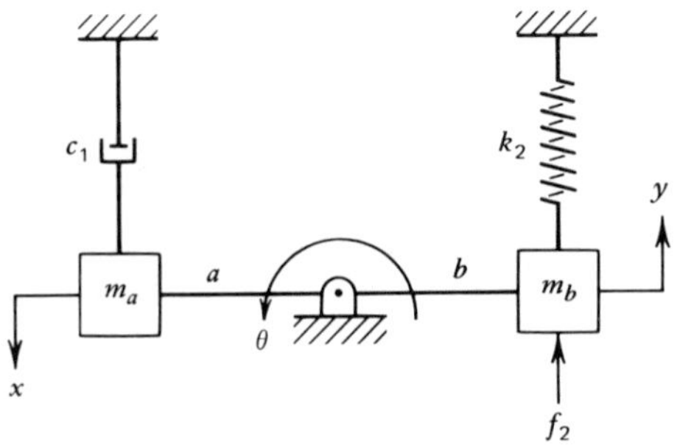


**Figure 2.14** **Levered system.**

Thus the dynamic model is

$$m_e\ddot{x} = -c_e\dot{x} - k_e x + f_e \tag{2.3-40}$$

For a rotational model, all the parameters must be changed because none of them are located at the pivot. The equivalent values at the pivot are

$$I_e = m_a a^2 + m_b b^2$$

$$k_e = k_2 b^2$$

$$c_e = ca^2$$

$$T_e = f_2 b$$

The equivalent dynamic model is

$$I_e\ddot{\theta} = -c_e\dot{\theta} - k_e\theta + T_e \tag{2.3-41}$$

## 2.4 ELECTRICAL SYSTEMS

Our principal interest in electrical systems here is concerned with those devices used for measurement and control, instead of applications in power transmission or communications, for example. The theory of electrical devices can be divided into direct-current (dc) or alternating-current (ac) theory. We emphasize dc theory and applications here primarily because advances in permanent magnet technology have made it possible to use dc motors in many control system applications that formerly required ac motors. Thus the associated circuitry is now more often dc. Since a dc motor is an electromechanical system, the appropriate models are combinations of the rotational models of Section 2.3 and the electrical models of this section. The resulting models are of higher order and are thus deferred to Chapters Five and Six.

Another way to categorize electrical theory is by lumped- versus distributed-parameter models. A lumped (or network) model can be used when the size of the electrical element is small compared to the electrical signal's wavelength. Since electrical signals propagate at the speed of light, the wavelengths are on the order of a mile or more for most application frequencies. Thus network models can be used for most applications of interest here. For this reason we discuss the resistance, capacitance, and inductance relations in terms of lumped-parameter systems.

## Circuit Elements

Resistance is a property that opposes the flow of current. According to the wavelength criterion, the resistance of a relatively short wire can be considered lumped at a point and represented by the resistance element shown in Table 2.14. Its resistance may be added to that of any elements attached to the wire. The voltage-current relation for a lumped-resistance $R$ is

$$i = \frac{v}{R} \tag{2.4-1}$$

Values for typical resistors used as circuit elements range from $R = 10^3\Omega$ to $R = 10^7\Omega$.

Capacitance is the property that allows charge to be stored. The relation between

**TABLE 2.14 Units and Representations for Common Electrical Quantities**[a]

| Quantity and Symbol | Units | Representation |
|---|---|---|
| Voltage $v$ | volt (v) | Voltage source 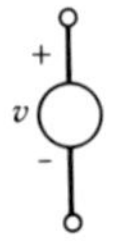 |
| Charge $Q$ | coulomb = N-m/volt | |
| Current $i$ | ampere (amp) = coulomb/sec | Current source  |
| Flux $\phi$ | volt-sec | |
| Resistance $R$ | ohm ($\Omega$) = volt/amp |   |
| Capacitance $C$ | farad (F) = coulomb/volt | 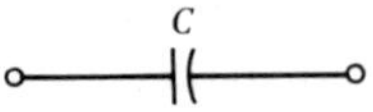  |
| Inductance $L$ | henry (H) = volt-sec/amp |   |
| Battery (voltage source plus internal resistance) | — | 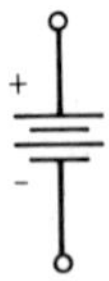 |
| Ground (zero voltage reference) | — |  |

[a] British Engineering and SI units are identical for electrical systems.

the stored charge and the voltage rise so produced is described by

$$v(t) = \frac{1}{C}\int_0^t i\,dt + \frac{Q_o}{C} \tag{2.4-2}$$

where $Q_o$ is the initial charge and the integral is the charge stored up to time $t$. A capacitor can be made by separating two conductors, such as metallic plates, with a nonconducting material such as an insulator or dielectric medium. Capacitors typically used have values between $C = 10^{-12}F$ and $10^{-4}F$.

A magnetic field (a flux) surrounds a moving charge or current. If the conductor of the current is coiled, the flux created by the current in one loop affects the adjacent loops. This flux is proportional to the time integral of the applied voltage, and the current is proportional to the flux. Thus

$$\phi \propto \int_0^t v\,dt$$

$$i \propto \phi$$

Combining these relations gives the voltage-current relation for an inductor.

$$i(t) = \frac{1}{L}\int_0^t v\,dt \tag{2.4-3}$$

The constant of proportionality is the inductance $L$. Most inductors have an inductance of less than $1H$.

The wire winding in a real inductor means that the element has resistance as well as inductance. Thus the real inductor can be modeled as an ideal inductor in series with an ideal resistor. Similar statements hold for real resistors, capacitors, and conductors. They can exhibit to some extent all three properties (resistance, capacitance, and inductance), and the appropriate circuit models might have to include these effects.

Circuit elements may also be classified as active or passive. Passive elements such as resistors, capacitors, or inductors are not sources of energy, although the last two can store it temporarily. The active elements are energy sources that drive the system. Their physical type can be chemical (batteries), mechanical (generators), thermal (thermocouples), and optical (solar cells).

Active elements are modeled as either ideal voltage sources or ideal current sources. Their circuit symbols are given in Table 2.14. An ideal voltage source supplies the desired voltage no matter how much current is drawn by the loading circuit. The opposite is true for a current source. Obviously no real source behaves exactly this way. For example, battery voltage drops because of heat generation as more current is drawn. If the current is small or constant in a given application, the battery can be treated as an ideal source. If not, the battery is frequently modeled as an ideal voltage source plus an internal resistance whose value can be obtained from the voltage-current curve for the battery.

## Circuit Laws

The causal relations summarized by the tetrahedron of state must be supplemented with other physical laws in order to model a given system. For electrical circuits, *Kirchhoff's current* and *voltage laws* are needed. The current law (or node law) is a statement of charge conservation. At a *node* (where two or more conductors are joined) the law states that the sum of the currents entering the node equals the sum of currents leaving the node. Recall that the positive direction of current flow (shown by an arrow) is opposite to the flow of negative charge.

Kirchhoff's voltage law (or loop law) states that the instantaneous algebraic sum of the voltages around any loop in a circuit is zero; that is, the sum of the voltage drops equals the sum of the voltage rises around the loop. The law is an expression of conservation of energy. The point in the circuit at which the voltage is zero is indicated by the *ground* symbol given in Table 2.14. It establishes the reference point against which voltages in other parts of the circuit are measured or specified. When we are interested only in relative voltage values and not absolute values, the ground symbol is often omitted.

## Applications

We now illustrate the preceding modeling concepts with several applications.

### *Example 2.9*

An example of a classical electrical circuit application is the dc generator shown in Figure 2.15, similar to that used on diesel locomotives. Diesel engine efficiency varies greatly with speed. This arrangement allows the diesel to be run at its optimum speed, while locomotive speed control is obtained by using a dc motor that is easily controlled and efficient over a wide speed range. A small voltage $v$ from the engineer's throttle is applied to the generator circuit. This voltage is amplified by the diesel turning an armature (a conductor wrapped around a ferromagnetic core) in the field

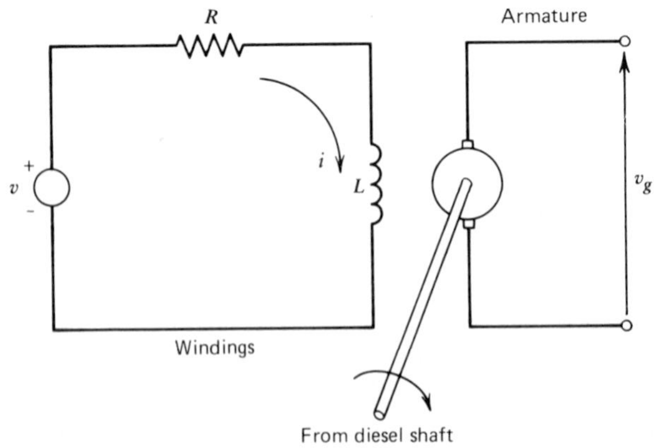


**Figure 2.15 Direct-current generator.**

created by the windings whose inductance is $L$. The voltage induced in the armature by its motion in the field is greater than the voltage $v$ because of the power supplied by the diesel. This amplified voltage $v_g$ is applied to the dc motor and is given by

$$v_g = K_g i \tag{2.4-4}$$

where $K_g$ is the generator constant that depends on the armature and winding construction. The winding resistance $R$ and inductance $L$ are assumed to be lumped at a point. Develop a model for the generated voltage $v_g$ to determine how long it takes $v_g$ to respond to a change in $v$.

The basic principles to be used are conservation of charge and of energy. For this circuit the former simply states that the current through each element is the same since no capacitive storage elements exist in the circuit. Application of Kirchhoff's voltage law to the winding's circuit, along with the constitutive relations (2.4-1) and (2.4-3), gives

$$v - iR - L\frac{di}{dt} = 0 \tag{2.4-5}$$

The desired model consists of this equation and (2.4-4). The energy is stored in the inductance compartment in the winding field. Thus an appropriate state variable is the current $i$. Given $v$, (2.4-5) can be solved with the methods of Chapter Three to obtain $i$, and thence $v_g$ from (2.4-4).

## *Example 2.10*

A circuit whose dynamics result from the capacitance-resistance causal path is the series $RC$ circuit shown in Figure 2.16 with a voltage source $v$. Develop a model for the voltage $y$ across the capacitor.

The voltage law states that

$$v - iR - y = 0$$

where the voltage $y$ across the capacitor $C$ is related to the current $i$ as follows.

$$y = \frac{1}{C}\int_0^t i\,dt + \frac{Q_0}{C}$$

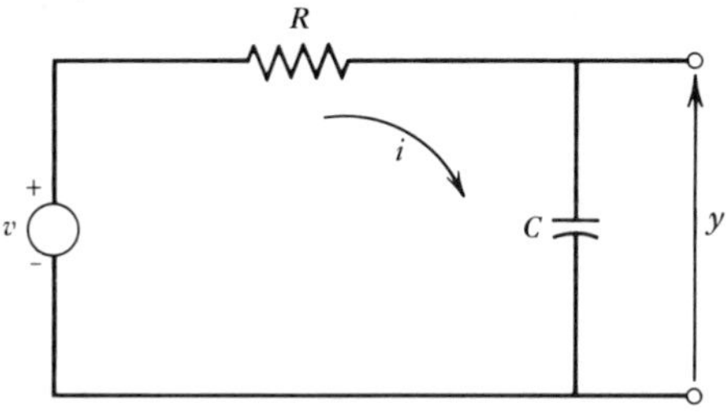


**Figure 2.16** Series *RC* circuit.

Differentiating this with respect to time and substituting for $i$ from the first equation gives

$$\frac{dy}{dt} = \frac{1}{C} i = \frac{v - y}{RC}$$

or

$$C \frac{dy}{dt} = -\frac{1}{R} y + \frac{1}{R} v \tag{2.4-6}$$

This circuit has many applications in filtering unwanted voltage variations from the input voltage $v$. The storage characteristics of the capacitor prevent rapid changes in the voltage $v$ from affecting the capacitor voltage $y$, just as the storage tank in Figure 2.1 dampens sudden changes in the input flow rate $q_i$. The model given by (2.4-6) is useful for determining the values of $R$ and $C$ in terms of the frequencies to be filtered out.

### *Example 2.11*

Another circuit with significant applications is the op amp circuit shown in Figure 2.17. Show that this device integrates the input voltage over time.

In Chapter One the op amp was introduced as a high-gain ($10^5$ to $10^9$) voltage amplifier. Thus the current drawn by the amplifier to produce the voltage $v_o$ is negligible and the voltage $v_1$ is nearly zero. Consequently the currents $i_1$ and $i_2$ are approximately equal, and

$$\frac{v_i - v_1}{R} = i_1 = i_2 = C \frac{d(v_1 - v_o)}{dt}$$

where $v_i$ is the input voltage. Since $v_1$ is approximately zero,

$$C \frac{dv_o}{dt} = -\frac{v_i}{R}$$

or

$$v_o(t) = -\frac{1}{RC} \int_0^t v_i \, dt \tag{2.4-7}$$

if the output voltage $v_o$ is initially zero. The output voltage can be seen to be proportional in magnitude to the integral of the input voltage, with a sign reversal. This simple circuit occurs in many useful devices for computing, signal generation, and control, some of which will be analyzed in later chapters.

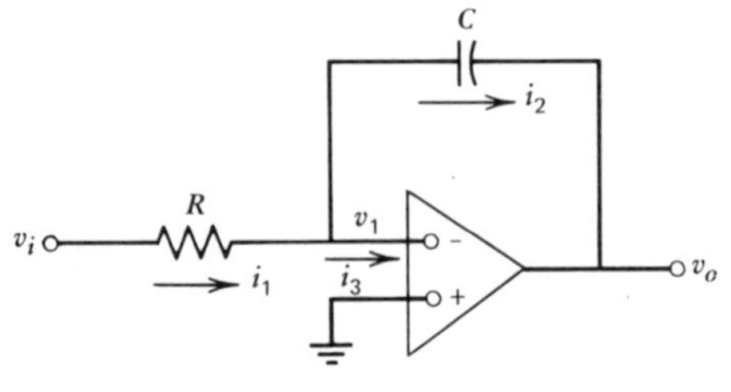


**Figure 2.17 Op-amp integrator.**

## Analogies

Different physical systems that can be represented by the same mathematical model are *analogous* systems. For example, comparison of the voltage-current relation for a resistor (2.4-1) with the laminar pipe-flow relation (2.2-7) shows that voltage drop is analogous to pressure drop, and that flow rate is analogous to current. The concept of analogous systems is useful for two reasons.

1. It allows measurements taken on one system to be extrapolated to make predictions concerning the behavior of an analogous system. For the analogy given above, current measurements taken on a properly arranged network of resistors can be used to predict the flow rates through an analogous fluid system. The assumption is that the system to be measured is easier to build than its analogous counterpart.
2. It allows insight developed for one system to be applied to an understanding of the analogous system.

The use of analogies as computational aids (the so-called *analog computer*) has rapidly declined as digital computers have increased in efficiency, availability, and flexibility. Op amp circuitry is now the most frequently used analog method and often provides a convenient way of analyzing dynamic problems. However, the technique is inconvenient and expensive unless one is in a situation requiring its frequent use. Appendix E provides more details for setting up the computer circuit.

The analogy between electrical systems and other system types is frequently used as a foundation for a modeling approach or philosophy. For some analysts, a mechanical system is more easily visualized as a set of resistors, capacitors, and inductors, while others prefer to work with the laws of the original physical system. Analysts should adopt the approach that is most comfortable for them. Analogies might help by providing a different viewpoint but are no substitute for understanding the physics of the particular situation. The assumptions made in order to develop a model can easily be forgotten if the modeler becomes engrossed in analyzing the "equivalent" system.

# 2.5 LIQUID-LEVEL SYSTEMS AND PIPE FLOW

Storage tanks and their connecting pipe networks are found in many industrial applications. The effort variable in such systems may be either pressure $p$ or "head" $h$, the liquid height producing hydrostatic pressure. The two are related by the fluid-weight density $\rho g$ as $p = \rho g h$. The fluid quantity can be either mass $m$ or volume $V$, with the rate variable selected as mass flow $q_m$ or volume flow rate $q$.

If volume $V$ is selected as quantity, the container's capacitance relation describes head as a function of volume. For a tank with vertical sides and bottom area $A$ (see Figure 2.18), this relation is $h = V/A$. Thus this tank's capacitance is $A$. If pressure is taken as the effort variable, the tank's capacitance relation is $p = \rho g\, V/A$, and the capacitance is $A/\rho g$. The term *compliance* is generally used instead of *capacitance* when pressure is the effort. The compliance of a fluid system is found by computing

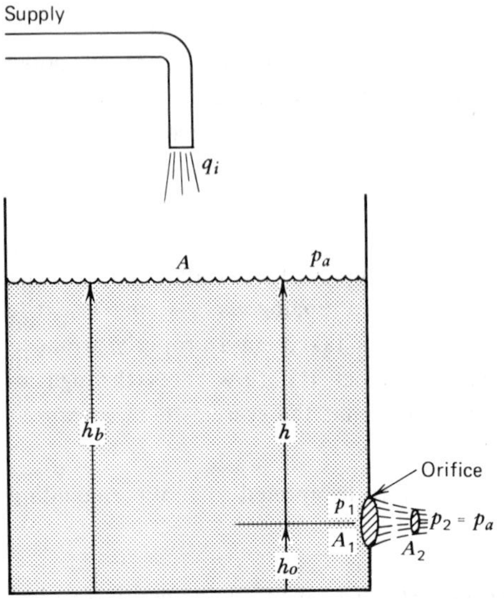


**Figure 2.18 Tank with an orifice.**

the pressure $p$ required to force a given quantity of fluid into the system. For incompressible fluids with volume as quantity, the compliance is $V/p$.

Fluid-flow resistance can result from orifices, wall friction in conduits such as pipes, and restrictions such as valves, elbows, and other fittings. The resistance relates the effort to the rate, and vice versa. Usually head and volume flow rate are selected as a pair for liquid systems. The conversion to mass flow is easily made via the fluid mass density. The resistance relation can be linear or nonlinear depending on flow conditions. Here we present some commonly used formulas with a minimum of fluid-flow theory because of space limitations. Detailed development can be found in most fluid mechanics texts. Our purpose here is to give the student some idea of how resistance calculations are made and a feeling for the circumstances in which the formulas are valid.

The inductance path is not discussed here because in most applications of interest to us, the fluid-inertia or kinetic-energy storage compartment is negligible compared to the capacitance compartment. Its inclusion leads to much more detailed models than the ones to be presented. Often such detail is not warranted.

Typical fluid mechanics units and conversion factors are given in Table 2.15. Some fluid properties of two common liquids, water and fuel oil, are given in Table 2.16.

## A Tank with an Orifice

Consider the tank shown in Figure 2.18 with vertical sides, bottom area $A$, and a circular orifice of area $A_1$. Assume that atmosphere pressure $p_a$ acts at the liquid surface and at the orifice outlet. The hydrostatic pressure resulting from the weight of a column of liquid of height $h$ is

$$p = \rho g h \tag{2.5-1}$$

TABLE 2.15 Typical Units in Fluid Mechanics

| Quantity and Symbol | British Engineering[a] | SI[a] | Conversion Factor[b] | Other Common Units[a] |
|---|---|---|---|---|
| Pressure $p$ | lb/ft² (psf) | N/m² (pascal) | 47.88 | lb/in.² (psi)<br>atmospheres (atm)[c]<br>millimeters of mercury (mm Hg)[d] |
| Volume flow rate $q$ | ft³/sec (cfs) | m³/sec | 0.02832 | gallons/minute (gpm)[e] |
| Mass density $\rho$[f] | slugs/ft³ | kg/m³ | 515.4 | – |
| Viscosity $\mu$ | lb-sec/ft² | N-sec/m² | 47.88 | Centipoise[g] |

[a] Other names or abbreviations are shown in parentheses.

[b] To convert from British Engineering Units to SI units, multiply the British units by this factor.

[c] 1 atmosphere = $1.01325 \times 10^5$ N/m² = 14.6959 psi = 760 mm Hg.

[d] 1 mm Hg at 0°C = 133.322 N/m².

[e] 1 U.S. liquid gallon = 0.13368 ft³.

[f] Weight density $\gamma = \rho g$ where $g = 32.174$ ft/sec² = 9.8066 m/sec².

[g] 1 centipoise = $10^{-3}$ N-sec/m².

with $\rho$ the liquid mass density and $g$ the acceleration due to gravity. The relation between the fluid velocity $v$ through the orifice and the pressure drop $(p_1 - p_2)$ across the orifice can be shown to be

$$v = \sqrt{\frac{2}{\rho}(p_1 - p_2)} \tag{2.5-2}$$

This is Torricelli's formula and dates from 1640.

For our case, $p_2 = p_a$, $p_1 = \rho g h + p_a$, and we obtain the following expression for the volume flow rate $q$ through the orifice.

$$q = A_1 v = A_1 \sqrt{2gh} \tag{2.5-3}$$

Experiments and advanced fluid-flow theory reveal that the jet area decreases to a minimum $A_2$ at a distance of about one orifice diameter and that the relation between

TABLE 2.16 Properties of Water and Fuel Oil

| Fluid | Mass Density $\rho$ (slugs/ft³) | Viscosity $\mu$ (lb-sec/ft²)[a] | Specific Gravity[b] |
|---|---|---|---|
| Water at 68°F | 1.94 | $2 \times 10^{-5}$ | 1 |
| Fuel oil at 68°F | 1.88 | $2 \times 10^{-2}$ | 0.97 |

[a] Viscosity is usually highly temperature dependent.

[b] Specific gravity is the ratio of the density to that of pure water at 4°C and a pressure of 1 atmosphere.

$A_2$ and $A_1$ is approximately given by $A_2 = 0.61A_1$, if we assume that the jet does not bend downward significantly. Also, frictional effects are not completely negligible. These factors lead to an improved equation for the volume discharge rate.

$$q = C_d A_1 \sqrt{2gh} \tag{2.5-4}$$

where $C_d$, the *discharge coefficient*, is approximately 0.60 for water if $h \geqslant 1.5$ ft and the orifice diameter $d \geqslant 0.2$ ft. Values for $C_d$ under other conditions and for different fluids are found in standard reference works (Reference 4, Chapter 8). This is a non-linear constitutive relation between the rate variable $q$ and the effort variable $h$.

The liquid mass in the tank is $\rho A h_b$. From conservation of mass, the rate of change of this mass must equal the difference between the mass inflow and outflow rates; that is,

$$\frac{d(\rho A h_b)}{dt} = \rho q_i - \rho q \tag{2.5-5}$$

For an incompressible fluid the density $\rho$ is constant and cancels from the equation. The orifice height $h_o$ is constant as is the tank area $A$. Thus

$$A \frac{dh_b}{dt} = A \frac{d(h + h_o)}{dt} = A \frac{dh}{dt} = q_i - q$$

Substituting (2.5-4) gives the final system model.

$$A \frac{dh}{dt} = q_i - C_d A_1 \sqrt{2gh} \tag{2.5-6}$$

## Pipe Flow

Industrial processes, such as those in a refinery, require the transport of large quantities of fluids such as liquid chemicals and steam. On the other hand, hydraulic and pneumatic control components use comparatively low flow rates in small-diameter lines. Both extremes involve the motion of a fluid in a conduit.

Assume that Newton's viscosity law (2.3-26) holds. For flow in which the average particle velocity is equal to the actual velocity (*laminar* flow), it can be shown that the volume flow rate through a level circular pipe is

$$q = \frac{\pi D^4}{128 \mu L}(p_1 - p_2) \tag{2.5-7}$$

where $D$ and $L$ are the pipe's diameter and length, and $\mu$ is the fluid's viscosity. The pressures at the pipe ends are $p_1$ and $p_2$. This is the *Hagen-Poiseuille law* (Reference 4). It can be generalized to the case where the pipe is not level by replacing $p_1$ with $(p_1 + \rho g h_1)$ and $p_2$ with $(p_2 + \rho g h_2)$, where $h_1$ and $h_2$ are the heights of the ends above the same arbitrary reference point. The Hagen-Poiseuille law strictly applies only to steady-state conditions in which the effects of the pipe entrance on the flow are negligible. However, the law has been found to give satisfactory answers in many situations involving a pressure drop that changes with time.

If the effort variable is taken to be pressure, (2.5-7) specifies the resistance relation between effort and rate.

$$R_L q = p_1 - p_2 \tag{2.5-8}$$

Comparison of the last two expressions shows that the resistance for laminar flow in a pipe is

$$R_L = \frac{128\mu L}{\pi D^4} \tag{2.5-9}$$

where the subscript $L$ has been included to denote laminar resistance.

Flow in which the average particle velocity does not equal the actual velocity is *turbulent*. A useful criterion for predicting the existence of turbulence is the *Reynolds number* $R_e$, the ratio of the fluid's inertial forces to the viscosity forces. For a circular pipe,

$$R_e = \frac{\rho \bar{v} D}{\mu} \tag{2.5-10}$$

where $\bar{v} = q/(\pi D^2/4)$, the average fluid velocity.

For $R_e > 2300$ the flow is often turbulent, while for $R_e < 2300$ laminar flow usually exists. The precise value of $R_e$ above which the flow becomes turbulent depends on, for example, the flow conditions at the pipe inlet. However, the criterion is useful as a rule of thumb.

The resistance relation for turbulent flow is nonlinear and of the form

$$R_T q^2 = p_1 - p_2 \tag{2.5-11}$$

where $R_T$ is the turbulent resistance. The resistance depends on the pipe length and diameter, and on fluid density. The equality in (2.5-11) is made possible by the introduction of the *friction factor* $f$, an empirically determined quantity. For turbulent pipe flow,

$$R_T = \frac{8fL\rho}{\pi^2 D^5} \tag{2.5-12}$$

The friction factor is a function of the Reynolds number. It is approximately constant for high Reynolds numbers. This asymptotic value of $f$ is a function of the pipe's diameter, material, and finish. Friction factor charts are readily available; for example, see Reference 4. Some typical values are given in Table 2.17.

Components such as valves, elbows, etc., resist flow and can be described by (2.5-11) where

$$R_T = \frac{8\rho C_L}{\pi^2 D^4} \tag{2.5-13}$$

with $D$ usually taken to be the diameter of the outlet side of the restriction. The dimensionless number $C_L$ is the *head-loss coefficient*, so named because the pressure drop caused by the restriction is equivalent to a decrease in the head driving the flow. Values of $C_L$ are available in manufacturer's literature or in handbooks (Reference 5). Values for common components are given in Table 2.17.

The orifice flow equations (2.5-2) and (2.5-4) can be combined to show that

TABLE 2.17 Friction Factors and Head-Loss Coefficients for Common Components

**Pipe Friction Factor $f$ (dimensionless)**[a]

| | $R_e$ | | | |
|---|---|---|---|---|
| | $10^4$ | $10^5$ | $10^6$ | $10^7$ |
| Commercial steel | | | | |
| 1 in. diameter | 0.032 | 0.025 | 0.023 | 0.023 |
| 10 in. diameter | 0.03 | 0.0195 | 0.005 | 0.0045 |
| Concrete (rough) | | | | |
| 10 in. diameter | 0.053 | 0.049 | 0.049 | 0.049 |
| 100 in. diameter | 0.034 | 0.023 | 0.021 | 0.020 |

**Head Loss Coefficient $C_L$ (dimensionless)**[b]

| | |
|---|---|
| Standard 90° elbow | 0.75 |
| Coupling | 0.04 |
| Gate valve | |
| Fully open | 0.70 |
| 1/4 open | 25.0 |
| Well-rounded pipe entrance | 0.04 |
| Sudden contraction | |
| Area ratio of 10% | 0.37 |
| Area ratio of 90% | 0.02 |

[a] From data given by Moody (L. F. Moody, "Friction Factors for Pipe Flow," *Transactions ASME*, Vol. 66, No. 8, 1944, pp. 671–684). Used with permission.

[b] Extracted from Table 5–19 in *The Chemical Engineer's Handbook*, R. Perry, ed., Fourth Edition, McGraw-Hill, New York, 1963. Used with permission.

(2.5-11) applies. Comparison of these expressions shows the orifice resistance to be

$$R_o = \frac{\rho}{2C_d^2A_1^2} \tag{2.5-14}$$

The resistance laws and formulas for laminar and turbulent pipe flow, orifice flow, and flow through restrictions are summarized in Table 2.18. The symbol shown is used to represent all types of fluid resistance.

## Applications

In liquid-level systems such as shown in Figure 2.19, energy is stored in two ways: as potential energy in the mass of liquid in the tank, and as kinetic energy in the mass of liquid flowing in the pipe. The transfer of energy between these compartments is opposed by the resistances, and energy is dissipated from the system by this process.

The development of any model involves a trade-off between accuracy and utility or, equivalently, between generality and simplicity. Thus some of the less

**TABLE 2.18 Fluid Resistance Formulas**

Resistance symbol

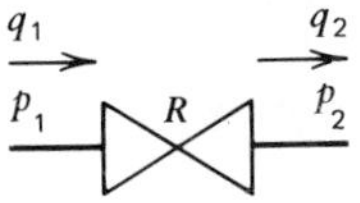


$$p = p_1 - p_2 = \text{pressure drop}$$
$$q = q_1 = q_2 = \text{flow rate}$$

Formulas (if $q$ is volume flow rate)[a]

Linear law $p = Rq$

Laminar pipe flow $R = R_L = \dfrac{128\mu L}{\pi D^4}$ (2.5-9)

Nonlinear law $p = Rq^2$

Orifice flow $R = R_o = \dfrac{\rho}{2C_d^2 A_1^2}$ (2.5-14)

Turbulent pipe flow $R = R_T = \dfrac{8fL\rho}{\pi^2 D^5}$ (2.5-12)

Turbulent flow in fittings $R = R_T = \dfrac{8\rho C_L}{\pi^2 D^4}$ (2.5-13)

[a] For mass flow rates, linear resistance is divided by mass density $\rho$; nonlinear resistance is divided by $\rho^2$.

important energy storage compartments might have to be ignored if the model is to be kept simple enough to be solved or evaluated conveniently. Here if the mass of liquid in the pipe is small enough or is flowing at a small enough velocity, the kinetic energy contained in it will be negligible compared to the potential energy in the tank. The dividing line between what is "small enough" and "not small enough" is frequently identifiable only with difficulty. The designer's experience, data from similar systems,

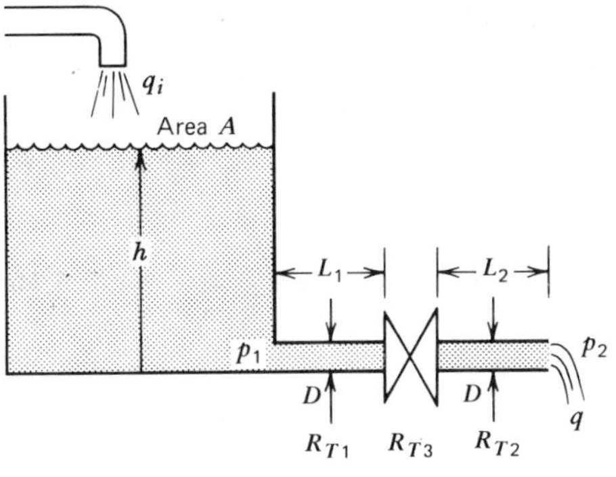


**Figure 2.19 Tank-pipe-valve system.**

and the degree of accuracy required of the model all play a role in determining when such approximations may be used. If the kinetic energy of the liquid is significant, more advanced fluid-flow theory is required. This is usually not the case for the scope of applications considered here, since the high-velocity applications usually involve low-density fluids such as air.

Our resistance formulas apply only beyond the entrance length $L_e$, which is the distance from the pipe entrance beyond which the velocity profile no longer changes with increasing distance. Beyond this point the flow is said to be fully developed. For laminar flow $L_e$ is given approximately by

$$L_e = 0.06DR_e \tag{2.5-15}$$

Since laminar flow can be expected only if $R_e < 2300$, $L_e$ might be as long as 138 pipe diameters. Of course, for small Reynolds numbers, $L_e$ is shorter. The increased mixing in turbulent flow causes the profile to develop faster so that $L_e$ is usually between 25 and 40 diameters for such flow. The smaller $L_e$ is relative to the pipe length, the better will be our resistance calculations. If $L_e$ is large, the resistance calculations require a partial differential equation, distributed-parameter model to describe the development of the velocity profile down the pipe length.

### *Example 2.12*

Develop a model for the behavior of the liquid height $h$ in the system shown in Figure 2.19. A component such as a valve has been inserted between the two lengths of pipe. Assume that turbulent flow exists and show that the total turbulent resistance is the sum of all element resistances.

Denote the pressure drop over the length of pipe 1 by $\Delta p_1$, that over pipe 2 by $\Delta p_2$, and that across the component by $\Delta p_3$. Then from (2.5-11),

$$R_{T1}q^2 = \Delta p_1$$

$$R_{T2}q^2 = \Delta p_2$$

$$R_{T3}q^2 = \Delta p_3$$

where the flow rate $q$ is the same through each element. The pipe resistances $R_{T1}$ and $R_{T2}$ are found from (2.5-12), and $R_{T3}$ from (2.5-13). The total pressure drop across all three elements is the sum of the drops across each element. Thus

$$p_1 - p_2 = R_{T1}q^2 + R_{T2}q^2 + R_{T3}q^2$$

or

$$p_1 - p_2 = R_T q^2$$

where

$$R_T = R_{T1} + R_{T2} + R_{T3} \tag{2.5-16}$$

The series law is therefore valid.

Since $p_1 - p_2 = \rho g h$ we have

$$q = \sqrt{\frac{\rho g h}{R_T}}$$

and from conservation of volume (mass),

$$A\frac{dh}{dt} = q_i - \sqrt{\frac{\rho g h}{R_T}} \tag{2.5-17}$$

*Example 2.13*

The reservoir system shown in Figure 2.20 is typical of some town water-supply arrangements. The large reservoir (a natural or artificial lake) supplies water to the smaller reservoir in which the water is treated for consumption. It is required to have a model of this system in order to find the necessary conduit diameters and to size the smaller reservoir based on an anticipated consumption rate schedule.

We assume that the lake is large enough that its level is relatively unaffected by the withdrawal of water to the lower reservoir. Thus the system input is taken to be a pressure source, in contrast to the flow-rate inputs seen thus far. The volume of fluid in the conduit is assumed to be much smaller than in the reservoirs, and its compressibility is neglected. Thus the only energy storage compartment of interest is the smaller reservoir. Note that substantial energy storage takes place in the lake, but the energy level does not change and hence is of interest only for its input pressure.

The next step is to select a horizontal datum to which all heights are referenced (see Figure 2.20). Mass (volume) conservation in the lower reservoir gives

$$A\frac{d(h_2 - z_2)}{dt} = A\frac{dh_2}{dt} = q_1 - q_o$$

The flow rate $q_1$ is determined from the laminar or turbulent resistance formulas, with the pressure drop across the conduit taken as $(p_1 - p_2)$. If laminar flow is assumed,

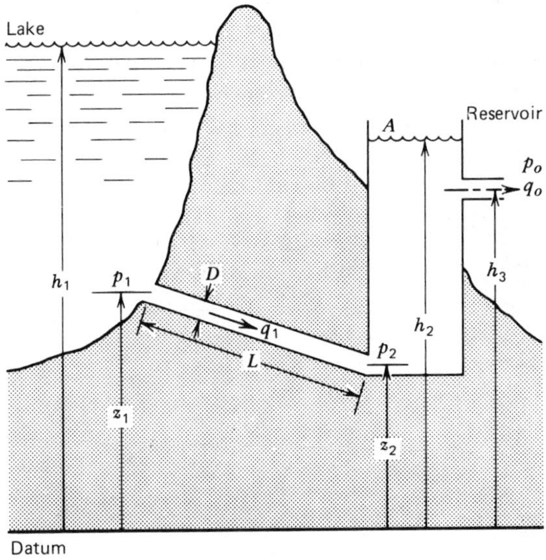


**Figure 2.20 Water-supply reservoir.**

the Hagen-Poiseuille law for an unlevel conduit can be used [(2.5-7) and following]. This gives

$$q_1 = \frac{\pi D^4}{128\mu L}(p_1 + \rho g z_1 - p_2 - \rho g z_2)$$

where $D$ is the conduit diameter and $L$ is its length. The inlet pressure is

$$p_1 = \rho g(h_1 - z_1)$$

Also,

$$p_2 = \rho g(h_2 - z_2)$$

so that

$$q_1 = \frac{\pi D^4 \rho g}{128\mu L}(h_1 - h_2)$$

The height difference $(h_1 - h_2)$ produces a hydrostatic pressure difference. Note that the difference $(z_1 - z_2)$ in the elevation of the conduit's ends does not influence the flow rate. The model is completed on specification of the demand flow rate $q_o$. After separating the given inputs from the unknown $h_2$, we obtain the final model.

$$A\frac{dh_2}{dt} + \frac{\pi D^4 \rho g}{128\mu L}h_2 = \frac{\pi D^4 \rho g}{128\mu L}h_1 - q_o \qquad (2.5\text{-}18)$$

This model is also a representation of a *surge tank* used to dampen fluctuations in flow rate from the reservoir to a hydraulic turbine used to generate electric power. The surge tank is an example of how a capacitance can be used to reduce the effects of rapid changes. This property is discussed in more detail in the next section.

## 2.6 HYDRAULIC DEVICES

A *hydraulic* device uses an incompressible liquid, such as oil, for its working medium. These devices are widely used with high pressures to obtain large forces in control systems, allowing large loads to be moved or high accelerations to be obtained. Here we analyze three devices commonly found in hydraulic systems: (1) *dampers* limit the velocity of mechanical elements; (2) *hydraulic servomotors* produce motion from a pressure source; (3) *accumulators* reduce fluctuations in pressure or flow rates.

### A Fluid Damper

A *damper* or *dashpot* is a mechanical device that exerts a force as a result of a velocity difference across it. One common design relying on fluid friction is shown in Figure 2.21

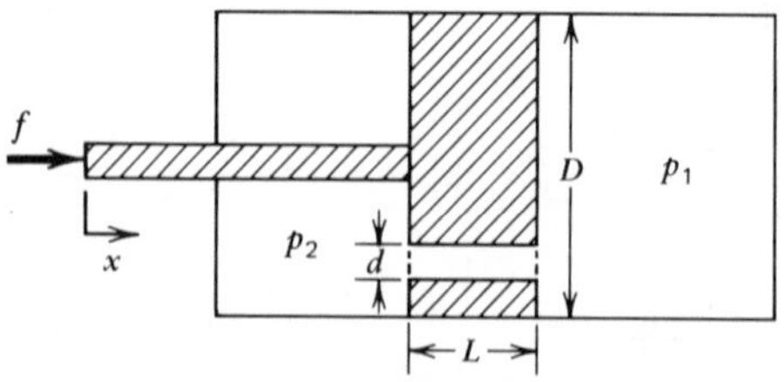


**Figure 2.21 Hydraulic damper.**

and is similar to that used in automotive shock absorbers. A piston of diameter $D$ and thickness $L$ has a cylindrical hole of diameter $d$ drilled through it. The housing is assumed to be stationary here, but the results can easily be generalized to include a movable housing. The piston rod extends out of the housing, which is sealed and filled with a viscous incompressible fluid, such as an oil. A force $f$ acting on the piston rod creates a pressure difference $(p_1 - p_2)$ across the piston such that if the acceleration is small,

$$f = A(p_1 - p_2) = A\,\Delta p \tag{2.6-1}$$

The net cross-sectional piston area is

$$A = \pi\left(\frac{D}{2}\right)^2 - \pi\left(\frac{d}{2}\right)^2$$

For a very viscous fluid at relatively small velocities, laminar flow can be assumed. If the length $L$ of the hole is much greater than the entrance length $L_e$, as is probably the case because $R_e$ is small, fully developed flow exists over most of the length, and the Hagen-Poiseuille law (2.5-7) can be applied to give

$$q = \frac{\pi d^4}{128\mu L}\Delta p = C_1\,\Delta p \tag{2.6-2}$$

The volume flow rate $q$ is also given by

$$q = A\frac{dx}{dt} \tag{2.6-3}$$

because the fluid is incompressible. Substitution of (2.6-1) and (2.6-2) into (2.6-3) shows that

$$f = c\frac{dx}{dt} \tag{2.6-4}$$

where the *damping coefficient* (see Table 2.10) is

$$c = \frac{A^2}{C_1} = 8\pi\mu L\left[\left(\frac{D}{d}\right)^2 - 1\right]^2 \tag{2.6-5}$$

From the principle of action and reaction, $f$ can also be considered to be the force exerted by the damper when moving at a velocity $dx/dt$. If the housing is also in motion, $dx/dt$ is taken to be the relative velocity between housing and piston.

If the relative velocity is large enough to produce turbulent flow in the passage, the nonlinear resistance formulas should be used. In this case the damper relation is the nonlinear expression given by (2.2-3) where

$$c = A^3 R_T \tag{2.6-6}$$

with $R_T$ the turbulent resistance and $v = dx/dt$.

Note that the fluid inertia, or kinetic energy storage compartment, has been assumed to be negligible, and that potential energy resulting from pressure difference acts against the fluid resistance to create the system's dynamics. A similar analysis can be performed for *air dampers* in which the working fluid is air. The common storm-door damper is such a device.

## The Hydraulic Servomotor

A hydraulic servomotor is shown in Figure 2.22. The *pilot valve* controls the flow rate of the working medium to the receiving unit (a piston). The pilot valve shown is known as a spool-type valve because of its shape. Fluid under pressure is available at the supply port. For the pilot valve position shown (the "line-on-line" position), both cylinder ports are blocked and no motion occurs. When the pilot valve is moved to the right, the fluid enters the right-hand piston chamber and pushes the piston to the left. The fluid displaced by this motion exits through the left-hand drain port. The action is reversed for a valve displacement to the left. Both drain ports are connected to a tank (a sump) from which the pump draws fluid to deliver to the supply port. The filters and accumulators necessary to clean the recirculated fluid and dampen pressure fluctuations are not shown.

Let $x$ denote the displacement of the pilot valve from its line-on-line position, and $y$ denote the displacement of the load and piston from their last position before the start of the motion. Note that a positive value of $y$ (to the left) results from a positive value of $x$ (to the right). The flow through the cylinder port uncovered by the pilot valve can be treated as flow through an orifice, and the orifice flow relations (2.5-11) and (2.5-14) can be applied. Let $p$ be the pressure drop across the orifice. Thus

$$q = C_d A \sqrt{2p/\rho} \tag{2.6-7}$$

where $q$ is the volume flow rate through the cylinder port, $A$ is the uncovered area of the port, $C_d$ is the discharge coefficient (which usually lies between 0.6 and 0.8 for this application), and $\rho$ is the mass density of the fluid. The area $A$ is equal to $wx$, where $w$ is the port width. If $C_d$, $\rho$, $p$, and $w$ are taken to be constant, (2.6-7) can be written as

$$q = Cx \tag{2.6-8}$$

Conservation of mass requires that the flow rate into the cylinder must equal the flow rate out. Therefore

$$\rho q = \rho A \frac{dy}{dt}$$

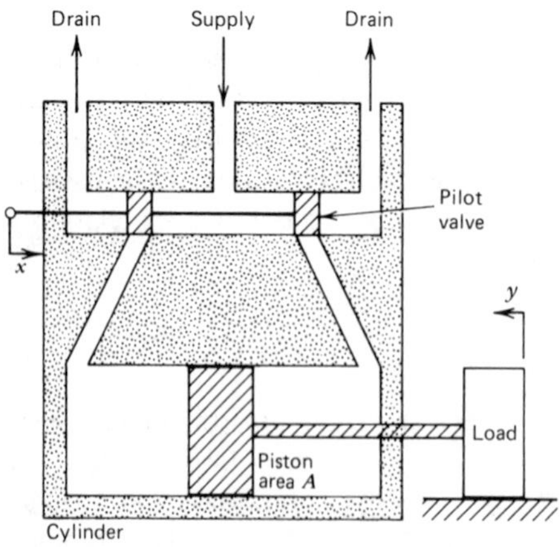


**Figure 2.22 Hydraulic servomotor.**

Combination of the last two equations gives the model for the servomotor.

$$\frac{dy}{dt} = \frac{C}{A} x \tag{2.6-9}$$

For high accelerations this model must be modified because it neglects the inertia of the load and piston. The piston force was assumed sufficient to move the load and is generated by the pressure difference across the piston. In Chapter Five a more sophisticated model will be developed by including these effects. However, (2.6-9) is sufficiently accurate for many applications.

## Accumulators

The compliance of fluid elements can have detrimental or beneficial effects. In hydraulic control systems operating with high force levels or at high speeds, the accuracy of the controller can be affected by the tendency of the nominally incompressible fluid medium to compress slightly. This compliance effect is greatly increased by the presence of bubbles in the hydraulic fluid, which makes the control action less positive and causes oscillations.

On the other hand, compliance can be useful in damping fluctuations or surges in pressure or flow rate, such as those resulting from reciprocating pumps and the addition or removal of other loads on a common pressure source. In such cases a compliance element called an accumulator is added to the system. This general term includes such different elements as a surge tank for liquid-level systems, a fluid column acting against an elastic element for hydraulic systems, and a bellows for gas systems. They all have the property of compliance, which allows them to store fluid during pressure peaks and to release fluid during low pressure. The result is to achieve more efficient and safe operation by damping out pressure pulses the way a flywheel smoothes the acceleration in a mechanical drive or an electrical capacitor smoothes or "filters" voltage ripples.

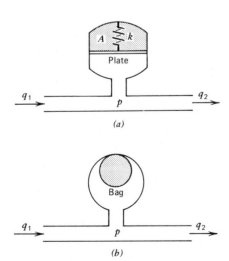

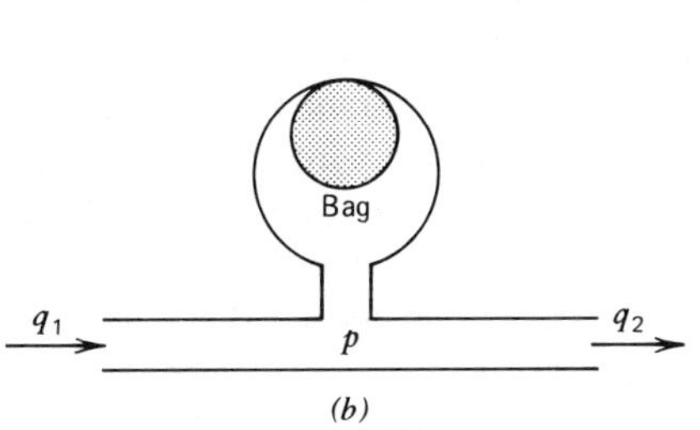


**Figure 2.23 Hydraulic accumulators. (*a*) Mechanical spring principle. (*b*) Compressible gas principle.**

Two typical arrangements for hydraulic systems are shown in Figure 2.23. In Figure 2.23*a* a plate and spring oppose the pressure of the liquid, and in Figure 2.23*b* the compliance is provided by a gas-filled bag that compresses as the liquid pressure increases. The simplest arrangement is a closed tube with air at the top. The air's compressibility acts against the liquid. This device is the common cure for the water-hammer effect in household plumbing.

All of the preceding devices utilize an equivalent elasticity to oppose the pressure, and thus they can be modeled similarly. If we take the displacement $x$ of the interface to be proportional to the force $f$ due to the liquid pressure, then from (2.2-17)

$$f = kx \tag{2.2-17}$$

where $k$ is the elastic constant of the device. Let $A$ be the area of the interface, $p$ the liquid pressure, and $q_1$, $q_2$ the volume flow rates into and out of the section. Neglect the pressure drop between the flow junction and the plate. From mass (volume) conservation,

$$\frac{d}{dt}(Ax) = A\frac{dx}{dt} = q_1 - q_2$$

Neglecting the inertia of the interface gives the force balance

$$pA = kx$$

The model for the dynamic behavior of the pressure as a function of the flow rates is obtained by combining the last two expressions.

$$\frac{A^2}{k}\frac{dp}{dt} = q_1 - q_2 \tag{2.6-10}$$

When the resistances upstream and downstream are defined, the flow rates $q_1$ and $q_2$ can be expressed as functions of the upstream and downstream pressures.

This accumulator's compliance is $V/p = A^2/k$. The compliance for gas-filled accumulators can be evaluated by using a gas model to determine the value for $k$. This technique is discussed in the following section.

## 2.7 PNEUMATIC ELEMENTS

While hydraulic devices use an incompressible liquid, the working medium in a *pneumatic* device is a compressible fluid such as air. Industrial control systems frequently have pneumatic components to provide forces greater than those available from electrical devices. Also, they afford more safety from fire hazards. The availability of air is an advantage for these devices because it may be exhausted to the atmosphere at the end of the device's work cycle, thus eliminating the need for return lines. On the other hand, the response of pneumatic systems is slower than that of hydraulic systems because of the compressibility of the working fluid.

The analysis of gas flow is more complicated than for liquid flow. For compressible fluids the mass and volume flow rates are not readily interchangeable. Since mass is conserved, pneumatic systems analysis uses mass flow rate, here denoted $q_m$ to distinguish from volume flow rate $q$. Because of the relatively low viscosity and density of gases, their flow is more likely to be turbulent. Finally, the possibility of supersonic (greater than the speed of sound) flow is more likely in gas systems.

Pressure is normally the only effort variable used for such systems. The kinetic energy of a gas is usually negligible, so the inductance relation is not employed. Typical units given in Table 2.15 also apply here. The nomenclature used here is identical to that used in Section 2.5 with one exception: here $q_m$ denotes *mass* flow rate.

## Thermodynamic Properties of Gases

Temperature, pressure, volume, and mass are functionally related for a gas. The model most often used to describe this relation is the *perfect gas law* (Reference 6), which describes all gases if the pressure is low enough and the temperature high enough. The law states that

$$pV = mR_gT \tag{2.7-1}$$

where $p$ is the absolute pressure of the gas with volume $V$, $m$ is the mass, $T$ its absolute temperature, and $R_g$ the *gas constant* that depends on the particular type of gas. The values of this constant and properties of air at standard temperature and pressure (STP) are given in Table 2.19. STP is defined as the following temperature and absolute pressure.

$$T = 0°\text{C} = 273\text{ K} = 32°\text{F} = 492°\text{R}$$

$$p = 1.0133 \times 10^5\text{ N/m}^2\text{ abs} = 14.7\text{ psia}$$

Gas does external work on its surroundings when its volume expands. If heat is added to the gas from its surroundings, some of this heat can do external work and the rest can increase the internal energy of the gas by raising its temperature. Consider the ratio of the amount of heat needed to raise the temperature of a unit mass of substance one degree, to the heat needed to raise a unit mass of water one degree at a specified temperature. This ratio is the *specific heat* of the substance at the specified temperature. Because the pressure and volume of a gas can change as its temperature changes, two specific heats are defined for a gas: one at constant pressure ($c_p$) and one at constant volume ($c_v$).

The amount of energy needed to raise the temperature of 1 kilogram of water from 14.5°C to 15.5°C is 4186 J. In the British Engineering system, the British thermal

**TABLE 2.19 Properties of Air at Standard Temperature and Pressure**

| | |
|---|---|
| Gas constant $R_g$[a] | 1715 ft-lb/slug-°R |
| Specific heats[b] | |
| $c_p$ | 7.72 Btu/slug-°R |
| $c_v$ | 5.5 Btu/slug-°R |
| Specific heat ratio | |
| $\gamma = c_p/c_v$ | 1.40 |
| Mass density $\rho$[c] | 0.0025 slug/ft$^3$ |
| Viscosity $\mu$[c] | $3.3 \times 10^{-7}$ lb-sec/ft$^2$ |

[a] To convert the gas constant to N-m/kg-K, multiply the British Units by 0.1672.

[b] To convert specific heat to J/kg-K, multiply the British units by 130.1.

[c] Conversion factors for mass density and viscosity are given in Table 2.15.

unit (Btu) can be considered to be the energy needed to raise 1 pound of water (1/32.174 of a slug in mass units) 1 degree Fahrenheit.

The perfect gas law allows us to solve for one of the variables $p$, $V$, $m$, or $T$ if the other three are given. We frequently do not know the values of three of the variables, and we need additional information. For a perfect gas this information is usually available in the form of a pressure-volume relation, or "process" relation. The following processes are possible. The subscripts 1 and 2 refer to the start and the end of the process. We assume the mass $m$ is constant.

1. *Constant-Pressure Process* ($p_1 = p_2$). The perfect gas law thus implies that $V_2/V_1 = T_2/T_1$. If the gas receives heat from the surroundings, some of it raises the temperature and some expands the volume.
2. *Constant-Volume Process* ($V_1 = V_2$). Here, $p_2/p_1 = T_2/T_1$. When heat is added to the gas, it merely raises the temperature since no external work is done in a constant-volume process.
3. *Constant-Temperature Process* ($T_1 = T_2$). Thus, $p_2/p_1 = V_1/V_2$. Any added heat does not increase the internal energy of the gas because of the constant temperature. It only does external work.
4. *Reversible Adiabatic (Isentropic) Process.* This process is described by the relation

$$p_1 V_1^\gamma = p_2 V_2^\gamma \tag{2.7-2}$$

where $\gamma = c_p/c_v$. *Adiabatic* means that no heat is transferred to or from the gas. *Reversible* means the gas and its surroundings can be returned to their original thermodynamic conditions. Since no heat is transferred, any external work done by the gas changes its internal energy by the same amount. Thus its temperature changes; that is,

$$W = mc_v(T_1 - T_2) \tag{2.7-3}$$

where $W$ is the external work. $W$ is positive if work is done on the surroundings.

5. *Polytropic Process.* The preceding four processes are only approximate models of reality. A real process can be more accurately modeled by properly choosing the exponent $n$ in the polytropic process

$$p\left(\frac{V}{m}\right)^n = \text{constant} \tag{2.7-4}$$

If the mass $m$ is constant, this general process reduces to the previous processes if $n$ is chosen as 0, $\infty$, 1 and $\gamma$, respectively, and if the perfect gas law is used.

## Pneumatic Resistance

As noted earlier, laminar flow rarely occurs in pneumatic devices, However, if the pipe flow is laminar and incompressible, the Hagen-Poiseuille formulas (2.5-7) and (2.5-9) can be used. For turbulent flow, the pipe-flow resistance can be found from (2.5-11) and (2.5-12). These formulations can be used only for incompressible flow.

Here we develop methods for determining pneumatic resistances when the flow is (1) compressible but subsonic, and (2) compressible and supersonic. The subject of compressible fluid flow is complex, and here we present only enough theory to permit the development of models sufficiently accurate for control system design. By making generous use of empirical constants such as the orifice discharge coefficient, we can keep the necessary theory to a minimum. (More detailed coverage of compressible flow can be found in Chapters 9 and 10 of Reference 4. More information on pneumatic as well as hydraulic systems is given in Reference 7.)

The theory of gas flow through an orifice provides the basis for modeling other pneumatic components. We assume that the perfect gas law applies and that the effects of viscosity (friction) and heat transfer are negligible; hence, an isentropic process is assumed. The orifice is shown in Figure 2.24*a*. Its cross section has an area $A$. The upstream absolute pressure is $p_1$, and the absolute back pressure $p_b$ is that pressure eventually achieved downstream from the orifice. Point 2 is that point in the orifice jet where the jet area is minimum.

The mass flow rate through the orifice is not a simple function of pressure drop. Its behavior is represented by Figure 2.24*b*, assuming that $p_1$ is held constant. For $p_b = p_1$, naturally no flow occurs (point *a* on Figure 2.24*b*). As $p_b$ is reduced the flow increases (point *b*). The pressure decreases along the length of the conduit (Figure 2.24*c*). For smaller values of $p_b$, a condition is eventually reached in which sonic velocity occurs at point 2. Thus it is impossible for any disturbances caused by further reductions in $p_b$ to be propagated upstream since the disturbance waves cannot travel faster than sound velocity. Since no disturbance wave can travel upstream, the flow can no longer change with further changes in the back pressure $p_b$. Thus as $p_b$ continues to decrease, the flow rate and pressure profile to the left of point 2 do not change. This is shown by points *d* and *e*.

The value of the back pressure corresponding to point *c* is the *critical* pressure $p_c$. The critical pressure for an ideal gas is (Reference 4, Chapter 10)

$$p_c = \left(\frac{2}{\gamma + 1}\right)^{\gamma/(\gamma-1)} p_1 \tag{2.7-5}$$

For air, $\gamma = 1.4$ and this gives

$$p_c = 0.528 p_1 \tag{2.7-6}$$

The results of chief interest to us are the expressions for the mass flow rate as functions of the pressures $p_1$ and $p_b$. Newton's laws, conservation of mass, and the isentropic process formula can be combined to show that the mass flow rate $q_m$ in the subsonic case ($p_b > p_c$) is

$$q_m = C_d A \sqrt{\frac{2}{R_g T_1} p_b (p_1 - p_b)} \tag{2.7-7}$$

where $C_d$ is an experimentally determined discharge coefficient that accounts for viscosity effects, and $T_1$ is the absolute temperature upstream from the orifice. For

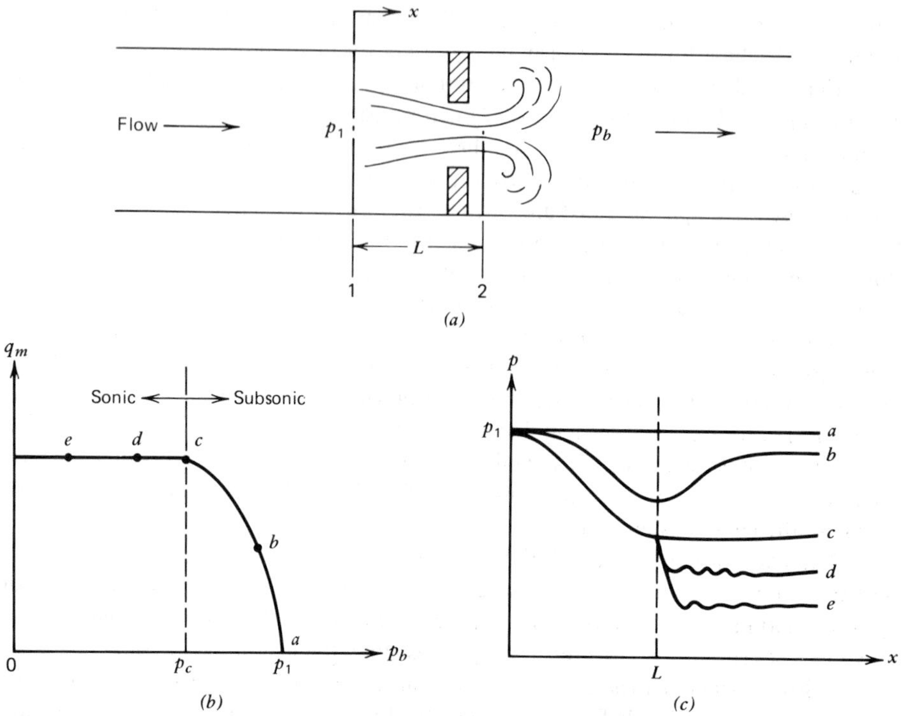


**Figure 2.24 Gas flow through an orifice. (*a*) Geometry. (*b*) Back-pressure–mass-flow rate relation. (*c*) Lengthwise pressure distribution.**

the sonic case ($p_b < p_c$),

$$q_m = C_d A \frac{p_1}{\sqrt{R_g T_1}} \sqrt{\gamma \left(\frac{2}{\gamma+1}\right)^{(\gamma+1)/(\gamma-1)}} \tag{2.7-8}$$

Note that the expression is independent of $p_b$ as long as $p_b < p_c$.

Valves and other flow restrictions in pneumatic systems can be modeled as an orifice by treating the term $C_d A$ as an equivalent orifice area, which can experimentally determined. The product $C_d A$ is the *effective cross-sectional area* of the component.

*Example 2.14*

A pneumatic valve was tested under the following conditions. The inlet and outlet absolute pressures were $p_1 = 1.5 \times 10^5$ N/m$^2$ and $p_b = 1.013 \times 10^5$ N/m$^2$ = atmospheric pressure. The inlet air temperature was $T_1 = 293$ K and the measured mass flow rate was $q_m = 0.1$ kg/sec. Estimate the effective cross-sectional area.

From (2.7-6) the critical pressure is $0.528\,(1.5 \times 10^5) = 0.79 \times 10^5$ N/m$^2$, which is less than $p_1$. Thus the subsonic flow condition is present and from (2.7-7) for air, we obtain

$$C_dA = \frac{0.1}{\sqrt{\dfrac{2}{287(293)} 1.013(1.5 - 1.013) \times 10^{10}}}$$

$$= 2.92 \times 10^{-4} \text{ m}^2$$

A typical condition in pneumatic systems is such that a small pressure drop exists across the component. Hence the flow is frequently subsonic. Also, the pressure changes often consist of small fluctuations about an average or steady-state constant pressure value. Under these circumstances the compressible flow resistance of pneumatic components can be modeled in the form of the turbulent resistance relation (2.5-11), written here as

$$R_T q_m^2 = \Delta p \tag{2.7-9}$$

where $\Delta p$ is the pressure drop across the component.

*Example 2.15*

Assume that the resistance of a pneumatic component is to be modeled as an orifice resistance. With reference to Figure 2.24, the input air pressure $p_1$ fluctuates about a steady-state pressure $p_s$ by a small amount $p_i$. Thus, $p_1 = p_s + p_i$. Similarly, the back pressure can be written as $p_b = p_s + p_o$, where $|p_o| \ll p_s$. Show that the resistance relation (2.7-9) holds. Find the value of $R_T$.

Because both $p_i$ and $p_o$ are small compared to $p_s$, $p_s + p_o > 0.528(p_s + p_i)$, and the flow is subsonic. From (2.7-7) and the given expressions for $p_1$ and $p_b$, we obtain

$$q_m = C_dA \sqrt{\frac{2}{R_g T_1}(p_s + p_o)(p_s + p_i - p_s - p_o)}$$

$$= C_dA \sqrt{\frac{2p_s}{R_g T_1}} \sqrt{1 + \frac{p_o}{p_s}} \sqrt{p_i - p_o}$$

Since $|p_o| \ll p_s$,

$$\sqrt{1 + \frac{p_o}{p_s}} \cong 1$$

and

$$q_m \cong C_dA \sqrt{\frac{2p_s}{R_g T_1}} \sqrt{p_i - p_o} \tag{2.7-10}$$

Comparison of this form with (2.7-9) shows that

$$\Delta p = p_i - p_o$$

$$R_T = \frac{R_g T_1}{2p_s C_d^2 A^2} \tag{2.7-11}$$

Note the similarity of this result to the expression (2.5-14) for the orifice resistance $R_o$, which is for *volume* flow rate. From the perfect gas law,

$$\frac{R_g T_1}{p_s} = \frac{V}{m} = \frac{1}{\rho}$$

and thus

$$R_T = \frac{1}{2\rho C_d^2 A^2} = \frac{1}{\rho^2} R_o \tag{2.7-12}$$

where $\rho$ is the density of the gas at pressure $p_s$ and temperature $T_1$. At this density the volume and mass flow rates are related by $q_m = \rho q$. Comparison of (2.7-9) and (2.5-11) shows that the resistance for mass flow equals the resistance for volume flow divided by the density squared. From (2.7-12), $R_T$ and $R_o$ satisfy this requirement. Hence they are equivalent. This is because the small pressure approximation used in Example 2.15 is equivalent to a linearization of the gas model about the reference pressure $p_s$, and thus about the reference density $\rho$. This has the effect of taking the gas to be an incompressible fluid, as was done in deriving (2.5-14).

### *Example 2.16*

Estimate the resistance of the pneumatic valve described in Example 2.14. Compare the estimate of flow rate given by (2.7-10) with the true value when $p_1 = 1.5 \times 10^5$ N/m$^2$.

The effective cross-sectional area of the valve was found to be $C_d A = 2.92 \times 10^{-4}$ m$^2$. From Table 2.19,

$$R_g = 1715(0.1672) = 286.7 \text{ N-m/kg-K}$$

With $T_1 = 293$ K and $p_s$ taken to be atmospheric pressure, (2.7-11) gives

$$R_T = \frac{(286.7)(293)}{2(1.013 \times 10^5)(2.92 \times 10^{-4})^2}$$

$$= 4.92 \times 10^6 \text{ N}^{-1} \text{ sec}^{-2}$$

From the given information, $p_i = (1.5 - 1.013) \times 10^5 = 0.487 \times 10^5$ N/m$^2$, and $p_o = 0$. Thus $\Delta p = p_i - p_o = 0.487 \times 10^5$, and

$$q_m = \sqrt{\frac{\Delta p}{R_T}} = 0.099 \text{ kg/sec}$$

The true (measured) value given previously is 0.1 kg/sec, for an error of 1%.

The analysis in Example 2.15 implicitly assumed that $p_i - p_o > 0$. Equation (2.7-10) gives an imaginary flow rate otherwise. If $p_i - p_o < 0$, the flow is reversed, and a similar analysis would result in an expression such as (2.7-10) with a sign change. These two cases can be treated with a single expression by using the *signed square root* function.

$$\text{SSR}(x) = \text{SIGN}(x)\sqrt{|x|} \tag{2.7-13}$$

where

$$\text{SIGN}(x) = \begin{cases} 1, & x > 0 \\ 0, & x = 0 \\ -1, & x < 0 \end{cases} \tag{2.7-14}$$

With this notation and (2.7-11), (2.7-10) can be generalized to

$$q_m \cong \frac{1}{\sqrt{R_T}} \text{SSR}(p_i - p_o) \tag{2.7-15}$$

## Pneumatic Capacitance

In pneumatic systems mass is the quantity variable and pressure the effort variable. Thus pneumatic capacitance is the relation between stored mass and pressure. Specifically, we take pneumatic capacitance to be the system's compliance as defined in Section 2.5; that is, capacitance $C$ (or compliance) is the ratio of the change in stored mass to the change in pressure, or

$$C = \frac{dm}{dp} \tag{2.7-16}$$

For a container of constant volume $V$ with a gas density $\rho$, this expression may be written as

$$C = \frac{d(\rho V)}{dp} = V\frac{d\rho}{dp} \tag{2.7-17}$$

If the gas undergoes a polytropic process,

$$p\left(\frac{V}{m}\right)^n = \frac{p}{\rho^n} = \text{constant} \tag{2.7-18}$$

and

$$\frac{d\rho}{dp} = \frac{\rho}{np} = \frac{m}{npV}$$

For a perfect gas, this shows the capacitance of the container to be

$$C = \frac{mV}{npV} = \frac{V}{nR_gT} \tag{2.7-19}$$

Note that the same container can have a different capacitance for different expansion processes, temperatures, and gases, since $C$ depends on $n$, $T$, and $R_g$.

*Example 2.17*

Find the capacitance of air in a rigid container with a volume of 0.5 $m^3$ for an isothermal process. Assume the air is initially at room temperature, 293 K.

The filling of the container can be modeled as an isothermal process if it occurs slowly enough to allow heat transfer to occur between the air and its surroundings. In this case, $n = 1$ and from (2.7-19),

$$C = \frac{0.5}{1(286.7)(293)} = 5.95 \times 10^{-6} \text{ kg-m}^2/\text{N}$$

## Modeling of Pneumatic Systems

Because fluid inertia is usually neglected in pneumatic analysis, the simplest model of such systems is a resistance-capacitance model for each mass storage element.

### *Example 2.18*

Air passes through a valve (modeled as an orifice) into a rigid container, as shown in Figure 2.25. Develop a dynamic model of the pressure change $p$ in the container as a function of the inlet pressure change $p_1$. Assume an isothermal process and that $p$ and $p_1$ are small variations from the steady-state pressure $p_s$.

From conservation of mass, the rate of mass increase in the container equals the mass flow rate through the valve. Thus

$$\frac{dm}{dt} = q_m$$

$$\frac{dm}{dt} = \frac{dm}{dp}\frac{dp}{dt} = C\frac{dp}{dt}$$

$$q = \frac{1}{\sqrt{R_T}} \text{SSR}(p_1 - p)$$

where $C$ and $R_T$ are given by (2.7-19) and (2.7-11), with $T = T_1$ and $n = 1$. The desired model is

$$\frac{V}{R_g T}\frac{dp}{dt} = C_d A \sqrt{\frac{2p_s}{R_g T}}\, \text{SSR}(p_1 - p)$$

or

$$\frac{dp}{dt} = \frac{C_d A}{V}\sqrt{2p_s R_g T}\, \text{SSR}(p_1 - p) \tag{2.7-20}$$

### *Example 2.19*

A pneumatic bellows, shown in Figure 2.26, is an expandable chamber usually made from corrugated copper because of this metal's good heat conduction and elastic

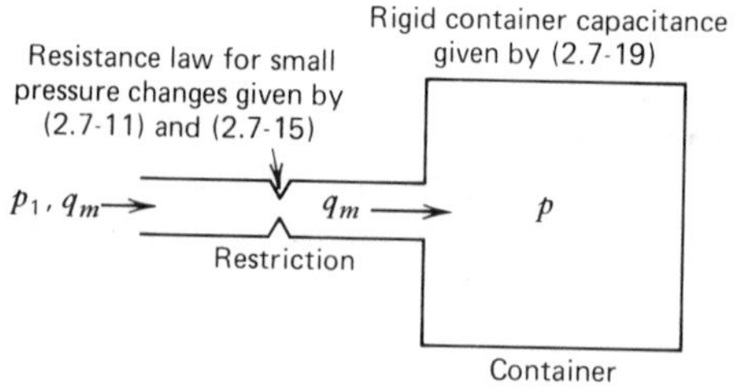


**Figure 2.25 Gas flow into a rigid container.**

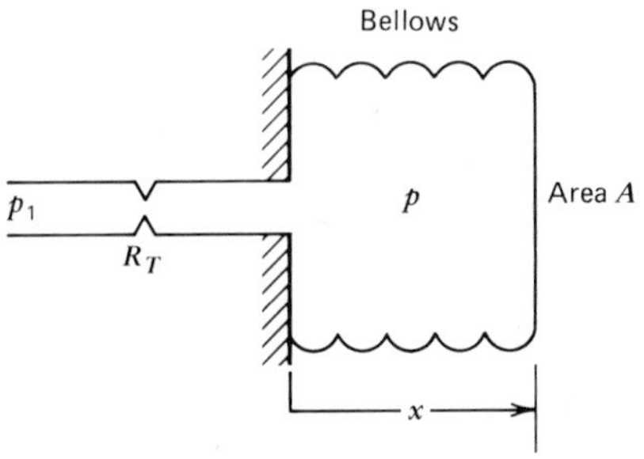


**Figure 2.26 Pneumatic bellows.**

properties. We model the elasticity of the bellows as a spring. The spring constant $k$ for the bellows can be determined from vendor's data or standard formulas. This device, when employed with a variable resistance $R_T$, is useful in pneumatic controllers as a feedback element to sense and control the pressure $p_1$. The displacement $x$ is transmitted by a linkage or beam balance to a pneumatic valve regulating the air supply. Develop a model for the dynamic behavior of $x$ with $p_1$ a given input.

We assume that the bellows expands or contracts slowly. Thus the product of its mass and acceleration is negligible, and a force balance gives

$$pA = kx \tag{2.7-21}$$

where the force exerted by the internal pressure is $pA$, $A$ is the area of the bellows, and $x$ is the displacement of the right-hand side (the left-hand side is fixed). The volume is $V = Ax$ and thus

$$pV = \frac{kx}{A}Ax = kx^2$$

The assumption of a slow process suggests an isothermal process. Thus the time derivative of the perfect gas law gives

$$\frac{d}{dt}(pV) = R_g T \frac{dm}{dt}$$

From the law for resistance (2.7-9)

$$\frac{dm}{dt} = q_m = \frac{1}{\sqrt{R_T}}\,\mathrm{SSR}(p_i - p)$$

Since

$$\frac{d(kx^2)}{dt} = 2kx\frac{dx}{dt}$$

these expressions give the following model.

$$2kx\frac{dx}{dt} = \frac{R_g T}{\sqrt{R_T}}\,\mathrm{SSR}\left(p_i - \frac{kx}{A}\right) \tag{2.7-22}$$

This is nonlinear because of the cross-product between $x$ and its derivative and because of the signed square-root relation.

The pneumatic capacitance of the bellows is found from (2.7-16) to be

$$C = \frac{dm}{dp} = \frac{V}{R_g T} = \frac{Ax}{R_g T}$$

which is a variable because it depends on $x$. This fact accounts for the complicated nature of (2.7-22) relative to the rigid-container model (2.7-20).

We frequently know the steady-state values of a system's variables but desire a dynamic model for small changes from these reference values, as in Example 2.15. In this case the system's static equations can be linearized individually using the Taylor series method given in Chapter One. The linearized relations can then be substituted into the describing differential equations to obtain the system model.

*Example 2.20*

Consider the pneumatic bellows in the previous example. Develop a linearized model for $x$.

Let $\delta x$, $\delta p_1$, $\delta p$, $\delta V$, and $\delta m$ represent the deviations of $x, p_1, p, V$, and $m$ from their steady-state values. Thus when the input deviation $\delta p_1$ remains zero, so do the other deviations. The linearized forms of the volume relation, force balance, and gas law are

$$\delta V = A\delta x$$

$$A\delta p = k\delta x$$

$$\delta V = \alpha_1 \delta p + \alpha_2 \delta m$$

where $\alpha_1$ and $\alpha_2$ are the linearization derivatives $\partial V/\partial p$ and $\partial V/\partial m$ evaluated at the reference condition. The linearized resistance relation gives

$$\delta q_m = \delta \dot{m} = \alpha_3 \delta p_1 - \alpha_4 \delta p$$

where $\alpha_3 = \alpha_4 = \partial q_m/\partial p_1$. Elimination of the intermediate variables gives

$$A\delta \dot{x} = \alpha_1 \frac{k}{A} \delta \dot{x} + \alpha_2 \delta \dot{m}$$

or

$$(A^2 - \alpha_1 k)\delta \dot{x} = \alpha_2 A \alpha_3 \left(\delta p_1 - \frac{k}{A} \delta x\right) \tag{2.7-33}$$

This is a linearized differential equation model of the form

$$\delta \dot{x} = a\delta x + b\delta p_1 \tag{2.7-24}$$

in terms of the unknown variable $\delta x$ and the given input $\delta p_1$.

## 2.8 THERMAL SYSTEMS

In thermal systems, energy is stored and transferred as heat. For this reason many energy conversion devices are examples of thermal systems, as are chemical processes where heat must be added or removed in order to maintain an optimal reaction temperature. Other examples occur in food processing and in environmental control in buildings.

Typical units are given in Table 2.20. The effort variable is temperature, or more

TABLE 2.20 Typical Units in Heat Transfer

| Quantity and Symbol | British Engineering | SI Metric | Conversion Factor[a] | Other Common Units |
|---|---|---|---|---|
| Heat energy $Q_h$ | ft-lb | J | 1.3557 | Btu,[b] calorie (cal)[c] |
| Heat flow rate $q_h$ | ft-lb/sec | W | 1.3557 | Btu/hr[b] |
| Thermal conductivity $k$ | ft-lb/sec-ft-°F | W/m-K | 8.0061 | Btu/hr-ft-°F[b] |
| Specific heat $c$, $c_p$, or $c_v$ | ft-lb/slug-°F | J/kg-K | 0.16721 | Btu/slug-°F[d] |
| Film (convection) coefficient $h$ | ft-lb/sec-ft$^2$-°F | W/m$^2$-K | 26.267 | Btu/hr-ft$^2$-°F[e] |

[a] To convert from British Engineering to SI units, multiply the British units by this factor.

[b] Conversion factors are given in Table 2.2.

[c] 1 cal = 4.18585 J.

[d] The conversion factor is given in Table 2.2.

[e] To convert $h$ from ft-lb/sec-ft$^2$-°F to Btu/hr-ft$^2$-°F, multiply by 4.6272.

precisely temperature difference. This can cause a heat transfer rate by one or more of the following resistance modes: conduction, convection, or radiation. The transfer of heat causes a change in the system's temperature. Thermal capacitance relates the system temperature to the amount of heat energy stored. It is the product of the system mass and its specific heat. No physical elements are known to follow a thermal inductance causal path.

In order to have a single lumped-parameter model, we must be able to assign a single temperature that is representative of the system. Thermal systems analysis is sometimes complicated by the fact that it is difficult to assign such a representative temperature to a body or fluid because of its complex shape or motion and resulting complex distribution of temperature throughout the system. Here we develop some guidelines as to when such an assignment can be made. When it cannot, several coupled lumped-parameter models or even a distributed-parameter model will be required.

Unlike current in an electrical circuit, the flow of heat often does not occur along a single, simple path, but might involve more than one mode. If the convective mode is present, the analysis must also consider a fluid system with its attendant difficulties. While the analysis of such complex phenomena is beyond the scope of this work, fortunately many analytical and empirical results are available for common situations, and we can use these to obtain the necessary coefficients for our thermal resistances.

## Heat Transfer

Heat transfer can be transferred in three ways: by *conduction* (diffusion through a substance), *convection* (fluid transport), and *radiation* (mostly infrared waves). The effort variable causing a heat flow is a temperature difference. The constitutive

**TABLE 2.21 Thermal Conductivity, Specific Heat, and Mass Density of Some Common Materials**

| Material[a] | Thermal Conductivity $k$ Btu/hr-ft-°F[b] | Specific Heat Btu/slug-°F | Mass Density $\rho$ slug/ft$^3$ [d] |
|---|---|---|---|
| Copper 32°F | 224 | 2.93 | 17.3 |
| Concrete 68°F | 0.65 | 6.8 | 4.04 |
| Fiber insulating board 70°F | 0.028 | – | 0.46 |
| Plate glass 70°F | 0.44 | 6 | 5.25 |
| Air 80°F, atmos. pressure | 0.01516 | 7.728 ($c_p$) | 0.0023 |
| Water 68°F | 0.345 | 32.14 | 1.942 |

[a] Values given for insulating board, concrete, and glass are typical, but can vary somewhat with composition.

[b] To convert thermal conductivity to *W/m-K*, multiply the British units by 1.7302.

[c] Conversion factor for specific heat is given in Table 2.19.

[d] Conversion factor for mass density is given in Table 2.15.

relation takes a different form for each of the three heat-transfer modes. The linear model for heat flow rate as a function of temperature difference is given by *Newton's law of cooling.*

$$Rq_h = \Delta T \tag{2.8-1}$$

The thermal resistance for convection is

$$R = \frac{1}{hA} \tag{2.8-2}$$

where $h$ is the *film coefficient* of the fluid-solid interface. For conduction, we will see that

$$R = \frac{d}{kA} \tag{2.8-3}$$

where $k$ is the *thermal* conductivity of the material, $A$ is the surface area, and $d$ the material thickness. Table 2.20 gives typical units for $h$ and $k$. Table 2.21 gives values of $k$ for some common materials.

Convective heat transfer is divided into two categories: forced convection – due, for example, to fluid pumped past a surface – and free or natural convection, due to motion produced by density differences in the fluid. The heat-transfer coefficient for convection is a complicated function of the fluid-flow characteristics especially. For most cases of practical importance the coefficient has been determined to acceptable accuracy, but a presentation of the results is lengthy and beyond the scope of this work. Standard references on heat transfer, such as Reference 8, contain this information for most cases.

Significant heat transfer can occur by radiation; the most notable example is solar energy. Thermal radiation produces heat when it strikes a surface capable of absorbing it. It can also be reflected or refracted, and all three mechanisms can occur at a single surface. When two bodies are in visual contact, a mutual exchange of energy occurs by emission and absorption. The *net* transfer of heat occurs from the warmer to the colder body. This rate depends on material properties, geometric factors affecting the portion of radiation emitted by one body and striking the other, and the amount of surface area involved. The net heat-transfer rate can be shown to be dependent on the body temperatures raised to the fourth power (a consequence of the *Stefan-Boltzmann* law).

$$q_h = \beta(T_1^4 - T_2^4) \tag{2.8-4}$$

The absolute body temperatures are $T_1$ and $T_2$, and $\beta$ is a factor incorporating the other effects. This factor is usually very small. Thus the effect of radiation heat transfer is usually negligible compared to conduction and convection effects, unless the temperature of one body is much greater than that of the other. Determination of $\beta$, like the convection coefficient, is too involved to consider here, but many results are available (see Reference 8).

## Some Simplifying Approximations

The thermal resistance for conduction through a plane wall given by (2.8-3) is an approximation that is valid only under certain conditions. Let us take a closer look at the problem. If we consider the wall to extend to infinity in both directions, then the heat flow is one-dimensional. If the wall material is homogeneous, the temperature gradient through the wall is constant under steady-state conditions (Figure 2.27*a*). *Fourier's law* of heat conduction states that the heat-transfer rate per unit area within a homogeneous substance is directly proportional to the negative temperature gradient. The proportionality constant is the thermal conductivity $k$. For the case shown in Figure 2.27*a*, the gradient is $(T_2 - T_1)/d$, and the heat-transfer rate is thus

$$q_h = -kA(T_2 - T_1)/d \tag{2.8-5}$$

where $A$ is the area in question. Comparing this with (2.8-1) shows that the thermal resistance is given by (2.8-3) with $\Delta T = T_1 - T_2$.

In transient conditions where the temperatures change with time, the temperature distribution is no longer linear. If the internal temperature gradients in the body are small, as would be the case for a large value of $k$, it is possible to treat the body as being at one uniform, average temperature and thus obtain a lumped- instead of distributed-parameter model. For solid bodies immersed in a fluid, a useful criterion for determining the validity of the uniform-temperature assumption is based on the *Biot* number, defined as

$$N_B = \frac{hL}{k} \tag{2.8-6}$$

where $L$ is the ratio of the volume to surface area of the body, and $h$ is the film coefficient. If the shape of the body resembles a plate, cylinder, or sphere, it may be considered to have a uniform temperature with a resulting error of less than 5% if $N_B$ is less than 0.1.

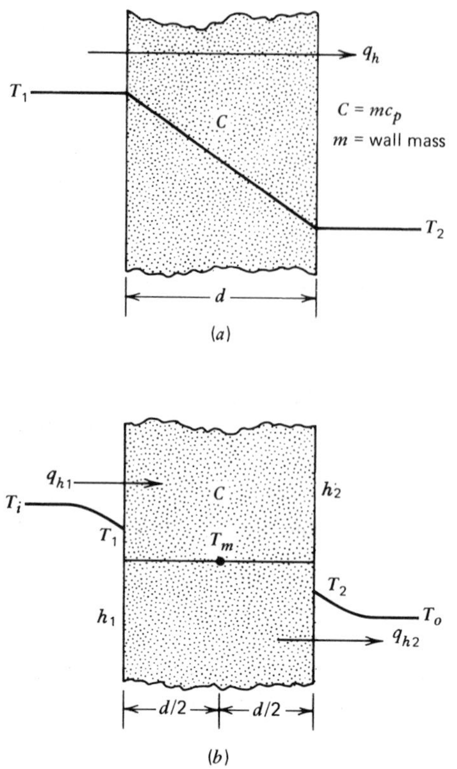


**Figure 2.27 One-dimensional heat transfer through a wall. (*a*) Steady-state temperature gradient. (*b*) Lumped approximation.**

## *Example 2.21*

A building wall made of cinder block has a $k$ value of 0.44 Btu/hr-ft-°F. The ratio $L$ is $Ad/2A$ or $d/2$. For a vertical wall in still air (an interior wall), $h$ is between 0.59 and 1.46 Btu/hr-ft$^2$-°F. For a vertical wall in a 15-mph wind, $h$ is approximately 6 (Reference 8). Can the wall temperature be considered uniform throughout?

For the interior wall, $N_B$ is greater than $0.59d/2(0.44)$, or $0.67d$. For an 8-in. wall, $N_B$ is 0.447, so that the preceding criterion does not enable us to treat the wall as a lumped mass with any assurance of accuracy, but we might still do so, especially if the time rates of change of the temperature inputs are small so that the temperature distribution in the wall is close to the steady-state distribution.

## *Example 2.22*

Consider a copper sphere 1 in. in diameter with $k = 212$ Btu/hr-ft-°F at 570°F, and $h = 5$ Btu/hr-ft$^2$-°F. Can its temperature be considered uniform?

Here $L$ is 1/6 and $N_B = 0.004$. According to the criterion we may treat the sphere with a lumped thermal analysis. But this example is typical of a thermal quenching application in which the sphere is dropped into the quenching liquid and thus the input is suddenly applied. Therefore $N_B$ should be much less than 0.1 to use a lumped analysis.

These examples point out that the rapidity of the input's variation must be taken into account in addition to the Biot number when considering a lumped analysis. The smaller the Biot number, or the slower the input variations, the more accurate will be the lumped (uniform temperature) model.

## A Wall Model

Conductive and convective resistance to heat transfer frequently occurs when one or more layers of solid materials are surrounded on both sides by fluid, such as shown in Figure 2.27*b*. We will use the figure to show how thermal capacitance is modeled and to show how thermal resistance is computed when both conduction and convection are present.

Assume that the wall material is homogeneous and that it is to be treated with a lumped approximation. The wall temperature is considered to be constant throughout and is denoted by $T_m$. The temperatures $T_1$ and $T_2$ are the temperatures in the surrounding fluid just outside the wall surface. There is a temperature gradient on each side of the wall across a thin film of fluid adhering to the wall. This film is known as the *thermal boundary layer*. The film coefficient $h$ is a measure of the conductivity of this layer. The temperatures outside of these layers are denoted by $T_i$ and $T_o$.

The thermal capacitance $C$ is the product of the wall mass times its specific heat. The capacitance of the boundary layer is generally negligible because of its small fluid mass. Thus, since no heat is stored in the layer, the heat-flow rate through the layer must equal the heat-flow rate from the surface to the mass at temperature $T_m$ considered to be located at the center of the wall. The length of this latter path is $d/2$. For the left-hand side this gives

$$q_{h1} = h_1 A (T_i - T_1) = \frac{kA}{d/2}(T_1 - T_m) \tag{2.8-7}$$

This allows us to solve for $T_1$ as a function of $T_i$ and $T_m$.

$$T_1 = \frac{2kT_m + dh_1 T_i}{2k + dh_1} \tag{2.8-8}$$

Similar equations can be developed for the right-hand flow $q_{h2}$ and temperature $T_2$.

An energy balance for the wall mass gives

$$C\frac{dT_m}{dt} = q_{h1} - q_{h2} \tag{2.8-9}$$

After $T_1$ and $T_2$ have been eliminated, this gives an equation for $T_m$ with $T_i$ and $T_o$ as inputs. Define

$$a_j = \frac{2kh_jA}{2k + dh_j}, \quad j = 1, 2 \tag{2.8-10}$$

and (2.8-9) becomes

$$C\frac{dT_m}{dt} = a_1 (T_i - T_m) - a_2 (T_m - T_o) \tag{2.8-11}$$

When the input temperatures are specified, $T_m$ can be determined as a function of time with the methods of Chapter Three.

### *Example 2.23*

Derive the series law for thermal resistances. Under what conditions is it valid?

If either the rate of change of $T_m$ or the wall capacitance $C$ is very small, the right-hand side of (2.8-11) can be taken to be zero and thus $q_{h1}$ equals $q_{h2}$. This relation can be manipulated to show that the wall temperature is

$$T_m = \frac{a_1 T_i + a_2 T_o}{a_1 + a_2} \tag{2.8-12}$$

and that the heat flow rate is

$$q_h = \frac{a_1 a_2}{a_1 + a_2}(T_i - T_o) \tag{2.8-13}$$

Thus the total resistance between $T_i$ and $T_o$ is $(a_1 + a_2)/a_1 a_2$. With (2.8-10) and some algebra, the total resistance can be shown to be

$$R = \left(\frac{1}{h_1} + \frac{d}{2k} + \frac{d}{2k} + \frac{1}{h_2}\right)\frac{1}{A} \tag{2.8-14}$$

This is the *series law* for thermal resistances. From the general definition of a series connection, the rate variable must be the same for all elements. Our wall model meets this test because all the elements have the same heat-flow rate under the stated assumptions.

Thermal elements that experience the same temperature difference are said to be in *parallel*, and their resistances combine in the same manner as parallel electrical resistances; that is, for three elements

$$\frac{1}{R} = \frac{1}{R_1} + \frac{1}{R_2} + \frac{1}{R_3} \tag{2.8-15}$$

The series and parallel laws for thermal systems are strictly true only under constant-temperature conditions or where capacitance is negligible. However, they are often utilized in transient analysis because of the difficulty of solving such problems without this assumption. The rapidity of the transients and the goals and constraints of the analysis must be considered in making this decision.

### *Example 2.24*

A certain radiator wall is made of copper with a conductivity $k = 218$ Btu/hr-ft-°F at 212°F. The wall is 1/8 in. thick and has circulating water on one side with a convection coefficient $h_1 = 400$ Btu/hr-ft$^2$-°F. Air is blown over the other side and has a coefficient $h_2 = 10$ Btu/hr-ft$^2$-°F. Find the resistance of the wall at steady state, on a square-foot basis. Which term contributes most to the resistance?

On a square-foot basis, $A = 1$ ft$^2$ and from (2.8-14) with $d = 1/8(12) = 1/96$ ft, the resistance is

$$R = \frac{1}{400} + \frac{1/96}{218} + \frac{1}{10}$$

$$= 0.0025 + 4.78 \times 10^{-5} + 0.1 = 0.1025 \frac{\text{hr}^\circ\text{F}}{\text{Btu}}$$

Thus the thermal boundary layer in the air contributes the most to the resistance. The metal wall is the smallest influence.

## A Solar Thermal Energy System

Heat loss can occur by means other than thermal resistance. For example, heated fluid can be pumped directly from a thermal storage element. This is the case in a solar heating unit with flat-plate collectors in which the working fluid may be air or water, perhaps with an antifreeze additive. The fluid is circulated through the collectors where it picks up energy that must be stored until needed. For air systems the storage element usually consists of rocks over which the hot air from the collector is circulated. Water systems use a tank of water for energy storage. Such a scheme is shown in Figure 2.28.

The collector efficiency is increased if the temperature of the water pumped into the collector is low. Thus it is taken from the bottom of the tank, near the inlet supplying cold water. Hot water for domestic use is drawn from the top. The heated water from the panels is circulated through a heat exchanger, and the cooled fluid is

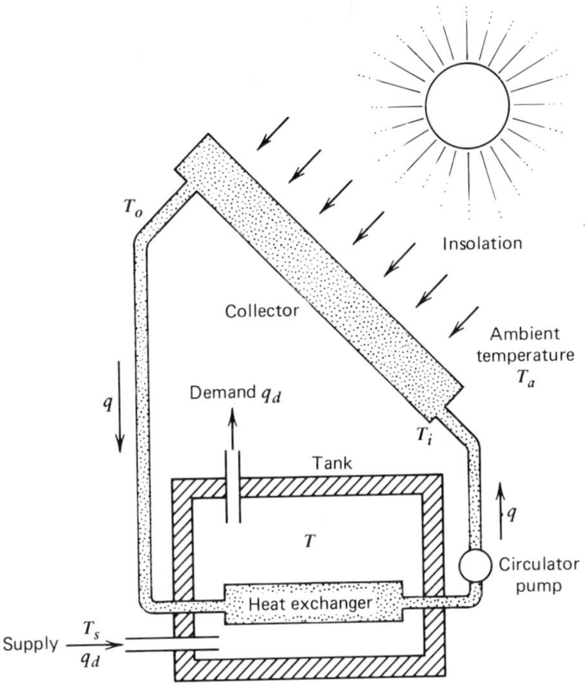


**Figure 2.28 Solar hot water system.**

returned to the panels. The heat exchanger prevents the mixing of the antifreeze solution with the tank water. Its location at the bottom increases the temperature difference and thus the heat transfer rate. The performance analysis of heat exchangers is an involved problem in fluid mechanics and heat transfer. For our purposes it is sufficient to employ the following input-output temperature relation (Reference 9).

$$T_o - T_i = \epsilon(T_o - T) \tag{2.8-16}$$

where $\epsilon$ is an effectiveness factor for the exchanger, $T_o$ is the temperature of the fluid entering the exchanger (out of the collector), $T_i$ is the temperature of the fluid emerging from the exchanger (and entering the collector), and $T$ is the temperature of the water in the tank. The exchanger can usually be designed to give $\epsilon = 0.9$, approximately.

This system has some interesting fluid circulation phenomena, but we will treat it solely as a thermal system by assuming that the tank water is completely mixed so that the tank temperature is uniform. When this is not true the tank is said to be *stratified*. It has been shown that a model with three layers of water at different temperatures (three thermal masses) is usually sufficient to treat the stratified tank problem (Reference 9). This results in three coupled differential equations. For now we will take the tank to be completely mixed.

Energy can also be stored in the walls of the tank; if significant, this adds another equation to the model (here we neglect wall storage). Let $q_d$ be the volume rate of hot water demand withdrawn from the tank. Thus, the volume rate of cold water supplied at temperature $T_s$ as makeup must also be $q_d$. Let $q$ be the volume flow rate circulated through the collectors at temperature $T_i$ and returned at $T_o$. The tank can be either buried in the ground or located in the basement. In most cases the heat loss through the tank walls is negligible, and a temperature drop of less than 0.5°F per hour is not uncommon. This effect will not be modeled here but can easily be included if desired by determining the appropriate thermal resistances.

Let $\rho_w$ and $c_{pw}$ be the mass density and specific heat of water. Then the thermal energy stored in the tank water is $\rho_w c_{pw} VT$, where $V$ is the tank volume. The rate of energy input to the fluid from the collector is a function of the fluid's temperature rise $(T_o - T_i)$ as it passes through the collector. This rate is given by $\rho q c_p(T_o - T_i)$, where $c_p$ and $\rho$ are the properties of the collector fluid. This term may also be considered the energy input rate to the tank from the heat exchanger, since the outlet temperature $T_i$ of the exchanger is determined by $T_o$ and $\epsilon$ through (2.8-16). The energy loss rate from the tank due to the demand is $\rho_w q_d c_{pw}(T - T_s)$. Thus an energy balance on the tank water shows that

$$c_{pw}\rho_w V\frac{dT}{dt} = \sigma\rho q c_p(T_o - T_i) - \rho_w q_d c_{pw}(T - T_s) \tag{2.8-17}$$

where $\sigma$ is a switching function representing a thermostatic controller used to turn the circulator pump on or off. We wish to stop collector circulation whenever the outlet temperature of the collector is less than the tank temperature. Otherwise, energy would be transferred to the collector from the tank. The simplest representation of $\sigma$ is

$$\sigma = \begin{cases} 0 & \text{if} \quad T_o \leqslant T \\ 1 & \text{if} \quad T_o > T \end{cases} \tag{2.8-18}$$

Equation (2.8-17) is not yet in a usable form because $T_o$ and $T_i$ are not given inputs. The appropriate input variables instead of $T_o$ and $T_i$ are

$S$ = solar insolation (energy) absorbed by the collector, per unit area, per unit time

$T_a$ = ambient (outside) temperature

The basic collector equation that relates these two variables to the energy delivered to the fluid is

$$\rho q c_p(T_o - T_i) = A_c F_R[S - U_c(T_i - T_a)] \tag{2.8-19}$$

where $A_c$ is the collector area, $U_c$ is its loss coefficient,* and $F_R$ is the collector heat removal factor. The last term on the right represents the heat loss to the ambient air. $F_R$ is an empirically based coefficient used to account for the fact that the temperature over the collector is greater than $T_i$, and thus the heat gain is less than that predicted by using $T_i$. Methods for determining $F_R$ and $S$ for given collector designs are given in Reference 9.

Equations (2.8-16) and (2.8-19) are linear algebraic equations that can be solved for $T_o$ and $T_i$ in terms of $T$, $S$, and $T_a$ in the form

$$T_o = b_1 T + b_2 S + b_3 T_a \tag{2.8-20}$$

$$T_i = b_4 T + b_5 S + b_6 T_a \tag{2.8-21}$$

and

$$\begin{aligned} T_o - T_i &= (b_1 - b_4)T + (b_2 - b_5)S + (b_3 - b_6)T_a \\ &= b_7 T + b_8 S + b_9 T_a \end{aligned} \tag{2.8-22}$$

where the coefficients $b_i$ depend on the model's parameters. Equation (2.8-22) can now be substituted in (2.8-17) to obtain the system model in closed form.

$$c_{pw}\rho_w V \frac{dT}{dt} = \sigma\rho q c_p(b_7 T + b_8 S + b_9 T_a) - \rho_w q_d c_{pw}(T - T_s) \tag{2.8-23}$$

Equation (2.8-20) is used to compute $T_o$ for use with the switching function $\sigma$.

Usually the collector flow rate $q$ is fixed according to the ratio of the collector area to tank volume. With given values for the model parameters and with $T_a$, $S$, and $q_d$ specified as functions of time, (2.8-23) can be solved for the tank temperature $T$ and the energy delivered to the demand as functions of time. This enables the system's efficiency to be evaluated. Techniques for solving (2.8-23) for $T$ as a function of time are covered in Chapter Three.

---

*Some writers use the loss coefficient instead of thermal resistance. The loss coefficient times the surface area equals the reciprocal of the area's thermal resistance. Here, $A_c U_c = 1/R_c$, where $R_c$ is the collector's resistance to heat loss to the ambient air.

*Example 2.25*

Obtain a closed-form model in SI units for the following system. Derivation of the collector parameters in SI units is given in Reference 9. We assume the user has specified parameters in British Engineering units. Neglect thermal losses through the tank walls. The given information is

Collector area $A_c = 2\ \text{m}^2$

Collector loss coefficient $U_c = 8\ \text{W/m}^2\text{-K}$

Heat removal factor $F_R = 0.834$ (dimensionless)

Collector volume flow rate $q = 0.06\ \text{m}^3/\text{hr}$

Collector fluid density $\rho = 1200\ \text{kg/m}^3$

Collector fluid specific heat $c_p = 6000\ \text{J/kg-K}$

Tank volume $V = 3.5\ \text{ft}^3$

Water supply temperature $T_s = 59°\text{F}$

Heat exchanger effectiveness $\epsilon = 0.9$

Since thermal systems tend to be relatively sluggish, the time unit is often taken to be an hour. Thus $U_c$ becomes

$$U_c = 8(3600) = 2.88 \times 10^4\ \text{J/hr-m}^2\text{-K}$$

From Tables 2.3, 2.15, 2.16 and 2.19 and the definition of a Btu, we obtain

$$\begin{aligned} V &= 3.5(0.02832) = 0.099\ \text{m}^3 \\ T_s &= \tfrac{5}{9}(59°\text{F} - 32) + 273 = 288\ \text{K} \\ c_{pw} &= 32.174\ \text{Btu/slug-}°\text{R} \\ &= 32.174\ (130.1) = 4185\ \text{J/kg-K} \\ \rho_w &= 1.94\ \text{slug/ft}^3 = 1.94(515.4) = 1000\ \text{kg/m}^3 \end{aligned}$$

With (2.8-16), (2.8-17), and (2.8-19), these values yield the following model, for temperature measured in degrees Kelvin.

$$4.14\frac{dT}{dt} = 4.32\sigma(T_o - T_i) - 41.85q_d(T - 288)$$

$$4.32 \times 10^5(T_o - T_i) = 1.668[S - 2.88 \times 10^4(T_i - T_a)]$$

$$T_o - T_i = 0.9(T_o - T)$$

The last two equations give the following solution for $T_o$ and $T_i$.

$$T_o = 0.878T + 4.238 \times 10^{-6}S + 0.1221T_a \qquad (2.8\text{-}24)$$

$$T_i = 0.9878T + 4.238 \times 10^{-7}S + 0.01221T_a \qquad (2.8\text{-}25)$$

Substituting these into the differential equation gives

$$4.14\frac{dT}{dt} = 4.32\sigma(-0.1098T + 3.814 \times 10^{-6}S + 0.1099T_a) - 41.85q_d(T - 288) \tag{2.8-26}$$

The switching function $\sigma$ is given by (2.8-18) using (2.8-24) and the solution for $T$ at any given time. The input variables here must be in the following units.

Ambient temperature $T_a$ = degrees K

Absorbed solar insolation $S$ = J/hr-m$^2$

Demand volume flow rate $q_d$ = m$^3$/hr

## 2.9 MODELS OF NONTECHNICAL SYSTEMS

Modeling and analysis techniques developed for engineering purposes are continually finding more applications in nontechnical fields such as biology, ecology, economics, and management. Many engineers with training in systems analysis are members of interdisciplinary teams solving problems in these areas. Such activities will increase because of the growing complexity of modern society's problems and systems. As a brief orientation to some of these exciting possibilities, we now explore a few simple models from several areas.

### Natural-Resource Management

The management of renewable and nonrenewable resources assumes greater importance as the world's population grows and becomes more advanced technologically, and as limited resources like petroleum become more scarce. In order to manage efficiently a renewable resource, such as an animal population or a wood-producing forest, a model of the resource's growth process is needed. The model now to be introduced is being utilized for fishery and forest management. For brevity we will treat only the fishery application.

The simplest realistic formulation of population growth states that the rate of growth is directly proportional to the size of the population, or

$$\frac{dy}{dt} = ry \tag{2.9-1}$$

where $y$ is the population size measured by number of individuals or by population mass (biomass). The growth coefficient $r$ is the per-capita growth rate if $y$ is measured in number of individuals, and it is the difference between the birth and death rates. It is assumed to be a constant here, although seasonal or other time-dependent effects have been modeled by letting $r$ be a function of time. Note that this formulation can be viewed as a result of conservation of mass. Population models such as this one have a long history. The English demographer and economist T. R. Malthus (1766–1834) studied this model, and the term *Malthusian growth* is used to denote the unlimited growth predicted by this model if $r$ is positive (see Chapter 3).

Many criticisms can be leveled at the Malthusian model, but it continues to find many applications. Births and deaths are discrete events, but for a large population the net effect of these processes resembles a continuous process (to be discussed in more detail when discrete-time models are considered in Chapter Four). It is unreasonable to expect that the growth rate $r$ will remain constant over a wide variety of conditions. In particular, as the population size increases, food or living space might become scarce, and the growth rate would be expected to diminish because the population would become less healthy on the average. This would increase the death rate and decrease the reproductive capability. The simplest way of modeling this effect is to assume that the net growth rate $r$ is a linear function of $y$, or

$$r = a - by \tag{2.9-2}$$

where $a$ and $b$ are positive constants. As $y$ approaches zero, $r$ approaches a constant value, as one would expect for a small population unhindered by environmental constraints. As $y$ increases, $r$ diminishes and eventually becomes zero, and then negative. This implies that there exists a constant population size, denoted by $K$, such that the population increases, remains constant, or decreases as $y$ is less than, equal to, or greater than $K$, respectively. The value of $K$ is the *carrying capacity* of the environment and can be found from (2.9-2) by setting $r$ to zero. This gives $K = a/b$. The model is frequently written in terms of $K$ as

$$\frac{dy}{dt} = r\left(1 - \frac{y}{K}\right)y \tag{2.9-3}$$

where $r = a$ and is introduced to emphasize that this model form approaches that of the Malthusian model for small $y$. Equation (2.9-3) is the well-known *logistic* growth model, whose solution is presented in Chapter Three. The model is associated with the names of Verhulst (1830) and Pearl and Reed (1920) who developed it separately.

In order to apply the logistic model to fishery management, a modification must be made. The minimum openings in a commercial fishnet are specified to allow immature fish to escape while trapping the larger fish. Thus only part of the population is subject to capture – called the *fishable stock*. If we assume that the time required for a fish to reach catchable size is short, we can take the size of the fishable stock to be $y$, the population size. The modification involves the introduction of a term describing the rate of capture or *harvest rate*. It is reasonable to assume that the harvest rate depends on the size of the fishable stock and on the amount of effort devoted to fishing for the species in question. The *fishing effort* $f$ is a measure of the efficiency of the fishing gear, the amount of time spent fishing, and the number of equivalent boats or units involved. Typical units are interesting: millions of hooks, boat days, or thousands of pots. The common assumption is that the harvest rate $h$ is directly proportional to the size of the fishable stock and the fishing effort $f$, and that

$$h = qfy \tag{2.9-4}$$

where $q$ is the *coefficient of catchability* whose value is determined by parameter-estimation techniques based on past fishing records.

Outflow rates from the fishable stock are the death and harvest rates, while the mass inflow results from birth. Conservation of mass applied to the fishable stock gives

$$\frac{dy}{dt} = r\left(1 - \frac{y}{K}\right)y - qfy \tag{2.9-5}$$

This model was first introduced in 1954 by the biologist M. B. Schaefer, and it is named for him. With the values of $r$, $K$, and $q$ estimated from past catch and effort data, the model can be used to compute the optimum value of fishing effort required to extract the maximum benefit from the fishery without driving the population to extinction.

## Biomedical Applications

Many engineers are employed in hospitals, research laboratories, and private industries in an area loosely termed *biomedical engineering.* Their activities deal with the application of fluid mechanics and thermodynamics to understanding the circulation, respiration, and temperature-regulation systems of the body; the use of circuit theory to analyze nerve behavior; and the development of instrumentation and diagnostic techniques such as electrocardiography and ultrasonics. A common thread through all these activities is the requirement to understand the dynamic behavior of systems with feedback properties. The importance of this need has been underscored by the addition of chapters on dynamic system theory to well-established textbooks on physiology (Reference 10) as well as ecology (Reference 11). A good introduction to the biomedical field is Reference 12. Here we present a commonly used model of the arterial system to show that even relatively simple models of the type developed in this chapter have useful applications in biomedical analysis.

A model is desired where blood pressure measurements can be used to estimate the volume of blood pumped in one heart pump-cycle (the stroke volume). A simple formulation of this problem conceives of the arteries as a system of connected elastic tubes with compliance (Reference 13, p. 399). Blood is pumped into the system from the left ventricle during *systole*. The arterial compliance absorbs some of this blood volume and releases it when the ventricle rests (*diastole*). The flow resistance $R$ is taken to result from the resistance of the capillaries, and the flow is assumed to be laminar. The compliance $C$ is the ratio of volume change to pressure change, so that

$$C\frac{dp}{dt} = q - \frac{1}{R}p \tag{2.9-6}$$

where $p$ is the blood pressure, $q$ is the ejection flow rate from the ventricle, and the discharge pressure in the capillaries is taken to be zero.

During diastole, $q = 0$ and the solution technique of Section 3.1 can be applied. During systole, $q = q_o =$ constant, and the solution given in Section 3.2 for a constant input can be used. With experimentally or analytically determined values for $C$ and $R$, measurements of $p(t)$ at various times can be used to estimate $q_o$. The stroke volume is then given by $V = q_o t_s$, where $t_s$ is the duration of systole.

## Socioeconomic Models

Energy flows in many media (thermal, electrical, etc.) are governed by a common set of physical laws, although widely differing notations and conventions often obscure this fact. Paynter's tetrahedron of state has helped to provide a unified view of such energetic systems. J. F. Forrester (References 14, 15, and 16) and H. T. Odum (Reference 17) have extended Paynter's concepts to socioeconomic and ecological systems in which the flows of interest are not only energy flows, but also flows of money, people, nutrients, etc. Actually, Odum reduces all of these flows to equivalent energy flows. Since we have already seen several models of natural populations, we focus here on a socioeconomic model.

Forrester's model structure consists of *levels, flow rates, decision functions*, and *information channels*. The levels are storage elements and can represent inventories, goods in transit, bank balances, factory space, or number of employees, for example. They are the present values resulting from the accumulated (integrated) difference between inflows and outflows. Flow rates define the instantaneous flows between the levels (the state variables). The decision functions are statements of policy that determine how the available information about levels leads to the present flow rates. As such, they determine how the contents of one level are transported to another level. The information channels are the basis for the decisions in that they provide information about the present level values to the decision functions.

The six networks (subsystems) commonly found in Forrester's models represent the dynamics of the following quantities.

1. Orders (not "commands" but "purchase orders").
2. Materials.
3. Money.
4. Personnel or population.
5. Capital equipment.
6. Information.

Levels within one network must all contain the same kind of quantity. In addition, any network may have subnetworks; for example, the material network may have lumber and concrete subnetworks. *Only* the information network can extend from a level in one network type to a decision function in any other network. Thus the information network is said to be in a superior position relative to the other types; it is viewed as forming the fundamental structure of the model. This contrasts with Odum's approach, which views energy as the superior network.

A prominent characteristic of Forrester's approach in his extensive use of *multipliers* in constitutive relations. For example (Reference 15), if we divide a city population into the following three groups:

1. Management and professionals.
2. Skilled laborers ($L$).
3. Unskilled workers ($U$).

a level equation is

$$\frac{dU}{dt} = U_a - U_d + L_s + U_b$$

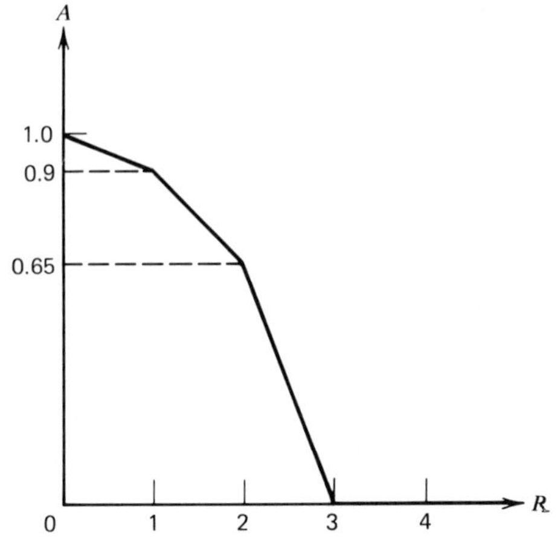


**Figure 2.29** **Attractiveness multiplier.**

where

$U_a$ = rate of arrival of $U$ into city

$U_d$ = rate of departure of $U$ from city

$L_s$ = net flow rate via social mobility between $L$ and $U$

$U_b$ = net rate of births minus deaths

The constitutive relation between $U_d$ and $U$ is

$$U_d = \frac{cU}{A}$$

where $c$ is a constant and $A$ is a multiplier representing the "attractiveness" of the city for the $U$-group. Forrester's method is to estimate the multiplier's value as a function of the other variables. For example, the multiplier $A$ is a function of $R$, the ratio of $U$-group families to $U$-group housing units. Figure 2.29 shows an example of $A$ as an estimated function of $R$.

Much of Forrester's work, and criticisms of it, deals with the manner in which the multiplier curves were obtained. Many of the curves simply represent best guesses. There is a great deal of controversy over the use of such models, especially for predicting the future of the world's economies and ecosystems. A sound understanding of the material in this and succeeding chapters is necessary to understand and to develop such models.

## 2.10 SUMMARY: MODELS OF COMPLEX SYSTEMS

For convenience we have segregated our coverage of physical systems into fluid (compressible and incompressible), thermal, mechanical, and electrical systems. Many designs contain a mixture of system types, and these subsystems can interact, often in subtle ways.

For example, Figure 2.30 represents a common configuration for a chemical

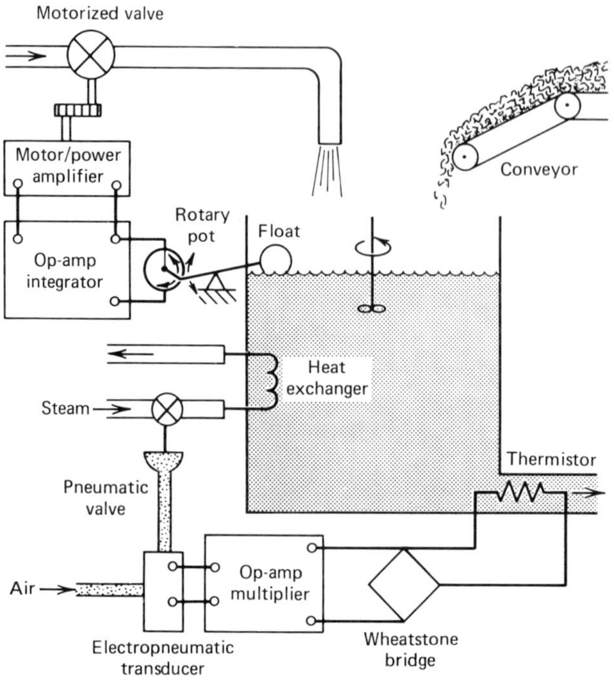


**Figure 2.30 Chemical processing system.**

processing system. A dry chemical is fed into the tank of liquid where it is dissolved and reacts with the liquid. The proper ratio of the two chemicals must be maintained. Thus the inflow rate of each is placed under feedback control. The volume in the tank is sensed by a float that moves a linkage proportionally to the tank height. The linkage in turn rotates a potentiometer (a variable resistance). This produces a voltage, which is compared to a reference voltage (not shown) indicative of the desired height. The operational amplifier (op amp) integrates this error signal voltage and opens or closes the motorized valve accordingly. The dynamics of the motor, gears, and load (valve) can be developed with the techniques of Sections 2.3. The device for controlling the feed rate of the dry chemical is not shown but would be a motor/amplifier connected to the conveyor drive rollers. The reasons for choosing an *integrating* op amp will be explored in Chapter Six.

The model for the tank height is linear if we assume that laminar flow exists and that the dry chemical is completely dissolved. The mixture is kept uniform by stirring, and its reaction rate is a function of the chemical concentration and the temperature. This is controlled by allowing steam to flow through a heat-exchanger coil. The model for the temperature of the tank mixture can be developed with an energy balance in the tank and the constitutive equation for the heat exchanger (Section 2.8). The steam flow is controlled by a pneumatic valve whose input air pressure is produced by a transducer that makes the pressure proportional to the error signal voltage generated by a Wheatstone Bridge, which is a device for computing differences in voltages. Here it is used to compare the voltage from a temperature-sensitive resistor (a thermistor) to a reference voltage indicative of the desired tank temperature.

The models of all these subsystems can be developed with the techniques of this chapter, and each would be a first-order model, probably a linear one. However, the total system model is of a higher order and is nonlinear. Let $y$ be the concentration of the reactant in the tank, $h$ the height, $C$ the tank surface area, and $T$ the reactant temperature. Then the total amount of reactant in the tank is $Chy$, and a material balance equation can be written for it. This equation is nonlinear because of the product of two of the state variables, $h$ and $y$. Also, *Arrhenius's law* states that the rate of the reaction is proportional to $Chy$ with the proportionality factor being a function of temperature. In this way, the heat-balance equation is coupled to the material-balance equation. The rate of heat generated by the reaction is proportional to the reaction rate and thus provides an additional coupling of the equations. If the thermistor or the pneumatic valve acts too slowly or too quickly, the action will result in additional coupling. The same can be said of the float, integrator, and motorized-valve dynamics. These interactions frequently require an overall supervisory control system to coordinate the individual controllers.

Even though coupling exists between the subsystems and the overall system model is of an order higher than one, it would be impossible to develop such a model without understanding how to describe each subsystem. Also, an analysis of the overall model is impossible without an ability to analyze a first-order model. We must start simply and work up the ladder of complexity. We will see that an analysis of each subsystem will enable us to decide whether or not its dynamics are such that they will cause significant interactions with the other subsystems. For example, if we can estimate the speed of response of the thermistor voltage to temperature changes, we can decide whether or not we must couple its behavior into the overall system model. Such analysis techniques for first-order models are the subject of the next chapter.

## REFERENCES

1. National Bureau of Standards, *NBSLD: A Computer Program for the Heating and Cooling Load Determination in Buildings*, NBS Building Science Series, Washington, D.C., 1976.

2. H. M. Paynter, *Analysis and Design of Engineering Systems*, MIT Press, Cambridge, 1961.

3. J. E. Shigley, *Simulation of Mechanical Systems*, McGraw-Hill, New York, 1967.

4. R. W. Fox and A. T. McDonald, *Introduction to Fluid Mechanics*, John Wiley, New York, 1978.

5. *Chemical Engineers' Handbook*, 4th edition, McGraw-Hill, New York, 1963.

6. G. J. Van Wylen and R. E. Sonntag, *Fundamentals of Classical Thermodynamics*, John Wiley, New York. 1977.

7. D. McCloy and H. R. Martin, *The Control of Fluid Power*, Longman, London, 1973.

8. American Society of Heating, Refrigeration and Air Conditioning Engineers, *ASHRAE Handbook of Fundamentals*, New York, 1972.

9. J. A. Duffie and W. A. Beckman, *Solar Thermal Energy Processes*, John Wiley, New York, 1974.

10. A. C. Guyton, *Textbook of Medical Physiology*, 4th edition, W. B. Saunders, Philadelphia, 1971.

11. E. P. Odum, *Ecology*, W. B. Saunders, Philadelphia, 1972.

12. J. H. Milsum, *Biological Control Systems Analysis*, McGraw-Hill, New York, 1966.

13. H. P. Schwan, ed., *Biological Engineering*, McGraw-Hill, New York, 1969.

14. J. W. Forrester, *Industrial Dynamics*, MIT Press, Cambridge, 1961.

15. J. W. Forrester, *Urban Dynamics*, MIT Press, Cambridge, 1968.

16. J. W. Forrester, *World Dynamics*, MIT Press, Cambridge, 1972.

17. H. T. Odum, *Environment, Power and Society*, Wiley-Interscience, New York, 1971.


## PROBLEMS

2.1 Derive the constitutive relations between the restoring force and displacement for the following systems (Figure P2.1). Do not assume small displacements.

(a) A pendulum with a "gravity spring."

(b) A cylinder with a "buoyancy spring." (Hint: Use Archimede's principle.)

(c) A wire of length $2L$ with an initial tension $T$.

2.2 A tank initially holds $v_o$ gallons of brine that contains $w$ pounds of salt. The input flow rate to the tank is $q_i$ gallons per minute of brine with a salt density of $\rho$ pounds of salt per gallon. The solution in the tank is well stirred and is pumped out at the *constant* rate of $q_o$ gallons/minute. Let $y(t)$ be the number of pounds of salt in the tank at time $t$. Develop the dynamic model for $y$ with $q_i$ and $q_o$ as the input variables.

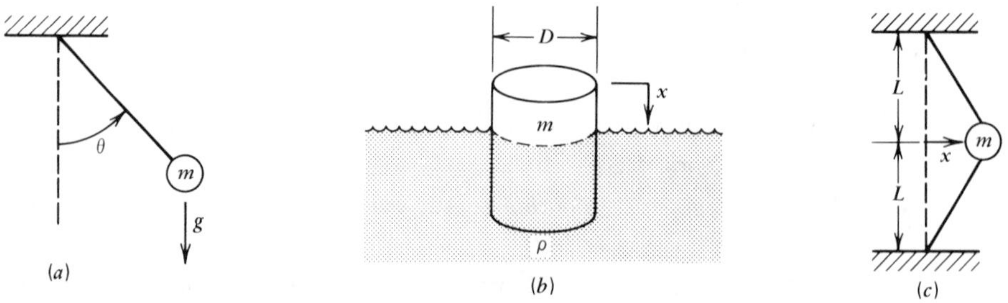


Figure P2.1

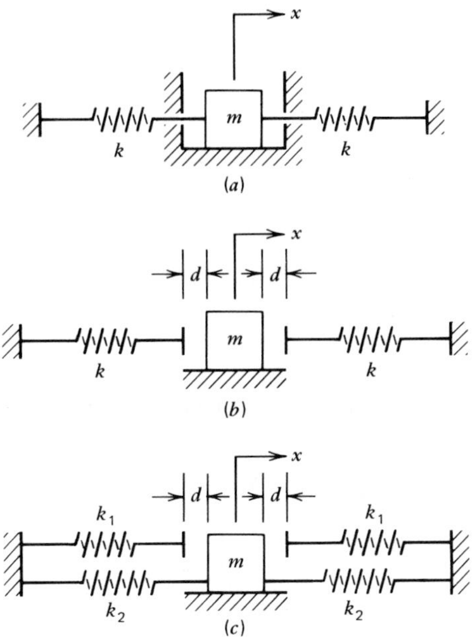


**Figure P2.4**

2.3 Derive the models for the mechanical systems shown in Figures 2.11*b* and 2.11*c*.

2.4 Plot the force exerted on the mass by the springs as a function of the displacement $x$ (Figure P2.4). Neglect friction.

2.5 Gear teeth have an elasticity that sometimes must be accounted for in a dynamic model. They also exhibit the phenomenon of *backlash* if the gears do not mesh tightly. Sketch the equivalent spring force of a gear tooth under these conditions, as a function of angular displacement.

2.6 The vibration of a motor mounted on the end of a cantilever beam can be modeled as a mass-spring system. The motor mass is 2 slugs and the beam mass is 1 slug. When the motor is placed on the beam it causes an additional static deflection of 0.01 feet. Find the equivalent mass $m$ and equivalent spring constant $k$.

2.7 A motor connected to a pinion gear of radius $R$ drives a load of mass $m$ on the rack (Figure P2.7). Neglect all masses except $m$. What is the energy-equivalent inertia due to $m$, as felt by the motor?

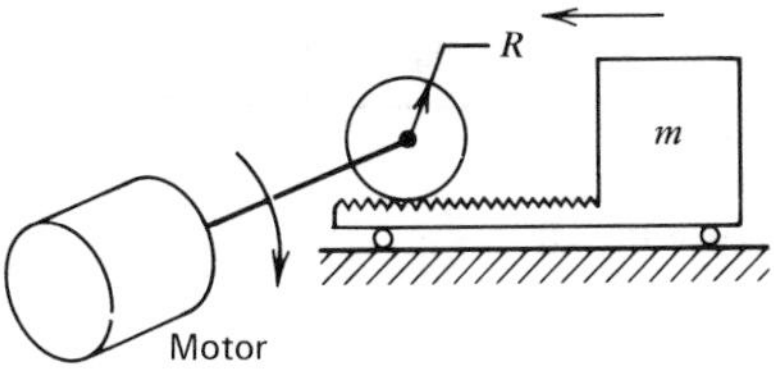


**Figure P2.7**

2.8 Find an expression for the torsional spring constant for the stepped shaft shown in Figure P2.8. (Hint: Are the two shaft segments in series or parallel?)

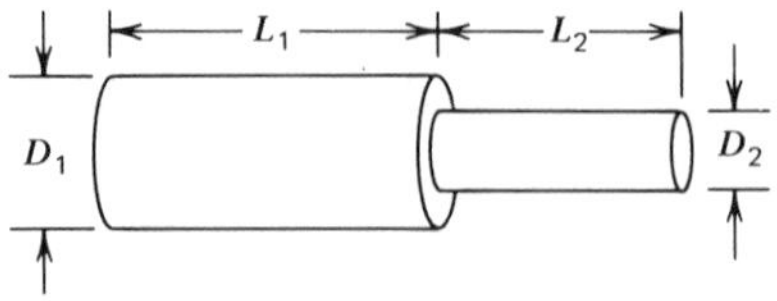


**Figure P2.8**

2.9 Determine a model for the levered systems shown in Figure 2.9, with the force $f$ as input. Assume small displacements and consider two cases.

**(a)** The output is $\theta$ and the lever is massless.

**(b)** The output is $\omega = \dot{\theta}$ and the lever has an inertia $I$ relative to the pivot.

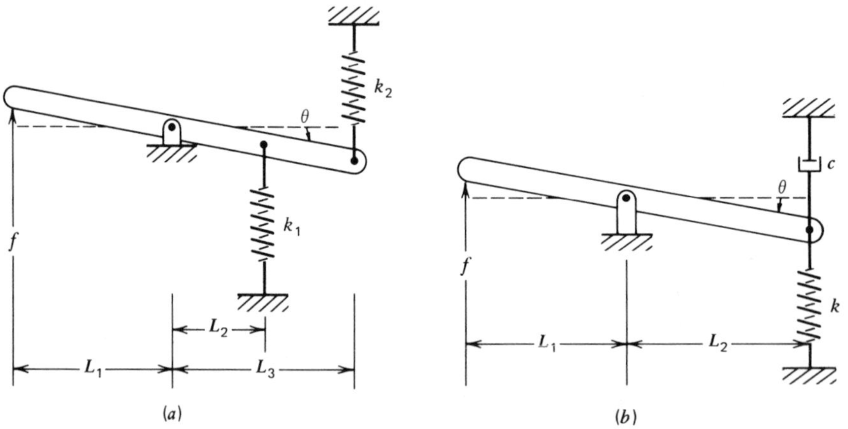


**Figure P2.9**

2.10 Derive models for the circuits shown in Figure P2.10. The input is the voltage $e_1$ and the output is the voltage $e_2$.

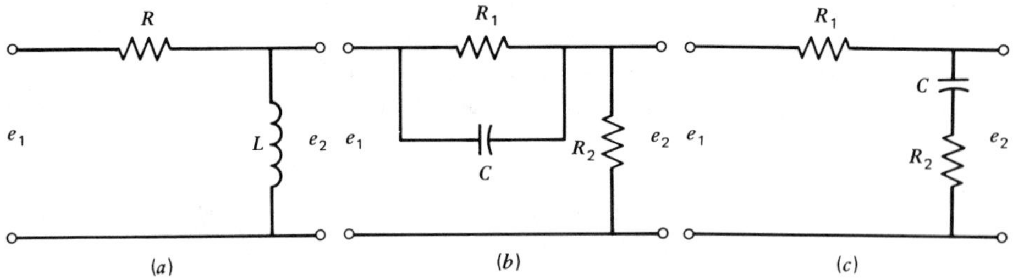


**Figure P2.10**

2.11 Derive models for the circuits shown in Figure P2.11. The input is the voltage $e$ and the output is the current $i$.

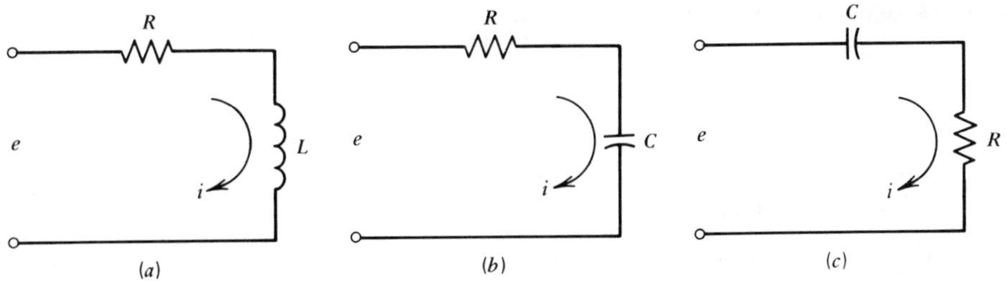


**Figure P2.11**

**2.12** Derive models for the op amp circuits shown in Figure P2.12. The input is the voltage $e_i$ and the output is the voltage $e_o$.

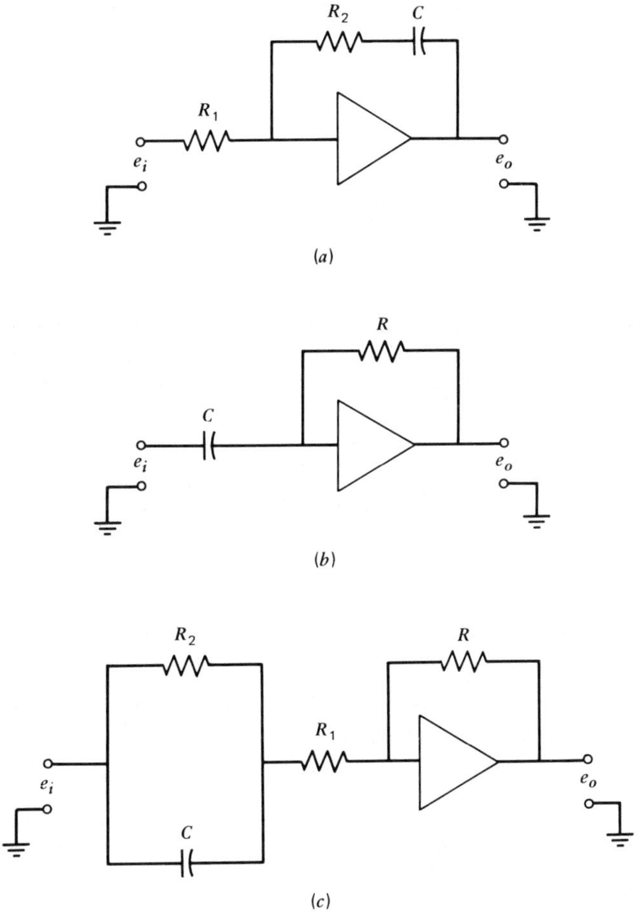


**Figure P2.12**

**2.13** A spherical water tank has a radius $r$, an air vent on top, and an orifice at the bottom (Figure P2.13)

(a) Find the constitutive relation between the liquid height $h$ and the liquid volume $V$.

(b) Develop a dynamic model for the height $h$.

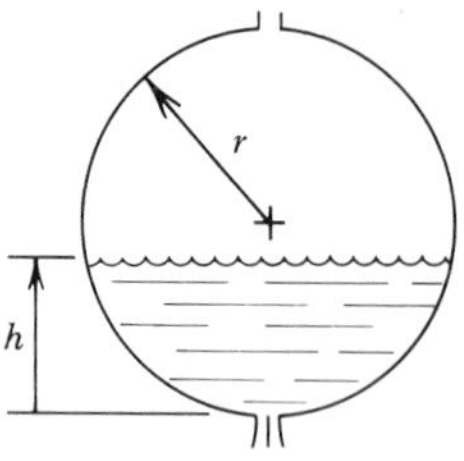

**Figure P2.13**

**2.14** Derive the nonlinear damping relations (2.2-3) and (2.6-6) when turbulent flow exists in the flow passage in Figure 2.21.

**2.15** Derive a model for the liquid height $h$ in a tank system whose input is the pressure $p_i$ (Figure P2.15).

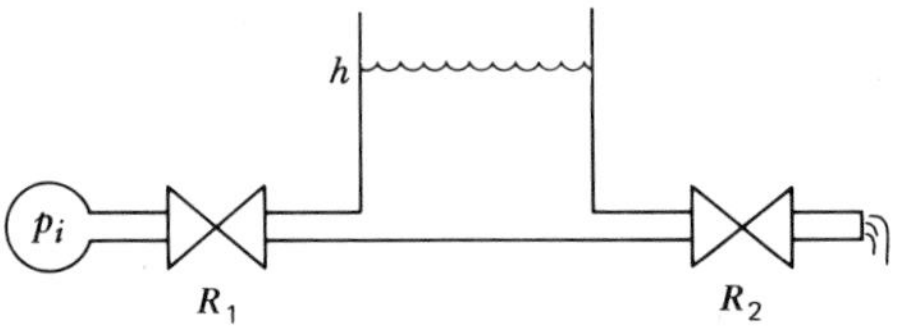

**Figure P2.15**

**2.16** A pneumatic restriction was tested under the following conditions. The inlet and outlet pressures (absolute) were $p_1 = 20$ psia and $p_b = 14.7$ psia. The inlet air temperature was $T_1 = 68°$F, and the measured mass flow rate was $q_m = 0.01$ slug/sec.

(a) Estimate the effective cross-sectional area $C_d A$.

(b) Estimate the turbulent resistance $R_T$ of the restriction.

**2.17** A rigid container has a volume of 10 ft$^3$. The air inside is initially at 68°F.

(a) Find the pneumatic capacitance of the container for an isothermal process.

(b) Develop a model for a system consisting of the container and the pneumatic inlet restriction analyzed in Problem 2.16.

**2.18** The pneumatic system shown in Figure P2.18 is a stack-type device used to control the output pressure $p_o$. The input pressures $p_r$ and $p_m$ represent reference and measurement pressures, respectively. The supply pressure is $p_s$.

Assume that the pressures are deviations from nominal values and develop an approximate model for the output pressure $p_o$. To do this, first develop a dynamic model relating $p_1$ to $p_o$, and then sum the pneumatic forces acting on the stem. Assume that the diaphragm and stem masses are negligible.

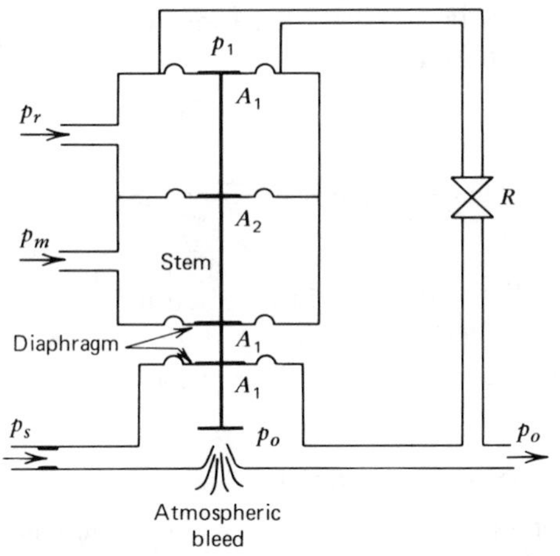


**Figure P2.18**

**2.19** A stepped copper shaft has the dimensions shown in Figure P2.19. Compute the thermal resistance and capacitance of each section of the shaft and of the total shaft.

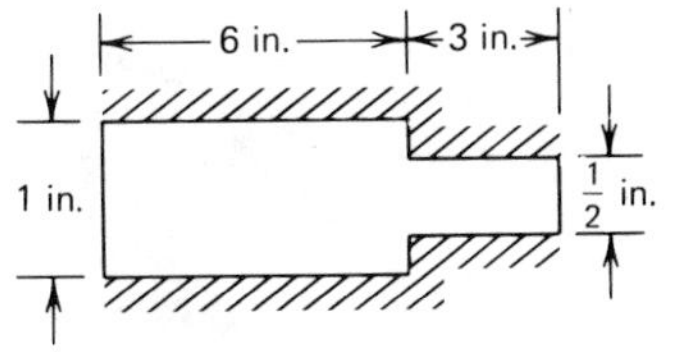


**Figure P2.19**

**2.20** Liquid at a temperature $T_i$ is pumped into a mixing tank at a volume flow rate $q$ (Figure P2.20). The container walls are perfectly insulated. The container volume is $V$ and the liquid within is assumed to be mixed and at a uniform temperature $T$. The liquid's specific heat and mass density are $c$ and $\rho$. Before the start of the process the inlet and outlet temperatures were $T_i = T = T_R$. Then $T_i$ is changed to $T_R + \Delta T_i$. Develop a model for $\Delta T = T - T_R$ with $\Delta T_i$ as the input.

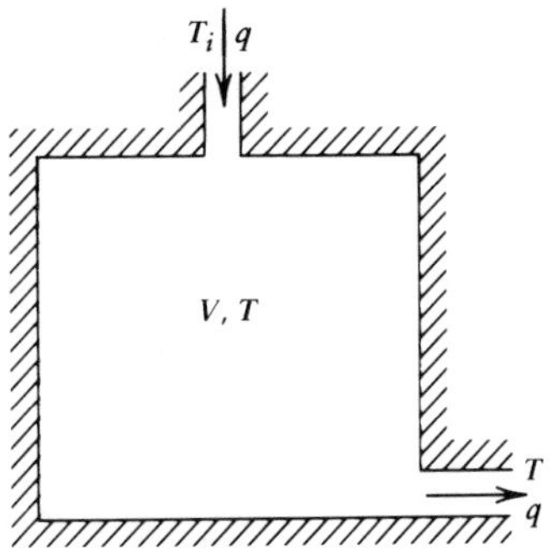


**Figure P2.20**

**2.21** A hot metal part is to be quenched in a water bath. Assume that the Biot number is small enough to allow a lumped-temperature model of the part. Denote its temperature by $T$, its capacitance by $C$, and its surface resistance by $R$. Assume that the bath is large enough so that its temperature remains constant. Derive a model for $T$.

**2.22** An electric heating element in an oven delivers a heat output rate of $q_i$. The object to be heated has a thermal capacitance $C$ and the oven walls have a resistance $R$. The temperature outside the oven is $T_o$.

(a) Assume that $R$ and $C$ are the only significant resistance and capacitance. Develop a model for $T$.

(b) Discuss the assumptions in (a) in light of the properties of the object in the oven (its mass and specific heat).

2.23 Consider the basic growth law (2.9-1). Modify the model to account for seasonal effects that produce a positive growth rate during the first half of the year, and a negative growth rate (a decay) during the second half of the year. Assume a sinusoidal variation in the rate.

2.24 Consider the Schaefer fishing model (2.9-5) for a situation in which $\dot{y} = 0$.

(a) Solve the resulting equation for the fishing yield $qfy$ under static conditions, and find the value of the fishing effort $f$ that maximizes the yield.

(b) Explain what happens if the effort is smaller or larger than the maximizing value.

# CHAPTER THREE

# Analysis of Continuous-Time Models

Once the system model has been developed, we can see what predictions it makes of the system's behavior under whatever conditions are of interest. In order to do this for a continuous-time, dynamic model, we must solve the model's differential equations. This chapter introduces the necessary solution methods. Those needed for discrete-time models are treated in Chapter Four.

Most of the methods needed to analyze dynamic models can be developed in terms of first-order models, the simplest case. Some types of behavior occur in higher-order models that are not found in the first-order case; these are discussed in Chapter Five. However, most of the differences lie in the increased algebraic complexity needed to analyze higher-order models, and not in the basic solution concepts. By concentrating on first-order models here, we do not obscure these fundamental concepts with algebraic manipulations.

Analytical methods for linear models are treated first, starting with simple input functions. As the complexity of the input and the resulting response are increased, we introduce appropriate methods to deal with the situation. We will see that linear models allow general solutions to be obtained for many types of input functions, without the need for numerical values for either the coefficients or the initial values. However, for nonlinear models, few closed-form solutions are available, and we must rely on either a linearized approximation or a numerical solution. We conclude the chapter with a treatment of these topics.

## 3.1 FREE RESPONSE OF THE LINEAR MODEL

The simplest dynamic model to analyze is a linear, first-order, constant-coefficient, ordinary differential equation without inputs. This lengthy string of modifiers refers to the following simple equation.

$$\frac{dy}{dt} = ry \tag{3.1-1}$$

where $r$ is a constant. There are several ways to solve this equation, but we wish to emphasize from the start the following useful fact.

> Any constant-coefficient, linear, ordinary differential equation or coupled set of such equations of any order, without inputs, can always be solved by assuming an exponential form for the solution.

With this in mind, the solution can be easily obtained. The general exponential form to be assumed for each unknown variable is

$$y(t) = Ae^{st} \tag{3.1-2}$$

where $A$ and $s$ are unknown constants. The time derivative of $y$ in this form is

$$\frac{dy}{dt} = sAe^{st}$$

Substitution of the last two relations into (3.1-1) gives

$$sAe^{st} = rAe^{st}$$

For the solution to be nontrivial (nonzero for arbitrary values of $t$), the constant $A$ must be nonzero and $s$ must be finite. Thus, $Ae^{st}$ does not equal zero, and we obtain

$$s = r \tag{3.1-3}$$

Equation (3.1-3) is known as the *characteristic* equation of the model and is a first-order polynomial equation in $s$. It has one solution, the *characteristic root.* Thus the solution of (3.1-1) is

$$y(t) = Ae^{rt} \tag{3.1-4}$$

for any nonzero $A$. In order to determine a value for $A$, an additional condition must be stipulated. This condition usually specifies the value of $y$ at time $t_0$, the start of the process, and is the *initial condition.* Evaluating the above equation at $t = t_0$ gives

$$A = y(t_0)e^{-rt_0}$$

and accordingly the solution is

$$y(t) = y(t_0)e^{-rt_0}\,e^{rt}$$

or

$$y(t) = y(t_0)e^{r(t-t_0)} \tag{3.1-5}$$

For autonomous models,* the origin of the time axis may be shifted so that $t_0$ may be taken to be zero without loss of generality. The solution is sketched in Figure 3.1 for positive, zero, and negative values of $r$.

If $r$ is negative, a new constant $\tau$ is usually introduced by the definition

$$\tau = -\frac{1}{r} \tag{3.1-6}$$

and the solution can be written as

$$y(t) = y(0)e^{-t/\tau} \tag{3.1-7}$$

The new parameter $\tau$ is the model's *time constant*, and it gives a convenient measure of the exponential decay curve. To see this, let $t = \tau$ in (3.1-7) to obtain

$$y(\tau) = y(0)e^{-1} \cong 0.37y(0) \tag{3.1-8}$$

After a time equal to one time constant has elapsed, $y$ has decayed to 37% of its initial value. Alternatively, we can say that $y$ has decayed by 63%. If $t = 4\tau$,

$$y(4\tau) = y(0)e^{-4} \cong 0.02y(0) \tag{3.1-9}$$

---

* The model $dy/dt = f(y, t)$ is *autonomous* when $f(y, t)$ is independent of $t$.

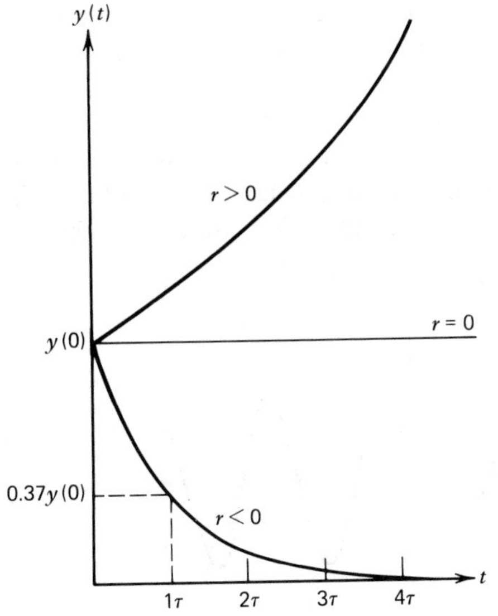


**Figure 3.1** **Free response of a linear first-order model.**

After a time equal to four time constants, $y$ has decayed to 2% of its initial value (or by 98%). Other times such as $t = 2\tau$, $3\tau$, etc., could be selected, but $\tau$ and $4\tau$ are the common choices because these times yield two points that, along with the initial value, provide sufficient spacing for sketching the curve. The results summarized by (3.1-8) and (3.1-9) are used frequently, and it is desirable to commit them to memory.

## Point Equilibrium and Stability

Often the information desired from a dynamic model simply concerns the existence of a *point equilibrium* or *equilibrium state* (or simply *equilibrium* in common usage) and its *stability*. A point equilibrium is a condition of no change in the model's variables. Mathematically it is determined by solving for the values of the variables that make *all* of the time derivatives identically zero. For (3.1-1) this implies that

$$0 = ry$$

If $r$ is zero, an infinite number of equilibrium values for $y$ exists. If $r$ is nonzero, the only point equilibrium is $y = 0$. Thus if $y$ is initially zero it will remain zero unless some cause (other than those already described by the model) displaces $y$ from its equilibrium value.

If $y$ is disturbed from equilibrium, a natural question to ask is: does $y$ return to its equilibrium? The answer is determined by the stability characteristics of the equilibrium. A *stable* equilibrium is one to which the model's variables return *and* remain, if slightly disturbed. The significance of the last phrase will become apparent later. Conversely, an *unstable* equilibrium is one from which the model's variables continue to recede, if slightly disturbed. There is a borderline situation, a *neutrally stable* equilibrium, defined to be one to which the model's variables do not return and remain, but from which they do *not* continue to recede.

These three cases can be illustrated by a ball rolling on a surface (Figure 3.2). In Figure 3.2*a*, if the ball is displaced but still within the valley ("slightly displaced"), it will roll back and forth around the bottom, transferring energy between kinetic and potential. If there is friction the ball finally comes to rest after the friction has dissipated all the energy. Thus the equilibrium state is the state in which the ball rests at the bottom, and it is stable if friction is present. A frictionless surface will allow the ball to oscillate forever about the bottom, and in this case the equilibrium

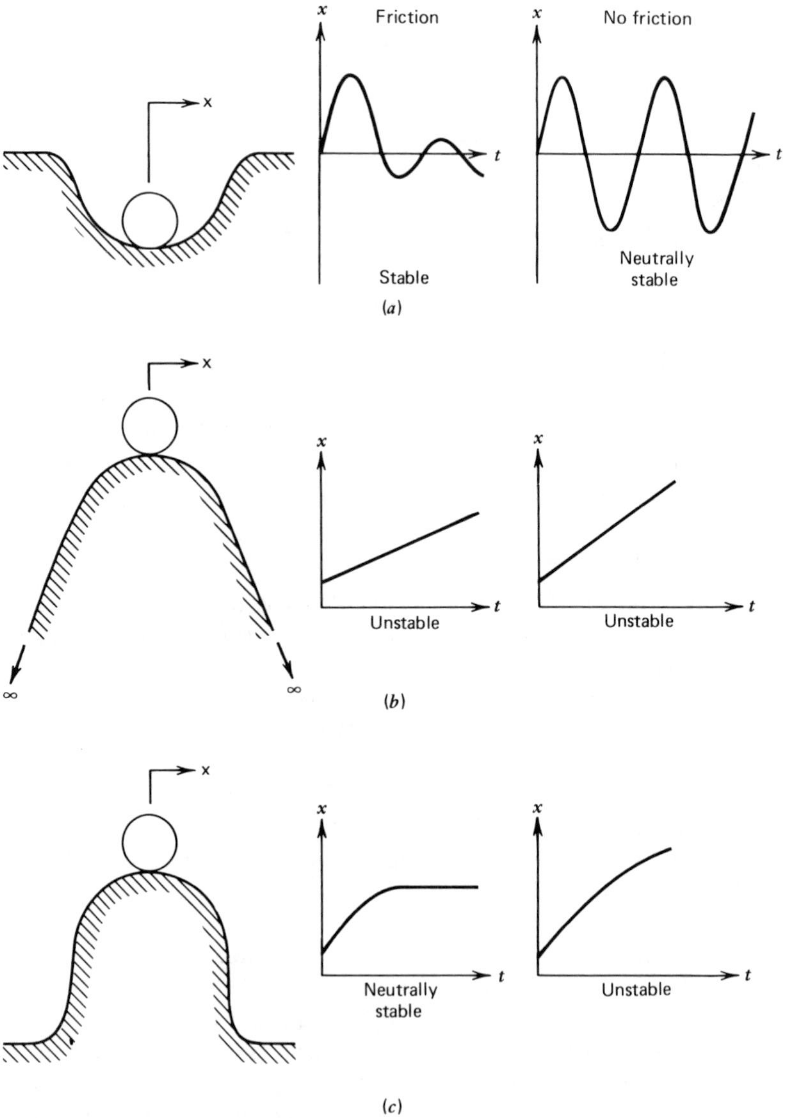


**Figure 3.2 Stability examples. (*a*) Ball in a hole. (*b*) Ball on an infinite hill. (*c*) Ball on a finite hill.**

is neutrally stable. If the ball is displaced so much that it now lies on the level surface, the equilibrium can no longer be considered stable even with friction.

The term *local stability* is sometimes used to emphasize the restriction to small displacements, while *global stability* implies stability with respect to all displacements. The situation with the ball placed on the level surface corresponds to a situation in which the original model's assumptions are no longer valid. Such would be the case for (3.1-1) as a population model, if the population size approached the carrying capacity of the environment (see Section 2.8).

For the ball precariously balanced on top of the hill (Figure 3.2*b*), this equilibrium is globally unstable (and thus locally unstable as well) for a hypothetical hill without a bottom, if the friction is not great enough to stop the ball. The absence of such a hill in reality points out that in the interpretation of instability, the assumptions of the model structure must be kept in mind. Thus in a population model of the form of (3.1-1) the equilibrium at $y = 0$ is globally unstable mathematically, but this does not mean that the population will eventually become infinite because the assumptions of the model (a constant growth coefficient) will eventually be violated as $y$ increases. For physical systems a locally unstable equilibrium usually means either that the equipment will fail as it recedes from equilibrium or that the model's assumptions are no longer valid as the system enters a new mode of behavior requiring a new model.

If the hill has a bottom consisting of a level surface (Figure 3.2*c*), the equilibrium is still locally unstable but now becomes neutrally stable in a global sense since the ball eventually comes to rest if there is friction. With no friction it remains globally unstable. Comparison of Figure 3.2*a* without friction and 3.2*c* with friction reveals that two types of neutral stability can occur. In the first type, the ball's position returns to equilibrium but does not remain. In the second type, the ball never returns but does not recede forever. Neutral stability of the first type (oscillatory behavior) is an abstraction that does not exist in physical systems unless a power source is present to sustain the oscillations. Otherwise, dissipative forces such as friction act to create a stable situation.

The first-order model has one characteristic root, and, as we will see, a second-order model has two roots, etc. The local stability properties of an equilibrium for a linear model are determined from the characteristic roots and by an equivalent but approximate process called *linearization* for a nonlinear model. A *linear* model is globally stable if it is locally stable, but the determination of global stability is frequently difficult for nonlinear models. This topic will be explored later in some detail, since the question of the stability of a control system is very important because the introduction of feedback can cause instability if the designer is not careful.

The characteristic root of (3.1-1) is given by (3.1-3), and from Figure 3.1 we can extract the following general statement.

> An equilibrium of the first-order linear model is globally stable if and only if its characteristic root $s$ is negative, and is neutrally stable if and only if $s$ is zero. Otherwise the equilibrium is unstable.

This statement will be later extended, with some modifications, to higher-order models. Thus the characteristic roots play a key role in the determination of local stability.

## Parameter Estimation

In many applications a system can be described by (3.1-1), but the parameter $r$ cannot be computed from basic principles. If past measurements of $y$ at various times are available, a value for $r$ can be estimated by the following technique. If the first data point is taken to be at $t_0 = 0$, the natural logarithm of both sides of (3.1-5) gives

$$\ln y(t) = \ln y(0) + rt \tag{3.1-10}$$

If the logarithms of the measurements of $y$ are plotted versus $t$, the transformed data should cluster about a straight line if (3.1-1) is an accurate model of the process (and if the measurement error is small!). This situation is shown in Figure 3.3 for a negative value of $r$. A straight line can be fitted by eye if great accuracy is not required and the value of $r$ is estimated from the slope of this line. If the points have some scatter, the method of *least squares* can be used to fit a line to the data. This method is developed in the problems at the end of the chapter. In either case, the line giving the best fit does not necessarily pass through the initial data point and should not be constrained to do so unless some compelling reason exists. With the estimated value of $r$, predictions can then be made about future system behavior.

*Example 3.1*

For the series $RC$ circuit model (2.4-6),

$$RC\frac{dy}{dt} + y = v$$

the characteristic root is $s = -1/RC$, and thus the time constant is $\tau = RC$. Typical values of $R$ and $C$ are 10,000 ohms and 1 microfarad (1 $\mu$f), respectively. These give a time constant of 0.01 sec. If the input voltage becomes zero but the circuit remains closed, the capacitor voltage decays by 98% in 0.04 sec.

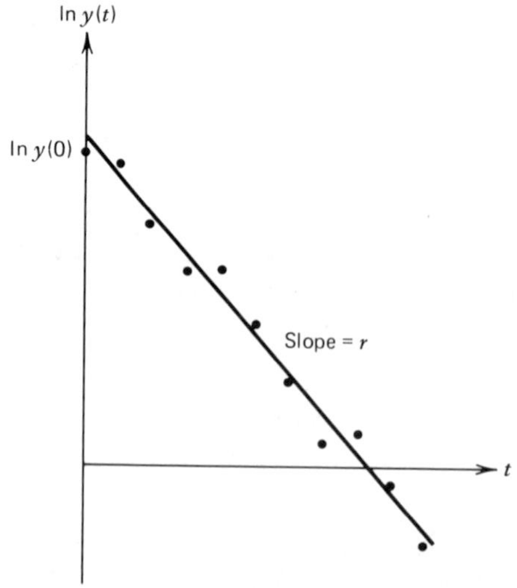


**Figure 3.3 Logarithmic transformation of data.**

*Example 3.2*

Compute the time constant for a tank-pipe system as shown in Figure 2.3$a$, assuming that the flow is laminar. The tank contains fuel oil at 70°F with a mass density $\rho$ of 1.82 slugs/ft$^3$ and a viscosity $\mu$ of 0.02 lb-sec/ft$^2$. The outlet pipe diameter $D$ is 1 in., and its length $L$ is 2 ft. The tank is 2 ft in diameter.

From conservation of mass we obtain the following model.

$$C\frac{dh}{dt} = -\frac{1}{R}\Delta p + q_i = -\frac{\rho g}{R}h + q_i \tag{3.1-11}$$

where the pressure drop $\Delta p$ due to the head of liquid is $\Delta p = \rho g h$. Here the tank area is $C = \pi$ ft$^2$. The laminar pipe resistance $R$ can be obtained from Table 2.17 or (2.5-9). The resistance is

$$R = \frac{128\mu L}{\pi D^4} = \frac{128(0.02)(2)}{\pi(1/12)^4} = 33794.4\,\frac{\text{lb-sec}}{\text{ft}^5}$$

and

$$\frac{\rho g}{R} = \frac{1.82(32.2)}{33794.4} = 0.001734\ \text{ft}^2/\text{sec}$$

Thus

$$\pi\frac{dh}{dt} = -0.001734h + q_i$$

and the time constant is $\tau = \pi/0.001734 = 1812$ sec $= 30.2$ min.

## 3.2 RESPONSE TO A STEP INPUT

All of the linear models developed in Chapter Two can be put into the following general form.

$$\dot{y} = ry + bv \tag{3.2-1}$$

where $v$ is the input. The solution obtained in Section 3.1 applies when no input is present – that is, when $v = 0$ – and this solution represents the intrinsic behavior of the system. Here we begin the analysis of (3.2-1) for the commonly encountered input functions of time.

The simplest of these perhaps is the *step function* depicted in Figure 3.4. As its name implies, it has the appearance of a single stair step. Before the initial time $t = 0$, the step function has a constant value, usually taken to be zero. At $t = 0$, $v$ jumps instantaneously to a new constant value, $M$, the magnitude, which it maintains thereafter. The step function is an approximate description of an input that can be switched on in a time interval that is very short compared to the time constant of the system. In the tank system of Figure 2.3$a$, we may model the input flow rate $q_i$ as a step if the inlet valve can be opened

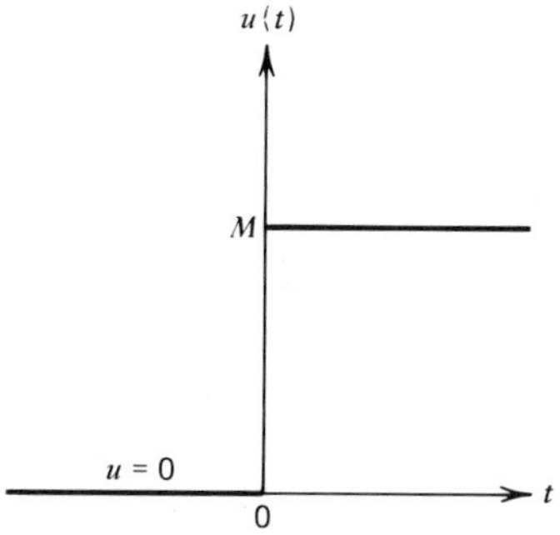


**Figure 3.4 Step input of magnitude $M$.**

much faster than the time it takes for significant changes in liquid height to occur. In Example 3.2, since $\tau = 30.2$ min, a valve-opening time of less than say, 1 min would seem to allow a step input approximation to be used with confidence. The magnitude $M$ would be the value of the flow rate if it is held constant.

A common notation for the *unit* step function ($M = 1$) is $u_s(t)$. Accordingly, the step with magnitude $M$ is denoted by $Mu_s(t)$.

## Solution for the Stable Case

There are several ways to obtain the closed form solution of (3.2-1) with a step input, if $r$ and $b$ are constants. The case where $r$ and $b$ are not constants is treated later. For now we will restrict ourselves to the stable case ($r < 0$), in which case (3.2-1) can be written as

$$\dot{y} = -\frac{1}{\tau}y + bv \tag{3.2-2}$$

The simplest way to proceed at this point is to employ some physical insight to obtain the form of the solution. If the input flow rate $q_i$ in Figure 2.3*a* is a step input, the tank height and thus the outflow rate will increase until the outflow rate equals the input rate, and the tank comes to an equilibrium state. Since this system is stable and its intrinsic behavior is given by (3.1-5), we assume that its response to a step input can be written in the form

$$y(t) = C_1 e^{-t/\tau} + C_2 \tag{3.2-3}$$

where $C_1$ and $C_2$ are undetermined constants. Note that as $t \to \infty, y \to C_2$ as we would expect. Substitution into (3.2-2) gives for $t > 0$

$$-\frac{1}{\tau}C_1 e^{-t/\tau} = -\frac{1}{\tau}(C_1 e^{-t/\tau} + C_2) + bM$$

This yields

$$C_2 = bM\tau$$

The last condition to be satisfied is the initial condition

$$y(0) = C_1 e^0 + bM\tau$$

or

$$C_1 = y(0) - bM\tau$$

This gives the step response

$$y(t) = [y(0) - bM\tau]e^{-t/\tau} + bM\tau \tag{3.2-4}$$

which is plotted in Figure 3.5 for positive values of $b$, $M$, and $y(0)$. Note once again the usefulness of the time constant in supplying convenient points for plotting. After a time equal to one time constant has elapsed, 37% of the difference between the initial value of $y$ and its final value still remains. After a four time-constant interval, only 2% of this difference remains. Thus for most cases of practical interest we may say that $y$ has reached its final value at $t = 4\tau$, and if the time required for the input to be "switched on" is much less than $\tau$, we can model the input as a step function.

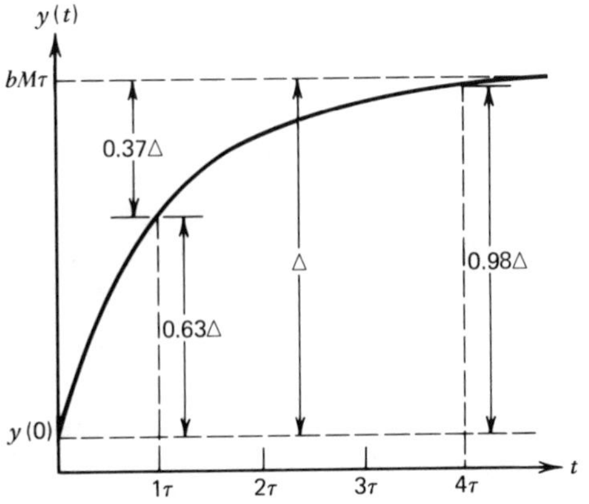


**Figure 3.5** **Step response of a first-order system.**

## Free and Forced Response

The solution given by (3.2-4) can be decomposed in several ways. We may think of the solution as being composed of two parts: one resulting from the "intrinsic" behavior of the system (the free response) and the other from the input (the forced response). Written this way the solution is

$$y(t) = \underbrace{y(0)e^{-t/\tau}}_{\text{free}} + \underbrace{bM\tau(1 - e^{-t/\tau})}_{\text{forced}}$$

The *principle of superposition* for a linear model implies that the *total* or *complete* response results from the sum of the free and the forced responses. This is borne out by the preceding equation. With this in mind we could have found the complete solution by adding the free response from (3.1-5) to the forced response given by (3.2-4) with $y(0) = 0$. This approach is sometimes useful for complicated (higher-order) models. In summary, the free response represents that part of the system behavior stimulated by the initial value of $y$, and the forced response represents the effect of the input.

## Transient and Steady -State Responses

We can also think of the response as being comprised of a term that eventually disappears (the *transient* response) and a term that remains (the *steady-state* response). For (3.2-4) the transient response is

$$[y(0) - bM\tau]e^{-t/\tau}$$

and the steady-state response is $bM\tau$. For a stable system the free response is always part of the transient response, and some of the forced response might appear in the transient response.

## Neutrally Stable and Unstable Cases

If $r$ is zero in (3.2-1), the solution given by (3.2-4) does not apply because $\tau$ is positive by definition. In this neutrally stable case, $\dot{y}$ is a constant for $t > 0$ and thus

$$y(t) = \int_0^t bM\,dt + y(0)$$

or

$$y(t) = bMt + y(0) \tag{3.2-5}$$

and $y$ continues to increase indefinitely in magnitude. Contrast this behavior with the step response for the stable case, in which $y$ approaches a constant. This difference results from the self-regulating (feedback) nature of the term $ry$ on the right-hand side of (3.2-1).

If $r$ is positive (unstable case), an approach similar to that used for the stable case produces the following solution.

$$y(t) = \left[y(0) + \frac{bM}{r}\right] e^{rt} - \frac{bM}{r} \tag{3.2-6}$$

The response increases exponentially.

### *Example 3.3*

(a) It is desired to find the motor torque necessary to drive a load inertia of 1.4 oz-in.-sec$^2$ at a constant speed of 150 rad/sec. The load's resisting torque is due entirely to dry friction and is measured to be 256 oz-in. The tentative motor choice has an armature inertia of 0.15 oz-in.-sec$^2$, a rotational loss coefficient of 7 oz-in./Krpm, and a maximum internal friction torque of 15 oz-in. These data are available from manufacturer's catalogs (for example, see Reference 1).

(b) Once the motor torque is determined, find how long it will take to reach the desired speed starting from rest.

(a) The inertias of the connecting shaft and gears are assumed to have been lumped with the load inertia, using the methods of Section 2.3. If we neglect the twisting of the shaft, we can also lump the armature and load inertias to get a model similar in form to that shown in Figure 2.12 where the applied torque $T_1$ is the motor torque $T_m$ minus the dry friction torque $T_F$. The damping coefficient $c$ is given by the rotational loss coefficient. Thus the needed values are

$$I = 1.4 + 0.15 = 1.55 \text{ oz-in.-sec}^2$$

$$T_1 = T_m - 256 - 15 = T_m - 271 \text{ oz-in.}$$

$$c = 7 \text{ oz-in./Krpm} = 0.07 \text{ oz-in./rad/sec}$$

At steady state the torques must balance, so that

$$T_m - 271 = 0.07(150)$$

where the torque on the right is the damping torque experienced at the design speed of 150 rad/sec. The required motor torque therefore is $T_m = 281.5$ oz-in.

(b) If this torque is suddenly applied and maintained, the step response may be used. The system time constant is

$$\tau = \frac{I}{c} = \frac{1.55}{0.07} = 22 \text{ sec}$$

and it will take approximately 88 sec to reach the desired speed if the motor torque is held constant. Later we will see that the torque of a dc motor does not remain constant when accelerating a load even if a constant motor voltage is applied. This is due to the coupling between the dynamics of the mechanical subsystem and the electrical subsystem. However, these results are accurate enough to be useful for finding the approximate motor size required.

## *Example 3.4*

Consider Example 3.2 with inflow to a 2-ft diameter tank. If the input flow rate of fuel oil $q_i$ is 20 gal/min, how long will it take to raise the oil level from 1 ft to 10 ft if the 1-in. outlet line remains open?

Since 7.48 gal = 1 ft$^3$, the input flow rate is 0.0443 ft$^3$/sec, and this is the magnitude of the step. The tank model (3.1-11), when put into the standard form of (3.2-2), gives

$$y = h, \qquad v = q_i$$

$$b = \frac{1}{C} = \frac{1}{\pi}, \quad \tau = \frac{RC}{\rho g} = 1812$$

Equation (3.2-4) can be transformed logarithmically to find the time required to reach 10 ft if $h(0) = 1$ ft.

$$bM\tau = \left(\frac{1}{3.14}\right)(0.0443)(1811) = 25.56 \text{ ft}$$

$$\ln\left(\frac{10 - 25.56}{1 - 25.56}\right) = -\frac{t}{1812}$$

The time required is

$$t = 827 \text{ sec} = 13.8 \text{ min}$$

A by-product of this analysis is the final height, 25.56 ft, that would be attained if the input flow rate were maintained constant at 20 gal/min.

## *Example 3.5*

For the previous example, how long would it take for the height to reach 10 ft if the outlet pipe were closed?

In this case the outflow term $-\rho gh/R$ in (3.1-11) is zero, and the neutrally stable case applies. From (3.2-5), the step response is

$$h(t) = \frac{1}{3.14}(0.0443)t + h(0)$$

With $h(t) = 10$ and $h(0) = 1$, the time required is

$$t = \frac{(10 - 1)(3.14)}{0.0443} = 636 \text{ sec} = 10.6 \text{ min}$$

# 3.3 LAPLACE TRANSFORM SOLUTION OF DIFFERENTIAL EQUATIONS

We will make extensive use of the Laplace transform in this and succeeding chapters. The advantage of the transform is that it converts linear differential equations into algebraic relations. With proper algebraic manipulation of the resulting quantities, the solution of the differential equation can be recovered in an orderly fashion by inverting the transformation process to obtain a function of time. Here we introduce this technique as applied to first-order models. In later chapters we will extend the method to higher-order cases.

In addition to providing a systematic solution procedure for differential equations, the Laplace transform allows us to develop a graphical representation of the system's dynamics in terms of block diagrams. Manipulation of such diagrams in Chapter One was limited to static elements. With the Laplace transform, the diagram concept can be extended to include dynamic elements. This is introduced in Section 3.4.

## The Laplace Transform

Appendix B is a self-contained presentation of the basic theory of the Laplace transform. Here we review only those properties of the transform that are needed to deal with first-order differential equations.

The Laplace transform $\mathscr{L}[y(t)]$ of a function $y(t)$ is defined to be

$$\mathscr{L}[y(t)] = \int_0^\infty y(t)e^{-st}\,dt \tag{3.3-1}$$

The integration removes $t$ as a variable, and the transform is thus a function only of the Laplace variable $s$, which may be a complex number. This integral with an infinite limit exists for most of the commonly encountered functions, if suitable restrictions are placed on $s$. An alternative notation is given by the use of the uppercase symbol to represent the transform of the corresponding lowercase symbol; that is,

$$Y(s) = \mathscr{L}[y(t)] \tag{3.3-2}$$

The variable $y(t)$ is assumed to be zero for negative time $t < 0$. For example, the unit step function $u_s(t)$ is such a function. If $y(t) = Mu_s(t)$, then its transform is

$$\mathscr{L}[y(t)] = \int_0^\infty Mu_s(t)e^{-st}\,dt = M\int_0^\infty e^{-st}\,dt = M\left.\frac{e^{-st}}{s}\right|_0^\infty = \frac{M}{s}$$

where we have assumed that the real part of $s$ is greater than zero so that the limit of $e^{-st}$ exists as $t \to \infty$. Similar considerations of the region of convergence of the integral apply for other functions of time. However, we need not concern ourselves with this here, since the transforms of all the common functions have been calculated and tabulated. Some other examples of the transform calculation are given in Appendix B. Table 3.1 is a short table of transforms; a more extensive table is given in Appendix B.

The transform of a derivative is of use. Applying integration by parts to the

TABLE 3.1 Laplace Transform Pairs

| $f(t),\ t \geqslant 0$ | $\mathscr{L}[f(t)] = F(s)$ |
|---|---|
| 1. $\delta(t)$ (unit impulse) | $1$ |
| 2. $u_s(t)$ (unit step) | $\frac{1}{s}$ |
| 3. $t$ | $\frac{1}{s^2}$ |
| 4. $t^n$ $(n > -1)$ | $\frac{n!}{s^{n+1}}$ |
| 5. $e^{-at}$ | $\frac{1}{s+a}$ |
| 6. $\sin bt$ | $\frac{b}{s^2+b^2}$ |
| 7. $\cos bt$ | $\frac{s}{s^2+b^2}$ |
| 8. $\frac{df}{dt}$ | $sF(s) - f(0)$ |
| 9. $\int_0^t f(t)\,dt$ | $\frac{1}{s}F(s)$ |
| 10. $f(t) = y(t - D)$ | $F(s) = e^{-sD}Y(s)$ |

definition of the transform, we obtain

$$\mathscr{L}\left(\frac{dy}{dt}\right) = \int_0^\infty \frac{dy}{dt}e^{-st}\,dt = y(t)e^{-st}\Big|_0^\infty + s\int_0^\infty y(t)e^{-st}\,dt$$
$$= s\mathscr{L}[y(t)] - y(0) \tag{3.3-3}$$

We have just seen how a multiplicative constant can be factored out of the integral. Also, the integral of a sum equals the sum of the integrals. These facts point out the *linearity property* of the transform; namely, that

$$\mathscr{L}[af_1(t) + bf_2(t)] = a\mathscr{L}[f_1(t)] + b\mathscr{L}[f_2(t)] \tag{3.3-4}$$

## Application to Differential Equations

The derivative and linearity properties can be used to solve the differential equation (3.2-1). If we multiply both sides of (3.2-1) by $\exp(-st)$ and then integrate over time from $t = 0$ to $t = \infty$, we obtain

$$\mathscr{L}(\dot{y}) = \mathscr{L}(ry + bv)$$
$$= r\mathscr{L}(y) + b\mathscr{L}(v)$$

where the second equality results from the linearity property of the transform. Using (3.3-3) and the alternate transform notation, we obtain

$$sY(s) - y(0) = rY(s) + bV(s) \tag{3.3-5}$$

where $V(s)$ is the transform of $v$. This equation is an algebraic equation for $Y(s)$ in terms of $V(s)$ and $y(0)$. The solution is

$$Y(s) = \frac{y(0)}{s-r} + \frac{b}{s-r}V(s) \tag{3.3-6}$$

The *inverse Laplace transform* $\mathscr{L}^{-1}[Y(s)]$ is that time function $y(t)$ whose transform is $Y(s)$. The inverse operation is also linear, and when applied to (3.3-6) it gives

$$y(t) = \mathscr{L}^{-1}\left[\frac{y(0)}{s-r}\right] + \mathscr{L}^{-1}\left[\frac{b}{s-r}V(s)\right] \tag{3.3-7}$$

From Table 3.1, it is seen that

$$\mathscr{L}^{-1}\left[\frac{y(0)}{s-r}\right] = y(0)e^{rt} \tag{3.3-8}$$

which is the free response. The forced response must therefore be given by

$$\mathscr{L}^{-1}\left[\frac{b}{s-r}V(s)\right] \tag{3.3-9}$$

This cannot be evaluated until $V(s)$ is specified.

*Example 3.6*

Compute the step response of (3.2-1).

If $v$ is a step of magnitude $M$, $V(s) = M/s$ from Table 3.1. For this case, the forced response is

$$\mathscr{L}^{-1}\left(\frac{b}{s-r}\frac{M}{s}\right)$$

This transform can be converted to a sum of simple transforms by a partial fraction expansion.

$$\frac{bM}{(s-r)s} = \frac{C_1}{s-r} + \frac{C_2}{s}$$

This is true only if

$$C_1 s + C_2(s-r) = bM$$

for arbitrary values of $s$. This implies that

$$C_1 + C_2 = 0$$

$$-rC_2 = bM$$

or

$$C_1 = -C_2 = \frac{bM}{r}$$

Thus the forced response is

$$\frac{bM}{r}e^{rt} - \frac{bM}{r}$$

The addition of the forced and free responses gives the previous results (3.2-6) or (3.2-4) for $r = -1/\tau$.

The form of a partial fraction expansion of a function depends on the roots of the function's denominator. When there are only a few roots, cross-multiplication by the lowest common denominator quickly produces a solution for the expansion's coefficients $C_1, C_2, \ldots$ For more complicated functions, the coefficients can be determined by the general-purpose formulas given in Appendix B.

*Example 3.7*

Find the forced response of the model $\dot{y} = ry + bv$ when the input is an exponential $v(t) = e^{at}$, $a \neq r$.

The response is given by

$$Y(s) = \frac{b}{s-r} V(s) = \frac{b}{s-r} \frac{1}{s-a}$$

But

$$\frac{b}{(s-r)(s-a)} = \frac{C_1}{s-r} + \frac{C_2}{s-a}$$

$$= \frac{b}{r-a}\left(\frac{1}{s-r} - \frac{1}{s-a}\right)$$

Thus

$$y(t) = \frac{b}{r-a}(e^{rt} - e^{at})$$

## Impulse Response

Besides the step function, the *pulse* function and its approximation, the *impulse*, appear quite often in the analysis and design of dynamic systems. In addition to being an analytically convenient approximation of an input applied for only a very short time, the impulse also is useful for estimating the system's parameters experimentally. The impulse is an abstraction that does not exist in the physical world but can be thought of as the limit of a rectangular pulse whose duration $T$ approaches zero while maintaining its *strength* $A$. The strength of an impulse or pulse is the area under its time curve (Figure 3.6).

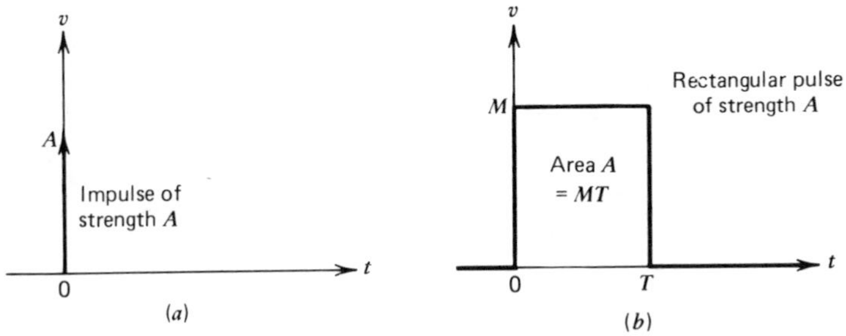


**Figure 3.6 Impulse and rectangular pulse. (*a*) Impulse of strength $A$. (*b*) Rectangular pulse. The impulse is the limit of the pulse as $T \to 0$ with $A$ held constant.**

The impulse response of the first-order model (3.2-1) can be obtained by the Laplace transform method. From Table 3.1, the transform of an impulse $v(t)$ of strength $A$ is

$$V(s) = A$$

From (3.3-6)

$$Y(s) = \frac{y(0)}{s-r} + \frac{b}{s-r}A = \frac{y(0)+bA}{s-r} \tag{3.3-10}$$

In the time domain this becomes

$$y(t) = [y(0)+bA]e^{rt} \tag{3.3-11}$$

Thus the impulse can be thought as being equivalent to an additional initial condition of magnitude $bA$.

## Pulse Response

The response due to a pulse can be obtained by using the step response (3.2-4) to find $y(T)$, which is then used as the initial condition for a zero-input solution. Alternately, the *shifting property* of Laplace transforms can be applied (Table 3.1). With this viewpoint the pulse in Figure 3.6*b* is taken to be composed of a step input of magnitude $M$ starting at $t = 0$, followed at $t = T$ by a step input of magnitude $(-M)$ (see Figure 3.7). The pulse input can now be expressed as follows.

$$v(t) = Mu_s(t) - Mu_s(t-T) \tag{3.3-12}$$

Its transform is

$$\begin{aligned} V(s) &= M\mathscr{L}[u_s(t)] - M\mathscr{L}[u_s(t-T)] \\ &= M\frac{1}{s} - Me^{-sT}\frac{1}{s} \\ &= \frac{M}{s}(1-e^{-sT}) \end{aligned} \tag{3.3-13}$$

Assume that the system is stable and that the initial condition is zero. The pulse response is found from (3.3-6) with $r = -1/\tau$ and a partial fraction expansion.

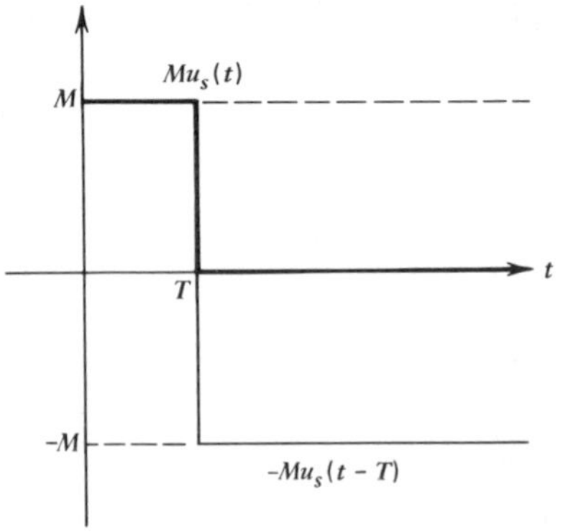


**Figure 3.7 Rectangular pulse as the superposition of two step functions.**

$$\begin{aligned} Y(s) &= \frac{\tau bM}{\tau s+1}\left(\frac{1-e^{-sT}}{s}\right) \\ &= \frac{C_1}{\tau s+1} + \frac{C_2}{s} - \frac{C_1}{\tau s+1}e^{-sT} - \frac{C_2}{s}e^{-sT} \end{aligned}$$

The transform has been expressed as the sum of elementary transforms. Cross-multiplication gives

$$C_1 s + C_2(\tau s+1) = \tau bM$$

or

$$C_1 = -\tau^2 bM$$

$$C_2 = \tau bM$$

In the time domain we obtain

$$y(t) = \frac{C_1}{\tau} e^{-t/\tau} + C_2 u_s(t) - \frac{C_1}{\tau} e^{-(t-T)/\tau} u_s(t-T) - C_2 u_s(t-T)$$

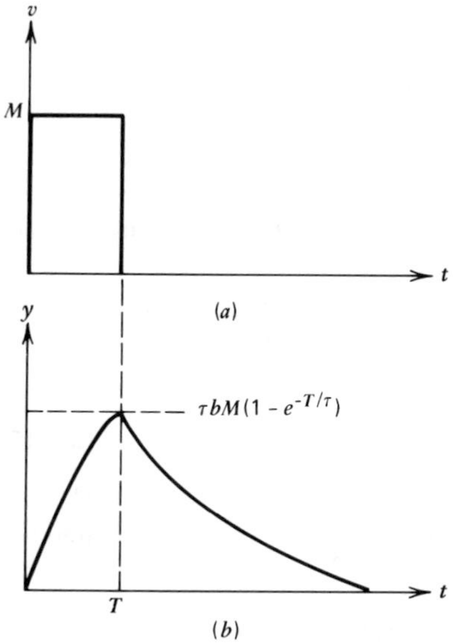


**Figure 3.8 Pulse response of a linear first-order system. (*a*) Pulse input. (*b*) Pulse reponse.**

For $0 < t < T$,

$$y(t) = \frac{C_1}{\tau} e^{-t/\tau} + C_2$$

$$= \tau bM - \tau bM\, e^{-t/\tau} \quad (3.3\text{-}14)$$

For $t \geqslant T$,

$$y(t) = \frac{C_1}{\tau} e^{-t/\tau} - \frac{C_1}{\tau} e^{-(t-T)/\tau} + C_2 - C_2$$

$$= -\tau bM(1 - e^{T/\tau})e^{-t/\tau} \quad (3.3\text{-}15)$$

This response is shown in Figure 3.8. The previous equation, when written in terms of the pulse strength $A = MT$, is

$$y(t) = -\frac{\tau bA}{T}(1 - e^{T/\tau})e^{-t/\tau} \quad (3.3\text{-}16)$$

If the strength $A$ is kept constant as $T$ approaches zero, L'Hôpital's rule gives

$$\lim_{T\to 0} y(t) = bA\, e^{-t/\tau}$$

This is the same as the impulse response when $y(0)$ is zero.

## Approximation of Pulse Response

A rule of thumb for determining when a pulse can be approximated by an impulse is obtained as follows. The Taylor series expansion for $\exp(T/\tau)$ is

$$e^{T/\tau} = 1 + \frac{T}{\tau} + \frac{1}{2}\left(\frac{T}{\tau}\right)^2 + \ldots + \frac{1}{n!}\left(\frac{T}{\tau}\right)^n + \ldots$$

If the first two terms in the series are retained and substituted into (3.3-16), the result is

$$y(t) = -\frac{\tau bA}{T}\left(1 - 1 - \frac{T}{\tau} - \ldots\right) e^{-t/T}$$

$$\cong bAe^{-t/\tau}$$

This is identical to the impulse response.

To see when it is justifiable to truncate the series in this way, consider the case where $T/\tau = 0.1$. Thus

$$e^{T/\tau} = 1 + 0.1 + 0.005 + 0.00017 + \ldots$$

For accuracy to the second decimal place, only the first two terms need be kept. This leads to the following guide, which can also be shown true for an *arbitrarily shaped* pulse.

*If the pulse duration is of the order of* $\tau/10$ *or less, the first-order system response is nearly the same as the impulse response.*

### *Example 3.8*

(a) Investigate the response of the hydraulic accumulator shown in Figure 3.9, for the situation in which the input flow rate $q_1$ jumps from 0.01 to 0.05 ft$^3$/sec, holds this value for 0.01 sec, and then returns to 0.01 ft$^3$/sec. Suppose that the flow rate $q_2$ out of the section is opposed by a linear resistance $R = 10^5$ lb-sec/ft$^5$. The accumulator plate is 1 in. in diameter so that $A = 0.0055$ ft$^2$. Assume that the spring constant is $k = 30$ lb/ft.

(b) Analyze the effectiveness of the accumulator in reducing the pressure peak.

(a) With the indicated assumptions, the accumulator model (2.6-10) becomes

$$\frac{A^2}{k}\dot{p} = q_1 - q_2 = q_1 - \frac{1}{R}(p - p_o) \tag{3.3-17}$$

where $R$ is the resistance and $p_o$ is the outlet or discharge pressure assumed to be constant. If the input flow rate $q_1$ is held constant at the value $q_{1ss}$, $p$ will approach a constant value $p_{ss}$ found from

$$0 = q_{1ss} - \frac{1}{R}(p_{ss} - p_o) \tag{3.3-18}$$

We can use this fact to simplify the analysis of the accumulator by introducing the new variables

$$y = p - p_{ss}$$

$$v = q_1 - q_{1ss}$$

$$\dot{y} = \dot{p} - \dot{p}_{ss} = \dot{p}$$

The model now is

$$\frac{A^2}{k}\dot{y} = v + q_{1ss} - \frac{1}{R}(y + p_{ss} - p_o)$$

$$= v - \frac{1}{R}y + q_{1ss} - \frac{1}{R}(p_{ss} - p_o)$$

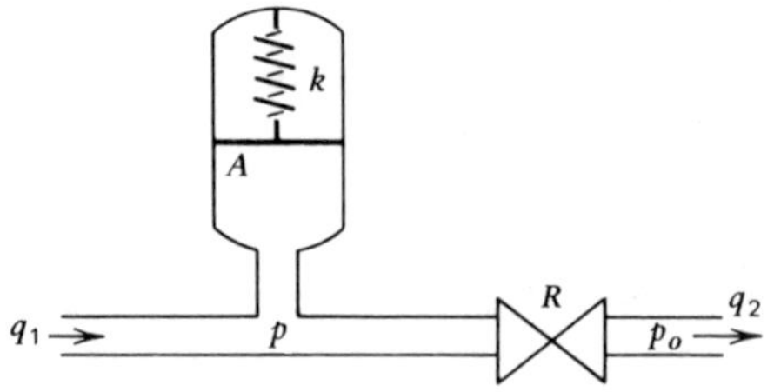


**Figure 3.9 Hydraulic accumulator with outlet resistance.**

or

$$\frac{A^2}{k}\dot{y} = v - \frac{1}{R}y \tag{3.3-19}$$

from (3.3-18). The introduction of the new variables $y$ and $v$ allows us to eliminate $p_o, q_{1ss}$, and $p_{ss}$, at least temporarily, from the analysis.

The given values of $A$ and $k$ result in a compliance $C$ of about

$$C = \frac{A^2}{k} = 10^{-6} \text{ ft}^5/\text{lb}$$

The fluid resistance $R$ is given as $10^5$ lb-sec/ft$^5$, and the time constant for the accumulator is $\tau = RC = 0.1$ sec. If $q_{1ss} = 0.01$ ft$^3$/sec, then $p_{ss} - p_o = 1000$ lb/ft$^2$ from (3.3-18).

From the given form of $q_1(t)$, the deviation $v$ is a pulse of magnitude $M = (0.05 - 0.01) = 0.04$ ft$^3$/sec, with a duration $T = 0.01$ sec. Since the duration is one-tenth of the time constant, the impulse response is approximately given by (3.3-11) with a zero initial condition and

$$MT = 0.04(0.01) = 4 \times 10^{-4}$$

Comparing (3.2-1) and (3.3-19) shows that $b = 1/C = 10^6$. Thus the response is

$$y(t) = 10^6(4 \times 10^{-4})e^{-10t} = 400e^{-10t}$$

This shows that the pressure in the section is

$$p(t) = y(t) + p_{ss} = 400e^{-400t} + 1000 + p_o$$

At the beginning of the pulse, the pressure quickly rises by 400 lb/ft$^2$, but decays back to 1000 lb/ft$^2$ after about 0.4 sec (4 time constants).

(b) It is instructive to compare this response to what would occur without the accumulator. In this case the section model is a static one obtained from (3.3-17) by setting the compliance to zero. The solution for $p$ is

$$p = Rq_2 + p_o = 10^5 q_2 + p_o$$

Under the stated conditions, the pressure $p$ increases immediately by $10^5(0.05 - 0.01) = 4000$ lb/ft$^2$, stays at its new value for the duration of the pulse, and drops by 4000 lb/ft$^2$ at the end of the pulse. Thus the pressure rise to 10 times larger than that endured with the accumulator.

## *Example 3.9*

The 1-in. copper sphere treated in Example 2.22 has a specific heat $c = 2.93$ Btu/slug-°F, a mass density $\rho = 17.33$ slugs/ft$^3$, a convection coefficient $h = 5$ Btu/hr-ft$^2$-°F, and a Biot number $N_B = 0.004$. Compute the sphere's temperature as a function of time if the temperature of the fluid flowing past the sphere suddenly drops from 570°F to 400°F for 5 minutes, and then returns to 570°F. Assume the sphere's temperature is initially 570°F.

Since $N_B < 0.1$, a lumped analysis is sufficient. The thermal capacitance $C$ is found from the product of the specific heat and the mass. The sphere's volume is 0.0003 ft$^3$.

$$C = 2.93(17.33)(0.0003) = 0.0154 \text{ Btu/°F}$$

The sphere's surface area is $A = 0.0218$ ft$^2$ and its surface resistance from (2.8-2) is

$$R = \frac{1}{hA} = 9.17 \text{ hr-}^\circ\text{F/Btu}$$

The lumped-parameter model for the sphere temperature $T$ is

$$C\dot{T} = -\frac{1}{R}(T - T_o) \tag{3.3-20}$$

where $T_o$ is the temperature of the surrounding fluid.

The sphere's time constant is $RC = 0.14$ hr $= 8.4$ min. Thus the pulse duration is too long to allow the impulse response to be used. After the pulse has disappeared, the forced response is given by (3.3-15). Once again it is convenient to transform variables. Since the sphere's temperature is initially 570°F, let $y = T - 570$, and $v = T_o - 570$. Equation (3.3-20) becomes

$$C\dot{y} = -\frac{1}{R}(y - v)$$

The constant $b$ in (3.2-1) is $1/RC$, and $v$ is a pulse of magnitude $M = -170$°F and duration $T = 5$ min $= 0.083$ hr. Since $y(0) = 0$, the forced response is also the complete response. From (3.3-14) the response is, for $t \leqslant 0.083$ hr,

$$y(t) = -170(1 - e^{-t/0.14}), \qquad t \leqslant 0.083 \text{ hr}$$

For $t \geqslant 0.083$ hr, it is, from (3.3-15),

$$\begin{aligned} y(t) &= 170(1 - e^{0.593})e^{-t/0.14} \\ &= -137.6e^{-t/0.14}, \qquad t \geqslant 0.083 \text{ hr} \end{aligned}$$

The actual sphere temperature is recovered from $T(t) = y(t) + 570$°F.

## Ramp Response

A *ramp function* is one whose first derivative $m$ with respect to time is constant, but not zero; that is, $v(t)$ is a ramp function if

$$v(t) = mt + c \tag{3.3-21}$$

Usually the constant $c$ is zero to describe a start-up process for a system initially at rest. The ramp response can be used to determine how well the system's output follows the start-up command. The *unit* ramp has a slope of unity ($m = 1$). For $c = 0$,

$$V(s) = \frac{m}{s^2} \tag{3.3-22}$$

For a zero initial condition, the response is given by (3.3-6) as

$$Y(s) = \frac{b}{s - r}\frac{m}{s^2} \tag{3.3-23}$$

This is our first encounter with a partial fraction expansion with repeated powers of $s$. It can be shown that the general expansion must include all powers of $s$ that appear in the original denominator (see Appendix B). Here the result is

$$Y(s) = \frac{C_1}{s} + \frac{C_2}{s^2} + \frac{C_3}{s-r}$$

Cross-multiplication by the least common denominator $s^2(s-r)$ gives, after grouping terms,

$$(C_1 + C_3)s^2 + (C_2 - rC_1)s - rC_2 = bm$$

Thus

$$C_1 + C_3 = 0$$

$$C_2 - rC_1 = 0$$

$$-rC_2 = bm$$

The solution is

$$C_1 = -\frac{bm}{r^2} = -C_3$$

$$C_2 = -\frac{bm}{r}$$

The ramp response is thus

$$y(t) = C_1 u_s(t) + C_2 t + C_3 e^{rt}$$

The case of most interest is the stable case, and for $t \geqslant 0$ the response is

$$y(t) = \tau bm(t - \tau + \tau e^{-t/\tau}) \tag{3.3-24}$$

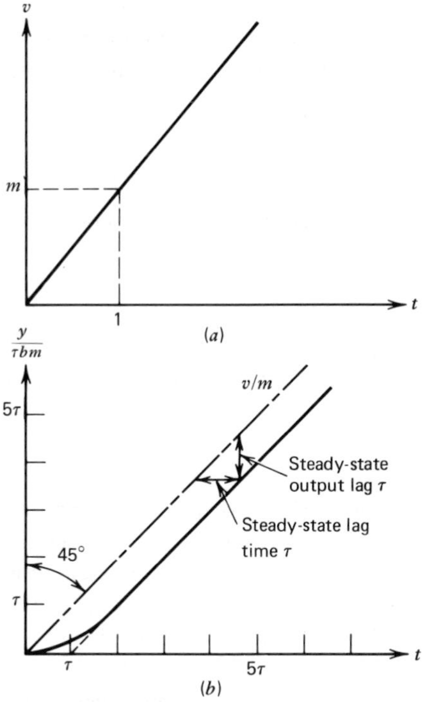


**Figure 3.10 Ramp response of a first-order linear system. (*a*) Ramp input. (*b*) Normalized ramp input and ramp response.**

This is shown in normalized form in Figure 3.10 along with the ramp input. The response approaches a straight line, which passes through the point $(0, \tau)$ and which is parallel to the normalized ramp input $v/m$, a unit ramp. In this steady-state condition the normalized output $y/\tau bm$, at any given time, is less than the normalized input by an amount $\tau$. Also, in this condition the normalized output at time $t$ equals what the normalized input was at time $(t - \tau)$. This phenomenon is the steady-state *lag time* and is another interpretation of the time constant.

The time constant also indicates how long it takes for the steady-state lag to become established. The difference between the normalized input and output is

$$\frac{v(t)}{m} - \frac{y(t)}{\tau bm} = \tau(1 - e^{-t/\tau})$$

From this it can be seen that the steady-state difference is $\tau$ (as $t \to \infty$). At $t = \tau$, the difference is $0.63\tau$; at $t = 4\tau$, the difference is $0.98\tau$.

## 3.4 TRANSFER FUNCTIONS AND BLOCK DIAGRAMS

Equation (3.3-6) contains much information about the system's behavior, as we have seen. Because the free and forced responses add to produce the total response, it is frequently convenient to assume that the initial condition $y(0)$ is zero. This allows us to focus our attention on the effects of the inputs. When the input analysis is completed, any free response due to a nonzero initial condition can be added to the result. This is a direct consequence of the superposition property of linear models.

We will be deeply involved in the analysis of systems made up of several components. Each component can influence the others by generating inputs to them. For example, the input flow rate to a tank might be the outlet flow rate from another tank. We therefore seek to establish an "input-output" format for analyzing the effect of each component on the overall behavior of the system. If this format can be developed in terms of algebraic relations, it will be to our advantage because the analysis task will be simplified. This algebraic framework is provided by (3.3-6).

Assume that $y(0) = 0$. Then (3.3-6) gives

$$\frac{Y(s)}{V(s)} = \frac{b}{s-r} = T(s) \tag{3.4-1}$$

Equation (3.4-1) describes the *transfer function* $T(s)$ of the system (3.2-1). The transfer function of a linear system is the ratio of the transform of the output to the transform of the input, with the initial conditions assumed to be zero. In other words, the transfer function is the ratio of the transforms of the forced response and the input.

### Block Diagram Elements

It is apparent that the transfer function describes that part of the forced response that results from the characteristics of the system itself. For example, the denominator of $T(s)$ in (3.4-1) is the characteristic polynomial of the system. This fact is of great significance in the analysis and design of linear systems.

The remainder of the forced response results from the characteristics of the input. The transform of the forced response is the product of the input transform and the transfer function; that is,

$$Y(s) = T(s)V(s) \tag{3.4-2}$$

This allows an algebraic and graphical representation of the cause-and-effect relations in a given system. The algebraic representation of the system's equations permits easier manipulation for analysis and design purposes, and the graphical representation allows the analyst to see the interaction between the system's components. The principal graphical representation to be employed here is the block diagram and is a visual display of the algebraic relations.

In Chapter One we saw a simple block diagram consisting of a multiplier block, repeated in Figure 3.11*a*. The block represents the algebraic relation between the Laplace transforms of the input $v$ and the output $y$. The arrows represent the transforms, and their direction indicates the cause-and-effect relation between the input and output. The *summer* and *comparator* symbols are used to represent addition and

| Type | Input-Output Relations: Time Domain | Input-Output Relations: Transform Domain | Symbol |
|---|---|---|---|
| (*a*) Multiplier | $y(t) = Kv(t)$ | $Y(s) = KV(s)$ | 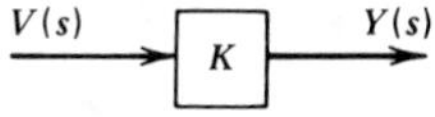 |
| (*b*) General transfer function | $y(t) = \mathscr{L}^{-1}[T(s)V(s)]$ | $Y(s) = T(s)V(s)$ | 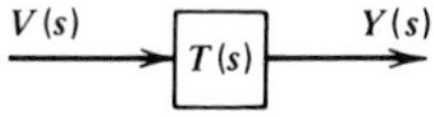 |
| (*c*) Summer | $y(t) = v_1(t) + v_2(t)$ | $Y(s) = V_1(s) + V_2(s)$ | 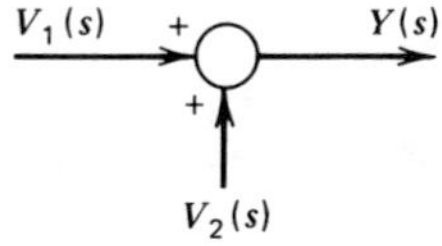 |
| (*d*) Comparator | $y(t) = v_1(t) - v_2(t)$ | $Y(s) = V_1(s) - V_2(s)$ | 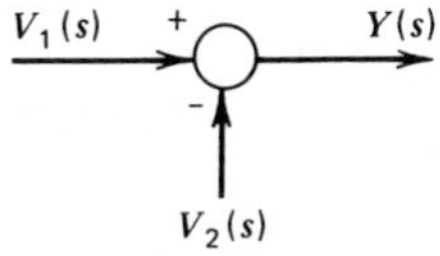 |
| (*e*) Takeoff point | $y(t) = v(t)$ | $Y(s) = V(s)$ | 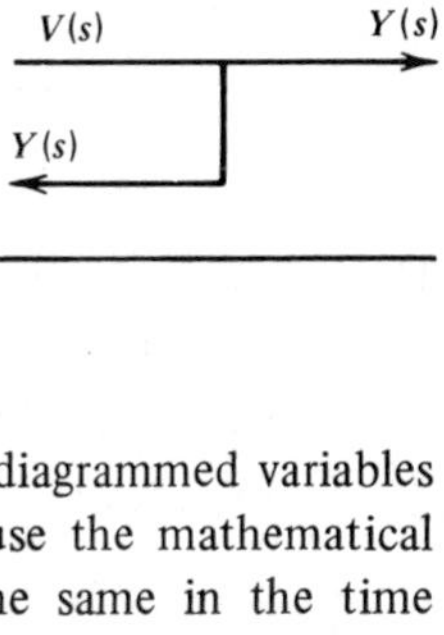 |

**Figure 3.11 Basic block diagram elements.**

subtraction. For these elements, as well as the multiplier, the diagrammed variables need not be transforms, but may be functions of time because the mathematical operations of addition, subtraction, and multiplication are the same in the time domain and the Laplace transform domain.

The concept of a transfer function is used to extend the notion of the multiplier block. If the input and output are related by a dynamic operation (integration or differentiation with respect to time), the time domain and transform domain relations are no longer identical. Only in the transform domain can such operations be represented by a multiplicative relation. The transfer function is the multiplication factor.

One must be careful to specify whether time domain or Laplace domain variables are used in the diagram. The reduction techniques to be developed for obtaining the overall system transfer function cannot be employed with time domain diagrams and can only be applied to block diagrams in which all the variables are transformed variables.

The *takeoff point* in Figure 3.11*e* is used to connect a variable with another part of the diagram and does *not* imply subtraction. A variable is taken to have the same value along the entire length of a line, until it is modified by one of the elements shown in parts (*a*) through (*d*) only of Figure 3.11.

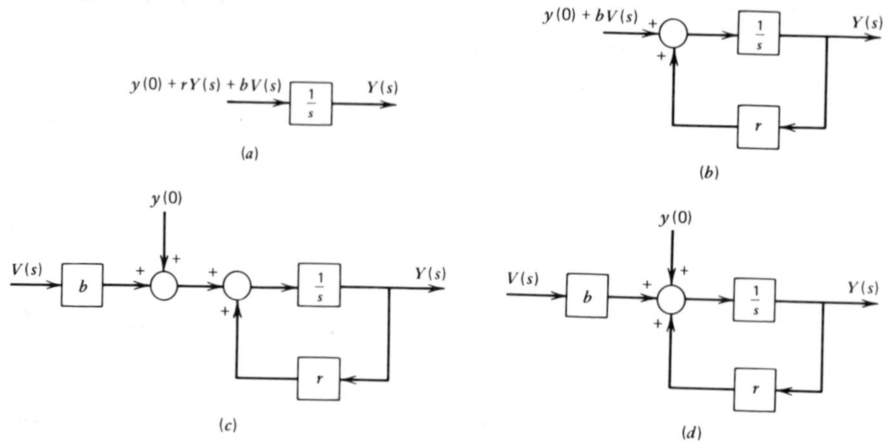


**Figure 3.12** **Construction of a block diagram. (*a*) Integrator representation. (*b*) and (*c*) Intermediate steps. (*d*) Final result.**

## Constructing a System Diagram

These concepts are more easily understood with the aid of an example. Equation (3.2-1) when transformed gives (3.3-5), which can be rearranged as

$$Y(s) = \frac{1}{s}\left[y(0) + rY(s) + bV(s)\right]$$

Thus $Y(s)$ can be treated as the output of a block with a transfer function, $1/s$, with the input to the block being the terms within the brackets. Figure 3.12 shows the step-by-step decomposition to obtain the final diagram. We now have the results of step (*a*). At this point it may seem strange to have $Y(s)$ in both the input and output. This is resolved by the use of a takeoff point, a multiplier, and a summer to create a *feedback* loop in step (*b*).

In step (*c*) the final form is obtained by using a multiplier. Another summer is used to separate the effects of $v$ and $y(0)$. Sometimes it is preferable to add or subtract more than two variables at a summer or comparator. This is permissible if a convention is established for assigning the (+) and (−) signs to the proper variables. *The convention used here is that the sign should appear clockwise from the variable's arrowhead.* Because of this more general usage, the terms *summer* and *comparator* are sometimes used interchangeably. The alternate diagram with the combined summers is shown in step (*d*).

At this point the objection might be raised that we have used an untransformed variable, $y(0)$, as an input to the diagram. The violation is only apparent, however, because $y(0)$ is considered to be the Laplace transform of an impulse function with a strength equal to $y(0)$ (see Table 3.1). This interpretation may be given *only* to *constants* that are inputs to the diagram. Constants cannot appear as internal values along a line. The common practice with these diagrams is to treat the initial conditions as zero and to add in their effect (the free response) at the conclusion of the analysis.

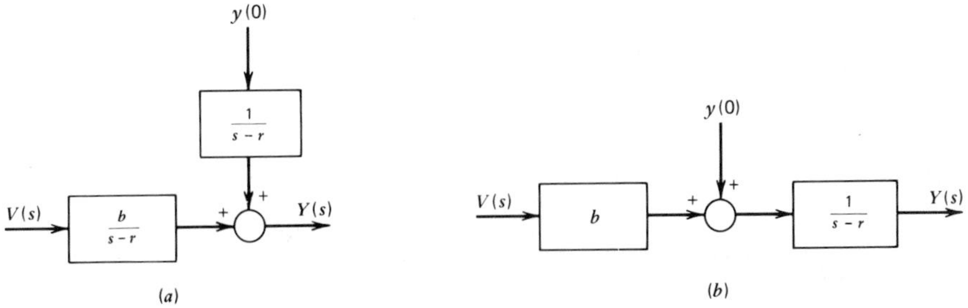


**Figure 3.13** Examples of reduced diagrams.

The emphasis of the diagram analysis is on the effect of the inputs on the system's behavior (the forced response).

## Block Diagram Reduction

If we had chosen to analyze (3.3-6) instead of (3.3-5), we would have obtained the diagram shown in Figure 3.13. Since the equations are equivalent, the diagrams must be equivalent as well. We can reduce the diagram in Figure 3.12*c* to that in either Figure 3.13*a* or 3.13*b* by using the rules of block diagram algebra. The form of the diagram obtained depends on how the transformed equation is arranged. The diagram manipulation rules simply require that the algebraic relations defined by the transformed equations be maintained. Any two diagram arrangements are equivalent if they correctly express the algebraic system equations and have the same input and output variables. However, one form may be more convenient for some purposes.

Figure 3.14 summarizes some of the common arrangements, along with an equivalent arrangement. The equivalence can easily be shown by working with the describing equations. In Figure 3.14*a* these are

$$W(s) = G_1(s)V(s)$$

$$Y(s) = G_2(s)W(s)$$

By eliminating the intermediate variable $W(s)$ we get

$$Y(s) = G_2(s)G_1(s)V(s)$$

The transfer function between $v$ and $y$ is $G_1(s)G_2(s)$. This reduction, called the *series* law, implies that we are not interested in analyzing the variable $w$ per se.

It is important to note that the series (or *cascade*) representation of two elements implies that the elements are nonloading; that is, the output $w$ of the first element is not changed by connecting it to the second element. In electrical terminology, this requires that the input impedance of the second element be infinite, which means that no power is being withdrawn from the first element. Of course, this situation is an idealization never achieved in practice. But the degree of loading should be determined before accepting the series representation. In electrical systems, nonloading is achieved by inserting an *isolation amplifier* with high impedance

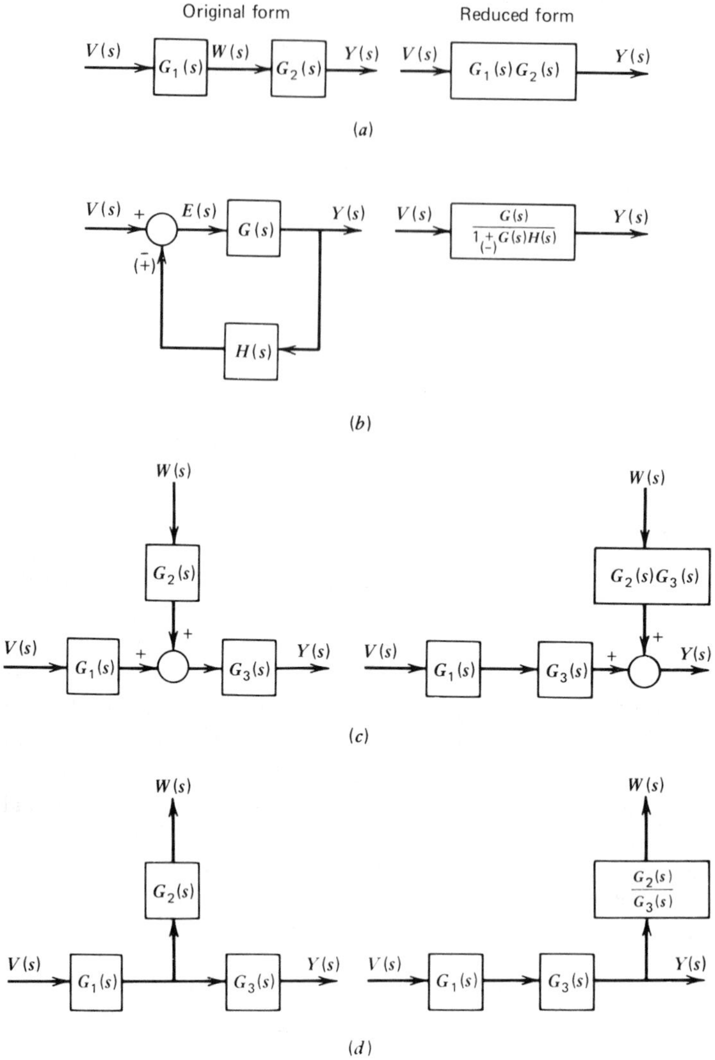


**Figure 3.14 Some common reduction formulas. (*a*) Series or cascaded elements. (*b*) Feedback loop. (*c*) Relocated summer. (*d*) Relocated takeoff point.**

between the two circuits in question. In cases where significant loading exists, the overall transfer function for the combined elements must be obtained from their equations rather than from the series law.

Another important reduction formula is illustrated in Figure 3.14*b*. In order to obtain the overall transfer function of a system, the feedback loops must each be reduced to an equivalent single block. The formula can be derived by writing the equations for the original diagram.

$$E(s) = V(s) \underset{(+)}{-} H(s)Y(s)$$

$$Y(s) = G(s)E(s)$$

where the sign in parentheses refers to the situation shown by parentheses in the diagram. Eliminating $E(s)$ gives

$$Y(s) = G(s)[V(s) \underset{(+)}{-} H(s)Y(s)]$$

or

$$Y(s) = \frac{G(s)}{1 \underset{(-)}{+} G(s)H(s)} V(s) \tag{3.4-3}$$

Note that if a comparator is used, $G(s)H(s)$ is added to 1 in the denominator. If a summer is used, $G(s)H(s)$ is subtracted from 1. Thus the sign in the denominator is always opposite to that in the feedback loop. The transfer function between $y$ and $v$ given in (3.4-3) is called the *closed-loop* transfer function because it incorporates the loop effects.

Figure 3.14*c* provides some more insight into the process of modifying diagrams. Follow the "flow" of the variable $V(s)$ through the original diagram. By the time its effect is felt on the output $Y(s)$, $V(s)$ has been multiplied by $G_1(s)$ and $G_2(s)$, and its sign has been preserved in passing through the summer. In the equivalent diagram, exactly the same operations occur to $V(s)$ before reaching the output, although in a different order. The same can be said of the flow of $W(s)$ through both diagrams. While this equivalence could have been shown from the system's equations, it seems easier to establish equivalence by tracing the variable's flow through the diagram. Figure 3.14*d* has two outputs, but the same technique can be applied to trace $V(s)$ through to each output.

### *Example 3.10*

Consider the diagram shown in Figure 3.15*a*, and find the transfer function between $y$ and $v$.

The first step in reducing the diagram to a single block is to separate the feedback loop from the paths containing the multipliers with the factors "$a$" and "$b$" (Figure 3.15*b*). Next reduce the loop to a single transfer function and do the same for the paths with the multipliers (Figure 3.15*c*). The last step is simply the combination of the two series elements (Figure 3.15*d*). It is standard practice to present transfer functions in the form of a numerator polynomial in $s$ divided by a denominator polynomial. This is the normalized form given in Figure 3.15*d*.

### *Example 3.11*

The series $RC$ circuit (Figure 2.16) has the model

$$RC\dot{y} + y = v$$

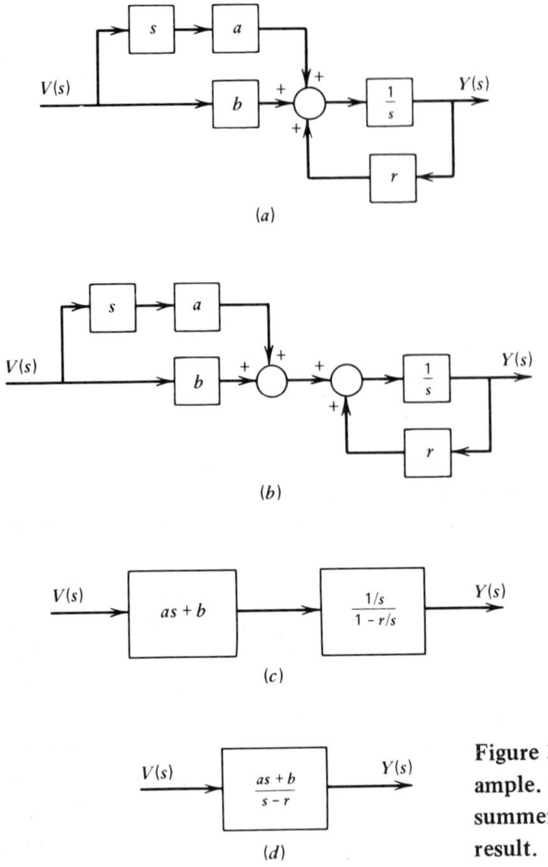


**Figure 3.15** Diagram reduction example. (*a*) Original diagram. (*b*) Relocated summer. (*c*) Loop reduction. (*d*) Final result.

where $v$ and $y$ are the input and output voltages. Its transfer function can be found from (3.4-1) with $r = -1/RC$ and $b = 1/RC$. It is

$$\frac{Y(s)}{V(s)} = \frac{1}{RCs + 1} \tag{3.4-4}$$

## Equations from Diagrams

System analysis is often facilitated by constructing a block diagram from the component's differential equations and using it to visualize the role of the component when it is connected with others in a system. In other applications, the block diagram is supplied without the system's equations, which must then be obtained for further analysis. Such is frequently the case when a number of elements, whose transfer functions are known, are connected together, and we need the equations describing the resulting system.

A simple example is shown in Figure 3.16*a*. An amplifier with a gain $K$ (the voltage amplification factor) is connected to a series $RC$ circuit. We assume that the

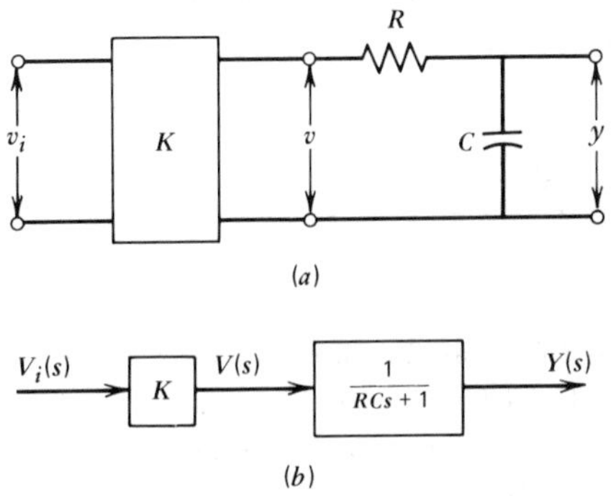


**Figure 3.16 Series *RC* circuit with an amplifier. (*a*) Circuit diagram. (*b*) Block diagram.**

amplifier is designed so that it is not loaded by the circuit. This means that the amplifier is capable of supplying the current drawn by the circuit, and the amplifier is said to have a low-output impedance. The external power supply to the amplifier is not shown. The input voltage $v_i$ is a signal voltage applied to the amplifier.

With the nonloading assumption we can use the schematic diagram to draw the block diagram (Figure 3.16*b*). The series blocks when combined give the transfer function between $v_i$ and the output voltage $y$.

$$\frac{Y(s)}{V_i(s)} = \frac{K}{RCs+1} \tag{3.4-5}$$

To recover the differential equation, the input $V_i(s)$ and the output $Y(s)$ are cross-multiplied by the numerator and denominator of the transfer function. This gives

$$(RC+1)Y(s) = KV_i(s)$$

or

$$RCsY(s) + Y(s) = KV_i(s)$$

A glance at the Laplace transform table (Table 3.1) reveals that this transformed equation is equivalent to

$$RC\dot{y} + y = Kv_i \tag{3.4-6}$$

which is the system's differential equation.

The general procedure for recovering the differential equation from the transfer function follows the same pattern as the preceding example. The rules of block diagram reduction are applied until the transfer function is obtained for each input-output pair. If an initial condition is present, as in Figure 3.12*d*, it can be set to zero to simplify the procedure, because the initial condition does not determine the differential equation. If $y(0)$ is set to zero in Figure 3.12*d*, you should be able to reduce the diagram and recover the original equation (3.2-1).

## Numerator Dynamics

You should also be able to verify easily that the system shown in Figure 3.15*d* has the following equation.

$$\dot{y} = ry + a\dot{v} + bv \tag{3.4-7}$$

This is a model of a system said to have *numerator dynamics* because of the dynamic operation indicated by $s$ in the numerator of the transfer function.

The existence of numerator dynamics has the effect of supplying an additional input to the system since the input's rate of change $\dot{v}$ appears in the differential equation. For example, if $v(t) = t$, then an increasing input signal appears due to $v$, as well as a constant input due to $\dot{v} = 1$. For systems with numerator dynamics, the response cannot be described just with the characteristic root. The value of the coefficient of $s$ in the numerator must also be considered.

First-order systems with numerator dynamics can have the following transfer functions.

$$T(s) = \frac{Ks}{s+d} \tag{3.4-8}$$

$$T(s) = \frac{K(s+c)}{s+d} \tag{3.4-9}$$

*Example 3.12*

The pneumatic component shown in Figure 3.17 finds applications in control systems for improving the response characteristics of the controller. A device used for this purpose is a *compensator*. Take the air to be incompressible and at constant temperature. Neglect the pneumatic capacitance of the chamber with pressure $p$ and find the transfer function relating $p$ to $p_i$.

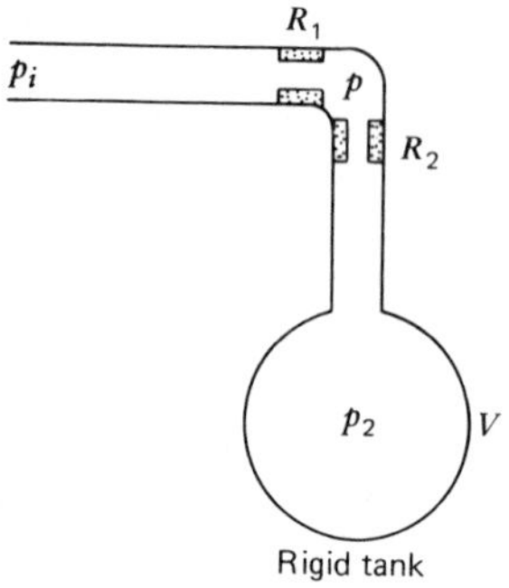


**Figure 3.17 Pneumatic compensator.**

Since the capacitance of the intermediate chamber is negligible, the mass flow rates through each resistance must be equal. Thus

$$q_m = \frac{p_i - p}{R_1} = \frac{p - p_2}{R_2} \tag{3.4-10}$$

The capacitance of the rigid tank is given by (2.7-19) as $C = V/nR_gT$. The mass flow rate $q_m$ into the tank equals the rate of change of mass in the tank, which is

$$\frac{dm}{dt} = C\frac{dp_2}{dt}$$

Therefore

$$\frac{p_i - p}{R_1} = C\frac{dp_2}{dt} \tag{3.4-11}$$

Solve (3.4-10) for $p_2$ and differentiate.

$$p_2 = p - \frac{R_2}{R_1}(p_i - p)$$

$$\frac{dp_2}{dt} = -\frac{R_2}{R_1}\frac{dp_i}{dt} + \left(1 + \frac{R_2}{R_1}\right)\frac{dp}{dt}$$

Substitution of this into (3.4-11) gives the result.

$$C\left(1 + \frac{R_2}{R_1}\right)\frac{dp}{dt} + \frac{1}{R_1}p = C\frac{R_2 dp_i}{R_1 dt} + \frac{p_i}{R_1}$$

or

$$\tau\frac{dp}{dt} + p = \tau_1\frac{dp_i}{dt} + p_i$$

where $\tau = R_1 C\left(1 + \dfrac{R_2}{R_1}\right)$ and $\tau_1 = R_2 C$.

The transfer function is

$$\frac{P(s)}{P_i(s)} = \frac{\tau_1 s + 1}{\tau s + 1} = K\frac{s+c}{s+d} \tag{3.4-12}$$

where $c = 1/\tau_1$, $d = 1/\tau$, and $K = \tau_1/\tau$.

## *Example 3.13*

Compute the unit step response for the models (3.4-8) and (3.4-9). Assume they are stable ($d > 0$). Consider three cases: $c = 0$, $c > d$, and $c < d$. Compare the response with the step response of the first-order model without numerator dynamics (3.2-1).

Let the input be $v$ and the output be $y$ so that

$$V(s) = \frac{1}{s}$$

$$Y(s) = T(s)\,V(s) = T(s)\,\frac{1}{s}$$

We assume that the initial value of $y$ is zero. From (3.4-9), we have

$$Y(s) = K\frac{s+c}{s+d}\,\frac{1}{s}$$

Since the factor $K$ is multiplicative, we can set $K = 1$ to simplify the analysis. The final results can then be multiplied by $K$. The expansion of $Y(s)$ is

$$Y(s) = \frac{C_1}{s} + \frac{C_2}{s+d} = \frac{1}{d}\left(\frac{c}{s} + \frac{d-c}{s+d}\right)$$

From Table 3.1 we obtain

$$y(t) = \frac{1}{d}\,[c + (d-c)\,e^{-dt}] \tag{3.4-13}$$

The case $c = 0$ represents the form (3.4-8), and its response is

$$y(t) = e^{-dt} \tag{3.4-14}$$

The response decays to zero from an initial value of unity (Figure 3.18*a*). Since we assumed that $y(0) = 0$, we can see that the numerator dynamics act on the step input to raise $y(t)$ instantaneously from $y(0) = 0$ to $y(0+) = 1$, where $t = 0+$ denotes a very short time after $t = 0$. Since this is impossible for a physical system, we must be careful in interpreting the response of models with numerator dynamics. A true step input is physically impossible also. Thus if we are interested in predicting the response, we must examine more closely either the model assumptions that led to the numerator dynamics or the assumptions that allowed us to model the input as a step.

The form (3.4-9) occurs when $c \neq 0$. For $c > d$, the response behaves much like that of the system without numerator dynamics, except that the value of $y$ is again instantaneously increased to 1. For $c < d$, the response decays to its steady state value of $c/d$ (Figures 3.18*b* and 3.18*c*).

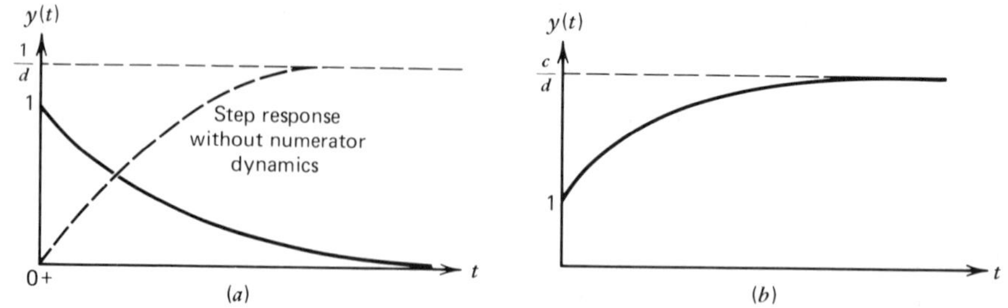


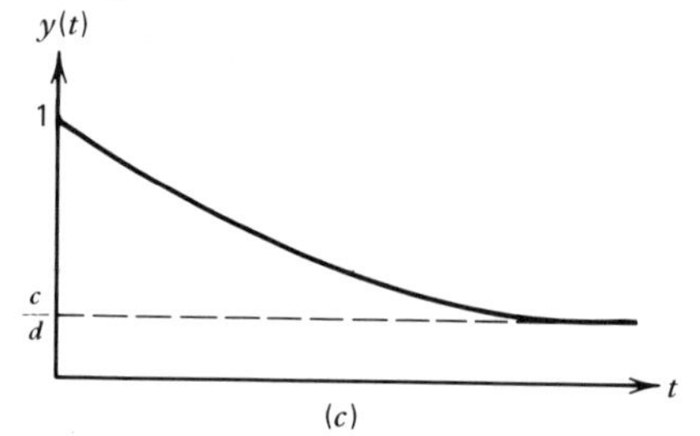


**Figure 3.18** Unit step response of a first-order model with numerator dynamics. (*a*) Response of model form (3.4-8). (*b*) Response of model form (3.4-9) for $c > d$. (*c*) Response of (3.4-9) for $c < d$.

## The Transfer Function and Impulse Response

Equation (3.3-10) shows that if the initial condition $y(0) = 0$, the unit impulse response is given by

$$Y(s) = \frac{b}{s - r}$$

or

$$y(t) = be^{rt}$$

But from (3.4-1)

$$T(s) = \frac{b}{s - r}$$

This shows that the Laplace transform of the unit impulse response is identical to the system's transfer function. In general, since $Y(s) = T(s)V(s)$, if $V(s) = 1$ (a unit impulse), then $Y(s) = T(s)$, as stated.

This important property allows the transfer function of a system to be determined experimentally by placing a pulse input on the system and measuring the ouput as a function of time. If the pulse duration is small compared to the system's time constant, the measured response will approximate the impulse response. The transfer function is found by computing the Laplace transform of the measured function by approximating it with a convenient functional form.

## The Effect of Dead Time on System Response

The shifting property of the Laplace transform can also be used to determine the response of a system with *dead time.* This occurs when an input results from a transporting fluid, for example. This is shown in Figure 3.19. A fluid with a temperature $y$ flows through a pipe. The fluid velocity is $u$ and is constant with time. The pipe length is $L$, so it takes a time $D = L/u$ for the fluid to move from one end to the

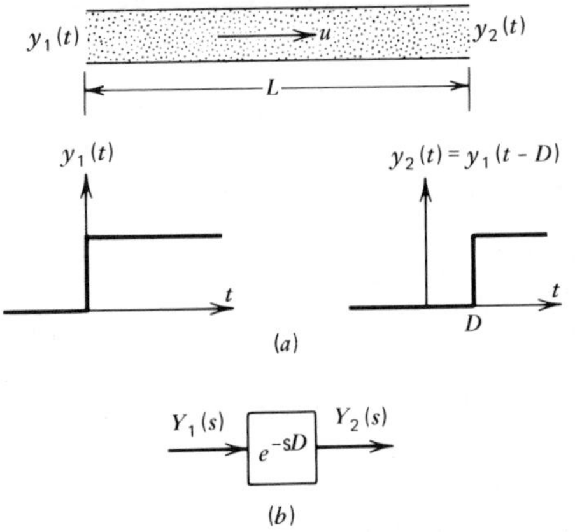


**Figure 3.19 Process with dead time $D = L/u$. (a) Time plots of the input and the response. (b) Block diagram.**

other. Let $y_1(t)$ denote the incoming fluid temperature and $y_2(t)$ denote the temperature of the fluid leaving the pipe.

Now suppose that the temperature of the incoming fluid suddenly increases. If this is modeled as a step function, the result is shown in Figure 3.19. If no heat energy is lost, then $y_2(t)$, the temperature at the output, is $y_1(t-D)$, where $y_1(t)$ is the temperature at the input. Thus a time $D$ later the output temperature suddenly increases.

A similar effect occurs for any change in $y_1(t)$; in general we may write

$$y_2(t) = y_1(t-D)$$

From the shifting theorem,

$$Y_2(s) = e^{-Ds}Y_1(s) \tag{3.4-15}$$

This result is shown in block diagram form in Figure 3.19. Figure 3.20 shows the general case of an element with dead time. In this case

$$Y(s) = e^{-Ds}G(s)V(s) \tag{3.4-16}$$

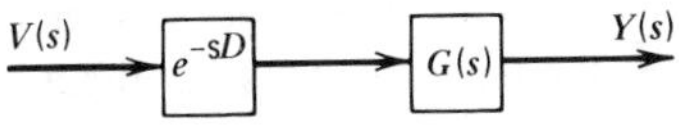


**Figure 3.20 System diagram with dead time.**

The presence of dead time means that the system does not have a characteristic equation of finite order. In fact, there are an infinite number of characteristic roots for a system with dead time. This can be seen by noting that the term $e^{-Ds}$ can be expanded in an infinite series as

$$e^{-Ds} = \frac{1}{e^{Ds}} = \frac{1}{1 + Ds + \dfrac{D^2s^2}{2!} - \cdots} \tag{3.4-17}$$

Methods for dealing with such models are presented in Chapter Eight.

## 3.5 THE INITIAL AND FINAL VALUE THEOREMS

In systems analysis the investigator often needs to know only the steady-state value of the output for a preliminary study. In such cases the *final value theorem* is frequently useful (Appendix B). If $y(t)$ and $dy/dt$ possess Laplace transforms, if the following limit exists, and if $y(t)$ approaches a definite value as $t \to \infty$, this theorem states that

$$y(\infty) = \lim_{s \to 0} [s\,Y(s)] \tag{3.5-1}$$

The derivation of this theorem is given in most texts on the theory of the Laplace transform. The theorem's conditions are satisfied if the system model is linear with constant coefficients (i.e. transformable so that $Y(s) = T(s)\,V(s)$) and if none of the roots of the denominator of $Y(s)$ are purely imaginary. The most common case in which the theorem fails is the case in which the input $v$ is a pure sinusoid. This introduces purely imaginary roots and a steady-state sinusoidal response. Thus $y(t)$ does not approach a definite value as $t \to \infty$.

The companion theorem is the *initial value theorem.*

$$y(0) = \lim_{s \to \infty} [s\,Y(s)] \tag{3.5-2}$$

The conditions under which the theorem is valid are that the limit and the transforms of $y$ and $dy/dt$ exist.

If the ramp response of $\dot{y} = ry + bv$ were not available, the final value theorem could have been used to find the steady-state difference between the input and output quickly. This is because, from (3.3-22) and (3.3-23) with $r = -1/\tau$, we obtain

$$\begin{aligned} v(\infty) - y(\infty) &= \lim_{s \to 0} s[V(s) - Y(s)] \\ &= \lim_{s \to 0} s\left(\frac{m}{s^2} - \frac{\tau b}{\tau s + 1}\frac{m}{s^2}\right) \\ &= \lim_{s \to 0} \frac{m}{s}\left(\frac{\tau s + 1 - \tau b}{\tau s + 1}\right) \\ &= \begin{cases} \infty, & \tau b \neq 1 \\ m\tau, & \tau b = 1 \end{cases} \end{aligned} \tag{3.5-3}$$

Thus the output does not follow the input unless $\tau b = 1$.

The initial and final value theorems are especially useful when the system is represented only in terms of its transfer function. The theorems can thus be directly applied without working through the solution of a differential equation.

### *Example 3.14*

(a) Derive formulas for the steady-state difference between the input $V(s)$ and the output $Y(s)$ for a unit step input and a unit ramp input. The system diagram is given in Figure 3.14*b*.

(b) Apply the results to the case where $G(s) = K/s$ and $H(s) = 1$ (an integrator with a feedback loop).

(a) The difference between $V(s)$ and $Y(s)$ is $E(s)$ in the figure, where

$$E(s) = V(s) - Y(s) = V(s) - G(s)E(s) \tag{3.5-4}$$

Solve for $E(s)$ in terms of $V(s)$.

$$E(s) = \frac{V(s)}{1 + G(s)}$$

The steady-state difference from the final value theorem is

$$e_{ss} = \lim_{s\to 0} sE(s) = \lim_{s\to 0} \frac{sV(s)}{1 + G(s)} \tag{3.5-5}$$

For a unit step input, $V(s) = 1/s$ and (3.5-5) gives

$$e_{ss} = \lim_{s\to 0} \frac{1}{1 + G(s)} = \frac{1}{1 + G(0)} \tag{3.5-6}$$

where $G(0) = \lim_{s\to 0} G(s)$.

For a unit ramp input, $V(s) = 1/s^2$ and

$$e_{ss} = \lim_{s\to 0} \frac{1}{s + sG(s)} = \frac{1}{\lim_{s\to 0} sG(s)} \tag{3.5-7}$$

(b) For $G(s) = K/s$, $G(0) = \lim_{s\to 0} K/s \to \infty$, and $e_{ss} \to 0$. Thus this particular system has a zero steady-state difference between the output and the step input.

For a unit ramp input, $\lim_{s\to 0} sG(s) = K$, and $e_{ss} = 1/K$, which is a finite but nonzero difference.

## 3.6 SIGNAL FLOW GRAPHS

An alternate graphical representation of the system's transformed equations is the *signal flow graph* in which variables are represented as *nodes*, and the operations on these variables are represented by directed line segments between the nodes. This contrasts with the block diagram's representation of variables as lines and operations as blocks. Because of this difference it is easier to find the transfer function of the system with a signal flow graph if the system has many loops. A general formula useful for reducing complicated graphs is *Mason's gain formula* (Appendix D). Here we will confine ourselves to a brief introduction to signal flow graphs.

The principal advantage of a block diagram is that the system's physical components (the elements performing the operations) are themselves represented by blocks indicating their operations, rather than by line segments. Thus the block diagram more closely resembles the physical structure of the system. The choice between the two representations is made on the basis of convenience.

The value of a variable at a node is obtained by *summing* the incoming signals at that node. This sum is transmitted to all line segments outgoing from the node.

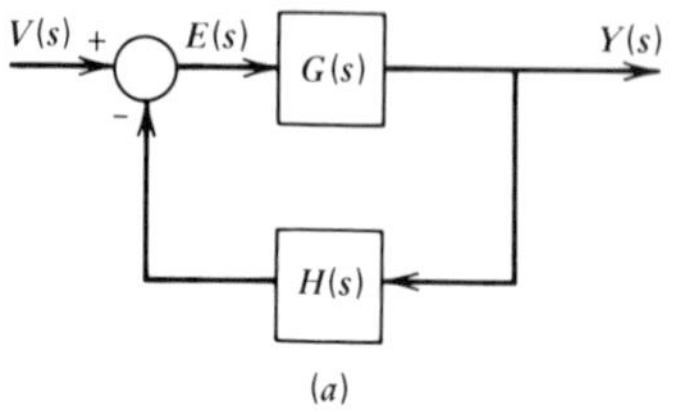


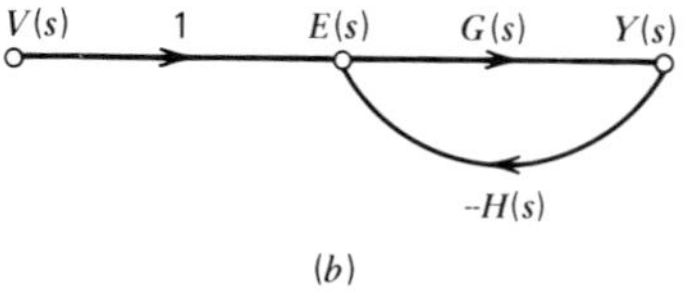


**Figure 3.21 Comparison of (*a*) Block diagram and (*b*) Signal flow graph.**

The directed line segments show the relation between the variables whose nodes are connected by the line, with the arrow indicating the direction of causality. Subtraction is indicated by the use of a negative sign with the operation denoted by the proper line segment. An example of a signal flow graph is shown in Figure 3.21 for the system shown in Figure 3.14*b*, repeated here for convenience. The variables are $V(s)$, $E(s)$, and $Y(s)$ and are shown as nodes (circles). The variable $E(s)$ is obtained by summing the incoming signals to get

$$\begin{aligned} E(s) &= 1V(s) + [-H(s)Y(s)] \\ &= V(s) - H(s)Y(s) \end{aligned}$$

The output $Y(s)$ is

$$Y(s) = G(s)E(s)$$

Combination of the last two equations results in the same transfer function that was obtained from the block diagram. With the basic properties of nodes and line segments we can develop rules similar to those given in Figure 3.14, with Mason's gain formula as one result.

As an additional example we display in Figure 3.22 the signal flow graph for the system represented by Figure 3.12*d* and (3.2-1). Note that the variable appearing at the middle node is $sY(s)$.

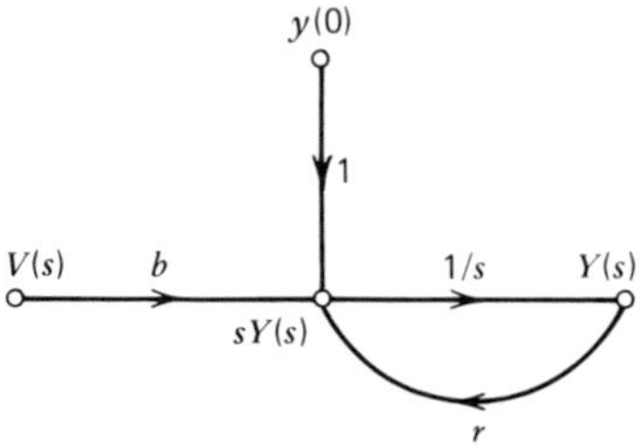


**Figure 3.22 Signal flow graph for the first-order linear system $\dot{y} = ry + bv$.**

Sometimes the rules of diagram manipulation are modified to allow time-domain notation to be used. For example, the system represented by Figures 3.12*d* and 3.22 can also be represented by a *simulation diagram* using time-domain notation. These are shown in Figure 3.23. The indefinite integration sign denotes that the incoming variable is integrated over time. Note that the initial condition now enters the diagram *after* the integration. This is easily seen by considering the following equation.

$$\dot{y} = w$$

Direct integration of both sides yields

$$y = \int w\,dt + y(0)$$

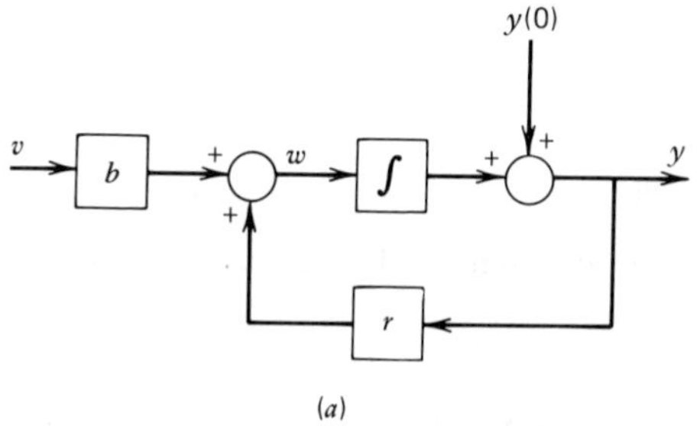


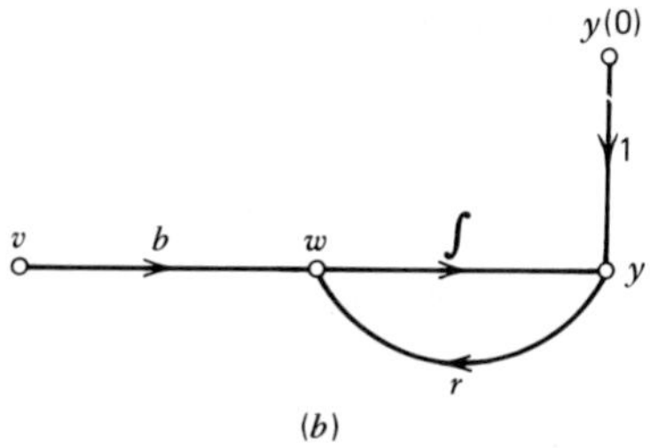


**Figure 3.23** System diagrams in the time domain. (*a*) Block diagram. (*b*) Signal flow graph.

The transform approach yields

$$Y(s) = \frac{1}{s}[W(s) + y(0)]$$

In the time-domain case the integration operation acts only on the input $w$, whereas in the transformed case the integration operator $1/s$ acts on the sum of $W(s)$ and $y(0)$.

## 3.7 IMPEDANCE AND TRANSFER FUNCTIONS

The transfer function is a useful vehicle for discussing the concept of impedance in more detail. In general, any transfer function between a rate variable as input and an effort variable as output is considered an impedance. In electrical systems, of course, this is the ratio of a voltage transform to a current transform and thus implies a current source. If the voltage output is measured at the terminals to which the driving current is applied, the impedance so obtained is the *driving-point* or *input impedance.* If the voltage is measured at another place in the circuit, the impedance obtained is a *transfer impedance* (because the effect of the input current has been transferred to another point). Impedance may be thought of as a generalized resistance, because it *impedes* current flow (ac or dc) just as a resistor *resists* dc current. Sometimes the term *admittance* is used. This is the reciprocal of impedance and is an indication of to what extent a circuit *admits* current flow.

The transfer functions of individual elements can be combined with series and parallel laws to find the transfer function or impedance at any point in the system. The impedances of three common electrical elements are found as follows. For the capacitor,

$$v(t) = \frac{1}{C}\int_0^t i\,dt$$

or

$$V(s) = \frac{1}{Cs}I(s) \tag{3.7-1}$$

and the impedance is

$$T(s) = \frac{1}{Cs}$$

For the inductor

$$v(t) = L\frac{di}{dt}$$

or

$$V(s) = LsI(s) \tag{3.7-2}$$

and its impedance is $T(s) = Ls$. Of course, the impedance of a resistor is $T(s) = R$.

## *Example 3.15*

The *lead compensator* network shown in Figure 3.24*a* is widely used to improve the performance of instruments and controllers. We wish to find the transfer function between the input voltage $v_i$ and the output voltage $v$.

Two impedances are in series if they have the same rate variable. If so, the total impedance is the sum of the individual impedances. If the impedances have the same effort difference across them, they are in parallel, and their impedances combine by the reciprocal rule

$$\frac{1}{T} = \frac{1}{T_1} + \frac{1}{T_2}$$

where $T$ is the total equivalent impedance. These two laws are the extensions to the dynamic case of the laws governing series and parallel static resistance elements.

It can be seen that $R_1$ and $C$ are in parallel. Thus their equivalent impedance is

$$\frac{1}{Z(s)} = \frac{1}{1/Cs} + \frac{1}{R_1}$$

or

$$Z(s) = \frac{R_1}{R_1Cs + 1}$$

The equivalent circuit is shown in Figure 3.24*b*. From this it can be seen that $Z$ and $R$ are in series. Thus

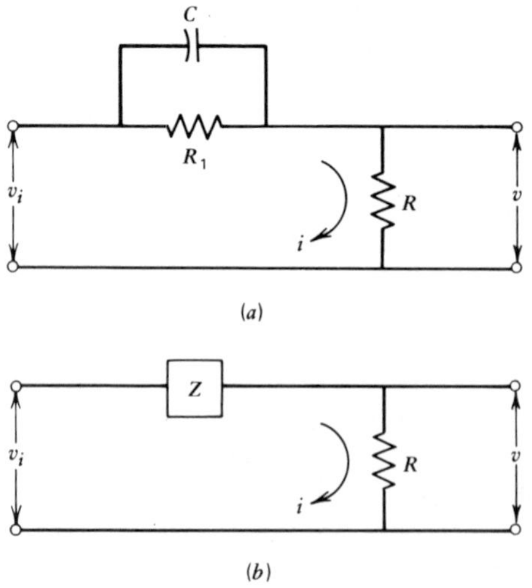


**Figure 3.24 Lead compensator. (*a*) Circuit. (*b*) Equivalent impedance representation.**

$$\frac{V_i(s)}{I(s)} = Z(s) + R$$

and

$$V(s) = RI(s)$$

Eliminating $I(s)$ from the last two relations yields the desired transfer function.

$$V_i(s) = \frac{V(s)}{R}[Z(s) + R]$$

or

$$T(s) = \frac{V(s)}{V_i(s)} = \frac{1}{\dfrac{Z(s)}{R} + 1}$$

$$= \frac{RR_1Cs + R}{RR_1Cs + R_1 + R} \tag{3.7-3}$$

The network is seen to be a first-order system with numerator dynamics. The differential equation is of the form given by (3.4-7).

Impedances and transfer functions in general can be used to develop a systematic algebraic procedure for determining network transfer functions for most system types when the network is a ladderlike arrangement of series impedances alternating with impedances *shunted* to ground like the resistor $R$ in Figure 3.24.

## 3.8 FREQUENCY RESPONSE

Periodic inputs occur extensively in both natural and artificial systems. An input $v(t)$ is periodic with a period $p$ if $v(t + p) = v(t)$ for all values of time $t$, where $p$ is a constant called the *period.* The earth's rotation and revolution about the sun produce periodic forcing functions in nature. Rotating unbalanced machinery produces such a force on the supporting structure. Internal combustion engines produce a periodic torque resulting from the engine's cycle, and reciprocating fluid pumps produce hydraulic and pneumatic pressures with one or more periodic components. Perhaps the most familiar example is the ac voltage. Many of these periodicities are sinusoidal, or nearly so, and we will begin the analysis with this type. In Section 3.9 it will be shown how the results for a sinusoidal input can be used to obtain the system response to a general periodic input by means of the *Fourier series* representation (Appendix A).

### Sinusoidal Response

Consider the general linear model

$$\dot{y} = ry + bv \tag{3.8-1}$$

where

$$v(t) = A \sin \omega t \tag{3.8-2}$$

with constant amplitude $A$ and frequency $\omega$. From Table 3.1, the transform of $v(t)$ is

$$V(s) = A\frac{\omega}{s^2+\omega^2} \tag{3.8-3}$$

Since $Y(s) = T(s)V(s)$, and $T(s) = b/(s-r)$, the forced response is given by

$$\begin{aligned} Y(s) &= \frac{b}{s-r}\frac{A\omega}{s^2+\omega^2} \\ &= bA\omega\frac{1}{s-r}\frac{1}{s-i\omega}\frac{1}{s+i\omega} \end{aligned} \tag{3.8-4}$$

where $i = \sqrt{-1}$. This can be represented as a partial fraction expansion of terms corresponding to the three distinct roots of the denominator.

$$Y(s) = \frac{C_1}{s-r} + \frac{C_2}{s-i\omega} + \frac{C_3}{s+i\omega}$$

Multiplication by the least common denominator and comparison of numerators shows that*

$$C_1 = \frac{bA\omega}{r^2+\omega^2}$$

$$C_2 = \frac{bA\omega}{(i\omega - r)(2i\omega)}$$

$$C_3 = \frac{bA\omega}{(-i\omega - r)(-2i\omega)}$$

It can be shown that $C_2$ and $C_3$ are complex conjugates, as they must be since $y$ is a real number. When the roots of the denominator are complex, an alternate expansion can be made to give real coefficients directly. For this case the expansion form is

$$Y(s) = \frac{C_1}{s-r} + \frac{C_4 s + C_5}{s^2+\omega^2} \tag{3.8-5}$$

where $C_1$ was given previously, and

$$C_4 = -C_1$$

$$C_5 = rC_4$$

Thus

$$Y(s) = C_1\left(\frac{1}{s-r} - \frac{s}{s^2+\omega^2} - \frac{r}{s^2+\omega^2}\right)$$

From Table 3.1

$$y(t) = C_1\left(e^{rt} - \cos\omega t - \frac{r}{\omega}\sin\omega t\right) \tag{3.8-6}$$

For the stable case $\left(r = -\dfrac{1}{\tau} < 0\right)$, the steady-state solution is

---

* The partial fraction expansion formulas given in Appendix B produce these results directly.

$$y(t) = -C_1\left(\cos\omega t + \frac{r}{\omega}\sin\omega t\right)$$

$$= -C_1\left(\cos\omega t - \frac{1}{\tau\omega}\sin\omega t\right) \tag{3.8-7}$$

which represents a sinusoidal oscillation at the same frequency as the input. In order to visualize this behavior more easily, use the identity

$$B\sin(\omega t + \phi) = B\sin\omega t\cos\phi + B\cos\omega t\sin\phi \tag{3.8-8}$$

Comparing this with (3.8-7) we see that

$$B\sin\phi = -C_1 = -\frac{bA\omega\tau^2}{1+\omega^2\tau^2} \tag{3.8-9}$$

$$B\cos\phi = \frac{C_1}{\omega\tau} \tag{3.8-10}$$

This implies that

$$\tan\phi = \frac{\sin\phi}{\cos\phi} = -\omega\tau \tag{3.8-11}$$

and

$$B^2(\sin^2\phi + \cos^2\phi) = B^2 = C_1^2\left(1 + \frac{1}{\omega^2\tau^2}\right)$$

By definition, $B$, $\tau$, $\omega$, and $A$ are positive, but $b$ may have any sign. Thus the solution for the amplitude is

$$B = \frac{\tau|b|A}{\sqrt{1+\omega^2\tau^2}} \tag{3.8-12}$$

The steady-state solution can now be written as

$$y(t) = B\sin(\omega t + \phi) \tag{3.8-13}$$

where

$$\phi = \tan^{-1}(-\omega\tau) \tag{3.8-14}$$

The quadrant of $\phi$ is determined as follows. If $b > 0$ then from (3.8-9) and (3.8-10), $\sin\phi < 0$ and $\cos\phi > 0$, and $\phi$ is thus in the fourth quadrant (270° to 360°). Similarly it follows that if $b < 0$, $\phi$ is in the second quadrant (90° to 180°). Note that the phase angle $\phi$ must be in radians when used in calculations with (3.8-13).*

These results can be used to find the response for related input types. For example, if $v = A\sin(\omega t + \psi)$, then the phase angle $\phi$ is simply increased by $\psi$. An important case in this category is the cosine input, since $\cos\omega t = \sin(\omega t + \pi/2)$.

The preceding results illustrate some important properties of linear systems.

---

* When formulas such as (3.8-14) are encountered, additional information is required to establish the proper quadrant since the arctangent is a double-valued function. Its principal value is defined to be in the range (−90°, 90°). Thus most calculators and computers will return an arctangent value in the fourth quadrant (0, −90°) if a negative number is entered, and a value in the first quadrant (0, 90°) if a positive number is entered.

The sinusoidal input $A \sin \omega t$ produces at steady state a sinusoidal output of the *same* frequency, but shifted in phase relative to the input. In addition, the amplitude of the output depends not only on the amplitude of the input and on the system's parameters, but also on the input frequency. The same is true of the phase shift; it is a function of frequency. These properties will allow us to develop some powerful methods for analysis and design.

## The Frequency Transfer Function

The solution given by (3.8-12) through (3.8-14) points out a useful application of the transfer function. Assume that the system is stable and write the transfer function $T(s) = b/(s - r)$ in terms of $\tau = -1/r$. This is

$$T(s) = \frac{\tau b}{\tau s + 1} \tag{3.8-15}$$

The transform of a sinusoidal steady-state output of frequency $\omega$ has the denominator $s^2 + \omega^2$. The roots are $s = \pm i\omega$; that is, the values of $s$ corresponding to sinusoidal oscillation of frequency $\omega$ are $s = \pm i\omega$. With this as a guide, let us see what the transfer function reveals when $s$ is replaced with $i\omega$.

$$T(i\omega) = \frac{\tau b}{\tau \omega i + 1} \tag{3.8-16}$$

This complex number is a function only of $\omega$ for fixed $\tau$ and $b$. It can be expressed in a more convenient form by multiplying through by the complex conjugate of the denominator.

$$\begin{aligned} T(i\omega) &= \frac{\tau b}{\tau \omega i + 1} \frac{1 - \tau \omega i}{1 - \tau \omega i} \\ &= \frac{\tau b (1 - \tau \omega i)}{1 + \tau^2 \omega^2} \end{aligned} \tag{3.8-17}$$

$T(i\omega)$ can be thought of as a *vector* in the complex plane where the vector's components are the real and imaginary parts of $T(i\omega)$. Figure 3.25 shows the case for $b > 0$. The vector rotates and its length changes as $\omega$ varies from 0 to $\infty$. The

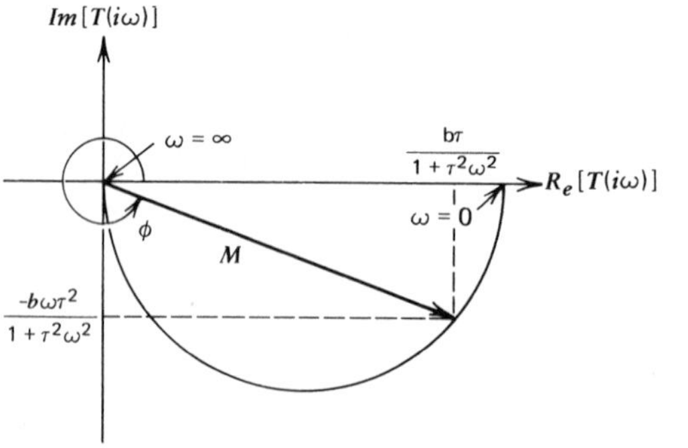


**Figure 3.25** Polar plot of a first-order linear system.

locus of the vector's tip is shown and is a semicircle in the fourth quadrant for positive $\omega$.

The transfer function with $s$ replaced by $i\omega$ is called the *frequency transfer function*, and its plot in vector form is the *polar plot.* The vector's magnitude $M$ and angle $\phi$ relative to the positive real axis are easily found from trigonometry to be

$$M(\omega) = \frac{\tau|b|}{\sqrt{1+\tau^2\omega^2}} \tag{3.8-18}$$

$$\phi(\omega) = \tan^{-1}(-\omega\tau) \tag{3.8-19}$$

where $M$ and $\phi$ are explicitly written as functions of $\omega$. Comparison of these expressions for $M(\omega)$ and $\phi(\omega)$ with those of $B$ and $\phi$ in (3.8-12) and (3.8-14) reveals the following useful fact (see Figure 3.26).

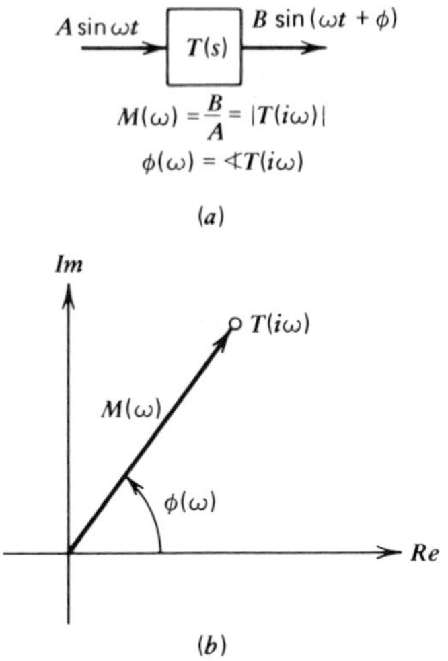


**Figure 3.26 Steady-state sinusoidal response of a linear system. (*a*) Input-output relation. (*b*) Polar representation of the frequency transfer function $T(i\omega)$.**

The magnitude $M$ of the frequency transfer function $T(i\omega)$ is the ratio of the sinusoidal steady-state output amplitude to the sinusoidal input amplitude. The phase shift of the output relative to the input is $\phi$, the angle of $T(i\omega)$ [the *argument* of $T(i\omega)$]. Both $M$ and $\phi$ are functions of the input frequency $\omega$.

In an alternate notation this means that

$$M(\omega) = |T(i\omega)| \tag{3.8-20}$$

$$\phi(\omega) = \measuredangle T(i\omega) \tag{3.8-21}$$

$$B = AM(\omega) \tag{3.8-22}$$

where $|T(i\omega)|$ and $\measuredangle T(i\omega)$ denote the magnitude and angle of the complex number $T(i\omega)$. Because of (3.8-22), $M$ is sometimes called the *amplitude ratio.*

## The Logarithmic Plots

Inspection of (3.8-18) reveals that the steady-state amplitude of the output decreases as the frequency of the input increases for $\tau \neq 0$. The larger $\tau$ is, the more the output amplitude decreases with frequency. At a high frequency the system's "inertia" prevents it from closely following the input. The larger $\tau$ is, the more sluggish is the system response. This also produces an increasing phase lag as $\omega$ increases.

The curves of $M$ and $\phi$ versus $\omega$ are difficult to sketch accurately. For this reason logarithmic plots are usually employed. Here the amplitude ratio $M$ is specified in *decibel* units, denoted db. The relation between a number $M$ and its decibel equivalent $m$ is

$$m = 20 \log M \qquad \text{db} \tag{3.8-23}$$

where the logarithm is to the base 10. For example, the number 10 corresponds to 20 db; the number 1 corresponds to 0 db; numbers less than 1 have negative decibel values. It is common practice to plot $m(\omega)$ in decibels versus $\log \omega$. Thus if $M$ is first converted to decibel units, semilog graph paper may be used. For easy reference, $\phi(\omega)$ is also plotted versus $\log \omega$.

Referring to (3.8-18), we have

$$\begin{aligned} m(\omega) &= 20 \log \frac{\tau|b|}{\sqrt{1+\tau^2\omega^2}} \\ &= 20 \log (\tau|b|) - 10 \log (1 + \tau^2\omega^2) \end{aligned} \tag{3.8-24}$$

To sketch the curve we approximate $m(\omega)$ in three frequency ranges. For $\omega\tau \ll 1$, $1 + \tau^2\omega^2 \cong 1$, and

$$\begin{aligned} m(\omega) &\cong 20 \log (\tau|b|) - 10 \log 1 \\ &= 20 \log (\tau|b|) \end{aligned} \tag{3.8-25}$$

Thus for $\omega \ll 1/\tau$, $m(\omega)$ is approximately constant. For $\omega\tau \gg 1$, $1 + \tau^2\omega^2 \cong \tau^2\omega^2$, and

$$\begin{aligned} m(\omega) &\cong 20 \log (\tau|b|) - 10 \log \tau^2\omega^2 \\ &= 20 \log (\tau|b|) - 20 \log \tau - 20 \log \omega \end{aligned} \tag{3.8-26}$$

This gives a straight line versus $\log \omega$. Its slope is $-20$ db/decade, where a *decade* is any 10:1 frequency range. At $\omega = 1/\tau$, this line gives $m(\omega) = 20 \log (\tau|b|)$. This is useful for plotting purposes but does not represent the true value of $m$ at that point. For $\omega = 1/\tau$, (3.8-24) gives

$$\begin{aligned} m(\omega) &= 20 \log (\tau|b|) - 10 \log 2 \\ &= 20 \log (\tau|b|) - 3.01 \end{aligned} \tag{3.8-27}$$

Thus at $\omega = 1/\tau$, $m(\omega)$ is 3.01 db below the low-frequency asymptote given by (3.8-25). The low-frequency and high-frequency asymptotes meet at $\omega = 1/\tau$, which is the *breakpoint* frequency.

The curve of $\phi$ versus $\omega$ is constructed as follows. For $\omega \ll 1/\tau$, (3.8-19) gives

$$\phi(\omega) \cong \tan^{-1}(0) = 0°$$

For $\omega = 1/\tau$,

$$\phi(\omega) = \tan^{-1}(-1) = -45°$$

and for $\omega \gg 1/\tau$,

$$\phi(\omega) \cong \tan^{-1}(-\infty) = -90°$$

Using these facts, the curve is easily sketched.

The curves for $T(s) = 1/(\tau s + 1)$ are shown in Figure 3.27. This corresponds to $\tau b = 1$ in (3.8-15). If $\tau b \neq 1$, the scale for the $m$ curve shown in the figure should be shifted by $20 \log (\tau|b|)$. The phase angle curve remains unchanged.

Also shown in the figure are the curves for $T(s) = \tau s + 1$. It is easy to show that these are the mirror images of the curves for $T(s) = 1/(\tau s + 1)$. We note that

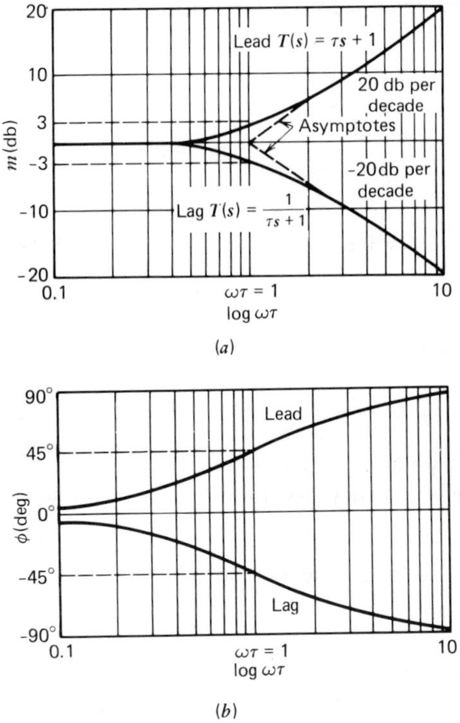


**Figure 3.27 Logarithmic frequency response plots for two first-order transfer functions. (*a*) Magnitude ratio *m* in decibels. (*b*) Phase angle.**

$$\left|\frac{1}{1 + i\omega\tau}\right| = \frac{1}{|1 + i\omega\tau|}$$

Thus

$$20 \log \left|\frac{1}{1 + i\omega\tau}\right| = -20 \log |1 + i\omega\tau|$$

Also

$$\measuredangle(1 + i\omega\tau) = \tan^{-1} \omega\tau$$

This angle is the negative of that given in (3.8-19).

## *Example 3.16*

A model of a *vibration isolator* is shown in Figure 3.28. Such a device is used to isolate the motion of point 1 from the input displacement $y_2$, which might be produced by a mechanism driving a lightweight tool at point 2. Point 1 represents a lightweight tool support that cannot be subjected to large deflections. Determine the values of $c$ and $K$ required to keep the amplitude of $y_1$ less than 10% of the amplitude of $y_2$. Assume that $y_2$ is

$$y_2(t) = 2 \sin(100t) \qquad \text{in.}$$

with time measured in seconds.

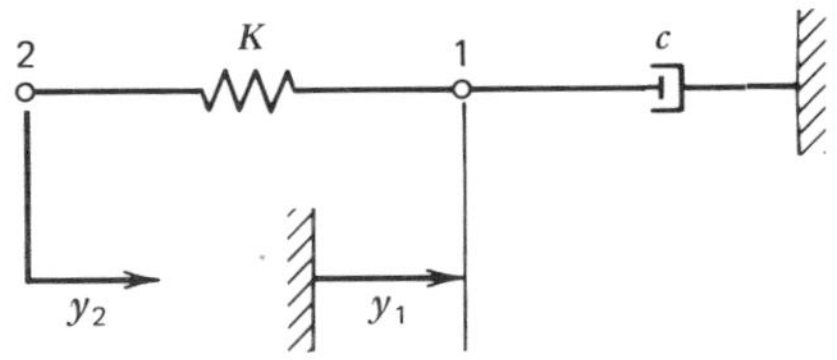


**Figure 3.28 Vibration isolator.**

The sum of the forces at point 1 must equal zero since the mass at this point is assumed to be zero. Thus

$$K(y_2 - y_1) - c\dot{y}_1 = 0$$

or

$$c\dot{y}_1 = -Ky_1 + Ky_2$$

The transfer function between $y_1$ and $y_2$ is

$$\frac{Y(s)}{Y_2(s)} = T(s) = \frac{K}{cs + K} \tag{3.8-28}$$

In terms of the notation for (3.8-15), $b = K/c$ and $\tau = c/K$. The amplitude requirement can be stated as

$$M \leqslant 0.1$$

In decibel units this gives

$$m \leqslant 20 \log (0.1) = -20$$

The requirement is met exactly if $m = -20$ db at $\omega = 100$ rad/sec. If we assume for now that the design point lies on the high-frequency asymptote given by (3.8-26) with $\omega = 100$, we have (since $b\tau = 1$)

$$\begin{aligned} m &= -20 = 20 \log 1 - 20 \log \tau - 20 \log 100 \\ &= -40 - 20 \log \tau \end{aligned}$$

The result is $\tau = 0.1$ sec. This gives $\omega\tau = 100(0.1) = 10$. From Figure 3.27 it can be seen that this point is accurately described by the high-frequency asymptote and our assumption is validated. The isolation requirement is met for any $K$ and $c$ such that $c/K = 0.1$ sec.

Other constraints must be used to specify the values of both $K$ and $c$. One such constraint might be the maximum force allowed to be transmitted to the mechanism causing the displacement $y_2$. Since we have designed the system so that point 1 is relatively motionless, the deflection of the spring $K$ is approximately equal to $y_2$, and the spring force is the force transmitted to the mechanism. This is $Ky_2 = K(2 \sin 100\,t)$. If this force must be less than, say, 50 lb, then $K$ must be less than 25 lb/in. This means that $c$ must be less than $K\tau = 25(0.1) = 2.5$ lb/in./sec. With this information, formulas such as the damper equation (Table 2.10) can be used to design or select the proper elements.

### *Example 3.17*

The presence of numerator dynamics in a first-order system can significantly alter the

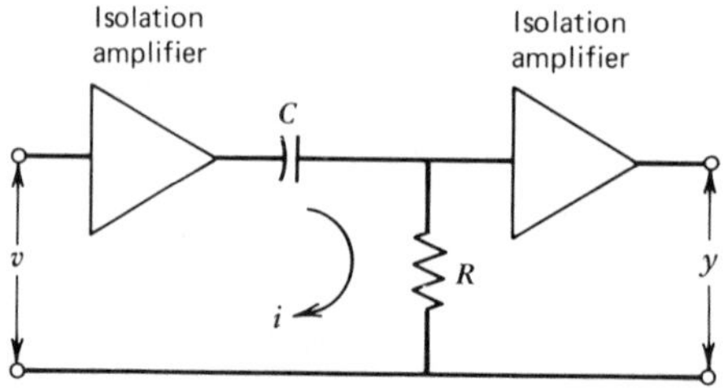


**Figure 3.29 High-pass filter circuit.**

system's frequency response. Consider again the series $RC$ circuit where the output voltage is now taken to be across the resistor (Figure 3.29). The two amplifiers serve to isolate the circuit from the loading effects of adjacent elements. The input impedance between the voltage $v$ and the current $i$ is found from the series law.

$$\frac{V(s)}{I(s)} = R + \frac{1}{Cs}$$

Thus

$$Y(s) = I(s)R = \frac{V(s)}{R + \frac{1}{Cs}} R$$

Determine the circuit's frequency response, and interpret its effect on the input.

The transfer function is

$$T(s) = \frac{Y(s)}{V(s)} = \frac{RCs}{RCs + 1} \tag{3.8-29}$$

Let $\tau = RC$ to obtain

$$T(i\omega) = \frac{i\omega\tau}{i\omega\tau + 1} = \frac{i\omega\tau(1 - i\omega\tau)}{(1 + i\omega\tau)(1 - i\omega\tau)}$$

$$= \frac{\omega^2\tau^2 + i\omega\tau}{1 + \omega^2\tau^2}$$

$$M(\omega) = \frac{\omega\tau}{1 + \omega^2\tau^2}\sqrt{\omega^2\tau^2 + 1} = \frac{\omega\tau}{\sqrt{1 + \omega^2\tau^2}}$$

The argument of the numerator of $T(i\omega)$ is

$$\measuredangle i\omega\tau = 90^\circ$$

and the argument of the denominator is

$$\measuredangle(1 + i\omega\tau) = \tan^{-1}\omega\tau$$

A property of a complex number is that its argument is the sum of the arguments in the numerator minus the sum of the arguments in the denominator (Appendix A). Here this gives

$$\phi = \measuredangle T(i\omega) = 90^\circ - \tan^{-1}\omega\tau$$

For $\omega\tau \ll 1$, $\phi \cong 90^\circ$ and $M(\omega) \cong \omega\tau$. Thus

$$m(\omega) \cong 20 \log \omega\tau = 20 \log \omega + 20 \log \tau, \qquad \omega\tau \ll 1$$

For low frequencies the log magnitude curve increases with a slope of 20 db/decade. For $\omega\tau \gg 1$,

$$m(\omega) \cong 20 \log 1 = 0$$

$$\phi(\omega) \cong 90^\circ - \tan^{-1}\infty = 0$$

For $\omega\tau = 1$,

$$m(\omega) = 20 \log \frac{1}{\sqrt{2}} = -3.01$$

$$\phi(\omega) = \tan^{-1} 1 = 45^\circ$$

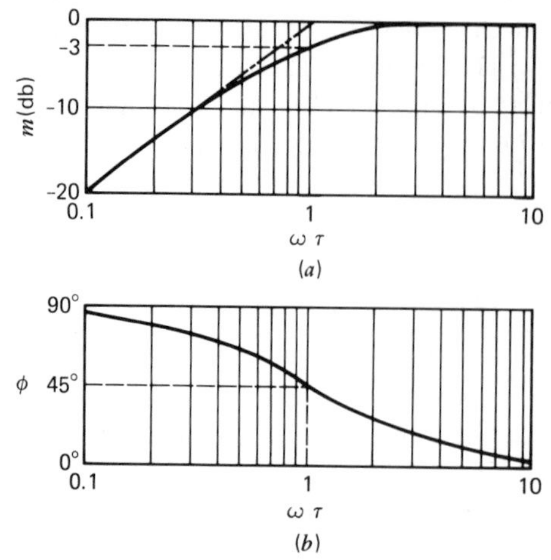


**Figure 3.30 Frequency response of the high-pass filter. (*a*) Magnitude plot. (*b*) Phase plot.**

The log magnitude and phase plots are shown in Figure 3.30. The circuit passes signals with frequencies above $\omega = 1/\tau$ with little attenuation and is called a *high-pass filter.* It is used to remove dc and low-frequency components from a signal. This is desirable when one wishes to study high-frequency components whose small amplitudes would be undiscernible in the presence of large-amplitude, low-frequency components. Such a circuit is incorporated into oscilloscopes for this reason. The voltage scale can then be selected so that the signal components to be studied will fill the screen.

## Experimental Determination of a Transfer Function

The frequency-response curves are an aid to developing a system model from test data. If a suitable apparatus can be devised to provide a sinusoidal input of adjustable frequency, then the system's output amplitude and phase shift relative to the input can be measured for various input frequencies. When these data are plotted on the logarithmic plot for a sufficient frequency range, it will be obvious whether or not the system can be modeled as a first-order linear system. If *both* the $m$ and $\phi$ data plots are of the forms shown in Figure 3.27, the parameters $\tau$ and $b$ can be estimated. This procedure is easiest for systems with electrical inputs and outputs, because for these systems variable-frequency oscillators and frequency-response analyzers are commonly available. Some of these can automatically sweep through a range of frequencies and plot the decibel and phase angle data. For nonelectrical outputs, such as a displacement, a suitable transducer can be used to produce an electrical measurement that can be analyzed. For nonelectrical inputs the task is more difficult but is frequently used nonetheless.

An advantage of frequency-response tests is that they often can be used on a device without interrupting its normal operations. The small-amplitude sinusoidal test signal is superimposed on the operating inputs, and the sinusoidal component of the output with the input frequency is subtracted out of the output measurements. Computer algorithms are available to do this.

## 3.9 BANDWIDTH AND PERIODIC INPUTS

The instantaneous power in an electrical circuit is $p = iv$ as we saw in Chapter Two. The transform of instantaneous power can be written as

$$P(s) = I(s)V(s) = T(s)I^2(s) \tag{3.9-1}$$

where $T(s)$ is the impedance $V(s)/I(s)$. If the input current is sinusoidal with an amplitude $I$ and frequency $\omega$, we can write the power as

$$p(t) = I^2|T(i\omega)| \sin \omega t \sin(\omega t + \phi) \tag{3.9-2}$$

since

$$v(t) = |T(i\omega)|I \sin(\omega t + \phi) \tag{3.9-3}$$

$$\phi = \measuredangle T(i\omega) \tag{3.9-4}$$

from the frequency-response properties of linear systems.

### Bandwidth

From (3.9-2) we see that the magnitude of the instantaneous power is a function of the input frequency $\omega$. This fact is used to establish another measure of system performance called *bandwidth.* Consider the series $RL$ circuit in Figure 3.31. Its impedance is

$$T(s) = \frac{V(s)}{I(s)} = R + Ls \tag{3.9-5}$$

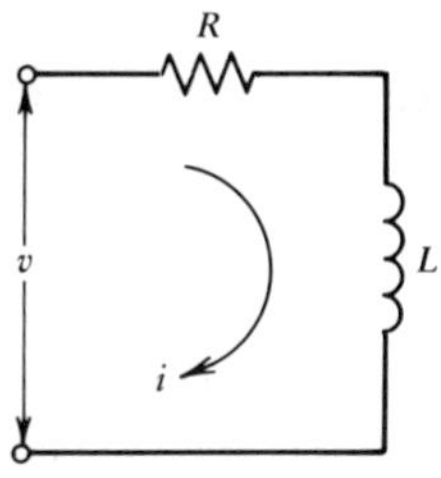

**Figure 3.31 Series $RL$ circuit.**

Thus

$$I(i\omega) = \frac{V(i\omega)/R}{1 + i\omega L/R}$$

At zero input frequency $\omega = 0$, the current attains its maximum value of

$$I_o = \frac{V(0)}{R}$$

Form the ratio

$$\left|\frac{I(i\omega)}{I_o}\right| = \left|\frac{1}{1 + i\omega L/R}\right|$$

$$= \frac{1}{\sqrt{1 + (\omega L/R)^2}}$$

$$= \frac{|T(0)|}{|T(i\omega)|}$$

When $\omega = R/L = \omega_c$, the current is 0.707 of its maximum value $I_o$, and the power consumed by the ciruit is one-half the maximum value at $\omega = 0$. At $\omega = \omega_c$, the preceding equation shows that

$$\frac{|T(0)|}{|T(i\omega_c)|} = 0.707$$

or

$$|T(i\omega_c)|_{\text{db}} = |T(0)|_{\text{db}} - 3.01 \text{ db} \tag{3.9-6}$$

That is, for input frequencies greater than $\omega_c$ (the *cutoff frequency*) the power delivered to the circuit is less than one-half its maximum value. Thus $\omega_c$ is termed the *half-power* frequency.

Equation (3.9-6) can be taken as the defining relation for the cutoff frequency $\omega_c$, the frequency at which the log magnitude curve drops 3.01 db below its peak value. In this example the peak occurred at $\omega = 0$, but this is not true for all systems (for example, see Figure 3.30). Also, there may be two cutoff frequencies; one to the left and one to the right of a peak in $T(i\omega)$.

The *bandwidth* is the frequency spread between the cutoff frequencies, or between $\omega = 0$ and the cutoff frequency, if the peak power occurs at $\omega = 0$. Figure 3.27 shows that $\omega_c$ is the corner frequency $1/\tau$ for $T(s) = 1/(\tau s + 1)$. Its bandwidth is $1/\tau$.

Bandwidth is a measure of the range of frequencies for which a significant portion of the system's input is felt by the output. The system filters out to a greater extent those input components whose frequencies lie outside the bandwidth. The bandwidth definition is arbitrarily based on the half-power points. Other definitions can be used but this is the most common. Since the first-order system's bandwidth is the reciprocal of its time constant, the bandwidth is seen to be an indication of the speed of response. The larger $\tau$ is, more sluggish is the response. A small bandwidth indicates a sluggish system, which poses a problem for the designer. For example, if the time constant of a pneumatic device is picked to be large enough to filter out unwanted pressure fluctuations, its response might be too slow for other purposes.

*Example 3.18*

For Example 3.3 of Section 3.2, the time constant of the motor-plus-load was $\tau = 22$ sec. The half-power frequency is $1/22 = 0.045$ rad/sec. Since the steady-state speed is 150 rad/sec, any sinusoidal torques resulting from misaligned bearings, for example, have frequencies well outside the system's bandwidth.

*Example 3.19*

Discuss the effects of a flywheel on a rotational system described by $I\dot{\omega} = T - b\omega$.

An example of an accumulator in a rotational system is a flywheel, which is simply an added inertia about the axis of rotation. Since $1/\tau = b/I$, any increase in $I$ will decrease the bandwidth of the system. With engines for which there is considerable torque variation over a cycle, such as with internal combustion engines, a flywheel is necessary to obtain a nearly constant speed.

## Response to General Periodic Inputs

The application of the sinusoidal input response is not limited to cases involving a single sinusoidal input. A basic theorem of analysis states that under some assumptions, which are generally satisfied in most practical applications, any periodic function can be expressed by a constant term plus an infinite series of sines and cosines with increasing frequencies. This theorem is the *Fourier theorem*, and its associated series is the *Fourier series.* It has the form

$$v(t) = a_0 + a_1 \cos\left(\frac{\pi t}{p}\right) + a_2 \cos\left(\frac{2\pi t}{p}\right) + \ldots + b_1 \sin\left(\frac{\pi t}{p}\right) + b_2 \sin\left(\frac{2\pi t}{p}\right) + \ldots \tag{3.9-7}$$

where $v(t)$ is the periodic function and $p$ is the *half-period* of $v(t)$. The constants $a_i$ and $b_i$ are determined by integration formulas applied to $v(t)$. Appendix A contains these formulas and summarizes the procedure required to construct the series.

In previous sections we have seen how to determine the steady-state response of a linear system when subjected to an input that is a constant (a step), a sine, or a cosine. When the input $v$ is periodic and expressed in the form of (3.9-7), the superposition principle states that the complete steady-state response is the sum of the steady-state responses due to each term in (3.9-7). Although this is an infinite series, in practice we have to deal only with a few of its terms, because those terms whose frequencies lie outside the system's bandwidth can be neglected as a result of the filtering property of the system.

### *Example 3.20*

For the vibration isolator shown in Figure 3.28, suppose that $c/K = \tau = 0.1$ sec, and that the displacement $y_2$ is no longer purely sinusoidal but consists of a sine for 0.5 sec followed by zero displacement for the second half of the period (see Figure 3.32). Such a motion might be produced by a rotating cam with a *dwell* (the zero displacement portion). Determine the steady-state response of $y_1$.

The Fourier series representation of $y_2(t)$ is, from Appendix A,

$$y_2(t) = \frac{1}{\pi} + \frac{1}{2}\sin 2\pi t - \frac{2}{\pi}\left(\frac{\cos 4\pi t}{1(3)} + \frac{\cos 8\pi t}{3(5)} + \frac{\cos 12\pi t}{5(7)} + \ldots\right) \tag{3.9-8}$$

For the system transfer function with $\tau = 0.1$ and $b = 1/\tau = 10$, the magnitude ratio $M$ from (3.8-18) is

$$M(\omega) = \frac{1}{\sqrt{1 + 0.01\omega^2}}$$

The decibel equivalent $m(\omega)$ may also be used and is convenient when working from a logarithmic plot. From (3.8-19) the phase shift is

$$\phi(\omega) = \tan^{-1}(-0.1\omega)$$

It is desired to find the steady-state motion $y$ of point 1 due to $y_2$. The first term in the series expansion for $y_2$ is a constant, $1/\pi = 0.318$. This corresponds to an input with zero frequency; at steady state it produces a response of $y_1 = 0.318$ since $M(0) = 1$. The second term in the series is a sine with an amplitude of 0.5 and a frequency of $2\pi$ rad/sec. Thus

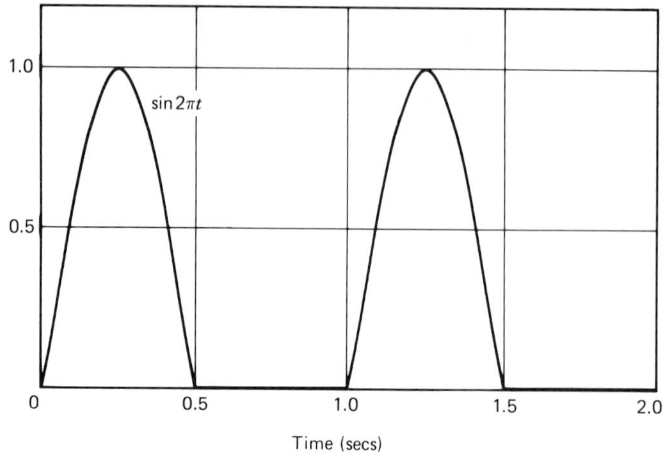


**Figure 3.32 Half-sine function.**

$$M(2\pi) = \frac{1}{\sqrt{1 + 0.01(2\pi)^2}} = 0.847$$

It produces a sinusoidal response of the same frequency with an amplitude of $0.847(0.5) = 0.424$, with a phase shift of $\phi(2\pi) = \tan^{-1}(-0.2\pi) = -0.56$ rad. These are the only terms within the bandwidth of the isolator (10 rad/sec). To demonstrate the filtering property and to illustrate how cosine terms are handled, we will treat the third term in the series.

Since $\cos 4\pi t = \sin(4\pi t + \pi/2)$, this term can be handled as a sine wave. The phase shift of $\pi/2$ radians is then added to $\phi(4\pi)$. Here

$$M(4\pi) = \frac{1}{\sqrt{1 + 0.01(4\pi)^2}} = 0.623$$

The amplitude of the response is $0.623(2/3\pi) = 0.132$, and the phase shift is $\phi(4\pi) + \pi/2 = -0.90 + \pi/2 = 0.67$ rad. From superposition the total steady-state response resulting from the three series terms retained is

$$y_1(t) = 0.318 + 0.423 \sin(2\pi t - 0.56) - 0.132 \sin(4\pi t + 0.67)$$

The input and the response are plotted in Figure 3.33. The difference between the input and output wave shapes results from the resistive or lag effect of the system, not from the omission of the higher-order terms in the series. To see this, we compute that part of the response amplitude resulting from the fourth term in the series. The amplitude contribution is $2M(8\pi)/15\pi = 0.016$, which is about 10% of the amplitude of the third term. The decreasing amplitude of the higher-order terms in the series for $y_2(t)$, when combined with the filtering property of the system, allows us to truncate the series when the desired accuracy has been achieved.

Before leaving this example, we point out that the maximum displacement of point 1 is 85% of the maximum displacement of the input. Thus the system's isolation characteristics for this input are quite different from those for the previous application

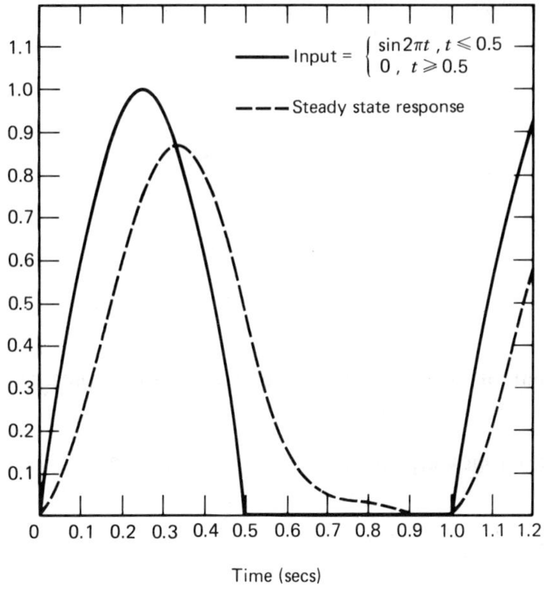


**Figure 3.33 Steady-state response of the isolator to the half-sine input.**

in which the input was a pure sine with a frequency of 100 rad/sec. Whenever the input frequency or wave shape is altered very much from its design value, the isolator cannot be expected to perform as well.

## 3.10 LINEAR SYSTEM RESPONSE TO GENERAL INPUTS

In the case of a sinusoidal or general periodic input to a linear model, we concentrated on the steady-state response and ignored the transient response. We now return to the problem of determining the complete response of the linear system,

$$\dot{y} = ry + bv \qquad (3.10\text{-}1)$$

The general solution can be developed by considering the input term $bv(t)$ in the model (3.10-1) to be composed of a series of rectangular pulses of width $\Delta\lambda$. Figure 3.34*a* shows one such pulse at some time $\lambda_i$ for an arbitrary input. The pulse may be considered as an impulse of strength $bv(\lambda_i)\,\Delta\lambda$ if $\Delta\lambda$ is small enough (see Section 3.3). The system's impulse response $(t - \lambda_i)$ time units later due to this input is

$$y(t) = bv(\lambda_i)\,\Delta\lambda\, e^{r(t-\lambda_i)}$$

This is shown in Figure 3.34*b* for a stable system. From the principle of superposition, the total response at time $t$ due to the input is the sum of all the responses due to pulses occurring before time $t$, or

$$y(t) = \sum_i e^{r(t-\lambda_i)} bv(\lambda_i)\,\Delta\lambda$$

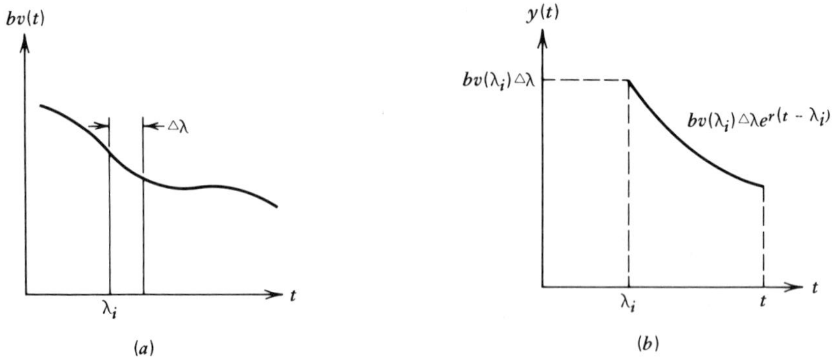


**Figure 3.34 Graphical development of general system response. (*a*) Input plot. (*b*) Response plot.**

In the limit the sum is replaced by the following integral as $\Delta\lambda$ approaches zero.

$$y(t) = \int_0^t e^{r(t-\lambda)} bv(\lambda)\, d\lambda$$

This is the forced response. The complete response is obtained by adding the free response, which gives

$$y(t) = y(0)e^{rt} + \int_0^t e^{r(t-\lambda)} bv(\lambda)\, d\lambda \tag{3.10-2}$$

This integral is the *superposition* or *convolution* integral. An alternate derivation of the solution uses the convolution property of the Laplace transform (Appendix B).

## Variable Coefficients

Equation (3.10-2) is valid regardless of the stability properties of the system, and it can be generalized to include the case of time-varying coefficients.* In this case, $r = r(t)$, $b = b(t)$, and the free response is of the form $\phi(t, t_0)y(t_0)$, where $\phi(t, t_0)$ is a time function that depends on the specific form of the function $r(t)$. The complete solution is

$$y(t) = \phi(t, t_0)y(t_0) + \int_{t_0}^t \phi(t, \lambda)b(\lambda)v(\lambda)\, d\lambda \tag{3.10-3}$$

where the initial time $t_0$ cannot be arbitrarily set to zero as before. The derivation of the form of $\phi(t, \lambda)$ can be found in most differential equation texts (Reference 3). The solution is

$$\phi(t, \lambda) = \exp\left[\int_\lambda^t r(\alpha)\, d\alpha\right] \tag{3.10-4}$$

---

* In the next section it will be seen that some transformations of nonlinear constant-coefficient models produce linear models with variable coefficients.

The correctness of this form can be shown by substituting it into the original equation (3.10-1).

From (3.10-4) we see that when $r$ is constant,

$$\phi(t, t_0) = \exp[r(t - t_0)] = \phi(t - t_0) \tag{3.10-5}$$

Thus with constant-coefficient models the origin of the time axis can be arbitrarily shifted, whereas with variable-coefficient models we cannot do so. The function $\phi(t, t_0)$ is known as the *transition function.* The name applies because $\phi(t, t_0)$ describes how the system state makes the transition from time $t$ to time $t_0$. The only other information required for the solution is that on the initial value and the input function. These enter the solution through multiplication by $\phi$.

*Example 3.21*

Find the transient response of the vibration absorber $\tau\dot{y}_1 + y_1 = y_2$ (Figure 3.28) to the half-sine input displacement (Figure 3.32).

The response can be determined with (3.10-2). Assume that $y_1(0) = 0$ and obtain

$$y_1(t) = \int_0^t e^{-(t-\lambda)/\tau} \frac{1}{\tau} y_2(\lambda)\, d\lambda$$

$$= \frac{1}{\tau} e^{-t/\tau} \int_0^t e^{\lambda/\tau} y_2(\lambda)\, d\lambda$$

If $t \leqslant 0.5$ sec, then $y_2(\lambda) = \sin 2\pi\lambda$ and

$$y_1(t) = \frac{1}{\tau} e^{-t/\tau} \int_0^t e^{\lambda/\tau} \sin 2\pi\lambda\, d\lambda$$

$$= \frac{1}{1 + \tau^2(2\pi)^2} (\sin 2\pi t - 2\pi\tau \cos 2\pi t + 2\pi\tau e^{-t/\tau}) \quad 0 \leqslant t \leqslant 0.5 \tag{3.10-6}$$

For $0.5 \leqslant t \leqslant 1.0$, $y_2(t) = 0$, and its contribution to the convolution integral is zero. The response is

$$y_1(t) = \frac{1}{\tau} e^{-t/\tau} \int_0^{0.5} e^{\lambda/\tau} \sin 2\pi\lambda\, d\lambda$$

$$= \frac{2\pi}{\tau} e^{-t/\tau} \left( \frac{e^{0.5/\tau} + 1}{1/\tau^2 + (2\pi)^2} \right) \quad 0.5 \leqslant t \leqslant 1.0 \tag{3.10-7}$$

Previously the isolator time constant was selected to be $\tau = 0.1$ sec. The complete solution for this case is shown in Figure 3.35 along with the input and the transient response for $\tau = 0.5$ for comparison. The transient response for $\tau = 0.1$ does not differ greatly from the steady-state response shown in Figure 3.33 because the period of the input is 10 times greater than the system's time constant. Thus the transient has disappeared quickly relative to the period of the input function. However, this is not the case when $\tau = 0.5$.

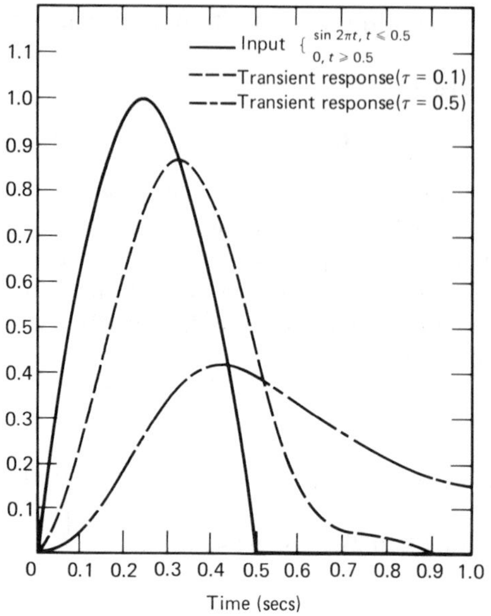


**Figure 3.35 Transient response of the vibration isolator to the half-sine input for $0 \leqslant t \leqslant 1.0$.**

## 3.11 NONLINEAR MODELS

Most of the linear models encountered thus far have resulted from some simplifying assumption. For example, a closer examination of the viscous damper model in Section 2.6 will reveal that the underlying physical process is inherently nonlinear and can be approximated by a linear relation only when the appropriate velocity is small. It is solely for reasons of mathematical simplicity that such approximations are made; as we will see, no general solution technique is available for nonlinear models.

The general form of a first-order *linear* ordinary differential equation is

$$a(t)\dot{y} + b(t)y + c(t) = 0 \tag{3.11-1}$$

Any first-order model not fitting this form is *nonlinear.* Thus the class of nonlinear models can be seen to be much larger than the linear class, and this indicates the difficulty in obtaining general nonlinear techniques. Unhappily for the analyst, prediction of the response of a nonlinear system is neither easy nor generally understood. Happily for the modeler, nonlinear models are not so restricted in their behavioral possibilities.

An important characteristic of nonlinear model response is that its form depends on the magnitude of the input and the initial conditions. For example, step inputs of different magnitudes might produce completely different forms of response for the same model. Also, the response might be oscillatory for one initial condition, whereas another might produce exponential behavior. This cannot occur in a linear model, whose response form (sinusoidal, exponential, etc.) is completely determined by the input form and the system's characteristic roots.

In this section we treat some of the elementary methods for dealing with simple nonlinear models. An extensive treatment is beyond the scope of this work, but the reader interested in further exploration of this fascinating subject is directed to References 3 and 4.

## Solution by Separation of Variables

The general first-order equation can be written as

$$\dot{y} = f(y, t) \tag{3.11-2}$$

where the inputs, if any, have been specified as functions of $y$ and $t$. The equation is said to be *separable* if it can be written as

$$f_1(t)g_1(y)\,dt + f_2(t)g_2(y)\,dy = 0 \tag{3.11-3}$$

Dividing by $f_2$ and $g_1$ produces

$$\frac{f_1(t)}{f_2(t)}dt = -\frac{g_2(y)}{g_1(y)}dy$$

An integration over time from 0 to $t$ corresponds to an integration over $y$ from $y(0)$ to $y(t)$. Thus

$$\int_0^t \frac{f_1(t)}{f_2(t)}dt = -\int_{y(0)}^{y(t)} \frac{g_2(y)}{g_1(y)}dy \tag{3.11-4}$$

If these integrals can be obtained in closed form, the solution of the equation can be obtained, at least in parametric form.

*Example 3.22*

The models of flow through an orifice or turbulent flow involve a square root relation (see Table 2.18). If there are no inputs, these models are of the form

$$\frac{dy}{dt} = -c\sqrt{y} \tag{3.11-5}$$

where $c$ is a constant. Separation of variables leads to

$$\frac{dy}{\sqrt{y}} = -c\,dt$$

The integral is

$$\int_{y(0)}^{y(t)} \frac{dy}{\sqrt{y}} = -c\int_0^t dt$$

or

$$2\sqrt{y}\,\Bigg|_{y(0)}^{y(t)} = 2\sqrt{y(t)} - 2\sqrt{y(0)} = -ct$$

It is possible to solve this equation for $y(t)$. However, care must be taken in dealing with the square root. As long as

$$t < \frac{2}{c}\sqrt{y(0)} \tag{3.11-6}$$

the right-hand side of the following equation is positive.

$$\sqrt{y(t)} = \sqrt{y(0)} - \frac{c}{2}t \tag{3.11-7}$$

If so, the solution can be written as

$$y(t) = \left(\sqrt{y(0)} - \frac{c}{2}t\right)^2 \tag{3.11-8}$$

This is valid for $y(0) > 0$ if (3.11-6) is satisfied. If $y(0) < 0$ the solution is a complex number, as can be deduced from (3.11-5). The borderline between these two cases occurs when $y(0) = 0$. From (3.11-5) we can see that a solution for this case is $y(t) = 0$.* These three solutions show that the form of the solution depends on the initial condition. If $y(0) > 0$, $y(t)$ decreases and equals zero at $t = 2\sqrt{y(0)}/c$. Equation (3.11-8) predicts that $y(t)$ then increases for $t > 2\sqrt{y(0)}/c$, but this is an incorrect interpretation because this equation is not a solution of the original differential equation if (3.11-6) is not satisfied.

Physical insight often helps to avoid acceptance of an erroneous solution. If (3.11-5) represents a tank with orifice flow, we know that $y$ (the liquid level) must remain at zero once the tank has emptied.

## Solution by Transformation

When the function $f(y, t)$ in (3.11-2) is nonlinear because the dependent variable $y$ is raised to a power, as in $\dot{y} = f(y^m, t)$, then the following transformation frequently reduces the equation to a linear form in $z$.

$$y = z^n \tag{3.11-9}$$

The derivative is

$$\dot{y} = nz^{n-1}\dot{z} \tag{3.11-10}$$

*Example 3.23*

In (3.11-5), $f(y^m, t) = -cy^{\frac{1}{2}}$, so $m = \frac{1}{2}$. From (3.11-9) and (3.11-10),

$$nz^{n-1}\dot{z} = -cz^{n/2}$$

or

$$\dot{z} = -\frac{c}{n}z^{1-n/2}, \quad n \neq 0$$

This equation is linear if $n = 2$. In this case,

$$\dot{z} = -\frac{c}{2}$$

and

$$z(t) = z(0) - \frac{c}{2}t$$

The solution in terms of $y(t)$ and $y(0)$ is easily recovered and is identical to (3.11-8).

*Example 3.24*

Consider the logistic growth equation (2.9-3), repeated here.

---

* Another solution when $y(0) = 0$ is $y(t) = c^2t^2/4$. For a discussion of the existence and uniqueness of solutions, see Reference 3.

$$\dot{y} = ry\left(1 - \frac{y}{K}\right) \tag{3.11-11}$$

The transformation given by (3.11-9) gives

$$\dot{z} = \frac{r}{n}z - \frac{r}{nK}z^{n+1}$$

This is linear if $n = -1$ and the solution is given by the step response of the linear model (3.2-4). In the present notation, this is

$$z(t) = \left[z(0) - \frac{1}{K}\right]e^{-rt} + \frac{1}{K} \tag{3.11-12}$$

Inversion of the transformation produces the solution in terms of the original variable.

$$y(t) = \frac{Ky(0)}{y(0) + [K - y(0)]\,e^{-rt}} \tag{3.11-13}$$

The solution's behavior for various initial conditions is illustrated in Figure 3.36, for $r$ and $K$ positive. The value $y = K$ is the carrying capacity of the environment, and it represents the stable equilibrium to which all solutions are attracted provided that $y(0) > 0$.

If $y(0) < 0$, the solution approaches $-\infty$. While a negative initial condition is not feasible for population growth models, this case serves as another example of how the form of a nonlinear equation's solution depends on the initial condition. The solution for $y(0) < 0$ cannot be obtained from (3.11-13) because the transformation from $z$ to $y$ becomes undefined when the denominator in this equation becomes zero. An examination of the original model (3.11-11) shows that $y$ approaches $-\infty$ for negative initial conditions.

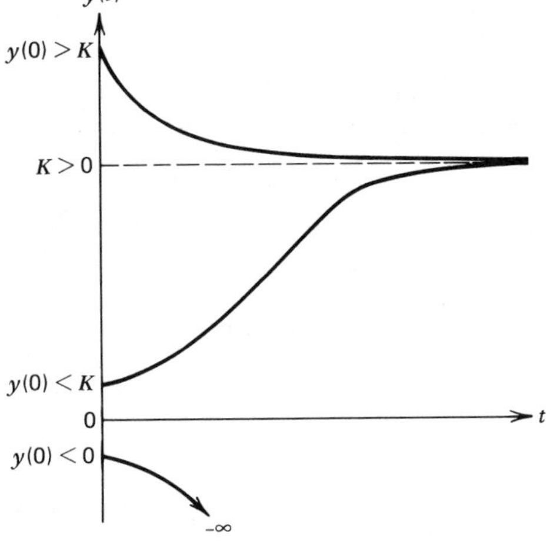


**Figure 3.36 Solutions of the logistic growth model.**

The logistic model is a special case of a larger class of equations described by the *generalized Riccati equation*, which is

$$\dot{y} + q(t)y + p(t)y^2 = w(t) \tag{3.11-14}$$

If $w(t)$ is identically zero, the equation may be reduced to linear form by means of the same transformation just employed; namely, $y = 1/z$. The resulting linear form is

$$\dot{z} = q(t)z + p(t) \tag{3.11-15}$$

whose solution is given in the form of (3.10-3). If $w(t)$ is not zero this transformation will not produce a linear equation. In this case, the transformation

$$y = \frac{\dot{z}}{p(t)z}$$

is used, but this results in a second-order linear equation. The solution of second-order equations is treated in Chapter Five.

## Piecewise Linear Models

Sometimes nonlinear models result from the inclusion of innocuous-looking effects for the sake of greater realism. A common example of this is the Coulomb friction shown in Figure 3.37. For now assume that the static and dynamic friction coefficients have the same value, denoted by $\mu$, and that an externally applied force $F$ acts on a mass $m$. Its equation of motion is

$$m\dot{v} = F \pm \mu N \tag{3.11-16}$$

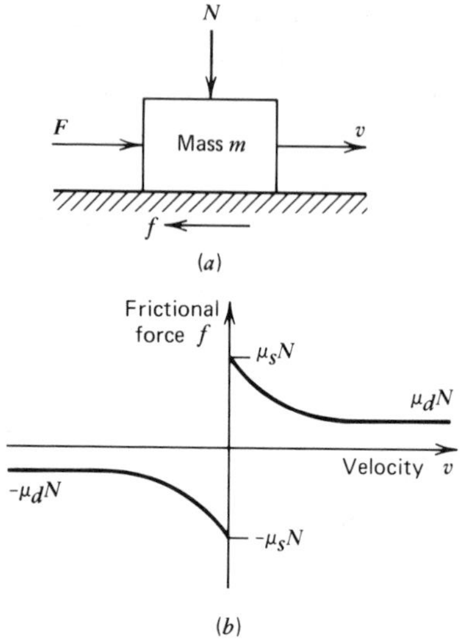


**Figure 3.37 Coulomb friction. (*a*) Mass sliding on a surface. (*b*) Constitutive relation.**

where $v$ is the velocity and $N$ is the normal force equal to the weight, mg. The plus sign is used if $v < 0$, and the minus sign is used if $v \geqslant 0$. Thus this model really represents two equations: one for $v < 0$ and one for $v \geqslant 0$. Each equation by itself is linear, but the total model is nonlinear because of the switching that occurs when $v$ changes sign. The model is said to be *piecewise linear.* Suppose that $F$ is sinusoidal with a frequency $\omega$ and amplitude $A$. Then

$$m\dot{v} = A \sin \omega t \pm \mu N$$

This is amenable to solution by direct integration. The result is

$$v(t) = v(t_1) - \frac{A}{m\omega}(\cos \omega t - \cos \omega t_1) \pm \mu N(t - t_1) \tag{3.11-17}$$

Given the inital velocity $v(t_1)$, (3.11-17) is solved using the appropriate sign until the velocity changes sign. The time when this occurs is denoted $t_2$, and $v(t_2)$ is the new initial condition when the procedure is repeated.

If the forcing function $F$ is such that $v$ changes sign several times, the procedure becomes tedious. The numerical methods presented in Section 3.13 can then be applied. If only a qualitative description of the response is needed, the graphic methods of Section 5.10 can be used.

## 3.12 LINEARIZATION OF DYNAMIC MODELS

Quite often, especially when dealing with higher-order systems, nonlinear equations are encountered that are intractable using the separation of variables method and for which no transformation is known to reduce the equation to linear form. However, it is often possible to approximate the nonlinear model with a linear one. We can then take advantage of the linear model's generality, in the sense that its response form is independent of the specific values of the initial conditions. This feature makes linear models extremely useful in system design, because with them the designer can focus on the structure of the system, and need not worry about the effects of different initial conditions that might be encountered when the system is in operation. This is especially true in the design of control systems, which must produce stable system behavior for a variety of initial conditions. The sacrifice incurred by a linear approximation is that its validity is limited to a certain range of operating conditions. However, we will see that if a feedback control system is operating properly, it preserves the validity of the assumption on which the linearization is based. We will attempt to shed some light on these concepts as well as on the procedures for developing linear approximations.

### Coordinate Translations in Dynamic Models

At this point it is appropriate to consider the effect of a coordinate system translation in a linear dynamic system. We have previously used this technique in the accumulator and sphere response examples (Examples 3.8 and 3.9). Suppose that the model

$$\dot{y} = ay + bv \tag{3.12-1}$$

is subjected to a constant input $v_e$. Then

$$\dot{y} = ay + bv_e$$

At steady state this will produce an output, denoted by $y_e$, such that $\dot{y}_e = 0$ and thus

$$y_e = -\frac{bv_e}{a} \tag{3.12-2}$$

The equilibrium state of the model is $y_e$ when the input is the constant $v_e$.

In many situations the input may change slightly from the reference or nominal value $v_e$; or the system may be subjected to a disturbance that suddenly changes $y$ from $y_e$ to a new value. In order to investigate the behavior of the system under these conditions, it is convenient to define a new set of variables $x$ and $u$ such that $x = y - y_e$ and $u = v - v_e$. To obtain the differential equation describing $x$ in terms of $u$, note that $\dot{x} = \dot{y} - \dot{y}_e = \dot{y}$, since $y_e$ is a constant. Substitution for $y$ and $v$ in (3.12-1) results in

$$\dot{y} = \dot{x} = a(x + y_e) + b(u + v_e)$$
$$= ax + bu + ay_e + bv_e$$

Use of (3.12-2) gives

$$\dot{x} = ax + bu \tag{3.12-3}$$

and we see that the form of the equation for $x$ and $u$ is the same as that for $y$ and $v$, because the equation is linear. Equation (3.12-3) is more convenient for analyzing deviations from the reference condition since the values of $y_e$ and $v_e$ do not appear explicitly. Note that the system's equilibrium state is now described by $x = 0$ when $u = 0$.

## Linearization by Taylor Series

The linearization method to be presented is based on the Taylor series expansion of a function of one or more variables (see Appendix A). As demonstrated in Chapter 1, the series can be used to approximate a general (differentiable) nonlinear function in the vicinity of a reference point of interest. By truncating the series after the first-order derivative terms, a straight line or plane is obtained that passes through the reference point. To be more specific, consider a static function of two variables $f(y, v)$. Denote the reference point by $(y_e, v_e)$. Then the Taylor series expansion of the function around the reference point is

$$f(y, v) = f(y_e, v_e) + \left(\frac{\partial f}{\partial y}\right)_e (y - y_e) + \left(\frac{\partial f}{\partial v}\right)_e (v - v_e)$$
$$+ \frac{1}{2!}\left(\frac{\partial^2 f}{\partial y^2}\right)_e (y - y_e)^2 + \left(\frac{\partial^2 f}{\partial y \partial v}\right)_e (y - y_e)(v - v_e)$$
$$+ \frac{1}{2!}\left(\frac{\partial^2 f}{\partial v^2}\right)_e (v - v_e)^2 + \dots \tag{3.12-4}$$

where $(\ \ )_e$ indicates that the term within the parentheses is evaluated at the point $(y_e, v_e)$. If $y$ and $v$ are sufficiently close to $y_e$ and $v_e$, then the series converges, and the second- and higher-order terms are smaller than the first-order terms. Thus the approximation is

$$\Delta f = ax + bu \tag{3.12-5}$$

where we have defined the following.

$$\Delta f = f(y, v) - f(y_e, v_e) \tag{3.12-6}$$

$$x = y - y_e \tag{3.12-7}$$

$$u = v - v_e \tag{3.12-8}$$

$$a = \left(\frac{\partial f}{\partial y}\right)_e \tag{3.12-9}$$

$$b = \left(\frac{\partial f}{\partial v}\right)_e \tag{3.12-10}$$

**TABLE 3.2 Taylor Series Linearization of a Dynamic Model**

| | |
|---|---|
| Nonlinear model | $\dot{y} = f(y, v), \quad v = v(t)$ |
| Reference solution | $\dot{y}_e = f(y_e, v_e)$ |
| | $\dot{y}_e = 0$ for a point equilibrium |
| Linearized model | $\dot{x} = ax + bu$ |
| | $x = y - y_e$ |
| | $u = v - v_e$ |
| | $a = \left(\dfrac{\partial f}{\partial y}\right)_e$ |
| | $b = \left(\dfrac{\partial f}{\partial v}\right)_e$ |

The quantities $\Delta f$, $x$, and $u$ represent the deviations or perturbations of $f$, $y$, and $v$ from their reference values. In terms of these deviations, the original nonlinear function $f(y, v)$ has been approximated by a plane in the $(x, u)$ coordinate system. Note that the slopes of this plane in the $x$ and $u$ directions equal the slopes of $f(y, v)$ in the $y$ and $v$ directions at the reference point, and that the values of these slopes depend on the specific reference point chosen. Different reference points will produce different planes, in general.

Now consider the nonlinear dynamic model

$$\dot{y} = f(y, v) \tag{3.12-11}$$

with the equilibrium values $y_e$, $v_e$. From the definitions given by (3.12-5) through (3.12-10), we can write the left- and right-hand sides of (3.12-11) as follows.

$$\dot{y} = \dot{x} + \dot{y}_e = \dot{x}$$
$$f(y, v) = f(y_e, v_e) + ax + bu$$

The latter equation is an approximation, of course. Since $\dot{y}_e = f(y_e, v_e) = 0$, we obtain from (3.12-11)

$$\dot{x} = ax + bu \tag{3.12-12}$$

This is the linearized approximation to the original equation (3.12-11). This procedure is summarized in Table 3.2.

The translation of the origin of a linear model to the equilibrium point results in the form (3.12-3), which is the same form as (3.12-12). However, in the first case the resulting model is exact, and in the second case it is only approximately correct for small values of $x$ and $u$.

*Example 3.25*

In Section 2.5 we saw that the height $h$ of liquid in a tank with an orifice can be described by (see Figure 3.38)

$$\dot{h} = \frac{1}{A_1} q - \frac{\alpha}{A_1}\sqrt{h} \tag{3.12-13}$$

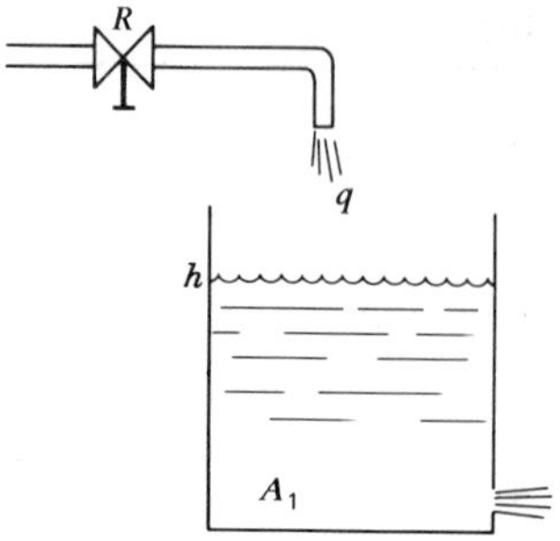


Figure 3.38 Tank with orifice.

$$\alpha = C_d A\sqrt{2g} \qquad (3.12\text{-}14)$$

Suppose the given parameter values are $A_1 = 1 = \alpha$. Obtain a linearized model in terms of the given equilibria at (a) $q_e = h_e = 1$, and (b) $q_e = 3$, $h_e = 9$.

For the given values, the model is

$$\dot{h} = q - \sqrt{h} = f(h, q) \qquad (3.12\text{-}15)$$

If the inflow rate is held constant at $q_e$, the height will eventually come to equilibrium at the value $h_e$ where

$$h_e = \left(\frac{q_e}{\alpha}\right)^2 \qquad (3.12\text{-}16)$$

From Table 3.2 the linearization coefficients are

$$a = \left(\frac{\partial f}{\partial h}\right)_e = -\tfrac{1}{2}h_e^{-\frac{1}{2}}$$

$$b = \left(\frac{\partial f}{\partial q}\right)_e = 1$$

and the variations are $x = h - h_e$ and $u = q - q_e$.

(a) For $h_e = q_e = 1$, $a = -\frac{1}{2}$, $b = 1$, and the linearized model is

$$\dot{x} = -\tfrac{1}{2}x + u$$

The time constant of the linearized model is 2.

(b) For $h_e = 9$, $q_e = 3$, we have $a = -1/6$, $b = 1$, and

$$\dot{x} = -\tfrac{1}{6}x + u$$

The time constant is 6. Note that the linearized model's time constant depends on the particular equilibrium chosen for the linearization.

An equation can be linearized around one of its solutions other than a point equilibrium. The subscript $e$ usually denotes an equilibrium value, but to avoid additional notation let $y_e(t)$ and $v_e(t)$ denote the functions about which the linearization is to be performed. The function $y_e(t)$ is the solution of $\dot{y} = f(y, v)$ when $v = v_e(t)$. Then $f(y_e, v_e, t)$ is no longer zero but a known function of time and equals $\dot{y}_e$. The same procedure as before results in (3.12-12) where the coefficients $a$ and $b$ are now known functions of time.

## Linearization Accuracy

The question that inevitably arises in a discussion of linearization is: how small must $x$ and $u$ be for the truncated Taylor series to be acceptable? There are two ways to frame an answer to this question.

Often the linearized model is obtained not for solution purposes, but for an indication of the local stability of an equilibrium. If the linearized equation is stable, then any slight perturbation will die out and the equilibrium is locally stable. On the other hand, linearization cannot be used to investigate the global stability of the equilibrium because the series truncation is not valid for large perturbations. If the linearized equation is unstable, the equilibrium is locally unstable at least. However, if the linearized equation is neutrally stable, no information is obtained about local stability because the second-order terms in the expansion are now critical. To see this, note that the coefficient $a$ in (3.12-12) is zero for neutral stability, and thus the first-order derivative term in the expansion is also zero. The remaining terms are thus significant and cannot be truncated. This invalidates the linearization, and a nonlinear analysis is required to resolve the issue.

Linearized models are widely used in the design of feedback control systems and are successful for the following reason. If the controller is designed to keep the system output near the desired equilibrium value, and if the design is based on a linearization around this value, then the controller will tend to keep the linearization accurate. This interesting property partly explains why models for controller design need not be as detailed as those required for other purposes.

## 3.13 NUMERICAL METHODS: AN INTRODUCTION

Despite the relatively powerful methods presented in this chapter, it is not always possible to obtain the closed-form solution of a dynamic model. In this section we introduce numerical methods for solving differential equations.* Our discussion will focus on first-order models, but the techniques are easily generalized to higher-order models for which the analytical methods of this chapter become difficult to apply because of the cumbersome algebra involved.

The essence of a numerical method is to convert the differential equation into an equivalent model composed of difference equations. In this form the model can be programmed on a calculator or digital computer. Numerical algorithms differ mostly as a result of the specific procedure used to obtain the difference equations. In general, as the accuracy of the approximation is increased, so is the complexity of the programming involved.

### The Euler Method

Consider the model

$$\frac{dy}{dt} = ry, \quad r = \text{constant} \tag{3.13-1}$$

From the definition of the derivative,

$$\frac{dy}{dt} = \lim_{\Delta t \to 0} \frac{y(t + \Delta t) - y(t)}{\Delta t}$$

---

* An alternate approach is to use analog simulation. This technique is covered in Appendix E.

If the time increment $\Delta t$ is chosen small enough, the derivative can be replaced by the approximate expression

$$\frac{dy}{dt} \cong \frac{y(t+\Delta t) - y(t)}{\Delta t} \tag{3.13-2}$$

Assume that the right-hand side of (3.13-1) remains constant over the time interval $(t, t + \Delta t)$ and replace (3.13-1) by the following approximation.

$$\frac{y(t+\Delta t) - y(t)}{\Delta t} = ry(t)$$

or

$$y(t + \Delta t) = y(t) + ry(t)\,\Delta t \tag{3.13-3}$$

The smaller $\Delta t$ is made, the more accurate are our two assumptions leading to (3.13-3). This technique for replacing a differential equation with a difference equation is the *Euler method.*

Equation (3.13-3) can be written in more convenient form as follows. At the initial time $t_0$, (3.13-3) gives:

$$y(t_0 + \Delta t) = y(t_0) + ry(t_0)\,\Delta t$$

Let $t_1 = t_0 + \Delta t$. Then

$$y(t_1 + \Delta t) = y(t_1) + ry(t_1)\,\Delta t$$

That is, (3.13-3) is applied successively at the instants $t_k$, where $t_{k+1} = t_k + \Delta t$, and we can write in general

$$y(t_{k+1}) = y(t_k) + ry(t_k)\,\Delta t \tag{3.13-4}$$

It is convenient to introduce a time index $k$ such that

$$y_k \triangleq y(t_k)$$

Thus (3.13-4) becomes

$$y_{k+1} = y_k + ry_k\,\Delta t = (1 + r\,\Delta t)\,y_k, \quad k = 0, 1, 2, \ldots \tag{3.13-5}$$

In this form it is easily seen that the continuous-time variable $y(t)$ has been represented by the discrete-time variable $y_k$. Equation (3.13-5) is a *recursion relation* or *difference equation.* It can be solved recursively (sequentially) at the instants $k = 0, 1, 2, \ldots$ starting with the initial value $y_0 = y(t_0)$.

In cases where the subscript notation is inconvenient, the variable $y_k$ is written as $y(k)$. Care must be taken not to confuse $y(k)$ with $y(t_k)$. For $k = 0, 1, 2, 3, \ldots,$ $y(k)$ represents $y(t)$ at the values of $t = t_0, t_1, t_2, \ldots$ In this notation, (3.13-5) becomes

$$y(k + 1) = (1 + r\,\Delta t)\,y(k), \quad k = 0, 1, 2, \ldots \tag{3.13-6}$$

With $r$ a constant we can easily solve the original differential equation (3.13-1), so we really have no need for (3.13-6). However, this is not the case for a general equation of the form

$$\dot{y} = f(y, v) \tag{3.13-7}$$

Assume the right-hand side to be constant over the interval $(t_k, t_{k+1})$ and equal to the value $f[y(t_k), v(t_k)]$. With the approximation (3.13-2), we have the Euler method for (3.13-7).

$$y(t_{k+1}) = y(t_k) + \Delta t\, f[y(t_k), v(t_k)] \tag{3.13-8}$$

This is easily programmed on a computer or calculator. The values of $t_0, y(t_0)$, $\Delta t$, and any parameters in the functions $f$ and $v$ must be read into the program, along with a stopping criterion. To be specific, suppose the model to be solved is

$$\dot{y} = 3ty^2 + v(t)$$

where $v(t)$ is a ramp function with a slope of 5: $v(t) = 5t$. The structure of a FORTRAN program to implement the Euler method for this equation would look like the following.

```
 read in: DT, Y0, T0, TMAX
 T = T0
 DEL = TMAX - T0
 Y = Y0
 IMAX = INT(DEL/DT) + 1
 DO 1 I = 1, IMAX
     T = T + DT
     V = 5*T
     Y = Y + DT*(3*T*Y*Y + V)
     print: T, Y
1 CONTINUE
 STOP
```

The statements for reading and printing depend on the particular machine. The variables DT, Y0, T0, and TMAX represent $\Delta t$, $y(t_0)$, $t_0$, and $t_{max}$, respectively. The time $t_{max}$ is the largest time at which the solution is desired.

The Euler method is the simplest algorithm for numerical solution of a differential equation, but it usually gives the least accurate results. We therefore will later consider some more accurate algorithms.

*Example 3.26*

Compare the exact solution and the Euler method using a step size $\Delta t = 0.1$ for

$$\dot{y} = ty, \quad y(0) = 1$$

where $0 \leqslant t \leqslant 1$.

The Euler algorithm (3.13-8) for this model is

$$y(t_{k+1}) = (1 + 0.1t_k)y(t_k) \tag{3.13-9}$$

For this case the exact solution can be obtained by separation of variables and direct integration. It is

$$y(t) = y(0)\exp(t^2/2)$$

The following table compares the results to four significant figures.

| | $y(t_k)$ | |
|---|---|---|
| $t_k$ | Exact | Euler |
| 0.1 | 1.005 | 1.000 |
| 0.2 | 1.020 | 1.010 |
| 0.3 | 1.046 | 1.030 |
| 0.4 | 1.083 | 1.061 |
| 0.5 | 1.133 | 1.104 |
| 0.6 | 1.197 | 1.159 |
| 0.7 | 1.278 | 1.228 |
| 0.8 | 1.377 | 1.314 |
| 0.9 | 1.499 | 1.419 |
| 1.0 | 1.649 | 1.547 |

The error in the Euler method increases with time. It is in error by 0.5% in the first step; by the tenth step, the error is 6.19%. The accuracy can be improved by using a smaller value of $\Delta t$. The step size $\Delta t = 0.1$ was chosen arbitrarily here for illustration of the mechanics of the method.

As with this example, the remaining chapter examples will present the results of several iterations so the student can test the methods. A hand-held calculator is usually sufficient for this purpose.

## Selecting a Step Size

Thus far we have not considered the effect of the "run time" of a numerical method. This is the time required for the software to perform the calculations, and it is frequently an important factor in selecting an algorithm. It is difficult to make a prior determination of the run time for a given algorithm. The complexity and length of a program are indicators. Also, obviously the smaller $\Delta t$ is, the more iterations will be required and the longer will be the total computer time required. The assembly and compilation times must also be considered in addition to run time on computers that utilize assemblers or compilers.

The process of selecting a proper step size $\Delta t$ begins with a consideration of the dynamics of the model and how fast the input changes with time. A trial value of $\Delta t$ is then selected that is small in relation to the above considerations. The trial solution for the response is obtained with this value. The step size is then reduced significantly (say by a factor of two) and the two responses are compared. If they agree to the number of significant figures specified by the analyst, the correct answer is taken to be the last solution. If not, $\Delta t$ is again reduced by a significant factor, and the process is repeated until the solution converges. Note that two values of $\Delta t$ that are close to each other will give similar solutions. Thus the method fails unless the successive $\Delta t$ values differ significantly.

In strictly analytical terms, as $\Delta t$ approaches zero, the solution of the difference equation approaches that of the corresponding differential equation. This is the

TABLE 3.3 Illustration of Round-off Error Effects

Model $\frac{dy}{dt} = -y, \quad y(0) = 1$

Solution $y(t) = e^{-t}$

| | Euler* | | True Solution* |
|---|---|---|---|
| | $\Delta t = 0.1$ | $\Delta t = 0.01$ | |
| $t$ | $y$ | $y$ | $y$ |
| 0 | 1 | 1 | 1 |
| 0.1 | 0.90 | 0.90 | 0.90 |
| 0.2 | 0.81 | 0.80 | 0.82 |
| 0.3 | 0.73 | 0.70 | 0.74 |
| 0.4 | 0.66 | 0.59 | 0.67 |
| 0.5 | 0.59 | 0.50 | 0.61 |

* Euler results were computed using only two significant figures. True solution has been rounded off to two figures.

reason for using a small $\Delta t$. The error produced when $\Delta t$ is not small enough to represent the differential equation accurately is termed *truncation error.* However, numerical considerations as well as computer time require that $\Delta t$ not be made too small. The reason is that the effect of *round-off error* is greater for smaller $\Delta t$ due to the larger number of iterations required. The round-off error is that error resulting from the inability of any digital device to represent a number by more than a finite number of binary bits or significant figures. For example, suppose we have a hypothetical device that is capable of representing numbers only to two significant figures. If we solve the equation

$$\frac{dy}{dt} = -y, \; y(0) = 1$$

with the Euler method, the algorithm is

$$y(k+1) = (1 - \Delta t)y(k)$$

The time constant $\tau$ is unity, and so we might try $\Delta t = 0.1$. The results are shown in Table 3.3 along with the true solution for comparison. Now suppose we use $\Delta t = 0.01$ to improve the accuracy. The displayed results indicate that instead of decreasing, the error has increased drastically due to the large number of iterations required relative to the number of significant figures carried by the machine.

For large $\Delta t$, the truncation error dominates, while for small $\Delta t$, the round-off error is most important. Our suggested method for selecting $\Delta t$ usually results in starting above the optimum $\Delta t$ and moving down. On most large machines in use today, the number of available significant figures is great enough that round-off error is usually not a problem unless one is solving high-order equations over very long times.

## 3.14 ADVANCED NUMERICAL METHODS

In light of the large variety of equation types that are nonlinear, it is no wonder that many different numerical methods exist for solving them. Some methods work well only for special classes of problems. Here we consider two methods that are generally useful. The first is a so-called predictor-corrector method based on the Euler method but with greater accuracy. It is easy to program and thus is suitable for calculators. The second method is the Runge-Kutta family of algorithms. Of the more advanced techniques, these are perhaps the most widely used in engineering applications. Runge-Kutta algorithms are widely available at computer installations and in software packages for more advanced calculators.

### Trapezoidal Integration

Some useful results can be obtained from the trapezoidal rule for integration. Consider the equation

$$\frac{dy}{dt} = v \tag{3.14-1}$$

The variable $y$ is the integral of $v$, or

$$y(t_{k+1}) = y(t_k) + \int_{t_k}^{t_{k+1}} v(\lambda)\,d\lambda \tag{3.14-2}$$

The trapezoidal formula is

$$\int_{t_k}^{t_{k+1}} v(\lambda)\,d\lambda = \frac{\Delta t}{2}\left[v(t_{k+1}) + v(t_k)\right] \tag{3.14-3}$$

where $\Delta t = t_{k+1} - t_k$. Substituting this into (3.14-2) and omitting the subscript notation gives

$$y(k+1) = y(k) + \frac{\Delta t}{2}\left[v(k+1) + v(k)\right] \tag{3.14-4}$$

where $y(k)$ stands for $y(t_k)$, etc. Equation (3.14-4) is the recursive algorithm that implements the integration in (3.14-2) via the trapezoidal formula and therefore solves (3.14-1) numerically.

### Predictor-Corrector Methods

The Euler method can have a serious deficiency in problems where the variables are rapidly changing because the method assumes the variables are constant over the time interval $\Delta t$. As a result, its truncation error can be shown to be of the order of $(\Delta t)^2$ (see Reference 5).

One way to improve the method would be to use a better approximation to the right-hand side of the model

$$\frac{dy}{dt} = f(y, v, t) \tag{3.14-5}$$

where $v$ is the input function. The Euler approximation is

$$y(t_{k+1}) = y(t_k) + \Delta t\, f[y(t_k), v(t_k), t_k] \tag{3.14-6}$$

Suppose instead we use the average of the right-hand side of (3.14-5) on the interval $(t_k, t_{k+1})$. This gives

$$y(t_{k+1}) = y(t_k) + \frac{\Delta t}{2}(f_k + f_{k+1}) \tag{3.14-7}$$

where

$$f_k = f[y(t_k), v(t_k), t_k] \tag{3.14-8}$$

with a similar definition for $f_{k+1}$. Equation (3.14-7) is equivalent to integrating (3.14-5) with the trapezoidal rule.

The difficulty with (3.14-7) is that $f_{k+1}$ cannot be evaluated until $y(t_{k+1})$ is known, but this is precisely the quantity being sought. A way out of this difficulty is to use the Euler formula (3.14-6) to obtain a preliminary estimate of $y(t_{k+1})$. This estimate is then used to compute $f_{k+1}$ for use in (3.14-7) to obtain the required estimate of $y(t_{k+1})$.

The notation can be changed to clarify the method. Let $x_{k+1}$ be the estimate of $y(t_{k+1})$ obtained from the Euler formula (3.14-6). Then, on omitting the $t_k$ notation from the other equations, we obtain the following description of the *predictor-corrector* process.

Euler predictor. $$x_{k+1} = y_k + \Delta t\, f(y_k, v_k, t_k) \tag{3.14-9}$$

Trapezoidal corrector. $$y_{k+1} = y_k + \frac{\Delta t}{2}[f(y_k, v_k, t_k) + f(x_{k+1}, v_{k+1}, t_{k+1})] \tag{3.14-10}$$

This algorithm is sometimes called the *modified Euler method.* However, note that any algorithm can be tried as a predictor or a corrector. Thus many methods can be classified as predictor-corrector, but we will limit our treatment to the modified Euler method. The truncation error for this method is of the order of $(\Delta t)^3$, a significant improvement over the $(\Delta t)^2$ error for the basic Euler method.

For purposes of comparison with the Runge-Kutta methods to follow, we can express the modified Euler method as

$$g_1 = \Delta t\, f(y_k, v_k, t_k) \tag{3.14-11}$$

$$g_2 = \Delta t\, f(y_k + g_1, v_{k+1}, t_k + \Delta t) \tag{3.14-12}$$

$$y_{k+1} = y_k + \tfrac{1}{2}(g_1 + g_2) \tag{3.14-13}$$

*Example 3.27*

Use the modified Euler method with $\Delta t = 0.1$ to solve the equation

$$\frac{dy}{dt} = ty, \quad y(0) = 1$$

up to time $t = 1.0$.

Here the equations for the algorithm become

$$x_{k+1} = y_k + \Delta t \, t_k \, y_k$$

$$y_{k+1} = y_k + \frac{\Delta t}{2}(t_k \, y_k + t_{k+1} x_{k+1})$$

On the first step, $t_0 = 0$, $k = 0$, $y_0 = 1$, and

$$x_1 = 1 + 0.1(0) = 1$$

$$y_1 = 1 + 0.05(0 + 0.1) = 1.005$$

On the second step, $t_1 = 0.1$, $k = 1$ and

$$x_2 = 1.005 + 0.1[0.1(1.005)] = 1.01505$$

$$y_2 = 1.005 + 0.05[0.1(1.005) + 0.2(1.01505)]$$

$$= 1.0201755$$

The results are shown in the following table rounded to six figures along with the analytical solution (see Example 3.26). It can be seen that the modified Euler method gives very accurate results for this problem. Its performance here is superior to the basic Euler method.

| | $y_k$ | |
|---|---|---|
| $t_k$ | **Exact** | **Modified Euler** |
| 0 | 1 | 1 |
| 0.1 | 1.00501 | 1.00500 |
| 0.2 | 1.02020 | 1.02018 |
| 0.3 | 1.04603 | 1.04599 |
| 0.4 | 1.08329 | 1.08322 |
| 0.5 | 1.13315 | 1.13305 |
| 0.6 | 1.19722 | 1.19707 |
| 0.7 | 1.27762 | 1.27739 |
| 0.8 | 1.37713 | 1.37677 |
| 0.9 | 1.49930 | 1.49876 |
| 1.0 | 1.64872 | 1.64788 |

Corrector equation (3.14-10) can also be used more than once on each $\Delta t$ interval in order to increase the accuracy. For example, the refined estimate $y_{k+1}$ from (3.14-10) can be used to replace $x_{k+1}$ in the right-hand side of that equation in order to get an improved estimate of $y_{k+1}$. This is repeated until the desired accuracy is achieved; then the next $\Delta t$ interval is treated the same way. However, this modification obviously increases the complexity of the programming.

## Runge-Kutta Methods

The Taylor series representation forms the basis of several methods for solving differential equations, including the Runge-Kutta methods. The Taylor series may be used to represent the solution $y(t + \Delta t)$ in terms of $y(t)$ and its derivatives as follows.

$$y(t + \Delta t) = y(t) + \Delta t\,\dot{y}(t) + \tfrac{1}{2}(\Delta t)^2 \ddot{y}(t) + \tfrac{1}{6}(\Delta t)^3 \dddot{y}(t) \ldots \qquad (3.14\text{-}14)$$

The required derivatives are calculated from the differential equation. For an equation of the form

$$\frac{dy}{dt} = f(t, y) \qquad (3.14\text{-}15)$$

these derivatives are

$$\dot{y}(t) = f(t, y)$$

$$\ddot{y}(t) = \frac{df}{dt}$$

$$\dddot{y}(t) = \frac{d^2 f}{dt^2}, \text{ etc.} \qquad (3.14\text{-}16)$$

where (3.14-15) is to be used to express the derivatives in terms of only $y(t)$. If these derivatives can be found, (3.14-14) can be used to march forward in time. In practice, the higher-order derivatives can be difficult to calculate, and the series (3.14-14) is truncated at some term. The number of terms kept in the series thus determines its accuracy. If terms up to and including the $n$th derivative of $y$ are retained, the truncation error at each step is of the order of the first term dropped – namely,

$$\frac{(\Delta t)^{n+1}}{(n + 1)!} \frac{d^{n+1} y}{dt^{n+1}} \qquad (3.14\text{-}17)$$

Sometimes the closed-form expression for the series can be recognized. For example, consider the equation

$$\dot{y} = ay^2, \; a < 0 \qquad (3.14\text{-}18)$$

The derivatives required for the series are

$$\ddot{y} = 2ay\dot{y} = 2a^2 y^3$$

$$\dddot{y} = 6a^2 y^2 \dot{y} = 6a^3 y^4$$

$$\vdots$$

$$\frac{d^n y}{dt^n} = (n + 1)a^n y^{n+1}$$

Substituting these into (3.14-14) gives

$$y(t + \Delta t) = y(t) + a\,\Delta t\,y^2(t) + (a\,\Delta t)^2 y^3(t) + (a\,\Delta t)^3 y^4(t) + \ldots + (a\,\Delta t)^n y^{n+1}(t) + \ldots$$

The closed-form expression can be recognized by multiplying by $a\,\Delta t$ to obtain

$$a\,\Delta t\, y(t+\Delta t) = \sum_{i=1}^{\infty} [a\,\Delta t\, y(t)]^i$$

$$= \frac{a\,\Delta t\, y(t)}{1-a\,\Delta t\, y(t)}$$

Thus

$$y(t+\Delta t) = \frac{y(t)}{1-a\,\Delta t\, y(t)} \tag{3.14-19}$$

Since we have not truncated the series, no approximation errors have been introduced, and (3.14-19) is exact for any $\Delta t$.

While few problems can be solved exactly with the Taylor series, the preceding example serves to illustrate the power of the method. Its chief difficulty is the need to compute the higher-order derivatives. For this reason, the Runge-Kutta methods were developed. These methods use several evaluations of the function $f(t, y)$ in a way that approximates the Taylor series. The number of terms in the series that is duplicated determines the order of the Runge-Kutta method. Thus a fourth-order Runge-Kutta algorithm duplicates the Taylor series through the term involving $(\Delta t)^4$.

## Development of the Second-Order Algorithm

We now demonstrate the development of the Runge-Kutta algorithms for the second-order case. For notational compactness, we replace the step size $\Delta t$ with $h$. Thus, $t_{k+1} = t_k + h$. Also, let the subscript $k$ denote evaluation at time $t_k$. Truncation of (3.14-14) beyond the second-order term and use of (3.14-5) gives

$$y_{k+1} = y_k + h\dot{y}_k + \tfrac{1}{2}h^2\ddot{y}_k$$

$$= y_k + hf_k + \tfrac{1}{2}h^2\left(\frac{df}{dt}\right)_k$$

But

$$\frac{df}{dt} = \frac{\partial f}{\partial t} + \frac{\partial f}{\partial y}\frac{dy}{dt} = \frac{\partial f}{\partial t} + \frac{\partial f}{\partial y} f$$

In simpler notation, $f_y$ denotes $\partial f/\partial y$, and $f_t$ denotes $\partial f/\partial t$. Thus

$$\frac{df}{dt} = f_t + f_y f$$

and

$$y_{k+1} = y_k + hf_k + \tfrac{1}{2}h^2(f_t + f_y f)_k \tag{3.14-20}$$

The second-order Runge-Kutta methods express $y_{k+1}$ as

$$y_{k+1} = y_k + w_1 g_1 + w_2 g_2 \tag{3.14-21}$$

where $w_1$ and $w_2$ are constant weighting factors, and

$$g_1 = hf(t_k, y_k) \tag{3.14-22}$$

$$g_2 = hf(t_k + \alpha h, y_k + \beta h f_k) \tag{3.14-23}$$

for constant $\alpha$ and $\beta$. We now compare (3.14-21) with (3.14-20). First expand $f$ in $g_2$ in a two-variable Taylor series.

$$g_2 = h[f_k + (f_t)_k \alpha h + (f_y)_k \beta h f_k + \ldots] \tag{3.14-24}$$

where the omitted terms are of the order of $h^2$. Substitution of this and $g_1$ into (3.14-21) and collection of terms give

$$y_{k+1} = y_k + h[w_1 f + w_2 f]_k + h^2 [w_1 \alpha f_t + w_2 \beta f f_y]_k$$

Comparison with (3.14-20) for like powers of $h$ shows that

$$w_1 + w_2 = 1 \tag{3.14-25}$$

$$w_1 \alpha = \tfrac{1}{2} \tag{3.14-26}$$

$$w_2 \beta = \tfrac{1}{2} \tag{3.14-27}$$

Thus the family of second-order Runge-Kutta algorithms is categorized by the parameters $(\alpha, \beta, w_1, w_2)$, one of which can be chosen independently. The choice $\alpha = 2/3$ minimizes the truncation error term (Reference 5). For $\alpha = 1$, the Runge-Kutta algorithm (3.14-21) through (3.14-23) corresponds to the trapezoidal integration rule if $f$ is a function only of $t$, and is the same as the predictor-corrector algorithm (3.14.9)–(3.14-10) for a general $f(y, t)$.

## Fourth-Order Algorithms

Second- and third-order Runge-Kutta algorithms have found extensive applications in the past. However, the fourth-order algorithm is now the most commonly used. Its derivation follows that same pattern as earlier, with all terms up to $h^4$ retained. The algebra required is extensive, and is given in Reference 6. Here we present the results. The algorithm is

$$y_{k+1} = y_k + w_1 g_1 + w_2 g_2 + w_3 g_3 + w_4 g_4 \tag{3.14-28}$$

$$\begin{aligned} g_1 &= hf(t_k, y_k) \\ g_2 &= hf(t_k + \alpha_1 h, y_k + \alpha_1 g_1) \\ g_3 &= hf[t_k + \alpha_2 h, y_k + \beta_2 g_2 + (\alpha_2 - \beta_2) g_1] \\ g_4 &= hf[t_k + \alpha_3 h, y_k + \beta_3 g_2 + \gamma_3 g_3 + (\alpha_3 - \beta_3 - \gamma_3) g_1] \end{aligned} \tag{3.14-29}$$

Comparison with the Taylor series yields eight equations for the ten parameters. Thus two parameters can be chosen in light of other considerations. Three common choices are as follows.

1. *Gill's Method.** This choice minimizes the number of memory locations required to implement the algorithm and thus is well suited for programmable calculators.

---

* Appendix C contains a FORTRAN subroutine that implements Gill's version of the fourth-order Runge-Kutta algorithm.

$$
\begin{aligned}
w_1 &= w_4 = 1/6 \\
w_2 &= (1 - 1/\sqrt{2})/3 \\
w_3 &= (1 + 1/\sqrt{2})/3 \\
\alpha_1 &= \alpha_2 = 1/2, \quad \alpha_3 = 1 \\
\beta_2 &= 1 - 1/\sqrt{2}, \quad \beta_3 = -1/\sqrt{2} \\
\gamma_3 &= 1 + 1/\sqrt{2}
\end{aligned} \tag{3.14-30}
$$

2. *Ralston's Method.* This choice minimizes a bound on the truncation error.

$$
\begin{aligned}
w_1 &= 0.17476028, \quad w_2 = -0.55148066 \\
w_3 &= 1.20553560, \quad w_4 = 0.17118478 \\
\alpha_1 &= 0.4, \quad \alpha_2 = 0.45573725 \\
\alpha_3 &= 1, \quad \gamma_3 = 3.83286476 \\
\beta_2 &= 0.15875964, \quad \beta_3 = -3.05096516
\end{aligned} \tag{3.14-31}
$$

3. *Classical Method.* This method reduces to Simpson's rule for integration if $f$ is a function only of $t$. Because the coefficients are computed from whole numbers, this method also requires less computer storage than the Ralston method. It is commercially available for programmable calculators.

$$
\begin{aligned}
w_1 &= w_4 = 1/6 \\
w_2 &= w_3 = 1/3 \\
\alpha_1 &= \alpha_2 = 1/2 \\
\beta_2 &= 1/2, \quad \beta_3 = 0 \\
\gamma_3 &= \alpha_3 = 1
\end{aligned} \tag{3.14-32}
$$

*Example 3.28*

Use the classical Runge-Kutta parameter values (3.14-32) with a step size $h = \Delta t = 0.1$ to solve

$$\dot{y} = ty, \quad y(0) = 1$$

up to $t = 1.0$.

The algorithm for the classical parameter values is

$$
\begin{aligned}
y_{k+1} &= y_k + \tfrac{1}{6}g_1 + \tfrac{1}{3}g_2 + \tfrac{1}{3}g_3 + \tfrac{1}{6}g_4 \\
g_1 &= hf(t_k, y_k) \\
g_2 &= hf(t_k + \tfrac{1}{2}h, y_k + \tfrac{1}{2}g_1) \\
g_3 &= hf[t_k + \tfrac{1}{2}h, y_k + \tfrac{1}{2}g_2 + (\tfrac{1}{2} - \tfrac{1}{2})g_1] \\
g_4 &= hf[t_k + h, y_k + 0g_2 + g_3 + (1 - 0 - 1)g_1]
\end{aligned}
$$

For the given problem $f = ty$. For the first iteration, $k = 0$, $t_0 = 0$, $y_0 = 1$, and

$$g_1 = 0.1(0) = 0$$

$$g_2 = 0.1(0 + 0.05)(1 + \tfrac{1}{2}0) = 0.005$$

$$g_3 = 0.1(0 + 0.05)(1 + \tfrac{1}{2}0.005) = 0.0050125$$

$$g_4 = 0.1(0 + 0.1)(1 + 0.0050125) = 0.010050125$$

Thus

$$y(0.1) = y_1 = 1 + \tfrac{1}{6}0 + \tfrac{1}{3}(0.005) + \tfrac{1}{3}(0.0050125) + \tfrac{1}{6}(0.010050125)$$

$$= 1.005012521$$

This answer agrees with the exact solution $\exp(t^2/2)$ out to the number of significant figures available on most pocket calculators. The result demonstrates the power of the Runge-Kutta method. The rest of the solution is as follows.

| $t$ | $y(k)$ |
|---|---|
| 0.1 | 1.005012521 |
| 0.2 | 1.02020134 |
| 0.3 | 1.046027859 |
| 0.4 | 1.08387065 |
| 0.5 | 1.133148446 |
| 0.6 | 1.197217347 |
| 0.7 | 1.277621279 |
| 0.8 | 1.377127695 |
| 0.9 | 1.499302362 |
| 1.0 | 1.648721007 |

The exact solution at $t = 1.0$, to ten figures, is 1.648721271. The numerical solution is correct to seven figures.

Numerical methods must be used carefully when the true solution changes rapidly. This can occur when either the free response or the forcing function is oscillatory. Another instance occurs when the time span for significant changes to occur in the free response is different from the time span for the forcing function. An example of this effect is given by the equation

$$\dot{y} = -y + 0.001e^{10t}, \; y(0) = 10 \tag{3.14-33}$$

Since the solution can be found with the analytical methods of this chapter, a numerical method is not needed. The problem is illustrative only. Its solution is

$$y(t) = 10e^{-t} + \frac{0.001}{11}(e^{10t} - e^{-t}) \tag{3.14-34}$$

The time constant is $\tau = 1$, and thus the free response decreases by 63% in one time unit. On the other hand, the forcing function $\exp(10t)$ increases by 22,026% in the same time. The existence of more than one time scale in the problem means that the step size $\Delta t$ must be selected with care. Also, for $0 \leqslant t \leqslant 0.9$, the free response dominates and $y$ decreases. For $t > 0.9$, the forcing function dominates, and $y$ increases. For $t < 0.9$, a step size of one-tenth of the time constant might be sufficient, but would be too large to use for $t > 0.9$. One approach would be to decrease the step size for $t > 0.9$. However, the Runge-Kutta method can deal with this solution using a constant step size if it is chosen small enough.

*Example 3.29*

Solve (3.14-33) using $\Delta t = 0.01$.

The classical Runge-Kutta algorithm gives the following solution. The exact solution is shown for comparison. The table presents the results of every twentieth iteration.

| $t$ | Runge-Kutta | Exact |
|---|---|---|
| 0.2 | 8.187904833 | 8.187904833 |
| 0.4 | 6.708102991 | 6.708102991 |
| 0.6 | 5.524741816 | 5.524741818 |
| 0.8 | 4.764244988 | 4.764245001 |
| 1.0 | 5.681167053 | 5.68116715 |
| 1.2 | 17.80780563 | 17.80780635 |
| 1.4 | 111.793615 | 111.7936204 |
| 1.6 | 809.8472175 | 809.8472567 |
| 1.8 | 5970.741384 | 5970.741674 |

The characteristics of the Runge-Kutta algorithms can be summarized as follows.

1. The Runge-Kutta algorithms do not require calculation of the higher derivatives of $y(t)$, as is required by the Taylor method. Calculation of these derivatives can be quite tedious, especially for sets of coupled differential equations of high order. Instead, the Runge-Kutta algorithms utilize the computation of $f(t, y)$ at various points. These algorithms can be easily generalized to handle higher-order equations and equation sets, as we will see in Chapter Five. On the other hand, error propagation in the Runge-Kutta method is not as easy to keep track of as in the Taylor method. One is advised to watch the $g_i$ values. If they differ drastically from each other, reduce the step size.
2. The Runge-Kutta algorithms are self-starting, and the step size can be changed easily between iterations. While we have not seen any methods for which this is not true, there are more accurate and faster methods that suffer those disadvantages. For example, higher-order predictor-corrector methods are not self-starting, and a Runge-Kutta method is frequently used to provide a starting solution for the predictor-corrector (Reference 6).
3. For some numerical methods, the difference equations used to approximate the original differential equation have more than one solution. One of these

solutions is the true one, while the others result from round-off errors. If these extraneous solutions grow with time (thus swamping the true solution), the method is said to be unstable. The Runge-Kutta methods have good stability characteristics in this respect.

4. The fourth-order Runge-Kutta method is probably the most widely used method for engineering applications. Its software is available for advanced calculators and most computers. Faster, more accurate methods are available, but these are most difficult to program (Reference 6). If great efficiency is required, such as with production codes to be run many times, one might consider a more advanced method. However, the Runge-Kutta method is suitable for most problems that involve a small number of runs.

## 3.15 SUMMARY

This chapter deals with the free and forced response of first-order continuous-time models. Several solution techniques are presented, and their relative merits depend on the nature of the input. The reader should become familiar with these techniques by means of the problems and should stress understanding of the nature of the model's behavior. In particular, note the role of the characteristic root (and the time constant for stable models) in determining the model's behavior for different input types. The following summary of the most important concepts in each section should help in this respect.

(3.1) The free response of the linear model given by (3.2-1) ($\dot{y} = ry + bv$) is the solution for a zero input and can be obtained by substituting an assumed exponential time function into the equation. The two unknown constants are found from the resulting characteristic (algebraic) equation and from the initial condition. The characteristic root $s = r$ determines the stability of the solution: $r > 0$ (unstable), $r = 0$ (neutrally stable), and $r < 0$ (stable). For the stable case, the time constant is $\tau = -1/r$ and is a measure of the rate of exponential decay of the solution. At $t = t_0 + \tau$, the solution is 37% of the initial value; at $t = t_0 + 4\tau$, it is 2% of the initial value.

(3.2) The response of $\dot{y} = ry + bv$ to a step input can be obtained by an assumed exponential plus constant solution form. The time constant is a measure of the time required to reach the final output value (63% after one time constant; 98% after four time constants). The superposition principle says that the total solution is the sum of the free response (dependent only on the initial condition) and the forced response (dependent only on the input). The solution can also be decomposed into a transient part (which disappears with time), and a steady-state part (which remains).

(3.3)–(3.7) Operator notation is introduced to provide algebraic representation of system models. The Laplace transform operator provides a general algebraic solution method for linear constant-coefficient models. A simple partial fraction expansion technique enables a systematic table lookup procedure to be used to recover the time-domain solution. The system's transfer function is the ratio of the output transform to the input transform, for zero initial conditions. Block diagram representations of the system are possible with the use of transfer function blocks for each subsystem. In addition to providing visualization of the system's dynamics, these diagrams can be reduced algebraically to find the overall system transfer function and differential equation. Thus the effect of new system components on the overall behavior is easily determined. An alternate representation, the signal flow graph, was presented and has some advantages when dealing with systems composed of many elements. The transfer function between a rate variable as input and an effort variable as output is an impedance, which may thus be thought of as a generalized resistance. It is useful in determining the power loading effects of connected elements and in combining dynamic elements in a model.

The impulse is an analytically convenient approximation of an input that is applied for only a very short time. If the input duration is less than one-tenth of the system's time constant, the impulse response may be used with considerable accuracy. The impulse response is equivalent to the free response with an adjusted initial condition. For longer-duration inputs, the pulse response is easily obtained with the shifting property of the Laplace transform.

The ramp response is a good description of a system's start-up behavior. The time constant is a measure of the difference between the input and output, as well as how long it takes for the steady-state lag to become established. The ramp response provides a good example of the use of the initial and final value theorems. Given the transfer function and the input transform, these theorems provide a quick indication of the initial and steady-state values of the output, under mild restrictions.

(3.8)–(3.9) The frequency response of a system describes its steady-state behavior resulting from periodic inputs. Later we will see that it is also useful in predicting transient behavior and stability properties. A sinusoidal input applied to a linear system produces a steady-state sinusoidal output of the same frequency, but with a different amplitude and a phase shift. The frequency transfer function is the transfer function with the Laplace variable $s$ replaced by $i\omega$, where $\omega$ is the input frequency. The magnitude $M$ of the frequency transfer function is the amplitude ratio between the input and output, and its argument is the phase shift $\phi$. Both are functions of $\omega$ and are usually plotted against $\log \omega$, with $M$ expressed in decibels. A polar plot is also useful sometimes. The logarithmic plots are easily sketched for the first-order system by using low- and high-frequency asymptotes, with the corner frequency given by $1/\tau$. The bandwidth is a measure of the filtering property of the system and is $1/\tau$ if no numerator dynamics are present. The Fourier series representation of a periodic function can be used with the system's filtering property to compute the general periodic response in terms of the sinusoidal response.

**(3.10)** Responses not falling into the preceding categories can be handled with the convolution integral and the transition function, as can variable-coefficient linear models. If the required integrals cannot be evaluated in closed form, the methods introduced in Sections 3.13 and 3.14 can be used to evaluate them numerically.

(3.11)–(3.12) Most of the preceding does not apply to nonlinear models. The solution form of nonlinear models can change drastically with the initial conditions or with the magnitude of the input. If they cannot be solved in closed form, say by separation of variables or transformation, numerical methods or reduction to an approximate linear form must be used. Linearization by Taylor series produces an approximate model in terms of the deviations of the state variables and inputs from their reference values. It is useful for stability analysis, when an approximate closed-form solution is needed, and for design of feedback control systems (whose performance reinforces the linearization's accuracy).

**(3.13)** When the analytical methods of the previous sections cannot be applied, the analyst must resort to a numerical solution. The Euler method was introduced and applied to a linear model possessing a variable coefficient. Selection of an adequate step size was analyzed in light of the opposing effects caused by round-off and truncation error.

**(3.14)** The wide variety of nonlinear models means that no one numerical method will be equally effective for all types of equations. We introduced the two most commonly used types: the predictor-corrector and the Runge-Kutta methods. They are sufficient for most applications of interest to us.

In succeeding chapters we will see that the techniques introduced here can be readily applied to higher-order models. More than one characteristic root will be involved, but the principle of superposition will allow us to treat each root separately (as an equivalent first-order system). The one exception occurs when complex roots exist; however, the techniques for these are similar. This is covered in Chapter Five. The techniques of this chapter are also similar to those needed for discrete-time models, the subject of the next chapter. It can probably be said that the chapter you have just finished is the most important of all.

## REFERENCES


1. Electrocraft Corporation, *DC Motors, Speed Controls, Servo Systems*, 5th edition, Hopkins, Minn., 1980.
2. R. J. Smith, *Circuits, Systems and Devices*, 3d edition, John Wiley, New York, 1976.
3. H. T. Davis, *Introduction to Nonlinear Differential and Integral Equations*, Dover, New York, 1962.
4. E. Kreysig, *Advanced Engineering Mathematics*, 3d edition, John Wiley, New York, 1972.
5. A. Ralston, *A First Course in Numerical Analysis*, McGraw-Hill, New York, 1965.
6. R. W. Hornbeck, *Numerical Methods*, Quantum Publishers, New York, 1975.


## PROBLEMS

**3.1** Given a set of measurements $x(t_i)$, $t_i$, we can select the two parameters $a$ and $b$ to obtain a straight line $x(t) = at + b$ that best fits the data according to the following least-squares criterion. We choose $a$ and $b$ to minimize $J$, the sum of the squares of the differences between the predicted and the measured values; that is,

$$J = \sum_{i=1}^{n} [at_i + b - x(t_i)]^2$$

where $n$ is the number of measurements. The criterion $J$ obviously can never be negative, and its minimum occurs when $\partial J/\partial a = \partial J/\partial b = 0$.

(a) Obtain the expressions for the values of $a$ and $b$ that minimize $J$.

(b) Find the straight line $x(t) = at + b$ that best fits the following data in the least-squares sense.

| $t_i$ | $x(t_i)$ |
|---|---|
| 0 | 0.1 |
| 1 | 1.8 |
| 2 | 3.9 |
| 3 | 6.1 |

3.2 (a) Apply the results of part a of Problem 3.1 to find the parameters $r$ and $y(0)$ that give the best fit for the free response $y(t) = y(0)e^{rt}$ of a first-order linear model.

(b) The liquid height $y(t)$ in a certain tank was measured at several times with no inflow applied. The data are given in the following table. For a linear outlet resistance the model is $\dot{y} = -y/RC$. Assume the measurements have some error, and estimate the value of the resistance $R$ with a least-squares approach. The tank area $C$ is 3 ft$^2$.

(c) Now suppose that $y(0)$ is known to be exactly 10.1 feet. How does this change the results of part (b)?

| $t_i$(sec) | $y(t_i)$ (ft) |
|---|---|
| 0 | 10.1 |
| 500 | 7.1 |
| 1000 | 5.0 |
| 1500 | 3.8 |
| 2000 | 2.6 |

3.3 A tank has the model $RC\dot{h} = -h + Rq_i$, with $C = 3$ ft$^2$ and $R = 700$ sec/ft$^2$. Suppose that the height $h$ is initially 5 ft when the inflow rate $q_i$ is suddenly turned on at a constant value of 0.05 ft$^3$/sec. Find $h(t)$.

3.4 (a) A motor torque increases with time until it reaches a constant value after 0.1 sec. When this torque is applied to a load with inertia $I = 0.007$ lb-ft-sec$^2$ and damping $c = 0.001$ lb-ft-sec/rad, is a step function an adequate representation of the torque input? Explain.

(b) If at $t = 0.2$ the torque is suddenly removed, can it be modeled as an impulse? Explain.

3.5 Use the Laplace transform method to obtain the forced response of the system $\dot{y} = -2y + v(t)$ for $t \geqslant 0$, for the following cases.

(a) $v(t) = t$

(b) $v(t) = t^2$

(c) $v(t) = te^{-t}$

(d) $$v(t) = \begin{cases} 4t, & 0 \leqslant t \leqslant 2 \\ -4t + 16, & 2 \leqslant t \leqslant 4 \\ 0, & t \geqslant 4 \end{cases}$$

(*Hint.* Use the ramp response and the shifting theorem.)

(e) $v(t) = t + 3$

3.6 A load inertia $I_2$ is driven through gears by a motor with inertia $I_1$. The gear ratio is 5:1 (the motor shaft has the greater speed). The gear and shaft inertias have been lumped into $I_1$ and $I_2$, respectively. The motor torque $T_1$ is 10 lb-ft, and the load torque due to viscous friction is $T_2 = 1.2\omega$ with $\omega$ in rad/sec (see Figure P3.6).

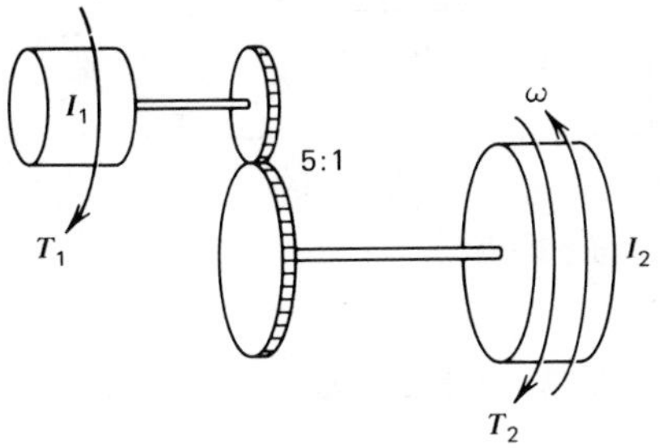


**Figure P3.6**

$$I_1 = 0.01, \quad I_2 = 0.5 \text{ sec}^2\text{-ft-lb/rad}$$

(a) What is the steady-state speed of the load $I_2$?

(b) Find the transfer function between $\omega$ and $T_1$.

(c) If $T_1$ is suddenly applied, approximately how long will it take for the speed to reach its steady-state value?

3.7 (a) The forced response of a linear system is the sum of the responses resulting from each input. Use this fact to reduce the diagram shown in Figure P3.7 in order to find the transfer functions relating $y$ to $v_1$ and $y$ to $v_2$.

(b) Find the differential equation model of the system.

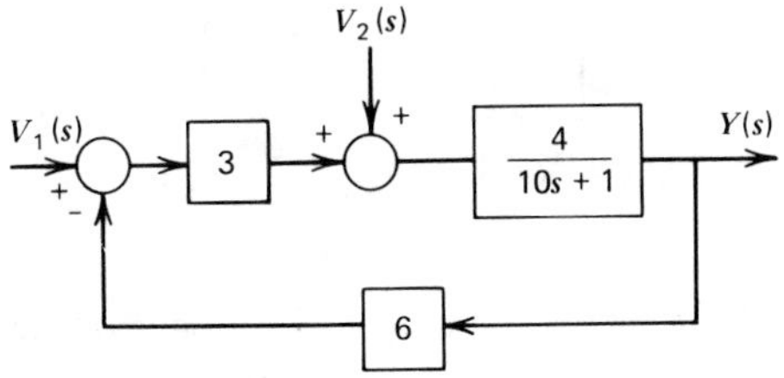


**Figure P3.7**

3.8 Construct the block diagrams and signal flow graphs of the models with numerator dynamics (3.4-8) and (3.4-9).

3.9 Use the impedance method to find the transfer functions for the circuits shown in Figure P3.9 for the indicated inputs and outputs.

(a) The input is $e$; the output is $i$.

(b) The input is $e_i$; the output is $e_o$. (This is a *lag* compensator.)

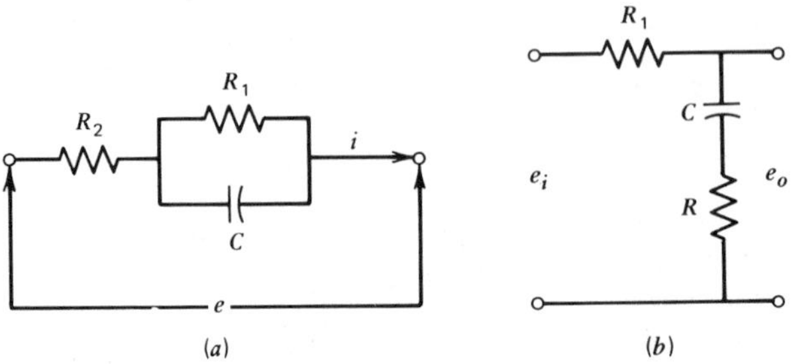


**Figure P3.9**

3.10 A feedback system with dead time is shown in Figure P3.10.

(a) Find the transfer function.

(b) The characteristic equation for this system has an infinite number of roots. Show this by writing the root as a complex number $s = a + ib$ and obtain the equations that must be solved to find $a$ and $b$.

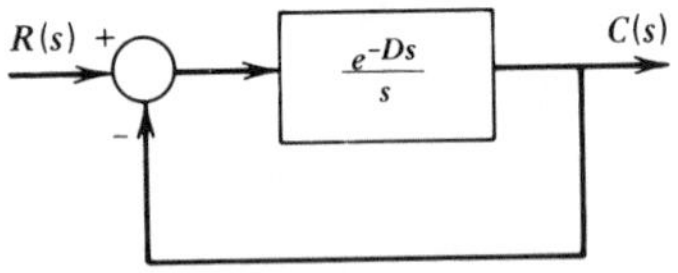


**Figure P3.10**

(c) For $D = 2$, set $b = 0$, and use Newton iteration to find the only real characteristic root (see Appendix A for the Newton iteration procedure for finding the roots of an equation). Estimate the system's time constant.

3.11 Use the final value theorem to compute the steady-state error $e$ between the the input and the output for the system shown in Figure P3.11, with the input functions as follows.

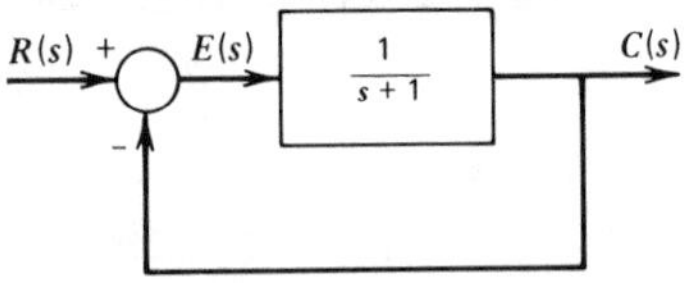


**Figure P3.11**

(a) $R(s) = 1/s$

(b) $R(s) = 1/s^2$

3.12 A thermometer is immersed in a liquid of temperature $v(t)$ °F. The temperature of the thermometer fluid is $y(t)$ °F. Time $t$ is in seconds. The thermal time constant of the thermometer wall is 100 sec, so that

$$100\dot{y} + y = v(t)$$

Suppose that $v(t)$ varies sinusoidally with an amplitude of 10°F and a frequency of 0.5 cycles per minute.

(a) Find the steady-state amplitude of $y$.

(b) Find the time lag in seconds between the occurrence of a peak in $v$ and a peak in $y$, at steady state.

**3.13** Given the model

$$0.1\dot{y} + y = v(t)$$

with the following Fourier series representation of the input

$$v(t) = 0.5 \sin 4t + 2 \sin 8t + 0.02 \sin 12t + 0.03 \sin 16t + \ldots$$

(a) What is the bandwidth of the system?

(b) Find an approximate description of the output $y(t)$ at steady state, using only those input components that lie within the bandwidth.

**3.14** A torque $T$ is applied to a load of inertia $I$. The damping is negligible so that $Is\Omega(s) = T(s)$, where $\omega$ is the speed of the load. For a sinusoidal torque $T(t) = A \sin \omega_f t$, plot the frequency response curves with $A/I = 2$ for a two-decade range centered at $\omega_f = 1$ rad/unit time.

**3.15** The following transfer functions result for the lead and lag compensator circuits when $R_1 = 2\ k\Omega$, $R = 1\ k\Omega$, and $C = 1\ \mu f$ (see Figures 3.24 and P3.9*b*). Construct the frequency response plots.

(a) $T(s) = \dfrac{1}{3}\dfrac{1 + 0.02s}{1 + 0.00667s}$

(b) $T(s) = \dfrac{1 + 0.01s}{1 + 0.03s}$

**3.16** Apply the initial value theorem to the two models with numerator dynamics, (3.4-8) and (3.4-9), for a unit step input and zero initial conditions. Explain the results.

**3.17** Repeat Problem 3.5 using the convolution integral (3.10-3) for zero initial conditions.

**3.18** The convolution integral (3.10-3) derived in Section 3.10 can also be derived from the convolution theorem of the Laplace transform (Appendix B). The theorem states that if $\mathscr{L}[v(t)] = V(s)$, $\mathscr{L}[g(t)] = G(s)$, and $\mathscr{L}[y(t)] = Y(s)$, then

$$y(t) = \mathscr{L}^{-1}[Y(s)] = \mathscr{L}^{-1}[G(s)V(s)] = \int_0^t v(\lambda)g(t-\lambda)\,d\lambda$$

$$= \int_0^t v(t-\lambda)g(\lambda)\,d\lambda$$

(a) Let $G(s)$ be the transfer function of the system $\dot{y} = ry + bv$. Then $g(t)$ is its impulse response. Use the convolution theorem to obtain (3.10-3).

(b) Calculation of inverse transforms is sometimes simplified with the convolution theorem if we can separate the transform into terms whose inverse transforms are known. Use this approach to find $\mathscr{L}^{-1}[1/(s+1)^2]$.

**3.19** The following model form results from a dilution problem like that developed in Problem 2.3. The variable $y$ represents the amount of salt in a tank of brine, as a function of time.

$$\dot{y} + \frac{2}{10+2t}y = 4$$

(a) Find the transition function.

(b) Find the response if $y(0) = 0$.

**3.20** The method of *variation of parameters* is useful for finding the response of linear models. Let the free response of a first-order model be written as $y(t) = cy_1(t)$, where $c$ is the constant determined by the initial condition. We attempt a solution for the forced response of the form

$$y(t) = f(t)y_1(t)$$

where $f(t)$ is found by substitution into the original differential equation.

(a) Find the free response of the model

$$\dot{y} + 2ty = t$$

(b) Use variation of parameters to find the complete response in terms of an arbitrary value of $y(0)$.

**3.21** The *Gompertz* growth model is

$$\dot{y} = ry\ln\frac{K}{y}$$

(a) Obtain its closed-form solution for $r, K > 0$.

(b) Compare its behavior with those of the exponential growth and logistic growth models (2.9-1) and (2.9-3).

**3.22** The following is a model of population size $y$ in which the growth is seasonally dependent. The mean growth rate is $R$, and the actual growth rate varies about $R$ with an amplitude $A$.

$$\dot{y} = [R + A \sin(2\pi t)]y, \quad t \text{ in years}$$

Solve this equation in closed form by

(a) Direct integration.

(b) Using the transformation $x = \ln y$.

3.23 The *Pella-Tomlinson* growth model is a generalization of the logistic growth equation. It is

$$\dot{y} = ry \left[1 - \left(\frac{y}{K}\right)^n\right]$$

where $r$, $K$, and $n$ are constants and $r, K > 0$. Solve this equation by

(a) Direct integration.

(b) Reduction to linear form with the transformation $z = y^{-m}$

3.24 Find the equilibria of the following model and obtain a linearized model for each of the first two nonnegative equilibria. Determine their stability properties.

$$\dot{y} = \sin y$$

3.25 Generalizations of the Schaefer fishery model treated in Section 2.9 are obtained by including the harvesting term $qfy$ with a growth model. Using the Pella-Tomlinson and the Gompertz models, we obtain

(1) $$\dot{y} = ry \left[1 - \left(\frac{y}{K}\right)^n\right] - qfy$$

(2) $$\dot{y} = ry \ln\left(\frac{K}{y}\right) - qfy$$

where $q$ is a constant and $f(t)$ is the fishing effort. In general it is not possible to solve these models in closed form unless the function $f(t)$ is very simple.

(a) Obtain the constant fishing effort $f_e$ that maximizes the equilibrium yield $Y_e = qf_e y_e$ for the models (1) and (2).

(b) Obtain a linearized dynamic model valid near the optimal equilibrium found in (a) for each model (1) and (2).

3.26 Use the Euler method to solve the equation $\dot{y} = y^2 + 1$, with $y(0) = 0$ for $0 \leqslant t \leqslant 1$ using (a) $\Delta t = 0.1$ and (b) $\Delta t = 0.01$. (c) Compare the results with the exact solution $y(t) = \tan t$.

3.27 Repeat Problem 3.26 using the modified Euler method.

3.28 Repeat Problem 3.26 using a version of the Runge-Kutta algorithm. Appendix C contains a program implementing Gill's version.

3.29 Write a computer program or calculator routine to solve the equation $\dot{y} = f(y, v)$, $v = v(t)$, using

(a) The Euler method.

(b) The modified Euler method.

3.30 A clue to the selection of a proper step size for a numerical solution can be obtained from the approximate speed of response of the system, if this is known, and from the rate of change of any input function. Consider the tank model

$$C\dot{h} = q_i - 0.0589\sqrt{h}$$

where $C = \pi$ ft$^2$ and time is in seconds. The model's solution for $q_i = 0$ can be found from the results of Example 3.22.

Suppose that the inflow rate is $q_i = 0.01t$ and $h(0) = 9$ ft. Choose a numerical algorithm and a step size and solve for $h(t)$ over the interval $0 \leqslant t \leqslant 100$. Solve the equation for at least two different step sizes to check the accuracy.

3.31 Consider the solar collector system of Example 2.25. Suppose the water temperature in the tank is initially 293 K at 7 AM.

(a) Use a numerical solution method to solve (2.8-26) for the water temperature in the tank as a function of time. The given input data are typical of a January day in the northern United States. The absorbed solar insolation $S$ includes the effect of the collector tilt angle.

| Time | $S$ kJ/hr-m$^2$ | $T_a$ K | $q_d$ m$^3$/hr |
|---|---|---|---|
| 7–8 AM | 0 | 274 | 0 |
| 8–9 AM | 850 | 274 | 0.05 |
| 9–10 AM | 1400 | 275 | 0.02 |
| 10–11 AM | 2200 | 277 | 0.01 |
| 11–12 AM | 3100 | 283 | 0 |
| 12–1 PM | 3100 | 283 | 0.01 |
| 1–2 PM | 2800 | 281 | 0.03 |
| 2–3 PM | 2200 | 281 | 0 |
| 3–4 PM | 1400 | 279 | 0.01 |
| 4–5 PM | 0 | 277 | 0.02 |

(b) The desired temperature of the water drawn from the tank is 120°F = 322 K. Find the backup energy that must be supplied to heat the water to this temperature for the given demand flow rate $q_d$.

# CHAPTER FOUR
# Discrete-Time Models and Sampled-Data Systems

We now come to the analysis of the second major category of models of interest to us: discrete-time models. These occur primarily, but not exclusively, in applications of digital computers for measurement and control. Usually the system being measured or controlled is a continuous-time system, but the discrete-time nature of the computer's sampling operations means that it is inconvenient to employ a continuous-time model. The discrete-time models of such applications are also called *sampled-data* models.

We will use the same approach in developing analytical methods for discrete-time models as was used for the continuous-time case. The emphasis of the chapter will be on first-order models so that the fundamental concepts are not lost in a jungle of algebraic manipulations. The results for linear models will be extended to the higher-order cases in a straightforward way in Chapter Five.

## 4.1 ORIGIN OF DISCRETE-TIME MODELS

Discrete-time models can arise from the description of systems that are inherently of a discrete-time nature, in the analysis of digital measurement and control systems, and in numerical solution methods for differential equations.

### Intrinsically Discrete Systems

In the models we have encountered in previous chapters, the independent variable, time, was continuous. This is the natural situation in physical systems. However, in some circumstances, particularly with nonphysical systems, it is more convenient to conceive of events as occurring at discrete instants. An imposed sequencing or scheduling of events naturally leads to a discrete formulation of the time variable. For example, a bank might compute interest for a savings account once per year on the amount in the account at the end of the year. If $y(k)$ represents the amount of money in the account at the beginning of the year $k$, then $y(1)$ is the amount initially deposited, $y(2)$ is the amount after the first interest payment is added, etc. If interest is computed every three months (quarterly), then $y(2)$ would still represent the amount after the first interest payment. That is, in discrete-time models we normally use integer values to represent the instants at which events occur. We will shortly see that such a formulation allows us to compute easily the total amount in the account after any specified number of interest periods.

As another example, the progress of a student through a college curriculum is indicated by the number of semesters completed. Suppose that $y(k)$ represents the number of mechanical engineering majors who have just finished semester $k$, where $k = 1$ represents the first semester of freshman year. Then $y(8)$ represents the number of graduating seniors. Note that the time index $k$ is quite distinct from physical time because the length of time between semesters is variable (due mainly to the summer recess). Such a model type is said to be event-oriented, not time-oriented. Although this representation ignores changes that occur between the discrete time points, such as a student who leaves at midsemester, the formulation would be useful to a registrar who needs to allocate classroom space for the following semester. A continuous-time model of student enrollment would in fact be too cumbersome for the registrar's purposes.

## Digital Systems

The widespread use of digital computers has had a profound impact on engineering practice and education. With the computational burden eased with computers, it is now possible to use many numerical analysis techniques whose potential was previously unrealized. Also, with the recent breakthroughs in circuit miniaturization, digital devices are now commonly used in experiments and industrial processes for data collection and system control. The use of computers and other digital devices in control systems is discussed in Chapter Six. In the present chapter we will present an introduction to some basic digital devices and the mathematical methods needed to analyze their behavior.

Digital devices can handle mathematical relations and operations only when expressed as a finite set of numbers, rather than as infinite-valued functions. Thus, any continuous measurement signal must be converted into a set of pulses by *sampling* – the process by which a continuous-time variable is measured at distinct, separated instants of time. The sequence of measurements replaces the smooth curve of the measured variable versus time, and the infinite set of numbers represented by the smooth curve is replaced by a finite set of numbers. This is done because a digital device cannot store a continuous signal. Each pulse amplitude is then rounded off to one of a finite number of levels depending on the characteristics of the machine. This is called *quantization*. Consequently, a digital device is one in which the signals are quantized in both time and amplitude. In an analog device the signals are analog; that is, they are continuous in time and are not quantized in amplitude.

Figures 4.1*a*, 4.1*b*, and 4.1*c* show the original analog signal, the sampling pulses, and the pulses after quantization for a simplified hypothetical machine with only four quantization levels (0, 1, 2, and 3). The sampling is done at times $t_0, t_1, t_2, \ldots$, and the index $k$ denotes these sampling times by $k = 0, 1, 2, \ldots$. The quantized signal is shown in Figure 4.1*c*, where we have assumed that any value in the interval $[0.5, 1.5)$ is rounded off to 1, etc. After quantization, the signal is *coded* by converting the quantization level of each pulse (0, 1, 2, or 3 here) into an equivalent binary number which the digital device can accept and store. Figure 4.1*d* shows the resulting binary number sequence for our example at the discrete times $k = 0, 1, 2, 3, 4, \ldots$. This is only one example of a binary coding scheme; several are in common use (Reference 1). The device that performs the sampling, quantization, and coding is an *analog-to-digital* (*A*/*D*) *converter*.

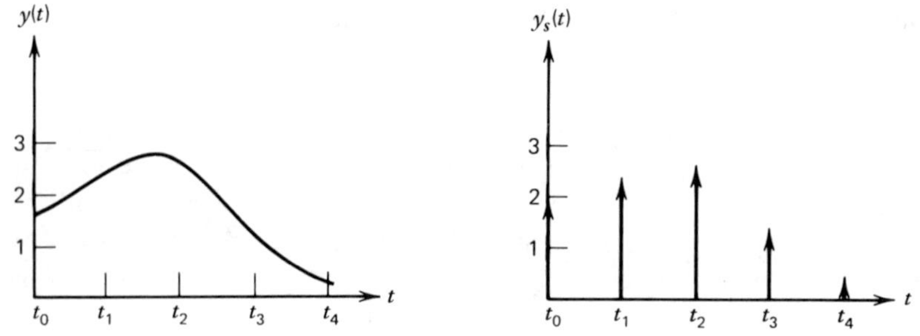


(a) (b)

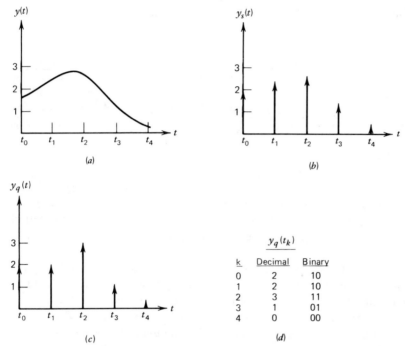

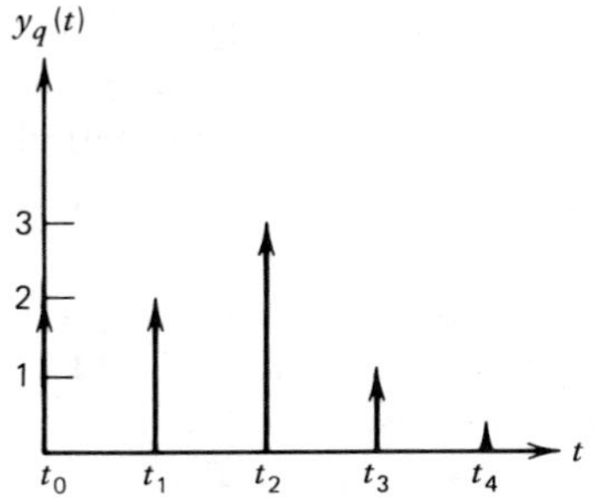


(c)

| | $y_q(t_k)$ | |
|---|---|---|
| k | Decimal | Binary |
| 0 | 2 | 10 |
| 1 | 2 | 10 |
| 2 | 3 | 11 |
| 3 | 1 | 01 |
| 4 | 0 | 00 |

(d)

**Figure 4.1 Sampling, quantization, and coding of an analog signal. (*a*) Original analog signal as a function of time. (*b*) Sampled signal. (*c*) Quantized signal for a hypothetical device with two bits. (*d*) Decimal-coded and binary-coded representations of the quantized signal.**

The number of binary digits carried by the machine is its *word length* and is obviously an important characteristic related to the device's resolution – the smallest change in the input signal that will produce a change in the output signal. The hypothetical machine in Figure 4.1 has two binary digits and thus four quantization levels. Thus, any change of the input over the interval [0.5, 1.5) produces no change in the output. With three binary digits, $2^3$ or 8, quantization levels can be had, and the resolution of the device would improve.

A/D converters in common use have a word length of 10 bits (a bit is one binary digit). This means that an input signal can be resolved to one part in $2^{10}$, or 1 in 1024. If the input signal has a range of 10 volts, the resolution is 10/1024, or approximately 0.01 volt. Thus the input must change by at least 0.01 volt in order to produce a change in the output. High-resolution devices are available with more bits, but their cost is higher and their speed of conversion is slower.

A concept pertinent to A/D conversion is *filtering*, a collection of techniques for removing the effects of noise and measurement error from a signal. For example, the series *RC* circuit in Chapter Three is used to filter noise out of voltage signals and is an analog filter. It will be shown in this chapter that it is possible to use an algorithm called a *digital filter*, which is stored in the software of the computer. Thus the difference between analog and digital filtering is that the latter is performed by a physical device (a circuit) inserted into the system specifically for the purpose of filtering, while digital filtering can be performed by a set of instructions stored in the computer.

When a digital computer is used for measurement and control the natural question to ask is: should we filter the measurement signals with an analog device before they are passed through an A/D converter, or should the A/D conversion be done on the noisy measurements with the filtering done afterward by a digital algorithm? Digital signals are more easily detected than analog signals and are more resistant to noise degradation in transmission and computing. Thus, since A/D conversion must eventually take place if a digital computer is involved, the trend is toward digital filtering in such applications.

In many digital system applications, it is necessary to convert binary outputs from the digital device into a usable form for the equipment being controlled (motors, heaters, etc.). This usually means a continuous or piecewise-continuous voltage. A *digital-to-analog (D/A) converter* performs this function. A common mathematical algorithm used in a D/A converter is the *zero-order data hold*. This will be discussed in Section 4.6.

The preceding discussion deals with hardware considerations. Another implication for digital systems concerns the proper form of models to be used in light of the discrete nature of digital operations versus the continuous nature of the physical devices and processes being measured and controlled by the computer. One view is that the engineer from the start should model the system to be measured or controlled by a set of discrete-time equations called difference equations. This model form is immediately compatible with the digital operations of the computer. The difficulty with this approach is that physical laws are not often stated in discrete-time form. For example, displacement and velocity are continuous variables, and Newton's laws of motion are naturally stated as differential equations. Thus the other view is that the initial system model should be in the form in which the appropriate physical laws are most conveniently and clearly stated. If this form is a set of differential equations, the numerical analysis techniques introduced in this chapter are used to convert them into a discrete-time formulation. This procedure mimics what actually happens in the physical system where sampling converts the continuous physical variables to discrete pulses. Because of this correspondence, here we will adopt the second viewpoint.

## Numerical Solution of Continuous-Time Models

Just as sampling is used to convert a continuous-time variable into a discrete-time variable, a similar process is used to convert a continuous-time model (a differential equation) into a discrete-time model. This form allows a solution to be obtained with a digital computer or calculator (inherently discrete-time devices). This technique is useful when a closed-form solution is not obtainable or when the assumptions required for the linearization of a nonlinear model are not satisfied.

Several algorithms for obtaining difference equations from differential equations were presented in Chapter Three. Recall for example, the Euler method applied to the model

$$\frac{dy}{dt} = ry, \qquad r = \text{constant} \tag{4.1-1}$$

If the time increment $\Delta t$ is chosen small enough, the derivative can be replaced by the approximate expression

$$\frac{dy}{dt} \cong \frac{y(t+\Delta t)-y(t)}{\Delta t} \tag{4.1-2}$$

Replacing the left-hand side of (4.1-1) with this expression, we obtain

$$y(t+\Delta t) = y(t) + ry(t)\,\Delta t \tag{4.1-3}$$

This can be written in several forms. If (4.1-2) is applied successively at the instants $t_k$, where $t_{k+1} = t_k + \Delta t$, we can write the model as

$$y(t_{k+1}) = (1 + r\,\Delta t)y(t_k) \tag{4.1-4}$$

or

$$y(k+1) = (1 + r\,\Delta t)y(k) \tag{4.1-5}$$

where $y(k)$ represents $y(t)$ evaluated at $t = t_k$, $k = 0, 1, 2, \ldots$.

While (4.1-5) can be solved recursively in a straightforward manner, it is tedious to do this if we wish to compute $y$ for a large value of $k$. Therefore, more sophisticated solution techniques are needed, and these are developed in this chapter.

## 4.2 FREE RESPONSE

Suppose that in our savings account example the bank compounds interest annually at 5%. If we have invested one dollar initially, $y(0) = \$1.00$, and after one year we would have accumulated an amount

$$y(1) = \$1.00 + 0.05(\$1.00) = \$1.05$$

At the end of the second year the amount would be

$$y(2) = \$1.05 + 0.05(\$1.05) = \$1.1025$$

In general an initial amount $y(0)$ invested at 5% will yield at the end of the first year an amount

$$y(1) = y(0) + 0.05y(0) = 1.05y(0)$$

and for the second year

$$y(2) = y(1) + 0.05y(1) = 1.05y(1)$$

From this it is easy to obtain the general formula

$$y(k) = y(k-1) + 0.05y(k-1)$$

or

$$y(k) = 1.05y(k-1), \qquad k = 1, 2, 3, \ldots \tag{4.2-1}$$

We can calculate the accumulated savings at any year $k$ either by repetitive multiplication by the factor 1.05 or by realizing that the process can be described by the formula

$$y(k) = (1.05)^k y(0) \tag{4.2-2}$$

This apparently is the solution of (4.2-1) with the initial condition $y(0)$. The form is convenient because it allows the amount in any year $k$ to be computed in one step. With this insight, we can hypothesize that the solution of the difference equation

$$y(k) = ay(k-1), \qquad k = 1, 2, 3, \ldots \tag{4.2-3}$$

is

$$y(k) = a^k y(0), \qquad k = 0, 1, 2, 3, \ldots \tag{4.2-4}$$

when $a$ is a constant. This is affirmed by noting that (4.2-4) implies that

$$y(k-1) = a^{k-1}y(0) = a^{-1}[a^k y(0)] = a^{-1}y(k)$$

which is equivalent to (4.2-3). Equation (4.2-4) also satisfies the initial condition at $k = 0$ since $a^0 = 1$.

Equation (4.2-3) represents a *linear first-order difference equation*. The term *linear* has the same meaning for difference equations as it does for differential equations. Similarly, a linear difference equation can be recognized as a linear relation between the state variables and inputs (if any). The *order* of a difference equation is the maximum difference between the time indices in the equation. Thus the equation

$$y(k) = 5y(k-1)y(k-2)$$

is *second-order* and nonlinear as a result of the cross-product term between $y(k-1)$ and $y(k-2)$.

The solution behavior of the difference equation (4.2-3) differs greatly from that of the corresponding differential equation (4.1-1). Seven types of behavior can arise from (4.2-4) depending on the value of $a$.

1. $a > 1$. The solution's magnitude grows with time and the solution keeps the sign of $y(0)$.
2. $a = 1$. The solution remains constant at $y(0)$.
3. $0 < a < 1$. The solution decays in magnitude and keeps the sign of $y(0)$.
4. $a = 0$. The solution jumps from $y(0)$ to zero at $k = 1$ and remains there.
5. $-1 < a < 0$. The magnitude decays but the solution alternates sign at each time step (an oscillation with a period of two time units).
6. $a = -1$. The magnitude remains constant at $y(0)$, but the solution alternates sign at each step.
7. $a < -1$. The magnitude grows and the sign of the solution alternates at each time step.

This behavior is summarized in Figure 4.2 for a positive $y(0)$. The smaller plots represent $y(k)$ versus $k$ for each of the seven numbered cases. The small circles represent the solution. The lines connecting the circles are visual aids only. The difference equation makes no statement about what happens between integer values of $k$.

## Stability

The *concepts* of point equilibrium and stability can be immediately transferred from the differential equation case (Section 3.1). However, the *determination* of a point equilibrium and its stability properties is done differently. Since a point equilibrium is a condition of no change, it is found for a differential equation by setting the time derivatives to zero. For a difference equation it is found by dropping all time indices;

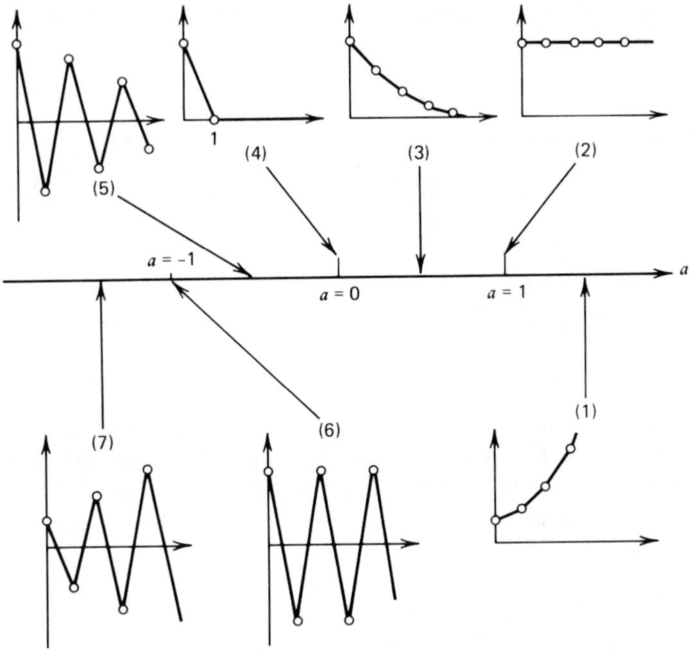


**Figure 4.2 Free response of the linear first-order equation (4.2-4) for various values of the coefficient $a$. Numbered cases refer to the discussion in the text.**

that is, by setting $y(k) = y(k-1) = y_e$ for all $k \geqslant 0$, where $y_e$ is the equilibrium value. With (4.2-3) this implies that

$$y_e = ay_e$$

That is satisfied if either

$$1 = a$$

or

$$y_e = 0$$

Thus if $a \neq 1$, the only point equilibrium possible for (4.2-3) is $y_e = 0$. If $a = 1$, any value of $y$ is an equilibrium. This can be seen in Figure 4.2, case (2).

The criterion for stability of the point equilibrium can be ascertained from Figure 4.2. For an equilibrium at $y = 0$, the only cases in which the solution approaches and remains at equilibrium are the cases where $-1 < a < 1$. In other words, the model is *stable* if $|a| < 1$, *unstable* if $|a| > 1$, and *neutrally stable* if $|a| = 1$. Contrast this with the criterion for (4.1-1). For that equation, instability occurs if $r > 0$, stability if $r < 0$, and neutral stability if $r = 0$.

## Relation to the Continuous-Time Model

It is apparent from Figure 4.2 that simple difference equations produce types of behavior not found in simple differential equations. In (4.1-1) no oscillations are possible. But since it is possible to derive a difference equation from a differential equation – with the Euler method, for example – then their two solutions should be related somehow. For comparison, the solution of (4.1-1) is repeated here.

$$y(t) = y(0)e^{rt} \tag{4.2-5}$$

If time $t_k$ corresponds to the discrete time index $k$, then from (4.2-4) and (4.2-5)

$$y(t_k) = y(0)e^{rt_k} \tag{4.2-6}$$

and

$$y(t_k) = y(k) = a^k y(0)$$

Comparison of these two expressions gives

$$e^{rt_k} = a^k$$

or

$$rt_k = \ln a^k = k \ln a \tag{4.2-7}$$

This indicates that a comparison between the two models can be made only if $a > 0$, since $\ln a$ is undefined for $a \leqslant 0$. This is not unexpected because oscillatory behavior occurs in the discrete-time model only if $a < 0$. Equation (4.2-7) can be used to find the value of $r$ required to make a continuous-time process equivalent to one in discrete time, and vice versa.

*Example 4.1*

One bank offers a savings plan with interest compounded *continuously*, while another offers a plan with 5% interest compounded *annually*. What rate must be used for continuous compounding to make it equivalent to 5% annual compounding? For annual compounding at 5%, how much is an investment worth after 10 years?

For annual compounding at 5%, $a = 1.05$ and

$$y(k) = 1.05y(k-1)$$

where $k$ measures the number of years of investment. For continuous compounding the growth of the investment is described by (4.1-1) where $r$ is the instantaneous rate of growth (the interest rate). Let $t_k$ measure the number of years of investment. Then $k = 1$ corresponds to $t_k = 1$, and from (4.2-7),

$$r = \ln 1.05 = 0.04879 = 4.879\%$$

Thus, continuous compounding at 4.879% is equivalent to 5% compounded annually. After $k = 10$ years, the initial investment $y(0)$ has grown to $y(10) = (1.05)^{10}y(0) = 1.62889y(0)$. The investment has increased by approximately 163%.

It is now clear that the form $z^k$ plays the same fundamental solution role in linear difference equations that the form $e^{st}$ plays in linear differential equations. We will see that this also holds true for higher-order equations, and thus a firm grasp of the first-order model's behavior will greatly promote understanding of behavior in higher-order models. The differences and similarities of the two first-order models are summarized in Table 4.1.

The form of the first-order equation given by (4.1-5) is equivalent to (4.2-3) with $a = 1 + r\Delta t$ and the index $k$ starting from 1 instead of 0. In general the linear model given by (4.2-3) is equivalent to

**TABLE 4.1 Comparison of Differential and Difference Equations**

| Characteristic | Differential Equation $\dot{y} = ry$ | Difference Equation $y(k) = ay(k-1)$ |
|---|---|---|
| 1. Solution | $y(t) = y(0)e^{rt}$ | $y(k) = y(0)a^k$ |
| 2. Solution behavior | No oscillation | Oscillations of period two if $a < 0$ |
| 3. Stability | | |
| Stable if | $r < 0$ | $\|a\| < 1$ |
| Neutrally stable if | $r = 0$ | $\|a\| = 1$ |
| Unstable if | $r > 0$ | $\|a\| > 1$ |
| 4. Time constant (time to decay by 63%) | $t = \tau = -1/r,\ r < 0$ | $k = -1/\ln \|a\|,\ \|a\| < 1$ |
| 5. Relation between the two models | (a) time $t_k$ corresponds to time $k$<br>(b) $rt_k = k \ln a$ if $a > 0$.<br>(c) no relation if $a \leqslant 0$. | |

$$y(k+1) = ay(k) \qquad k = 0, 1, 2, \ldots$$

Both equations have the same solution: (4.2-4).

## Linearization and Nonlinear Models

Nonlinear difference equations are convenient model forms in many applications. For example, when population growth occurs at discrete times and when generations are completely nonoverlapping, a difference equation model is more appropriate than a differential equation. These phenomena occur in some insect populations where the prior generation dies off before the next is born from eggs laid the previous year. In this case the difference equation analog of the logistic growth equation in Section 2.9 is commonly used. This is

$$y(k+1) = y(k) \exp\left\{r\left[1 - \frac{y(k)}{K}\right]\right\} \tag{4.2-8}$$

with $r$ being the instantaneous growth rate and $K$ the carrying capacity of the environment. Note that when $y \ll K$, (4.2-8) gives the solution of the basic growth equation (4.1-1) at the discrete time points (see Table 4.1).

Few analytical solutions exist for nonlinear difference equations. Sometimes the nonlinear equation can be solved by transforming it into a linear one. For example, the equation

$$y(k+1) = \frac{y(k)}{1 + y(k)} \tag{4.2-9}$$

can be converted to

$$x(k+1) = x(k) + 1 \tag{4.2-10}$$

by the transformation

$$x(k) = \frac{1}{y(k)} \tag{4.2-11}$$

The solution of (4.2-10) can be found easily by mathematical induction to be

$$x(k) = x(0) + k \tag{4.2-12}$$

which gives

$$y(k) = \frac{y(0)}{1 + ky(0)} \tag{4.2-13}$$

Unfortunately most common nonlinear models have no known analytical solution, and we must resort to either step-by-step iteration (if the parameter values and initial conditions are specified numerically) or approximate methods of analysis. The technique of linearization introduced in Chapter Three can also be used for difference equations as an approximate method of analysis. The general first-order equation can be written as

$$y(k+1) = f[y(k)] \tag{4.2-14}$$

If this equation has an equilibrium solution $y_e$, then

$$y_e(k+1) = y_e(k)$$

and

$$y_e = f(y_e) \tag{4.2-15}$$

If $y_e$ is unknown, (4.2-15) must be solved. This might require a numerical method, since $f(y_e)$ is nonlinear (see Appendix A for one such method). Once $y_e$ is known, we expand $f[y(k)]$ in a Taylor series expansion, and keep only the first-order term as follows,

$$f[y(k)] = f(y_e) + \left[\frac{\partial f}{\partial y(k)}\right]_e [y(k) - y_e] + \ldots \tag{4.2-16}$$

where the derivative is evaluated at the equilibrium point $y_e$. Upon substituting this expression into the right-hand side of (4.2-14), and using (4.2-15), we get

$$y(k+1) - y_e = \left[\frac{\partial f}{\partial y(k)}\right]_e [y(k) - y_e]$$

Let

$$x(k) = y(k) - y_e \tag{4.2-17}$$

$$a = \left[\frac{\partial f}{\partial y(k)}\right]_e \tag{4.2-18}$$

to obtain

$$x(k+1) = ax(k) \tag{4.2-19}$$

This is a linear difference equation whose solution is given as before.

$$x(k) = x(0)a^k \tag{4.2-20}$$

Equation (4.2-19) is an approximation to the original equation (4.2-14) in the vicinity of the equilibrium point. As such it can be used to determine the local stability of the model, as was done in Section 3.8. From the solution (4.2-20) we see that the equilibrium is locally stable if and only if $|a| < 1$, and unstable if $|a| > 1$, in a local

**TABLE 4.2 Linearization of a Nonlinear Difference Equation**

| | |
|---|---|
| Nonlinear equation | $y(k+1) = f[y(k), v(k)]$ |
| Equilibrium condition | $y_e = f(y_e, v_e)$, $v_e$ given |
| Linearized approximation | $x(k+1) = ax(k) + bu(k)$ |
| | $x(k) = y(k) - y_e$ |
| | $u(k) = v(k) - v_e$ |
| | $a = \left[\dfrac{\partial f}{\partial y(k)}\right]_e$ |
| | $b = \left[\dfrac{\partial f}{\partial v(k)}\right]_e$ |
| Local stability | Stable if $\lvert a \rvert < 1$ |
| | Unstable if $\lvert a \rvert > 1$ |
| | Neutral stability unable to be predicted from linearized model |

sense. The technique says nothing about the neutral or the global stability of the model. Table 4.2 summarizes the method. The results for a model with an input variable $v(k)$ can be found in a manner similar to that in Section 3.12.

*Example 4.2*

Find a linearized approximation to the model (4.2-8) and investigate its local stability properties, for the nonzero equilibrium.

The equilibrium solutions are given by

$$y_e = y_e \exp\left[r\left(1 - \frac{y_e}{K}\right)\right] \tag{4.2-21}$$

One solution is $y_e = 0$, which is of no interest in a population model. The other is $y_e = K$, and is the one of interest. From (4.2-18) we have

$$a = \left\{\left[1 - \frac{r}{K}y(k)\right] \exp\left[r\left(1 - \frac{y(k)}{K}\right)\right]\right\}_e \tag{4.2-22}$$

For $y_e = K$, this gives $a = 1 - r$. Thus the equilibrium is stable in a local sense if and only if $0 < r < 2$.

## Nonlinear Behavior

The dynamic behavior of even simple first-order nonlinear difference equations can be suprisingly varied and includes phenomena not seen in first-order differential equations. For example, the behavior of the model (4.2-8) for $r > 2$ is very complex, as pointed out by May (Reference 2). We have seen that the equilibrium at $y_e = K$ is locally stable if $0 < r < 2$, and Figure 4.3*a* shows the results of a recursive solution for $r = 1.8$ and $y(0)/K = 0.1$. For $2 < r < 2.526$ the solution settles into an oscillation of period

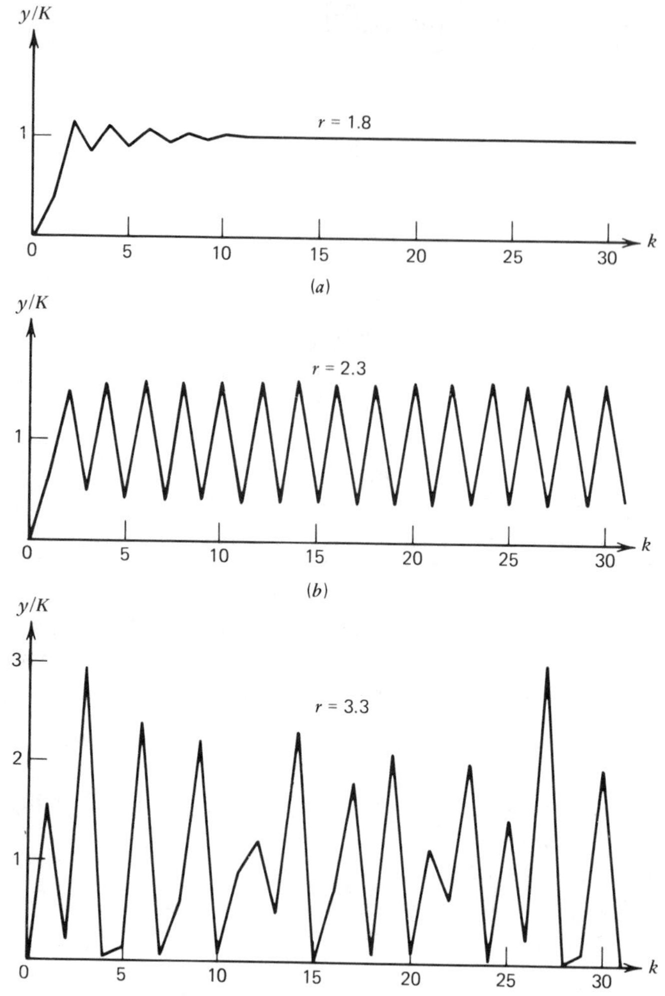


**Figure 4.3 Dynamic behavior of the population density $y/K$ given by (4.2-8). (*a*) Stable equilibrium point. (*b*) Stable two-point cycle. (*c*) Chaotic behavior. (Here the behavior of the solution depends on the initial condition.)**

two (Figure 4.3*b*). For $r > 2.526$, oscillations of period four, eight, sixteen, etc., appear as $r$ is increased. There is a limiting value of $r$ above which "chaotic" behavior occurs where there exists an uncountable number of initial conditions $y(0)$ for which the solution does not ultimately settle into any cycle. The solution has the appearance of a random sequence; hence the term *chaotic* (Figure 4.3*c*). Also, slightly different initial conditions can drastically change the appearance of the solution. For (4.2-8) May has shown that chaotic behavior occurs when $r > 2.692$.

Analytic techniques for predicting what parameter values result in chaotic behavior are beyond the scope of this text. Much of the work is in the preliminary stage, and there is great potential for new discoveries. What is important to realize is that even a simple deterministic nonlinear model can possess complex behavior. Stated another way, simple nonlinear models do not necessarily possess simple behavior. Thus complex processes might indeed be able to be represented by a simple model. On the other hand, because chaotic solutions resemble random sequences, it might be difficult to distinguish the deterministic component from the random effects in any data taken from such a process.

## 4.3 SAMPLING

Sampling extracts a discrete-time signal from a continuous-time signal. If the sampling frequency is not selected properly, the resulting sample sequence will not accurately represent the original signal. Fortunately the proper frequency is readily determined in many cases via the sampling theorem, to be presented shortly.

### Impulse Representation of Sampling

Sampling can be represented by the opening and closing of a switch, as shown in Figure 4.4. In digital systems this is accomplished electronically, not mechanically. Let $y(t)$ be the continuous-time signal and $y^*(t)$ the discrete-time signal resulting from sampling $y(t)$ briefly every $T$ seconds. *Uniform sampling* occurs when the sampling period $T$ is constant. Only this case will be treated here.

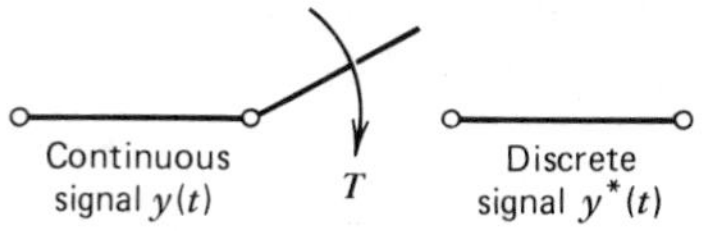


**Figure 4.4 Symbol for a sampler with a sampling period $T$.**

Figure 4.5 illustrates the sampling process. Figure 4.5*a* shows the signal to be sampled. In Figure 4.5*b* the duration $\Delta$ represents the length of time the sampling switch remains closed. The circuitry in the sampler is designed to take this duration into account, and thus the sampler output corresponds to an average value of $y$ over the interval $\Delta$ (Reference 1). One representation of the sampling process considers the sampler outputs to be impulses with a strength equal to the sampled value of $y$ at the sampling times. This is shown in Figure 4.5*c*.

Let $\delta(t)$ be the unit impulse function. Then $\delta(t-kT)$ is a unit impulse occurring at time $kT$. The sampled sequence, denoted $y^*(t)$, can be represented by a train of impulses occurring at the sampling times $kT$, each having a strength equal to the sample value at that time; that is,

$$y^*(t) = y(0)\delta(t) + y(T)\delta(t-T) + y(2T)\delta(t-2T) + \dots$$

$$= \sum_{i=0}^{N} y(iT)\delta(t-iT) \tag{4.3-1}$$

where $N$ is the number of samples.

Only the values of $y(t)$ at the sampling instants are retained by the digital memory. Thus the continuous time variable $t$ is no longer needed and can be replaced with the discrete index $k$, where $t = kT$. The times $t = 0, 1, 2, \dots$ corresponds to $k = 0, 1, 2, \dots$, and (4.3-1) can be written as

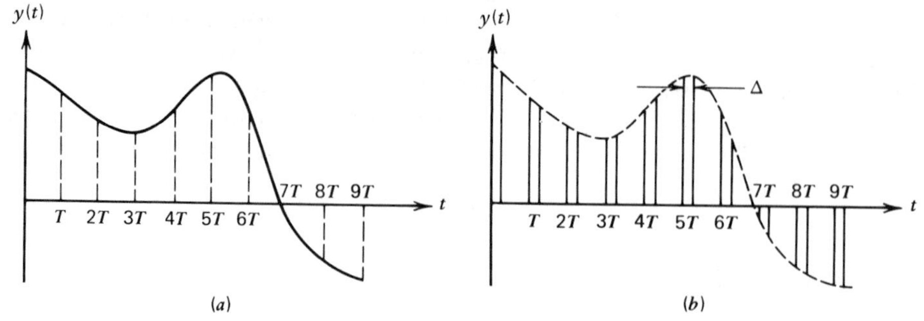


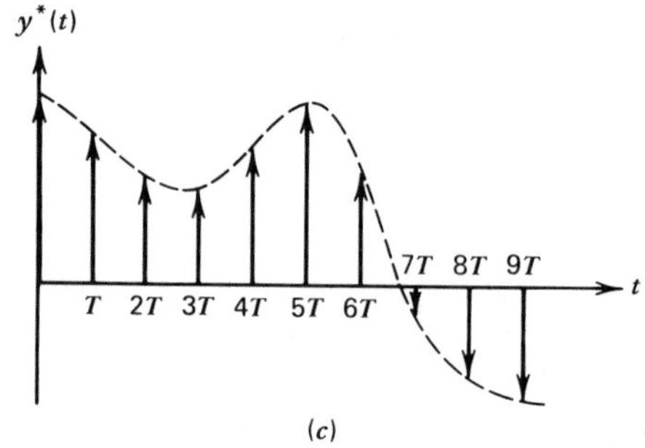


**Figure 4.5** Sampling process. (*a*) The original signal. (*b*) Illustration of the sampling duration Δ. (*c*) Impulse representation of the sampled signal.

$$y^*(kT) = \sum_{i=0}^{N} y(iT)\delta(kT - iT) \tag{4.3-2}$$

In this form the sampling period $T$ is obviously superfluous once it is specified, and we can simply use the index $k$, as was done with the Euler method (4.1-4) and (4.1-5). In this notation,

$$y_k^* = y^*(kT)$$

$$\delta_{k-i} = \delta(kT - iT)$$

and we have

$$y_k^* = \sum_{i=0}^{N} y_i \delta_{k-i}$$

If the subscripts are cumbersome the preceding is replaced by

$$y^*(k) = \sum_{i=0}^{N} y(i)\delta(k - i) \tag{4.3-3}$$

where it is understood that $y(k)$ represents $y_k$, which in turn represents $y(kT)$.

## Frequency Content of Signals

The proper value of the sampling period depends on the nature of the signal being sampled. This is easily shown with a sinusoid of period $P$ (Figure 4.6). If the sampling period $T$ is slightly greater than the half-period $P/2$, the case shown in the figure, it is possible to miss completely one lobe of the sinusoid. If $T < P/2$, each lobe will always

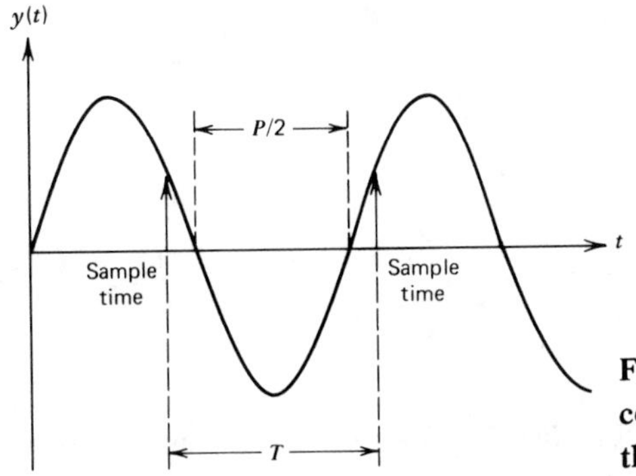


**Figure 4.6 Sampling a sinusoid. The adjacent samples are far enough apart to miss the oscillation in $y(t)$.**

be sampled at least once, and the oscillation will be detected. Stated in terms of frequencies, the sampling frequency $1/T$ must be at least twice the sinusoidal frequency $1/P$.

The issue is not as clear when the signal does not consist of sinusoids. If the signal $y(t)$ is periodic with period $P$, it can be represented by a Fourier series (Appendix A) as

$$y(t) = \sum_{n=0}^{\infty} A_n \sin\left(\frac{n\pi t}{P} + \phi_n\right) \tag{4.3-4}$$

where $A_n \geqslant 0$ is the amplitude of the $n$th harmonic and $\phi_n$ is its phase angle. The plot of $A_n$ versus the frequency $n\pi/P$ is called the *spectrum* of the signal, in analogy with the light spectrum used in physical optics. For example, the half-sine function shown in Figure 3.32 has the Fourier series

$$\begin{aligned} y(t) &= \frac{1}{\pi} + \frac{1}{2}\sin 2\pi t - \frac{2}{3\pi}\cos 4\pi t - \frac{2}{15\pi}\cos 8\pi t - \frac{2}{35\pi}\cos 12\pi t + \ldots \\ &= \frac{1}{\pi} + \frac{1}{2}\sin 2\pi t + \frac{2}{3\pi}\sin\left(4\pi t + \frac{3\pi}{2}\right) \\ &\quad + \frac{2}{15\pi}\sin\left(8\pi t + \frac{3\pi}{2}\right) + \frac{2}{35\pi}\sin\left(12\pi t + \frac{3\pi}{2}\right) + \ldots \end{aligned}$$

The spectrum of this signal is shown in Figure 4.7. Notice how the amplitudes decrease with frequency for the higher frequencies. This characteristic is typical of physical signals because of "inertia" effects.

The spectrum of a periodic signal provides a convenient representation of the relative importance of each frequency component. Thus it would be useful to have a similar representation for signals that are not periodic (aperiodic). This is provided by the *Fourier transform* of the signal and is defined to be

$$\mathscr{F}[y(t)] = Y(\omega) = \int_{-\infty}^{\infty} y(t)\, e^{-i\omega t}\, dt \tag{4.3-5}$$

The inverse transform is

$$\mathscr{F}^{-1}[Y(\omega)] = y(t) = \frac{1}{2\pi}\int_{-\infty}^{\infty} Y(\omega)\, e^{i\omega t}\, dt \tag{4.3-6}$$

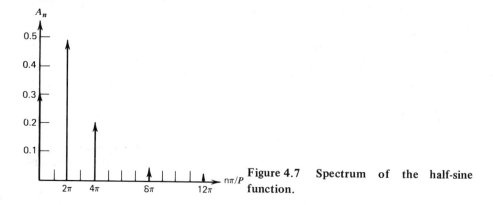

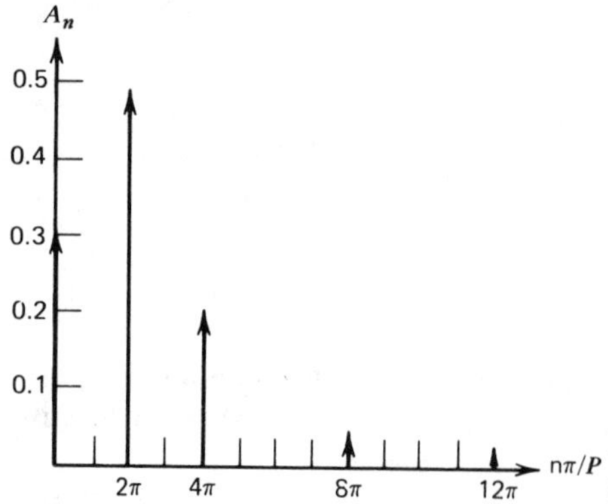


**Figure 4.7 Spectrum of the half-sine function.**

Note the similarity to the Laplace transform. Here the integral is taken over all time rather than only over positive time. The transform variable $\omega$ is a real number and is taken to be circular frequency. If $y(t) = 0$ for $t \leqslant 0$ and if the region of convergence for the Laplace transform includes the imaginary axis, then

$$Y(\omega) = [Y(s)]_{s=i\omega} \tag{4.3-7}$$

where $Y(s)$ is the Laplace transform of $y(t)$.

The Euler formula relating the complex exponential in (4.3-5) to $\sin \omega t$ and $\cos \omega t$ can be used to present a heuristic derivation of the Fourier transform pair from the Fourier series formulas for $A_n$ (Appendix A). The essence of the argument is to allow the period $P$ to approach infinity while the separation in signal frequencies is allowed to approach zero. In the limit the summation in (4.3-4) approaches the integral in (4.3-6). The details are given in Reference 3, a good source of information dealing with sampling and digital filtering. The important point here is that the magnitude $Y(\omega)$ of the Fourier transform can also be used as a measure of the signal's frequency content, for aperiodic signals, just as $A_n$ is used for periodic signals. The plot of $|Y(\omega)|$ versus $\omega$ is the spectrum of $y(t)$.

The energy content of a sinusoid is proportional to the square of its amplitude. The energy contained in a general aperiodic signal is found by integrating the square of the amplitude $|Y(\omega)|$ over all frequencies, and is

$$\int_0^\infty |Y(\omega)|^2 \, d\omega$$

*Example 4.3*

Compute the Fourier transform of the signal

$$y(t) = \begin{cases} 0, & t < 0 \\ e^{-t/\tau}, & t \geqslant 0 \end{cases}$$

Find its spectrum and the frequency $\omega_u$ for which 99% of the signal's energy lies in the range $0 \leqslant \omega \leqslant \omega_u$.

From (4.3-6)

$$Y(\omega) = \int_{-\infty}^{0} 0 \, dt + \int_0^\infty e^{-t/\tau} e^{-i\omega t} \, dt$$

$$= \int_0^\infty e^{-[(1/\tau)+i\omega]t}\, dt = \left. \frac{-e^{-[(1/\tau)+i\omega]t}}{(1/\tau)+i\omega} \right|_0^\infty$$

$$= \frac{\tau}{1 + i\omega\tau}$$

This is the same form as the frequency transfer function of the first-order system with a time constant $\tau$. This is given by (3.8-16) with $b = 1$. The result could also have been obtained from (4.3-7) with $Y(s) = \tau/(\tau s + 1)$. The magnitude is

$$|Y(\omega)| = \frac{\tau}{\sqrt{1 + \tau^2\omega^2}}$$

This is plotted in Figure 3.27*a* in logarithmic form. The spectrum is flatter for smaller values of $\tau$ and falls off with $\omega$ more rapidly for larger values of $\tau$. Thus a rapidly decaying exponential has a higher frequency content than one that decays slowly.

The upper frequency $\omega_u$ is found from

$$\int_0^{\omega_u} |Y(\omega)|^2\, d\omega = 0.99 \int_0^\infty |Y(\omega)|^2\, d\omega$$

With the given expression for $Y(\omega)$, this becomes

$$\int_0^{\omega_u} \frac{\tau^2}{1+\tau^2\omega^2}\, d\omega = 0.99 \int_0^\infty \frac{\tau^2}{1+\tau^2\omega^2}\, d\omega$$

$$\tan^{-1}(\tau\omega_u) - \tan^{-1}(0) = 0.99[\tan^{-1}(\infty) - \tan^{-1}(0)]$$

or

$$\omega_u = \frac{1}{\tau}\tan\left(\frac{0.99\pi}{2}\right) = \frac{63.657}{\tau}\ \frac{\text{radians}}{\text{unit time}}$$

If $\tau = 1$ sec, 99% of the exponential signal's energy lies in a frequency band of $0 < \omega < 63.657$ rad/sec.

## Aliasing

Now that we are able to determine the frequency content of a signal, we need to know what effects the higher frequencies have on the sampling process. Uniform sampling of the sinusoid

$$y(t) = \sin(\omega_1 t + \phi)$$

produces the sequence

$$y(kT) = \sin(k\omega_1 T + \phi)$$

No loss of information is incurred in sampling if $T$ is selected so that $0 \leqslant \omega_1 \leqslant \pi/T$, as shown earlier. If we sample the following signals,

$$x(t) = \sin\left[\left(\omega_1 + \frac{2\pi n}{T}\right)t + \phi\right]$$

$$z(t) = \sin\left[\left(\frac{2\pi n}{T} - \omega_1\right)t + \pi - \phi\right]$$

the resulting sequences can be seen to be identical to $y(kT)$, after applying several trigonometric identities. These are

$$x(kT) = \sin\left[\left(\omega_1 + \frac{2\pi n}{T}\right) kT + \phi\right]$$

$$= \sin(k\,\omega_1 T + \phi)$$

$$z(kT) = \sin\left[\left(\frac{2\pi n}{T} - \omega_1\right) kT - \pi + \phi\right]$$

$$= \sin(k\,\omega_1 T + \phi)$$

for *any* positive integer $n$. The signals $y(t)$ and $x(t)$ *differ* in circular frequency by an integral multiple of $2\pi/T$. Similarly, the *sum* of the frequencies of $y(t)$ and $z(t)$ is an integral multiple of $2\pi/T$.

We conclude that uniform sampling cannot distinguish between two sinusoidal signals when their circular frequencies have a sum or difference equal to $2\pi n/T$, where $n$ is any positive integer. This means that the only effective frequency range for uniform sampling is $0 \leqslant \omega \leqslant \pi/T$. That is, the signal being sampled must have no circular frequency content above $\pi/T$, if uniform sampling is not to distort the signal's information. The frequency $\pi/T$ (radians per unit time) is called the *Nyquist frequency* or the *folding frequency*, because all frequencies in the signal are folded into the interval $0 \leqslant \omega \leqslant \pi/T$ by uniform sampling (Reference 1). This phenomenon is called *aliasing*. It is typically seen in motion pictures of a rotating spoked wheel or aircraft propeller. As it rotates faster it appears to slow down and then stop, or even rotate backward. The sampling process produced by the picture frames "aliases" the high rotation speed into the lower-frequency interval defined by the Nyquist frequency.

## The Sampling Theorem

The concept of aliasing leads directly to the sampling theorem.

### *Sampling Theorem*

A continuous-time signal $y(t)$ can be reconstructed from its uniformly sampled values $y(kT)$ if the sampling period $T$ satisfies

$$T \leqslant \frac{\pi}{\omega_u}$$

where $\omega_u$ is the highest frequency contained in the signal; that is,

$$Y(\omega) = 0, \qquad \omega > \omega_u$$

where $Y(\omega)$ is the Fourier transform of $y(t)$.

With the principal exception of a pure sinusoid, most physical signals have no finite upper frequency $\omega_u$. Their spectra $|Y(\omega)|$ approach zero only as $\omega \to \infty$. In such cases we estimate $\omega_u$ by finding the frequency range containing most of the signal's energy, as in Example 4.3. With a conservative engineering design philosophy, a safety factor between two and ten is applied to determine the sampling rate. This factor is necessary also because we do not have in practice an infinite sequence of impulses as required by the sampling theorem. Because of aliasing, a low-pass filter, called a *guard* filter, is inserted before the sampler to eliminate frequencies above the Nyquist frequency $\pi/T$. By definition, these frequencies do not contribute significantly to the signal's energy. Thus they are equivalent to noise in the system and should be filtered out anyway.

*Example 4.4*

Compute the uniform sampling period required for the signal $e^{-t/\tau}$.

From Example 4.3, the upper frequency for the 99% energy criterion was $\omega_u = 63.657/\tau$. From the sampling theorem,

$$T \leqslant \frac{\pi}{\omega_u} = \frac{\pi\tau}{63.657} = 0.049\tau$$

With a safety factor of ten, the sampling period is $T = 0.0049\tau$. The guard filter would be designed to filter out frequencies above $\pi/T$.

In many applications the Fourier transform integral in (4.3-5) is impossible to evaluate in closed form. Also, the signal might be given in tabular or graphical form. In such cases we may compute the integral numerically. With standard integration algorithms like the trapezoidal rule, the number of arithmetic operations goes as $N^2$, where $2N$ is the number of data points or subintervals. Most computer installations support a version of the Fast Fourier Transform (FFT) algorithm, which is a numerical method requiring $N \log N$ arithmetic operations. Thus for large $N$, the savings in computation time and accuracy is quite large (Reference 3).

## 4.4 THE z TRANSFORM

The impulse representation of sampling provides a convenient way of introducing the $z$ transform. The first three terms of (4.3-2) are

$$y^*(t) = y(0)\delta(t) + y(T)\delta(t-T) + y(2T)\delta(t-2T) + \dots \qquad (4.4\text{-}1)$$

If both sides are transformed with the Laplace transform we get

$$Y^*(s) = y(0) + y(T)e^{-Ts} + y(2T)e^{-2Ts} + \dots$$

where the *shifting property* of the Laplace transform has been used. From Table B.1 in Appendix B, this property states that for two functions $g(t)$ and $f(t)$, where

$$g(t) = \begin{cases} f(t-a) & t \geqslant a \\ 0 & t < a \end{cases}$$

the transform of $g(t)$ is

$$\mathscr{L}[g(t)] = e^{-as}\mathscr{L}[f(t)]$$

Here $f(t) = \delta(t)$ and $\mathscr{L}[\delta(t)] = 1$.

Make the definition

$$z = e^{Ts} \tag{4.4-2}$$

and write $Y^*(s)$ as a function of $z$.

$$Y^*(z) = y(0) + y(T)\frac{1}{z} + y(2T)\frac{1}{z^2} + \dots \tag{4.4-3}$$

The transformed variable $Y^*(z)$ is the $z$ *transform* of the function $y^*(t)$. The transformation is written as $\mathscr{Z}[y(t)]$, and is a power series in $1/z$. Comparison of (4.4-1) and (4.4-3) shows that $1/z$ is a *delay* operator representing a time delay $T$. Also, $1/z^2$ represents a delay $2T$, etc. With this interpretation we can recover a sampled sequence from its $z$ transform. For example, the transform

$$Y^*(z) = 5 + \frac{3}{z} + \frac{1}{z^2} + \frac{-1}{z^3} + \frac{-4}{z^4} + \frac{-5}{z^5} + \dots$$

corresponds to the sample values $y(t) = 5, 3, 1, -1, -4, -5, \dots$ for $t = 0, T, 2T, 3T, 4T, 5T, \dots$.

## Commonly Occurring Transforms

If $y^*(t)$ is a unit impulse at $t = 0$, then $y(0) = 1$ and $y(it) = 0$ for $i \neq 0$. The $z$ transform of this function is 1, and is identical to its Laplace transform. Unfortunately, the simple relationship between a sampled sequence and its $z$ transform series does not always provide a convenient form to handle because the series is usually infinite. What is needed is a closed-form expression for the transform. Fortunately, these are available for commonly occurring functions like the step, ramp, and exponential. Here we take all signals to be zero for $t < 0$ and thus confine our attention to the one-sided $z$ transform. If $y(t)$ is nonzero for $t < 0$, the two-sided transform is needed (Reference 4).

### *Example 4.5*

Find the $z$ transform of the sequence $y = \{1, a, a^2, a^3, \dots\}$ with $y = 0$ for $t < 0$.

The sequence values are the coefficients of the transform's series expression; that is,

$$Y(z) = 1 + \frac{a}{z} + \frac{a^2}{z^2} + \frac{a^3}{z^3} + \dots = \sum_{k=0}^{\infty} a^k z^{-k} \tag{4.4-4}$$

This is the *geometric* series, and it is shown in most calculus texts that its sum can be expressed as

$$Y(z) = \frac{1}{1 - az^{-1}} = \frac{z}{z - a} \tag{4.4-5}$$

if $|z| > |a|$. The series is said to have a radius of absolute convergence of $|z| = |a|$.

If $a = 1$ the preceding sequence corresponds to the sample sequence of a unit step function that starts just before sampling begins. The $z$ transform of this function is

$$Y(z) = \frac{z}{z - 1} \tag{4.4-6}$$

Similarly the exponential $y(t) = \exp(-bt)$ has the transform

$$Y(z) = 1 + \frac{e^{-bT}}{z} + \frac{e^{-2bT}}{z^2} + \ldots \tag{4.4-7}$$

This is the form of (4.4-4) with $a = \exp(-bT)$. Thus from (4.4-5)

$$\mathscr{Z}(e^{-bt}) = \frac{z}{z - e^{-bT}}$$

*Example 4.6*

Compute the $z$ transform of the sampled unit ramp, $y(t) = t$.

The sampled series representation of the unit ramp is

$$\begin{aligned} Y(z) &= 0 + \frac{T}{z} + \frac{2T}{z^2} + \ldots \\ &= Tz\left(\frac{1}{z^2} + \frac{2}{z^3} + \ldots\right) \\ &= -Tz\frac{d}{dz}\left(\frac{1}{z} + \frac{1}{z^2} + \ldots\right) \end{aligned}$$

With (4.4-6) this can be written as

$$Y(z) = -Tz\frac{d}{dz}\left(\frac{z}{z-1}\right) = \frac{Tz}{(z-1)^2} \tag{4.4-8}$$

The method of differentiation used to obtain the preceding transform can be generalized. In summation notation the $z$ transform is defined as

$$\mathscr{Z}[y(kT)] = Y(z) = \sum_{k=0}^{\infty} y(kT)z^{-k} \tag{4.4-9}$$

or equivalently as

$$\mathscr{Z}[y(k)] = Y(z) = \sum_{k=0}^{\infty} y(k)z^{-k} \tag{4.4-10}$$

Differentiation and multiplication by $z$ gives

$$\begin{aligned} -z\frac{d}{dz}Y(z) &= \sum_{k=0}^{\infty} ky(kT)z^{-k} \\ &= \mathscr{Z}[ky(kT)] \end{aligned}$$

The resulting property is

$$\mathscr{Z}[ky(kT)] = -z\frac{d}{dz}Y(z) \tag{4.4-11}$$

with a similar result obtained for the form given by (4.4-10). This property can be used to obtain new $z$ transform pairs.

**TABLE 4.3 Laplace and $z$ Transforms for Sampled Functions**

| | $y(t)$ $t \geq 0$ | $Y(s) = \int_0^\infty y(t)e^{-st}\,dt$ | $Y(z) = \sum_{k=0}^{\infty} y(kT)z^{-k}$ |
|---|---|---|---|
| 1. | $1$ | $\frac{1}{s}$ | $\frac{z}{z-1}$ |
| 2. | $t$ | $\frac{1}{s^2}$ | $\frac{zT}{(z-1)^2}$ |
| 3. | $\frac{t^2}{2}$ | $\frac{1}{s^3}$ | $\frac{z(z+1)T^2}{2(z-1)^3}$ |
| 4. | $e^{-at}$ | $\frac{1}{s+a}$ | $\frac{z}{z-e^{-aT}}$ |
| 5. | $te^{-at}$ | $\frac{1}{(s+a)^2}$ | $\frac{zTe^{-aT}}{(z-e^{-aT})^2}$ |
| 6. | $\sin \omega t$ | $\frac{\omega}{s^2+\omega^2}$ | $\frac{z \sin \omega T}{z^2 - 2z \cos \omega T + 1}$ |
| 7. | $\cos \omega t$ | $\frac{s}{s^2+\omega^2}$ | $\frac{z(z-\cos \omega T)}{z^2 - 2z \cos \omega T + 1}$ |
| 8. | $e^{-at} \sin \omega t$ | $\frac{\omega}{(s+a)^2+\omega^2}$ | $\frac{ze^{-aT} \sin \omega T}{z^2 - 2ze^{-aT} \cos \omega T + e^{-2aT}}$ |
| 9. | $e^{-at} \cos \omega t$ | $\frac{s+a}{(s+a)^2+\omega^2}$ | $\frac{z^2 - 2e^{-aT} \cos \omega T}{z^2 - 2ze^{-aT} \cos \omega T + e^{-2aT}}$ |

Table 4.3 lists the $z$ transforms of common sampled functions along with their Laplace transforms. The $z$ transforms are obtained by manipulating series summations in a manner similar to that used in the previous examples. The time period $T$ does not appear in the transform if the time indexing is referenced solely by $k$, as in (4.4-10). This was the case in Example 4.5. The conversion between the two forms is easily done by letting $T = 1$ in the sampled transform. For example, $y(k) = k$ represents the unit ramp expressed without reference to an absolute sampling time. Its transform is $z/(z-1)^2$, which can be obtained either from (4.4-10) or from (4.4-8) with $T = 1$. Note that this does not imply that the sampling period actually is unity. For convenience, this alternate form is shown in Table 4.4.

## 4.5 THE TRANSFER FUNCTION AND SYSTEM RESPONSE

Our primary interest in the $z$ transform is that it helps us to compute the forced response of difference equation models and allows a transfer function description of

**TABLE 4.4 z Transform Pairs**

| | $y(k)$ for $k \geqslant 0$ | $Y(z) = \sum_{k=0}^{\infty} y(k)z^{-k}$ |
|---|---|---|
| 1. | $1$ | $\dfrac{z}{z-1}$ |
| 2. | $a^k$ | $\dfrac{z}{z-a}$ |
| 3. | $k$ | $\dfrac{z}{(z-1)^2}$ |
| 4. | $k^2$ | $\dfrac{z(z+1)}{(z-1)^2}$ |
| 5. | $ka^k$ | $\dfrac{az}{(z-a)^2}$ |
| 6. | $a^k \sin \omega k$ | $\dfrac{az \sin \omega}{z^2 - 2az \cos \omega + a^2}$ |
| 7. | $a^k \cos \omega k$ | $\dfrac{z(z - a \cos \omega)}{z^2 - 2az \cos \omega + a^2}$ |

discrete-time linear systems. Both of these applications can be demonstrated with the following first-order model.

$$y[(k+1)T] = ay(kT) + bu(kT) \tag{4.5-1}$$

If $T$ is specified, this is more simply expressed as

$$y(k+1) = ay(k) + bu(k) \tag{4.5-2}$$

where $u(k)$ is a specified input or forcing function; $a$ and $b$ are constants.

It is easily established from the definition (4.4-10) that the $z$ transform is a linear operator. Therefore, transforming both sides of (4.5-2) gives

$$\begin{aligned} \mathscr{Z}[y(k+1)] &= \mathscr{Z}[ay(k) + bu(k)] \\ &= a\mathscr{Z}[y(k)] + b\mathscr{Z}[u(k)] \end{aligned} \tag{4.5-3}$$

Equation (4.4-10) also implies that

$$\begin{aligned} \mathscr{Z}[y(k+1)] &= y(1) + y(2)z^{-1} + y(3)z^{-2} + \dots \\ &= z[y(0) + y(1)z^{-1} + y(2)z^{-2} + \dots] - zy(0) \end{aligned}$$

or

$$\mathscr{Z}[y(k+1)] = zY(z) - zy(0) \tag{4.5-4}$$

This is the *left-shifting property* of the transform and is analogous to the Laplace transform of a derivative. Application of the shifting property to (4.5-3) gives

$$zY(z) - zy(0) = aY(z) + bU(z)$$

or

$$Y(z) = \frac{z}{z-a}y(0) + \frac{b}{z-a}U(z) \tag{4.5-5}$$

This shows that the complete solution consists of the sum of two terms: a free response and a forced response. As with the first-order linear differential equation, the free response depends only on the initial condition, while the forced response results from the input function. Thus, the two effects can be treated separately.

With the linearity property and the results of Example 4.5, the free response is found to be

$$\mathscr{X}^{-1}\left[\frac{z}{z-a}y(0)\right] = y(0)a^k \tag{4.5-6}$$

where $\mathscr{X}^{-1}$ represents the inverse transformation. We have seen this before in (4.2-4).

## The Transfer Function

If $y(0)$ is zero, the transform of the solution is

$$Y(z) = \frac{b}{z-a}U(z)$$

The ratio $Y(z)/U(z)$ of the output and input transforms is the transfer function of the system. It is also called the *pulse transfer function* or *discrete transfer function* to distinguish it from its Laplace transform counterpart for continuous-time systems.

The transfer function of (4.5-2) is

$$\frac{Y(z)}{U(z)} = \frac{b}{z-a} \tag{4.5-7}$$

The characteristic root of the model is the root of the denominator of the transfer function after it is normalized to a ratio of two polynomials in $z$. Here the root is $z = a$, and it tells much about the system behavior, as can be seen from Figure 4.2.

### *Example 4.7*

Investigate the behavior of the system for which $a = 0$, $b = 1$, and $y(0) = 0$. Interpret its transfer function.

The transfer function is $1/z$ and

$$Y(z) = \frac{1}{z}U(z) \tag{4.5-8}$$

The original difference equation shows that

$$y(k+1) = u(k) \tag{4.5-9}$$

This model represents a simple time shift of the input such that $y(1) = u(0)$, $y(2) = u(1)$, etc. In this sense the transfer function $1/z$ represents a shift of one unit forward in time.

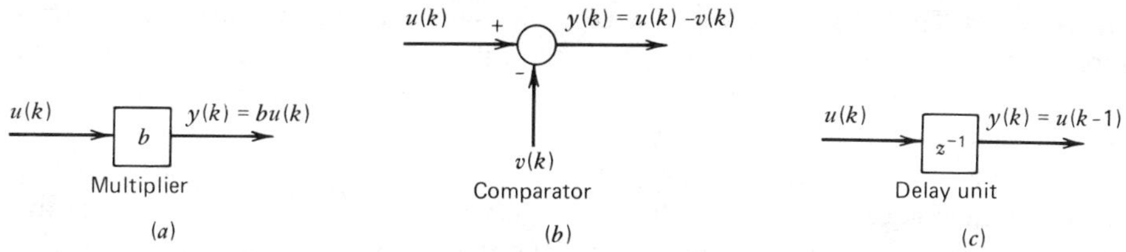


**Figure 4.8 Three basic elements of block diagrams for discrete-time systems. (*a*) Multiplier. (*b*) Comparator. (*c*) Delay unit.**

## System Diagrams

One advantage of the transfer function concept is that its input-output structure provides a basis for a graphic description of system behavior, such as a block diagram. The three basic elements of block diagrams for discrete-time systems are shown in Figure 4.8. The multiplier and comparator are identical to their continuous-time counterparts. The new element in the *delay* unit, which delays the input signal by one time unit to produce the output. From Example 4.7 the transfer function of the delay unit is seen to be $1/z$. In discrete-time diagrams the delay unit plays a central role similar to that played by the integrator element whose transfer function is $1/s$.

These elements are sufficient to describe the first-order system (4.5-2), as shown in Figure 4.9*a*. The corresponding signal flow diagram is shown in Figure 4.9*b*. For clarity the time variables have been shown instead of the transformed variables required by the strict interpretation of the diagram. The block diagram with transformed variables is shown in Figure 4.9*c*.

The discrete-time system equations become algebraic equations after being

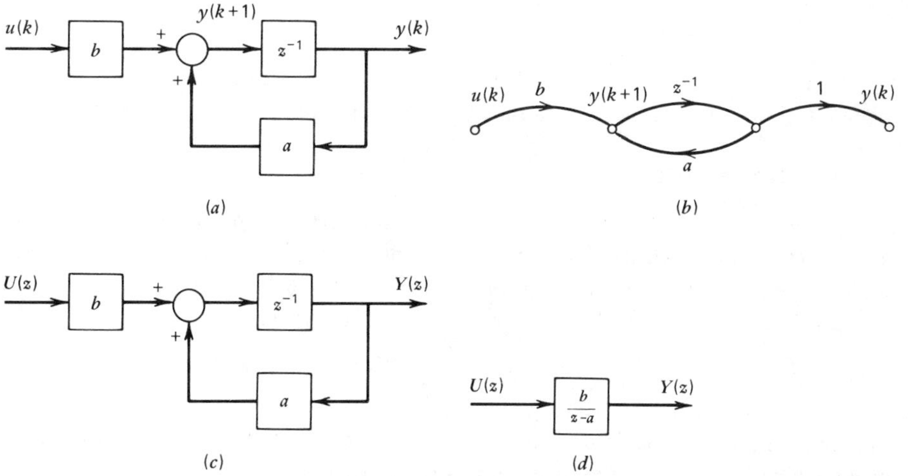


**Figure 4.9 Representations of a first-order discrete-time system. (*a*) Block diagram in the time domain. (*b*) Signal flow graph in the time domain. (*c*) Block diagram in terms of *z*-transforms. (*d*) Reduced block diagram showing the system transfer function.**

transformed. Then they obey the same rules of diagram algebra as Laplace-transformed continuous-time equations. Figure 3.9 contains the common reduction formulas. For example, the loop reduction formula in Figure 3.14*b* can be applied to Figure 4.9*c* to obtain the transfer function given by (4.5-7).

A pulse transfer function for an analog system can be found by approximating its differential equation model with a difference equation representation such as that given by the Euler method. However, several different approximations are in common use, and each will give a different pulse transfer function. This topic is covered in more detail in Section 4.9.

*Example 4.8*

Use the Euler method to find the pulse transfer function of the series $RC$ circuit shown in Figure 2.16. The input voltage is denoted here by $u(t)$, and the output voltage by $y(t)$.

The differential equation relating $y$ to $u$ is

$$RC\frac{dy}{dt} = u - y \tag{4.5-10}$$

The Euler approximation assumes the right-hand side of this equation is constant over the step size, denoted here by $T$. Using this approximation and the expression (4.1-2) for $dy/dt$, we obtain

$$y(k+1) = y(k) + \frac{T}{RC}[u(k) - y(k)] \tag{4.5-11}$$

For zero initial conditions, the $z$ transform of this equation is

$$zY(z) = \left(1 - \frac{T}{RC}\right)Y(z) + \frac{1}{RC}U(z)$$

or

$$T(z) = \frac{Y(z)}{U(z)} = \frac{b}{z + b - 1} \tag{4.5-12}$$

where $b = T/RC$. This is the desired pulse transfer function.

## System Response

The system transfer function is useful in computing the response to a given input. In most cases a partial fraction expansion is required to decompose the transformed function into a set of simpler terms whose inverse transforms are given in Table 4.4. For functions with a simple root, the transform is of the form

$$\frac{Cz}{z + d}$$

Thus the expansion is made in terms of similar expressions. This is more easily done by expanding $Y(z)/z$ as a series of terms of the form $1/(z + d)$.

*Example 4.9*

Find the impulse response of (4.5-2) for $y(0) = 0$ and an impulse of strength $A$.

For the impulse, $U(z) = A$ from Table 4.4, and from (4.5-7),

$$Y(z) = \frac{b}{z-a} A = \frac{bz}{z-a} \frac{A}{z}$$

The expansion of $Y(z)/z$ is

$$\frac{Y(z)}{z} = \frac{b}{z-a} \frac{A}{z} = \frac{C_1}{z-a} + \frac{C_2}{z}$$

Cross-multiplication gives

$$C_1 z + C_2(z-a) = bA$$

or

$$(C_1 + C_2)z - aC_2 = bA$$

From

$$C_1 + C_2 = 0$$

$$-aC_2 = bA$$

the solution is

$$C_1 = -C_2 = \frac{bA}{a}$$

Thus

$$Y(z) = \frac{bA}{a}\left(\frac{z}{z-a} - 1\right)$$

From the transform table we obtain

$$y(k) = \frac{bA}{a}[a^k - \delta(k)]$$

This satisfies the initial condition: $y(0) = 0$ because $\delta(0) = 1$. For $k > 0$ the solution reduces to

$$y(k) = \frac{bA}{a} a^k = bAa^{k-1} \tag{4.5-13}$$

This is equivalent to the free response with $y(0) = bA/a$. However, if we consider the start of the problem to be at $k = 1$, the impulse response is equivalent to the free response with initial condition $y(1) = bA$. With this interpretation, the result resembles that for the continuous-time model.

### *Example 4.10*

Compute the unit step response of (4.5-2) if $y(0) = 0$.

Here $u(k) = \{1, 1, 1, \ldots\}$ for $k \geqslant 0$, and from Table 4.4 and (4.5-7)

$$U(z) = \frac{z}{z-1}$$

$$Y(z) = \frac{b}{z-a} \frac{z}{z-1}$$

The expansion is

$$\frac{Y(z)}{z} = \frac{b}{(z-a)(z-1)} = \frac{b}{a-1}\left(\frac{z}{z-a} - \frac{z}{z-1}\right)$$

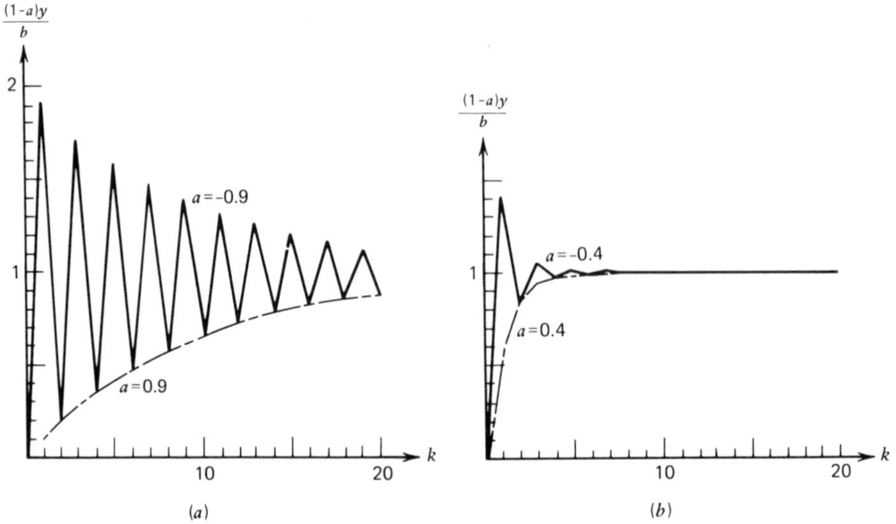


**Figure 4.10 Normalized step response of the first-order system (4.5-2). The response for $a > 0$ provides the envelope of the oscillatory response for $a < 0$.**

From the transform table we obtain

$$y(k) = \frac{b}{a-1}(a^k - 1) = \frac{b}{1-a}(1-a^k) \tag{4.5-14}$$

The response normalized by the factor $b/(1-a)$ is shown in Figure 4.10 for stable values of the root $z = a$. Note the occurrence of oscillations for $a < 0$.

*Example 4.11*

Compute the ramp response of (4.5-2) with $y(0) = 0$.

For the ramp function with a slope $m$,

$$u(k) = mk$$

$$U(z) = \frac{mz}{(z-1)^2}$$

and

$$\frac{Y(z)}{z} = \frac{b}{z-a}\frac{m}{(z-1)^2}$$

Thus

$$\frac{Y(z)}{z} = \frac{b}{z-a}\frac{m}{(z-1)^2} = \frac{C_1}{z-a} + \frac{C_2}{(z-1)^2} + \frac{C_3}{z-1}$$

where the expansion for repeated roots is performed similarly to that in Section 3.3.

The result is

$$C_1 = \frac{mb}{(a-1)^2} = -C_3$$

$$C_2 = (1-a)C_1$$

and

$$Y(z) = \frac{mb}{(a-1)^2}\left[\frac{z}{z-a} + \frac{(1-a)z}{(z-1)^2} - \frac{z}{z-1}\right]$$

The response is

$$y(k) = \frac{mb}{(a-1)^2}\,[a^k + (1-a)k - 1] \tag{4.5-15}$$

If uniform sampling with a period $T$ is applied to the *unit* ramp function, $u(t) = t$, the sampled sequence is $u(kT) = kT$. This is not a *unit* ramp in terms of the time index $k$ unless $T = 1$. When speaking of unit functions such as the unit ramp, the time variables ($t$ or $k$) should be specified.

## Final Value Theorem

As with continuous-time systems, the transfer function provides a convenient way of computing the steady-state value of the output by means of the final value theorem. To derive this theorem for a discrete-time system, note that

$$\mathscr{Z}[y(k+1)] = zY(z) - zy(0) = \sum_{k=0}^{\infty} y(k+1)z^{-k}$$

$$\mathscr{Z}[y(k)] = Y(z) = \sum_{k=0}^{\infty} y(k)z^{-k}$$

Subtracting the second equation for the first gives

$$(z-1)Y(z) - zy(0) = \sum_{k=0}^{\infty} [y(k+1) - y(k)]z^{-k} = y(\infty) - y(0)$$

Thus

$$\frac{(z-1)}{z}Y(z) = y(0) + \frac{y(\infty) - y(0)}{z}$$

and

$$\lim_{z \to 1} \frac{z-1}{z} Y(z) = y(\infty)$$

The theorem is valid only when the preceding limit exists. This occurs when the system is stable and the input is such that the output can approach a definite value. Mathematically, these conditions are satisfied if the product $(z-1)Y(z)/z$ has no poles $z$ such that $|z| \geqslant 1$. In this case the product is said to be *analytic* for $|z| \geqslant 1$.

This and other properties of the $z$ transform are listed in Table 4.5.

# 4.6 SAMPLED-DATA SYSTEMS

In most engineering applications there is at least one component of the system that is analog in nature. The motion of a mechanical element is an example and thus can be described by differential equations. If a digital device is used to measure or control

TABLE 4.5 Properties of the $z$ Transform

| Property | $y(t)$ or $y(k)$ | $\mathscr{Z}[y(t)]$ or $\mathscr{Z}[y(k)]$ |
|---|---|---|
| 1. Linearity | $ay(t) + bx(t)$ | $aY(z) + bX(z)$ |
| 2. Right-shifting | $y(t - mT)$<br>$y(k - m)$ | $z^{-m}Y(z)$ |
| 3. Left-shifting | (a) $y(t + T)$<br>$y(k + 1)$ | $zY(z) - zy(0)$ |
| | (b) $y(t + 2T)$<br>$y(k + 2)$ | $z^2Y(z) - z^2y(0) - zy(T)$<br>$z^2Y(z) - z^2y(0) - zy(1)$ |
| | (c) $y(t + mT)$ | $z^mY(z) - \sum_{i=0}^{m-1} y(iT)z^{m-i}$ |
| | $y(k + m)$ | $z^mY(z) - \sum_{i=0}^{m-1} y(i)z^{m-i}$ |
| 4. Differentiation | $ty(t)$ | $-Tz\frac{d}{dz}[Y(z)]$ |
| | $ky(k)$ | $-z\frac{d}{dz}[Y(z)]$ |
| 5. Convolution | $\sum_{i=0}^{k} x(iT)y(kT - iT)$ | $X(z)Y(z)$ |
| 6. Summation | $\sum_{i=0}^{k} y(iT)$ | $\frac{z}{z-1}Y(z)$ |
| 7. Multiplication by an exponential | (a) $e^{-at}y(t)$ | $Y(ze^{aT})$ |
| | (b) $a^ky(k)$ | $Y\left(\frac{z}{a}\right)$ |
| 8. Initial value theorem | $y(0) = \lim_{\|z\| \to \infty} Y(z)$ | |
| 9. Final value theorem | $y(\infty) = \lim_{z \to 1} \frac{(z-1)}{z} Y(z)$<br>if $\frac{(z-1)}{z} Y(z)$ is analytic for $\|z\| \geq 1$ | |

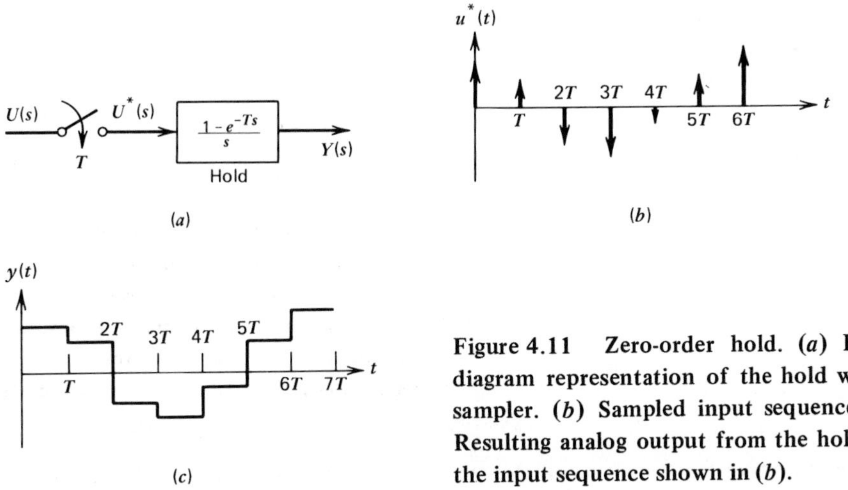


**Figure 4.11 Zero-order hold. (*a*) Block diagram representation of the hold with a sampler. (*b*) Sampled input sequence. (*c*) Resulting analog output from the hold for the input sequence shown in (*b*).**

the mechanical element, the resulting system is a discrete-continuous hybrid. Such systems are called *sampled-data* systems. The mathematical modeling of sampled-data systems must be treated with special care because of their mixed nature. Here we present pertinent modeling and analysis techniques for such applications.

## Zero-Order Hold Circuit

The binary form of the output from the digital device is first converted to a sequence of short voltage pulses. However, it usually is not possible to drive a load such as a motor with short pulses. In order to deliver sufficient energy, the pulse amplitude might have to be so large as to be infeasible to be generated. Also, large voltage pulses might saturate or even damage the system being driven.

The solution to this problem is to smooth the output pulses to produce a signal in analog (continuous) form. Thus the digital-to-analog converter performs two functions: first, generation of the output pulses from the digital representation produced by the machine, and second, conversion of these pulses to analog form.

The simplest way of converting an impulse sequence into a continuous signal is to hold the value of the impulse until the next one arrives. Since the impulses are really short-duration pulses, the net effect is to extend their duration to $T$, the sampling period (Figure 4.11). The device that accomplishes this is called a *zero-order hold* or *boxcar hold*, from the output signal shape. The term *zero-order* refers to the zero-order polynomial used to extrapolate between the sampling times. A first-order hold uses a first-order polynomial (a straight line with nonzero slope) for extrapolation. The zero-order hold is by far the easiest to construct physically and to program, and it is the most widely used.

If the input to the hold is a unit impulse, its resulting output will reveal the transfer function. For such an input at $t = 0$, the output is a pulse with a unit amplitude from $t = 0$ to $t = T$. The Laplace transform of this pulse is obtained with the shifting property and is

$$G(s) = \frac{1}{s}(1 - e^{-Ts}) \tag{4.6-1}$$

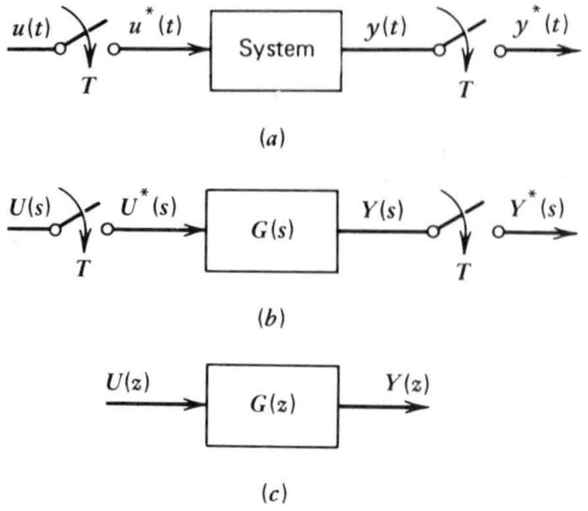


**Figure 4.12** Equivalent representations of a sampled-data system. (*a*) Time-domain diagram. (*b*) *s*-domain diagram. (*c*) *z*-domain diagram (automatically implies the existence of an input sampler and of either a real or a fictitious output sampler.)

This is the transfer function of the zero-order hold. Referring to Figure 4.11*a* we see that

$$Y(s) = G(s)U^*(s) \tag{4.6-2}$$

Each impulse at the input to the zero-order hold is converted into a rectangular pulse of width $T$ and a height $u(kT)$ equal to the sample value at that time. If these pulses are applied to an element whose time constants are large compared to $T$, the pulses can be considered to be impulses with a strength $Tu(kT)$. Thus the hold not only converts the impulses into analog form but also supplies a gain $T$.

## Pulse Transfer Functions of Analog Elements

When discrete-time signals are used with a hold to drive an analog device such as a motor, the block diagram representation naturally consists of a mixture of discrete elements like samplers and continuous elements representing the transfer functions of the analog devices. The alternative to this hybrid diagram is to describe all the system elements by difference equations and pulse transfer functions, and to use the discrete-time block diagram elements shown in Figure 4.8. For example, the differential equation describing the dynamics of a motor and its load might be converted to a difference equation by the Euler method, and its discrete-time transfer function obtained as in Section 4.5. This alternate approach avoids a mixed description of the system but has the disadvantages associated with approximations required to obtain the describing difference equations. The optimal choice is not always clear. In this section we present the mixed description.

We know that for the system shown in Figure 4.12*b*

$$Y(s) = G(s)U^*(s)$$

and for Figure 4.12*c*

$$Y(z) = G(z)U(z)$$

The question to be answered is: given $G(s)$, what is $G(z)$? Since the transfer functions $G(s)$ and $G(z)$ represent the unit impulse response $g(t)$ in the $s$ and $z$ domains,

$$g(t) = \mathscr{L}^{-1}[G(s)] \tag{4.6-3}$$

and

$$G(z) = \mathscr{Z}[g(t)] \tag{4.6-4}$$

Thus $G(z)$ is obtained from the $z$ transform of $g(t)$, which can be found from $G(s)$.

Table 4.3 has been constructed to facilitate this procedure. Commonly encountered time functions are given there along with their Laplace and $z$ transforms. These entries give $G(s)$ and $G(z)$ for the corresponding $g(t)$. Thus, given $G(s)$, the appropriate $G(z)$ is found on the same line in the table. For more complicated forms not found in the table, $G(s)$ must be written as a partial fraction expansion. The preceding procedure is then applied to each term in the expansion.

### *Example 4.12*

Find the pulse transfer function and difference equation for the series $RC$ circuit shown in Figure 4.13*a*. The input voltage $u(t)$ is sampled before being applied to the circuit. The output is the voltage $y$. The block diagram is shown in Figure 4.13*b*.

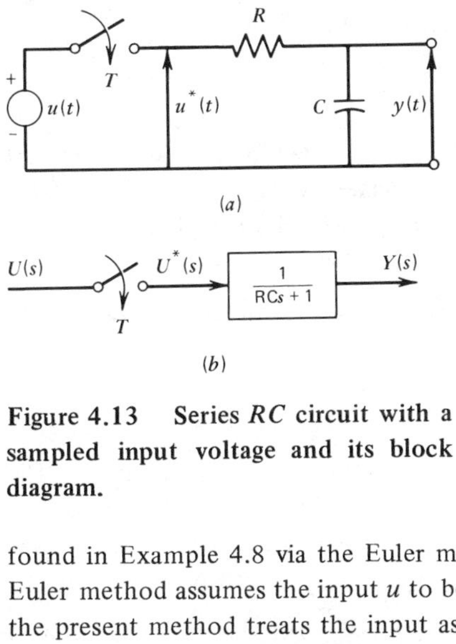


**Figure 4.13 Series $RC$ circuit with a sampled input voltage and its block diagram.**

From the diagram and Figure 4.12*b* we see that

$$G(s) = \frac{1}{RCs+1} = \frac{1/RC}{s+1/RC} \tag{4.6-5}$$

From Table 4.3, No. 5,

$$G(z) = \frac{1}{RC}\frac{z}{z-e^{-T/RC}} = \frac{Y(z)}{U(z)} \tag{4.6-6}$$

and

$$y(k+1) - e^{-T/RC}y(k) = \frac{1}{RC}u(k+1) \tag{4.6-7}$$

This pulse transfer function and difference equation are quite different from those found in Example 4.8 via the Euler method. One reason for the difference is that the Euler method assumes the input $u$ to be constant over the sampling interval $T$, whereas the present method treats the input as a train of impulses. The second reason is that the $z$-transform entries $G(z)$ in Table 4.3 represent the exact solution for the impulse response corresponding to $G(s)$. Equation (4.6-7) represents this solution for $y(t_{k+1})$ with initial condition $y(t_k)$ and an impulse input $u(t_{k+1})$ occurring at time $t_{k+1}$ (not at $t_k$). It involves no approximation for the derivative, as does the Euler method.

The diagrams in Figures 4.12*a* and 4.12*b* both have samplers at their outputs, whereas that in Figure 4.13*b* does not. Although the output voltage of the $RC$ circuit in Figure 4.13*b* is analog, its pulse transfer function relates the output to the input only at the sample times. That is, imagine a "fictitious" output sampler that is synchronized with the input sampler. The output of this fictitious sampler is $y^*(t)$ and is the output that can be computed from the pulse transfer function. Thus Figure 4.12 portrays the general situation, in that the output samplers shown there may be real or fictitious, depending on the application.

If the sampler is fictitious (i.e., an analog output), and if it is desired to compute the output values between the sample times, a more complicated analysis is required.

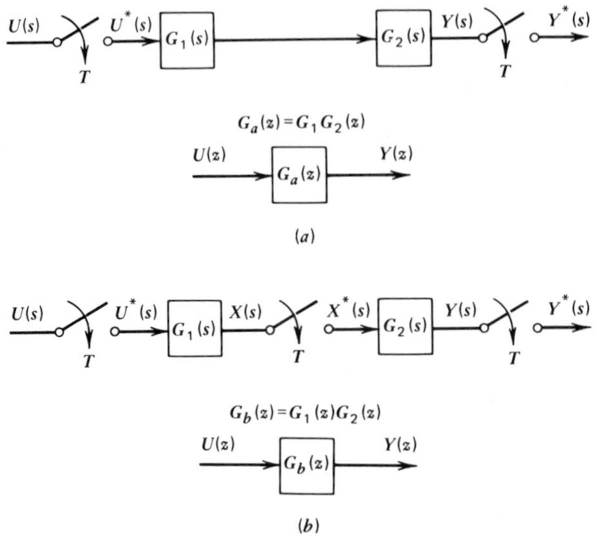


**Figure 4.14 Transfer functions of cascaded elements. (*a*) Analog elements not separated by a sampler and their cascaded pulse-transfer function $G_a(z)$. (*b*) Analog elements separated by a sampler and their cascaded pulse-transfer function $G_b(z)$.**

The essence of one technique is to choose the fictitious sampling period to be a submultiple of the real period $T$.

## Cascaded Elements

Care is needed in obtaining the pulse transfer function for cascaded elements. The pulse transfer function between $U$ and $Y$ in Figure 4.14*a* is not the same as that between $U$ and $Y$ in Figure 4.14*b*. If no sampler is present between elements, the usual cascade formula applies. Thus for Figure 4.14*a*,

$$Y(s) = G_1(s)G_2(s)U^*(s) = G_a(s)U^*(s) \tag{4.6-8}$$

and

$$Y(z) = G_a(z)U(z)$$

where

$$G_a(z) = \mathscr{Z}[g_a(t)]$$

$$g_a(t) = \mathscr{L}^{-1}[G_1(s)G_2(s)] \tag{4.6-9}$$

For Figure 4.14*b*,

$$Y(s) = G_2(s)X^*(s)$$

$$X(s) = G_1(s)U^*(s)$$

$$X(z) = G_1(z)U(z)$$

$$Y(z) = G_2(z)X(z) = G_2(z)G_1(z)U(z) = G_b(z)U(z) \tag{4.6-10}$$

where

$$G_i(z) = \mathscr{Z}[g_i(t)] \qquad i = 1, 2 \tag{4.6-11}$$

$$g_i(t) = \mathscr{L}^{-1}[G_i(s)] \qquad i = 1, 2$$

So the pulse transfer functions $G_a(z)$ and $G_b(z)$ are not equal in general because $G_a(z)$ does not equal $G_2(z)G_1(z)$ in general. This is most easily proved by a counter-example.

*Example 4.13*

Refer to Figure 4.14. Let $G_1(s) = 1/(s + c)$ and $G_2(s) = 1/(s + d)$. Find $G_a(z)$ and $G_b(z)$.

(a) Find $G_a(z)$. From (4.6-8)

$$G_a(s) = \frac{1}{s+c}\frac{1}{s+d}$$

$$= \frac{1}{d-c}\left(\frac{1}{s+c} - \frac{1}{s+d}\right)$$

Table 4.3 gives

$$G_a(z) = \frac{1}{d-c}\left(\frac{z}{z-e^{-cT}} - \frac{z}{z-e^{-dT}}\right)$$

$$= \frac{z}{d-c}\frac{e^{-cT}-e^{-dT}}{(z-e^{-cT})(z-e^{-dT})} \qquad \textbf{(4.6-12)}$$

(b) Find $G_b(z)$. From (4.6-10),

$$G_b(z) = G_1(z)G_2(z)$$

where from Table 4.3,

$$G_1(z) = \frac{z}{z-e^{-cT}}$$

$$G_2(z) = \frac{z}{z-e^{-dT}}$$

Thus

$$G_b(z) = \frac{z^2}{(z-e^{-cT})(z-e^{-dT})} \qquad \textbf{(4.6-13)}$$

$G_a(z)$ is obviously not equal to $G_b(z)$.

In light of the preceding observation, we now introduce a new notation for use with cascaded elements. In Figure 4.14*a*, $G_a(z)$ is found by operating on the product $G_1(s)G_2(s)$. We will denote this operation by

$$G_a(z) = G_1G_2(z) = \mathscr{Z}\{\mathscr{L}^{-1}[G_1(s)G_2(s)]\} \qquad (4.6\text{-}14)$$

Note that $G_1G_2(z) = G_2G_1(z)$. In this notation the $z$ operator is meant to apply to the product, and Example 4.13 shows that in general

$$G_1(z)G_2(z) \neq G_1G_2(z) \qquad (4.6\text{-}15)$$

Thus it is important to look for a sampler between elements (the preceding analysis assumes that all samplers are synchronized with the same period).

We can use these results to obtain the transfer function for a system with a zero-order hold. If the hold element is cascaded with an analog element, with no sampler in between, write the transfer function product as

$$G(s) = G_1(s)G_2(s) \qquad (4.6\text{-}16)$$

where $G_1(s) = 1 - e^{-Ts}$ and $G_2(s)$ is the remainder of $G(s)$. This gives

$$G(s) = G_2(s) - e^{-Ts}G_2(s)$$

and

$$g(t) = g_2(t) - g_2(t-T)$$

From the third property in Table 4.5,

$$\begin{aligned} G(z) &= G_2(z) - z^{-1}G_2(z) \\ &= (1 - z^{-1})G_2(z) = \frac{z-1}{z}G_2(z) \end{aligned} \tag{4.6-17}$$

This identity simplifies the analysis.

*Example 4.14*

The series $RC$ circuit is now shown in Figure 4.15 with a sample and hold acting on the input voltage. Find the pulse transfer function and difference equation.

From (4.6-16)

$$G_2(s) = \frac{1}{s}\frac{1/RC}{s + 1/RC} = \frac{1}{s} - \frac{1}{s + 1/RC}$$

and

$$\begin{aligned} G_2(z) &= \frac{z}{z-1} - \frac{z}{z - e^{-T/RC}} \\ &= \frac{z(1 - e^{-T/RC})}{(z-1)(z - e^{-T/RC})} \end{aligned}$$

The required transfer function is obtained from (4.6-17).

$$G(z) = \frac{z-1}{z}G_2(z) = \frac{1 - e^{-T/RC}}{z - e^{-T/RC}} = \frac{Y(z)}{U(z)} \tag{4.6-18}$$

This implies that

$$y(k+1) = e^{-T/RC}y(k) + (1 - e^{-T/RC})u(k) \tag{4.6-19}$$

Because of the zero-order hold's effect, (4.6-19) represents the pulse response of the differential equation (4.5-10) for the circuit, just as (4.6-7) represents the impulse response. Even with a piecewise-constant input, (4.6-19) still differs from (4.5-11) because the latter involves a derivative approximation.

## Block Diagram Algebra for Sampled-Data Systems

Given the block diagram in terms of Laplace transforms, the usual rules of block diagram algebra can be applied to reduce the diagram and obtain the pulse transfer

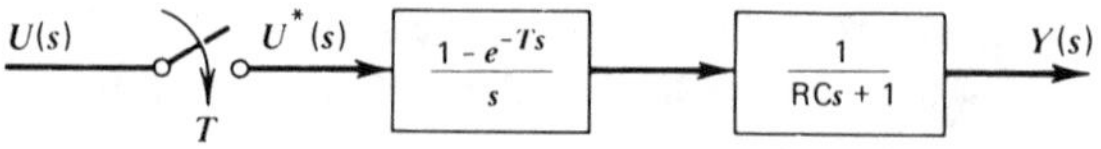


**Figure 4.15** Series *RC* circuit block diagram with a sample and hold acting on the input voltage.

function. Here one must be careful whenever a sampler is present within a loop. The procedure is to check all signal paths. If a signal passes from one element to another without going through a sampler, the transfer functions for these elements must be combined according to (4.6-14). The algebraic relations are then written in terms of $z$ transformed variables and the reduction carried out.

For example, consider the diagram shown in Figure 4.16*a*. The signal $E^*(s)$ passes through $G(s)$ to $H(s)$ without encountering a sampler. Thus

$$B(z) = GH(z)E^*(z)$$

Because the $z$ transform specifies quantities only at the sample times,

$$E^*(z) = E(z)$$

Also, from the diagram,

$$C(z) = G(z)E^*(z)$$

$$E(z) = R(z) - B(z)$$

Combining these, we obtain

$$E(z) = R(z) - GH(z)E(z)$$

$$E(z) = \frac{R(z)}{1 + GH(z)}$$

$$C(z) = G(z)\frac{R(z)}{1 + GH(z)}$$

The transfer function is

$$T(z) = \frac{C(z)}{R(z)} = \frac{G(z)}{1 + GH(z)} \tag{4.6-20}$$

Note that the presence of the output sampler outside the loop did not affect the transfer function.

Now consider Figure 4.16*b*, where a sampler has been inserted in the loop between $G(s)$ and $H(s)$. In this case

$$B(z) = G(z)H(z)E^*(z)$$

and the rest of the equations are the same as before. The resulting algebra gives

$$T(z) = \frac{C(z)}{R(z)} = \frac{G(z)}{1 + G(z)H(z)} \tag{4.6-21}$$

Note that this differs from (4.6-20). Related forms are given in Figure 4.16*c*, 4.16*d*, and 4.16*e*.

When no sampling is performed on the error signal $e(t)$, the preceding procedure does not yield a pulse transfer function that is a ratio of the output $C(z)$ to the input $R(z)$. Figure 4.16*f* shows an example of this. From the diagram,

$$C(s) = G(s)[R(s) - H(s)C^*(s)]$$

$$= G(s)R(s) - G(s)H(s)C^*(s)$$

Thus

$$C(z) = GR(z) - GH(z)C^*(z)$$

But

$$C^*(z) = C(z)$$

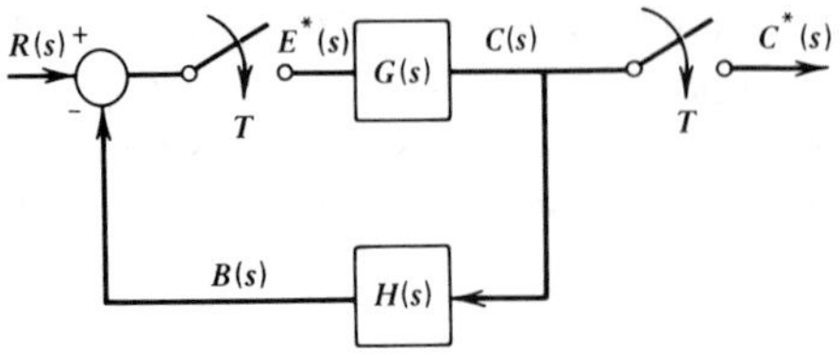


$$C(z) = \frac{G(z)R(z)}{1 + GH(z)}$$

(a)

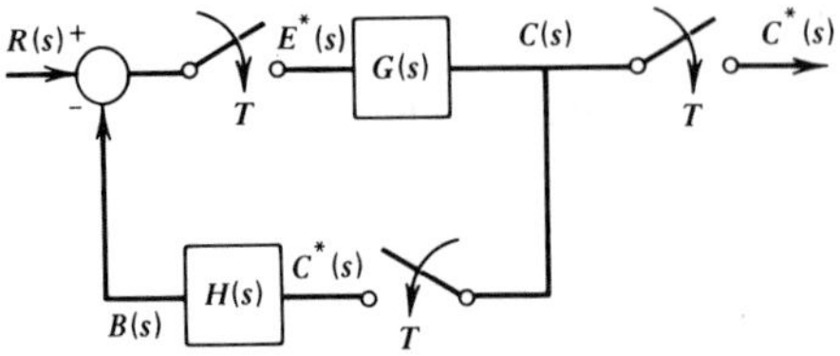


$$C(z) = \frac{G(z)R(z)}{1 + G(z)H(z)}$$

(b)

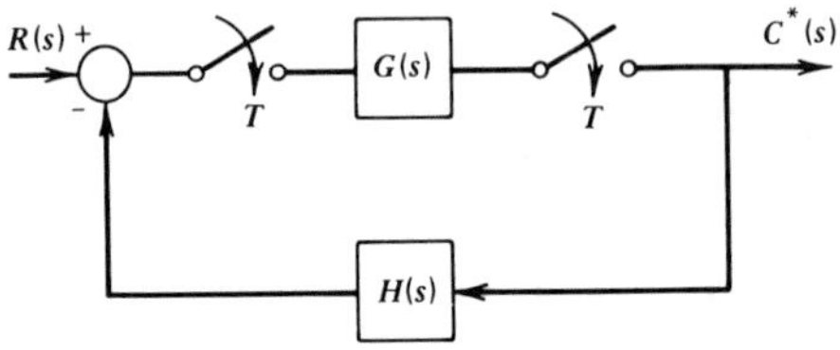


$$C(z) = \frac{G(z)R(z)}{1 + G(z)H(z)}$$

(c)

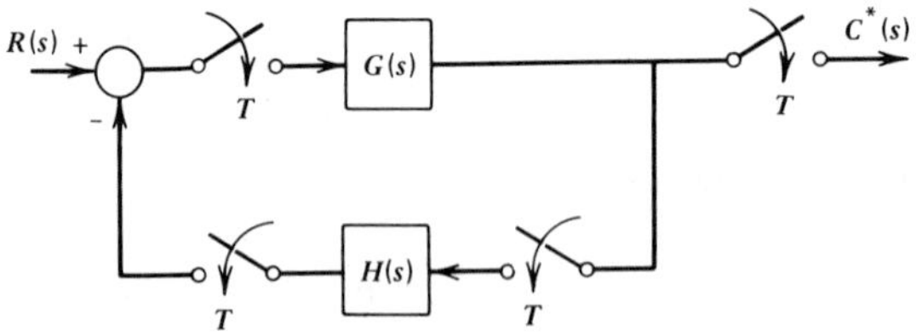


$$C(z) = \frac{G(z)R(z)}{1 + G(z)H(z)}$$

(d)

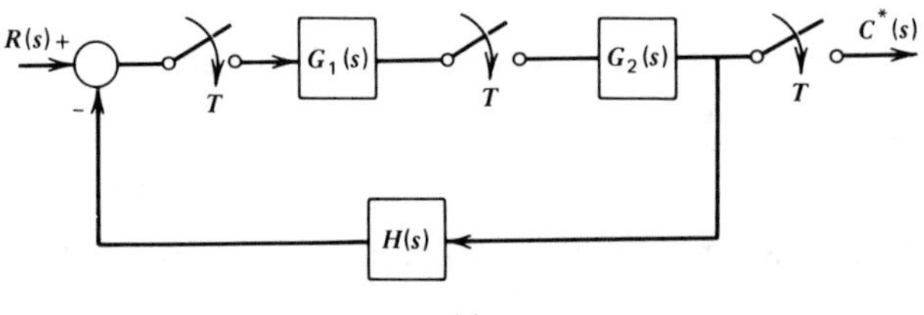


$$C(z) = \frac{G_1(z)G_2(z)R(z)}{1 + G_1(z)G_2H(z)}$$

(e)

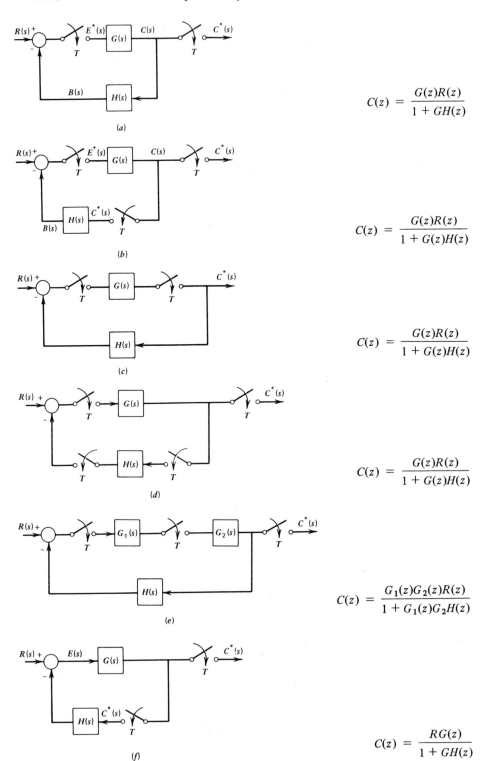

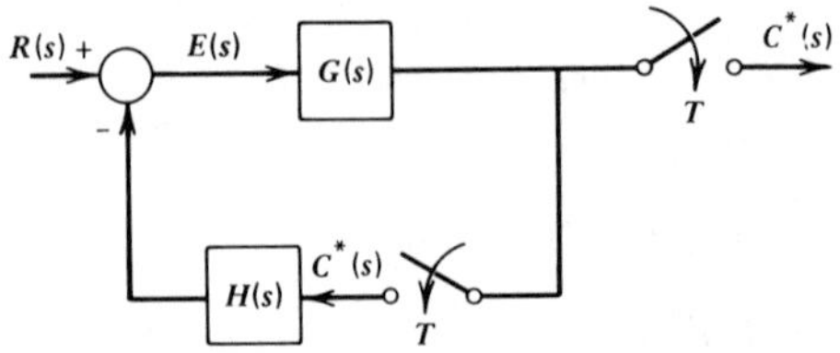


$$C(z) = \frac{RG(z)}{1 + GH(z)}$$

(f)

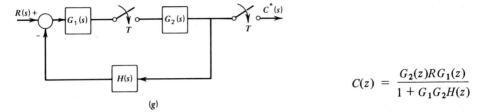

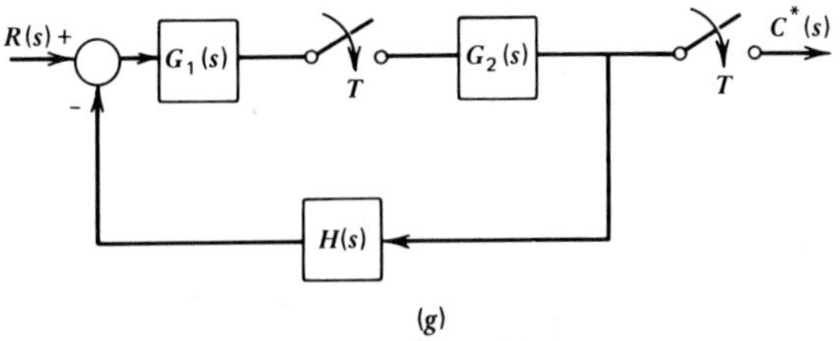


$$C(z) = \frac{G_2(z)RG_1(z)}{1 + G_1G_2H(z)}$$

**Figure 4.16 Output relations for various loop and sampler configurations.**

from which we obtain

$$C(z) = \frac{GR(z)}{1 + GH(z)} \tag{4.6-22}$$

A related form is given in Figure 4.16*g*.

Note that any starred (sampled) Laplace quantity, such as $C^*(s)$ in the preceding example, is *not* combined in product form in the manner of (4.6-14). The only variables to which (4.6-14) applies are those that have never been subjected to sampling [such as $R(s)$ in the preceding example] or that have passed through a dynamic operation after being sampled (the dynamic operation actually creates a new, unsampled variable).

It is not permissible to create a pulse transfer function for the preceding example by inserting a fictitious sampler at the input, because this would change the physics of the situation represented by the diagram (the analog input would be replaced by a train of impulses). Fictitious samplers are permissible only at the output, because they are simply a means of selecting the values of the output at the times of interest to us – namely, the sample times.

The introduction of sampling can destabilize a system, as shown by the following example.

### *Example 4.15*

Compare the stability properties of the system shown in Figure 4.17, with and without a sample-and-hold on the error signal. Assume $K > 0$.

The analog system in Figure 4.17*a* has the transfer function

$$T(s) = \frac{K}{s + K + 3}$$

This system is stable for all $K > 0$. For the system in Figure 4.17*b*,

$$G(s) = (1 - e^{-sT})\frac{K}{s(s+3)} = (1 - e^{-sT})\frac{K}{3}\left(\frac{1}{s} - \frac{1}{s+3}\right)$$

$$G(z) = \frac{K}{3}\frac{z-1}{z}\left(\frac{z}{z-1} - \frac{z}{z - e^{-3T}}\right) = \frac{K}{3}\frac{1 - e^{-3T}}{z - e^{-3T}}$$

$$GH(z) = G(z)$$

Thus its transfer function is

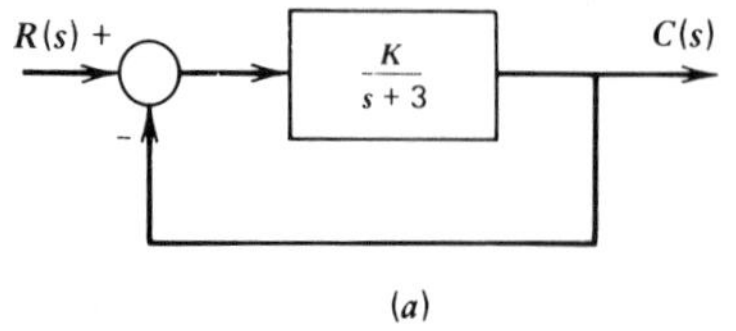


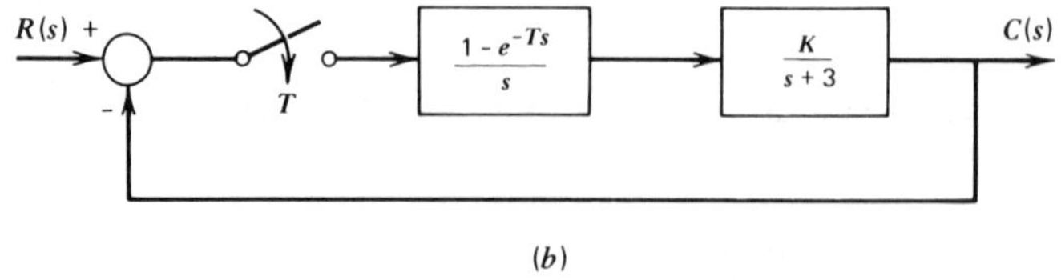


**Figure 4.17 Instability caused by sampling. (*a*) Original analog system. (*b*) Analog system with a sample and hold inserted.**

$$T(z) = \frac{K}{3} \frac{1 - e^{-3T}}{z - e^{-3T} + \dfrac{K}{3}(1 - e^{-3T})}$$

For an unstable system, $|z| > 1$. This occurs if either

$$e^{-3T} > \frac{K+3}{K+3} = 1$$

or

$$e^{-3T} < \frac{K-3}{K+3}$$

The first condition is never true for $T > 0$, and we now consider the implications of the second condition. As the sampling frequency is increased ($T \to 0$), the stability of the system is improved (its behavior approaches that of the analog system). For example, with $T = 1$, the system is unstable for $K > 3.315$. For $T = 0.5$, instability occurs if $K > 4.723$.

## 4.7 DIGITAL FILTERING: TIME-DOMAIN ANALYSIS

Over the years technological developments have caused an evolution in the meaning of the term *filter*. The primitive concept referred to a physical device that removes high-frequency components from a signal. These components are referred to generally as "noise". An example of such a device is the low-pass *RC* circuit treated in Section 3.4. The signal variable is not always electrical, however. For example, an accumulator (Section 2.6) filters out high-frequency pressure fluctuations. With the improvement and miniaturization of op amp circuitry it became possible to build devices that operated on sensor signals in a closed-loop manner to reduce the uncertainties inherent in the measurements. In the sense that sensor measurement uncertainties constituted "noise" in the measurement signal, these devices came to be called filters because they removed the noise to yield a better estimate of the measured variable. One of the most powerful and popular algorithms for designing such circuits is the *Kalman filter* algorithm, named after its developer.

Filtering took on a new meaning with the introduction of digital devices. Now the term *filter* also includes any algorithm or formula that acts on a sequence of input *numbers* to produce another sequence. Such an algorithm is called a *digital filter*. Its function is the same as that of an analog filter; namely, to remove unwanted components from a signal or number sequence. Perhaps the most common example is the running average of a sequence of numbers $u(k)$. The two-point average is expressed by the formula

$$\hat{y}(k+1) = \tfrac{1}{2}[u(k+1) + u(k)] \tag{4.7-1}$$

where $\hat{y}$ is the estimate.

Modern systems theory terminology distinguishes between *filtering*, *smoothing*, and *prediction*. If the independent variable is time, smoothing refers to the process of obtaining the best estimate of a previous value of the measured variable, using past measurements up to the present time. A common example of a smoothing algorithm is the least-squares criterion for fitting a curve to data. Filtering is the process of estimating the current value of the variable, and prediction refers to estimating a future value. The term *estimation* refers to all three processes. The techniques of digital filtering can be extended to develop methods for smoothing and prediction.

## Filter Design in the Time-Domain

The two-point running average formula given by (4.7-1) can be analyzed by difference equation techniques. It contains the implicit assumption that $y$, the value to be estimated, is a constant and that the measurements $u(k)$ fluctuate about the true value of $y$. If the fluctuation amplitude is large, an average over a longer string of data would be used. The $n$-point average is

$$\hat{y}(k+1) = \frac{1}{n}[u(k+1) + u(k) + u(k-1) + \ldots + u(k-n+2)] \tag{4.7-2}$$

The larger $n$ is, the more digital storage is required. This might be a critical consideration for a small special-purpose device.

Equation (4.7-2) is an example of a *nonrecursive* filter. This implies that the data $u(k)$ must be processed in batch form and that the algorithm does not take advantage of previous calculations as new data arrive. However, the algorithm can be rearranged to alleviate this problem somewhat by noting that

$$\hat{y}(k) = \frac{1}{n}[u(k) + u(k-1) + \ldots + u(k-n+1)] \tag{4.7-3}$$

Thus

$$u(k) + u(k-1) + \ldots + u(k-n+1) = n\hat{y}(k) - u(k-n+1)$$

Substituting this into (4.7-2) gives

$$\hat{y}(k+1) = \hat{y}(k) + \frac{1}{n}[u(k+1) - u(k-n+1)] \tag{4.7-4}$$

This says that the estimate of $y$ at time $(k+1)$ consists of the sum of the previous estimate and a correction term proportional to the difference between the latest measurement and the measurement $n$ time periods earlier. The proportional factor

$1/n$ weights the correction term relative to the previous estimate $\hat{y}(k)$. If $n$ is large, the filter places more emphasis on the previous estimate. This makes sense because this estimate results from a long string of data for large $n$ and presumably contains more information than is in the two measurements, $u(k+1)$ and $u(k-n+1)$. This form of the running average algorithm is said to be *recursive* because it uses the previous average to reduce the number of required calculations.

While the number of computations required by (4.7-4) is less than those for (4.7-2), the same number of data points must be stored even though (4.7-4) uses only two at a time. If the sampling rate is high, the number of storage locations ($n$) will be large. Thus if we can find a filter algorithm that utilizes only the current measurement and the previous estimate, we will have considerably reduced both the storage requirements and the number of required calculations. With the additional insight that linear equations are easier to work with, we are led to try the following filter form.

$$\hat{y}(k+1) = w_1\hat{y}(k) + w_2u(k+1) \tag{4.7-5}$$

where $w_1$ and $w_2$ are weighting factors measuring the relative importance of the previous estimate and the current measurement $u(k+1)$.

Up to now we have been considering the measurement of a constant $y$. A more general case is the one in which the quantity to be estimated is a variable $y$ described by the linear model.

$$y(k+1) = ay(k) \tag{4.2-3}$$

It is emphasized that the variable $y$, which presumably represents a physical process, usually cannot be described exactly by such a simple model. On the other hand, our measurements of $y$ are not exact either. The coefficients $w_1$ and $w_2$ can be chosen to express the relative degrees of confidence we have in our model of the process and our measurements. If we believe our model completely and do not trust our measurements, we choose $w_1 = a$ and $w_2 = 0$. This choice tells the filter not to use the measurements at all and to estimate $y(k+1)$ using the model given by (4.2-3). The other extreme is complete faith in our measurements and none in the model. This leads to the choice $w_1 = 0$ and $w_2 = 1$. The current estimate of $y$ is then taken to be the value of the last measurement.

The preceding interpretation of the filter weights suggests the following rearrangement of the algorithm.

$$\hat{y}(k+1) = a\hat{y}(k) + K[u(k+1) - a\hat{y}(k)] \tag{4.7-6}$$

where $w_1 = a(1-K)$ and $w_2 = K$. The first term on the right of the equation is the prediction of the current value of $y$ based on the model (4.2-3). The second term is a correction term consisting of the difference between the current measurement and the current value of $y$ as predicted by the model. This difference is weighted by the filter gain $K$. From the two extreme cases discussed previously we see that a choice of $K = 0$ reflects complete faith in the model, and of $K = 1$, complete faith in the measurement.

The filter's dynamics are revealed by its transfer function. The $z$ transform of (4.7-6) is

$$z\hat{Y}(z) - z\hat{y}(0) = a\hat{Y}(z) + KzU(z) - Kzu(0) - aK\hat{Y}(z) \tag{4.7-7}$$

or

$$\hat{Y}(z) = \frac{z[\hat{y}(0) - Ku(0)]}{z - a + aK} + \frac{Kz}{z - a + aK} U(z) \tag{4.7-8}$$

The denominator of the transfer function has the single root $z = a(1 - K)$, and the filter estimate is inherently nonoscillatory if $a(1 - K) > 0$. [An oscillating data sequence $u(k)$ could cause oscillations, however.] From the preceding discussion we see that $K$ lies in the range $0 < K < 1$. Thus no inherent oscillation of the estimate occurs unless $a < 0$. This is acceptable because $a < 0$ implies that the process being estimated is itself oscillatory. If the quantity to be estimated is modeled as a constant, then $a = 1$ and the filter is stable if $0 \leqslant K < 1$.

*Example 4.16*

Assume a process $y(t)$ is to be modeled as the decaying exponential, $y(0) \exp(-t/\tau)$, where $\tau$ is assumed to be known at least approximately. The variable $y$ might represent the temperature of an object being cooled. Develop a linear recursive filter to estimate $y(t)$ using only the current measurement.

The given process is continuous-time, so a suitable sampling period $T$ must be selected. The sampling theorem states that $T = 0.049\tau$ is sufficient for this process. With a safety factor of ten, we use $T = 0.0049\tau$. The discrete-time model of the process is

$$y[(k+1)T] = y(kT)e^{-T/\tau} = y(kT)e^{-0.0049} = 0.9951y(kT)$$

With the filter form given by (4.7-6), $a = 0.9951$, and the desired algorithm is

$$\hat{y}(k+1) = 0.9951\hat{y}(k) + K[u(k+1) - 0.9951\hat{y}(k)]$$

where the time between discrete instants is $T$, and $u(k)$ is the error-contaminated measurement of the process variable $y(k)$. (If the measurements did not contain error, we would not need a filter.) Our choice of a value for $K$ depends on how confident we are of (1) the form of the process model (exponential), (2) our knowledge of the model parameter $\tau$, and (3) our measurements.

The filter algorithm given by (4.7-6) requires an initial value $\hat{y}(0)$ in order to start. Frequently this value taken to be the value given by the first measurement $u(1)$; that is, $\hat{y}(0) = u(1)$. In view of the additional requirement of choosing a value for the gain $K$, one might wonder if this filter has any advantages over traditional estimation methods such as averaging or least-squares techniques. If $y$ is taken to be a constant, an estimate can be obtained by averaging the measurements; that is,

$$\hat{y}(k) = \frac{1}{k} \sum_{i=1}^{k} u(i) \tag{4.7-9}$$

When the next measurement becomes available at time $k + 1$, the new estimate is

$$\hat{y}(k+1) = \frac{1}{k+1} \sum_{i=1}^{k+1} u(i)$$

$$= \frac{k}{k+1} \left[ \frac{1}{k} \sum_{i=1}^{k} u(i) \right] + \frac{1}{k+1} u(k+1)$$

or

$$\hat{y}(k+1) = \frac{k}{k+1}\hat{y}(k) + \frac{1}{k+1}u(k+1) \tag{4.7-10}$$

This is the recursive form of the simple averaging process. It eliminates the need to store all the past measurements. Note that in this case a starting value of $\hat{y}$ at $k = 0$ need not be provided because the coefficient of $\hat{y}(k)$ vanishes when $k = 0$. The algorithm can be rewritten as

$$\hat{y}(k+1) = \hat{y}(k) + \frac{1}{k+1}[u(k+1) - \hat{y}(k)] \tag{4.7-11}$$

In this form a resemblance can be seen to the filter form given by (4.7-6) with $a = 1$. However, here the filter gain $1/(k+1)$ is time-varying. The measurement is weighted less and less as time goes on.

### *Example 4.17*

Compare the two methods (4.7-6) and (4.7-11) of estimating a constant whose true value is 1, for the case in which the initial measurement is $u(1) = 0$, and succeeding measurements are perfect $[u(k) = 1, k > 1]$.

(a) Equation (4.7-6). Since this filter is formulated with $u(k+1)$ only for $k \geqslant 0$, $u(0)$ does not enter, and it may be taken to be zero along with $u(k)$ for $k < 0$. Thus the measurement sequence is a unit step function delayed by two time units. From the shifting property,

$$U(z) = z^{-2}\left(\frac{z}{z-1}\right) = \frac{1}{z(z-1)}$$

Equation (4.7-7) with $a = 1$ and $u(0) = 0$ gives

$$\hat{Y}(z) = \frac{z\hat{y}(0)}{z-1+K} + \frac{Kz}{(z-1+K)}\frac{1}{(z-1)z}$$

The expansion gives

$$\frac{\hat{Y}(z)}{z} = \frac{\hat{y}(0)}{z-1+K} + \frac{K}{(1-K)z} + \frac{1}{(K-1)(z-1+K)} + \frac{1}{z-1}$$

Thus

$$\hat{y}(k) = y(0)(1-K)^k + \frac{K}{1-K}\delta(k) + \frac{1}{K-1}(1-K)^k + 1$$

If we take the first measurement $u(1)$ as a guess for $y(0)$, then $\hat{y}(0) = 0$ and for $k \geqslant 1$,

$$\hat{y}(k) = 1 - (1-K)^{k-1}$$

If we have equal confidence in our measurements and our assumption that $y$ is constant, then we would choose $K = 0.5$. In this case the estimate $\hat{y}$ is within 10% of the true value after the fifth measurement. This response is tabulated in the accompanying table.

(b) Equation (4.7-10). No initial condition is needed here and we can solve the equation recursively. (The $z$ transform cannot be used because of the time-varying coefficient.) The results are tabulated in the table. The estimate is within 10% of the true value after the tenth measurement.

| | | $\hat{y}(k)$ | |
|---|---|---|---|
| Time $k$ | $u(k)$ | (4.7-6) with $\hat{y}(0) = 0$ and $K = 0.5$ | (4.7-10) |
| 1 | 0 | 0 | 0 |
| 2 | 1 | 0.5 | 0.5 |
| 3 | 1 | 0.75 | 0.6667 |
| 4 | 1 | 0.885 | 0.75 |
| 5 | 1 | 0.9375 | 0.8 |
| 6 | 1 | 0.9688 | 0.8333 |
| 7 | 1 | 0.9844 | 0.8571 |
| 8 | 1 | 0.9922 | 0.875 |
| 9 | 1 | 0.9961 | 0.8888 |
| 10 | 1 | 0.9980 | 0.9 |

Note that both methods require a model of the process being estimated. The averaging algorithm implicitly assumes that $y$ is constant. Similarly, estimation techniques based on least-squares criteria also require a process model. For example, if we fit a straight line to the data, we are assuming that the process can be modeled by a straight line. The so-called *weighted* least-squares method also requires a selection of weighting factors for every measurement as indicators of their reliability.

The filter algorithm given by (4.7-6) requires also a selection of values for the gain $K$ and initial condition $\hat{y}(0)$. In spite of this, the filter has advantages in that the rate of convergence to the true value can be faster than for the other methods. The algorithm can be extended to higher-order models, and systematic methods are available for determining the filter gains in terms of the statistical properties of the process model uncertainties and measurement errors (see Reference 5).

## 4.8 FREQUENCY RESPONSE OF DISCRETE SYSTEMS

The response of a discrete-time system to a sinusoidal input is determined in a manner similar to that used for continuous-time systems. If a sinusoid of circular frequency $\omega$ and amplitude $A$ is sampled at a sampling period $T$, the resulting sequence is

$$u(k) = A \sin k\omega T \tag{4.8-1}$$

If this sequence is applied as an input to a stable system whose transfer function is $T(z)$, the steady-state output is

$$y(k) = B \sin (k\omega T + \phi) \tag{4.8-2}$$

where

$$M = \frac{B}{A} = |T(e^{i\omega T})| \tag{4.8-3}$$

$$\phi = \measuredangle T(e^{i\omega T}) \tag{4.8-4}$$

Thus the steady-state output is also sinusoidal with the same frequency $\omega$. The amplitude ratio $M$ and phase shift $\phi$ are found by substituting $z = \exp(i\omega T)$ in the transfer function. [Recall that the substitution $s = i\omega$ is made in $T(s)$ for continuous-time systems]. As in Chapter Three, polar or rectangular plots can be constructed for $M$ and $\phi$ as functions of $\omega$.

Both $M$ and $\phi$ are periodic functions of $\omega$ with period $2\pi/T$. This is true because $\exp(i\omega T)$ is periodic in $\omega$ with period $2\pi/T$, as can be shown from the Euler identity relating $\exp(i\omega T)$ to $\sin \omega T$ and $\cos \omega T$. Also, from the sampling theorem, the highest significant frequency can be no greater than $\pi/T$, if $T$ has been chosen correctly. Thus we need to consider the behavior of $M$ and $\phi$ only over the frequency range $0 \leqslant \omega \leqslant \pi/T$.

*Example 4.18*

Investigate the frequency response characteristics of the filter algorithm given by (4.7-6).

The transfer function from (4.7-8) is

$$T(z) = \frac{Kz}{z - a + aK} \tag{4.8-5}$$

Substitution of $z = \exp(i\omega T)$ and use of Euler's identity gives

$$T(e^{i\omega T}) = \frac{Ke^{i\omega T}}{e^{i\omega T} - a + ak}$$

$$= \frac{K}{1 + a(K-1)e^{-i\omega T}}$$

$$= \frac{K}{1 + a(K-1)\cos \omega T - ia(K-1)\sin \omega T}$$

After some manipulation the magnitude and phase angle can be shown to be

$$M = |T(e^{i\omega T})| = K[a^2(K-1)^2 + 2a(K-1)\cos \omega T + 1]^{-1/2} \tag{4.8-6}$$

$$\phi = \tan^{-1}\left[\frac{a(K-1)\sin \omega T}{a(K-1)\cos \omega T + 1}\right]$$

Two cases are possible: $a \geqslant 0$ and $a < 0$ (we assume here that $0 \leqslant K \leqslant 1$). The frequency range of interest is $0 \leqslant \omega \leqslant \pi/T$. The values of $M$ at the two extreme frequencies are also of interest. At $\omega = 0$,

$$M = \frac{K}{|a(K-1)+1|}$$

and at $\omega = \pi/T$,

$$M = \frac{K}{|a(K-1)-1|}$$

If $a > 0$ the maximum value of $M$ occurs at $\omega = 0$ and the minimum at $\omega = \pi/T$. The opposite is true for $a < 0$. Thus the algorithm defined by (4.7-6) is a *low-pass* filter if $a > 0$, and a *high-pass* filter if $a < 0$. This is consistent with our previous interpre-

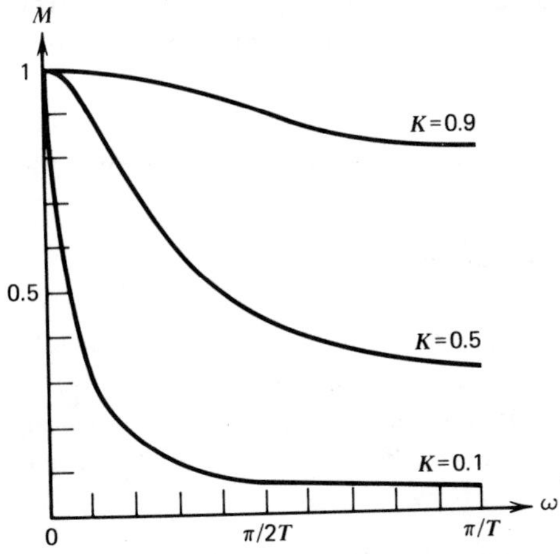


**Figure 4.18** **Magnitude ratio $M$ versus circular frequency $\omega$ for the digital filter (4.8-5) with $a = 1$.**

tation of the algorithm. For $a < 0$ the filter is trying to estimate the value of a variable assumed to be alternating in sign at every time step (and hence possessing a circular frequency of $\omega = \pi/T$). We would expect the filter to "look for" such fluctuations and to pass the high-frequency components.

Figure 4.18 shows the magnitude of ratio $M$ plotted versus frequency for the case $a = 1$ and several values of $K$. For $K = 0.9$ very little attentuation occurs at any frequency. This is the case in which heavy weight is given to the measurement. For $K = 0.1$, heavy weight is given to the model; the filter algorithm expects to be estimating a constant and therefore tends to ignore measurement fluctuations with higher frequencies. From Table 4.1 the time constant for the filter is $1/\ln(1 - K)$. The smaller value of $K$ produces a larger time constant and hence a more sluggish system. For small $K$ the filter estimate generally takes longer to converge to the true value, but also does a better job of filtering out high-frequency components resulting from, for example, measurement noise.

The general linear algorithm given by

$$y(k + 1) = ay(k) + bu(k + 1) \tag{4.8-8}$$

is a low-pass filter if $a > 0$, and a high-pass filter if $a < 0$. Define the system bandwidth in the same way as for continuous-time systems; that is, the bandwidth is the range of frequencies for which the magnitude ratio $M$ satisfies

$$M \geqslant \frac{M_{max}}{\sqrt{2}} \tag{4.8-9}$$

where $M_{max}$ is the maximum value of $M$. In general, as the designer reduces the filter's bandwidth to suppress measurement noise, the response of the filter becomes sluggish. If a satisfactory compromise cannot be reached, a higher-order algorithm often works.

*Example 4.19*

Write the software required to implement a digital filter for estimating an exponentially decaying voltage whose time constant is thought to be approximately 0.5 sec. The electrical instrumentation has line noise due to nearby equipment operating on 60-Hz ac current.

From the results of Example 4.4 we require a sampling period of $T = 0.0049(0.5) = 0.00245$ sec. Thus the highest circular frequency the digital filter will see is

$$\omega = \frac{\pi}{T} = 1282 \text{ rad/sec}$$

Since 60 Hz is 377 rad/sec, the noise frequency is seen by the filter and the bandwidth should be made less than this frequency. For the filter form given in Example 4.16, in terms of the notation of (4.7-6), $a = 0.9951$. Using (4.8-6) and the bandwidth definition (4.8-9), we find that the noise frequency lies outside the bandwidth if $K \leqslant 0.58$. For $K = 0.58$, the filter algorithm is

$$\hat{y}(k+1) = 0.9951\hat{y}(k) + 0.58[u(k+1) - 0.9951\hat{y}(k)]$$

or

$$\hat{y}(k+1) = 0.4179\hat{y}(k) + 0.58u(k+1) \tag{4.8-10}$$

The general outline of a FORTRAN program implementing this filter is shown below. The first measurement is used to initialize the filter estimate. The measurements will be read by the program every $2.45 \times 10^{-3}$ sec. Thus the time for the program to execute one cycle must be less than this time, if the filter is to run in "real time."

```
  I = 0
1 READ U
  IF(I.EQ.0)  YI = U
  I = I + 1
  Y = 0.4179 * YI + 0.58 * U
  OUTPUT Y
  YI = Y
  GO TO 1
```

The frequency-response curve used to design this filter is a representation of the filter's steady-state response and does not include the effect of transients. That is, the filter might perform well at steady state but give poor estimates before the steady state is reached. An estimate of how long this takes is provided by the time constant. Referring to Table 4.1, the time constant is

$$k = -\frac{1}{\ln 0.4179} = 1.15 \text{ steps}$$

Thus the transients will be essentially gone after $4(1.15) \cong 5$ steps, or $5(2.45 \times 10^{-3}) = 0.012$ sec. This time is short compared to the estimated time constant of the input (0.5 sec).

## 4.9 INVERSE OPERATORS: THE TUSTIN METHOD

Numerical solution of a differential equation requires that it be converted to discrete-time form. One way of doing this is to replace the time derivatives by an equivalent finite-difference expression. The Euler method does this by expressing the derivative as follows.

$$\frac{dy}{dt} = \frac{y(t+\Delta t) - y(t)}{\Delta t} \tag{4.9-1}$$

Substitution of this into the equation

$$\frac{dy}{dt} = f(y, t) \tag{4.9-2}$$

yields the difference equation

$$y(t_{k+1}) = y(t_k) + f[y(t_k), t_k]\,\Delta t \tag{4.9-3}$$

This is the Euler recursion relation.

A better approximation to the derivative is the *Tustin* approximation. This is based on the trapezoidal rule presented in Section 3.14. For the equation

$$\frac{dy}{dt} = v \tag{4.9-4}$$

the trapezoidal formula results in the approximate integration rule

$$y(k+1) = y(k) + \frac{\Delta t}{2}[v(k+1) + v(k)] \tag{4.9-5}$$

The pulse transfer function between $y$ and $v$ is found from (4.9-5) as follows. For zero initial conditions,

$$zY(z) = Y(z) + \frac{\Delta t}{2}(z+1)V(z)$$

Thus

$$Y(z) = \frac{\Delta t}{2}\frac{1+z^{-1}}{1-z^{-1}}V(z) \tag{4.9-6}$$

Equation (4.9-6) states that the transfer function between the integral $y$ as output and the integrand $v$ as input is

$$\frac{Y(z)}{V(z)} = I(z) = \frac{\Delta t}{2}\frac{1+z^{-1}}{1+z^{-1}} \tag{4.9-7}$$

for the trapezoidal formula. On the other hand, since differentiation and integration are inverse operations, (4.9-6) also says that the transfer function between the derivative $v$ and the integral $y$ as input is

$$\frac{V(z)}{Y(z)} = D(z) = \frac{1}{I(z)} = \frac{2}{\Delta t}\frac{1-z^{-1}}{1+z^{-1}} \tag{4.9-8}$$

Equation (4.9-8) is the Tustin approximation for the $z$-transform of $v$, the derivative of $y$, or

$$\mathscr{Z}\left(\frac{dy}{dt}\right) = V(z) = D(z)Y(z) \tag{4.9-9}$$

The Tustin approximation is used as follows. The $z$ transform of $\dot{y} = ry + by$ is

$$\mathscr{Z}\left(\frac{dy}{dt}\right) = r\mathscr{Z}(y) + b\mathscr{Z}(v)$$

Using (4.9-8) and (4.9-9), we get

$$\frac{2}{\Delta t}\frac{1-z^{-1}}{1+z^{-1}}Y(z) = rY(z) + bV(z)$$

or

$$\left(\frac{2}{\Delta t} - r\right) zY(z) = \left(\frac{2}{\Delta t} + r\right) Y(z) + b(z+1)V(z)$$

The corresponding difference equation is

$$\left(\frac{2}{\Delta t} - r\right) y(k+1) = \left(\frac{2}{\Delta t} + r\right) y(k) + bv(k+1) + bv(k)$$

or

$$y(k+1) = \frac{2+r\Delta t}{2-r\Delta t}y(k) + \frac{b\,\Delta t}{2-r\Delta t}[v(k+1)+v(k)] \tag{4.9-10}$$

Equation (4.9-10) is the recursion algorithm that implements the numerical integration of $\dot{y} = ry + bv$ via the trapezoidal rule. Even though the differential equation is linear, the recursion algorithm is useful when the input $v(t)$ is not Laplace transformable, or when $v(t)$ is given only at discrete instants.

The Tustin approximation is easily generalized to higher-order models. From (4.9-9), $V(z)$ is the transform of the derivative of $y$. Therefore, the transform of the second derivative of $y$ is the transform of the derivative of $v$; that is,

$$\mathscr{Z}\left(\frac{d^2y}{dt^2}\right) = D(z)V(z) = [D(z)]^2Y(z) \tag{4.9-11}$$

In general, the Tustin approximation for the $n$th derivative is

$$\mathscr{Z}\left(\frac{d^ny}{dt^n}\right) = [D(z)]^nY(z) \tag{4.9-12}$$

with $D(z)$ given by (4.9-8).

*Example 4.20*

Use the Tustin method to derive a difference equation approximation for

$$\frac{d^2y}{dt^2} + 5\frac{dy}{dt} + 3y = v$$

Take the transform of both sides and use (4.9-8) and (4.9-11) to obtain

$$\{[D(z)]^2 + 5D(z) + 3\}Y(z) = V(z)$$

$$\left[\frac{4}{(\Delta t)^2}\frac{1-2z^{-1}+z^{-2}}{1+2z^{-1}+z^{-2}} + \frac{10}{\Delta t}\frac{1-z^{-1}}{1+z^{-1}} + 3\right] Y(z) = V(z)$$

For convenience let $a = \Delta t$, multiply by $(1+z^{-1})^2a^2z^2$, and collect terms to obtain

$$[(3a^2+10a+4)z^2 + (6a^2-8)z + (3a^2-10a+4)]\,Y(z) = a^2(z^2+2z+1)V(z)$$

The corresponding difference equation is

$$c_2 y(k+2) + c_1 y(k+1) + c_0 y(k) = a^2 v(k+2) + 2a^2 v(k+1) + a^2 v(k) \tag{4.9-13}$$

*Where*

$$c_2 = 3a^2 + 10a + 4$$
$$c_1 = 6a^2 - 8$$
$$c_0 = 3a^2 - 10a + 4$$
$$a = \Delta t$$

The response characteristics of this second-order model are treated in the next chapter.

Other approximations to the transform of $dy/dt$ can be developed by applying other integration formulas to (4.9-4) with a procedure similar to that we have used for the trapezoidal rule. The staircase approximation to the integral results in the Euler method; Simpson's rule results in still another derivative approximation, and so on. However, the Tustin approximation is one of the most commonly used methods.

## 4.10 SUMMARY

Analysis of discrete-time models resembles that of continuous-time models in many ways. The Laplace and $z$-transforms allow linear models to be reduced to algebraic problems in terms of the $s$ and $z$ variables, respectively, and the concept of a transfer function can be applied to both model types. The response can be obtained by partial fraction expansion and inversion of the transform. The only difference occurs when we return to the time domain. For example, the unit step function corresponds to $1/s$ in the Laplace domain but to $z/(z-1)$ in the $z$ domain. The stability regions reflect this difference [$|z| < 1$ versus $Re(s) < 0$]. Thus facility for continuous-time analysis is easily transferred to discrete-time analysis, and vice versa.

Important applications of discrete-time modeling and analysis occur where an analog system is coupled to a digital device to produce a sampled-data system. No natural sampling period exists in such cases, and the designer must select it unless it has been specified by some other source. Of course, no comparable decision must be made in continuous-time systems. Also, the location of the samplers in the system must be carefully considered before reducing the block diagram to find the pulse transfer function. This is because cascaded elements combine differently depending on whether or not a sampler is located between them. We have also seen that a transfer function will not exist for configurations that do not contain an input sampler (see Figures 4.16*f* and 4.16*g*). Finally, the presence of sampling with some type of hold, usually a zero-order hold, can alter the stability characteristics of the system. This phenomenon is not seen in continuous-time systems.

There are several advantages to using a digital device instead of an analog element. To illustrate these, assume that the application is the implementation of a filter of some sort (low pass, bandpass, or high pass).

- Digital implementation readily allows the use of time-varying coefficients – in the simple averaging process (4.7-11), for example.
- Algorithms can be implemented digitally even when analog implementation is impossible because of unrealizable component values (for example, extremely large capacitance values).
- Nonlinear algorithms can easily be implemented in digital form. Design of nonlinear analog elements is often difficult.
- Greater accuracy is possible with digital systems.
- Digital systems are less sensitive to environmental factors like humidity, temperature, pressure, and noise.

On the other hand, digital systems have some disadvantages.

- Digital systems are active and thus require power (analog filters like the *RC* circuit are often passive).
- Input signals with a wide range of values cannot be handled unless the number of bits is sufficient for accurate quantization.
- Processing is limited to signals whose highest frequency content is less than the logic circuitry speed.

For these reasons one finds applications for both analog and digital systems.

## REFERENCES


1. J. B. Peatman, *The Design of Digital Systems*, McGraw-Hill, New York, 1972.
2. R. M. May, "Biological Populations with Nonoverlapping Generations: Stable Points, Stable Cycles, and Chaos," *Science*, Vol. 186, pp. 645–647, 1974.
3. R. W. Hamming, *Digital Filters*, Prentice Hall, Englewood Cliffs, N.J., 1977.
4. H. J. Blinchikoff and A. I. Zverev, *Filtering in the Time and Frequency Domains*, John Wiley, New York, 1976.
5. A. Gelb, ed., *Applied Optimal Estimation*, MIT Press, Cambridge, 1974.


## PROBLEMS

**4.1** Numerical algorithms preprogrammed into a calculator's memory are often in the form of difference equations. An example of this is the equation

$$x(k) = \tfrac{1}{2}\left[x(k-1) + \frac{u}{x(k-1)}\right]$$

where $u$ is a constant (for example, a number entered from the keyboard)

y Ball, Plays

Nebraska
uman Relations Training

Illinois at Chicago

ry Education

al Estate School

Center

Medical Transcription

(a) What function does this equation implement? (*Hint.* Find the equilibrium solution.)

(b) Simulate the algorithm for a value of $u$, say $u = 4$, with several initial values $x(0)$, and investigate its dynamics.

4.2 One way to compute the integral

$$x(t) = \int_0^t u(\lambda)\, d\lambda$$

is to convert it into a difference equation. Let

$$x(kT) = \int_0^{kT} u(\lambda)\, d\lambda$$

where $T$ is the discretization interval.

(a) Use the rectangular approximation

$$\int_{kT-T}^{kT} u(\lambda)\, d\lambda = Tu(kT - T)$$

to derive a difference equation for $x(kT)$ in terms of $x(kT - T)$ and $u(kT - T)$. What is the initial condition?

(b) Suppose that $u(t) = t$ and we wish to find $x(t = 1)$. Select an interval $T$ and simulate the equation found in (a) to obtain $x(1)$. Compare with the exact value.

4.3 Suppose that we apply a piecewise-constant torque $m(kT)$ to an inertia $I$. Use the step response of the continuous-time model $I\dot{\omega} = m$ to obtain a difference equation model of the speed $\omega$ at the discrete times $0, 2T, 3T, \ldots$.

4.4 Use the convolution integral (3.10-3) for the model $\dot{y} = ry + bv$ to obtain a difference equation model for the response $y(t)$ at the discrete times $t = 0, T, 2T, \ldots, kT, \ldots$, given that the input $v(t)$ is piecewise constant and changes only at the times $t = kT, k = 1, 2, \ldots$.

4.5 A difference equation used to model the dynamics of insect populations is the Hassel-May-Varley model

$$y(k + 1) = \lambda y(k)[1 + ay(k)]^{-b}$$

where $\lambda$, $a$, and $b$ are positive.

(a) Find the equilibrium solution of the population density $y$.

(b) Linearize the model and discuss its stability.

4.6 Determine the Fourier series representation of the function shown in Figure P4.6 and plot its spectrum.

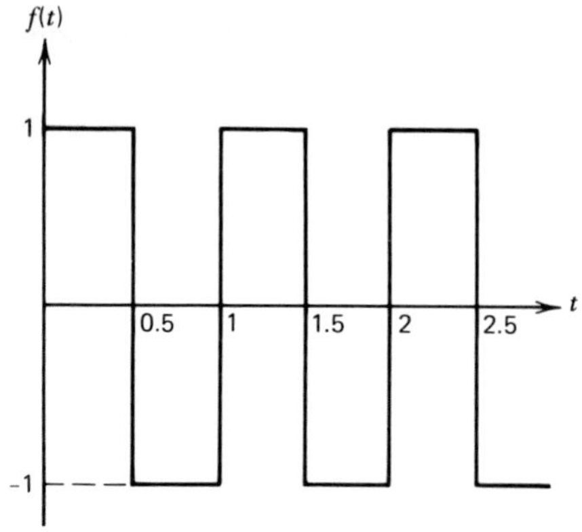


**Figure P4.6**

**4.7** For the signal whose spectrum was found in Problem 4.6, compute the frequency range that contains 99% of the signal's energy and estimate the required sampling frequency.

**4.8** Compute the $z$-transforms for the following functions.

(a) $\frac{1}{2}t^2$

(b) $e^{-at}$

(c) $te^{-at}$

(d) $\sin \omega t$

(e) $\cos \omega t$

(f) $e^{-at} \sin \omega t$

(g) A unit amplitude pulse of duration $L$.

**4.9** Use the Euler method to compute the pulse transfer function of the models with dead time.

(a) $\dot{y} = ay(t) + by(t-D) + u(t)$

(b) $\dot{y} = ay(t) + bu(t-D)$

Assume that the dead time $D$ is an integral multiple of the sampling period $T$, so that $D = nT$, $n =$ integer.

**4.10** Use the Euler method to find the pulse transfer function of the model with numerator dynamics.

$$\dot{y} = ay + b\dot{u} + cu$$

**4.11** Compute the forced response of the system $y(k+1) = 0.5y(k) + u(k)$ for $u(k)$ as

(a) $k^2$

(b) $ka^k$

(c) $a^k \sin \omega k$

(d) $a^k \cos \omega k$

**4.12** Use the convolution summation (Table 4.5) to compute the forced response of the model $y(k+1) = 0.1y(k) + u(k)$ for $0 \leqslant k \leqslant 5$, where

| $k$ | $u(k)$ |
|---|---|
| 0 | 2 |
| 1 | 3 |
| 2 | 4 |
| 3 | 2 |
| 4 | 1 |

**4.13** Write a calculator or computer program to simulate the response of the model $y(k+1) = ay(k) + u(k)$, with $u(k)$ a given set of values.

**4.14** Draw the block diagram and signal flow graph for the model

$$y(k+1) = ay(k) + bu(k) + cu(k+1)$$

**4.15** Compute the step response of the model

$$y(k+1) = 0.1y(k) + bu(k) + cu(k+1)$$

**4.16** Use the final value theorem to compute the steady-state error $e$ between the input and the output for the system shown in Figure P4.16 with the input functions

(a) $r(k) = 1$

(b) $r(k) = k$

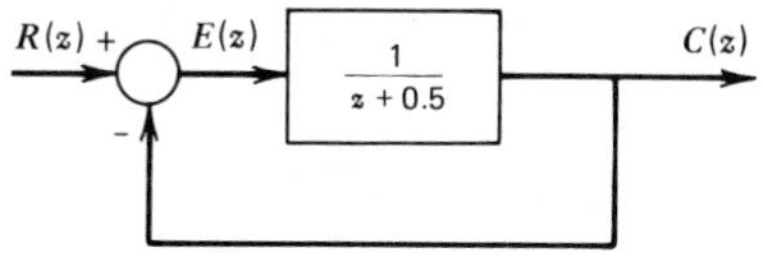


**Figure P4.16**

**4.17** Find the pulse transfer function and difference equation for the following models where $u$ is a sampled input (no hold is present).

(a) $\dot{y} = ay(t) + bu(t-D), \qquad D = nT$

(b) $\dot{y} = ay + b\dot{u} + cu$

**4.18** Find the pulse transfer function and difference equation for the following systems when a sample and hold is applied to the input $u$.

(a) $G(s) = \dfrac{Y(s)}{U(s)} = K\dfrac{s+c}{s+a}$

(b) $G(s) = \dfrac{Y(s)}{U(s)} = K\dfrac{s}{s+a}$

**4.19** Derive the relations given in Figure 4.16*c*, 4.16*d*, 4.16*e*, and 4.16*g*.

**4.20** Determine the stability properties of the system with negative unity feedback around the element $G(s)$, where

$$G(s) = \frac{K}{s+5}$$

(a) with, and (b) without a sample-and-hold on the error signal.

**4.21** The relationship between the Laplace variable $s$ and the $z$ variable is $z = e^{sT}$. Suppose that a first-order continuous-time system is required to have a time constant in the range $2 \leqslant \tau \leqslant 5$. Find the corresponding range for the characteristic root $z$ that a sampled data system must have to meet this requirement. Consider two cases

(a) $T = 1$.

(b) $T = 0.1$.

**4.22** (a) Consider the analog element shown in Figure P4.22*a*. We wish to use this element to create a system with a time constant of 0.5. To do this we place a gain $K$ in series with the element and a unit feedback loop around the combination. This is shown in Figure P4.22*b*. Find the value of $K$ required to give a time constant of 0.5.

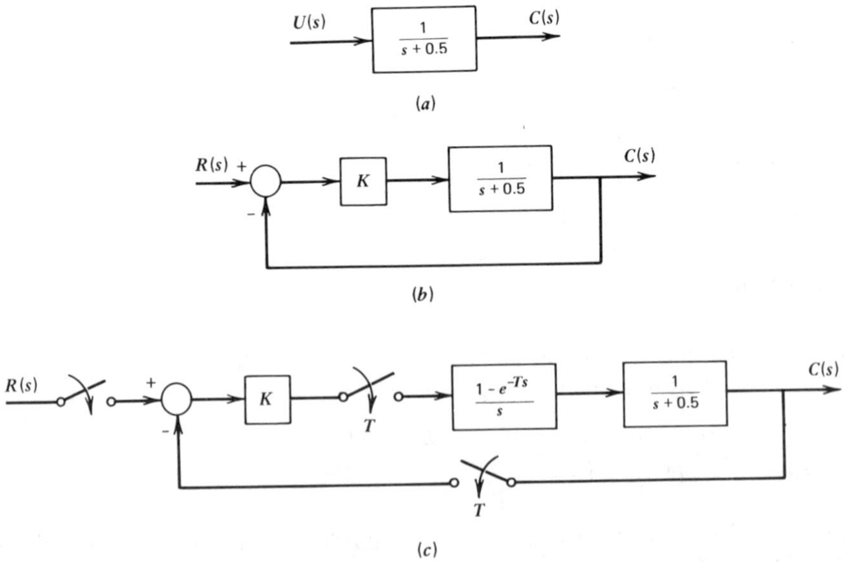


Figure P4.22

(b) We now wish to implement the design of part a in digital form. This is shown in Figure P4.22*c*. Find the closed-loop transfer function $T(z) = C(z)/R(z)$. Determine the equivalent root location for $z$ that corresponds to a time constant of 0.5, using a sample period of $T = 0.4$. Determine the value of $K$ that will give this response.

(c) For $T = 0.4$, compute the system's unit step response for the values of $K$ found in parts a and b. Which value of $K$ best meets the specification of a time constant of 0.5?

**4.23** (a) Repeat parts b and c of Problem 4.22 using $T = 0.2$.

(b) Compare the two design methods for digital systems.

(1) Calculation of $K$ from $s$-plane specifications, followed by discretization.

(2) Discretization and calculation of $K$ from equivalent $z$-plane root locations.

**4.24** Use a subroutine to generate random numbers, and write a program to simulate the behavior of the two-point averaging filter (4.7-1).

**4.25** Write a program to simulate the behavior of the filter in Example 4.16 for a set of exponentially decaying data contaminated with a random component. Investigate the choice of $K$ and the filter's initial condition $\hat{y}(0)$.

**4.26** Plot the frequency response curve for (4.7-8) for $a = -1$ and $K = 0.1, 0.5$, and 0.9.

**4.27** Redo the filter design of Example 4.19 for the case in which the voltage time constant is 1 sec instead of 0.5 sec.

**4.28** Simulate the behavior of the filter in Example 4.19 for the case in which the measured voltage is $u(t) = 10e^{-2t} + 0.5 \sin(377t)$.

**4.29** Use the Tustin method to derive a difference equation approximation for

$$\dot{y} = ry + bv + c\dot{v}$$

# CHAPTER FIVE

# Analysis of Higher-Order Systems

Our aims in this chapter are twofold: (1) to extend the first-order results of Chapters Two, Three, and Four to the second-order case, and (2) to introduce concepts and methods needed to handle high-order systems.

The primary difference between first-order and second-order linear models is that the characteristic roots of second-order models can be complex numbers. This leads to types of behavior not seen in first-order models, and sometimes requires modified algorithms in order to handle quantities with real and imaginary parts. The second-order case contains all of the possible types of behavior that can occur in linear systems. This is because a characteristic equation with real coefficients must have roots that fall into one of the three categories.

1. Real and distinct.
2. Distinct but complex conjugate pairs.
3. Repeated.

The first case was treated in Chapters Three and Four. The second and third cases will be covered in this chapter.

Once we understand the behavior generated by each of the three root cases, the superposition principle will allow us to apply the results to a linear system of any order. For example, the response of a sixth-order system with a complex conjugate pair of roots, two real repeated roots, and two real distinct roots can be analyzed by determining that part of the response resulting from each root type. The linear combination of each response pattern will give the total system response.

With this insight, classical solution methods using the Laplace transform, for example, can be applied to a linear model of any order, in theory at least. However, in practice, once the system order exceeds three, the algebraic manipulations required for many purposes become cumbersome. For such cases it is helpful to modify the solution procedures so that a digital computer can be used to provide computational assistance. If the model equations are expressed in terms of vector-matrix notation, this provides a convenient structure for computer applications. Here we introduce this notation and use it to develop some helpful methods. The topic of matrix methods is continued in Chapter Nine at an advanced level.

The chapter is structured as follows. We first consider linear continuous-time models. The origins and examples of higher-order models are treated first, followed by a discussion of the various forms they can take. The free response is presented along with a powerful stability analysis tool, the Routh-Hurwitz criterion. The forced response is then obtained for several common types of input functions. Nonlinear analysis and numerical techniques conclude the continuous-time material. The origins and analysis of discrete-time models conclude the chapter; the latter parallel the continuous-time case. We emphasize the importance of interpreting performance specifications in the continuous-time domain in terms of equivalent locations for the characteristic roots in the $z$-plane. This is of importance in sampled-data systems in which the object to be controlled is an analog element.

## 5.1 ORIGIN OF HIGHER-ORDER MODELS

Most physical elements are either zero-order (static), first-order, or second-order. Thus system models of order greater than one can occur if the system contains at least one second-order element or at least two first-order elements. Mechanical elements with significant mass and having a force input and displacement output are inherently second-order elements because Newton's law requires that two integrations must be performed on the force input to obtain the displacement. Similarly, electrical elements having inductance and capacitance will be second-order because they are governed by both integral causality paths on the tetrahedron of state. Liquid-level, hydraulic, pneumatic, and thermal elements are usually first-order unless inertia effects are included (in which case they are really examples of mechanical elements). However, when more than one such element occurs in a system, the system order is greater than one.

Development of higher-order models follows the approach outlined in Chapter Two. Often our systems will be of mixed type, such as with electromechanical systems, and the modeling techniques for electrical and mechanical elements both apply. We now present some examples of how such higher-order models are obtained.

### Liquid-Level Systems

*Example 5.1*

Develop a model for the time behavior of the heights $h_1$ and $h_2$ in the tank system shown in Figure 5.1$a$. The input variable is the volume flow rate $q$. Assume a linear resistance relation. The bottom areas of the tanks are $A_1$ and $A_2$.

Conservation of volume is equivalent to conservation of mass for incompressible fluids. Thus, for tank 1,

$$A_1 \frac{dh_1}{dt} = q - q_1$$

$$q_1 = \frac{1}{R_1}(h_1 - h_2)$$

For tank 2,

$$A_2 \frac{dh_2}{dt} = q_1 - q_o$$

$$q_o = \frac{1}{R_2} h_2$$

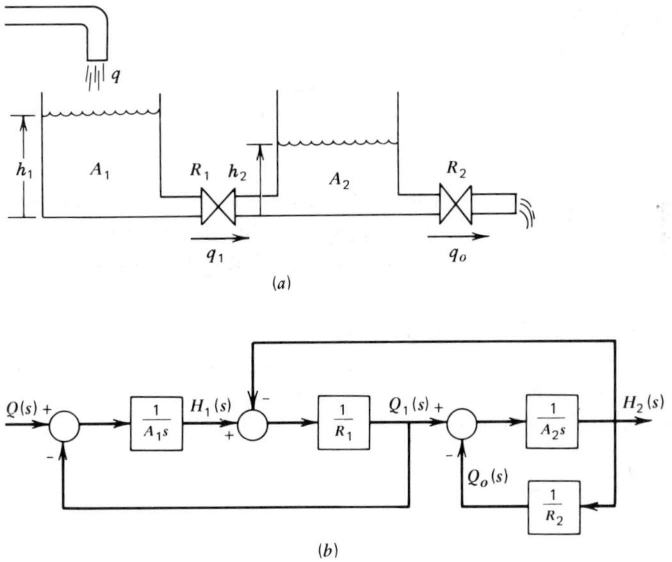


**Figure 5.1** System of two coupled first-order liquid-level elements and its block diagram.

Substituting for $q_1$ and $q_o$ gives the desired model.

$$A_1 \frac{dh_1}{dt} = q - \frac{1}{R_1}(h_1 - h_2) \tag{5.1-1}$$

$$A_2 \frac{dh_2}{dt} = \frac{1}{R_1}(h_1 - h_2) - \frac{1}{R_2} h_2 \tag{5.1-2}$$

The block diagram for the two-tank system can be easily obtained by applying the Laplace transform to the last two equations for zero initial conditions. Division by $A_1 s$ and $A_2 s$ gives

$$H_1(s) = \frac{1}{A_1 s}\left\{Q(s) - \frac{1}{R_1}[H_1(s) - H_2(s)]\right\} \tag{5.1-3}$$

$$H_2(s) = \frac{1}{A_2 s}\left\{\frac{1}{R_1}[H_1(s) - H_2(s)] - \frac{1}{R_2}H_2(s)\right\} \tag{5.1-4}$$

In this form $H_1$ and $H_2$ are recognized as being generated as the outputs of integrators with gains $1/A_1$ and $1/A_2$. The inputs to the integrators are the quantities within the braces. These quantities are generated from system inputs and feedback loops. The resulting diagram is shown in Figure 5.1*b*.

## Thermal Systems

### *Example 5.2*

The oven shown in Figure 5.2 has a heater with appreciable capacitance $C_1$. The other capacitance is that of the oven air $C_2$. The corresponding temperatures are $T_1$ and $T_2$,

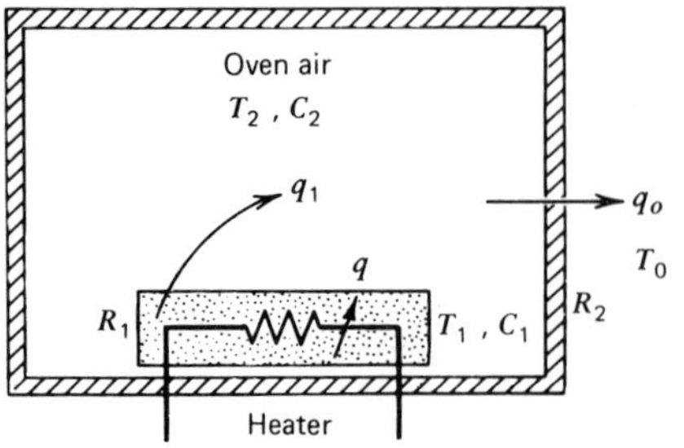


**Figure 5.2 Two thermal masses: an oven with significant heater capacitance $C_1$ (oven wall has negligible capacitance).**

and the outside temperature is $T_o$. The thermal resistance of the heater-air interface is $R_1$; that of the oven wall is $R_2$. Develop a model for $T_1$ and $T_2$, with input $q$, the heat flow rate delivered to the heater mass.

Conservation of energy for the heater gives

$$C_1 \frac{dT_1}{dt} = q - q_1$$

$$q_1 = \frac{1}{R_1}(T_1 - T_2)$$

For the oven air,

$$C_2 \frac{dT_2}{dt} = q_1 - q_o$$

$$q_o = \frac{1}{R_2}(T_2 - T_o)$$

Combining these we obtain

$$C_1 \frac{dT_1}{dt} = q - \frac{1}{R_1}(T_1 - T_2) \quad (5.1\text{-}5)$$

$$C_2 \frac{dT_2}{dt} = \frac{1}{R_1}(T_1 - T_2) - \frac{1}{R_2}(T_2 - T_o) \quad (5.1\text{-}6)$$

Note the similarity of these equations (5.1-1) and (5.1-2). This is a consequence of the fact that both systems consist of two first-order elements coupled in an identical manner by their resistance elements. The corresponding electrical network is shown in Figure 5.3. The current input $i$ is analogous to the flow rate $q$ in the two previous examples. The voltages $(v_1, v_2)$ are analogous to the heights $(h_1, h_2)$ and the temperatures $(T_1, T_2)$. The input voltage $v_o$ is analogous to the outside temperature $T_o$. Thus we see that $T_o$ could also be considered an input variable for the oven problem. No analogous quantity appears in the tank system, because we assumed the outlet pressure to be zero. If such a pressure $p_o$ exists, it would appear as an equivalent head $h_o$ in the model, where $h_o = p_o/\rho g$.

## A Direct-Current Motor System

Figure 5.4 shows an electromechanical system consisting of an armature-controlled dc motor driving a load inertia. The armature is the rotating part of the motor. It

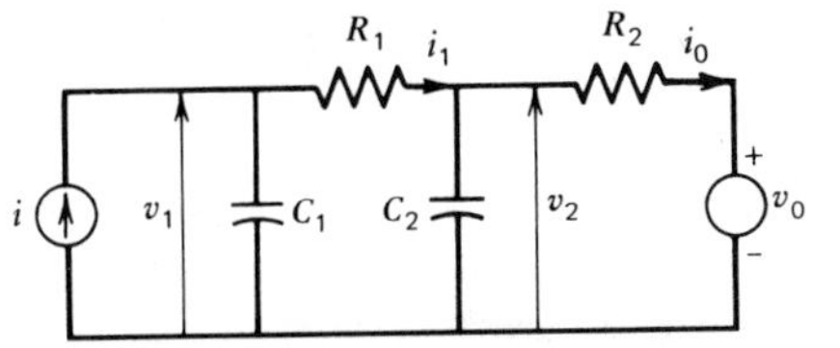


**Figure 5.3 Second-order electrical circuit analogous to the systems shown in Figures 5.1 and 5.2.**

usually consists of a wire conductor wrapped around an iron core. This winding has an inductance $L$, shown in the figure. The resistance $R$ in the figure represents the lumped value of the armature resistance and any external resistance deliberately introduced to change the motor's behavior.

The armature is surrounded by a magnetic field. The reaction of this field with the armature current produces a torque that causes the armature to rotate. If the armature voltage $v$ is used to control the motor, the motor is said to be *armature-controlled.* In this case the field is produced by an electromagnet supplied with a constant voltage or by a permanent magnet. It is now possible to manufacture permanent magnets of high field intensity and armatures of low inertia so that motors with a high torque-to-inertia ratio are available. This has opened up new areas of application for dc motors in control systems. This is fortunate because the speed of a dc motor is easier to control than that of an ac motor. Also, the analysis and design of the associated dc circuitry is simpler than for ac systems.

## *Example 5.3*

Develop a model for the armature-controlled system shown in Figure 5.4, with the armature voltage $v$ as input and the load speed $\omega$ as output.

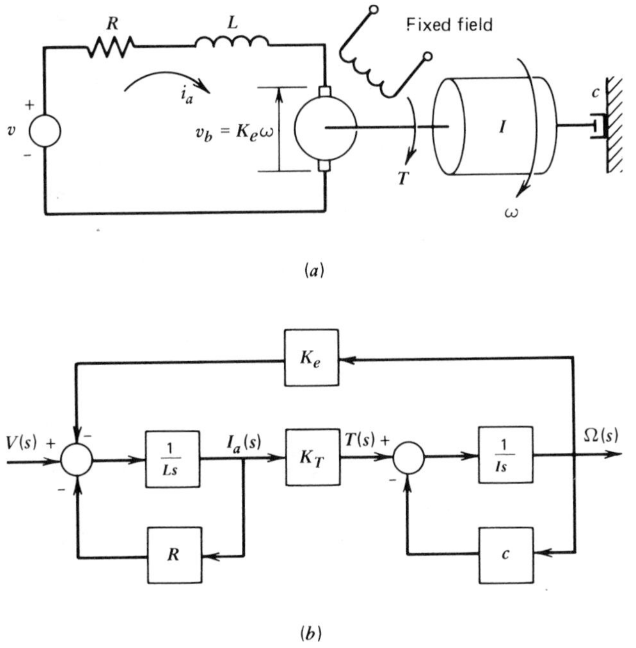


**Figure 5.4 Armature-controlled dc motor with a load and its block diagram.**

This motor type produces a torque $T$ that is proportional to the armature current $i_a$.

$$T = K_T i_a \tag{5.1-7}$$

The torque constant $K_T$ depends on the strength of the field and other details of the motor's construction. The *user* of such motors (as opposed to the motor *designer*) can obtain values of $K_T$ for a specific motor from the manufacturer's literature.

The motion of a current-carrying conductor in a field produces a voltage in the conductor that opposes the current. This voltage is called the *back emf* (electromotive force). Its magnitude is proportional to the speed and is given by

$$e_b = K_e \omega \tag{5.1-8}$$

The value of the constant $K_e$ is found with other product data (it is sometimes called the *voltage constant*). Equations (5.1-7) and (5.1-8) are constitutive relations for the motor.

The back emf is a voltage drop in the armature circuit. Thus Kirchhoff's voltage law gives

$$v = i_a R + L \frac{di_a}{dt} + K_e \omega \tag{5.1-9}$$

From Newton's law applied to the inertia $I$,

$$\begin{aligned} I \frac{d\omega}{dt} &= T - c\omega \\ &= K_T i_a - c\omega \end{aligned} \tag{5.1-10}$$

These two equations constitute the system model. They are coupled because of the back emf term. This coupling is shown graphically by the block diagram (Figure 5.4*b*), in which the outer feedback loop represents the influence of the speed $\omega$ on the current $i_a$.

The diagram is obtained by solving (5.1-9) for the derivative $di_a/dt$ and transforming. The result is

$$I_a(s) = \frac{1}{Ls}[V(s) - RI_a(s) - K_e \Omega(s)]$$

$$\Omega(s) = \frac{1}{Is}[K_T I_a(s) - c\Omega(s)]$$

These relations define the input-output relations for integrators with gains $1/L$ and $1/I$, respectively.

## An Electrical Circuit

*Example 5.4*

Develop a model for the circuit shown in Figure 5.5*a*. The input is the voltage $v$; the output is the voltage $v_o$ across the capacitance.

There are two loops in this circuit, and Kirchhoff's voltage law must be applied to each. Thus

$$v = i_1 R_1 + i_2 R_2$$

$$i_2 R_2 = L \frac{di_o}{dt} + v_o$$

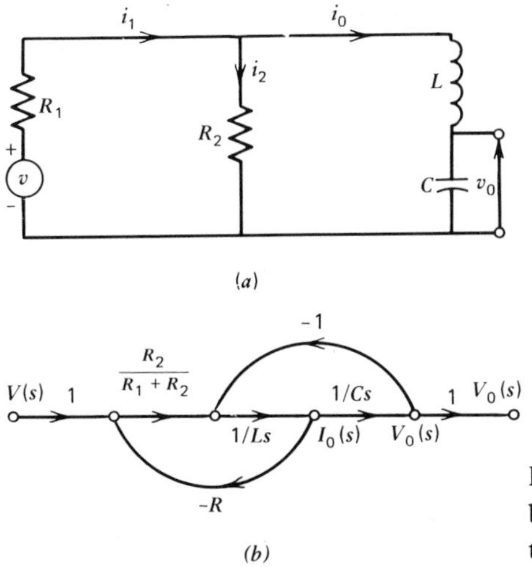


**Figure 5.5 Second-order circuit produced by coupling a capacitance and an inductance, and its signal flow graph.**

From conservation of charge,

$$i_1 = i_o + i_2$$

The output voltage is

$$v_o = \frac{1}{C}\int_0^t i_o \, dt \tag{5.1-11}$$

To reduce these equations, keep in mind that the state variables are $i_o$ (for the inductor) and $v_o$ (for the capacitor). Since $v$ is an input, the other variables ($i_1$ and $i_2$) are extraneous and may be eliminated with algebra. The result is

$$L\frac{di_o}{dt} = \frac{R_2}{R_1 + R_2}(v - R_1 i_o) - v_o \tag{5.1-12}$$

$$C\frac{dv_o}{dt} = i_o \tag{5.1-13}$$

where the last equation results from differentiating the capacitor relation (5.1-11). The signal flow graph is shown in Figure 5.5*b*. It was obtained in the same manner as the previous block diagrams.

## A Mechanical System With Elasticity and Damping

### *Example 5.5*

Many mechanical systems can be represented as an equivalent mass-spring-damper system, such as in Figure 5.6*a*. Derive the equation of motion for the displacement, with the applied force $f$ as the input.

Recall that the constitutive relation for a linear spring states that the spring's restoring force is directly proportional to the displacement from its free length $L$. If the mass $m$ is attached to the spring and allowed to settle to its equilibrium position, the spring will be stretched by an amount $\Delta$ from its free length. The force balance

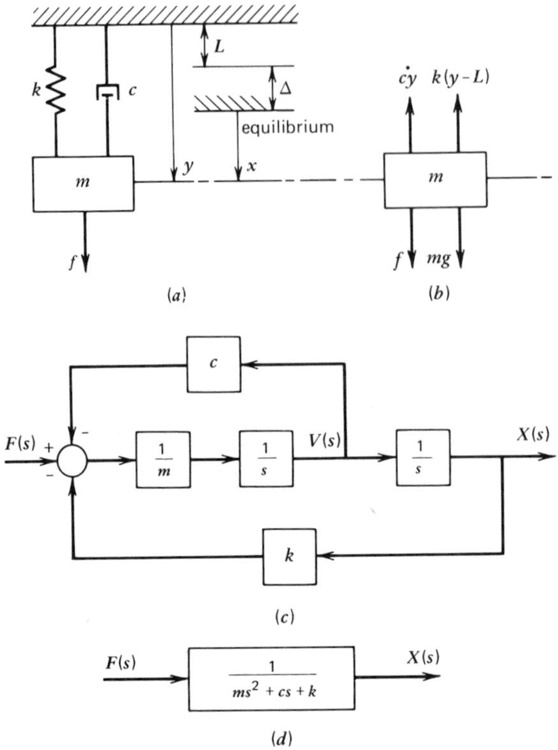


**Figure 5.6 Second-order model for mechanical vibration. (*a*) Coordinates for mass, spring, and damper elements. (*b*) Free-body diagram for (*a*). (*c*) State-variable block diagram for (*a*). (*d*) Reduced block diagram showing the transfer function between $x$ and $f$.**

at this equilibrium is (if $f = 0$)

$$mg = k\Delta \tag{5.1-14}$$

where $mg$ is the gravity force and $k\Delta$ is the spring force. This relation determines $\Delta$.

Now consider the dynamic situation with $f \neq 0$. The damper force is $c\dot{y}$ and the spring force is $k(y - L)$. Newton's second law states that

$$m\ddot{y} = f - c\dot{y} - k(y - L) + mg \tag{5.1-15}$$

The model can be simplified by using the coordinate $x$, as defined in Figure 5.6*a*. The required relations are

$$y - L = x + \Delta$$

$$\dot{y} = \dot{x}$$

$$\ddot{y} = \ddot{x}$$

In terms of $x$, we have

$$m\ddot{x} = f - c\dot{x} - k(x + \Delta) + mg$$

In light of (5.1-14), the model becomes

$$m\ddot{x} = f - c\dot{x} - kx \tag{5.1-16}$$

In this form the gravity term cancels the spring force term $k\Delta$. This simplification is commonly used in vibration models; when the coordinate origin is located at the equilibrium position of the mass, the static force terms do not appear in the resulting dynamic equation. This coordinate choice is often used but is sometimes not clearly stated. It is important to remember that the model form (5.1-16) applies only when $x$ is measured from the equilibrium resulting when $f = 0$.

## State-Variable Representation

Equation (5.1-16) is a single second-order equation, whereas the models encountered thus far in this chapter consist of a pair of coupled first-order equations. Equation (5.1-16) can be written in first-order form by using the velocity $v$ as the second variable. The following set is equivalent to (5.1-16).

$$\dot{x} = v \tag{5.1-17}$$

$$m\dot{v} = f - cv - kx \tag{5.1-18}$$

State variables for this system are $x$ and $v$, since they can be used to describe the potential energy (PE) and kinetic energy (KE).

$$\text{KE} = \tfrac{1}{2}mv^2$$

$$\text{PE} = \tfrac{1}{2}k(x + \Delta)^2 - mg(x + \Delta)$$

A review of the previous examples will show that all of the models can be written as coupled first-order equations in terms of the system's state variables. This holds true in general; each state variable produces a single first-order equation with it as the dependent variable and time as the independent variable. This is so because of the path of integral causality associated with each state variable.

A system with $n$ state variables can therefore be described by a set of $n$ coupled first-order equations. The equivalence of (5.1-16) with (5.1-17) and (5.1-18) hints that such a system can also be described by a single $n$th-order equation. We will see in the next section that this is true, and we will develop methods for obtaining one description from another.

The block diagram of the state-variable representation is shown in Figure 5.6*c*, while that of the reduced equation is given in Figure 5.6*d*. The latter diagram can be obtained directly by transforming (5.1-16), or indirectly by reducing Figure 5.6*c* using block diagram algebra. This reduction is shown in Figure 5.7. In Figure 5.7*a* a separate inner loop has been created by using another comparator. The inner loop is then reduced via the loop reduction formula given in Chapter Three. The resulting transfer function is $1/(s + c)m$ and is shown in Figure 5.7*b*. Application of the series law and the loop reduction formula once more will give the reduced diagram shown in Figure 5.6*d*.

The reduced diagram shows that the transfer function between $x$ and $f$ is

$$T(s) = \frac{X(s)}{F(s)} = \frac{1}{ms^2 + cs + k} \tag{5.1-19}$$

Note that this could have also been obtained directly from the differential equation description. The diagram reduction method is more tedious, and therefore is usually applied only when the describing equations are not given.

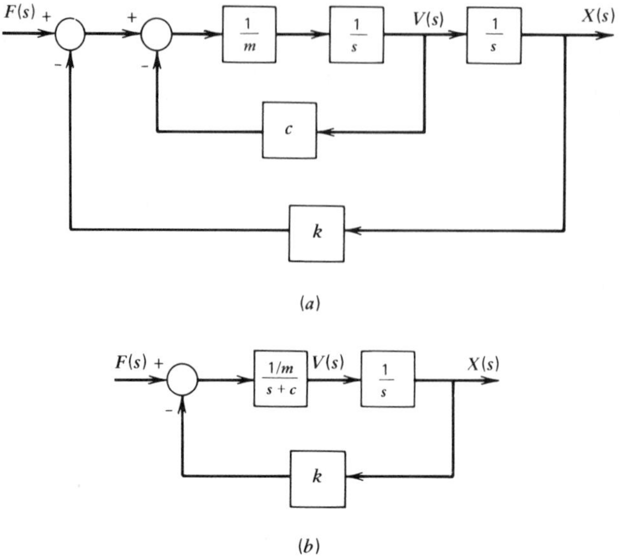


**Figure 5.7 Reduction of the block diagram for the mechanical vibration model. (*a*) Introduction of a new comparator to create a separate inner loop. (*b*) Result of reducing the inner loop created in (*a*).**

## A Hydromechanical System

Machine tools are an example of an application of the preceding model. The applied force $f$ can be supplied by a hydraulic servomotor (Figure 5.8*a*). The mass $m$ represents that of a cutting tool and the power piston, while $k$ represents the combined effects of the elasticity naturally present in the structure and that introduced by the designer to achieve proper performance. A similar statement applies to the damping $c$. The valve displacement $z$ is generated by a control system in order to move the tool through its prescribed motion.

In terms of the notation of Figure 5.8, the analysis in Section 2.6 showed that

$$\dot{x} = \frac{C}{A} z \tag{5.1-20}$$

where $A$ is the power piston area, and $C$ is a constant that relates the volume flow rate $q$ through the orifice to the displacement $z$. That analysis neglected the inertia of the load and piston. If the term $m\ddot{x}$ is large, the model (5.1-20) becomes inaccurate. Here we extend the analysis to cover this case.

The spool valve shown in Figure 5.8 has two *lands.* If the width of the land is greater than the port width, the valve is said to be *overlapped.* In this case a dead zone exists in which a slight change in the displacement $z$ produces no power piston motion. Such dead zones create control difficulties and are avoided by designing the valve to be *underlapped* (the land width is less the port width). For such valves there will be a small flow opening even when the valve is in the neutral position ($z = 0$). This gives it a higher sensitivity than an overlapped valve.

Let $p_s$ and $p_o$ denote the supply and outlet pressures, respectively. Then the

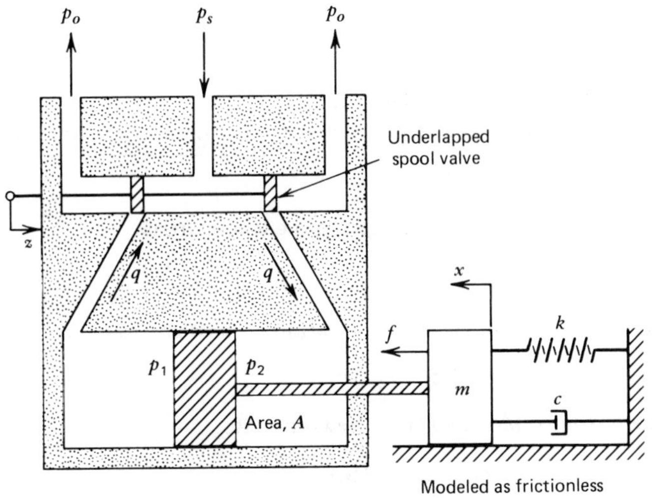


**Figure 5.8 Hydraulic servomotor with a load.**

pressure drop from inlet to outlet is $p_s - p_o$, and this must equal the sum of the drops across both valve openings and the power piston. That is,

$$p_s - p_o = 2\Delta p_v + \Delta p$$

where the drop across the valve openings is

$$\Delta p_v = p_s - p_2 = p_1 - p_o$$

and the drop across the piston is

$$\Delta p = p_2 - p_1$$

Therefore

$$\Delta p_v = \tfrac{1}{2}(p_s - p_o - \Delta p)$$

If we take $p_s$ and $p_o$ to be constant, we see that $\Delta p_v$ varies only if $\Delta p$ changes. The derivation in Section 2.6 treated $\Delta p_v$ as a constant.

Thus the variables $z$ and $\Delta p$ determine the volume flow rate, as

$$q = f(z, \Delta p)$$

This relation is nonlinear. For the reference equilibrium condition ($z = 0$, $\Delta p = 0$, $q = 0$), a linearization gives

$$q = C_1 z - C_2 \Delta p \tag{5.1-21}$$

where the variations from equilibrium are simply $z$, $\Delta p$, and $q$. The linearization constants are available from theoretical and experimental results (Reference 1). The constant $C_1$ is identical to $C$ in (5.1-20), and both $C_1$ and $C_2$ are positive. The effect of the underlapping shows up in the nonzero flow predicted from (5.1-21) when $z = 0$. The pressure drop $\Delta p$ is caused by the reaction forces of the load mass, elasticity, and damping, and by the supply pressure difference $(p_s - p_o)$.

*Example 5.6*

(a) Derive a model of the system shown in Figure 5.8*a*. Take into account the inertia effects of the load and assume the valve is underlapped.

(b) Show that the model reduces to (5.1-20) when $c = k = 0$ and either of the following approximations is valid.

1. $m/A \to 0$ (large piston force compared to the load inertia)
2. $C_2 \to 0$ ($q$ independent of $\Delta p$)

(a) Assume an incompressible fluid. Then conservation of mass and (5.1-21) give

$$A\dot{x} = q = C_1 z - C_2 \Delta p$$

The force generated by the piston is $A\Delta p$, and from Newton's law,

$$m\ddot{x} = -c\dot{x} - kx + A\Delta p$$

where $x$ is measured from the equilibrium point of the mass $m$. Substitute $\Delta p$ from the second equation to obtain

$$A\dot{x} = C_1 z - C_2\left(\frac{m}{A}\ddot{x} + \frac{c}{A}\dot{x} + \frac{k}{A}x\right)$$

or

$$\frac{C_2 m}{A}\ddot{x} + \left(\frac{cC_2}{A} + A\right)\dot{x} + \frac{C_2 k}{A}x = C_1 z \tag{5.1-22}$$

This is the desired model with $z$ as the input and $x$ as the output.

(b) To show its equivalence with (5.1-20), transform (5.1-22) and find the transfer function.

$$T(s) = \frac{X(s)}{Z(s)} = \frac{C_1}{\dfrac{C_2 m}{A}s^2 + \left(\dfrac{cC_2}{A} + A\right)s + \dfrac{C_2 k}{A}} \tag{5.1-23}$$

For $c = k = 0$, this becomes

$$T(s) = \frac{C_1}{\dfrac{C_2 m}{A}s^2 + As}$$

When either $m/A \to 0$ or $C_2 \to 0$, we obtain

$$T(s) = \frac{C_1}{As}$$

This is equivalent to the differential equation (5.1-20) since $C_1 = C$.

The preceding analysis neglects leakage around the power piston as well as compressibility of the fluid. For some applications involving very large forces, these can be important effects, but the details of their analysis are beyond our scope here (see Reference 1).

## 5.2 FORMS OF THE CONTINUOUS-TIME MODEL

We have just seen that the application of physical laws can produce a second-order model in the form of a single equation or two first-order equations. The forms are equivalent, but each has its own advantages for analyzing the system's behavior. Here we discuss both forms in general.

## State-Variable Form

The general form of the second-order linear model in terms of the state variables $x_1$ and $x_2$ is

$$\dot{x}_1 = a_{11}x_1 + a_{12}x_2 + b_{11}u_1 + b_{12}u_2 \tag{5.2-1}$$

$$\dot{x}_2 = a_{21}x_1 + a_{22}x_2 + b_{21}u_1 + b_{22}u_2 \tag{5.2-2}$$

where we have assumed that two inputs $(u_1, u_2)$ act on the system. This is the *state-variable* form of the model. If the outputs $(y_1, y_2)$ are linear combinations of the state variables and the inputs, then

$$y_1 = c_{11}x_1 + c_{12}x_2 + d_{11}u_1 + d_{12}u_2 \tag{5.2-3}$$

$$y_2 = c_{21}x_1 + c_{22}x_2 + d_{21}u_1 + d_{22}u_2 \tag{5.2-4}$$

If the coefficients are constant, these equations can be Laplace-transformed. The system diagram can then be readily drawn using two integrator elements whose outputs are $x_1$ and $x_2$, and whose inputs are the right-hand sides of (5.2-1) and (5.2-2). The system outputs are then constructed from (5.2-3) and (5.2-4). This diagram is called the *state diagram* because it is constructed from the state-variable form of the model and thus graphically shows how the state variables are generated.

### *Example* 5.7

Draw the state diagram for the following model.

$$\dot{x}_1 = -4x_1 + 2x_2 + 4u_1$$

$$\dot{x}_2 = -3x_2 + 5u_2$$

The output variables are

$$y_1 = x_1 + x_2 + 3u_1$$

$$y_2 = x_2$$

Assuming zero initial conditions, we transform the model and divide the first two equations by $s$ to obtain

$$X_1(s) = \frac{1}{s}[-4X_1(s) + 2X_2(s) + 4U_1(s)]$$

$$X_2(s) = \frac{1}{s}[-3X_2(s) + 5U_1(s)]$$

$$Y_2(s) = X_2(s)$$

Using the guidelines for the inputs and outputs of the two integrators, we obtain the diagram in Figure 5.9.

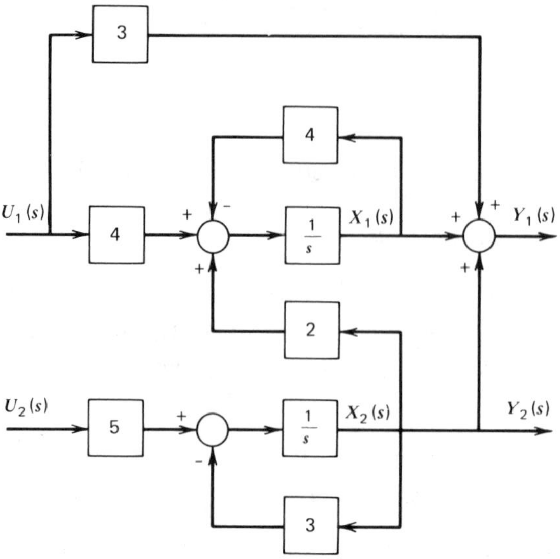


Figure 5.9 State diagram for Example 5.7.

## Transfer Functions

A transfer function is a relation between an input and an output. When there are two inputs and two outputs, there are four possible input-output pairs, and therefore four possible transfer functions for the system. These can be found either by reducing the system diagram or by algebraic manipulation of the transformed equations. If the diagram is in state-variable form, usually the easiest way is to obtain the state equations and then use the algebraic method. These equations are obtained by choosing the state variables as the outputs of the integrators. The right-hand side of each state equation is given by the input to the integrator.

Transform (5.2-1) through (5.2-4) and collect terms to obtain

$$(s-a_{11})X_1(s)-a_{12}X_2(s) = x_1(0)+b_{11}U_1(s)+b_{12}U_2(s) \tag{5.2-5}$$

$$-a_{21}X_1(s)+(s-a_{22})X_2(s) = x_2(0)+b_{21}U_1(s)+b_{22}U_2(s) \tag{5.2-6}$$

$$Y_1(s) = c_{11}X_1(s)+c_{12}X_2(s)+d_{11}U_1(s)+d_{12}U_2(s) \tag{5.2-7}$$

$$Y_2(s) = c_{21}X_1(s)+c_{22}X_2(s)+d_{21}U_1(s)+d_{22}U_2(s) \tag{5.2-8}$$

The transfer function describes the effect of a given input on a given output. It does not describe the effects of other inputs or initial conditions on that output. Therefore the other inputs and all initial conditions are set to zero when deriving a transfer function. That is, the transfer function between the output $y_i$ and the input $u_j$ is

$$T_{ij}(s) = \left.\frac{Y_i(s)}{U_j(s)}\right|_{\substack{U_k(s)=0,\ k\neq j \\ x_k(0)=0,\ \text{for all } k}} \tag{5.2-9}$$

*Example 5.8*

Find the transfer function $T_{12}(s)$ for the system given in Example 5.7.

The desired transfer function is defined as

$$T_{12}(s) = \left.\frac{Y_1(s)}{U_2(s)}\right|_{\substack{U_1(s)=0 \\ x_1(0)=x_2(0)=0}}$$

Comparing the state equations for the given model with the general form given by (5.2-1) through (5.2-4), we see that

$$a_{11} = -4, a_{12} = 2, a_{21} = 0, a_{22} = -3$$

$$b_{11} = 4, b_{12} = 0 = b_{21}, b_{22} = 5$$

$$c_{11} = 1 = c_{12}, c_{21} = 0, c_{22} = 1$$

$$d_{11} = 3, d_{12} = d_{21} = d_{22} = 0$$

Therefore from (5.2-5) and (5.2-6), with $x_1(0) = x_2(0) = U_1(s) = 0$,

$$(s+4)X_1(s) - 2X_2(s) = 0$$

$$(s+3)X_2(s) = 5U_2(s)$$

Solve for $X_1(s)$ and $X_2(s)$.

$$X_1(s) = \frac{5}{s+3}U_2(s)$$

$$X_2(s) = \frac{2}{s+4}X_2(s) = \frac{10}{(s+4)(s+3)}U_2(s)$$

Substitute these into (5.2-7) with $U_1(s) = 0$. The result is

$$Y_1(s) = X_1(s) + X_2(s) = \left[\frac{10}{(s+4)(s+3)} + \frac{5}{s+3}\right]U_2(s) = \frac{5(s+6)}{(s+4)(s+3)}U_2(s)$$

The transfer function is

$$T_{12}(s) = \frac{5(s+6)}{(s+4)(s+3)}$$

A similar procedure will yield any of the transfer functions. We could obtain the general results from (5.2-5) through (5.2-8), but the expressions are cumbersome. Instead, we will present the results compactly with the aid of matrix notation (Appendix A). This approach will also allow us to generalize the results to higher-order models.

## Matrix Form of the State Equations

We can form a two-dimensional column vector $\mathbf{x}$ with the two state variables $x_1$ and $x_2$; that is,

$$\mathbf{x} = \begin{bmatrix} x_1 \\ x_2 \end{bmatrix} \tag{5.2-10}$$

The vector $\mathbf{x}$ is the *state vector.* Similarly the *input vector* is

$$\mathbf{u} = \begin{bmatrix} u_1 \\ u_2 \end{bmatrix} \tag{5.2-11}$$

and the *output vector* is

$$\mathbf{y} = \begin{bmatrix} y_1 \\ y_2 \end{bmatrix} \tag{5.2-12}$$

From the rules of matrix-vector multiplication, the state equations and the output equations (5.2-1) through (5.2-4) can be written as

$$\begin{bmatrix} \dot{x}_1 \\ \dot{x}_2 \end{bmatrix} = \begin{bmatrix} a_{11} & a_{12} \\ a_{21} & a_{22} \end{bmatrix} \begin{bmatrix} x_1 \\ x_2 \end{bmatrix} + \begin{bmatrix} b_{11} & b_{12} \\ b_{21} & b_{22} \end{bmatrix} \begin{bmatrix} u_1 \\ u_2 \end{bmatrix}$$

$$\begin{bmatrix} y_1 \\ y_2 \end{bmatrix} = \begin{bmatrix} c_{11} & c_{12} \\ c_{21} & c_{22} \end{bmatrix} \begin{bmatrix} x_1 \\ x_2 \end{bmatrix} + \begin{bmatrix} d_{11} & d_{12} \\ d_{21} & d_{22} \end{bmatrix} \begin{bmatrix} u_1 \\ u_2 \end{bmatrix}$$

or

$$\dot{\mathbf{x}} = \mathbf{A}\mathbf{x} + \mathbf{B}\mathbf{u} \tag{5.2-13}$$

$$\mathbf{y} = \mathbf{C}\mathbf{x} + \mathbf{D}\mathbf{u} \tag{5.2-14}$$

where

$$\dot{\mathbf{x}} = \begin{bmatrix} \dot{\mathbf{x}}_1 \\ \dot{\mathbf{x}}_2 \end{bmatrix} \tag{5.2-15}$$

$$\mathbf{A} = [a_{ij}] \tag{5.2-16}$$

$$\mathbf{B} = [b_{ij}] \tag{5.2-17}$$

$$\mathbf{C} = [c_{ij}] \tag{5.2-18}$$

$$\mathbf{D} = [d_{ij}] \tag{5.2-19}$$

for $i = 1, 2$ and $j = 1, 2$. The matrix $\mathbf{A}$ is the *system matrix*; $\mathbf{B}$ is the *control* or *input matrix*; $\mathbf{C}$ and $\mathbf{D}$ are the *output matrices.*

Note that the derivative (or integral) of a vector or matrix is the vector or matrix of the derivatives (or integrals) of the elements. Therefore, since the Laplace transform is an integration operator, the transform of a vector is the vector consisting of the transforms of the elements; namely,

$$\mathbf{X}(s) = \begin{bmatrix} X_1(s) \\ X_2(s) \end{bmatrix} \tag{5.2-20}$$

Therefore

$$\mathscr{L}(\dot{\mathbf{x}}) = \begin{bmatrix} sX_1(s) - x_1(0) \\ sX_2(s) - x_2(0) \end{bmatrix} = s\mathbf{X}(s) - \mathbf{x}(0) \tag{5.2-21}$$

Transforming (5.2-13) and (5.2-14) and using this result, we obtain

$$s\mathbf{X}(s) - \mathbf{x}(0) = \mathbf{A}\mathbf{X}(s) + \mathbf{B}\mathbf{U}(s) \tag{5.2-22}$$

$$\mathbf{Y}(s) = \mathbf{C}\mathbf{X}(s) + \mathbf{D}\mathbf{U}(s) \tag{5.2-23}$$

The first of these can be arranged as

$$(s\mathbf{I} - \mathbf{A})\mathbf{X}(s) = \mathbf{x}(0) + \mathbf{B}\mathbf{U}(s) \tag{5.2-24}$$

where the identity matrix $\mathbf{I}$ has been used to allow $\mathbf{X}(s)$ to be factored out of the expression. The matrix

$$\mathbf{R}(s) = s\mathbf{I} - \mathbf{A} \tag{5.2-25}$$

is the *resolvent matrix*. Its properties will tell us much about the system's behavior.

The transfer function is defined for zero initial conditions. With $\mathbf{x}(0) = \mathbf{0}$, (5.2-24) gives

$$\mathbf{X}(s) = (s\mathbf{I} - \mathbf{A})^{-1}\mathbf{B}\mathbf{U}(s) \tag{5.2-26}$$

where the resolvent matrix inverse $(s\mathbf{I} - \mathbf{A})^{-1}$ is assumed to exist. Substitute this into (5.2-23) and factor out $\mathbf{U}(s)$ to obtain

$$\mathbf{Y}(s) = [\mathbf{C}(s\mathbf{I} - \mathbf{A})^{-1}\mathbf{B} + \mathbf{D}]\,\mathbf{U}(s) \tag{5.2-27}$$

The matrix multiplying $\mathbf{U}(s)$ is the *transfer function matrix* $\mathbf{T}(s)$.

$$\mathbf{T}(s) = \mathbf{C}(s\mathbf{I} - \mathbf{A})^{-1}\mathbf{B} + \mathbf{D} \tag{5.2-28}$$

The elements of $\mathbf{T}(s)$ are the transfer functions as defined by (5.2-9); (5.2-28) provides an alternate way of computing these elements.

### *Example 5.9*

Use (5.2-28) to find $T_{12}(s)$ for the model given in Example 5.7.

For this model,

$$\mathbf{A} = \begin{bmatrix} -4 & 2 \\ 0 & -3 \end{bmatrix} \qquad \mathbf{B} = \begin{bmatrix} 4 & 0 \\ 0 & 5 \end{bmatrix}$$

$$\mathbf{C} = \begin{bmatrix} 1 & 1 \\ 0 & 1 \end{bmatrix} \qquad \mathbf{D} = \begin{bmatrix} 3 & 0 \\ 0 & 0 \end{bmatrix}$$

Thus,

$$s\mathbf{I} - \mathbf{A} = s\begin{bmatrix} 1 & 0 \\ 0 & 1 \end{bmatrix} - \begin{bmatrix} -4 & 2 \\ 0 & -3 \end{bmatrix} = \begin{bmatrix} (s+4) & -2 \\ 0 & (s+3) \end{bmatrix}$$

As shown in Appendix A, the inverse of a $(2 \times 2)$ matrix $\mathbf{M}$ is

$$\mathbf{M}^{-1} = \frac{1}{m_{11}m_{22} - m_{12}m_{21}}\begin{bmatrix} m_{22} & -m_{12} \\ -m_{21} & m_{11} \end{bmatrix} \tag{5.2-29}$$

Letting $\mathbf{M} = s\mathbf{I} - \mathbf{A}$, we see that

$$(s\mathbf{I} - \mathbf{A})^{-1} = \frac{1}{(s+4)(s+3)}\begin{bmatrix} (s+3) & 2 \\ 0 & (s+4) \end{bmatrix} = \begin{bmatrix} \dfrac{1}{s+4} & \dfrac{2}{(s+4)(s+3)} \\ 0 & \dfrac{1}{s+3} \end{bmatrix}$$

and

$$\mathbf{T}(s) = \begin{bmatrix} 1 & 1 \\ 0 & 1 \end{bmatrix}\begin{bmatrix} \dfrac{1}{s+4} & \dfrac{2}{(s+4)(s+3)} \\ 0 & \dfrac{1}{s+3} \end{bmatrix}\begin{bmatrix} 4 & 0 \\ 0 & 5 \end{bmatrix} + \begin{bmatrix} 3 & 0 \\ 0 & 0 \end{bmatrix}$$

The transfer functions are obtained by carrying out the indicated multiplications. They are

$$T_{11}(s) = \frac{4}{s+4} + 3 = \frac{3s+16}{s+4}$$

$$T_{12}(s) = \frac{10}{(s+4)(s+3)} + \frac{5}{s+3} = \frac{5(s+6)}{(s+4)(s+3)}$$

$$T_{21}(s) = 0$$

$$T_{22}(s) = \frac{5}{s+3}$$

The answer for $T_{12}(s)$ agrees with that found in the previous example, as it should.

All of the transfer functions are obtained at once with the matrix method. It is cumbersome when the inverse of the resolvent matrix is difficult to form. If only some of the transfer functions are required, the algebraic method used in Example 5.8 is sometimes easier to apply.

The matrix form of the state equations is valid for a model of any order, although we have used only second-order examples. An $n$th-order model will have $n$ state variables, and the state vector $x$ will have $n$ elements. Equations (5.2-13) and (5.2-14) represent this general case. The power of this notation is demonstrated by the fact that the general transfer function relation (5.2-28) was obtained for a system of any order by using the matrix form of the state equations. A second advantage with the matrix form is that it allows general-purpose computer programs to be developed for analyzing systems of any order. The numerical methods of Section 3.13 and 3.14 are easily generalized by using the state vector form. All that must be changed in such programs are the dimension statements for the vectors and matrices. A third advantage of (5.2-13) is that it has the appearance of a first-order model, and therefore the first-order techniques of Chapter Three can be mimicked at least formally, keeping in mind the requirements of the matrix algebra. For example, the Laplace transformation of (5.2-13) to obtain (5.2-26) follows very closely the transfer function development of Chapter Three for the first-order model.

The state vector notation also allows a compact description of the system in graphical form, called the state-vector diagram. This can be either a block diagram or a signal flow graph. Both are shown in Figure 5.10 for the system described by (5.2-13) and (5.2-14). The reduced transfer function diagram is shown in Figure 5.10*c*. In the vector diagram, the signals (now multiple or vector quantities) are shown with a double-lined arrow.

## Reduced Form

Despite the generality of the matrix form, a single higher-order equation in terms of only one output is often easier to deal with. This form is often used when we are interested in the behavior of only one of the variables. The resulting single equation of higher order is the *reduced form* of the model.

A transfer function exists for each input-output pair. Therefore a differential equation relating each input-output pair can be derived from the transfer function.

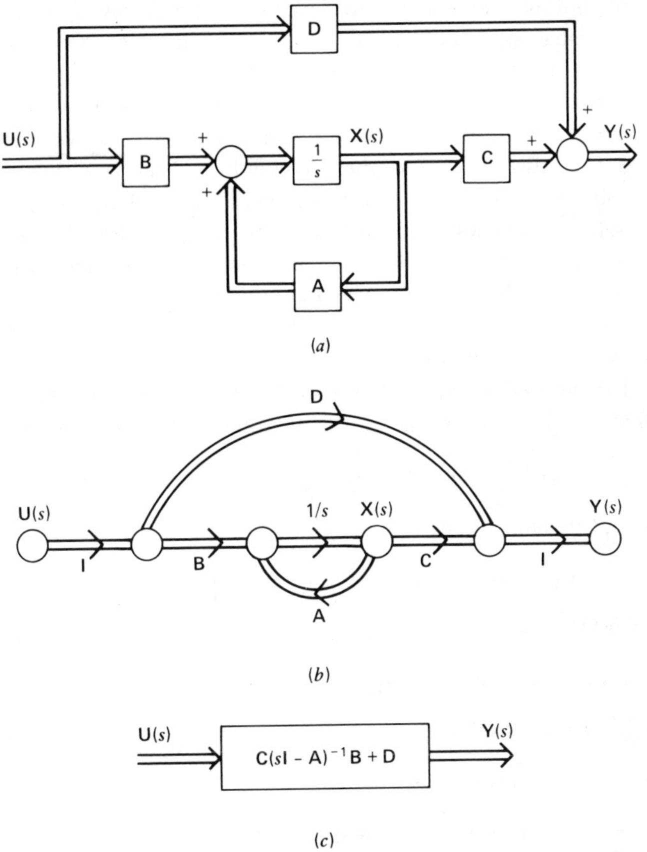


**Figure 5.10 Graphical representations of the state vector model: $\dot{\mathbf{x}} = \mathbf{Ax} + \mathbf{Bu}$, $\mathbf{y} = \mathbf{Cx} + \mathbf{Du}$. (*a*) Block diagram for the Laplace transform variables. (*b*) Signal flow graph. (*c*) Reduced diagram showing the transfer function matrix between y and u.**

For example, the transfer function $T_{12}(s)$ from Example 5.9 is

$$T_{12}(s) = \frac{Y_1(s)}{U_2(s)} = \frac{5(s+6)}{(s+4)(s+3)}$$

Cross-multiplication gives

$$(s^2 + 7s + 12)Y_1(s) = 5(s+6)U_2(s)$$

The corresponding differential equation is

$$\ddot{y}_1 + 7\dot{y}_1 + 12y_1 = 5\dot{u}_2 + 30u_2$$

Three other equations can be derived in a similar manner from the remaining three transfer functions of this example problem.

It was previously stated that only a single second-order equation would be required to describe a system with two state variables, but now we see that four second-order equations result from four transfer functions. This apparent contradiction is easily resolved by noting that the output variables **y** are simply a matter of choice on the part of the analyst. They represent the combination of state and input variables of interest. As such, there is an unlimited number of possible combinations, and therefore an unlimited number of possible output variables and differential equations required to relate them to the input variables. Thus far we have used two output variables for illustration, but there could have been a different number. The important point is that only the state variables, and not the output variables, constitute the set of variables necessary and sufficient to describe the dynamics of the system, given the input variables as functions of time. Thus the differential equations in terms of the output variables do not necessarily form a complete description of the system's dynamics. To obtain such a description, we need the differential equations for the state variables.

Suppose therefore that we define the output vector **y** to be the state vector **x**. Then $\mathbf{C} = \mathbf{I}$, and $\mathbf{D} = \mathbf{0}$ in (5.2-14), and from (5.2-28) the transfer function between **u** and **x** is

$$\mathbf{T}(s) = (s\mathbf{I} - \mathbf{A})^{-1}\mathbf{B} \tag{5.2-30}$$

For the second-order system, this can be written as

$$\mathbf{T}(s) = \frac{1}{\Delta(s)}\begin{bmatrix} (s-a_{22}) & a_{12} \\ a_{21} & (s-a_{11}) \end{bmatrix}\begin{bmatrix} b_{11} & b_{12} \\ b_{21} & b_{22} \end{bmatrix} \tag{5.2-31}$$

where $\Delta(s)$ is the determinant of the resolvent matrix.

$$\Delta(s) = |\mathbf{R}(s)| = |s\mathbf{I} - \mathbf{A}| \tag{5.2-32}$$

or

$$\Delta(s) = (s-a_{11})(s-a_{22}) - a_{12}a_{21}$$
$$= s^2 - (a_{11} + a_{22})s + a_{11}a_{22} - a_{12}a_{21} \tag{5.2-33}$$

The transfer functions for $x_1$ are

$$T_{11}(s) = \frac{b_{11}(s-a_{22}) + a_{12}b_{21}}{\Delta(s)} \tag{5.2-34}$$

$$T_{12}(s) = \frac{b_{12}(s-a_{22}) + a_{12}b_{22}}{\Delta(s)} \tag{5.2-35}$$

where

$$X_1(s) = T_{11}(s)U_1(s) + T_{12}(s)U_2(s) \tag{5.2-36}$$

Multiply by $\Delta(s)$ to obtain

$$\Delta(s)X_1(s) = N_{11}(s)U_1(s) + N_{12}(s)U_2(s) \tag{5.2-37}$$

where $N_{11}$ and $N_{12}$ are the numerators of $T_{11}$ and $T_{12}$. Insert the transfer function expressions, collect terms, and revert to the time domain to obtain the following differential equation for $x_1$.

$$\ddot{x}_1 - (a_{11} + a_{22})\dot{x}_1 + (a_{11}a_{22} - a_{12}a_{21})x_1 = b_{11}\dot{u}_1 + (a_{12}b_{21} - a_{22}b_{11})u_1$$
$$+ b_{12}\dot{u}_2 + (a_{12}b_{22} - a_{22}b_{12})u_2 \tag{5.2-38}$$

This equation is a complete description of the dynamics of the system described by (5.2-1) and (5.2-2).

Applying a similar procedure to the following relations will give the differential equation for $x_2$.

$$X_2(s) = T_{21}(s)U_1(s) + T_{22}(s)U_2(s) \tag{5.2-39}$$

$$\Delta(s)X_2(s) = N_{21}(s)U_1(s) + N_{22}(s)U_2(s) \tag{5.2-40}$$

$$T_{21}(s) = \frac{a_{21}b_{11} + (s - a_{11})b_{21}}{\Delta(s)} \tag{5.2-41}$$

$$T_{22}(s) = \frac{a_{21}b_{12} + (s - a_{11})b_{22}}{\Delta(s)} \tag{5.2-42}$$

The resulting differential equation for $x_2$ is also a complete description of the system's dynamics, and it is equivalent to (5.2-38). Its left-hand side will be identical to that of (5.2-38), with $x_1$ replaced by $x_2$. This is so because $\Delta(s)$ multiplies each variable in the respective equations, (5.2-37) and (5.2-40). The significance of this fact will become clear shortly.

*Example 5.10*

Suppose we are interested only in the height $h_2$ in the second tank in Figure 5.1. Find its differential equation for the situation in which both tank areas have the value $A$ and both resistances have the value $R$.

For this system the state variables can be chosen as $x_1 = h_1$ and $x_2 = h_2$. The input is $u_1 = q$, and the appropriate matrices are obtained from (5.1-1) and (5.1-2) with $R_1 = R_2 = R$ and $A_1 = A_2 = A$.

$$A = \begin{bmatrix} \dfrac{-1}{R_1A_1} & \dfrac{1}{R_1A_1} \\ \dfrac{1}{R_1A_2} & \dfrac{-1}{A_2}\left(\dfrac{1}{R_1} + \dfrac{1}{R_2}\right) \end{bmatrix} = \frac{1}{RA}\begin{bmatrix} -1 & 1 \\ 1 & -2 \end{bmatrix}$$

$$B = \begin{bmatrix} \dfrac{1}{A_1} \\ 0 \end{bmatrix} = \begin{bmatrix} \dfrac{1}{A} \\ 0 \end{bmatrix}$$

Then

$$\Delta(s) = s^2 + \frac{3}{RA}s + \left(\frac{1}{RA}\right)^2$$

$$T_{21}(s) = \frac{1}{RA^2\Delta(s)}$$

Note that $b_{21}$ and $b_{22}$ are taken to be zero since there is no second input $u_2$. The solution for $x_2$ is obtained from (5.2-39). [Note that this is equivalent to the algebraic solution of (5.1-3) and (5.1-4).]

$$X_2(s) = \frac{U_1(s)}{RA^2\Delta(s)}$$

or

$$(R^2A^2s^2 + 3RAs + 1)X_2(s) = RU_1(s) \tag{5.2-43}$$

With $x_2 = h_2$ and $u_1 = q$, this gives

$$R^2A^2\ddot{h}_2 + 3RA\dot{h}_2 + h_2 = Rq \tag{5.2-44}$$

The reduced form of the model in terms of a single-state variable is most easily accomplished by the preceding transfer function method. It cannot be used when the differential equations have variable coefficients or are nonlinear. In such cases one or more equations must be differentiated and substituted into the remaining equations.

*Example 5.11*

Reduce the following nonlinear model to one equation in terms of $x_1$.

$$\dot{x}_1 = x_2$$

$$\dot{x}_2 = -2x_2^2 - x_1 + u$$

Differentiate the first equation with respect to time and substitute $\dot{x}_2$ from the second equation.

$$\ddot{x}_1 = \dot{x}_2 = -2x_2^2 - x_1 + u$$

Substitute $x_2$ from the first equation to obtain the desired result.

$$\ddot{x}_1 = -2\dot{x}_1^2 - x_1 + u$$

## State Equations from Transfer Functions

In many applications the transfer functions of each system element are given, and the overall system transfer functions are obtained by diagram reduction. It is often desirable to obtain a state variable representation for the system. This is easily done if the diagram consists only of integrators, multipliers, and comparators. In this case the outputs of the integrators can be chosen as the state variables, as we have seen. Since the diagram can be arranged in more than one way, more than one state model can be obtained. However, if the diagram contains more complicated operational elements, this method cannot be used directly.

We now illustrate a method for using the transfer function to rearrange the diagram so that state variables can be identified. The order of the system, and therefore the number of state variables required, can be found by examining the denominator of the transfer functions. Note that

$$(s\mathbf{I} - \mathbf{A})^{-1} = \frac{1}{|s\mathbf{I} - \mathbf{A}|} \operatorname{adj}(s\mathbf{I} - \mathbf{A}) \tag{5.2-45}$$

and (5.2-28) becomes

$$\mathbf{T}(s) = \frac{\mathbf{C} \operatorname{adj}(s\mathbf{I} - \mathbf{A})\mathbf{B} + \mathbf{D}|s\mathbf{I} - \mathbf{A}|}{|s\mathbf{I} - \mathbf{A}|} \tag{5.2-46}$$

This shows that the denominators of all the transfer functions are the same and are given by the determinant $|s\mathbf{I} - \mathbf{A}|$, since the matrices $\mathbf{A}, \mathbf{B}, \mathbf{C}$, and $\mathbf{D}$ contain no $s$-operator terms. This determinant is an $n$th-order polynomial in $s$ if $\mathbf{A}$ is an $(n \times n)$

matrix. Therefore, the system is of order $n$, and $n$ state variables are required. Physical considerations (integral causality) can be used to show that the order of the numerator of (5.2-46) is equal to or less than $n$.

To illustrate how a state model can be derived from a transfer function, consider the following first-order model with numerator dynamics.

$$T(s) = \frac{Y(s)}{U(s)} = \frac{b_0 s + b_1}{s + a_1} \tag{5.2-47}$$

where the input is $u$ and the output is $y$. Cross-multiplying and reverting to the time domain, we obtain

$$\dot{y} + a_1 y = b_0 \dot{u} + b_1 u \tag{5.2-48}$$

An alternate way of obtaining the differential equation is to divide the numerator and denominator of (5.2-47) by $s$.

$$T(s) = \frac{b_0 + b_1/s}{1 + a_1/s} = \frac{Y(s)}{U(s)} \tag{5.2-49}$$

The essence of the technique is to obtain a 1 in the denominator, which is then used to isolate $Y(s)$. Cross-multiplication gives

$$\begin{aligned} Y(s) &= -\frac{a_1}{s} Y(s) + b_0 U(s) + \frac{b_1}{s} U(s) \\ &= \frac{1}{s}[b_1 U(s) - a_1 Y(s)] + b_0 U(s) \end{aligned}$$

The term multiplying $1/s$ is the input to an integrator; the integrator's output can be selected as a state variable $x$. Thus

$$\begin{aligned} Y(s) &= X(s) + b_0 U(s) \\ X(s) &= \frac{1}{s}[b_1 U(s) - a_1 Y(s)] \\ &= \frac{1}{s}[b_1 U(s) - a_1 X(s) - a_1 b_0 U(s)] \end{aligned}$$

or

$$\dot{x} = -a_1 x + (b_1 - a_1 b_0) u \tag{5.2-50}$$

$$y = x + b_0 u \tag{5.2-51}$$

The signal flow graph is shown in Figure 5.11. Compare (5.2-48) with (5.2-50), and note that the derivative of the input does not appear in the latter form.

Now consider the second-order model for the mechanical vibration problem. Its transfer function is given by (5.1-19).

$$T(s) = \frac{X(s)}{F(s)} = \frac{1}{ms^2 + cs + k} \tag{5.1-19}$$

Divide by $ms^2$ to obtain a unity term.

$$T(s) = \frac{X(s)}{F(s)} = \frac{1/ms^2}{1 + c/ms + k/ms^2}$$

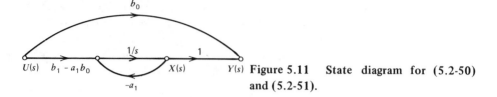

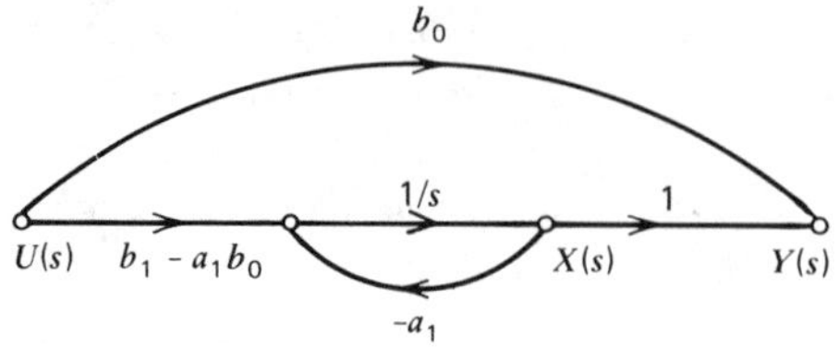


**Figure 5.11** **State diagram for (5.2-50) and (5.2-51).**

Cross-multiply and rearrange to identify the integration operators.

$$X(s) = -\frac{c}{ms}X(s) - \frac{k}{ms^2}X(s) + \frac{1}{ms^2}F(s)$$

$$= \frac{1}{s}\left\{-\frac{c}{m}X(s) + \frac{1}{s}\left[\frac{1}{m}F(s) - \frac{k}{m}X(s)\right]\right\} \tag{5.2-52}$$

Defining the state variables as the outputs of the integrators, we see that

$$X_1(s) = X(s) = \frac{1}{s}\left[-\frac{c}{m}X(s) + X_2(s)\right]$$

$$X_2(s) = \frac{1}{s}\left[\frac{1}{m}F(s) - \frac{k}{m}X(s)\right]$$

or

$$X_1(s) = \frac{1}{s}\left[-\frac{c}{m}X_1(s) + X_2(s)\right]$$

$$X_2(s) = \frac{1}{s}\left[\frac{1}{m}F(s) - \frac{k}{m}X_1(s)\right]$$

Figure 5.12*a* is the signal flow graph derived from these relations. The corresponding state equations in standard form are

$$\begin{bmatrix}\dot{x}_1\\ \dot{x}_2\end{bmatrix} = \begin{bmatrix}-\frac{c}{m} & 1\\ -\frac{k}{m} & 0\end{bmatrix}\begin{bmatrix}x_1\\ x_2\end{bmatrix} + \begin{bmatrix}0\\ \frac{1}{m}\end{bmatrix}u \tag{5.2-53}$$

$$y = [1 \quad 0]\begin{bmatrix}x_1\\ x_2\end{bmatrix} \tag{5.2-54}$$

where $u = f$ and $y = x = x_1$.

Equation (5.2-52) can also be arranged as follows. Let

$$X_1(s) = m\,X(s)$$

$$X_2(s) = s\,X_1(s)$$

Then

$$X_2(s) = \frac{1}{s}\left[F(s) - \frac{c}{m}X_2(s) - \frac{k}{m}X_1(s)\right]$$

These relations give the diagram shown in Figure 5.12*b*. The resulting state model

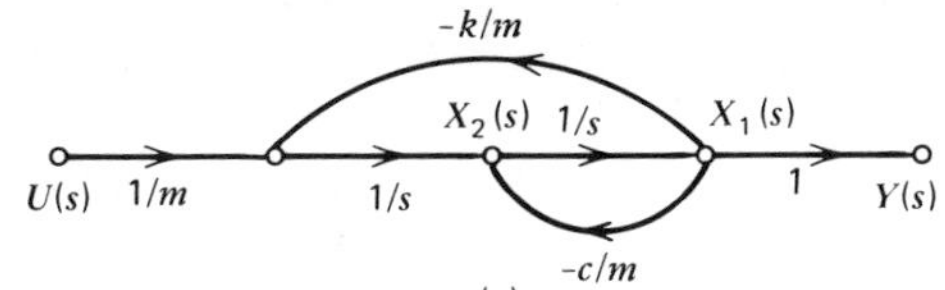


(a)

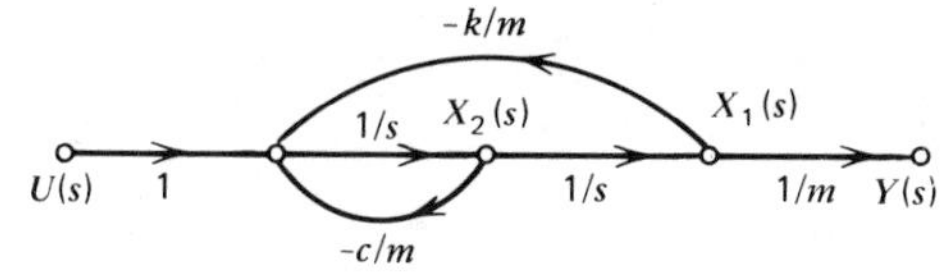


(b)

**Figure 5.12 State diagrams for the vibration model (5.1-19). (*a*) Diagram for the state model (5.2-53) and (5.2-54); (*b*) Diagram for the state model (5.2-55) and (5.2-56).**

with $u = f$ and $y = x$ is

$$\begin{bmatrix} \dot{x}_1 \\ \dot{x}_2 \end{bmatrix} = \begin{bmatrix} 0 & 1 \\ -\dfrac{k}{m} & -\dfrac{c}{m} \end{bmatrix} \begin{bmatrix} x_1 \\ x_2 \end{bmatrix} + \begin{bmatrix} 0 \\ 1 \end{bmatrix} u \tag{5.2-55}$$

$$y = \begin{bmatrix} \dfrac{1}{m} & 0 \end{bmatrix} \begin{bmatrix} x_1 \\ x_2 \end{bmatrix} \tag{5.2-56}$$

Note that the state variables obtained by this technique do not always have straightforward physical interpretations. With the original choice of state variables in (5.1-17) and (5.1-18), $x_1$ is displacement $x$ and $x_2$ is velocity $v$. For (5.2-53), $x_1$ is the displacement, but $x_2$ is the integral of the difference between the applied force $f$ and the spring force $kx$, divided by the mass $m$ ($\dot{x}_2$ is the acceleration due to this force difference). In (5.2-55), $x_2$ is the momentum $mv$, while $x_1$ is an artificial variable that is proportional to displacement ($x_1 = mx$). Convenient physical interpretations of the state variables are sometimes sacrificed to obtain special forms of the **A**, **B**, **C**, or **D** matrices that are useful for analysis. For example, note that the matrix **A** for (5.2-53) has only zeroes and ones in the second column, whereas that for (5.2-55) has only these terms in the first row. The significance of special forms will be discussed in Chapters Nine and Ten.

## 5.3 FREE RESPONSE

The free response of second-order linear continuous-time models is more varied than that of first-order models, because oscillatory as well as exponential functions can occur. Here we will study this behavior using the reduced form of the model.

### Solution from the Reduced Form

The mass-spring-damper system shown in Figure 5.6 is often used to illustrate the free response of a second-order system. Its model (5.1-16) can be written as

$$m\ddot{x} + c\dot{x} + kx = f \tag{5.3-1}$$

The free response can be obtained by Laplace transformation, but here we will obtain it by using the trial solution

$$x(t) = Ae^{st} \tag{5.3-2}$$

where $A$ and $s$ are constants to be determined. This approach is used to emphasize that the free response of any linear constant-coefficient equation can always be obtained by exponential substitution.

Differentiation of (5.3-2) twice gives

$$\dot{x} = sAe^{st}$$

$$\ddot{x} = s^2Ae^{st}$$

Substitute these into (5.3-1) with $f = 0$, and collect terms to get

$$(ms^2 + cs + k)Ae^{st} = 0$$

As with the first-order model in Section 3.1, a general solution is possible only if $Ae^{st} \neq 0$. Thus

$$ms^2 + cs + k = 0 \tag{5.3-3}$$

This is the model's *characteristic equation.* It gives the following solution for the unknown constant $s$.

$$s = \frac{-c \pm \sqrt{c^2 - 4mk}}{2m} \tag{5.3-4}$$

We immediately see that two characteristic roots occur. Each root generates a solution of the form of (5.3-2). If the roots are distinct, the free response is a linear combination of these two forms. Let $s_1$ and $s_2$ denote the two roots. The free response is

$$x(t) = A_1e^{s_1t} + A_2e^{s_2t} \tag{5.3-5}$$

It is easy to show that this solves (5.3-1) with $f = 0$, by computing $\dot{x}$ and $\ddot{x}$, substituting these expressions into (5.3-1) and noting that both $s_1$ and $s_2$ satisfy (5.3-3).

Two state variables are required to describe this system's dynamics. Therefore the initial values of these variables must be specified in order for the solution to be completely determined. This means that the two constants $A_1$ and $A_2$ are determined by the two initial conditions. If the values of $x(t)$ and $\dot{x}(t)$ are given at $t = 0$, then from (5.3-5),

$$x(0) = A_1e^0 + A_2e^0 = A_1 + A_2 \tag{5.3-6}$$

Differentiating (5.3-1) and evaluating at $t = 0$, we get

$$\dot{x}(0) = s_1A_1 + s_2A_2 \tag{5.3-7}$$

The solution for $A_1$ and $A_2$ in terms of $x(0)$ and $\dot{x}(0)$ is

$$A_1 = \frac{\dot{x}(0) - s_2x(0)}{s_1 - s_2} \tag{5.3-8}$$

$$A_2 = \frac{s_1x(0) - \dot{x}(0)}{s_1 - s_2} \tag{5.3-9}$$

*Example 5.12*

Find the free response of (5.3-1) for $m = 1$, $c = 5$, and $k = 4$. The initial conditions are $x(0) = 1$ and $\dot{x}(0) = 5$.

The characteristic roots from (5.3-4) are $s_1 = -1$ and $s_2 = -4$. From (5.3-8) and (5.3-9), $A_1 = 3$, $A_2 = -2$, and (5.3-5) gives

$$x(t) = 3e^{-t} - 2e^{-4t}$$

## Oscillatory Solutions

The solution given by (5.3-5) is convenient to use only when the charactistic roots are real and distinct. If the roots are complex, the behavior of $x(t)$ is difficult to visualize from (5.3-5). We now rearrange the solution into a more useful form for the case of complex roots.

Complex roots of (5.3-3) occur if and only if $c^2 - 4mk < 0$. If so, the roots are complex conjugates. For now, assume that the real part of the roots is negative and write the roots as

$$s_1 = -a + ib \tag{5.3-10}$$

$$s_2 = -a - ib \tag{5.3-11}$$

where

$$a = \frac{c}{2m} \tag{5.3-12}$$

$$b = \frac{\sqrt{4mk - c^2}}{2m} \tag{5.3-13}$$

For this case, $b > 0$ if $m > 0$. The solution (5.3-5) can be written as

$$\begin{aligned} x(t) &= A_1 e^{(-a+ib)t} + A_2 e^{(-a-ib)t} \\ &= e^{-at}(A_1 e^{ibt} + A_2 e^{-ibt}) \end{aligned}$$

Euler's identity for the complex exponential shows that

$$e^{ibt} = \cos bt + i \sin bt \tag{5.3-14}$$

$$e^{-ibt} = \cos bt - i \sin bt \tag{5.3-15}$$

Thus

$$x(t) = e^{-at}[(A_1 + A_2)\cos bt + i(A_1 - A_2)\sin bt] \tag{5.3-16}$$

Substitution of $s_1$ and $s_2$ into the expressions for $A_1$ and $A_2$ shows that

$$A_1 = \frac{x(0)}{2} - \frac{\dot{x}(0) + ax(0)}{2b} i \tag{5.3-17}$$

and that $A_2$ is the complex conjugate of $A_1$. Write $A_1$ as $A_R + A_I i$. Then

$$A_1 + A_2 = 2A_R$$

$$i(A_1 - A_2) = -2A_I$$

and

$$x(t) = e^{-at}(2A_R \cos bt - 2A_I \sin bt) \tag{5.3-18}$$

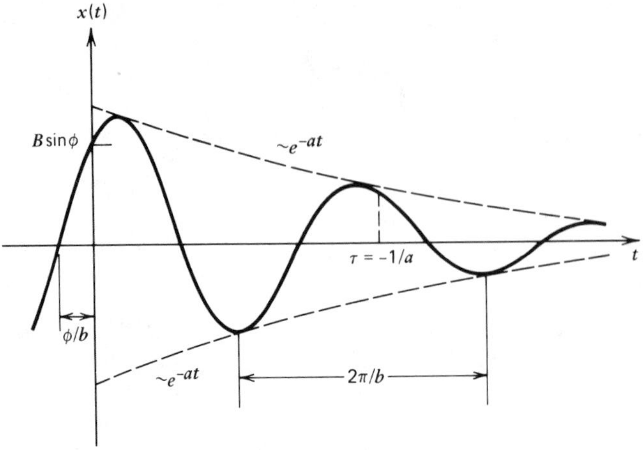


**Figure 5.13** Free response of the second-order model with complex roots $x(t) = Be^{-at}\sin(bt + \phi)$.

where

$$A_R = \frac{x(0)}{2} \tag{5.3-19}$$

$$A_I = -\frac{\dot{x}(0) + ax(0)}{2b} \tag{5.3-20}$$

It is now apparent from (5.3-18) that the free response consists of exponentially decaying oscillations with a frequency of $b$ radians per unit time. The precise nature of the oscillation can be displayed by writing (5.3-18) as a sine wave with a phase shift. Use the identity

$$B\sin(bt + \phi) = B\cos\phi\sin bt + B\sin\phi\cos bt \tag{5.3-21}$$

and compare with the term in parentheses in (5.3-18). This shows that

$$B\sin\phi = 2A_R \tag{5.3-22}$$

$$B\cos\phi = -2A_I \tag{5.3-23}$$

We can define $B$ to be positive and absorb any negative signs with the phase angle $\phi$. Thus

$$B^2\sin^2\phi + B^2\cos^2\phi = 4(A_R^2 + A_I^2)$$

or

$$B = 2\sqrt{A_R^2 + A_I^2} \tag{5.3-24}$$

since $\sin^2\phi + \cos^2\phi = 1$. With $B$ found from this expression, $\phi$ is computed from (5.3-22) with (5.3-23) used to determine the quadrant of $\phi$.

The free response for complex roots is thus given by

$$x(t) = Be^{-at}\sin(bt + \phi) \tag{5.3-25}$$

This is illustrated in Figure 5.13. The sinusoidal oscillation has a frequency $b$ (radians/unit time) and therefore a period of $2\pi/b$. The amplitude of oscillation decays exponentially; that is, the oscillation is bracketed on top and bottom by envelopes that are proportional to $e^{-at}$. These envelopes have a time constant of $\tau = 1/a$. Thus the amplitude of the next oscillation occurring after $t = 1/a$ will be less than 37% of the peak amplitude. For $t > 4/a$, the amplitudes are less than 2% of the peak.

## *Example 5.13*

Solve (5.3-1) for $m = 1$, $c = 3$, $k = 8.5$, and $f = 0$. The initial conditions are $x(0) = 1$ and $\dot{x}(0) = 5$.

The characteristic roots are

$$s = \frac{-3 \pm \sqrt{9 - 34}}{2} = -1.5 \pm 2.5i$$

and therefore $a = 1.5$ and $b = 2.5$. The constants $A_R$, $A_I$, $B$, and $\phi$ can be found from (5.3-19), (5.3-20), and (5.3-22) through (5.3-24). However, we now show that it is sufficient simply to remember the solution form (5.3-25). From this we see that

$$\dot{x}(t) = -aBe^{-at}\sin(bt + \phi) + bBe^{-at}\cos(bt + \phi) \tag{5.3-26}$$

and

$$x(0) = B\sin\phi \tag{5.3-27}$$

$$\dot{x}(0) = -aB\sin\phi + bB\cos\phi \tag{5.3-28}$$

Substitute $B\sin\phi$ into the last equation to obtain

$$B\cos\phi = \frac{\dot{x}(0) + ax(0)}{b} \tag{5.3-29}$$

The present numerical values give

$$B\sin\phi = 1$$

$$B\cos\phi = \frac{5 + 1.5}{2.5} = 2.6$$

Since $B$ is defined to be positive, these relations imply that $\sin\phi > 0$ and $\cos\phi > 0$. Thus $\phi$ is in the first quadrant and

$$\tan\phi = \frac{\sin\phi}{\cos\phi} = \frac{1}{2.6} = 0.3846$$

$$\phi = 21.04^\circ = 0.367 \text{ radians}$$

Finally,

$$B = \frac{1}{\sin\phi} = 2.786$$

and the solution is

$$x(t) = 2.786e^{-1.5t}\sin(2.5t + 0.367) \tag{5.3-30}$$

The time constant is

$$\tau = \frac{1}{1.5} = 0.667 \text{ time units}$$

and the oscillation will have essentially disappeared after 4(0.667) = 2.67 time units.

## Behavior with Repeated Roots

When the two roots of the characteristic equation are repeated (equal), the combination given by (5.3-5) does not consist of two linearly independent functions. In this case it can be shown that the proper combination is

$$x(t) = A_1 e^{s_1 t} + tA_2 e^{s_2 t} \tag{5.3-31}$$

where $s_1 = s_2$. This situation occurs for the vibration model (5.3-1) when $c^2 - 4mk = 0$. In this case, $s_1 = -c/2m$.

If the repeated roots are negative, the free response (5.3-31) decays with time, despite the presence of $t$ as a multiplier, because $e^{s_1 t}$ approaches zero faster than $t$ becomes infinite. To see this, replace the exponential by its series representation. This series contains powers of $t$ greater than one.

A peak can occur before the solution begins to decay, depending on the relative values of $A_1$ and $A_2$, which are found from the initial conditions as before.

$$A_1 = x(0) \tag{5.3-32}$$

$$A_2 = \dot{x}(0) - s_1 x(0) \tag{5.3-33}$$

# 5.4 THE CHARACTERISTIC EQUATION

We saw in Section 5.2 that the reduced forms of the second-order model are (5.2-37) for the state variable $x_1$ and (5.2-40) for $x_2$. Comparison of these forms with (5.3-3) shows that the characteristic equation for both reduced forms is $\Delta(s) = 0$, where $\Delta(s)$ is given by (5.2-33). Thus both forms have the same characteristic roots. For example, if the free response for $x_1$ is a damped sinusoid, the free response for $x_2$ will be of the same form, with the same frequency and time constant. Only the amplitudes and phase angles will differ.

We will now present some convenient criteria to use in determining the system's stability, its time constant, oscillatory behavior, and frequency of oscillation, if any. Because we have used the vibration model as our example, we will use its form of the characteristic equation as the reference.

$$\Delta(s) = ms^2 + cs + k \tag{5.4-1}$$

Note that this is equivalent to the form (5.2-33), but enjoys the advantage of simple notation. The coefficients $m$, $c$, and $k$ can be readily obtained from the differential equation of the reduced form.

## Stability

We have now treated all three situations that can arise for the free response of the linear second-order model with constant coefficients. With $f = 0$ in the vibration model (5.3-1), the only possible equilibrium is $x = 0$ if $k \neq 0$.* From the definition

---

* If $k = 0$ there exists an infinite number of equilibrium positions, but in this case the model (5.3-1) can be reduced to a first-order equation in terms of the speed $v = \dot{x}$.

of stability given in Chapter Three, it can be seen that the solution $x(t)$ must approach zero as $t$ becomes infinite, in order for the equilibrium to be stable. This occurs in the real, distinct-roots case and the repeated-roots case only if both roots are negative. In the case of complex roots, (5.3-25) shows that the real part of the roots must be negative. We can summarize all three cases by stating that the system is stable if and only if both roots have negative real parts. (For the real-root cases, the imaginary parts are considered to be zero.)

Neutral, or limited, stability occurs for this second-order system in the complex-roots case if the real part is zero. The solution is then a constant-amplitude sinusoidal oscillation about the equilibrium. Neutral stability also occurs if one root is negative and the other zero. Since two repeated zero roots are impossible here, we can generalize this result to state that the equilibrium is neutrally stable when at least one root has a zero real part and the other root does not have a positive real part.

Conversely, it can be seen from (5.3-5) that only one root need be positive for $x(t)$ to become infinite. For the complex-roots case, this happens if the real part is positive. Thus for all root cases, if at least one root has a positive real part, the equilibrium is unstable. As is the case for many of the results of this chapter, these observations hold true for systems of higher order.

Discussion of a second-order system automatically implies that the coefficient $m$ in (5.3-1) is nonzero. With this restriction, an analysis of the root solution (5.3-4) will show that both roots have negative real parts if and only if $m$, $c$, and $k$ are positive. If either $c$ or $k$ is zero, the equilibrium is neutrally stable.

## The Damping Ratio

The behavior of the preceding free response for the stable case can be conveniently characterized by the *damping ratio* $\zeta$ (sometimes called the *damping factor*). For the characteristic equation (5.4-1), this is defined as

$$\zeta = \frac{c}{2\sqrt{mk}} \tag{5.4-2}$$

Repeated roots occur if $c^2 - 4mk = 0$; that is, if $c = 2\sqrt{mk}$. This value of the damping constant is the *critical damping constant*, and when $c$ has this value the system is said to be critically damped. If the actual damping constant is greater than $2\sqrt{mk}$, two real distinct roots exist, and the system is *overdamped.* If $c < 2\sqrt{mk}$, complex roots occur, and the system is *underdamped.* The damping ratio is thus seen to be the ratio of the actual damping constant $c$ to the critical value. For a critically damped system, $\zeta = 1$. Exponential behavior occurs if $\zeta > 1$, and oscillations exist for $\zeta < 1$.

The damping ratio is used with the reduced model form as a quick check for oscillatory behavior. For example, the model (5.2-44) for the height $h_2$ has the following damping ratio.

$$\zeta = \frac{3RA}{2\sqrt{R^2A^2}} = \frac{3}{2} > 1$$

Since $\zeta > 1$, no oscillations will occur in the free response regardless of the initial values of the liquid heights.

## Natural and Damped Frequencies of Oscillation

When there is no damping, the characteristic roots are purely imaginary. The imaginary part, and therefore the frequency of oscillation, for this case is $b = \sqrt{k/m}$. This frequency is termed the *undamped natural frequency*, or simply the *natural frequency*, and is used as a reference value denoted by $\omega_n$. Thus

$$\omega_n = \sqrt{\frac{k}{m}} \tag{5.4-3}$$

We can write the characteristic equation in terms of the parameters $\zeta$ and $\omega_n$. First divide (5.4-1) by $m$ and use the fact that $2\zeta\omega_n = c/m$. The equation becomes

$$s^2 + 2\zeta\omega_n s + \omega_n^2 = 0 \tag{5.4-4}$$

and the roots are

$$s = -\zeta\omega_n \pm i\omega_n\sqrt{1-\zeta^2} \tag{5.4-5}$$

Comparison with (5.3-10) shows that $a = \zeta\omega_n$ and $b = \omega_n\sqrt{1-\zeta^2}$. Since the time constant $\tau$ is $1/a$,

$$\tau = \frac{1}{\zeta\omega_n} \tag{5.4-6}$$

The frequency of oscillation is sometimes called the *damped natural frequency*, or simply *damped frequency* $\omega_d$ to distinguish it from $\omega_n$.

$$\omega_d = \omega_n\sqrt{1-\zeta^2}. \tag{5.4-7}$$

For the underdamped case ($\zeta < 1$), we see that $\omega_d < \omega_n$.

## Graphical Interpretation

The preceding relations can be represented graphically by plotting the location of the roots (5.4-5) in the complex plane (Figure 5.14), in which the imaginary part is plotted versus the real part. This plot is said to be in the $s$-plane. Since the roots are conjugate, we consider only the upper root.

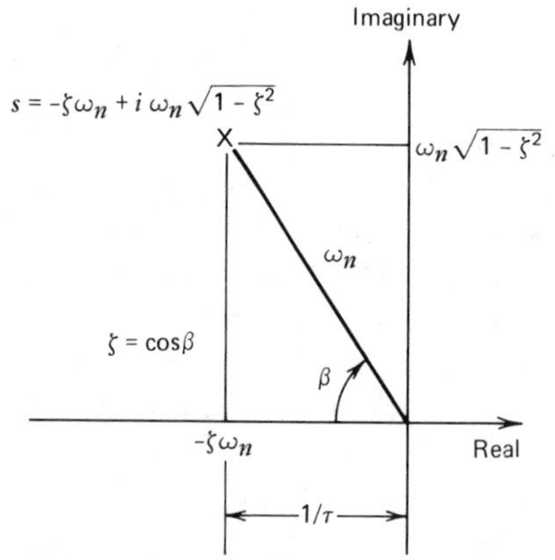


**Figure 5.14 Location of the upper complex root in terms of the parameters $\zeta, \tau, \omega_n, \omega_d$.**

From the stability criterion we see that a root lying in the right-half plane results in an unstable system. The parameters $\zeta$, $\omega_n$, $\omega_d$, and $\tau$ are normally used to describe stable systems only, and so we will assume for now that all roots lie in the left-hand plane.

The lengths of two sides of the right triangle shown in Figure 5.14 are $\zeta\omega_n$ and $\omega_n\sqrt{1-\zeta^2}$. Thus the hypoteneuse is of length $\omega_n$. It makes an angle $\beta$ with the negative real axis, and

$$\cos\beta = \zeta \tag{5.4-8}$$

Therefore all roots lying on the circumference of a given circle centered on the

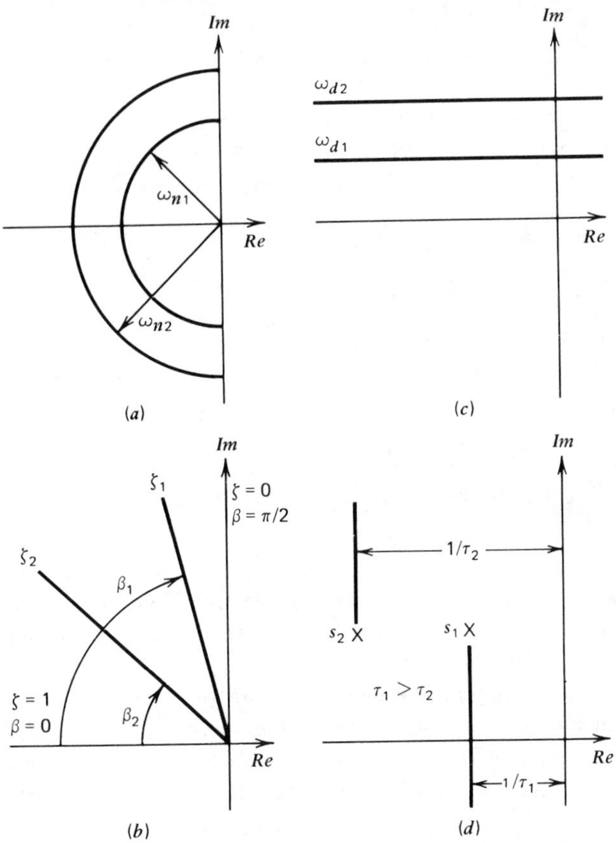


**Figure 5.15** Graphical representation of the parameters $\zeta$, $\tau$, $\omega_n$, and $\omega_d$ in the complex plane. (*a*) Roots with the same natural frequency $\omega_n$ lie on the same circle. (*b*) Roots with the same damping ratio $\zeta$ lie on the same line through the origin. (*c*) All roots on a given line parallel to the real axis have the same damped natural frequency $\omega_d$. (*d*) All roots on a given line parallel to the imaginary axis have the same time constant $\tau$. The root lying the farthest to the right is the dominant root.

origin are associated with the same undamped natural frequency $\omega_n$. Figure 5.15*a* illustrates this for two different frequencies, $\omega_{n1}$ and $\omega_{n2}$. From (5.4-8) we see that all roots lying on the same line passing through the origin are associated with the same damping ratio (Figure 5.15*b*). The limiting values of $\zeta$ correspond to the imaginary axis ($\zeta = 0$) and the negative real axis ($\zeta = 1$). Roots lying on a given line parallel to the real axis all give the same damped natural frequency (Figure 5.15*c*).

## The Dominant-Root Concept

Since $\tau = 1/\zeta\omega_n$, the distance from the root to the imaginary axis equals the reciprocal of the time constant for that root. All roots lying on a given vertical line have the same time constant, and the greater the distance of the line from the imaginary axis, the smaller the time constant (Figure 5.15*d*); this leads to the concept of the *dominant* *root.* For a given characteristic equation, this is the root that lies the farthest to the right in the $s$-plane. Therefore, if the system is stable, the dominant root is the root with the largest time constant.

The dominant-root concept allows us to simplify the analysis of a given system by focusing on the root that plays the most important role in the system's dynamics. For example, if the two roots are $s_1 = -20$ and $s_2 = -2$, the free response is of the form

$$x(t) = A_1 e^{-20t} + A_2 e^{-2t}$$

For time measured in seconds, the first exponential has a time constant $\tau_1 = 1/20$ sec, while the second exponential's time constant is $\tau_2 = 1/2$ sec. The first exponential will have essentially disappeared after 4/20 sec, while the second exponential takes about $4/2 = 2$ sec to disappear. Unless the constant $A_1$ is very much larger than $A_2$, after about 0.2 sec the solution is given approximately by

$$x(t) \simeq A_2 e^{-2t}$$

The root $s_2 = -2$ is the dominant root. The term in the free response corresponding to the dominant root remains nonzero longer than the other terms.

The usefulness of the dominant root in the preceding example is that it allows us to estimate the response time for the system. This is the time constant of the dominant root. Obviously, the greater the separation between the dominant root and the other roots, the better the dominant-root approximation.

The dominant behavior can sometimes be given by more than one root – for example, by repeated roots or a complex conjugate pair. While this provides no useful information for a second-order system, it will prove to be useful for higher-order systems. For example, a fourth-order system will have four roots. Suppose these roots are $-10 \pm i5$ and $-2 \pm i4$. As we will see later, the free response is of the form

$$x(t) = B_1 e^{-10t} \sin(5t + \phi_1) + B_2 e^{-2t} \sin(4t + \phi)$$

From Figure 5.16 we see that the dominant-root pair is $-2 \pm i4$, and after about 4/10 of a time unit, $x(t)$ is given approximately by the second sinusoid. The time constant, frequencies, and damping ratio of the system are estimated from those of the dominant root. These are shown in Figure 5.16.

We have so far devoted much effort to achieve a thorough understanding of first- and second-order models. One reason for this is that in many practical problems, the physical system can be approximated by a first- or second-order model because of the dominant-root concept.

## 5.5 THE ROUTH- HURWITZ STABILITY CRITERION

Perhaps the most important question to be answered by a dynamic model concerns the stability of the system under study. If the proposed system is predicted to be unstable, the designer immediately seeks to modify the system to make it stable. This will be especially true in the design of feedback control systems.

Thus it is convenient to have available a criterion that gives a yes-or-no answer to the stability question. Slightly different criteria were developed separately by Hurwitz in 1895 and by Routh in 1905 for the purpose of predicting the stability

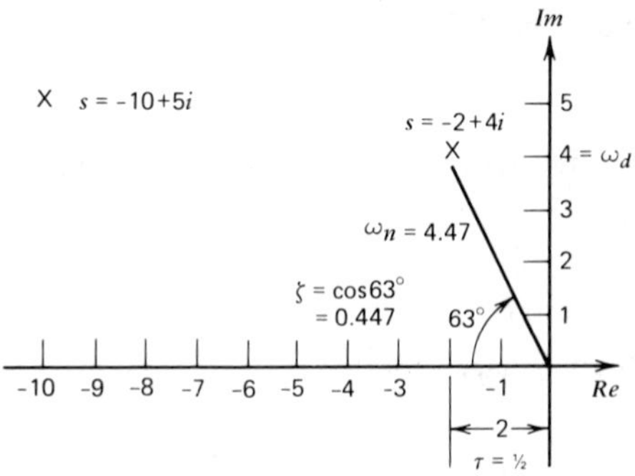


**Figure 5.16** Dominant-root concept applied to a fourth-order system. An estimate of the system's behavior is given by the parameters $\zeta$, $\tau$, $\omega_n$, and $\omega_d$ of the dominant root $s = -2 + 4i$. (The two conjugate roots lying in the lower half plane are not shown.)

of rotation of a body about a given axis. Their criteria furnish the same information and are summarized here for the second- and third-order cases. In Appendix F we state the results for the general case without proof (which is lengthy).

Recall that the stability properties of a linear system are determined by the roots of its characteristic equation. If any root lies in the right half of the complex plane, the system is unstable. The Routh-Hurwitz criterion tells us how many roots lie in the right-half plane. By shifting the origin of the complex plane to the left, we will see that the criterion can also be used to determine the number of roots that lie to the right of any given vertical line. This enables us to estimate the location and type of the dominant root or root pair and thus to obtain the approximate time constant, oscillation frequency, and damping ratio for the system without the necessity of solving for the characteristic roots. Finally, the Routh-Hurwitz criterion sometimes gives enough information to enable us to determine the location of all the roots. While closed-form solutions exist for the roots of polynomials up to fourth order, these formulas are cumbersome for third- and fourth-order equations, and the Routh-Hurwitz criterion often gives enough information with much less effort.

## The Second-Order Case

Since the quadratic formula is relatively simple, it is not difficult to use it to show that the characteristic equation

$$b_2 s^2 + b_1 s + b_0 = 0 \tag{5.5-1}$$

represents a stable system if and only if $b_2$, $b_1$, and $b_0$ all have the same sign. If $b_0 = 0$, the system is neutrally stable, since one root is $s = 0$. If $b_1 = 0$ and $b_0/b_2 > 0$, the system is again neutrally stable with two purely imaginary roots. (Of course, if $b_2 = 0$ the system is no longer second-order.)

## The Third-Order Case

Third-order models occur quite frequently, and the Routh-Hurwitz results are simply stated. Consider the equation

$$b_3 s^3 + b_2 s^2 + b_1 s + b_0 = 0 \tag{5.5-2}$$

For simplicity, assume that we have normalized the equation so that $b_3 > 0$. The system is stable if and only if all of the following conditions are satisfied.

$$\left.\begin{aligned} b_2 > 0, \quad b_1 > 0, \quad b_0 > 0 \\ b_1 b_2 > b_0 b_3 \end{aligned}\right\} \tag{5.5-3}$$

The neutrally stable cases occur if (5.5-3) is satisfied except for the following.

1. $b_0 = 0$ (a root exists at $s = 0$).
2. $b_2 = 0$ (a pair of purely imaginary roots).
3. $b_1 b_2 - b_0 b_3 = 0$ (a pair of purely imaginary roots).

*Example 5.14*

A certain system has the characteristic equation

$$s^3 + 9s^2 + 26s + K = 0 \tag{5.5-4}$$

Find the range of $K$ values for which the system will be stable.

From (5.5-3) it is necessary and sufficient that

$$K > 0$$

$$9(26) - K > 0$$

Thus the system is stable if and only if $0 < K < 234$.

*Example 5.15*

For the characteristic equation (5.5-4), find the value of $K$ required so that the dominant time constant is no larger than $\frac{1}{2}$.

The time constant requirement means that no root can lie to the right of $s = -2$. Translate the origin of the $s$-plane to $s = -2$ by substituting $s = p - 2$ into (5.5-4). This gives

$$p^3 + 3p^2 + 2p + K - 24 = 0 \tag{5.5-5}$$

We can apply the Routh-Hurwitz criterion to this equation to determine when all roots $p$ have negative real parts (and thus when all $s$ roots lie to the left of $s = -2$). From (5.5-3) this occurs when

$$K - 24 > 0$$

$$3(2) - (K - 24) > 0$$

Or $24 < K < 30$.

Even more information about the system's behavior (damping ratio, etc.) can be extracted from the Routh-Hurwitz criterion (this is explored in more detail in Appendix F, Example F.3).

## 5.6 THE TRANSITION MATRIX

We have obtained the free response of the second-order model from its reduced form. When we encounter higher-order models later, it will not be convenient to use the reduced form. For this reason we now introduce a matrix method for obtaining the free response.

In Section 5.2 we used the Laplace transformation of the state equation $\dot{\mathbf{x}} =$

$\mathbf{Ax} + \mathbf{Bu}$ to obtain

$$(s\mathbf{I} - \mathbf{A})\mathbf{X}(s) = \mathbf{x}(0) + \mathbf{BU}(s) \tag{5.6-1}$$

For the free-response calculation we can take $u(t)$ to be zero. This gives

$$(s\mathbf{I} - \mathbf{A})\mathbf{X}(s) = \mathbf{x}(0) \tag{5.6-2}$$

or

$$\mathbf{X}(s) = (s\mathbf{I} - \mathbf{A})^{-1}\mathbf{x}(0) \tag{5.6-3}$$

This says that the inverse of the resolvent matrix can be considered to be the transfer function matrix relating the free response $\mathbf{x}(t)$ to the initial condition $\mathbf{x}(0)$. Since $\mathbf{x}(0)$ is a vector of constants, (5.6-3) shows the free response to be of the form

$$\mathbf{x}(t) = \boldsymbol{\varphi}(t)\mathbf{x}(0) \tag{5.6-4}$$

where the matrix $\boldsymbol{\varphi}(t)$ is given by

$$\boldsymbol{\varphi}(t) = \mathscr{L}^{-1}[(s\mathbf{I} - \mathbf{A})^{-1}] \tag{5.6-5}$$

The matrix $\boldsymbol{\varphi}(t)$ is the *transition matrix* and is analogous to the transition function for the first-order model (see Section 3.10).

One advantage of expressing the free response in the form (5.6-4) is that the effects of the initial conditions are clearly distinguished from the intrinsic or natural behavior of the system as expressed by $\boldsymbol{\varphi}(t)$. This is especially useful when we wish to evaluate $\mathbf{x}(t)$ numerically at one or more values of $t$ for a variety of initial conditions.

Equation (5.6-4) represents only one of several ways by which the transition matrix can be computed. Other methods, some of which are suitable for implementation on a digital computer, are presented in Chapter Nine. Here we illustrate the use of (5.6-4) with an example.

### *Example 5.16*

Compute the transition matrix given the state equations

$$\dot{x}_1 = x_2$$

$$\dot{x}_2 = -4x_1 - 5x_2$$

Transforming the state equations gives

$$sX_1(s) - x_1(0) = X_2(s)$$

$$sX_2(s) - x_2(0) = -4X_1(s) - 5X_2(s)$$

or

$$sX_1(s) - X_2(s) = x_1(0)$$

$$4X_1(s) + (s+5)X_2(s) = x_2(0)$$

This is (5.6-2) in component form. These equations can be solved by Cramer's rule, by substitution or by matrix inversion. With the last method we have

$$s\mathbf{I} - \mathbf{A} = \begin{bmatrix} s & -1 \\ 4 & (s+5) \end{bmatrix}$$

and

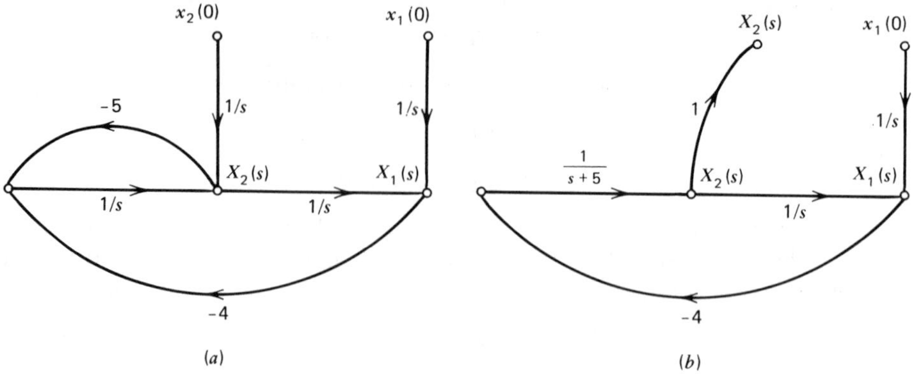


**Figure 5.17** (*a*) State diagram for Example 5.17. (*b*) Partially reduced diagram with $x_2(0) = 0$.

$$\Phi(s) = (sI - A)^{-1} = \frac{1}{s(s+5)+4}\begin{bmatrix}(s+5) & 1 \\ -4 & s\end{bmatrix}$$

$$= \frac{1}{(s+1)(s+4)}\begin{bmatrix}(s+5) & 1 \\ -4 & s\end{bmatrix} \quad (5.6\text{-}6)$$

Thus with a partial fraction expansion,

$$\Phi_{11}(s) = \frac{(s+5)}{(s+1)(s+4)} = \frac{4}{3}\frac{1}{s+1} - \frac{1}{3}\frac{1}{s+4}$$

or

$$\phi_{11}(t) = \tfrac{4}{3}e^{-t} - \tfrac{1}{3}e^{-4t}$$

Performing the same procedure with the other three entries in the matrix, we obtain

$$\phi_{12}(t) = \tfrac{1}{3}e^{-t} - \tfrac{1}{3}e^{-4t}$$

$$\phi_{21}(t) = \tfrac{4}{3}e^{-4t} - \tfrac{4}{3}e^{-t}$$

$$\phi_{22}(t) = \tfrac{4}{3}e^{-4t} - \tfrac{1}{3}e^{-t}$$

### *Example 5.17*

Compute the transition matrix for the system of the previous example, given the state diagram shown in Figure 5.17*a*.

This approach considers the elements of $\Phi(s)$ to be the transfer functions between the two outputs $X_1(s)$ and $X_2(s)$, and the two inputs $x_1(0)$ and $x_2(0)$. Set $x_2(0) = 0$ temporarily and reduce the diagram to find the transfer functions between $X_1(s)$, $X_2(s)$ and $x_1(0)$. From Figure 5.17*b*, the result is

$$\Phi_{11}(s) = \frac{X_1(s)}{x_1(0)} = \frac{s+5}{s(s+5)+4}$$

$$\Phi_{21}(s) = \frac{X_2(s)}{x_1(0)} = \frac{-4}{s(s+5)+4}$$

**TABLE 5.1 The Transition Matrix for Second-Order Systems**

State model $\begin{bmatrix}\dot{x}_1 \\ \dot{x}_2\end{bmatrix} = \begin{bmatrix}a_{11} & a_{12} \\ a_{21} & a_{22}\end{bmatrix}\begin{bmatrix}x_1 \\ x_2\end{bmatrix}$, $a_{ij}$ constant

Characteristic equation $s^2 - (a_{11} + a_{22})s + a_{11}a_{22} - a_{12}a_{21} = 0$

Transition matrix $\boldsymbol{\varphi}(t) = [\phi_{ij}(t)]$

$$\mathbf{x}(t) = \boldsymbol{\varphi}(t)\mathbf{x}(0)$$

*Case 1*

Real, distinct roots $s = -a, -b$

$$\phi_{11}(t) = \frac{1}{b-a}[(b + a_{22})e^{-bt} - (a + a_{22})e^{-at}]$$

$$\phi_{12}(t) = \frac{a_{12}}{b-a}(e^{-at} - e^{-bt})$$

$$\phi_{21}(t) = \frac{a_{21}}{b-a}(e^{-at} - e^{-bt})$$

$$\phi_{22}(t) = \frac{1}{b-a}[(b + a_{11})e^{-bt} - (a + a_{11})e^{-at}]$$

*Case 2*

Complex conjugate roots $s = -a \pm ib$

$$\phi_{11}(t) = e^{-at}\left[\cos bt - \frac{(a_{22} + a)}{b}\sin bt\right]$$

$$\phi_{12}(t) = \frac{a_{12}}{b}e^{-at}\sin bt$$

$$\phi_{21}(t) = \frac{a_{21}}{b}e^{-at}\sin bt$$

$$\phi_{22}(t) = e^{-at}\left[\cos bt - \frac{(a_{11} + a)}{b}\sin bt\right]$$

These are the entries in the first column of $\mathbf{\Phi}(s)$, as given by (5.6-6). Similarly, with $x_1(0) = 0$, we can reduce the diagram to obtain $\Phi_{12}(s)$ and $\Phi_{22}(s)$. The rest of the procedure follows as in Example 5.16 in order to obtain $\boldsymbol{\varphi}(t)$.

The Laplace transform procedure can be readily applied to the general second-order case. These results are given in Table 5.1 for real distinct roots and complex roots.

## Calculation of the Forced Response From $\phi(t)$

If $\mathbf{u}(t) \neq 0$, the resulting response can be found from (5.6-1) as follows.

$$\mathbf{X}(s) = (s\mathbf{I} - \mathbf{A})^{-1}\mathbf{x}(0) + (s\mathbf{I} - \mathbf{A})^{-1}\mathbf{B}\mathbf{U}(s) \tag{5.6-7}$$

The scalar form of the convolution theorem (Problem 3.18) can be extended to the vector case as follows. If $\mathscr{L}[\mathbf{v}(t)] = \mathbf{V}(s)$ and $\mathscr{L}[\boldsymbol{\varphi}(t)] = \boldsymbol{\Phi}(s)$, the theorem states that

$$\mathscr{L}^{-1}[\boldsymbol{\Phi}(s)\mathbf{V}(s)] = \int_0^t \boldsymbol{\varphi}(t-\lambda)\mathbf{v}(\lambda)\,d\lambda \tag{5.6-8}$$

If we take $\mathbf{v}(t) = \mathbf{B}\mathbf{u}(t)$, then

$$\mathscr{L}^{-1}[\boldsymbol{\Phi}(s)\mathbf{B}\mathbf{U}(s)] = \int_0^t \boldsymbol{\varphi}(t-\lambda)\mathbf{B}\mathbf{u}(\lambda)\,d\lambda \tag{5.6-9}$$

Since $\boldsymbol{\varphi}(t) = \mathscr{L}^{-1}[(s\mathbf{I} - \mathbf{A})^{-1}]$, (5.6-9) must be the forced response. Therefore, the complete response from (5.6-7) is

$$\mathbf{x}(t) = \boldsymbol{\varphi}(t)\mathbf{x}(0) + \int_0^t \boldsymbol{\varphi}(t-\lambda)\mathbf{B}\mathbf{u}(\lambda)\,d\lambda \tag{5.6-10}$$

This is the vector equivalent of the scalar form (3.10-1) in which $\phi(t) = e^{rt}$.

Equation (5.6-10) provides the basis for several useful computational procedures. For example, if the input vector $\mathbf{u}$ is a vector of step functions, then $\mathbf{u} = \mathbf{p}$, a constant vector for $t \geqslant 0$. Because $\mathbf{B}$ is constant,

$$\int_0^t \boldsymbol{\varphi}(t-\lambda)\mathbf{B}\mathbf{u}(\lambda)\,d\lambda = \int_0^t \boldsymbol{\varphi}(t-\lambda)\,d\lambda\,\mathbf{B}\mathbf{p} \tag{5.6-11}$$

If $\boldsymbol{\varphi}(t)$ is known in analytical form, the preceding integral can be evaluated in closed form. For example, the system of Example 5.16 resulted in a $(4 \times 4)$ matrix $\boldsymbol{\varphi}(t)$. The integral of a matrix is the matrix of the integrals of the elements. For this example,

$$\int \boldsymbol{\varphi}(t-\lambda)\,d\lambda = \begin{bmatrix} \int \phi_{11}(t-\lambda)\,d\lambda & \int \phi_{12}(t-\lambda)\,d\lambda \\ \int \phi_{21}(t-\lambda)\,d\lambda & \int \phi_{22}(t-\lambda)\,d\lambda \end{bmatrix}$$

For $\phi_{11}$,

$$\begin{aligned} \int_0^t \phi_{11}(t-\lambda)\,d\lambda &= \frac{1}{3}\int_0^t (4e^{-(t-\lambda)} - e^{-4(t-\lambda)})\,d\lambda \\ &= \tfrac{4}{3}e^{-t}\int_0^t e^{\lambda}\,d\lambda - \tfrac{1}{3}e^{-4t}\int_0^t e^{4\lambda}\,d\lambda \\ &= \tfrac{5}{4} - \tfrac{4}{3}e^{-t} + \tfrac{1}{12}e^{-4t} \end{aligned}$$

The other elements are evaluated in the same way. Since $\boldsymbol{\varphi}(t)$ depends only on $\mathbf{A}$ and not on $\mathbf{B}$, $\mathbf{u}$, or $\mathbf{x}(0)$, these results can be used to evaluate the response for any $\mathbf{B}$, any step function, and any set of initial conditions $\mathbf{x}(0)$.

The evaluation of the integrals of the $\boldsymbol{\varphi}(t)$ elements is tedious for high-order systems. However, with the use of computer subroutines for numerical evaluation of the integrals and for matrix multiplication, we can develop a powerful program for analyzing the response of linear systems of high order. More detailed discussion of these methods and results for other input types are given in Section 9.3.

## 5.7 STEP RESPONSE

The step function models any rapid change in the input from one constant level to another. In this section we treat the step response of second-order systems. In later sections the impulse, ramp, and sine response will be covered.

### Substitution Method

Consider the following model.

$$m\ddot{x} + c\dot{x} + kx = du(t) \tag{5.7-1}$$

where $u(t)$ is the input and $m, c, k, d$ are constants. If this model is stable, any constant input $u$ will produce a steady-state response such that $\ddot{x} = \dot{x} = 0$ and

$$kx = du \tag{5.7-2}$$

Let $d$ absorb the magnitude of $u(t)$, and consider $u(t)$ to be a unit step input, so that $u(t) = 0$, $t < 0$, and $u(t) = 1$, $t \geqslant 0$. Then from (5.7-2) the response at steady state is $x = d/k$.

Before steady state is reached, a transient term exists that is of the same form as the free response. If the model is underdamped, this term is of the form $Be^{-at}\sin(bt + \phi)$, where the characteristic roots have been written as $s = -a \pm ib$. We therefore try a solution in the form of the sum of the transient and steady-state forms; that is

$$x(t) = Be^{-at}\sin(bt + \phi) + \frac{d}{k} \tag{5.7-3}$$

Substitution of this trial form into (5.7-1) with $u(t) = 1$ will verify its correctness. The constants $B$ and $\phi$ are determined from the initial conditions *after* the steady-state term $d/k$ has been included. Because

$$\dot{x}(t) = -aBe^{-at}\sin(bt + \phi) + bBe^{-at}\cos(bt + \phi)$$

we obtain

$$x(0) = B\sin\phi + \frac{d}{k}$$

$$x(0) = -Ba\sin\phi + bB\cos\phi$$

or

$$B\sin\phi = x(0) - \frac{d}{k} \tag{5.7-4}$$

$$B\cos\phi = \frac{a}{bk}[kx(0) - d] + \frac{1}{b}\dot{x}(0) \tag{5.7-5}$$

These can be solved for $B$ and $\phi$ in a manner similar to that used to find the free response. For the case of zero initial conditions, we get

$$x(t) = \frac{d}{k}\left[\frac{\sqrt{a^2+b^2}}{b} e^{-at}\sin(bt+\phi)+1\right] \tag{5.7-6}$$

$$\tan\phi = \frac{b}{a} \tag{5.7-7}$$

where $\phi$ is in the third quadrant (remember we are assuming $a > 0$, $b > 0$). In terms of the parameters $\zeta$, $\omega_n$, this solution becomes

$$x(t) = \frac{d}{k}\left[\frac{1}{\sqrt{1-\zeta^2}} e^{-\zeta\omega_n t}\sin(\omega_n\sqrt{1-\zeta^2}\,t+\phi)+1\right] \tag{5.7-8}$$

$$\tan\phi = \frac{\sqrt{1-\zeta^2}}{\zeta} \tag{5.7-9}$$

where $\phi$ is in the third quadrant. If the initial conditions are not zero, we can simply add the free response to the result.

For an overdamped or critically damped system, the method is essentially the same. A function of the form of the free response is added to the steady-state term. For an overdamped case, this gives,

$$x(t) = A_1 e^{s_1 t} + A_2 e^{s_2 t} + \frac{d}{k}$$

where $s_1$ and $s_2$ are the distinct negative roots. For zero initial conditions this gives

$$x(t) = \frac{d}{k}\left(\frac{s_2}{s_1-s_2} e^{s_1 t} - \frac{s_1}{s_1-s_2} e^{s_2 t} + 1\right) \tag{5.7-10}$$

The result for the critically damped case is left to the student.

The step response is illustrated in Figure 5.18 with a plot of the normalized response variable $kx/d$ as a function of the normalized time variable $\omega_n t$. The plot gives the response for four values of the damping ratio, with $\omega_n$ held constant. When $\zeta > 1$, the response is sluggish and does not overshoot the steady-state value. As $\zeta$ is decreased the speed of response increases. The critically damped case, $\zeta = 1$, is the case in which the steady-state value is reached most quickly but without oscillation. Instrument mechanisms are frequently designed to be critically damped for this reason. When a voltage to be measured is applied to a voltmeter, it acts like a step input to the mechanism. We would like the needle of the meter to approach the indicated position as quickly as possible but without oscillating around it. This is the critically damped case.

As $\zeta$ is decreased below 1, the response overshoots and oscillates about the final value. The smaller $\zeta$ is, the larger the overshoot and the longer it takes for the oscillations to die out. There are design applications in which we wish the response to be near its final value as quickly as possible, with some oscillation tolerated. (For example, a radar antenna might need to be pointed quickly in the general direction of a target.) As $\zeta$ is decreased to zero (no damping), the oscillations never die out.

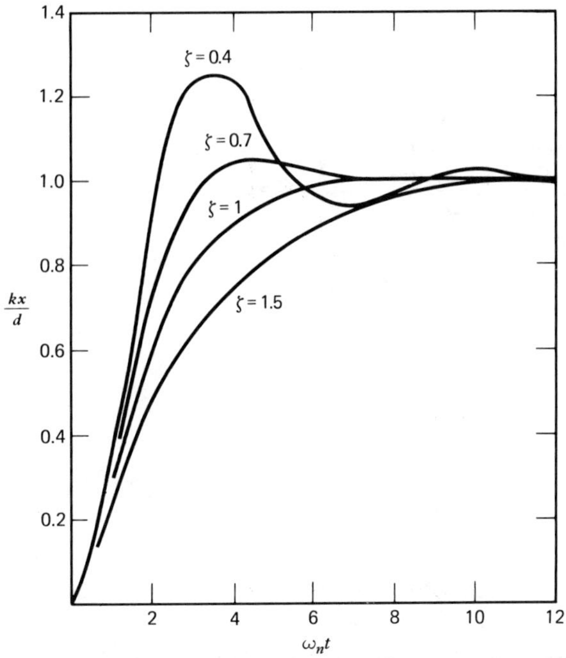


**Figure 5.18** **Normalized unit step response of the model** $m\ddot{x} + c\dot{x} + kx = du(t)$.

## Transient-Response Specifications

Performance criteria for the transient response of a dynamic system are frequently stated in terms of the parameters shown in Figure 5.19 for a typical step response of an underdamped system (not necessarily a second-order system). The *maximum overshoot* is the maximum deviation of the output $x$ above its steady-state value $x_{ss}$. It is sometimes expressed as a percentage of the final value. Since the maximum overshoot increases with decreasing $\zeta$, it is sometimes used as an indicator of the relative stability of the system. The *peak time* $t_p$ is the time at which the maximum overshoot occurs. The *settling time* $t_s$ is the time required for the oscillations to stay within some specified small percentage of the final value. The most common values used are 2% and 5%. The latter choice is shown in the figure. If the final value of the response differs from some desired value, a steady-state error exists.

The *rise time* $t_r$ can be defined as the time required for the output to rise from 10% to 90% of its final value. However, no agreement exists on this definition. Sometimes the rise time is taken to be the time required for the response to reach the final value for the first time. Other definitions are also in use. Finally, the *delay time* $t_d$ is the time required for the response to reach 50% of its final value.

These parameters are relatively easy to obtain from an experimentally determined step-response plot. However, if they are to be determined in analytical from a differential equation model, the task is difficult for models of order greater than two. Here we obtain expressions for these quantities from the second-order step response given by (5.7-8).

Setting the derivative of (5.7-8) equal to zero gives expressions for both the

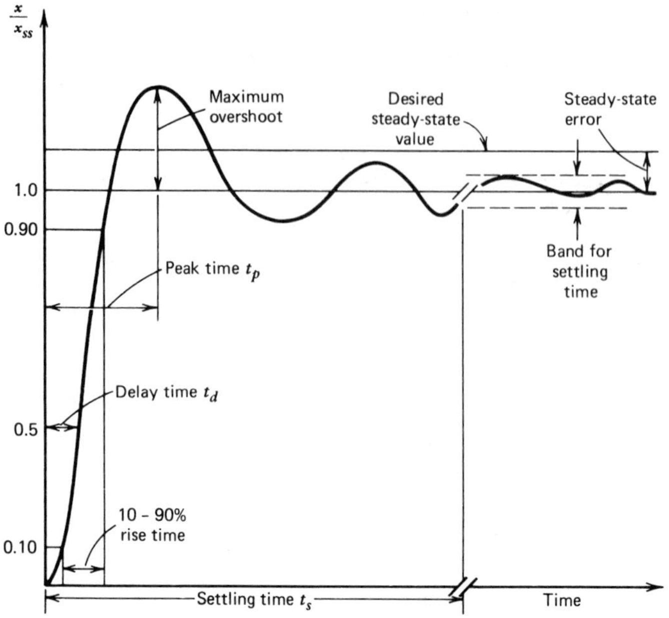


**Figure 5.19 Transient performance specifications based on step response.**

maximum overshoot and the peak time $t_p$. After some trigonometric manipulation, the result is

$$\frac{dx}{dt} = \frac{d}{k}\left(\frac{\omega_n}{\sqrt{1-\zeta^2}}\, e^{-\zeta\omega_n t}\sin \omega_n\sqrt{1-\zeta^2}\, t\right) = 0$$

This gives, for $t < \infty$,

$$\omega_n\sqrt{1-\zeta^2}\, t = n\pi, \quad n = 0, 1, 2, \ldots$$

The times at which extreme values of the oscillations occur are thus

$$t = \frac{n\pi}{\omega_n\sqrt{1-\zeta^2}} \tag{5.7-11}$$

The odd values of $n$ give the times of overshoots, and the even values correspond to the times of undershoots. The maximum overshoot occurs when $n = 1$. Thus,

$$t_p = \frac{\pi}{\omega_n\sqrt{1-\zeta^2}} \tag{5.7-12}$$

The magnitudes of the overshoots and undershoots are found by substituting (5.7-11) into (5.7-10). After some manipulation, the result is

$$x(t)|_{\text{extremum}} = \frac{d}{k}\,[1 + (-1)^{n-1}\, e^{-n\pi\zeta/\sqrt{1-\zeta^2}}] \tag{5.7-13}$$

The maximum overshoot is found when $n = 1$.

$$\text{maximum overshoot} = x_{\max} - x_{ss} = \frac{d}{k}\, e^{-\pi\zeta/\sqrt{1-\zeta^2}} \tag{5.7-14}$$

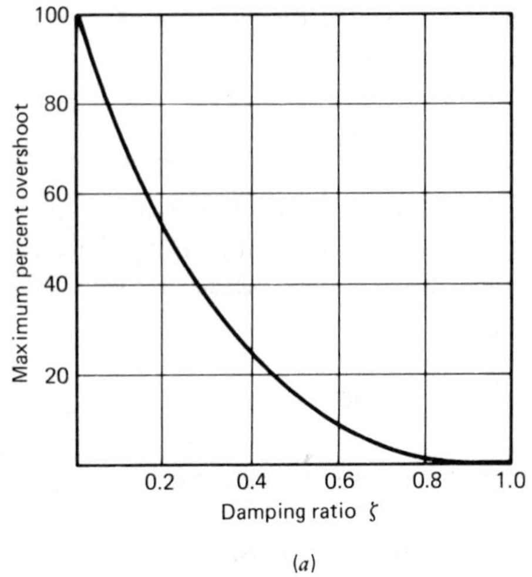


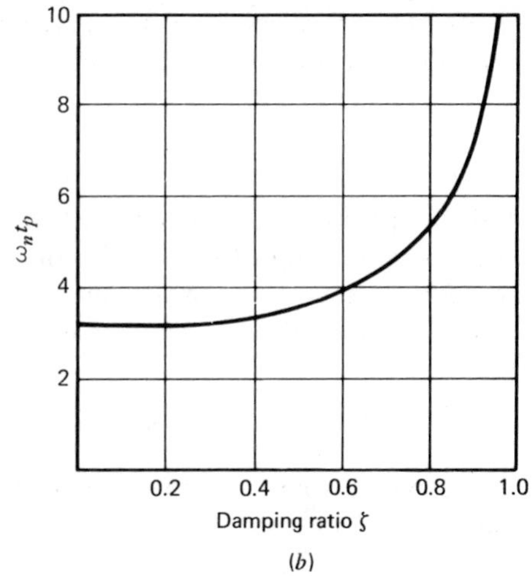


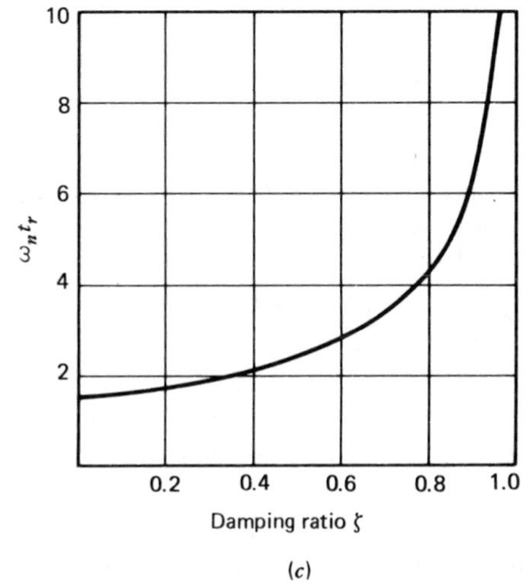


**Figure 5.20 Transient response specifications as functions of the damping ratio $\zeta$. (*a*) Maximum percent overshoot. (*b*) Normalized peak time $\omega_n t_p$. (*c*) Normalized 100% rise time $\omega_n t_r$.**

The preceding expressions show that the maximum overshoot and the peak time are functions only of the damping ratio $\zeta$ for a second-order system. The percent overshoot is

$$\text{percent maximum overshoot} = \frac{x_{\max} - x_{ss}}{x_{ss}} 100$$

$$= 100 e^{-\pi\zeta/\sqrt{1-\zeta^2}} \tag{5.7-15}$$

This is shown graphically in Figure 5.20*a*. The normalized peak time $\omega_n t_p$ is plotted versus $\zeta$ in Figure 5.20*b*.

Analytical expressions for the delay time, the rise time, and the settling time are difficult to obtain. For the delay time, set $x = 0.5x_{ss} = 0.5d/k$ in (5.7-8) to get

$$e^{-\zeta\omega_n t_d} \sin(\omega_n\sqrt{1-\zeta^2}\, t_d + \phi) = -0.5\sqrt{1-\zeta^2} \tag{5.7-16}$$

where $\phi$ is given by (5.7-9). For a given $\zeta$ and $\omega_n$, $t_d$ can be obtained by a numerical procedure such as Newton's method (Appendix A). This is easily done on a calculator, especially if the following straight-line approximation is used as a starting guess.

$$t_d \simeq \frac{1 + 0.7\zeta}{\omega_n}, \quad 0 \leqslant \zeta \leqslant 1 \tag{5.7-17}$$

A similar procedure can be applied to find the rise time $t_r$. In this case two equations must be solved, one for the 10% time and one for the 90% time. The difference between these times is the rise time. These calculations are made easier by using the following straight-line approximation.

$$t_r\big|_{10,90\%} \simeq \frac{0.8 + 2.5\zeta}{\omega_n}, \quad 0 \leqslant \zeta \leqslant 1 \tag{5.7-18}$$

This approximation was obtained by plotting the results of many such computer solutions.

If we choose the alternate definition of rise time – that is, the first time at which the final value is reached – then the solution for $t_r$ is easier to obtain. Set $x = x_{ss} = d/k$ in (5.7-8) to get

$$e^{-\zeta\omega_n t} \sin(\omega_n\sqrt{1-\zeta^2}\, t + \phi) = 0$$

This implies that for $t < \infty$,

$$\omega_n\sqrt{1-\zeta^2}\, t + \phi = n\pi, \quad n = 0, 1, 2, \ldots \tag{5.7-19}$$

For $t_r > 0$, $n = 2$, since $\phi$ is in the third quadrant. Thus

$$t_r\big|_{100\%} = \frac{2\pi - \phi}{\omega_n\sqrt{1-\zeta^2}} \tag{5.7-20}$$

where $\phi$ is given by (5.7-9). The rise time is inversely proportional to the natural frequency $\omega_n$ for a given value of $\zeta$. A plot of the normalized rise time $\omega_n t_r$ versus $\zeta$ is given in Figure 5.20*c*.

In order to express the settling time in terms of the parameters $\zeta$ and $\omega_n$, we can use the fact that the exponential term in the solution (5.7-8) provides the envelopes of the oscillations. These envelopes are found by setting the sine term to $\pm 1$ in (5.7-8). The magnitude of the difference between each envelope and the final value $d/k$ is

$$\frac{d}{k}\frac{e^{-\zeta\omega_n t}}{\sqrt{1-\zeta^2}}$$

Both envelopes are within 5% of the final value when

$$\frac{e^{-\zeta\omega_n t}}{\sqrt{1-\zeta^2}} \leqslant 0.05$$

The 5% settling time can be found from the preceding expression. It can also be approximated by noting that $e^{-3} \cong 0.05$, and using the formula

$$t_s\big|_{5\%} \cong \frac{3}{\zeta\omega_n} \tag{5.7-21}$$

The 2% settling time is found in a similar way by recalling that $e^{-4} \cong 0.02$. Thus $t_s$ is approximately four time constants.

$$t_s|_{2\%} \cong \frac{4}{\zeta\omega_n} \tag{5.7-22}$$

*Example 5.18*

Typical parameter values for an armature-controlled dc motor of the type shown in Figure 5.4 are

| | |
|---|---|
| $K_e = 0.199$ volts/rad/sec | $R = 0.43$ ohms |
| $K_T = 26.9$ oz-in./amp | $c = 0.07$ in.-oz/rad/sec |
| $L = 0.0021$ henrys | $I = 0.1$ oz-in.-sec$^2$ |

**(a)** Find the differential equation relating the load speed $\omega$ to the applied voltage $v$.

**(b)** Find $\zeta$, $\tau$, $\omega_n$, and $\omega_d$ for this system.

**(c)** A step input voltage of $v = 20$ volts is applied when the system is initially at rest. Evaluate the step response.

(a) The governing state equations for the state variables $\omega$ and $i_a$ (the armature current) were obtained in Section 5.1. Transforming (5.1-9) and (5.1-10) for zero initial conditions gives

$$V(s) = (R + Ls)I_a(s) + K_e\Omega(s)$$

$$(Is + c)\Omega(s) = K_T I_a(s)$$

Eliminate $I_a(s)$ by substituting from the second equation.

$$V(s) = \frac{(R + Ls)(Is + c)}{K_T}\Omega(s) + K_e\Omega(s)$$

This becomes

$$[LIs^2 + (RI + cL)s + cR + K_eK_T]\,\Omega(s) = K_T V(s)$$

The differential equation is

$$LI\ddot{\omega} + (RI + cL)\dot{\omega} + (cR + K_eK_T)\omega = K_T v(t)$$

The electrical units (ohm, volt, henry, amp, sec) are independent of the mechanical units (oz, in., rad, sec) except for the time unit. Thus, because seconds are used in both systems, we need not convert any of the units; they are all compatible. With the given data, we get

$$0.00021\ddot{\omega} + 0.04315\dot{\omega} + 5.3832\omega = 26.9v(t)$$

where $\omega$ is in rad/sec and $v(t)$ is in volts.

(b) Comparing the preceding with the general form (5.7-1) and using the formulas (5.3-35), (5.3-36), (5.3-39), and (5.3-40), we obtain

| | |
|---|---|
| $\zeta = 0.6417$ | $\omega_n = 160.1$ rad/sec |
| $\omega_d = 122.8$ rad/sec | $\tau = 0.0097$ sec |

(c) Since $\zeta < 1$, the system is underdamped, and oscillations will occur when a step voltage is applied. The solution (5.7-8) and (5.7-9) applies with $d = 26.9(20) = 538$. (Recall that (5.7-8) is for a unit step input). This gives the following response.

$$\omega(t) = 99.944[1.3038e^{-102.7t}\sin(122.8t + 4.016) + 1]$$

where the phase angle is in radians. The steady-state speed of the load is 99.944 rad/sec. From (5.7-12), the time of the maximum overshoot is $t_p = 0.0255$ sec. The maximum speed is 107.9 rad/sec. Thus, the percent overshoot is approximately 8%.

The delay time can be found approximately from (5.7-17). This gives

$$t_d = \frac{1 + 0.7(0.6421)}{160.1} = 0.0091 \text{ sec}$$

The delay time here is approximately one-tenth of the time constant. The 100% rise time is found from (5.7-20) to be

$$t_r|_{100\%} = \frac{2\pi - 4.016}{122.7} = 0.0185 \text{ sec}$$

The approximation to the 5% settling time is

$$t_s|_{5\%} = \frac{3}{102.8} = 0.0292 \text{ sec}$$

and the 2% settling time is approximately

$$t_s|_{2\%} = 0.0362 \text{ sec}$$

## Response Calculation From the Laplace Transform

The substitution method just presented allows the step response to be determined quickly for the second-order model (5.7-1). However, for higher-order systems, or for more complicated input functions, this method relies too heavily on memorized solution forms to be generally useful. When a complex conjugate root pair occurs in such problems, the following Laplace transform technique is useful.

Assume that the Laplace transformation of the system model results in the following expression for the transform of the response $X(s)$

$$X(s) = \frac{P(s)}{Q(s)} \quad (5.7\text{-}23)$$

where $Q(s)$ is a quadratic polynomial of the form

$$Q(s) = (s + a)^2 + b^2$$

Thus the roots are $s = -a \pm ib$. The term $P(s)$ represents the remaining terms in $X(s)$. A partial fraction expansion of $X(s)$ for the quadratic roots gives

$$X(s) = \frac{C_1}{s + a - ib} + \frac{C_2}{s + a + ib} \quad (5.7\text{-}24)$$

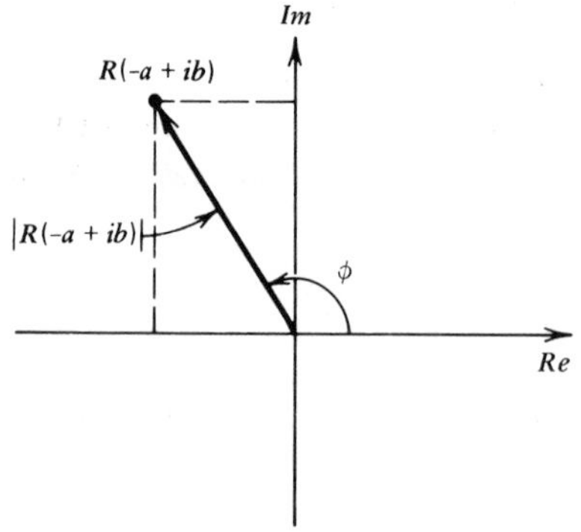


**Figure 5.21** Polar representation of the complex factor $R(-a + ib)$.

where

$$C_1 = \lim_{s\to -a+ib}\left[\frac{P(s)}{Q(s)}(s+a-ib)\right] \tag{5.7-25}$$

$$C_2 = \lim_{s\to -a-ib}\left[\frac{P(s)}{Q(s)}(s+a+ib)\right] \tag{5.7-26}$$

After the factor is canceled, $C_1$ has the form

$$\begin{aligned} C_1 &= \lim_{s\to -a+ib}\left[\frac{P(s)}{s+a+ib}\right] \\ &= \frac{1}{2ib}R(-a+ib) \end{aligned} \tag{5.7-27}$$

where $R$ is a function of the complex number $-a+ib$, and

$$R(-a+ib) = \lim_{s\to -a+ib}[P(s)] \tag{5.7-28}$$

Since $R(-a+ib)$ can be complex in general, it can be expressed as a magnitude and a phase angle, as shown in Figure 5.21.

$$R(-a+ib) = |R(-a+ib)|\,e^{i\phi}$$

Thus

$$C_1 = \frac{1}{2ib}|R(-a+ib)|\,e^{i\phi}$$

and since $C_2$ is the conjugate of $C_1$,

$$C_2 = -\frac{1}{2ib}|R(-a+ib)|\,e^{-i\phi}$$

Reverting to the time domain from (5.7-24), we obtain

$$\begin{aligned} x(t) &= C_1 e^{-at}e^{ibt} + C_2 e^{-at}e^{-ibt} \\ &= \frac{1}{b}|R(-a+ib)|\,e^{-at}\,\frac{e^{ibt}e^{i\phi} - e^{-ibt}e^{-i\phi}}{2i} \\ &= \frac{1}{b}|R(-a+ib)|\,e^{-at}\sin(bt+\phi) \end{aligned} \tag{5.7-29}$$

where we have used the identity

$$\sin\theta = \frac{e^{i\theta} - e^{-i\theta}}{2i}$$

$$\theta = bt + \phi$$

The phase angle is given by

$$\phi = \measuredangle R(-a+ib) \tag{5.7-30}$$

The preceding expressions can be used to determine that part of the free or forced response resulting from a pair of complex roots. If the system is of order greater than two, or if the input function introduces more roots, then (5.7-29) gives only part of the response. The rest of the response is found by doing a complete partial-fraction expansion for all the roots and adding all of the resulting response terms. The method will be illustrated by the following example and others to follow in Section 5.8.

*Example 5.19*

(a) Obtain the unit step response for the following model where the derivative of the input $u$ also affects the response. (This is a case of numerator dynamics.)

$$m\ddot{x} + c\dot{x} + kx = du + g\dot{u}$$

Assume that $g$ is constant and that $\zeta < 1$.

(b) Compare the response of part a with that for $g = 0$.

(a) With zero initial conditions, the transformation gives

$$(ms^2 + cs + k)X(s) = (d + gs)U(s) = \frac{d + gs}{s}$$

The effect of the derivative input is seen to be like that resulting from nonzero initial conditions such that $m\dot{x}(0) + cx(0) = g$. Solving for $X(s)$ and using the parameters $\zeta$ and $\omega_n$, we get

$$X(s) = \frac{d\omega_n^2}{zk} \frac{s + z}{(s^2 + 2\zeta\omega_n s + \omega_n^2)s}$$

where we have defined $z = d/g$. Since $P(s)$ is defined to contain everything in $X(s)$ except the quadratic factors,

$$P(s) = \frac{d\omega_n^2}{zk} \frac{s + z}{s}$$

Therefore, with $a = \zeta\omega_n$ and $b = \omega_n\sqrt{1 - \zeta^2}$,

$$R(-a + ib) = \lim_{s \to -a+ib} \left[\frac{d\omega_n^2}{zk} \frac{(s + z)}{s}\right]$$

$$= \frac{d\omega_n^2}{zk} \left(\frac{z - a + ib}{-a + ib}\right) \left(\frac{-a - ib}{-a - ib}\right)$$

$$= \frac{d\omega_n^2}{zk} \frac{a^2 + b^2 - az - ibz}{a^2 + b^2}$$

$$= \frac{d\omega_n}{zk} (\omega_n - \zeta z - iz\sqrt{1 - \zeta^2})$$

The magnitude and phase angle are

$$|R(-a + ib)| = \frac{d\omega_n}{zk} \sqrt{(\omega_n - \zeta z)^2 + z^2(1 - \zeta^2)}$$

$$= \frac{d\omega_n}{zk} \sqrt{\omega_n^2 - 2\zeta\omega_n z + z^2}$$

$$\phi = \measuredangle R(-a + ib) = \tan^{-1}\left(\frac{z\sqrt{1 - \zeta^2}}{\omega_n - \zeta z}\right), \quad \omega_n - \zeta z > 0$$

or

$$\phi = \pi + \tan^{-1}\left(\frac{z\sqrt{1-\zeta^2}}{\zeta z - \omega_n}\right), \quad \omega_n - \zeta z < 0 \tag{5.7-31}$$

or

$$\phi = \frac{\pi}{2}, \quad \omega_n - \zeta z = 0$$

That part of the response resulting from the quadratic roots is found from (5.7-29) using the preceding expressions for $R(-a+ib)$ and $\phi$. The complete expansion of $X(s)$ here also must include the effect of the third root, $s = 0$. The expansion is

$$X(s) = \frac{C_1}{s+a-ib} + \frac{C_2}{s+a+ib} + \frac{C_3}{s}$$

where

$$C_3 = \lim_{s\to 0}[X(s)s] = \frac{d}{k}$$

The root $s = 0$ contributes the term $d/k$ to the response. Adding the contributions for each root gives the total solution

$$x(t) = \frac{d}{k}\left[\frac{1}{z}\sqrt{\frac{\omega_n^2 - 2\zeta\omega_n z + z^2}{1-\zeta^2}}\, e^{-\zeta\omega_n t}\sin(\omega_n\sqrt{1-\zeta^2}\, t + \phi) + 1\right] \tag{5.7-32}$$

where $\phi$ was given previously.

(b) The larger $g$ is, the more significant is the effect of the derivative of the input. To see exactly what the effect is, compare the solution for part a to (5.7-8), which is the solution for the case $g = 0$. Both solutions have the same frequency and steady-state value; however, the amplitude of oscillation and the phase angle are both different. The ratio of the amplitude of (5.7-32) to that of (5.7-8) is

$$\frac{1}{z}\sqrt{\omega_n^2 - 2\omega_n z + z^2} = \sqrt{\left(\frac{\omega_n}{z}\right)^2 - \frac{2\omega_n}{z} + 1}$$

As $g$ increases, $z = d/g \to 0$, and the amplitude for the case with numerator dynamics becomes very large compared to the case $g = 0$. As $g \to 0$, $z \to \infty$, and the amplitudes become equal, as expected.

As the numerator zero approaches the imaginary axis (i.e., as $z \to 0$), the phase angle $\phi \to 0$. Physically this means that $g$ is so large that the response follows the step input very quickly. For $g \neq 0$, a plot of the response versus $\omega_n t$ would show that it passes the final value earlier than when $g = 0$ and that the overshoot is greater. The effect is much like that of decreasing $\zeta$ while holding $\omega_n$ constant (see Figure 5.19).

## 5.8 IMPULSE AND RAMP RESPONSE

As discussed in Chapter Three, an impulse input represents an input function of arbitrary shape, but whose duration is small compared to the dominant time constant of the system. This approximation greatly simplifies the mathematics. The ramp function, on the other hand, is not necessarily an approximation for the real input; it simply represents the situation in which the input function increases linearly with time. Here we obtain the response of both the impulse and ramp functions by means of the Laplace transform.

## Impulse Response

Consider the second-order model

$$m\ddot{x} + c\dot{x} + kx = du(t) \tag{5.8-1}$$

If the input $u(t)$ is zero, the transform gives

$$(ms^2 + cs + k)X(s) = (ms + c)x(0) + m\dot{x}(0)$$

However, if the initial conditions are zero and the input is a unit impulse $[U(s) = 1]$, then

$$(ms^2 + cs + k)X(s) = d$$

Comparison of the last two expressions shows that the response to a unit impulse is equivalent to the free response for the special set of initial conditions such that

$$d = (ms + c)x(0) + m\dot{x}(0)$$

or

$$x(0) = 0$$

$$m\dot{x}(0) = d$$

For a mass-spring damper system, this means that if the mass is started at the equilibrium $x(0) = 0$ with a velocity $\dot{x}(0) = d/m$, the resulting free response is equivalent to the unit impulse response for zero initial conditions. This corresponds to the impulse-momentum principle of Newtonian mechanics, which says that the change in momentum, here $m\dot{x}(0)$, must equal the impulse, which is the time integral of the applied force.

## Ramp Response

The response of (5.8-1) to a unit ramp input $[U(s) = 1/s^2]$ can be easily obtained with the transform method. If the characteristic roots are complex, the approach developed in Section 5.7 is useful. Consider the underdamped case first. The solution for $X(s)$ is

$$X(s) = \frac{d\omega_n^2/k}{s^2(s^2 + 2\zeta\omega_n s + \omega_n^2)}$$

where from (5.7-23) and following, we see that

$$P(s) = \frac{d\omega_n^2}{ks^2}$$

From (5.7-28), (5.7-30), and the relations $a = \zeta\omega_n$, $b = \omega_n\sqrt{1-\zeta^2}$,

$$R(-a+ib) = \frac{d\omega_n^2}{k}\frac{1}{(-a+ib)^2} = \frac{d\omega_n^2}{k}\frac{2\zeta^2 - 1 + i2\zeta\sqrt{1-\zeta^2}}{\omega_n^2}$$

$$\phi = \begin{cases} \tan^{-1}\dfrac{2\zeta\sqrt{1-\zeta^2}}{2\zeta^2-1}, & \text{if } 2\zeta^2 - 1 \geqslant 0 \\ \tan^{-1}\dfrac{2\zeta\sqrt{1-\zeta^2}}{2\zeta^2-1} + \pi, & \text{if } 2\zeta^2 - 1 < 0 \end{cases} \tag{5.8-2}$$

$$|R(-a+ib)| = \frac{d}{k}$$

From (5.7-29), the part of the solution due to the quadratic roots is

$$\frac{d}{k\omega_n\sqrt{1-\zeta^2}}e^{-\zeta\omega_n t}\sin(\omega_n\sqrt{1-\zeta^2}\,t+\phi)$$

The complete expansion of $X(s)$ includes the repeated roots at $s = 0$ and is the form

$$X(s) = \frac{C_1}{s-a-ib} + \frac{C_2}{s-a+ib} + \frac{C_3}{s^2} + \frac{C_4}{s}$$

where

$$C_3 = [X(s)s^2]|_{s=0} = \frac{d}{k}$$

$$C_4 = \left\{\frac{d}{ds}[X(s)s^2]\right\}\Bigg|_{s=0} = -\frac{2\zeta d}{k\omega_n}$$

The total solution is

$$x(t) = \frac{d}{k}\left[t - \frac{2\zeta}{\omega_n} + \frac{e^{-\zeta\omega_n t}}{\omega_n\sqrt{1-\zeta^2}}\sin(\omega_n\sqrt{1-\zeta^2}\,t+\phi)\right] \tag{5.8-3}$$

where $\phi$ is given by (5.8-2).

For the overdamped case the solution is cumbersome when presented in terms of $\zeta$ and $\omega_n$. Let the two real, distinct roots be $s_1$ and $s_2$. The transform of (5.8-1) with $U(s) = 1/s^2$ gives

$$X(s) = \frac{d/m}{s^2\left(s^2 + \frac{c}{m}s + \frac{k}{m}\right)} = \frac{d/m}{s^2(s-s_1)(s-s_2)}$$

$$= \frac{C_1}{s-s_1} + \frac{C_2}{s-s_2} + \frac{C_3}{s^2} + \frac{C_4}{s}$$

The coefficients are found by the usual way.

$$C_1 = \lim_{s\to s_1}[X(s)(s-s_1)] = \frac{d}{ms_1^2(s_1-s_2)}$$

$$C_2 = \lim_{s\to s_2}[X(s)(s-s_2)] = \frac{d}{ms_2^2(s_2-s_1)}$$

$$C_3 = \lim_{s\to 0}[X(s)s^2] = \frac{d}{ms_1s_2}$$

$$C_4 = \lim_{s\to 0}\frac{d}{ds}[X(s)s^2] = \frac{d(s_1+s_2)}{ms_1s_2} \tag{5.8-4}$$

The solution is

$$x(t) = C_1e^{s_1t} + C_2e^{s_2t} + C_3t + C_4 \tag{5.8-5}$$

This process can be repeated for the critically damped case, in which $s_2 = s_1$. The result is

$$x(t) = \frac{d}{ms_1^2}\left[e^{s_1t}\left(t - \frac{2}{s_1}\right) + t + \frac{2}{s_1}\right] \tag{5.8-6}$$

The difference between the ramp input and the response always changes with time unless $d = k$. To show this, define $e(t) = u(t) - x(t)$. Then

$$\begin{aligned} E(s) &= U(s) - X(s) \\ &= \frac{1}{s^2} - \frac{d}{s^2(ms^2 + cs + k)} \\ &= \frac{1}{s^2}\left(\frac{ms^2 + cs + k - d}{ms^2 + cs + k}\right) \end{aligned}$$

For stable characteristic roots, the final value theorem can be used to show that at steady state the difference is finite if $k = d$ and infinite otherwise. If $k = d$, $E(s)$ becomes

$$E(s) = \frac{1}{s}\left(\frac{ms + c}{ms^2 + cs + k}\right)$$

and

$$e_{ss} = \frac{c}{k}$$

If $c/k$ is positive, the response lags the input; that is, $x(t) < u(t)$ at steady state. The transient approach to steady state is oscillatory if $\zeta < 1$ and nonoscillatory if $\zeta \geqslant 1$. These responses are illustrated in Figure 5.22. If $k \neq d$, the response and input plots are not parallel at steady state.

## 5.9 FREQUENCY RESPONSE

In Chapter Three we obtained the response of a first-order system to a sinusoidal input. It was shown that a stable system has a steady-state sinusoidal output when subjected to a sinusoidal input. The output has the same frequency as the input but is shifted in phase. The phase and the ratio of the amplitude of the output to that of the input depend on the input frequency $\omega$ as well as the system's parameters. Specifically, with reference to Figure 5.23$a$, we define the amplitude ratio (or magnitude ratio) $M$ to be

$$M(\omega) = \frac{B(\omega)}{A} \tag{5.9-1}$$

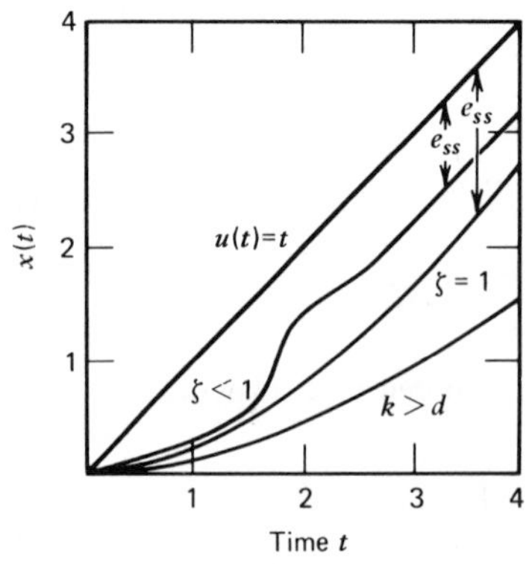


**Figure 5.22** Unit ramp response of the second-order system

$$T(s) = \frac{d}{ms^2 + cs + k}$$

Actual time for transients to disappear is approximately four time constants. (For the case shown $\tau \cong 1$.)

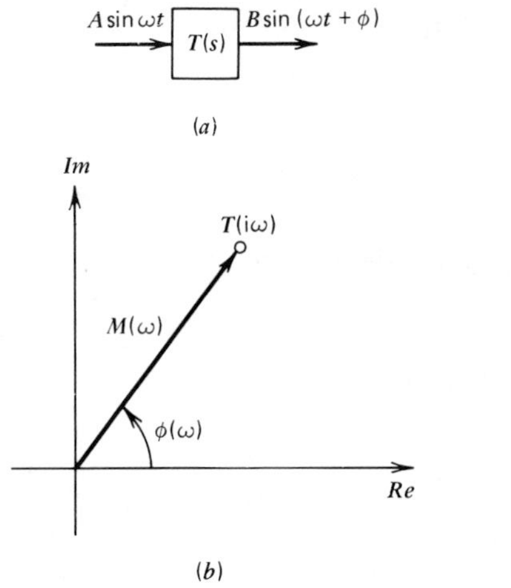


**Figure 5.23 Steady-state sinusoidal response of a linear system. (*a*) Input-output relations $M = B/A = T(i\omega)$; $\phi = \measuredangle T(i\omega)$. (*b*) Polar representation of $T(i\omega) = M\measuredangle\phi$.**

where $B(\omega)$ is the amplitude of the output and is a function of $\omega$ for a constant input amplitude $A$. Thus $M$ is also a function of $\omega$. If the system transfer function is $T(s)$, then from the results of Chapter Three,

$$M(\omega) = |T(i\omega)| \tag{5.9-2}$$

$$\phi(\omega) = \measuredangle T(i\omega) \tag{5.9-3}$$

With $i\omega$ substituted for $s$, $T(i\omega)$ can be considered as a vector with a magnitude $M(\omega)$ and an angle $\phi(\omega)$ relative to the positive real axis (Figure 5.23*b*). A plot of the tip of this vector as $\omega$ varies from 0 to $\infty$ is the polar plot of the system.

An alternate graphic representation of $T(i\omega)$ consists of the log-magnitude and log-phase plots, in which $M(\omega)$ and $\phi(\omega)$ are plotted versus $\log \omega$. Usually $M(\omega)$ is plotted in decibel units, as $m(\omega)$, where

$$m(\omega) = 20 \log M(\omega) \tag{5.9-4}$$

Figure 5.24 reproduces the results of Chapter Three for the simple lag transfer function $1/(\tau s + 1)$ and the simple lead transfer function $(\tau s + 1)$. Both plots have a zero (db) magnitude ratio for small frequencies ($\omega \ll 1/\tau$). At the corner frequency $\omega = 1/\tau$, $m(\omega)$ for the lag system breaks down with a slope of $-20$ db/decade, while that for the lead system breaks up with a slope of 20 db/decade. Both plots can be sketched with the high- and low-frequency asymptotic approximations, with a $\pm$ 3 db correction at the corner frequency. The phase angles of both systems are zero at low frequency while that for the lag system passes through $-45°$ at the corner frequency and approaches $-90°$ for high frequencies. The phase angle plot for the lead system is the mirror image of the lag plot.

Numerical values for $m(\omega)$ and $\phi(\omega)$ for any $\omega$ can easily be obtained with a

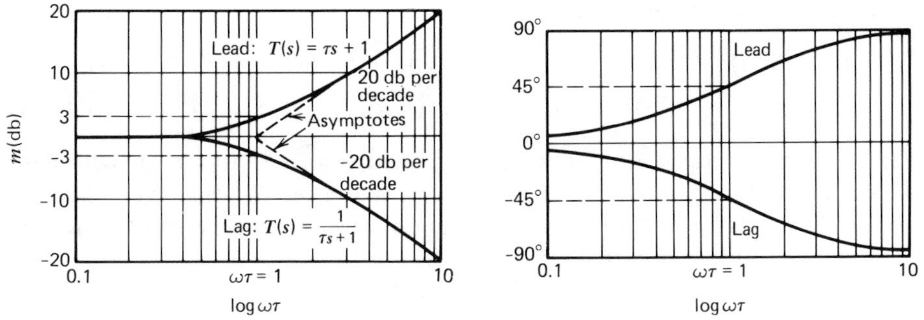


**Figure 5.24 Frequency-response plots for a simple lead and a simple lag. (*a*) Attenuation plots. (*b*) Phase-angle plots.**

calculator. The required expressions for the lead transfer function $T(s) = \tau s + 1$ are

$$m(\omega) = 20 \log |\tau\omega i + 1| = 20 \log \sqrt{\tau^2\omega^2 + 1}$$
$$= 10 \log (\tau^2\omega^2 + 1) \quad (5.9\text{-}5)$$
$$\phi(\omega) = \measuredangle(1 + \tau\omega i) = \tan^{-1} \omega\tau \quad (5.9\text{-}6)$$

The values of $m(\omega)$ and $\phi(\omega)$ for the simple lag are the negative of the preceding expressions.

## Frequency Response of an Overdamped System

Consider the second-order model

$$m\ddot{x} + c\dot{x} + kx = du(t)$$

Its transfer function is

$$T(s) = \frac{d}{ms^2 + cs + k}$$

If the system is overdamped, both roots are real and distinct, and we can write $T(s)$ as

$$T(s) = \frac{d/k}{\frac{m}{k}s^2 + \frac{c}{k}s + 1} = \frac{d/k}{(\tau_1 s + 1)(\tau_2 s + 1)} \quad (5.9\text{-}7)$$

where $\tau_1$ and $\tau_2$ are the time constants of the roots. Substituting $s = i\omega$ gives

$$T(i\omega) = \frac{d/k}{(\tau_1\omega i + 1)(\tau_2\omega i + 1)} \quad (5.9\text{-}8)$$

In general, if a complex number $T(i\omega)$ consists of products and ratios of complex factors, such that

$$T(i\omega) = K\frac{N_1(i\omega)N_2(i\omega)\ldots}{D_1(i\omega)D_2(i\omega)\ldots} \quad (5.9\text{-}9)$$

where $K$ is a real constant, then from the properties of complex numbers

$$|T(i\omega)| = \frac{|K|\,|N_1(i\omega)|\,|N_2(i\omega)|\ldots}{|D_1(i\omega)|\,|D_2(i\omega)|\ldots} \quad (5.9\text{-}10)$$

In decibel units, this implies that

$$m(\omega) = 20 \log |T(i\omega)| = 20 \log |K| + 20 \log |N_1(i\omega)| + 20 \log |N_2(i\omega)| + \ldots$$
$$- 20 \log |D_1(i\omega)| - 20 \log |D_2(i\omega)| - \ldots \quad (5.9\text{-}11)$$

That is, when expressed in logarithmic units, multiplicative factors in the numerator are summed, while those in the denominator are subtracted. We can use this principle graphically to add or subtract the contribution of each term in the transfer function to obtain the plot for the overall system transfer function.

Also, for the form (5.9-7), the phase angle is

$$\phi(\omega) = \measuredangle T(i\omega) = \measuredangle K + \measuredangle N_1(i\omega) + \measuredangle N_2(i\omega) + \ldots$$
$$- \measuredangle D_1(i\omega) - \measuredangle D_2(i\omega) - \ldots \quad (5.9\text{-}12)$$

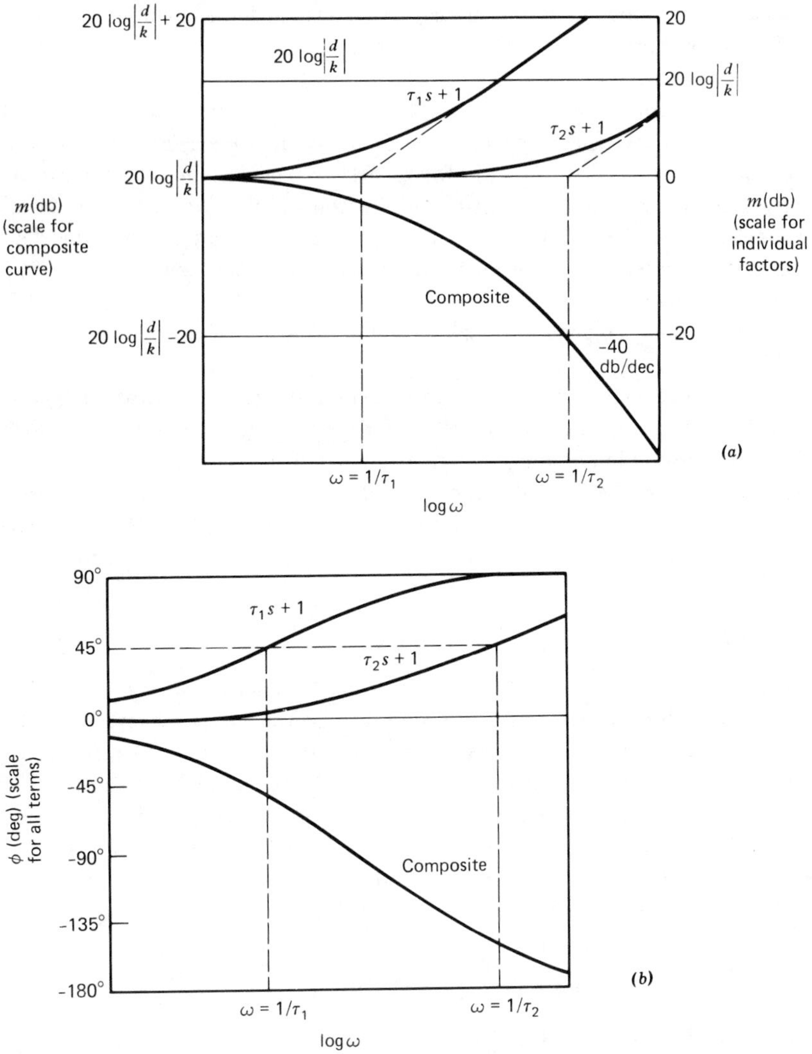


**Figure 5.25 Frequency-response plots for the overdamped system**

$$T(s) = \frac{d/k}{(\tau_1 s + 1)(\tau_2 s + 1)}$$

The phase angles of multiplicative factors in the numerator are summed, while those in the denominator are subtracted. This allows us to build the composite phase angle plot from the plots for each factor.

Returning to (5.9-8) we see that

$$m(\omega) = 20 \log |T(i\omega)| = 20 \log \left|\frac{d}{k}\right| - 20 \log |\tau_1 \omega i + 1| - 20 \log |\tau_2 \omega i + 1| \quad (5.9\text{-}13)$$

$$\phi(\omega) = \measuredangle \frac{d}{k} - \measuredangle (\tau_1 \omega i + 1) - \measuredangle (\tau_2 \omega i + 1) \quad (5.9\text{–}14)$$

Thus the magnitude ratio plot in db consists of a constant term, $20\log|d/k|$, minus the sum of the plots for two first-order lead terms. The corner frequencies are $\omega_1 = 1/\tau_1$ and $\omega_2 = 1/\tau_2$. Assume that $\tau_1 > \tau_2$. Then for $1/\tau_1 < \omega < 1/\tau_2$, the slope is approximately $-20$ db/decade. For $\omega > 1/\tau_2$, the contribution of the term $(\tau_2\omega i + 1)$ is significant. This causes the slope to decrease by an additional 20 db/decade, to produce a net slope of $-40$ db/decade for $\omega > 1/\tau_2$. The rest of the plot can be sketched by graphically adding or subtracting the contribution of each term for as many frequencies as desired. Usually the asymptotic approximations with the 3 db corrections at the corner frequencies will produce a plot of sufficient accuracy. The result is shown in Figure 5.25*a* for $d > k$. The phase angle plot shown in Figure 5.25*b* is produced in a similar manner by making use of (5.9-14). Note that if $d/k > 0$, $\measuredangle(d/k) = 0°$.

Note that the general shape of the curve can be deduced without specific knowledge of the constants $d$, $k$, $\tau_1$, and $\tau_2$, except for the information that $\tau_1 > \tau_2$. Such a general sketch is useful in obtaining an understanding of the system's behavior and is often used to determine the allowable range of values for one or more of the system's parameters. If more accurate numerical information is required, (5.9-2) and (5.9-3) can be evaluated directly with a calculator.

### *Example 5.20*

A simplified representation of a vehicle's front wheel system consisting of the tire, springs, and shock absorber is shown in Figure 5.26*a*. The spring $k_1$ represents the elasticity of the tire, while $k$ represents that of the shock absorber. Determine the frequency response of the tire displacement $x$ to a road surface displacement $y$. Assume that the tire does not leave the road and that the inertia of the car body is so large that the body can be considered as a fixed support. The tire weighs 30 lb, and the other constants are $k = 12{,}000$ lb/ft, $k_1 = 2400$ lb/ft, and $c = 360$ lb-sec/ft.

Let the tire displacement $x$ be measured from a suitable static equilibrium position. Then Newton's law gives the following model.

$$m\ddot{x} + c\dot{x} + kx = k_1(y - x)$$

For the given values, $m = 30/g = 0.933$ slugs, and

$$0.933\ddot{x} + 360\dot{x} + 14400x = 2400y$$

or

$$\ddot{x} + 386\dot{x} + 15434x = 2572y$$

The transfer function between $y$ as input and $x$ as output is

$$T(s) = \frac{2572}{s^2 + 386s + 15434}$$

The damping factor is $\zeta = 1.55$, and the roots are $s_1 = -45.3$ and $s_2 = -341$. The corresponding time constants are $\tau_1 = 0.022$ sec and $\tau_2 = 0.0029$ sec. Thus the transfer function can be written as

$$T(s) = \frac{2572}{(s + 45.3)(s + 341)} = \frac{0.1665}{(0.022s + 1)(0.0029s + 1)}$$

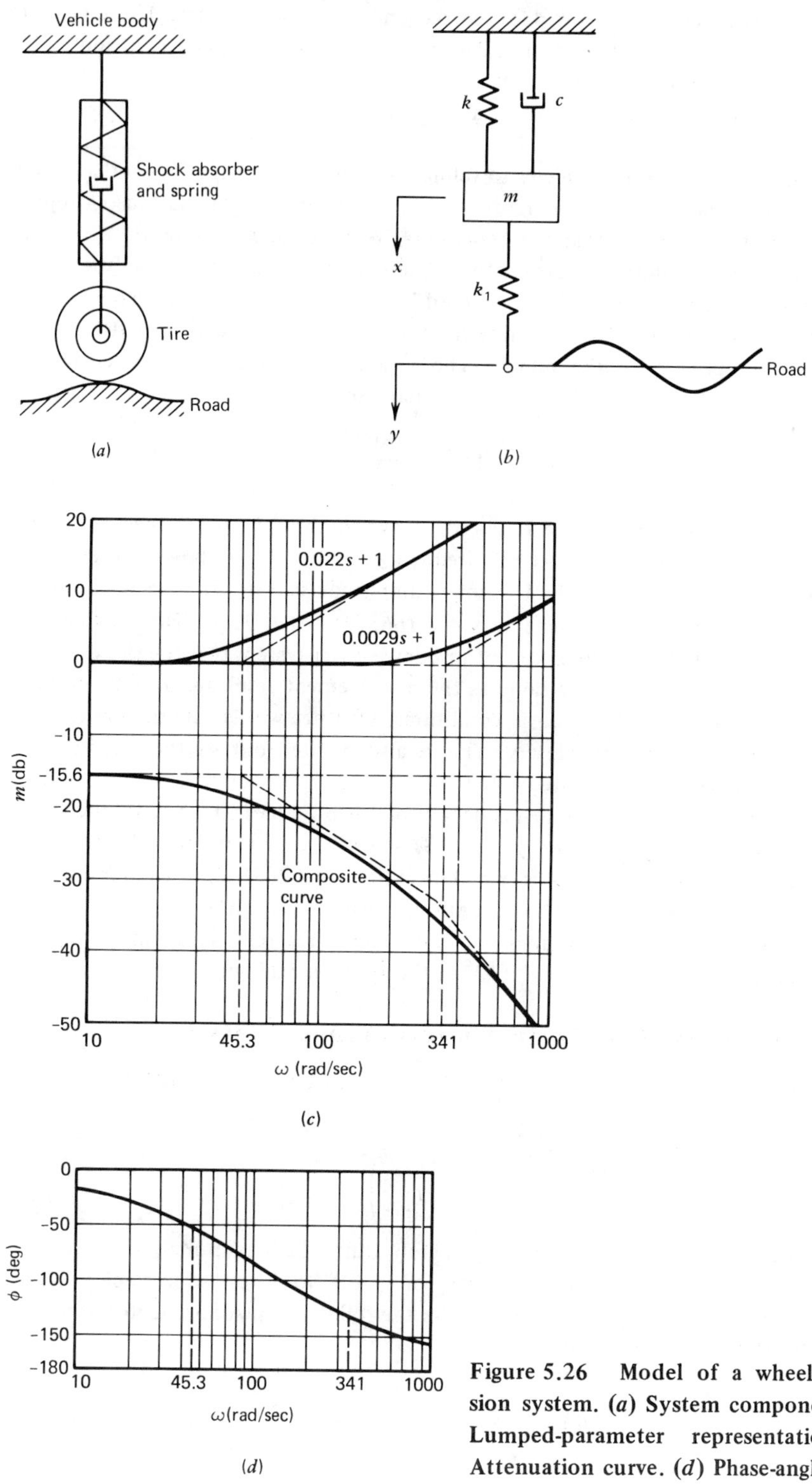


**Figure 5.26** Model of a wheel suspension system. (*a*) System components. (*b*) Lumped-parameter representation. (*c*) Attenuation curve. (*d*) Phase-angle curve.

The corner frequencies are $\omega_1 = 45.3$ rad/sec and $\omega_2 = 341$ rad/sec. The constant term 0.1665 shifts the $m(\omega)$ curve by $20 \log 0.1665 = -15.6$ db, and contributes $0^\circ$ to the phase angle plot. It is best to sketch the $m(\omega)$ plot first without the shift of $-15.6$ db, and then simply adjust the scale by this amount after the sketch is made. The result is shown in Figure 5.26*c*.

Measurements made on vehicles traveling on typical roads have shown that some construction techniques result in periodic wave patterns in the road surface. A representative wavelength is 50 ft and a representative road surface amplitude is 3 in. Ripples with a much shorter wavelength can also appear, but usually only on dirt roads, where they give the road a "washboard" appearance and have a smaller amplitude – say, 0.5 in. In order to determine the forcing frequency of these patterns, we must know the speed $V$ of the vehicle. The period $\tau$ is related to the wavelength $L$ and $V$ by $\tau = L/V$. The forcing frequency in radians/sec is

$$\omega = 2\pi \frac{1}{\tau} = \frac{2\pi V}{L}$$

For a speed of 60 mph = 88 ft/sec, the wavelength of 50 ft produces a forcing frequency of $\omega = 11.06$ rad/sec. At this frequency, Figure 5.26*c* shows that $m(11) \cong -15.6$ db. Thus $M(11) = 0.1665$. The input amplitude is 3 in. or 0.25 ft. Thus the amplitude of the tire displacement is $x = 0.1665(3) = 0.4995$ in. The forcing frequency is so low relative to the speed of response of the system that the tire displacement amplitude is essentially the same as the displacement resulting from an applied displacement of $y = 3$ in. under static conditions. In other words, the tire has time to follow the input motion very closely. This is also pointed out by the fact that the phase shift is almost zero at this frequency.

The washboard effect with a wavelength of 1 ft produces a forcing frequency of $\omega = 553$ rad/sec. At this frequency, $m(553) = -42.5$ db. Thus $M(553) = 0.0075$, and the resulting tire amplitude is $x = 0.0075(0.5) = 0.0038$ in. The amplitude is much smaller than for the 50-ft wavelength primarily because the forcing frequency is so high that the wheel system does not have time to respond. This is also indicated by the large negative phase shift of $-143^\circ$ at this frequency (Figure 5.26*d*).

## Frequency Response of an Underdamped System

If the transfer function given by (5.9-7) has complex conjugate roots, it can be expressed as

$$T(s) = \frac{d/k}{\dfrac{m}{k}s^2 + \dfrac{c}{k}s + 1} = \frac{d/k}{\left(\dfrac{s}{\omega_n}\right)^2 + 2\zeta\dfrac{s}{\omega_n} + 1} \tag{5.9-15}$$

We have seen that the constant term $d/k$ merely shifts the magnitude ratio plot up or down by a fixed amount and adds either $0^\circ$ or $-180^\circ$ to the phase angle plot. Therefore, for now let us take $d/k = 1$ and consider the following quadratic factor, obtained from (5.9-15) by replacing $s$ with $i\omega$.

$$T(i\omega) = \frac{1}{\left(\dfrac{i\omega}{\omega_n}\right)^2 + \dfrac{2\zeta}{\omega_n}\omega i + 1} = \frac{1}{1 - \left(\dfrac{\omega}{\omega_n}\right)^2 + \dfrac{2\zeta\omega}{\omega_n}i} \tag{5.9-16}$$

The magnitude ratio is

$$
\begin{aligned}
m(\omega) &= 20\log\left|\frac{1}{1-\left(\dfrac{\omega}{\omega_n}\right)^2+\dfrac{2\zeta\omega}{\omega_n}i}\right| \\
&= -20\log\sqrt{\left(1-\frac{\omega^2}{\omega_n^2}\right)^2+\left(\frac{2\zeta\omega}{\omega_n}\right)^2} \\
&= -10\log\left[\left(1-\frac{\omega^2}{\omega_n^2}\right)^2+\left(\frac{2\zeta\omega}{\omega_n}\right)^2\right]
\end{aligned}
\tag{5.9-17}
$$

The asymptotic approximations are as follows. For $\omega \ll \omega_n$,

$$m(\omega) \cong -20\log 1 = 0$$

For $\omega \gg \omega_n$,

$$
\begin{aligned}
m(\omega) &\cong -20\log\sqrt{\frac{\omega^4}{\omega_n^4}+4\zeta^2\frac{\omega^2}{\omega_n^2}} \\
&\cong -20\log\sqrt{\frac{\omega^4}{\omega_n^4}} \\
&= -40\log\frac{\omega}{\omega_n}
\end{aligned}
$$

Thus for low frequencies, the curve is horizontal at $m = 0$, while for high frequencies it has a slope of $-40$ db/decade, just as in the overdamped case. The high-frequency and low-frequency asymptotes intersect at the corner frequency $\omega = \omega_n$.

The underdamped case differs from the overdamped case in the vicinity of the corner frequency. To see this, examine $M(\omega)$.

$$M(\omega) = \frac{1}{\sqrt{\left(1-\dfrac{\omega^2}{\omega_n^2}\right)^2+\left(\dfrac{2\zeta\omega}{\omega_n}\right)^2}} \tag{5.9-18}$$

This has a maximum value when the denominator has a minimum. Setting the derivative of the denominator with respect to $\omega$ equal to zero shows that the maximum $M(\omega)$ occurs at $\omega = \omega_n\sqrt{1-2\zeta^2}$. This frequency is the *resonant frequency* $\omega_r$. The peak of $M(\omega)$ exists only when the term under the radical is positive; that is, when $\zeta \leqslant 0.707$. Thus

$$\omega_r = \omega_n\sqrt{1-2\zeta^2} \qquad 0 \leqslant \zeta \leqslant 0.707 \tag{5.9-19}$$

The value of the peak $M_p$ is found by substituting $\omega_r$ into $M(\omega)$. This gives

$$M_p = M(\omega_r) = \frac{1}{2\zeta\sqrt{1-\zeta^2}}, \qquad 0 \leqslant \zeta \leqslant 0.707 \tag{5.9-20}$$

If $\zeta > 0.707$, no peak exists, and the maximum value of $M$ occurs at $\omega = 0$ where $M = 1$. Note that as $\zeta \to 0$, $\omega_r \to \omega_n$, and $M_p \to \infty$. For an undamped system the resonant frequency is the natural frequency $\omega_n$.

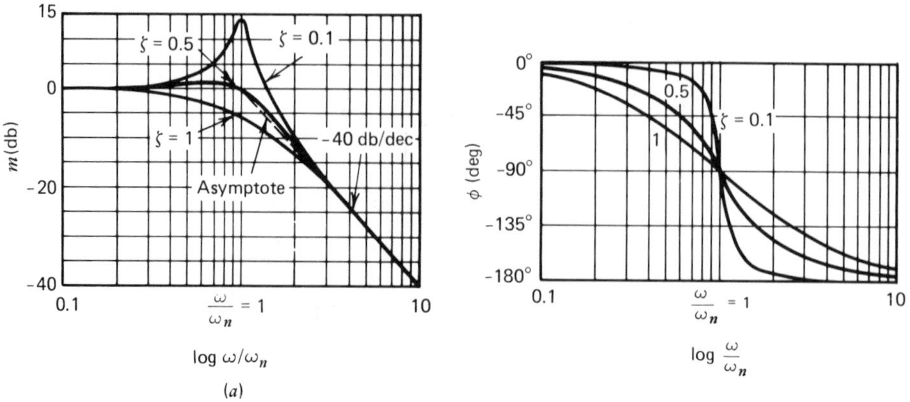


**Figure 5.27 Frequency-response plots for the underdamped system**

$$T(s) = \frac{d/k}{\left(\frac{s}{\omega_n}\right)^2 + \frac{2\zeta s}{\omega_n} + 1}$$

A plot of $m(\omega)$ versus $\log \omega$ is shown in Figure 5.27*a* for several values of $\zeta$. Note that the correction to the asymptotic approximations in the vicinity of the corner frequency depends on the value of $\zeta$. The peak value in decibels is

$$m_p = m(\omega_r) = -20 \log (2\zeta\sqrt{1-\zeta^2}) \tag{5.9-21}$$

At $\omega = \omega_n$,

$$m(\omega_n) = -20 \log 2\zeta \tag{5.9-22}$$

The curve can be sketched more accurately by repeated evaluation of (5.9-18) for values of $\omega$ near $\omega_n$.

The phase angle plot is obtained in a similar manner. From the additive property for angles (5.9-12) we see that for (5.9-16),

$$\phi(\omega) = -\measuredangle \left[1 - \left(\frac{\omega}{\omega_n}\right)^2 + \frac{2\zeta\omega}{\omega_n} i\right]$$

Thus

$$\tan \phi(\omega) = -\left[\frac{\dfrac{2\zeta\omega}{\omega_n}}{1 - \left(\dfrac{\omega}{\omega_n}\right)^2}\right] \tag{5.9-23}$$

where $\phi(\omega)$ is in the 3rd or 4th quadrant. For $\omega \ll \omega_n$,

$$\phi(\omega) \cong -\tan^{-1} 0 = 0°$$

For $\omega \gg \omega_n$,

$$\phi(\omega) \cong -180°$$

At the corner frequency,

$$\phi(\omega_n) = -\tan^{-1} \infty = -90°$$

This result is independent of $\zeta$. The curve is skew-symmetric about the inflection point at $\phi = -90°$ for all values of $\zeta$. The rest of the plot can be sketched by evaluating (5.9-23) at various values of $\omega$. The plot is shown for several values of $\zeta$ in Figure 5.27*b*. At the resonant frequency,

$$\phi(\omega_r) = -\tan^{-1} \frac{\sqrt{1-2\zeta^2}}{\zeta} \tag{5.9-24}$$

## Rotating Unbalance in Machinery

A common cause of sinusoidal forcing in machines is the unbalance that exists to some extent in every rotating machine. The unbalance is caused by the center of mass of the rotating part not coinciding with the center of rotation. Let $M$ be the total mass of the machine and $m$ the rotating mass causing the unbalance. Consider the entire unbalanced mass $m$ to be lumped at its center of mass, a distance $R$ from the center of rotation. This distance is the *eccentricity*. Figure 5.28*a* shows this situation. The main mass is thus $(M-m)$ and is assumed to be constrained to allow only vertical motion.

The motion of the unbalanced mass $m$ will consist of the vector combination of its motion relative to the main mass $(M-m)$ and the motion of the main mass. For a constant speed of rotation $\omega_R$, the rotation produces a radial acceleration of $m$ equal to $R\omega_R{}^2$. This causes a force to be exerted on the bearings at the center of rotation. This force has a magnitude $mR\omega_R{}^2$ and is directed radially outward. The vertical component of this unbalance force is, from Figure 5.28*b*,

$$f = mR\omega_R{}^2 \sin \omega_R t \tag{5.9-25}$$

*Example 5.21*

Alternating-current motors are usually designed to run at a constant speed, typically either 1750 or 3500 rpm. One such motor for a power tool weighs 20 lb and is to be mounted on a steel cantilever beam as shown in Figure 5.29*a*. Static force calculations and space considerations suggest that a beam 6 in. long, 4 in. wide, and 3/8 in. thick would be suitable. The rotating part of the motor weighs 10 lb and has an eccentricity

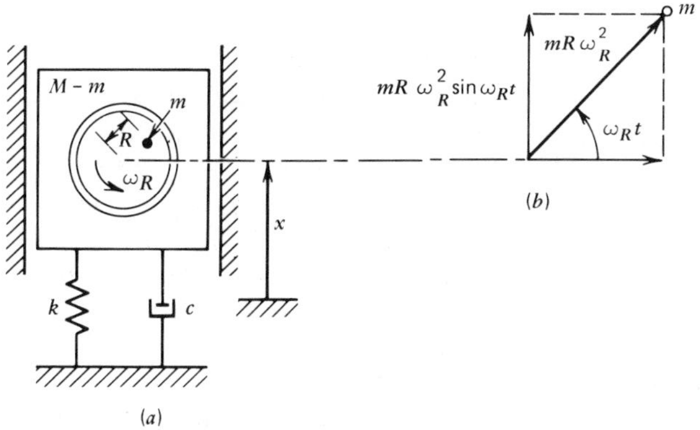


**Figure 5.28 Machine with rotating unbalance. (*a*) System components. (*b*) Force diagram.**

of 0.001 ft. (This distance can be determined by standard balancing methods or by applying the analysis in this example in reverse, using a vibration test.) The damping ratio for such beams is difficult to determine but is usually very small, say $\zeta \leqslant 0.1$. Estimate the amplitude of vibration of the beam at steady state.

For the equivalent system shown in Figure 5.29*b*, we have from Tables 2.7, 2.8, and 2.11, the following spring constant and equivalent mass (after converting inches to feet).

$$k = \frac{Ewh^3}{4L^3} = \frac{(4.32 \times 10^9)(0.333)(0.03125)^3}{4(0.5)^3}$$

$$= 8.78 \times 10^4 \text{ lb/ft}$$

$$m_e = \frac{20}{32.17} + 0.23(15.2)(0.333)(0.5) = 0.640 \text{ slugs}$$

The unbalanced mass is $m = 10/32.17 = 0.311$ slugs. The model for the system is

$$m_e\ddot{x} + c\dot{x} + kx = f(t) = mR\omega_R^2 \sin \omega_R t$$

This gives the transfer function

$$T(s) = \frac{X(s)}{F(s)} = \frac{1}{m_e s^2 + cs + k} = \frac{1/k}{\dfrac{m_e}{k}s^2 + \dfrac{c}{k}s + 1}$$

Thus

$$T(i\omega) = \frac{1/k}{1 - \left(\dfrac{\omega}{\omega_n}\right)^2 + \dfrac{2\zeta\omega}{\omega_n}i}$$

where $1/k = 1.139 \times 10^{-5}$ and $\omega_n = \sqrt{k/m_e} = 370$ rad/sec $= 3537$ rpm. Comparing this transfer function with (5.9-15) we see that $d/k \neq 1$, and we must include this factor in our calculations. This factor shifts the $m(\omega)$ curve by $-98.9$ db. Alternately, we may use the response plot shown in Figure 5.27*a* and multiply the result by $1/k$.

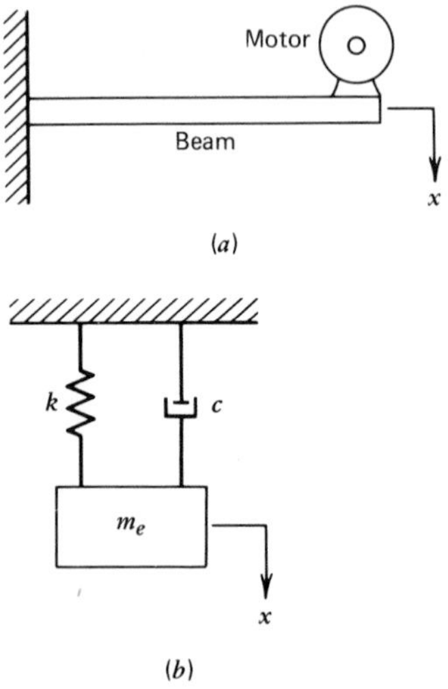


**Figure 5.29 Motor vibration problem of Example 5.21. (*a*) Motor supported by a cantilever beam. (*b*) Lumped-parameter model.**

The motor speed of 1750 rpm is 183 rad/sec. This gives $\omega_R/\omega_n = 0.485$. From the plot [or from (5.9-17)] we see that $m = 2.37$ db, which corresponds to a magnitude ratio of 1.314. Multiplying by the $1/k$ factor gives $(1.314)1.139 \times 10^{-5} = 1.496 \times 10^{-5}$. The amplitude of the forcing function is $mR\omega_R^2 = 0.311(0.001)(183)^2 = 10.4$ lb. Thus the steady-state amplitude is $10.4(1.496 \times 10^{-5}) = 15.6 \times 10^{-5}$ ft.

On the downward oscillation the total amplitude as measured from horizontal is the preceding value plus the static deflection, or $15.6 \times 10^{-5} + 20/87800 = 3.84 \times$

$10^{-4}$ ft. If this deflection would cause too much stress in the beam, one or more of the following changes could be made in the design.

1. Add a damper to the system.
2. Reduce the unbalance in the motor.
3. Increase the separation between the forcing frequency and the natural frequency either by selecting a beam with more stiffness (a larger $k$), by reducing the mass of the system, or by adding mass to shift $\omega_n$ far to the left of $\omega_R$ on the plot.

The preceding results are not very sensitive to the assumed value of $\zeta = 0.1$. If $\zeta = 0.05$ or 0.2, the calculated amplitude of vibration would be $15.7 \times 10^{-5}$ ft or $15.2 \times 10^{-5}$ ft, respectively. However, if we had used a motor with a speed of 3500 rpm = 366 rad/sec, this choice would put the forcing frequency very close to the natural frequency. In this region the assumed value of $\zeta$ would be critical in the amplitude calculation. In practice, such a design would be avoided. That is, in vibration analysis the most important quantity to know is the natural frequency $\omega_n$. If the damping is slight, the resonant frequency is near $\omega_n$. If $\omega_n$ is designed so that it is not close to the forcing frequency, it is not necessary to know the precise amount of damping.

## The Effect of Numerator Dynamics

The second-order models we have seen so far in this section have not had the $s$ operator in the numerator of their transfer functions. Consequently, the magnitude ratio is small at high frequencies. However, the introduction of numerator dynamics can produce a large magnitude ratio at high frequencies. This effect can be used to advantage – in instrument design, for example. The instrument shown in Figure 5.30 illustrates this point. With proper selection of the natural frequency of the device, it can be used either as a *vibrometer* to measure the amplitude of a sinusoidal displacement $y$ or as an *accelerometer* to measure the amplitude of the acceleration $\ddot{y}$, which is also sinusoidal. When used to measure ground motion from an earthquake, for example, the instrument is commonly referred to as a seismograph.

The mass displacement $x$ and the support displacement $y$ are relative to an inertial reference, with $x = 0$ corresponding to the equilibrium position of $m$ when $y = 0$. With the potentiometer arrangement shown, the voltage $v$ is porportional to

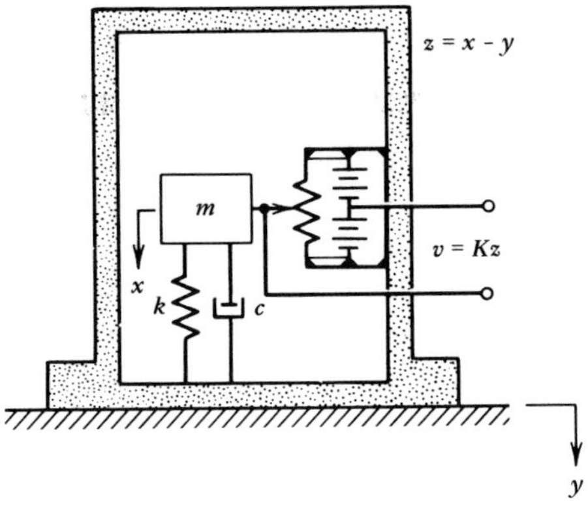


**Figure 5.30** **Vibration instrument.**

the relative displacement $z$ between the support and the mass $m$, where $z = x - y$. Newton's law gives

$$m\ddot{x} = -c(\dot{x} - \dot{y}) - k(x - y) = 0$$

In terms of $z$, this becomes

$$m\ddot{z} + c\dot{z} + kz = -m\ddot{y} \tag{5.9-26}$$

The transfer function between the input $y$ and the output $z$ is

$$T(s) = \frac{Z(s)}{Y(s)} = \frac{-ms^2}{ms^2 + cs + k} = \frac{-s^2/\omega_n^2}{\dfrac{s^2}{\omega_n^2} + \dfrac{2\zeta s}{\omega_n} + 1} \tag{5.9-27}$$

Substituting $s = i\omega$ and referring to (5.9-11), we see that the numerator gives the following contribution to the log magnitude ratio

$$20 \log |N_i(i\omega)| = 20 \log \left| \left(\frac{i\omega}{\omega_n}\right)^2 \right| = 40 \log \frac{\omega}{\omega_n}$$

This term contributes 0 db to the net curve at the corner frequency $\omega = \omega_n$, and it increases the slope by 40 db/decade over all frequencies. Thus at low frequencies the slope of $m(\omega)$ is 40 db/decade, and at high frequencies the slope is zero. The plot is sketched in Figure 5.31.

For a vibrometer this plot shows that the device's natural frequency $\omega_n$ must be selected so that $\omega \gg \omega_n$, where $\omega$ is the oscillation frequency of the displacement to be measured $[y(t) = A \sin \omega t]$. For $\omega \gg \omega_n$,

$$|T(i\omega)| \cong 40 \log \frac{\omega}{\omega_n} - 40 \log \frac{\omega}{\omega_n} = 0 \text{ db}$$

and thus $|z| \cong |y| = A$, as desired. The voltage $v$ is directly proportional to $A$ in this case. The physical explanation for this result is the fact that the mass $m$ cannot respond to high-frequency input displacements. Its displacement $x$ therefore remains fixed, and the motion $z$ directly indicates the motion $y$.

To design a specific vibrometer, we must know the lower bound of the input displacement frequency $\omega$. The frequency $\omega_n = \sqrt{k/m}$ is then made much smaller than this bound by selecting a large mass and a "soft" spring (small $k$). However, these choices are governed by constraints on the allowable deflections. For example, a very soft spring will have a large distance between the free length and the equilibrium positions.

An accelerometer can be obtained by using the lower end of the frequency

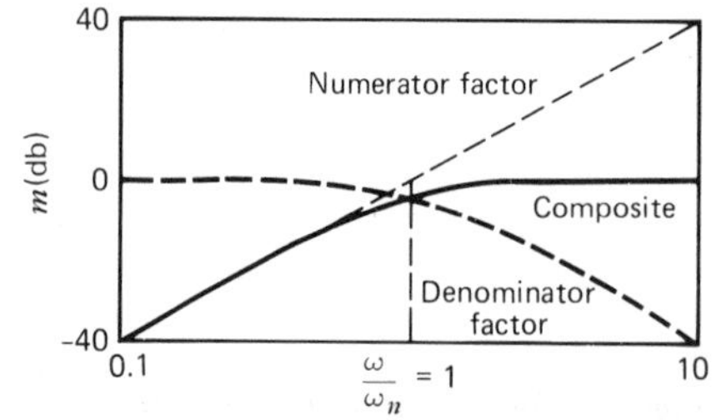


**Figure 5.31 Attenuation curve for the vibration instrument shown in Figure 5.30 (assuming that $\zeta = 1$).**

range; that is, select $\omega_n \gg \omega$, or equivalently, for $s$ near zero, (5.9-27) gives

$$T(s) \cong -\frac{s^2}{\omega_n^2} = \frac{Z(s)}{Y(s)}$$

or

$$Z(s) \cong -\frac{1}{\omega_n^2} s^2 Y(s)$$

The term $s^2 Y(s)$ represents the transform of $\ddot{y}$, so the output of the accelerometer is

$$|z| \cong \frac{1}{\omega_n^2} |\ddot{y}| = \frac{\omega^2}{\omega_n^2} A$$

With $\omega_n$ chosen large (using a small mass and a "stiff" spring), the input acceleration amplitude $\omega^2 A$ can be determined from $z$ (or $v$).

## Polar Plots

In Chapter Three the polar plot was introduced as an alternative way of presenting the magnitude and phase information contained in the frequency transfer function $T(i\omega)$. In this plot the vector representation of $T(i\omega)$ is used, with $M(\omega)$ as the length of the vector and $\phi(\omega)$ as its angle. The principles used to obtain the rectangular plots can be readily applied to sketch the polar plot. For example, consider the underdamped second-order system specified by (5.9-15). From Figure 5.27*b*, we see that the phase angle varies from 0° at $\omega = 0$ to $-180°$ at $\omega = \infty$. Thus the vector $T(i\omega)$ moves from the fourth quadrant to the third quadrant as $\omega$ increases. Figure 5.27*a* shows that if $\zeta \leqslant 0.707$, the length of the vector increases with $\omega$ until the resonant frequency $\omega_r$ is reached. For $\omega > \omega_r$, the vector length decreases. For $\zeta > 0.707$, the maximum vector length occurs at $\omega = 0$. When $\omega = \omega_n$, the phase angle is $-90°$. Therefore the locus of the vector tip intersects the imaginary axis when $\omega = \omega_n$. These observations enable us to obtain the plot shown in Figure 5.32. As $\zeta$ increases, one root becomes dominant, and the system's behavior approaches that of a first-order system. For large $\zeta$, the polar plot approaches a semicircle like that for a first-order system.

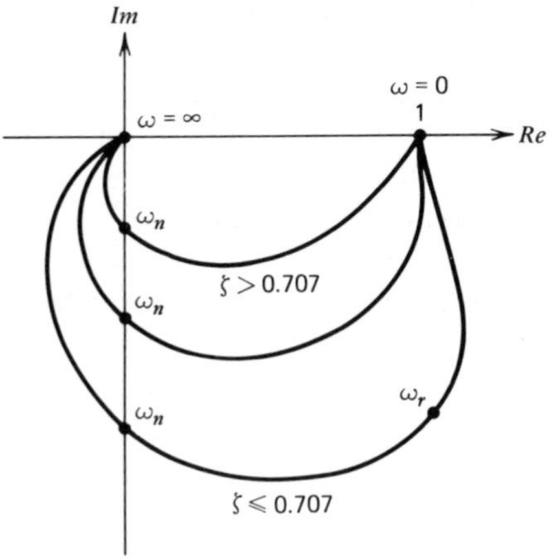


**Figure 5.32** Polar plots for the second-order system

$$T(s) = \frac{d/k}{\left(\frac{s}{\omega_n}\right)^2 + \frac{2\zeta s}{\omega_n} + 1}$$

Polar plots for other second-order system types can be obtained in a similar manner, although a variety of possible shapes exists.

## 5.10 NONLINEAR HIGHER-ORDER SYSTEMS

The analysis of nonlinear second-order and higher-order systems differs somewhat from that of first-order systems. The closed-form solution techniques discussed in Chapter Three do not work as often for second-order cases, because two integrations are now required instead of one. This makes evaluation of the resulting integrals very difficult. However, many such systems can be analyzed, in a restricted sense at least, by linearization or by the numerical solution methods of Chapter Three. These techniques can be extended with slight modification to higher-order models, as will be seen. We now also introduce another approach: phase-plane analysis.

### The Phase Plane

Phase-plane analysis was developed as a convenient graphical way of presenting the dynamics of a mechanical system. The *phase variables* used in this plot were originally taken to be the displacement and velocity of the mass in question. However, any convenient choice of a pair of state variables can be employed. The *phase plot* is a plot of one state variable versus the other, with time as a parameter on the resulting curve. For given initial values of the state variables, one curve is generated. This is the *trajectory* of the system's behavior. When the trajectories are sketched for many different sets of initial conditions, the resulting plot provides a compact and easily interpreted summary of the system's response.

The phase-plane concept can be applied to both linear and nonlinear models and to both first-order and second-order models. Consider the nonlinear logistic growth model presented in Section 2.8.

$$\dot{y} = ry\left(1 - \frac{y}{K}\right) \tag{5.10-1}$$

A plot of $\dot{y}$ versus $y$ is shown in Figure 5.33$a$ for $y \geqslant 0$. For first-order models with unique solutions, only one trajectory is possible. The maximum value of $\dot{y}$ is seen to occur at $y = K/2$. The plot is interpreted as follows. For a given initial value $y(0)$, the arrows on the plot indicate the behavior of the system for increasing time. The direction of the arrows can be found from the sign of $\dot{y}$ in each quadrant. From (5.10-1) we see that $\dot{y} > 0$ if $0 < y < K$. Thus if $y(0) < K$, $y(t)$ increases until it reaches the equilibrium at $y = K$. If $y(0) > K$, $y(t)$ decreases until the same equilibrium is reached. This result is clearly shown by the plot.

The linearized analysis can be demonstrated on this plot. We approximate the plot near the equilibrium $y = K$ by a straight line passing through the point $y = K$, $\dot{y} = 0$, with a slope equal to the slope of the phase plot at that point (Figure 5.33$b$). This slope is

$$\left.\frac{\partial \dot{y}}{\partial y}\right|_{y=K} = \left.\left(r - 2\frac{r}{K}y\right)\right|_{y=K} = -r$$

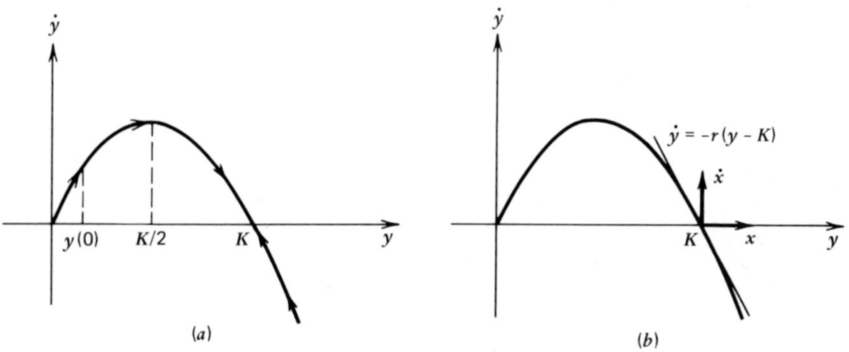


**Figure 5.33** (*a*) **Phase-plane plot of the logistic equation.** (*b*) **Graphical representation of linearization. The straight line is tangent to the phase plot at** $y = K$.

The straight line is given by

$$\dot{y} = -r(y - K)$$

If the coordinate system is shifted to the point $\dot{y} = 0, y = K$, and the new coordinates denoted $(\dot{x}, x)$, this line can be expressed as

$$\dot{x} = -rx \tag{5.10-2}$$

since $x = y - K$ and $\dot{x} = \dot{y}$. Equation (5.10-2) is the linearized approximation to (5.10-1). It tells us that the equilibrium is stable if $r > 0$. This result is also plain from the phase plot. Equation (5.10-2) also gives us an estimate of the time constant near equilibrium.

Linear second-order undamped systems are typified by the mechanical oscillator shown in Figure 5.34*a*. If the supporting surface is modeled as frictionless, the model is the same as that developed in Example 2.4, Section 2.3.

$$m\ddot{x} + kx = 0 \tag{5.10-3}$$

where $x = 0$ is the equilibrium position. It was seen that this equation can be integrated to yield

$$m\frac{v^2}{2} + k\frac{x^2}{2} = \text{constant} \tag{5.10-4}$$

where $v$ is the velocity $\dot{x}$. The constant on the right is determined by the initial con-

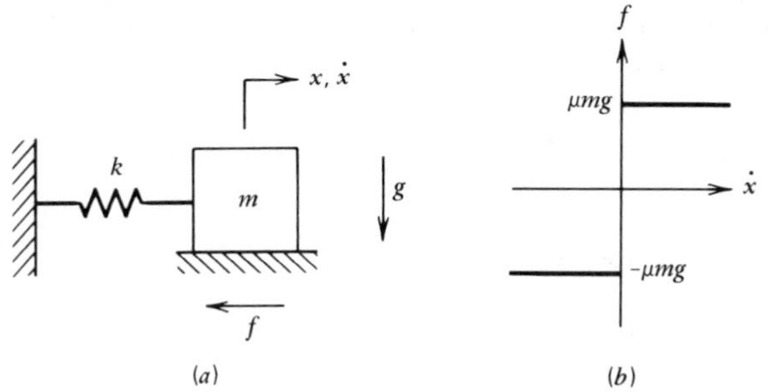


**Figure 5.34** (*a*) **Mass-spring system with friction.** (*b*) **Coulomb friction force as a function of velocity.**

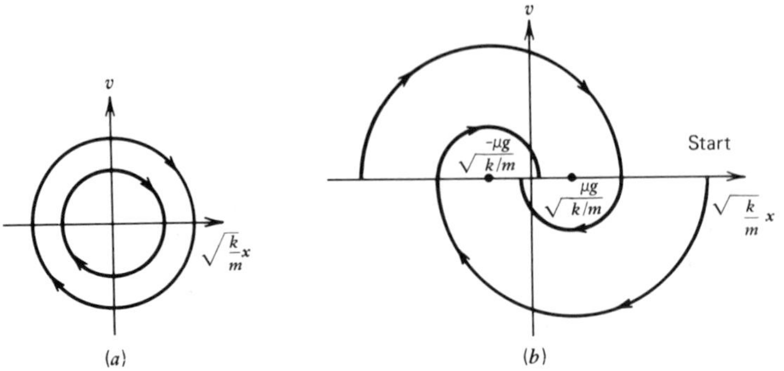


**Figure 5.35 Phase-plane plots for the mass-spring system. (*a*) Without friction. (*b*) With Coulomb friction.**

ditions and represents the initial energy of the system. Equation (5.10-4) describes a family of ellipses in the phase plane ($v$ versus $x$) and can be rearranged as follows.

$$v^2 + \left(\sqrt{\frac{k}{m}}x\right)^2 = C \tag{5.10-5}$$

where $C$ is a constant. In this form we can recognize a family of circles in terms of the modified phase variables $\left(v, \sqrt{\frac{k}{m}}x\right)$, Figure 5.35*a*. The radii of the circles are given by $R = \sqrt{C}$. The direction of motion can be determined from (5.10-3). When $x \neq 0$, the action of the spring pulls the mass back to $x = 0$. This determines the sign of $v$ as a function of $x$. Note that the phase plot clearly shows the neutral stability properties of this system. The trajectories never pass through the point (0, 0), nor do they continue to recede indefinitely from this point.

If Coulomb friction occurs on the supporting surface, the model takes the form

$$m\ddot{x} + kx = -\mu mg, \qquad \dot{x} > 0 \tag{5.10-6}$$

$$m\ddot{x} + kx = \mu mg, \qquad \dot{x} < 0 \tag{5.10-7}$$

where the friction force is given in Figure 5.34*b*. This model is piecewise linear, and the phase plane is particularly useful for such models. Reduce the preceding model to a single equation by defining a new displacement $y$, as

$$y = x + \frac{\mu mg}{k}\frac{|\dot{x}|}{\dot{x}}$$

The model now becomes

$$m\ddot{y} + ky = 0 \tag{5.10-8}$$

This is of the same form as (5.10-4), and thus

$$v^2 + \left(\sqrt{\frac{k}{m}}y\right)^2 = C \tag{5.10-9}$$

Substituting back for $x$, we get

$$v^2 + \frac{k}{m}\left(x + \frac{\mu mg}{k}\right)^2 = C_1, \quad v > 0 \tag{5.10-10}$$

$$v^2 + \frac{k}{m}\left(x - \frac{\mu mg}{k}\right)^2 = C_2, \quad v < 0 \tag{5.10-11}$$

In terms of the coordinates $v, \left(\sqrt{\frac{k}{m}}x\right)$, these equations describe a family of circles centered at $v = 0$, $x = -\mu mg/k$ for $v > 0$, and at $v = 0, x = \mu mg/k$ for $v < 0$. Now the trajectories approach the equilibrium at the origin, because the friction dissipates the energy.

We can easily use the phase plot to obtain information about the solution as a function of time. Suppose the motion is started by displacing the mass a distance $x(0) = \sqrt{\frac{m}{k}}\alpha$ and releasing it with zero velocity. This starting point is indicated in Figure 5.35*b*. The system follows the trajectory shown. When the velocity is again zero, on the opposite side of the plot, we see that the magnitude of the normalized displacement is

$$\left|\frac{k}{m}x\right| = \alpha - \frac{2\mu g}{\sqrt{k/m}}$$

or

$$|x| = x(0) - \frac{2\mu mg}{k} \tag{5.10-12}$$

Thus the reduction in displacement amplitude is $2\mu mg/k$ for a half-cycle; extending the analysis to the upper half plane, we can show that the amplitude reduction is $4\mu mg/k$ for a full cycle. The decay rate of the amplitude is therefore linear in time, as compared to exponential decay for the viscous damping case.

Sketching a phase plot is facilitated by expressing the model in state variable form. Consider the second-order system

$$\frac{dx_1}{dt} = f_1(x_1, x_2) \tag{5.10-13}$$

$$\frac{dx_2}{dt} = f_2(x_1, x_2) \tag{5.10-14}$$

Divide the second equation by the first and cancel the $dt$ terms to obtain

$$\frac{dx_2}{dx_1} = \frac{f_2(x_1, x_2)}{f_1(x_1, x_2)} \tag{5.10-15}$$

With $x_1$ now the independent variable, this differential equation describes the phase plane trajectories. It can be solved analytically in some cases, or numerically. For a complete classification of trajectory types and a discussion of numerical methods, see Reference 2.

## Linearization of the Reduced Form

The linearization technique presented in Chapter Two for first-order models is based on a truncated Taylor series representation of the model's nonlinearity and is equivalent to the straight-line approximation used in Figure 5.33*b*. The method can be easily extended to higher-order models, either by first linearizing the function producing the nonlinearity and then developing the differential equations, or by developing the equations first and then linearizing them. The first approach is sometimes more convenient, but one must be careful that the linearization is performed for values of the state and input variables that correspond to an equilibrium solution of the system's equations. The second approach automatically accounts for this.

The first approach was illustrated in Example 2.20, in which a linearized pneumatic bellows model was developed. We consider now a second-order example. The pendulum shown in Figure 5.36*a* can be considered as a representation of an unbalanced rotating load shown in Figure 5.36*b*. The mass $m$ represents the unbalanced mass lumped at its center of mass. We assume that a torque $T$ is applied to the system about the fixed axis of rotation. This torque would be produced by a control motor trying to position the load at some desired angle.

The mass moment of inertia of the pendulum as given in Table 2.5 is $I = mL^2$ for a massless rod. The gravity moment about the axis is $mgL \sin \theta$. This is the nonlinear term. If we are interested in motion about $\theta = 0$, the Taylor series expansion of $\sin \theta$ near $\theta = 0$ gives $\sin \theta \cong \theta$ for $\theta$ in radians. Thus the gravity moment is approximately $mgL\theta$, and Newton's law gives the model

$$mL^2\ddot{\theta} = T - mgL\theta \tag{5.10-16}$$

The reference point $\theta = 0$ is also an equilibrium solution of the model (5.10-16) if $T = 0$. The free response is like that of the mass-spring system, with a natural frequency of $\omega_n = \sqrt{g/L}$ for small amplitudes.

If we are interested in keeping the mass near the vertical position, we use the expansion of $\sin \theta$ near $\theta = \pi$; that is,

$$\sin\theta \cong \sin\pi + \left.\frac{\partial \sin\theta}{\partial\theta}\right|_{\theta=\pi} (\theta - \pi) = \pi - \theta$$

The gravity movement is approximately $mgL(\pi - \theta)$, and

$$mL^2\ddot{\theta} = T - mgL(\pi - \theta) \tag{5.10-17}$$

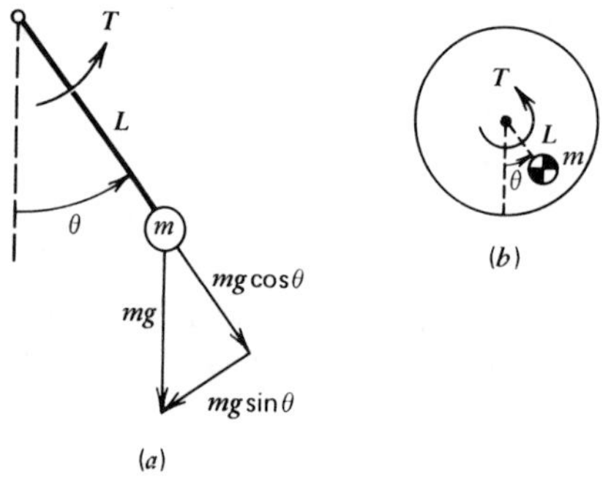


**Figure 5.36** **(*a*) Pendulum with an applied torque *T*. (*b*) Equivalent model of an unbalanced load driven by a torque *T*.**

Again, the linearization reference point $\theta = \pi$ is also an equilibrium solution of the model (5.10-17) if $T = 0$. Here it is convenient to introduce the deviation variable $x = \theta - \pi$ to obtain

$$mL^2\ddot{x} = T + mgLx \tag{5.10-18}$$

The equilibrium at $x = 0$ is seen to be unstable.

## Linearization of the Vector Form

The second approach to linearization requires that the describing differential equation be written first, the equilibrium solutions then obtained, and finally the linearization performed. This process is generally stated in terms of the vector form of the state equations. Assume that the state and input vectors are $\mathbf{z}$ and $\mathbf{v}$, where $\mathbf{z}$ is $(n \times 1)$ and $\mathbf{v}$ is $(m \times 1)$. The state equations are

$$\dot{\mathbf{z}} = \mathbf{f}(\mathbf{z}, \mathbf{v}) \tag{5.10-19}$$

For a constant input vector $\mathbf{v}_e$, (5.10-19) determines the equilibrium values of $\mathbf{z}$, denoted $\mathbf{z}_e$, by

$$\mathbf{0} = \mathbf{f}(\mathbf{z}_e, \mathbf{v}_e) \tag{5.10-20}$$

With $\mathbf{z}_e$ and $\mathbf{v}_e$ known, we expand $\mathbf{f}(\mathbf{z}, \mathbf{v})$ in a multivariable Taylor series about $(\mathbf{z}_e, \mathbf{v}_e)$ and keep only the first-order terms.

$$\dot{\mathbf{z}} = \mathbf{f}(\mathbf{z}, \mathbf{v}) \cong \overbrace{\mathbf{f}(\mathbf{z}_e, \mathbf{v}_e)}^{0} + \left[\frac{\partial \mathbf{f}}{\partial \mathbf{z}}\right]_e (\mathbf{z} - \mathbf{z}_e) + \left[\frac{\partial \mathbf{f}}{\partial \mathbf{v}}\right]_e (\mathbf{v} - \mathbf{v}_e) \tag{5.10-21}$$

where the partial derivative matrices are defined as follows. The subscript $e$ means that all derivatives are evaluated at the equilibrium $(\mathbf{z}_e, \mathbf{v}_e)$. The matrices are

$$\left[\frac{\partial \mathbf{f}}{\partial \mathbf{z}}\right] = \left[\frac{\partial f_i}{\partial z_j}\right] = \begin{bmatrix} \dfrac{\partial f_1}{\partial z_1} & \dfrac{\partial f_1}{\partial z_2} & \cdots \\ \dfrac{\partial f_2}{\partial z_1} & \dfrac{\partial f_2}{\partial z_2} & \cdots \\ \cdots & \cdots & \cdots \end{bmatrix} \tag{5.10-22}$$

for $i = 1, 2, \ldots, n$, $j = 1, 2, \ldots, n$; and

$$\left[\frac{\partial \mathbf{f}}{\partial \mathbf{v}}\right] = \left[\frac{\partial f_i}{\partial v_j}\right] = \begin{bmatrix} \dfrac{\partial f_1}{\partial v_1} & \dfrac{\partial f_1}{\partial v_2} & \cdots \\ \dfrac{\partial f_2}{\partial v_1} & \dfrac{\partial f_2}{\partial v_2} & \cdots \\ \cdots & \cdots & \cdots \end{bmatrix} \tag{5.10-23}$$

for $i = 1, 2, \ldots, n$, $j = 1, 2, \ldots, m$.

Use (5.10-20) to eliminate the first term in the series (5.10-21), and define the deviations as

$$\mathbf{x} = \mathbf{z} - \mathbf{z}_e \tag{5.10-24}$$

$$\mathbf{u} = \mathbf{v} - \mathbf{v}_e \tag{5.10-25}$$

Since $\dot{\mathbf{x}} = \dot{\mathbf{z}}$, (5.10-21) gives the desired linearized model,

$$\dot{\mathbf{x}} = \mathbf{A}\mathbf{x} + \mathbf{B}\mathbf{u} \tag{5.10-26}$$

where $\mathbf{A} = [a_{ij}]$, $\mathbf{B} = [b_{ij}]$, and

$$a_{ij} = \left[\frac{\partial f_i}{\partial z_j}\right]_e, \qquad b_{ij} = \left[\frac{\partial f_i}{\partial v_j}\right]_e \tag{5.10-27}$$

The matrix $\mathbf{A}$ is $(n \times n)$, and $\mathbf{B}$ is $(n \times m)$.

If a nonlinear output equation is encountered, say in the form

$$\mathbf{w} = \mathbf{g}(\mathbf{z}, \mathbf{v}) \tag{5.10-28}$$

this is linearized in a similar fashion to obtain the linear output equation $\mathbf{y} = \mathbf{C}\mathbf{x} + \mathbf{D}\mathbf{u}$, where

$$\mathbf{C} = \left[\frac{\partial g_i}{\partial z_j}\right]_e, \qquad \mathbf{D} = \left[\frac{\partial g_i}{\partial v_j}\right]_e \tag{5.10-29}$$

$$\mathbf{y} = \mathbf{w} - \mathbf{w}_e$$

For an original output vector $\mathbf{w}$ of dimension $(p \times 1)$, $\mathbf{C}$ is $(p \times n)$ and $\mathbf{D}$ is $(m \times n)$.

*Example 5.22*

Use the vector method to obtain a linearized model for the system shown in Figure 5.36 for the equilibrium $\theta_e = 3\pi/4$, $T_e = 0.707\, mgL$. Evaluate the stability.

Newton's law gives the reduced form

$$mL^2\ddot{\theta} = T - mgL \sin\theta \tag{5.10-30}$$

To put this into the state variable form (5.10-19), let $z_1 = \theta$, $z_2 = \dot{\theta}$, and $v = T$. Then

$$\left.\begin{aligned} \dot{z}_1 &= z_2 \\ \dot{z}_2 &= \frac{v}{mL^2} - \frac{g}{L}\sin z_1 \end{aligned}\right\} \tag{5.10-31}$$

From (5.10-27) we obtain

$$A = \begin{bmatrix} 0 & 1 \\ \dfrac{0.707g}{L} & 0 \end{bmatrix}, \qquad B = \begin{bmatrix} 0 \\ \dfrac{1}{mL^2} \end{bmatrix}$$

With $x_1 = z_1 - 3\pi/4$, $x_2 = z_2$ and $u = v - 0.707\, mgL$, the linearized state equations are

$$\left.\begin{aligned} \dot{x}_1 &= x_2 \\ \dot{x}_2 &= \frac{0.707g}{L}x_1 + \frac{1}{mL^2}u \end{aligned}\right\} \tag{5.10-32}$$

The characteristic equation is $mL^2s^2 - 0.707\, mgL = 0$, and this equilibrium is thus unstable.

## Numerical Methods

The numerical methods developed in Chapter Three for solving nonlinear first-order differential equations are easily adapted to higher-order equations expressed in state variable form. This form consists of a set of coupled first-order equations, such as

$$\frac{d\mathbf{y}}{dt} = \mathbf{f}(\mathbf{y}, \mathbf{v}) \tag{5.10-33}$$

where $\mathbf{v}$ is an input function of time.

Application of the Euler method this equation gives

$$\mathbf{y}(t_{k+1}) = \mathbf{y}(t_k) + \Delta t\, \mathbf{f}[\mathbf{y}(t_k), \mathbf{v}(t_k)] \tag{5.10-34}$$

The predictor-corrector method applied to (5.10-33) gives

$$\text{Predictor.} \quad \mathbf{x}_{k+1} = \mathbf{y}_k + \Delta t\, \mathbf{f}(\mathbf{y}_k, \mathbf{v}_k) \tag{5.10-35}$$

$$\text{Corrector.} \quad \mathbf{y}_{k+1} = \mathbf{y}_k + \frac{\Delta t}{2}[\mathbf{f}(\mathbf{y}_k, \mathbf{v}_k) + \mathbf{f}(\mathbf{x}_{k+1}, \mathbf{v}_{k+1})] \tag{5.10-36}$$

This is identical to the scalar form of the algorithm, except for the vector notation.

The extension of the fourth-order Runge-Kutta algorithms to the vector case is just as easily made. All of the comments made in Chapter Three regarding selection of algorithms, choice of step size, etc., apply to higher-order models. A FORTRAN program that implements Gill's fourth-order Runge-Kutta algorithm is given in Appendix C. This algorithm is sufficient to handle problems of the type encountered in this text.

## 5.11 DISCRETE-TIME MODELS OF HIGHER ORDER

Most of the analytical techniques developed in this chapter can be applied with slight modification to linear discrete-time models because the Laplace transform and $z$-transform operators reduce their respective model types to a problem in linear algebra. Because of this similarity an extensive discussion will not be presented here; instead we will outline the techniques while emphasizing the similarities and pointing out those areas where differences occur. First, we present a few applications.

### Finite-Difference Approximations

Finite-difference approximations for derivatives are used to convert differential equations into difference equations suitable for numerical solutions. The Tustin approximation introduced in Section 4.9 is an example. For a second-order derivative the approximation with a time step $T$ is

$$\mathscr{Z}\left(\frac{d^2y}{dt^2}\right) = [D(z)]^2 Y(z)$$

where

$$D(z) = \frac{2}{T}\frac{1 - z^{-1}}{1 + z^{-1}}$$

In Section 4.9 it was shown that this approximation converts the equation

$$\ddot{y} + 5y + 3\dot{y} = v$$

into the difference equation

$$a_2 y(k+2) + a_1 y(k+1) + a_0 y(k) = T^2 v(k+2) + 2T^2 v(k+1) + T^2 v(k) \tag{5.11-1}$$

where

$$a_2 = 3T^2 + 10T + 4$$
$$a_1 = 6T^2 - 8$$
$$a_0 = 3T^2 - 10T + 4$$

The behavior of such an approximation – for example, its stability properties – must be understood before it can be used to advantage.

## Digital Filters

Numerical algorithms used for digital filtering and processing of data often require a model of higher order than the first-order algorithms treated in Sections 4.7 and 4.8. For example, the frequency response of the low-pass filter given in Section 4.8 can be improved by cascading more than one such filter. For $n$ filters in cascade the overall filter transfer function is

$$T(z) = \left(\frac{Kz}{z - a + aK}\right)^n \tag{5.11-2}$$

Cascading allows the constants $a$ and $K$ to be selected with more freedom to improve the signal rejection properties (a sharper falloff with frequency) while preserving the transient response characteristics. Similar statements apply to the design of high-pass and band-pass filters, and higher-order models are common in such applications.

The following second-order filter is easily derived and is useful for estimating displacement and velocity, given only displacement measurements contaminated with error. Define the following variables.

$u(k)$ = displacement measurement at time $k$

$p(k)$ = prediction of the displacement at time $k$, after processing the measurement $u(k-1)$

$y(k)$ = estimate of the displacement at time $k$, after processing the measurement $u(k)$

$v(k)$ = estimate of the velocity at time $k$, after processing $k$ measurements

The filter consists of three stages. The first is a prediction of the current displacement at time $k$ based on the estimates obtained from the previous measurement at time $k-1$. Noting that velocity is the derivative of displacement, we apply the Euler approximation to obtain

$$p(k) = y(k-1) + Tv(k-1)$$

where $T$ is the time between measurements. The second stage corrects this prediction using the difference between the measurement and the prediction at time $k$; that is,

$$y(k) = p(k) + a[u(k) - p(k)]$$

where $a$ is a weighting factor that indicates the reliability of the measurement (see Section 4.7 for a discussion of this concept). The final stage of the filter is a correction of the velocity estimate, with a weighted velocity error correction term.

$$v(k) = v(k-1) + b\left[\frac{u(k) - v(k-1)}{T} - \frac{p(k) - v(k-1)}{T}\right]$$

Using the first equation to eliminate $p(k)$, we obtain the filter algorithm.

$$y(k) = (1-a)y(k-1) + (1-a)Tv(k-1) + au(k) \tag{5.11-3}$$

$$v(k) = -\frac{b}{T}y(k-1) + (1-b)v(k-1) + \frac{b}{T}u(k) \tag{5.11-4}$$

The performance of this filter will be analyzed in Section 5.12.

## Sampled-Data Systems

The design of digital control systems sometimes requires a model of a dc motor whose input voltage is produced by a sample-and-hold device. Assume the motor is field-controlled. Thus no back emf is present.* For now we neglect the armature inductance of the motor, and lump its inertia with that of the load. If viscous friction is present and the output is angular displacement, the resulting model is shown in Figure 5.37. From the property stated by (4.6-17) for a zero-order hold,

$$G(z) = \frac{z-1}{z}G_2(z) = \frac{\Theta(z)}{E(z)}$$

where

$$G_2(s) = \frac{K_m}{s^2(Is+c)} = \frac{K}{s^2(s+b)}$$

The new constants are $K = K_m/I, b = c/I$. Expand $G_2(s)$ to compute $G_2(z)$ as follows.

$$G_2(s) = \frac{K}{b}\left[\frac{1}{s^2} - \frac{1}{b}\left(\frac{1}{s} - \frac{1}{s+b}\right)\right]$$

From Table 4.3,

$$G_2(z) = \frac{K}{b}\left[\frac{zT}{(z-1)^2} - \frac{1}{b}\left(\frac{z}{z-1} - \frac{z}{z-a}\right)\right]$$

where $a = e^{-bT}$. The overall transfer function is

$$G(z) = \frac{K}{b^2}\,\frac{(bT-1+a)z + 1 - a - bTa}{(z-1)(z-a)} \tag{5.11-5}$$

The denominator indicates a second-order model.

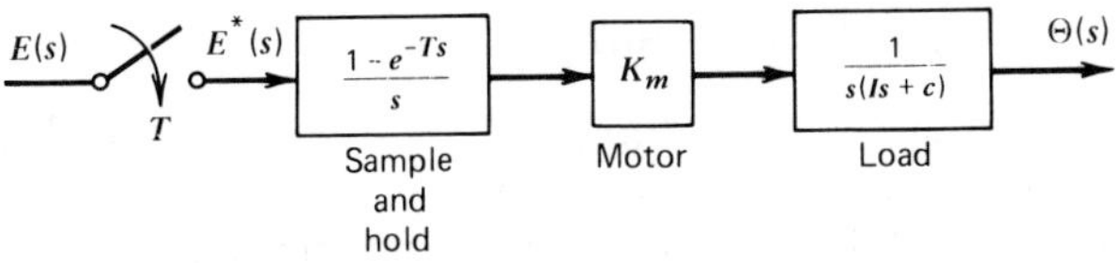


**Figure 5.37 Block diagram of a dc motor-amplifier system with a sampled voltage input and an angular displacement as the output.**

---

* More discussion of field-controlled motors is given in Problem 5.6.

## Model Forms

The preceding examples illustrate the various forms in which second-order discrete models can appear. The finite-difference model (5.11-1) is in the reduced form analogous to that for continuous-time models. The digital filter given by (5.11-3) and (5.11-4) is in state variable form, with the estimates of the displacement and the velocity as the state variables. Finally, (5.11-5) is a transfer function model. The techniques for converting from one model form to another are similar to those developed for the continuous-time case. For example, to construct a state model from a transfer function, use the same procedure as in Section 5.2, but with $1/s$ replaced with $1/z$. Consider a discrete model analogous to (5.1-19).

$$\begin{aligned} T(z) &= \frac{X(z)}{U(z)} = \frac{1}{az^2 + bz + c} \\ &= \frac{1/az^2}{1 + b/az + c/az^2} \end{aligned} \tag{5.11-6}$$

Cross-multiply and rearrange to obtain

$$X(z) = \frac{1}{z}\left\{-\frac{b}{a}X(z) + \frac{1}{z}\left[\frac{1}{a}U(z) - \frac{c}{a}X(z)\right]\right\}$$

Define $X_1(z) = X(z)$, $X_2(z) = [U(z) - cX(z)]/a$. This gives the state equations

$$\left.\begin{aligned} x_1(k+1) &= -\frac{b}{a}x_1(k) + x_2(k) \\ x_2(k+1) &= -\frac{c}{a}x_1(k) + \frac{1}{a}u(k) \end{aligned}\right\} \tag{5.11-7}$$

To obtain a reduced model, cross-multiply (5.11-6) directly and revert to the time domain. The result is

$$ax(k+2) + bx(k+1) + cx(k) = u(k) \tag{5.11-8}$$

The process is identical to that for Laplace transforms.

The matrix forms of the linear state equations and output relations are

$$\mathbf{x}(k+1) = \mathbf{Px}(k) + \mathbf{Qu}(k) \tag{5.11-9}$$

$$\mathbf{y}(k) = \mathbf{Cx}(k) + \mathbf{Du}(k) \tag{5.11-10}$$

where we have used **P** and **Q** instead of **A** and **B** so that we can distinguish between them later. The dimensions of the vector-matrix quantities are the same as those in Section 5.2. The $z$-transform applied to (5.11-9) and (5.11-10) gives

$$\begin{aligned} z\mathbf{X}(z) - z\mathbf{x}(0) &= \mathbf{PX}(z) + \mathbf{QU}(z) \\ \mathbf{Y}(z) &= \mathbf{CX}(z) + \mathbf{DU}(z) \end{aligned}$$

or

$$(z\mathbf{I}-\mathbf{P})\mathbf{X}(z) = z\mathbf{x}(0)+\mathbf{Q}\mathbf{U}(z) \qquad (5.11\text{-}11)$$

$$\mathbf{Y}(z) = \mathbf{T}(z)\mathbf{U}(z) \qquad (5.11\text{-}12)$$

$$\mathbf{T}(z) = \mathbf{C}(z\mathbf{I}-\mathbf{P})^{-1}\mathbf{Q}+\mathbf{D} \qquad (5.11\text{-}13)$$

where $\mathbf{T}(z)$ is the transfer function matrix.

We see from (5.11-11) that the characteristic equation is

$$|z\mathbf{I}-\mathbf{P}| = 0 \qquad (5.11\text{-}14)$$

In the next section we use these results to predict the free and forced response.

If the index $k$ is shifted by one in (5.11-3) and (5.11-4), the other standard state variable form results. This is

$$\mathbf{x}(k+1) = \mathbf{P}\mathbf{x}(k)+\mathbf{Q}\mathbf{u}(k+1) \qquad (5.11\text{-}15)$$

A similar procedure can be used to find the transfer function. However, the characteristic equation is still given by (5.11-14).

## 5.12 DISCRETE-TIME RESPONSE

The characteristic equation for the second-order case can be written as

$$\alpha z^2+\beta z+\gamma = 0 \qquad (5.12\text{-}1)$$

As should be expected, different types of response occur depending on whether the roots of this equation are real or complex. In the first-order case treated in Chapter Four, we saw that the basic solution form is

$$x(k) = Az^k$$

For two distinct roots $z_1$ and $z_2$, the free response of (5.11-8) is

$$x(k) = A_1z_1^k+A_2z_2^k \qquad (5.12\text{-}2)$$

where $A_1$ and $A_2$ depend on two initial conditions, say $x(0)$ and $x(1)$. The concept of a dominant root applies in discrete time as well. For example, if $|z_1|>|z_2|$, the $z_1$ term dominates the behavior. If either root is negative, the solution oscillates. If any root has a magnitude greater than unity, the system is unstable.

For repeated roots the solution form becomes

$$x(k) = A_1z_1^k+kA_2z_1^k \qquad (5.12\text{-}3)$$

### Complex Roots

We have seen that oscillations can occur in the free response of a first-order discrete system, but these oscillations are restricted to a period of two time steps because they are caused by a negative number raised to increasing integer powers. Free oscillations with a period different from two are possible only for higher-order models with complex roots. Assume that these are of the form $z=-a\pm ib$. In this case (5.12-2) is not convenient and we seek a simpler form. This is obtained by using *DeMoivre's theorem* for a complex number raised to a power. The complex number $r=z_1=$

$-a \pm ib$ can be written in terms of its magnitude and phase angle as

$$r = |r|\, e^{i\theta}$$

Thus

$$r^k = |r|^k e^{ik\theta}$$
$$= |r|^k (\cos k\theta + i \sin k\theta) \qquad (5.12\text{-}4)$$

where in this case, $|r| = \sqrt{a^2 + b^2}$ and $\tan\theta = -b/a$. For $z_2 = -a - ib$, a conjugate relation is the result. Substituting these expressions into (5.12-2) we find that

$$x(k) = |r|^k \left[(A_1 + A_2)\cos k\theta + i(A_1 - A_2)\sin k\theta\right]$$

From the same logic as in the continuous-time case, the result is

$$x(k) = |r|^k B \sin(k\theta + \phi) \qquad (5.12\text{-}5)$$

where $r$ is the magnitude of the upper root $z_1 = -a + ib$, $\theta$ is its phase angle, and the constants $B$ and $\phi$ depend on the initial conditions $x(0)$ and $x(1)$.

The response for the complex-roots case has a frequency of $\theta$ radians/time step. The model is seen to be stable if $|r| < 1$, neutrally stable if $|r| = 1$, and unstable otherwise. This completes all the cases that can occur, and the general stability criterion for linear discrete-time models can now be stated in terms of a unit circle in the $z$ plane (Figure 5.38). For an $n$th-order system, if any one of the $n$ roots lies outside the unit circle, the system is unstable. A root lies outside the circle if its magnitude is greater than one.

The roots of the denominator of the transfer function (5.11-5) are $z = 1$ and $z = a$. Thus the sampled-data motor system is neutrally stable since one root lies on the circumference of the unit circle while the other lies within the circle ($a < 1$ if $bT > 0$). The root $z = 1$ produces a constant term proportional to $1^k$ in the free response, while the root $z = a$ produces the term $a^k$, which decays with time.

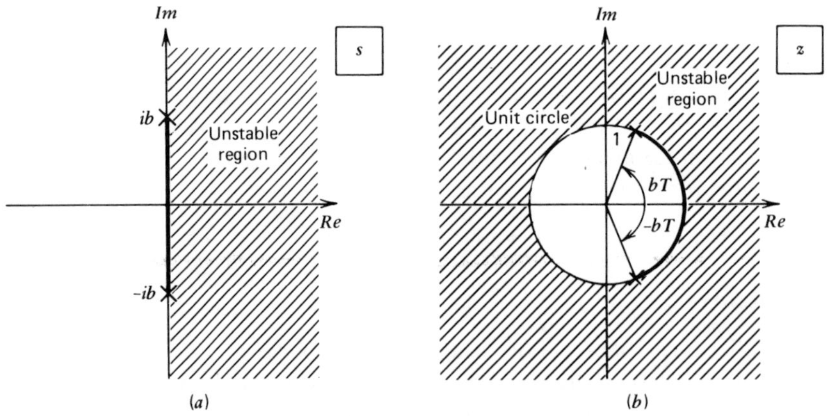


**Figure 5.38 Stability criteria for continuous and discrete time systems. (*a*) In the *s*-plane, the unstable region is the entire right-half plane. (*b*) In the *z*-plane, the unstable region is that outside of the unit circle at the origin. Equivalent locations of a pair of neutrally stable roots are also shown.**

## A Stability Test

The Routh-Hurwitz criterion provides a convenient stability test that can be applied to the characteristic polynomial without solving for the roots. Originally developed to detect the occurrence of a root in the right half of the $s$ plane, it can be converted for use with discrete models by employing the following transformation.

$$z = \frac{s+1}{s-1} \tag{5.12-6}$$

This transformation maps the inside of the unit circle in the $z$ plane onto the entire left half of the $s$ plane. Substitute $z$ from (5.12-6) into the characteristic equation in terms of $z$. This gives a polynomial in $s$ to which the Routh-Hurwitz criterion can be applied. The number of sign changes in the first column of the array indicates the number of roots that lie outside the unit circle.

*Example 5.23*

Determine the stability conditions for the characteristic equation

$$\alpha z^2 + \beta z + \gamma = 0$$

With (5.12-6), the characteristic equation becomes

$$\alpha\left(\frac{s+1}{s-1}\right)^2 + \beta\left(\frac{s+1}{s-1}\right) + \gamma = 0$$

or

$$(\alpha + \beta + \gamma)s^2 + 2(\alpha - \gamma)s + (\alpha - \beta + \gamma) = 0$$

The result of the Routh-Hurwitz criterion says that no root lies in the right half of the $s$ plane if $(\alpha + \beta + \gamma)$, $(\alpha - \gamma)$, and $(\alpha - \beta + \gamma)$ all have the same sign. Thus the second-order discrete-time system is stable if and only if

$$\frac{\alpha - \gamma}{\alpha + \beta + \gamma} > 0 \tag{5.12-7}$$

and

$$\frac{\alpha - \beta + \gamma}{\alpha + \beta + \gamma} > 0 \tag{5.12-8}$$

For the Tustin approximation algorithm (5.11-1), these results give

$$\alpha + \beta + \gamma = 12T^2$$

$$\alpha - \gamma = 20T$$

$$\alpha - \beta + \gamma = 16$$

Therefore the algorithm is stable for all positive values of the step size $T$.

The displacement-velocity filter given by (5.11-3) and (5.11-4) has the following system matrix $\mathbf{P}$ [see (5.11-15)].

$$\mathbf{P} = \begin{bmatrix} (1-a) & (1-a)\,\Delta t \\ -\dfrac{b}{\Delta t} & (1-b) \end{bmatrix}$$

From $|z\mathbf{I} - \mathbf{P}| = 0$ we see that

$$z^2 + (a + b - 2)z + 1 - a = 0$$

and

$$\alpha + \beta + \gamma = b$$

$$\alpha - \beta + \gamma = 4 - 2a - b$$

$$\alpha - \gamma = a$$

For $a > 0$ and $b > 0$, the filter will be stable if and only if $4 - 2a - b > 0$. If this stability condition is not satisfied, the displacement and velocity estimates $y(k)$ and $v(k)$ will never settle down to constant values. Thus the effects of the initial (and probably incorrect) estimates of displacement and velocity will never be eliminated, regardless of the accuracy of the measurements $u(k)$.

## Transient Performance Specifications

The Laplace transform of a sampled time function in Chapter Four led naturally to the definition of the variable $z$ as

$$z = e^{sT} \tag{5.12-9}$$

This formula is useful in relating the behavior of a time function as specified by its roots in the $s$ plane to the location of the corresponding roots in the $z$ plane. In Sections 5.4 and 5.7 we saw how transient behavior can be characterized by the damping ratio, natural frequency, and time constant of the dominant root in the $s$ plane. Therefore (5.12-9) gives the root locations in the $z$ plane required to produce the same transient behavior. The transformation (5.12-9) is not one to one. If $z = e^{s_1 T}$, then $z = e^{s_2 T}$ also, where

$$s_2 = s_1 + i\,\frac{2\pi n}{T}$$

and $n$ is integer. However, if guard filters are used to prevent aliasing, we need not be concerned with the solutions for $n > 0$.

Consider a stable root pair $s = -a \pm ib$. The corresponding $z$ values are

$$z = e^{-aT} e^{\pm ibT} = e^{-aT}(\cos bT \pm i \sin bT) \tag{5.12-10}$$

We have seen that purely imaginary roots $s = \pm ib$ correspond to $z$ roots on the unit circle $z = \cos bT \pm i \sin bT$ (Figure 5.38). Horizontal lines of constant frequency and variable time constant in the $s$ plane correspond to radial lines in the $z$ plane (Figure 5.39*a*). These make an angle of $\theta = \pm bT$ with the positive real axis. For small time constants, the $z$ roots lie close to the origin (these roots are denoted by O in both plots). As the time constant increases, the roots move in the direction of the arrows until the stability limit is reached (denoted by X). From this we see that the *dominant* root in the $z$ plane is that root lying closest (radially) to the unit circle.

If the time constant is held fixed and the damped frequency varied, the $z$ roots move in a circle of radius $e^{-aT}$ (Figure 5.39*b*). For a stable system ($a > 0$), the circle is within the unit circle.

To see the result of a root moving along a radial line of fixed damping ratio $\zeta = \cos\beta$, write the upper root as

$$s = -a + ib = -\zeta\omega_n + i\omega_n\sqrt{1 - \zeta^2}$$

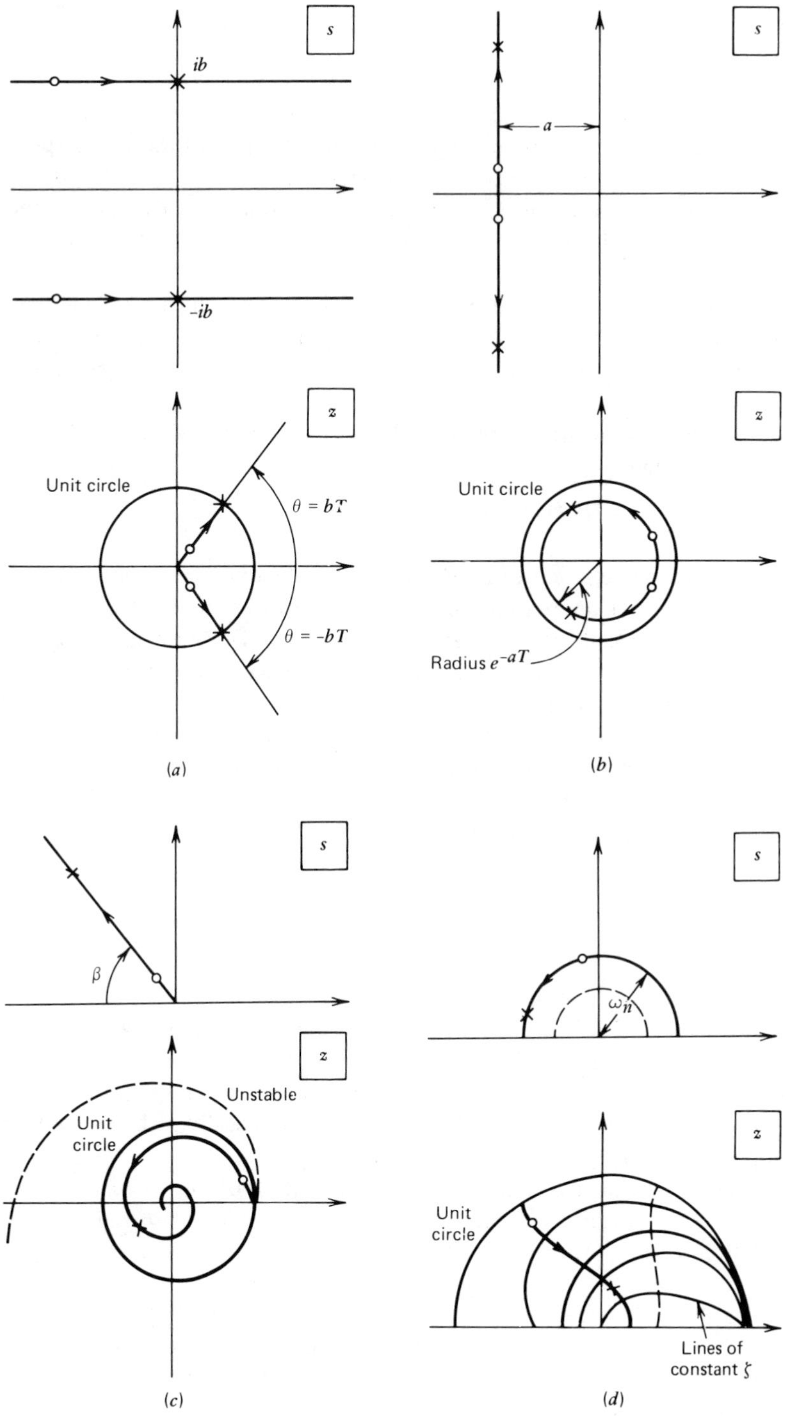


Figure 5.39 Equivalent root paths in the $s$ plane and $z$ plane. ($a$) Roots with the same oscillation frequency $b$. ($b$) Roots with the same time constant $\tau = 1/a$. ($c$) Roots with the same damping ratio $\zeta$. ($d$) Roots with the same natural frequency $\omega_n$.

The $z$ root is

$$z = e^{-\zeta\omega_n T} e^{i\omega_n T\sqrt{1-\zeta^2}} \tag{5.12-11}$$

For fixed $\zeta$ and $T$, the $s$ root moves out from the origin as $\omega_n$ is increased; the $z$ root rotates counterclockwise with decaying amplitude. The result is a logarithmic spiral (Figure 5.39*c*). The lower root $s = -a - ib$ produces a clockwise spiral that is not shown. As damping is decreased, $\beta$ increases to 90° (the neutrally stable case), and the spiral becomes the unit circle. For $\beta \geqslant 90°$, the spiral opens outward (the unstable case).

Curves of constant $\omega_n$ are concentric circles in the $s$ plane. In the $z$ plane they are lines perpendicular to the spirals for constant $\zeta$ (Figure 5.39*d*). As the $s$ root moves counterclockwise on its circle, the corresponding $z$ root moves toward the origin.

To illustrate how the relations are applied, consider a system described by the following transfer function.

$$T(z) = \frac{z + p}{\alpha z^2 + \beta z + \gamma}$$

This could be the transfer function of a sampled-data system whose output is a continuous function of time. Thus it would be natural to describe the desired transient response in terms of the specifications given in Section 5.7. The sampling time $T$ will appear in some if not all of the parameters $\alpha, \beta, \gamma$, and $p$. We assume that these can be selected to some extent by the designer.

Suppose that the step response of the preceding system is to have the following properties: (1) maximum percent overshoot $\leqslant 10\%$; (2) the 100% rise time $\leqslant 5$; and (3) the 2% settling time $\leqslant 20$. From the appropriate relations in Section 5.7 we can see that these are equivalent to the following.

1. $\zeta \geqslant 0.6$
2. $\omega_n \geqslant 3/5$
3. $\zeta\omega_n \geqslant 4/20$

We note in passing that the results are for a second-order system without any finite zeroes, whereas the transfer function has a zero at $z = -p$. In Example 5.19 the effect of such a zero was shown to increase the overshoot. We temporarily neglect this effect here; the resulting design should be checked analytically or by simulation to see if an adjustment is necessary.

The stability requirement is implicit. For $\zeta \geqslant 0.6$ both roots of the denominator of $T(z)$ must be within the region enclosed by the spirals shown in Figure 5.40*a*. These can be plotted from (5.12-11) for a fixed $T$ with $\zeta = 0.6$ and $\omega_n$ as a parameter. The lower spiral results from the other conjugate root for $\zeta = 0.6$. For $\omega_n \geqslant 3/5$, the $z$ roots must lie to the left of the curve obtained from (5.12-11) with $\omega_n = 3/5$ and $\zeta$ as a parameter (Figure 5.40*b*). Finally, the third specification is a restriction on the real part of the $s$ roots. This requires that the $z$ roots lie within a circle of radius $R = \exp(-4T/20)$ (Figure 5.40*c*).

The design is achieved by selecting a suitable sampling period $T$, sketching the boundaries, and choosing the design parameters to place the $z$ roots within the region

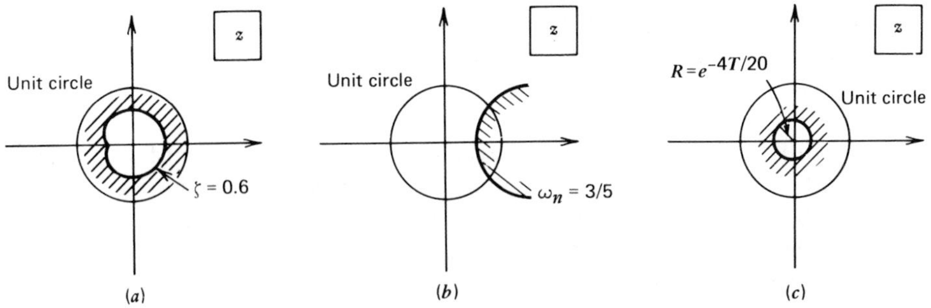


**Figure 5.40 Transient response specifications in the $z$ plane. ($a$) Overshoot constraint. ($b$) Rise-time constraint. ($c$) Settling-time constraint.**

allowed by all the boundaries. A technique that greatly eases this task is the root locus method to be presented in Chapter Eight. Consequently, we will return to this topic in more detail.

### Forced Response

The forced response of discrete-time models can be computed by direct substitution, if the solution form is known, or by the systematic method of the $z$ transform, as in Section 4.5. The only difference occurs when complex characteristic roots exist. In this case, the computations can be simplified by the introduction of a complex factor similar to the factor $R(-a + ib)$ of Section 5.7 (5.7-28) and by the use of DeMoivre's theorem (5.12-4).

For example, consider a single-input–single-output second-order system of the form

$$\alpha x(k+2) + \beta x(k+1) + \gamma x(k) = \delta u(k) \tag{5.12-12}$$

with complex conjugate roots $z = -a \pm ib = |r|e^{\pm i\theta}$. The unit step response is of the form

$$x(k) = A + B|r|^k \sin(k\theta + \phi) \tag{5.12-13}$$

The unknown constants are $A$, $B$, and $\phi$. These can be found by substituting $x(k)$ into (5.12-13) and using the initial conditions $x(0)$ and $x(1)$.

## 5.13 SUMMARY

We have now seen all the possible types of dynamic response that can occur in a linear system. The superposition property states that the free response is the linear combination of the response terms generated by each characteristic root. Thus even if the model is of third order or higher, its behavior can be understood in terms of the response properties of first- and second-order systems. This is because the characteristic roots must be either (1) real and distinct, (2) distinct but complex conjugate pairs, or (3) repeated. All three cases can occur in second-order models. For an approximate analysis the dominant-root concept can be used to reduce the system to an equivalent first- or second-order system.

In practice, however, the best method for analyzing linear systems depends on the system's order. For third order or less, Laplace transform analysis of the reduced model form is perhaps the most useful. For systems consisting of several coupled first- and second-order elements, the effort required to combine the elements' equations into a single equation of high order is probably not justified. However, the state equations can be readily obtained for each element and combined to produce the vector-matrix representation of the complete system. Laplace transform methods can then be applied to the vector state equations. When the resulting algebra is too cumbersome, the state equations provide a convenient form for numerical analysis.

In the next chapter we will begin to apply these methods to the design and analysis of control systems.

## REFERENCES


1. D. McCloy and H. Martin, *The Control of Fluid Power*, Longman, London, London, 1973.
2. Y. Takahashi, M. Rabins, and D. Auslander, *Control*, Addison-Wesley, Reading, Mass., 1970.


## PROBLEMS

5.1 Draw the block diagrams or signal flow graphs for the systems shown in Figure P5.1 and find their transfer function either from the differential equations or by reducing the diagrams.

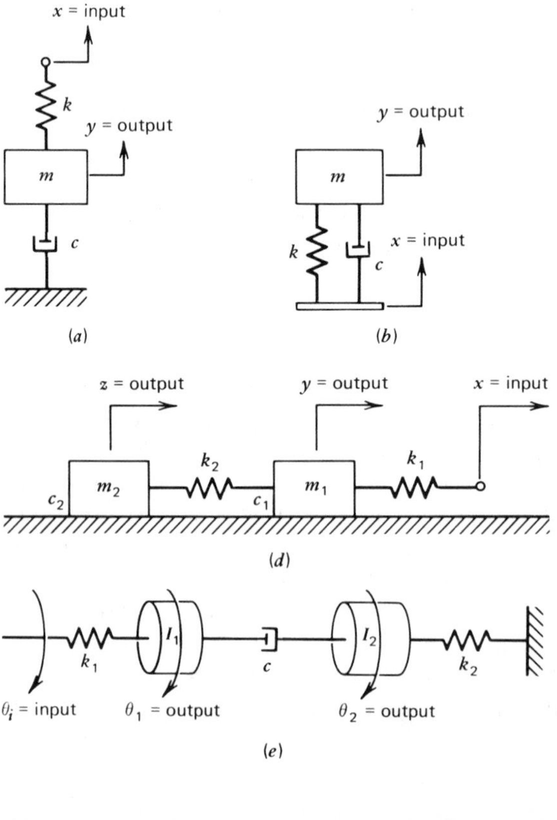

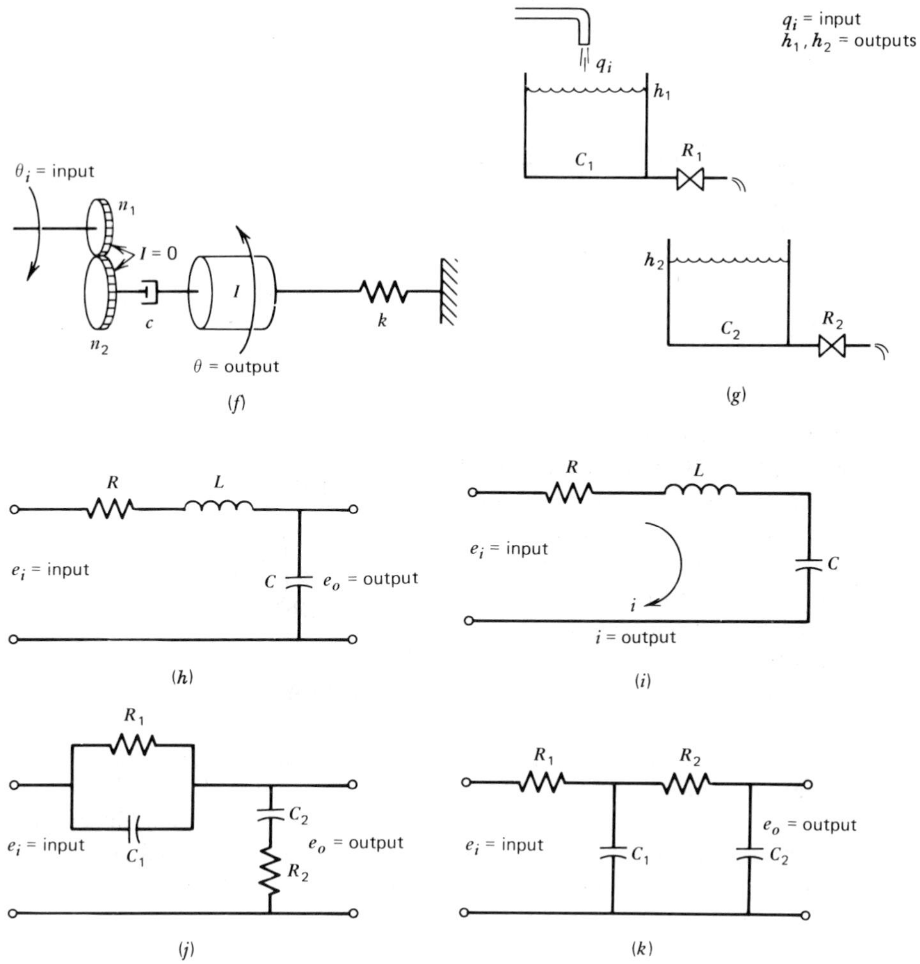


**Figure P5.1**

5.2 Derive the equivalent inertia formula for the lead screw given in Table 2.5. Use kinetic energy equivalence to do this.

5.3 The derivation of the dc motor model in Example 5.3 neglected the elasticity of the motor-load shaft. Figure P5.3 shows a model that includes this elasticity, denoted by its equivalent spring constant $k$. The motor inertia is $I_1$ and the load inertia is $I_2$.

(a) Find the transfer function with $\theta_2$ as output and $v$ as input.

(b) Reduce the results of part a to the rigid-shaft case ($k \to \infty$, $I_1 + I_2 \to I$), and compare with the transfer function obtained in Example 5.3.

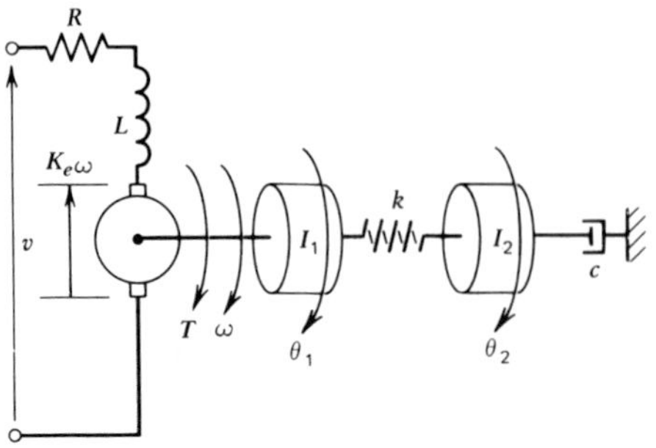


Figure P5.3

5.4 A thermometer can be represented as two thermal masses with thermal capacitances $C_1$ and $C_2$ (Figure P5.4). The total thermal resistances of the inner and outer glass surfaces are $R_1$ and $R_2$, respectively. Develop a model for the temperatures $T_1$ and $T_2$ with the external temperature $T_o$ as input.

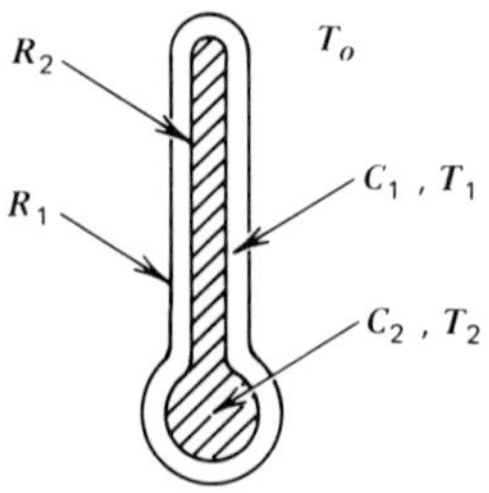


Figure P5.4

5.5 The torque constant for an armature-controlled dc motor relates the armature current $i$ to the motor torque $T$ as $T = K_t i$. The constant $K_e$ relates the back emf $e_b$ to the motor speed $\omega$ as $e_b = K_e\omega$. Establish the relationship between $K_t$ and $K_e$ by equating the two expressions for the power developed by the motor. These expressions are $P = e_b i_a$ (watts) and $P = \frac{1}{550} T\omega$ (hp). The units of $K_t$ are lb-ft/amp and those of $K_e$ are volt/rad/sec.

5.6 The dc motor treated in Example 5.3 is armature-controlled. Motor control can also be obtained by varying the voltage applied to the field windings and keeping the armature current constant (see Figure P5.6). In this case the motor torque is proportional to the field current, so that $T = K_m i_f$, where $K_m$ is the motor constant.

Develop the transfer function $\Omega(s)/E_f(s)$ for the field-controlled dc motor. Explain why no back emf occurs.

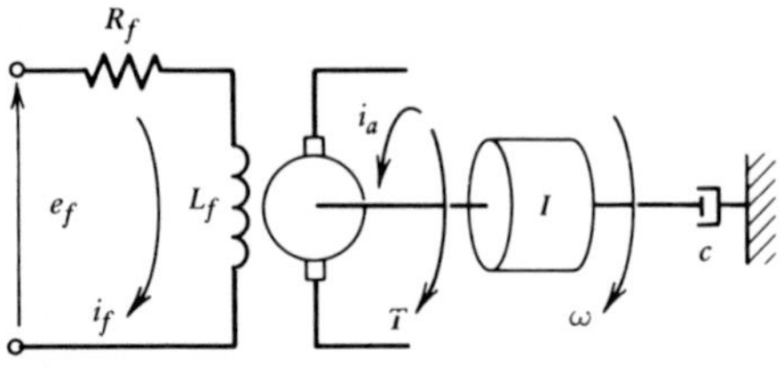


Figure P5.6

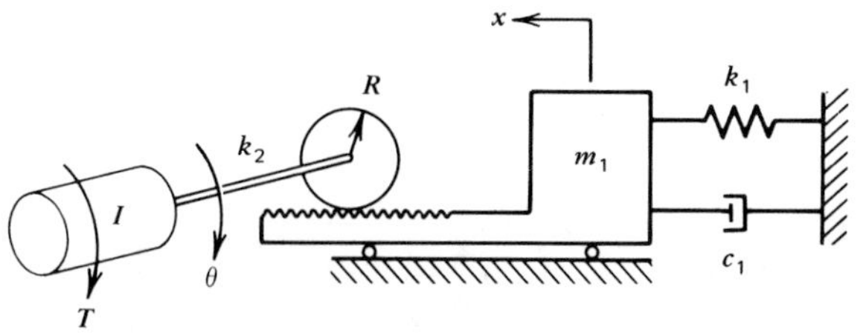


**Figure P5.7**

5.7 The system shown in Figure P5.7 might represent a machine tool slide driven through a rack and pinion by a motor, or an electrohydraulic actuator in which the mass $m_1$ represents the spool valve (see Problem 5.8).

The gear shaft is supported by frictionless bearings and thus can only rotate. Assume that the only significant masses in the system are the motor with inertia $I$ and the slide with mass $m_1$. Obtain the system's differential equations with the motor torque $T$ as the input.

5.8 The system of the previous problem is now used to move the spool valve of a hydraulic servomotor, as shown in Figure P5.8. Develop a transfer function model of the system with the load displacement $y$ as the output and the electric motor torque $T$ as input. Use the underlapped valve model developed in Section 5.1 and the model developed in Problem 5.7 for the following special cases: (a) $I = 0$, and (b) $I = m_1 = 0$.

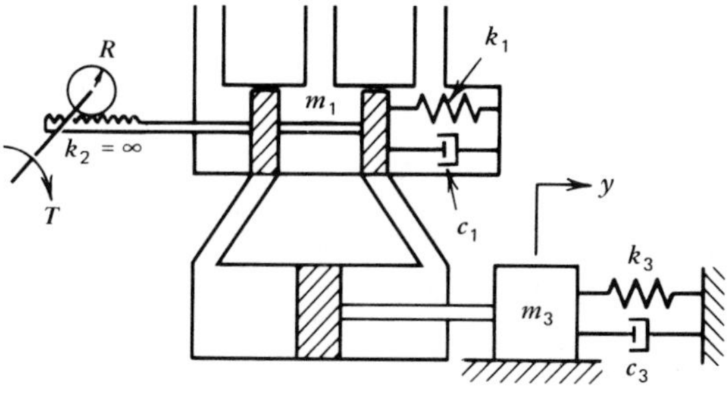


**Figure P5.8**

5.9 Figure P5.9 is a diagram of a speed-control system in which the dc motor voltage $e_1$ is generated by a dc generator driven by a prime mover. This system has been used on locomotives where the prime mover is a diesel engine that operates most efficiently at one speed. The efficiency of the electric motor is not as sensitive to speed and thus can be used to drive the locomotive. The motor voltage $e_1$ is varied by changing the generator input voltage $e$. The voltage $e_1$ is related to the generator field current $i_f$ by $e_1 = K_1 i_f$.

Draw the system's block diagram and find the transfer function relating the speed $\omega$ to the voltage $e$.

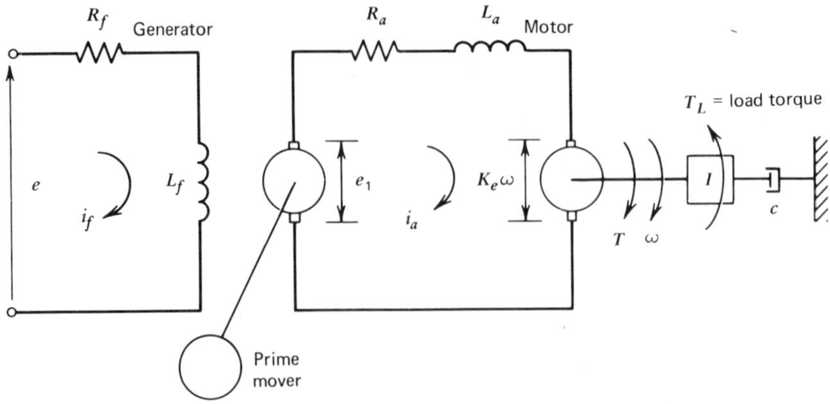


Figure P5.9

**5.10** For the tank shown in Figure P5.10, brine containing $w$ lb of salt/ft$^3$ flows in at a rate of $p$ ft$^3$/min. The resulting mixture flows out at a rate of $q$ ft$^3$/min. Assume that the tank mixture is kept uniform by stirring. Thus the salt density in the outflow is the same as that in the tank; namely, $v$ lb/ft$^3$. Let $y$ be the total amount of salt in the tank, in pounds. Neglect the volume of the outflow pipe. The input variables $w, p$ and the parameters $R, C$ are given.

(a) Find differential equations for $y$ and $h$ in terms of the given quantities.

(b) Suppose that the input flow rate $p$ has been held at the constant value $p_o$ long enough for the height to come to equilibrium at $h_o$. At a later time, say $t = 0$, the input brine density $w$ is set to zero, but the flow rate is maintained at $p_o$. Find the resulting salt density $v$ as a function of time, in terms of its value $v(0)$ at $t = 0$.

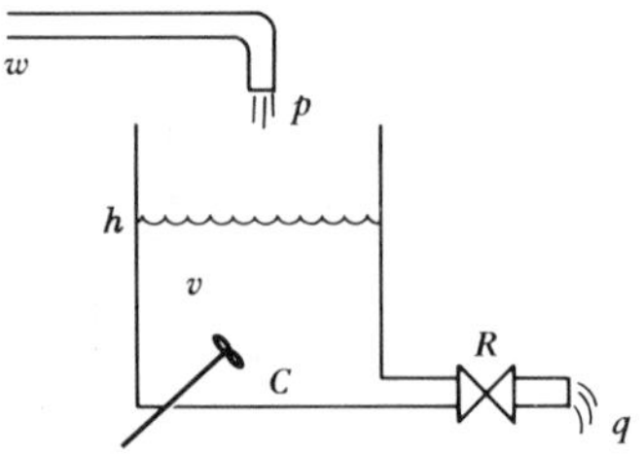


Figure P5.10

**5.11** For the tank shown in Figure P5.11, liquid flows in at a temperature $T_i$ and the volume flow rate $q_i$. Assume the outlet flow is laminar. Let $T$ be the temperature of the liquid in the tank and outflow. Assume that the only way heat is lost from the tank liquid is by the outflow. The liquid's mass density is $\rho$ and its specific heat is $c$.

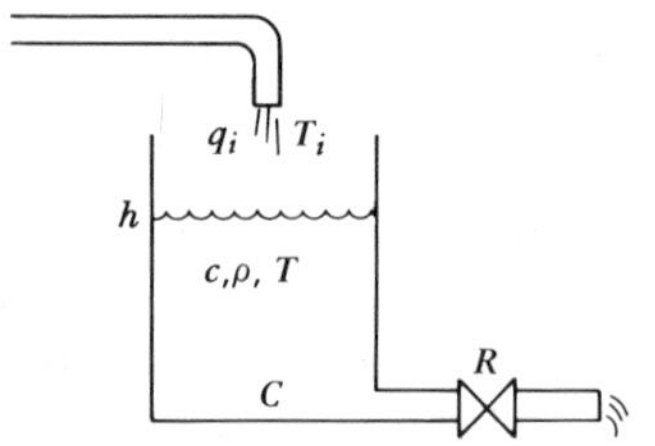


Figure P5.11

Derive the differential equations for the height $h$ and the temperature $T$, with $q_i$ and $T_i$ as the input variables.

**5.12** Given the following model

$$\dot{x}_1 = -5x_1 + 3x_2 + 2u_1$$

$$\dot{x}_2 = -4x_2 + 6u_2$$

and the output equations

$$y_1 = x_1 + 3x_2 + 2u_1$$

$$y_2 = x_2$$

(a) Draw the state diagram.

(b) Find the characteristic roots.

**5.13** With the state diagram shown in Figure P5.13, find the following.

(a) The state and output equations.

(b) The transfer function matrix by diagram reduction and by formula (5.2-28).

(c) The reduced model forms for each output.

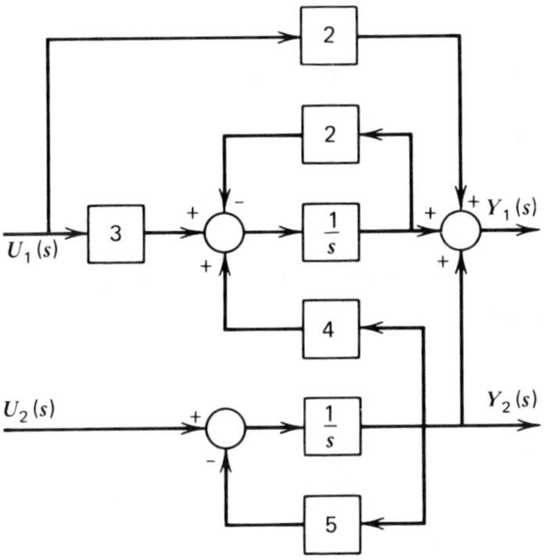


**Figure P5.13**

**5.14** Find the characteristic roots for the following model.

$$\dot{x}_1 = x_2$$

$$\dot{x}_2 = x_3$$

$$\dot{x}_3 = x_2 - 2x_3 + u$$

**5.15** Given the following transfer functions, find a state variable model for each.

(a) $T(s) = \dfrac{s+2}{s^2+4s+3}$

(b) $T(s) = \dfrac{1}{s^3+6s^2+4s+6}$

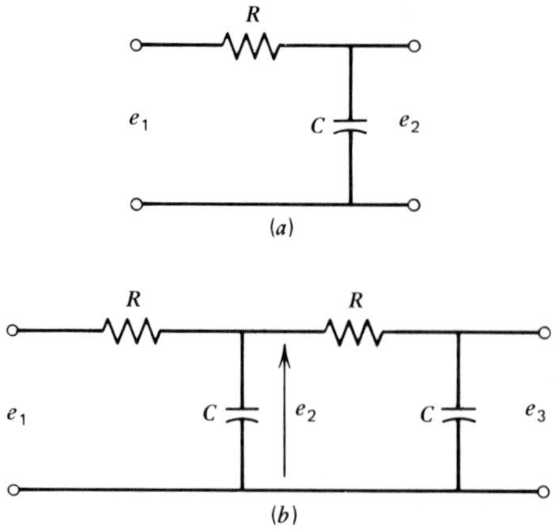


**Figure P5.16**

**5.16** The transfer function of the series $RC$ circuit shown in Figure P5.16 is

$$\frac{E_2(s)}{E_1(s)} = \frac{1}{RCs + 1}$$

Is the transfer function of the network shown in Figure P5.16*b* equal to

$$\frac{E_3(s)}{E_1(s)} = \frac{E_3(s)E_2(s)}{E_2(s)E_1(s)} = \left(\frac{1}{RCs + 1}\right)^2$$

Explain.

**5.17** Find the free response and the step response (for zero initial conditions) of the following models, by direct substitution and by Laplace transform. In part a the oscillatory component should be in the form of a sine function with a phase shift.

(a) $\ddot{x} + 4\dot{x} + 8x = 2, \quad x(0) = 0, \ \dot{x}(0) = 1$

(b) $\ddot{x} + 8\dot{x} + 12x = 2, \ x(0) = 0, \ \dot{x}(0) = 1$

**5.18** Find the ramp response of the following model by direct substitution and by Laplace transform.

$$\ddot{x} + 8\dot{x} + 12x = 6t, \quad x(0) = 1, \quad \dot{x}(0) = 0$$

**5.19** Determine $\zeta$, $\omega_n$, $\tau$, and $\omega_d$ for the dominant root for each of the following sets of characteristic roots.

(a) $s = -2, \ -3 \pm i$

(b) $s = -3, \ -2 \pm 2i$

**5.20** Given the system matrix

$$\mathbf{A} = \begin{bmatrix} r & 1 \\ -5 & 2 \end{bmatrix}$$

(a) Find the values of the parameter $r$ for which the system is

(1) Stable.

(2) Neutrally stable.

(3) Unstable.

(b) For the stable case, for what values of $r$ is the system

(1) Underdamped?

(2) Overdamped?

**5.21** Use the Routh-Hurwitz criterion (Appendix F) to determine the number of roots that have positive real parts.

(a) $s^4 + 2s^3 + 6s^2 + 8s + 8 = 0$

(b) $s^5 + 2s^4 + 2s^3 + 4s^2 + s + 1 = 0$

**5.22** Use the Routh-Hurwitz criterion to determine the number of roots that have positive real parts, and find the root locations by means of the auxiliary equation.

(a) $s^3 + 2s^2 + 5s + 24 = 0$

(b) $s^3 + 2s^2 + 4s + 8 = 0$

(c) $s^4 + 5s^3 + 7s + 5s + 6 = 0$

**5.23** Use the Routh-Hurwitz criterion to determine the allowable range of $K$ so that the systems with the following characteristic equations will be stable.

(a) $s^3 + as^2 + Ks + b = 0$ ($a, b$ known constants)

(b) $s^3 + as^2 + bs + K = 0$ ($a, b$ known constants)

(c) $s(s + 2)(s^2 + s + 1) + K = 0$

(d) $s(2s + 1)(3s + 1) + K(s + 1) = 0$

(e) $s^3 + 3s^2 + 3s + 1 + K = 0$

(f) $s^4 + 6s^3 + 11s^2 + 6s + K = 0$

**5.24** A certain system has the characteristic equation

$$s(Ts + 1) + K = 0$$

It is desired that all the roots lie to the left of the line $s = -b$ to guarantee a time constant no larger than $\tau = 1/b$. Use the Routh-Hurwitz criterion to determine the value of $K$ and $T$ so that both roots meet this requirement.

**5.25** Determine the range of $K$ values that will give a time constant of less than $\frac{1}{2}$ for the following characteristic equation.

$$s^3 + 10s^2 + 31s + K = 0$$

**5.26** Write a computer or calculator program to obtain the exact roots for the following.

(a) Quadratic polynomial.

(b) Cubic polynomial (see a mathematical handbook for the solution formulas).

5.27 Given the system matrix

$$\mathbf{A} = \begin{bmatrix} 0 & 1 \\ -k & -c \end{bmatrix}$$

where $k > 0$ and $c > 0$, use the Laplace transform method to compute the entries of the transition matrix for the following

(a) The underdamped case.

(b) The overdamped case.

5.28 Given the system matrix

$$\mathbf{A} = \begin{bmatrix} -\alpha & \beta \\ -\beta & -\alpha \end{bmatrix}$$

where $\beta > 0$, use the Laplace transform method to compute the entries of the transition matrix.

5.29 Compute the step response of the model

$$m\ddot{x} + c\dot{x} + kx = u(t)$$

assuming that the characteristic roots are repeated: $s = -a, -a$.

5.30 Consider the system

$$m\ddot{x} + c\dot{x} + kx = b\dot{u} + u, \quad x(0) = \dot{x}(0) = 0$$

where $u(t)$ is a unit step function. Use the initial value theorem to determine the values of $x(0+)$ and $\dot{x}(0+)$ in the following two cases. Explain the results.

(a) $b = 0$

(b) $b = 1$

5.31 Find the response of the following models for a unit step input (all initial conditions are zero).

(a) $\ddot{x} + 7\dot{x} + 12x = \dot{u} + 6u$

(b) $\ddot{x} + 6\dot{x} + 9 = \dot{u} + 9u$

(c) $\ddot{x} + 2\dot{x} + 5x = \dot{u} + u$

(d) $\dddot{x} + 6\ddot{x} + 11\dot{x} + 6x = u$ [*Hint.* $s^3 + 6s^2 + 11s + 6 = (s + 1)(s + 2)(s + 3)$].

(e) $\ddot{x} + 3\dot{x} + 2x = \dot{u}$

5.32 Derive (5.8-6).

5.33 Plot the frequency response curves for the following.

(a) $$T(s) = \frac{5}{(10s + 1)(4s + 1)}$$

(b) $$T(s) = \frac{4}{s^2 + 10s + 100}$$

(c) $$T(s) = \frac{4}{s(s^2 + 10s + 100)}$$

(d) $$T(s) = \frac{10}{s^2(s + 1)}$$

(e) $$T(s) = \frac{s}{(2s + 1)(5s + 1)}$$

(f) $$T(s) = \frac{s^2}{(2s + 1)(5s + 1)}$$

5.34 For the plots found in Problem 5.33, find the resonance frequency and bandwidth, if applicable.

5.35 (a) Write three computer or calculator subroutines that can compute the magnitude $m$ (db) and phase angle $\phi$ for the following factors, with the frequency $\omega$ and the appropriate parameters $\tau$, $\zeta$, $\omega_n$ as inputs to the subroutines.

(1) $s$

(2) $\tau s + 1$

(3) $\left(\frac{s}{\omega_n}\right)^2 + \frac{2\zeta}{\omega_n} s + 1, \quad \zeta < 1$

(b) Write a program using the subroutines in part a to compute $m$ and $\phi$ over a range of frequencies for a general transfer function composed of any number of the three types of factors.

5.36 A mass-spring-damper system is described by the model

$$m\ddot{x} + c\dot{x} + kx = f(t)$$

*Where*

$m$ = 0.05 slugs

$c$ = 0.4 lb/ft/sec

$k$ = 5 lb/ft

$f(t)$ = externally applied force (lb), shown in Figure P5.36.

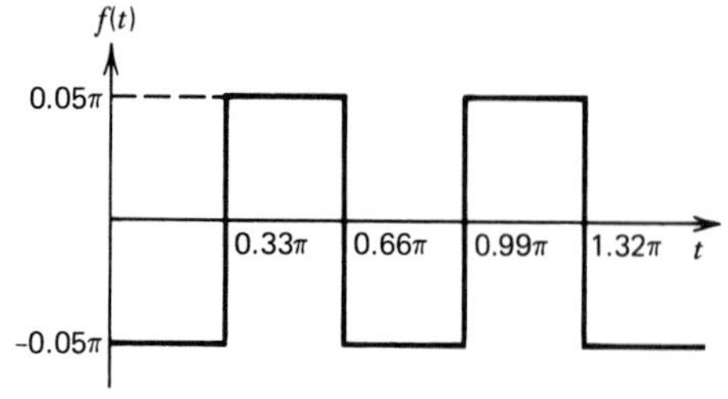


**Figure P5.36**

The forcing function can be expanded in a Fourier series as follows (see Appendix A).

$$f(t) = -0.2\left(\sin 3t + \tfrac{1}{3}\sin 9t + \tfrac{1}{5}\sin 15t + \tfrac{1}{7}\sin 21t + \ldots + \frac{1}{n}\sin 3nt + \ldots\right) \qquad n \text{ odd}$$

(a) Find the bandwidth of this system.

(b) Find the steady-state response $x(t)$ by considering only those components of the $f(t)$ expansion that lie within the bandwidth of the system.

**5.37** Figure P5.37 shows an application of a *dynamic vibration absorber.* The absorber consists of a mass $m_2$ and elastic element $k_2$ that are connected to the main mass $m_1$. If the forcing function $f$ is sinusoidal, proper selection of $m_2$ and $k_2$ can reduce the amplitude of motion of $m_1$ to zero. The technique is very effective and has been widely used.

(a) Derive the frequency transfer function for the displacement $x_1$ and $x_2$ with the force $f$ as the input. The expressions are simplified if we let $r_1 = \omega/\omega_{n1}$, $r_2 = \omega/\omega_{n2}$, $b = \omega_{n2}/\omega_{n1}$, and $\mu = m_2/m_1$, where

$$\omega_{n1} = \sqrt{k_1/m_1}, \; \omega_{n2} = \sqrt{k_2/m_2}$$

(b) What value of $r_2$ will make $|X_1(i\omega)| = 0$? This result gives the design condition for the absorber.

(c) For the value of $r_2$ found in part b, what is $|X_2(i\omega)|$? Assume that the forcing function is $f(t) = F_0 \sin \omega t$.

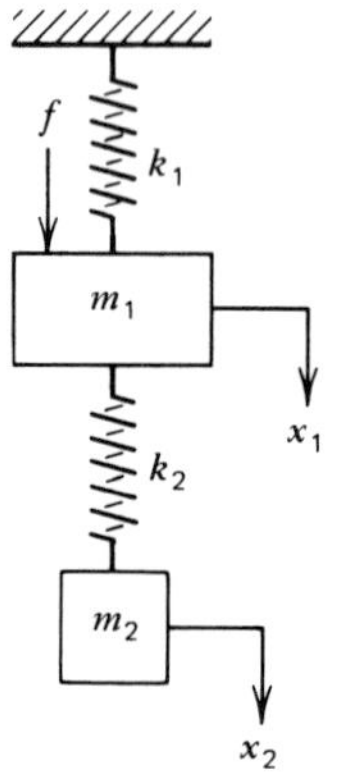


**Figure P5.37**

**5.38** (a) A certain machine with supports has an experimentally determined natural frequency of 3.43 Hz. It will be subjected to a rotating unbalance force with an amplitude of 3 lb and a frequency of 3 Hz. Design a vibration absorber for this machine. The available clearance for the absorber's motion is 1 in.

(b) What will happen if the forcing frequency is much different from the design value, say due to a startup process?

5.39 Sketch the phase plane plots for the following models.

(a) The Gompertz growth law (see Problem 3.21).

(b) The Pella-Tomlinson growth law with $n = 2$ (see Problem 3.23).

(c) A model of liquid height in a tank with an orifice and no inflow.

$$\dot{h} = -\sqrt{h}$$

(d) $\ddot{x} + (\dot{x})^2 = 0$

(e) $\ddot{x} + x = 0$

5.40 Obtain a linearized set of equations for the model developed in part a of Problem 5.10. Assume the nominal values of $y, h, w$, and $p$ are constants.

5.41 Models of the Lotka-Volterra type have been often used to model the dynamics of two species, one a prey (or host) and the other a predator (or parasite). One specific form is

$$\frac{dH}{dt} = (a_1 - c_1 H - b_1 P)H$$

$$\frac{dP}{dt} = (-a_2 + b_2 H)P$$

where the constants are positive or zero. The classic Lotka-Volterra model results when $c_1 = 0$. The host population density is $H$; that of the parasite is $P$.

(a) Discuss the significance of each term in the model.

(b) Find the positive equilibrium solution and obtain the linearized model.

(c) Evaluate the stability of the equilibrium for

(1) $c_1 \neq 0$

(2) $c_1 = 0$

5.42 An alternate host-parasite model is that of Leslie and Gower.

$$\frac{dH}{dt} = (a_1 - c_1 P)H$$

$$\frac{dP}{dt} = \left(a_2 - c_2 \frac{P}{H}\right)P$$

(a) Repeat parts a and b of Problem 5.41.

(b) Evaluate the stability. Will oscillations occur?

(c) What is the effect of $c_1$ and $c_2$ on the linearized dynamics?

5.43 Some relay controllers produce a force proportional to the input displacement, up to a certain point at which they saturate and produce a constant force. This

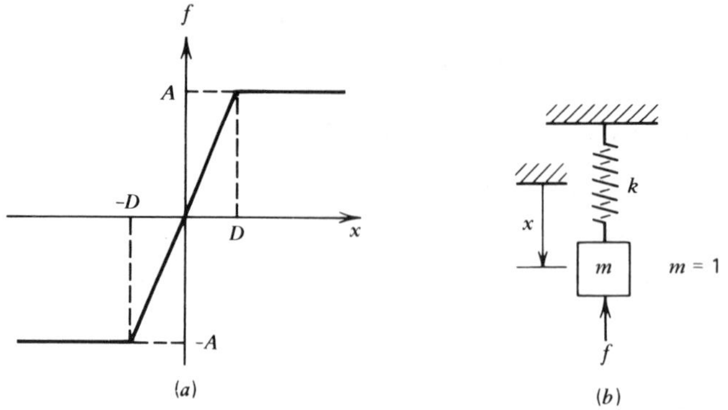


Figure P5.43

is shown in Figure P5.43 where $f$ is the relay force, $x$ the displacement, $A$ the saturation level of $f$, and $D$ the displacement at which saturation occurs. Suppose that such a relay is used to regulate the displacement of an undamped oscillatory system, as shown.

(a) Obtain the phase plane equations $x_2$ versus $x_1$, where $x_1$ is the displacement and $x_2$ the velocity of the mass.

(b) Assume that the initial displacement $x_1(0)$ is zero. Find the smallest initial velocity $x_2(0)$ that will result in saturation.

(c) With $x_1(0) = 0$ and $x_2(0)$ large enough to cause saturation, find the time it takes for the relay to saturate the first time.

5.44 Van der Pol's equation is a nonlinear model for some oscillatory processes. It is

$$\ddot{y} - b(1 - y^2)\dot{y} + ay = 0$$

(a) Put the model in state variable form and linearize it.

(b) Let $a = 1$ and use a numerical technique to find $y(t)$ for

(1) $b = 0.1, \quad y(0) = \dot{y}(0) = 1, \quad 0 \leqslant t \leqslant 25$

(2) $b = 0.1, \quad y(0) = \dot{y}(0) = 3, \quad 0 \leqslant t \leqslant 25$

(3) $b = 3, \quad y(0) = \dot{y}(0) = 1, \quad 0 \leqslant t \leqslant 25$

(c) Compare the results of part b with the linearized behavior.

5.45 A bandpass filter centered on the frequency $\omega_c$ can be obtained by cascading two first-order filters with roots at $z = r\exp(\pm i\omega_c T)$, where $r$ is a parameter. The filter transfer function is

$$T(z) = \frac{bz^2}{(z - re^{ia})(z - re^{-ia})}$$

where $a = \omega_c T$.

(a) What restrictions must be placed on $r$ for the filter to be stable?

(b) What value must the parameter $b$ have in order for the filter gain to be unity at $\omega = \omega_c$?

**5.46** Consider the system shown in Figure P5.46. Find the transfer function $C(z)/R(z)$ and determine the range of values of $K$ for which the system will be stable for the following cases.

(a) $G_1(s) = K, \quad G_2(s) = \dfrac{1}{s(s+4)}$ for $T = 0.25$ and $T = 0.025$.

(b) $G_1(s) = \dfrac{K}{s+1}, \quad G_2(s) = \dfrac{1}{s+2}$ for $T = 0.1$

(c) $G_1(s) = K, \quad G_2(s) = \dfrac{1}{s^2 + 2s + 5}$ for $T = 0.1$

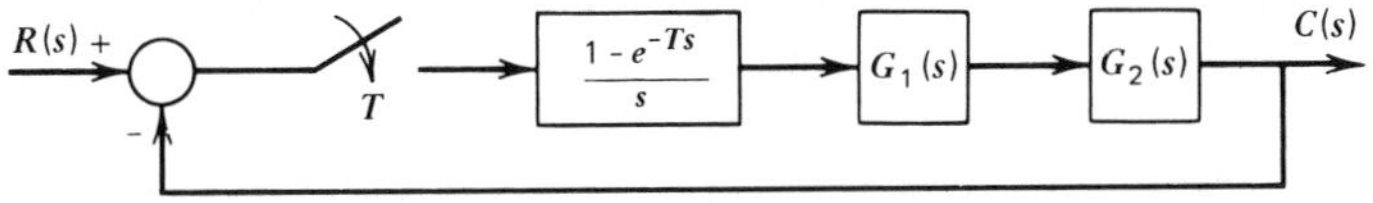


**Figure P5.46**

**5.47** Consider the armature-controlled dc motor treated in Example 5.18.

(a) For the parameter values given in the example, find the pulse transfer function between the motor speed $\omega$ and the motor voltage $v$. Assume that the voltage $v$ is applied through a sample and zero-order hold.

(b) The characteristic roots of the analog system are $s = -102.7 \pm i122.8$. Let the sampling period be $T = 0.001$ sec. Find the corresponding characteristic $z$ roots and compare.

(c) Let $T = 0.0001$ sec. Repeat part b.

**5.48** Put the transfer function models found in Problems 5.46 and 5.47 into state variable form.

**5.49** Given the transfer function

$$\frac{C(z)}{R(z)} = T(z) = \frac{z+3}{2z(z^2-4)}$$

obtain a state variable model.

**5.50** For the following transfer functions, $T(z) = C(z)/R(z)$. Find the free response. The initial conditions are $c(0) = 1, c(1) = 2$.

(a) $T(z) = \dfrac{1}{(z-0.5)(z+0.3)}$

(b) $T(z) = \dfrac{1}{(z-0.5)^2(z-0.1)}$

(c) $T(z) = \dfrac{z+0.2}{z^2+0.2z+0.26}$

(d) $T(z) = \dfrac{1}{z^2+0.2z+0.26}$

**5.51** Develop a procedure to handle complex roots for discrete-time systems, using a factor similar to $R(-a+ib)$ in (5.7-28).

**5.52** Use the $z$ transform to compute the unit step response for the systems whose transfer functions are given in Problem 5.50, parts a, c, and d. Assume zero initial conditions.

**5.53** Use the $z$ transform to compute the unit ramp response for the systems whose transfer functions are given in Problem 5.50, parts a and c.

**5.54** Consider the displacement-velocity filter, (5.11-3) and (5.11-4), for the situation in which the velocity $v(t) = 10$ ft/sec, and the initial displacement is $y(0) = 10$ ft. That is, $y(t) = 10 + 10t$. Assume that the measurements are perfect and that the filter's estimates are initialized to zero. Assume that the sampling period is $T = 0.1$ sec.

(a) Suppose that the filter's behavior in the continuous-time domain is specified to be critically damped $\zeta = 1$. Find the requirements on the filter's coefficient $a$ in terms of $b$ in order to achieve this. The filter must be stable.

(b) Compute the response of the filter in terms of $b$, given the previously specified inputs $y(t)$ and $v(t)$. Discuss the performance of the filter and the effect of $b$ on this performance.

**5.55** Plot the locations in the $z$ plane that correspond to the following $s$-plane locations. Assume that $T = 0.01$.

(a) $s = -a \pm i3, \quad 0 \leqslant a \leqslant 2$

(b) $s = -a \pm i, \quad 0 \leqslant a \leqslant 2$

(c) $s = -2 \pm ib, \quad 0 \leqslant b \leqslant 2$

(d) $s = -3 \pm ib, \quad 0 \leqslant b \leqslant 2$

(e) $\zeta = 0.707, \quad 0.1 \leqslant \tau \leqslant 1$

(f) $\omega_n = 2, \quad 0.5 \leqslant \zeta \leqslant 1$

(g) $\omega_n = 1, \quad 0.5 \leqslant \zeta \leqslant 1$

5.56 The discrete-time version of the Lotka-Volterra model (see Problem 5.41) is

$$H(k+1) = H(k)\exp\left[r_1 - c_1 H(k) - b_1 P(k)\right]$$

$$P(k+1) = P(k)\exp\left[-r_2 + b_2 H(k)\right]$$

(a) Find the equilibrium solution and linearize the model.

(b) Let $c_1 = 0$ and analyze the stability of the equilibrium. Compare the results with those for the continuous-time case.

# CHAPTER SIX

# Feedback-Control Systems

We now come to the third and final topic of this work: the control of dynamic systems. The modeling and analysis techniques developed in the previous chapters can be readily applied to the design of control systems. Our emphasis on linear and linearized models is justified by the fact that feedback improves linearity and acts to keep the linearization accurate by preventing large deviations from the reference operating condition.

We begin our study of feedback control with an introduction to its history, some examples, and its terminology (Sections 6.1 and 6.2). Sensors are required to provide the measurements necessary for feedback, and actuators are needed to supply the control forces. We present an overview of the devices available for these purposes (Sections 6.3 and 6.4). The heart of the controller is the element used to implement the control logic, called the *control law*, and in Sections 6.5, 6.6, and 6.7 we introduce and analyze the most common laws in use today. Finally, we discuss the design of devices used to implement the control law physically. These can be categorized as analog electronic (Section 6.8), pneumatic (Section 6.9), hydraulic (Section 6.10), and digital (Sections 6.11 and 6.12). In the latter case we develop additional algorithms that take advantage of the features of digital computers used as controllers.

## 6.1 FEEDBACK CONTROL: CONCEPTS, HISTORY AND APPLICATIONS

Some of the basic concepts of feedback and its uses were introduced in Chapter One as applied to static systems. It was shown that feedback can improve a system's linearity and sensitivity to parameter variations. The associated cost of this improvement is a decrease in amplification or gain. In this and the remaining chapters we demonstrate that these properties of feedback also apply to the dynamic case, and we show how to use these properties to design systems for controlling variables commonly found in machines and processes.

A control system may be defined as a system whose purpose is to regulate or adjust the flow of energy in some desired manner. A *closed-loop* or *feedback-control system* uses measurements of the output to modify the system's actions in order to achieve the intended goal. An *open-loop* control system does not use feedback, but adjusts the flow of energy according to a prescribed schedule. Examples of open-loop systems are cam-driven elements and timers. The term *automatic control system* is sometimes used to distinguish a system that acts without outside intervention from a manually controlled system that requires regular supervision by an operator. Thus an automatic control system can be either open or closed loop.

Open-loop controllers are relatively inexpensive but are limited to situations where events are quite predictable. For example, a timer can be used to switch a building's lights on at night and off in the morning. However, if a particular day is very dark the system will not give adequate performance. If a light-level sensor is added to the controller, it becomes a closed-loop system. The challenging design problems lie in the area of feedback controllers, to which we will henceforth limit our attention.

In addition to improvement of linearity and parameter sensitivity, feedback in a control system enables the controller to respond to changes in commands and to disturbances acting on the system. Perhaps the most familiar control system is the thermostat for controlling the temperature inside a house. The command is the desired temperature setting. The thermostat measures the actual temperature, compares it with the command temperature, and turns the furnace on or off depending on the difference between the actual and command temperatures. A disturbance to the system is anything that changes the actual inside temperature, other than the heating system. A change in outside temperature or the opening of doors are examples of disturbances. By sensing the change in inside temperature, the thermostat takes action to keep the temperature near its desired value.

It is helpful to consider some simple historical examples of control systems at this point.

## History of Feedback Control

Examples of feedback-control systems have been identified dating back to the third century B.C. Ktesibios, a Greek living in Alexandria, is credited with building a self-regulating flow device for use in water clocks (Reference 1). Such a clock uses a float indicator to mark the passage of time (Figure 6.1). Its accuracy is dependent on how constant is the inflow rate of water to the tank. This flow rate can change if the supply pressure changes. Ktesibios introduced a secondary tank with a float and valve tapered to match the opening of the supply line. As long as the water height in this secondary tank remains constant, the flow rate into the main water clock tank will not change. Suppose that the supply flow rate increases. This raises the level in the secondary tank,

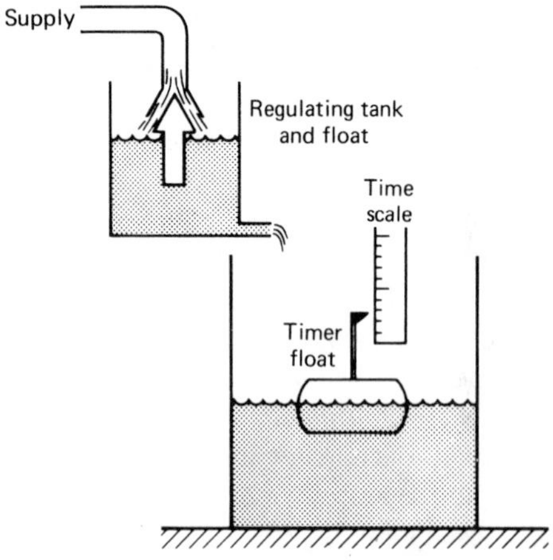


**Figure 6.1 Ktesibios' water clock.**

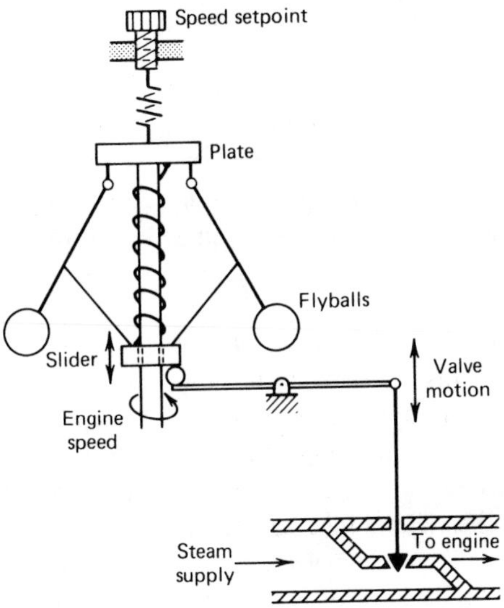


**Figure 6.2 James Watt's flyball governor for speed control of a steam engine.**

and the float tends to close off the supply line and decrease the flow. The level then drops and the supply flow increases. The sensitivity of the system to changes in supply pressure depends in part on the area of the secondary tank. Rapid fluctuations in supply pressure will have less effect on the water clock if this area is made larger.

Little progress was made in control systems during the next 2000 years. Several examples of such systems appear in windmill designs used by the Dutch in the fifteenth century. The windmills were made to pump water and grind grain more efficiently by the use of an auxiliary propeller at a right angle to the main blades. This was connected to a gear train that turned the main blades so they always faced into the wind.

The arrival of the machine age was accompanied by a large increase in the number of control system designs. The development of the steam engine led to the requirement for a speed-control device to maintain constant speed in the presence of changes in load torque or steam pressure. In 1788, James Watt of Glasgow developed his now famous flyball governor for this purpose (Figure 6.2). Watt took the principle of sensing speed with the centrifugal pendulum of Thomas Mead and used it in a feedback loop on a steam engine (Reference 2). As the motor speed increases, the flyballs move outward and pull the slider upward. The upward motion of the slider closes the steam valve, thus causing the engine to slow down. If the engine speed is too slow, the spring force overcomes that due to the flyballs, and the slider moves down to open the steam valve. The desired speed can be set by moving the plate to change the compression in the spring.

The principle of the flyball governor is still used for speed-control applications. Typically a hydraulic servomotor is connected to the slider to provide the high forces required to move large supply valves.

Analytical tools for control problems were first developed by J. C. Maxwell in 1868 for application to telescope position control, but a general understanding of the principle of feedback control was not developed until the twentieth century.

From 1900 to 1940, significant developments occurred in large-scale power generation, aeronautics, chemical industries, and electronics. The development of the vacuum-tube amplifier in particular resulted in many analytical techniques for the design of feedback systems. The names of Black, Bode, and Nyquist of the Bell Telephone Laboratories are responsible for many of these principles, especially those involving frequency response.

The advent of World War II gave additional impetus to the design of new types of control systems. The rapid increase in aircraft speeds made the manually controlled antiaircraft gun obsolete. The perfection of the radar-controlled gun was a major engineering advancement of the period because it represented the first application to a mechanical system of the mathematical methods developed for electronic amplifiers. One might say that control theory as a science first appeared at this time because a set of general principles were formulated that could be applied to any type of system.

Norbert Wiener, who helped develop the radar-controlled gun, contributed greatly to the emerging science with analytical techniques and by enlarging its purview to include applications other than to machines. He coined the term *cybernetics* to describe the study of control and communication in humans, animals, and machines (Reference 3). The term is derived from the Greek word for the person who steers a ship (the ship's controller). Although we have described a control system as one that regulates the flow of energy, the concepts can be applied to systems involving the flow of information, money, or other quantities. Control systems are found in abundance in biological and ecological systems. Also, feedback system analysis and design techniques have been applied to economic and social systems. References for these applications are found in Chapter Two.

Following World War II, system design techniques were consolidated primarily around transfer-function methods based on the Laplace transform. Significant advances in applications were made in the areas of aircraft and missile guidance systems and nuclear power.

By the late 1950s, the digital computer had begun to be applied to the design of control systems and to be used as a control element itself. Rediscovery of the state-space point of view, originally developed in classical mechanics, was made at this time. The 1960s were dominated by theoretical developments based on this approach. Initially, difficulty was encountered in applying digital computers as controllers, for many reasons, and the typical hardware during this decade consisted of pneumatic and hydraulic devices and operational amplifiers. The most significant applications were probably in the space program.

At the start of the 1970s it was realized that the so-called modern control theory based on the state-space approach could not entirely replace the "classical" theory based on frequency response and the Laplace transform. The modern control engineer should have a working knowledge of both approaches. In terms of hardware, the decade witnessed the widespread use of digital controllers, first as large-scale machines, then as minicomputers, and finally as microcomputers.

The 1980s should see the microcomputer continue to replace other devices as control elements, although there will always be a need for operational amplifiers, pneumatics, and hydraulics because of simplicity, reliability, and force requirements. Increasing demands for fuel economy will require improved control systems for aircraft and automobile engines, as well as other types of power plants. Classical control theory, with its emphasis on single-input, single-output systems, is not capable of

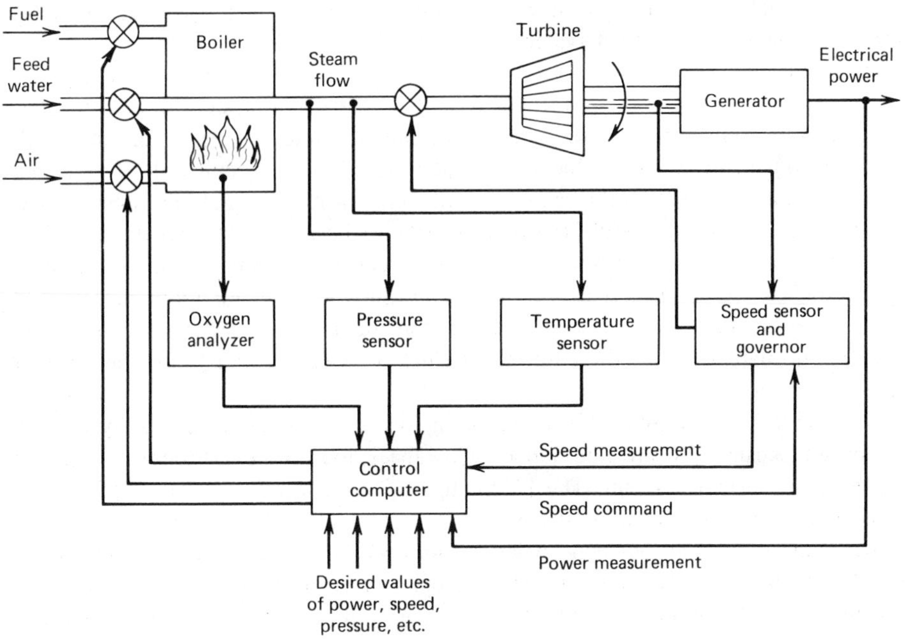


**Figure 6.3 Computer control system for a boiler-generator. Each important variable requires its own controller. The interaction between variables calls for coordinated control of all loops.**

dealing with some of these applications because several variables must be controlled, as shown in Figure 6.3.

In the future we can expect significant applications of automatic control systems in the space program because of the expense of sending people into space. The use of robots in this area as well as in industrial situations will become more widespread. Robotics has much to contribute in areas where the task is hazardous or in which human response time is not fast enough. High-speed transportation systems on land, sea, and in the air will require better control systems for improved passenger safety and comfort. Also, lighter-weight structures can be used if improved controllers are available to deal with the problems caused by the flexibility of such structures. With more stringent requirements for safety and efficiency in all areas, there will be much to do in both the theory and the practice of feedback-control systems.

## The Role of Modeling and Analysis

The modeling techniques of the previous chapters are required for the successful design of control systems. Transients due to disturbances and changes in the command input make it essential that a dynamic model be developed for the system to be controlled. The structure of the resulting control system will be heavily dependent on the form of this model. Thus the model should contain the most important behavior of the system, but no more than is necessary to achieve a satisfactory design. Quite often a first- or second-order model in terms of the system's dominant roots is satisfactory.

The analytical techniques developed thus far are extremely useful in predicting the performance of a control system. The quantifiers of transient performance – such as the time constant, damping ratio, and natural frequency – enable the designer to adjust the parameters of the controller to suit the particular system to be controlled. For example, consider the problem of balancing a cylindrical object on end in the palm of your hand. Such an object could be a pencil or a broom handle. The form of the mathematical model is the same for both, but the moments of inertia have different values. The broom handle is easier to balance because of its higher inertia. Our hand, eyes, and brain constitute the control system in this example, and the rapidity with which we respond to the object's motion must be faster for the pencil than for the broom handle. If we were to design a controller to replace the hand-eye-brain system, the parameters of the controller (which we will call *gains*) must be different for each object.

Analytical techniques for determining the stability of a control system are important because the introduction of a feedback loop around a stable system can result in an unstable system. The parameters of the controller must be selected to avoid this, and they depend on the system to be controlled. If we respond too quickly in balancing the pencil, our hand will overshoot too far and cause the pencil to topple too far in the opposite direction. The system will be unstable. On the other hand, the broom handle is more forgiving of fast (or slow) response because of its higher inertia. In general, we will see that instability is more of a problem in control systems with higher performance specifications.

The previous example shows that stability and speed of response can be conflicting objectives. In addition, control system accuracy can be another competing requirement. Simple control systems with loose performance specifications can sometimes be designed by trial and observation because of the self-correcting nature of feedback. However, the performance requirements and expense of modern systems can be quite high, and an analytical approach is usually necessary to design a control system with desirable accuracy, speed of response, and stability characteristics.

## 6.2 CONTROL SYSTEM STRUCTURE

The electromechanical position control system shown in Figure 6.4 illustrates the structure of a typical control system. A load with an inertia $I$ is to be positioned at some desired angle $\theta_r$. A dc motor is provided for this purpose. The system contains viscous damping, and a disturbance torque $T_d$ acts on the load, in addition to the motor torque $T$. The origin of the disturbance torque depends on the particular application. If the load to be positioned is a radar antenna, for example, wind gusts produce a torque whose magnitude and time of occurrence are unknown to some extent. If the motor is to control the position of a valve in a flow line (such as a damper in a forced-hot-air heating system), the disturbance torque would result from a change in fluid forces caused by a change in supply pressure in the line. In both examples, the effects of Coulomb friction could also be modeled as a disturbance, in the sense that the magnitude of such a torque is difficult to predict (coefficients of friction are not easily computed and are usually not constants).

Because of the disturbance the angular position $\theta$ of the load will not necessarily

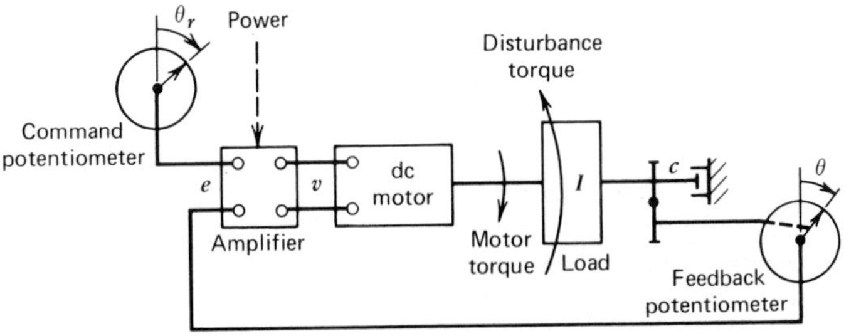


**Figure 6.4 Position-control system using a dc motor.**

equal the desired value $\theta_r$. For this reason a potentiometer is used to measure the displacement $\theta$. A potentiometer consists of a wire-wound resistor with a sliding electrical contact, or wiper. As shown in Figure 6.5, when the wiper rotates the resistance between the wiper contact and ground changes. This produces a voltage at the wiper that is a function of its angular displacement. If the resistance of the winding is uniform, and if the wiper circuit draws negligible current, the wiper voltage is proportional to the angular displacement, so that $e = K\theta$. A simple circuit analysis can be used to show that $K = V/2\pi$, where $V$ is the constant voltage supplied to the potentiometer. Potentiometers are also available for applications in which the displacement is a translation, not a rotation.

The potentiometer voltage representing the controlled position $\theta$ is compared to the voltage generated by the command potentiometer. This device enables the operator to dial in the desired angle $\theta_r$. The amplifier sees the difference $e$ between the two potentiometer voltages. The basic function of the amplifier is to increase the small error voltage $e$ up to the voltage level required by the motor and to supply enough current required by the motor to drive the load. In addition, the amplifier may shape the voltage signal in certain ways to improve the performance of the system. We will return to this aspect later.

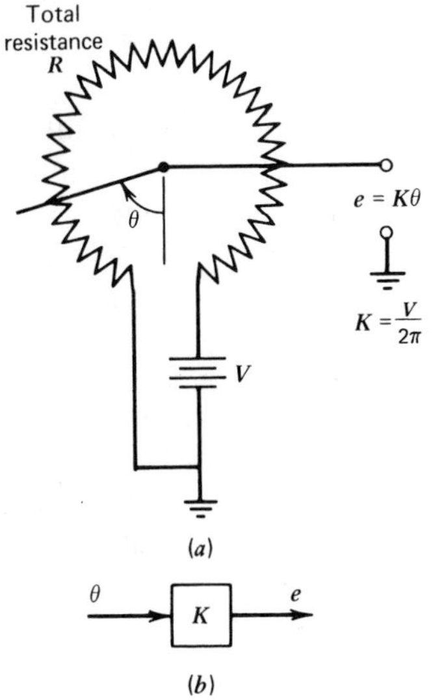


**Figure 6.5 Rotary potentiometer and its block diagram.**

The system operates as follows. If both potentiometers have the same constant $K$, the error voltage is

$$e = K(\theta_r - \theta)$$

When $\theta$ does not equal $\theta_r$, a nonzero voltage $e$ appears at the input terminals of the amplifier. Assume for now that the amplifier produces a voltage $v$ proportional to $e$, and with the same sign; that is, $v = K_a e$, where $K_a$ is the amplifier gain. Then if $\theta_r > \theta$, $e$ and $v$ are positive and the motor produces a torque $T$ that increases $\theta$. This continues until $e = 0$ and therefore $\theta = \theta_r$. Similar events occur if $\theta_r < \theta$. When the

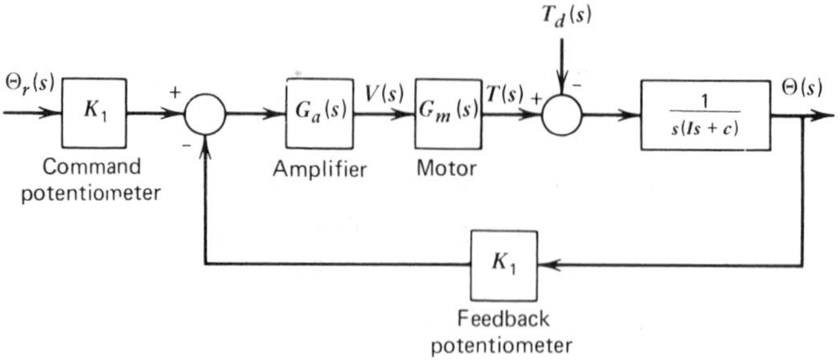


**Figure 6.6 Block diagram of the dc position-control system shown in Figure 6.4.**

action of a disturbance torque causes $\theta$ to deviate from $\theta_r$, the same process acts to restore $\theta$ to its desired value. Thus the control system is seen to provide two basic functions: (1) to respond to a command input that specifies a new desired value for the controlled variable, and (2) to keep the controlled variable near the desired value in spite of disturbances. The presence of the feedback loop, with which the measurement of the controlled variable is used to alter the operation of the motor, is seen to be vital to both functions.

A block diagram of this system will help to analyze its performance. This is shown in Figure 6.6, where the transfer functions for the amplifier and motor have been left in general form so that we can consider various models later. It is important to recall at this time some assumptions of block diagram representations. Cascaded elements must be "nonloading." For example, the motor cannot directly influence the voltage $v$ coming from the amplifier (except by way of the feedback loop). Thus the motor must have a high input impedance. Similarly, the potentiometer derivation assumes that the wiper current is negligible; that is, the amplifier must have a high input impedance. If these assumptions are not valid, the entire subsystem must be represented by a single transfer function derived from basic principles, rather than from diagram reduction. Refer to Chapter Three for more discussion of this point.

The power supplies required for the potentiometers and the amplifier are not shown in block diagrams of control system logic because they do not contribute to the control logic. However, their existence cannot be ignored. The motor's torque-current relation is $T = K_t i$. Therefore, as the load on the motor is increased, it calls for more current from the amplifier. If the available current is limited (which is true for all real systems), we must take this limitation into account before accepting any controller designs as final.

Some additional modeling approximations might also be contained in the diagram. We have taken the motor, damping and disturbance torques to be acting on the inertia $I$ of the load. This implies that the inertia of the motor has been lumped into that of the load. In reality, the damping torque, for example, might result from electromagnetic effects within the motor itself, and the motor torque acts directly on the rotor. But the approximation is usually a good one, especially when the shaft connecting the motor and load is stiff.

## A Standard Diagram

The electromechanical positioning system fits the general structure of a control system

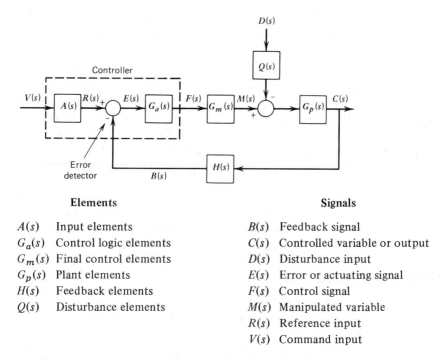

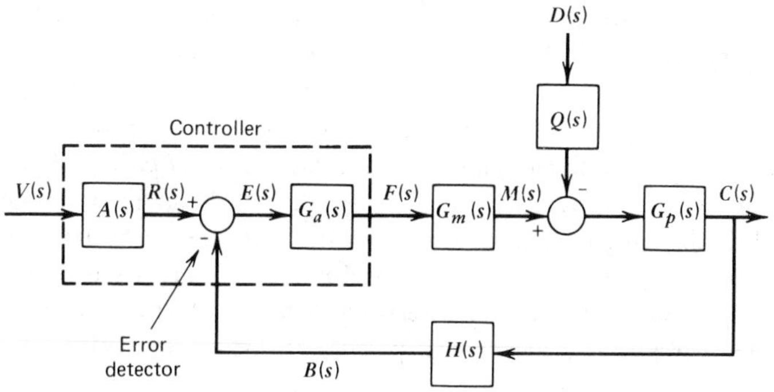


| Elements | | Signals | |
|---|---|---|---|
| $A(s)$ | Input elements | $B(s)$ | Feedback signal |
| $G_a(s)$ | Control logic elements | $C(s)$ | Controlled variable or output |
| $G_m(s)$ | Final control elements | $D(s)$ | Disturbance input |
| $G_p(s)$ | Plant elements | $E(s)$ | Error or actuating signal |
| $H(s)$ | Feedback elements | $F(s)$ | Control signal |
| $Q(s)$ | Disturbance elements | $M(s)$ | Manipulated variable |
| | | $R(s)$ | Reference input |
| | | $V(s)$ | Command input |

**Figure 6.7 Terminology and basic structure of a feedback-control system.**

(Figure 6.7). This figure also gives some standard terminology. Not all systems can be forced into this format, but it serves as a reference for discussion.

The controller is generally thought of as a logic element that compares the command signal with the measurement of the output and decides what should be done. The input and feedback elements are transducers for converting one type of signal into another type. The potentiometer converts displacement into voltage. This allows the error detector directly to compare two signals of the same type (e.g., two voltages). Not all functions show up as separate physical elements. The error detector in Figure 6.4 is simply the input terminals of the amplifier. The control logic elements produce the control signal, which is sent to the *final control elements*. These are the devices that develop enough torque, pressure, heat, etc., to influence the elements under control. Thus the final control elements are the "muscle" of the system, while the control logic elements are the "brain." In this chapter we are primarily concerned with the design of the logic to be used by this brain.

The object to be controlled is the *plant*. The *manipulated variable* is generated by the final control elements for this purpose. The disturbance input also acts on the plant. This is an input over which the designer has no control, and perhaps for which little information is available as to the magnitude, functional form, or time of occurrence. The disturbance can be a random input, such as wind gusts on a radar antenna, or deterministic, such as Coulomb friction effects. In the latter case, we can include the friction force in the system model by using a nominal value for the coefficient of friction. The disturbance input would then be the deviation of the friction force from this estimated value and would represent the uncertainty in our estimate.

Several control system classifications can be made with reference to Figure 6.7. A *regulator* is a control system in which the controlled variable is to be kept constant in spite of disturbances. An example is the temperature-control system for a house. Once the desired temperature is set, say at 68°F, the controller is to keep the room temperature near this value. The command input for a regulator is its *set point*. On the other hand, a *follow-up system* is supposed to keep the controlled variable near the command value, which is changing with time. An example of a follow-up system is a machine tool in which a cutting head must trace a specific path in order to shape the product properly. This is also an example of a *servomechanism*, which is a control system whose controlled variable is a mechanical position, velocity, or acceleration. The thermostat system is not a servomechanism, but is a *process-control system*, where the controlled variable describes a thermodynamic process. Typically such variables are temperature, pressure, flow rate, liquid level, chemical concentration, etc.

The following examples illustrate these concepts and provide practice in setting up the description of the control system.

## A Temperature Controller

The oven described in Example 5.2 is now considered as a candidate for control. If we neglect the thermal capacitance of the heater, $C_1 = 0$, and the model is

$$C_2 \frac{dT_2}{dt} = q - \frac{1}{R_2}(T_2 - T_o) \tag{6.2-1}$$

A control system for this process is shown in Figure 6.8. A power amplifier is used to supply current to the resistance-type heater. The input to the amplifier consists of the voltage from the command potentiometer, which represents the desired value of the temperature $T_2$ and the voltage from the temperature sensor. These voltages are $K_1 T_r$ and $K_2 T_2$, respectively.

Suppose the ambient temperature $T_o$ is reasonably constant at 70°F. In order for the oven temperature $T_2$ to be held constant at some desired value, say 200°F, the heat input rate from the heater must equal that lost through the resistance $R_2$; that is, for $T_2 = T_r = 200$°F, the steady-state heater output must be

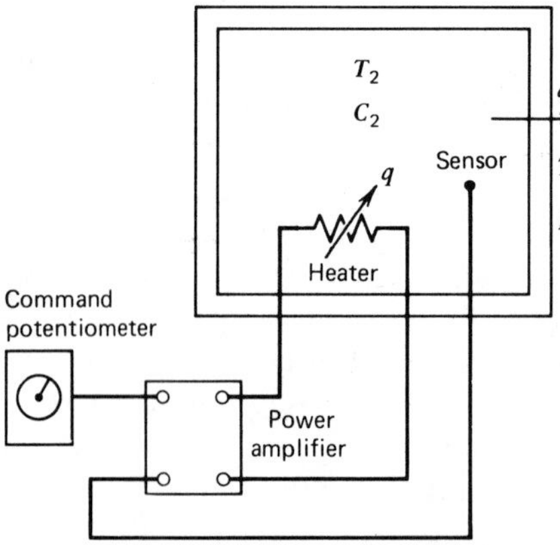


**Figure 6.8 Electrical temperature-control system.**

$$q_{ss} = \frac{1}{R_2}(200^\circ - 70^\circ) = \frac{130^\circ}{R_2}$$

The amplifier circuit must be constructed so that the potentiometer setting will produce this heat flow when $T_r = T_2 = 200°\text{F}$.

The second function of the amplifier circuit is to compare the voltage from the potentiometer with that from the temperature sensor and to modify the heater output by an amount $m$. We can thus express the heater output as

$$q = q_{ss} + m \tag{6.2-2}$$

If the rate $m$ can be made proportional to the difference between the potentiometer voltage and the sensor voltage, then

$$m = K_a e \tag{6.2-3}$$

$$e = K_1 T_r - K_2 T_2 = K_2(T_r - T_2) \tag{6.2-4}$$

where $K_a$ is a gain associated with the amplifier, and the scale factor of the potentiometer has been chosen so that $K_1 = K_2$.

It is convenient and standard practice to choose the block-diagram variables as deviations from reference values corresponding to a zero-error equilibrium operating condition. Let us choose the temperatures 200° and 70° as our references for the oven and ambient temperatures, and define the following deviations.

$$x = T_2 - 200^\circ$$

$$v = T_o - 70^\circ$$

The deviation $m = q - q_{ss}$ is the corresponding variable for heat rate, and the governing differential equation becomes

$$C_2 \frac{dx}{dt} = m - \frac{1}{R_2}(x - v) \tag{6.2-5}$$

A change in ambient temperature from 70° is represented by a nonzero value for the disturbance variable $v$. On the other hand, to allow the possibility of selecting a new oven temperature, define a set point deviation $u$ as

$$u = T_r - 200^\circ$$

In this case, $e = K_2(u - x)$; the diagram is shown in Figure 6.9. Note that the diagram does not show the equilibrium reference values, but only shows the deviations from these values. If the new oven temperature is desired to be 250°, the input $u$ is a step function with a magnitude of $250^\circ - 200^\circ = 50^\circ$.

The selection of deviation variables was not explicitly made in the previous position-control system. This is because any angular position of the load is an equilibrium operating condition when the motor is turned off and no disturbance torque is acting. In this case the reference motor torque and voltage are zero, and the variables $T$ and $v$ can be considered to be deviations from these zero reference values. If the nominal disturbance torque were not zero, the reference motor torque and voltage would be nonzero as well, and deviations would be defined from these values.

The primary reason for employing deviation variables is that they readily allow

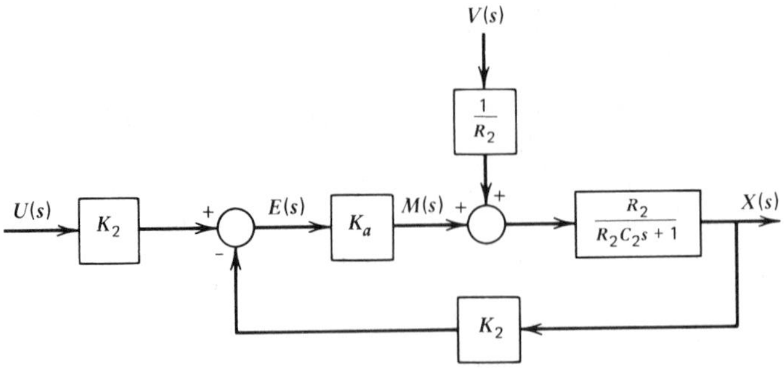


**Figure 6.9 Block diagram of the temperature-control system shown in Figure 6.8.**

a linearized model to be used to describe the system. Physical systems are inherently nonlinear, but most controllers are the result of a successful application of linear systems analysis. This is because the principal function of the controller is to eliminate deviations from some desired operating value. The next example demonstrates this point.

## A Liquid-Level Controller

A simplified illustration of a liquid-level controller commonly found in industry is shown in Figure 6.10. The liquid is supplied under pressure $p$ to the control valve. The

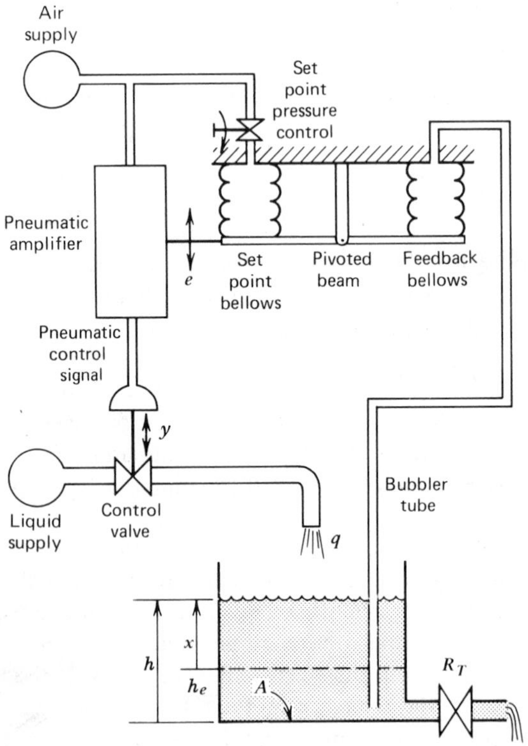


**Figure 6.10 Pneumatic liquid-level control system.**

valve motion is produced by a diaphragm with pneumatic pressure on one side and a resisting spring on the other. The air pressure is produc d by a pneumatic amplifier whose output is affected by the motion of a beam with a bellows at each end (this amplifier is discussed in Section 6.4). A bubbler tube filled with air indicates the liquid level by sensing the hydrostatic pressure in the tank. The air pressure in the tube is connected to the feedback bellows. The set point of the controller indicates the desired liquid level. This is set by changing the pressure in the set point bellows via an adjustable restriction in the supply line. The system is balanced when the level is at the desired value and the feedback pressure equals the set point pressure. If the level is below the desired value, the beam pivots counterclockwise. The amplifier and valve are designed to respond to this motion by increasing the flow rate $q$. The flow $q$ is a function of $p$ and the valve motion $y$.

Two possible disturbances to this system are changes in the supply pressures of the air and the liquid. Here we consider a change in the liquid supply pressure $p$. Assume that the outlet restriction is nonlinear. Then the governing equations are

$$A\frac{dh}{dt} = q - \sqrt{\frac{h}{R_T}} \tag{6.2-6}$$

$$q = f(p, y) \tag{6.2-7}$$

Let the level $h_e$ denote the nominal equilibrium level, $q_e$ the flow rate required to maintain this level, $p_e$ the nominal supply pressure, and $y_e$ the nominal valve position. Define the deviations

$$x = h - h_e$$

$$z = y - y_e$$

$$v = p - p_e$$

$$w = q - q_e$$

Measure the beam displacement $e$ from its position at this equilibrium condition. This displacement is an indication of the system error, and the beam acts like an error detector. The desired level $h_r$ is not always $h_e$. Therefore define the set point deviation $u = h_r - h_e$.

Linearization of the tank and valve equation gives

$$A\frac{dx}{dt} = w - \frac{1}{R}x \tag{6.2-8}$$

$$\frac{1}{R} = \frac{1}{2A}\sqrt{\frac{1}{R_T h_e}} \tag{6.2-9}$$

$$w = K_1 z + K_2 v \tag{6.2-10}$$

$$K_1 = \left(\frac{\partial f}{\partial y}\right)_e, \qquad K_2 = \left(\frac{\partial f}{\partial p}\right)_e \tag{6.2-11}$$

The linearized resistance is $R$ and is positive. The constant $K_1$ is positive because of the assumed operation of the valve; $K_2$ is positive because an increase in supply pressure causes an increase in flow rate.

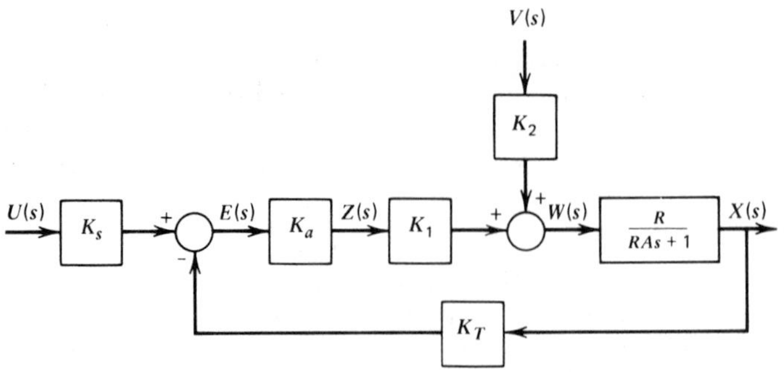


**Figure 6.11** **Block diagram of the liquid-level control system of Figure 6.10.**

The beam motion $e$ depends on the set point pressure and feedback pressure, which in turn depend on $h_r$ and $h$, respectively. Thus $e = g(h_r, h)$, and in linearized form

$$e = K_s u - K_T x \tag{6.2-12}$$

$$K_s = \left(\frac{\partial g}{\partial h_r}\right)_e, \qquad K_T = \left(\frac{\partial g}{\partial h}\right)_e \tag{6.2-13}$$

since $e$ is already a deviation. The set point and bubbler tube constants $K_s$ and $K_T$ depend on the beam geometry, bellows area and stiffness, and air supply pressure. The latter quantity can be absorbed into $K_s$ because we have assumed it is constant.

The valve displacement $y$ is a function of the beam motion, via the pneumatic amplifier. In linearized form this gives

$$z = K_a e \tag{6.2-14}$$

where $K_a$ is the linearization constant for the amplifier-valve combination.

The linearized description of the system is now complete, and the block diagram can be drawn (Figure 6.11). Caution must be observed when using this model. For a large enough change in set point or in supply pressure away from their nominal values $h_e$ and $p_e$, the linearized model is inaccurate, and the linearization constants should be recomputed at the new operating condition. However, in many cases the original linearized description is quite adequate.

## 6.3 TRANSDUCERS AND ERROR DETECTORS

The control system structure shown in Figure 6.7 indicates a need for physical devices to perform several types of functions. Here we present a brief overview of some available transducers and error detectors. Actuators and devices used to implement the control logic are discussed in Sections 6.4, 6.8, 6.9, and 6.10. Space limitations prevent a detailed discussion of these devices. The technology is rapidly changing, and the most up-to-date information is available only in manufacturers' literature. Some of these devices have been analyzed in previous examples; models for the remainder can be obtained by applying the techniques of Chapter Two or by consulting appropriate references (for example, see References 4 through 7).

## Displacement and Velocity Transducers

A *transducer* is a device that converts one type of signal into another type. An example is the potentiometer, which converts displacement into voltage. In addition to this conversion, the transducer can be used to make measurements. In such applications the term *sensor* is more appropriate. For our purposes we will assume that the sensor measurements are accurate. A detailed discussion of sensor errors is given in Reference 4. The design of control systems that compensate for imperfect measurements relies on filtering theory and is an advanced topic (Reference 8).

In addition to the potentiometer, displacement can also be measured electrically with a *linear variable differential transformer* (LVDT) or a *synchro*. An LVDT measures the linear displacement of a movable magnetic core through a primary winding and two secondary windings (Figure 6.12). An ac voltage is applied to the primary. The secondaries are connected together and also to a detector that measures the voltage and phase difference. A phase difference of 0° corresponds to a positive core displacement, while 180° indicates a negative displacement. The amount of displacement is indicated by the amplitude of the ac voltage in the secondary. The detector converts this information into a dc voltage $e_o$ such that $e_o = Kx$. The LVDT is sensitive to small displacements. Two of them can be wired together like the potentiometers in Section 6.2 to form an error detector.

A synchro is a rotary differential transformer, with angular displacement as either the input or output. They are often used in pairs (a *transmitter* and a *receiver*) where a remote indication of angular displacement is needed. When a transmitter is used with a synchro *control transformer*, two angular displacements can be measured and compared (Figure 6.13). The output voltage $e_o$ is approximately linear with angular difference within ± 70°, so that $e_o = K(\theta_1 - \theta_2)$.

Displacement measurements can be used to obtain forces and accelerations. For example, the displacement of a calibrated spring indicates the applied force. The accelerometer described in Section 5.9 is another example. Still another is the *strain gage* used for force measurement. It is based on the fact that the resistance of a fine wire changes as it is stretched. The change in resistance is detected by a circuit that can be calibrated to indicate the applied force.

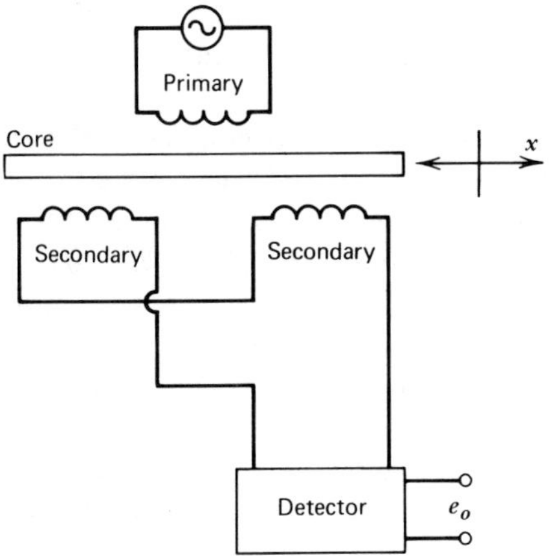


**Figure 6.12 Linear variable differential transformer (LVDT).**

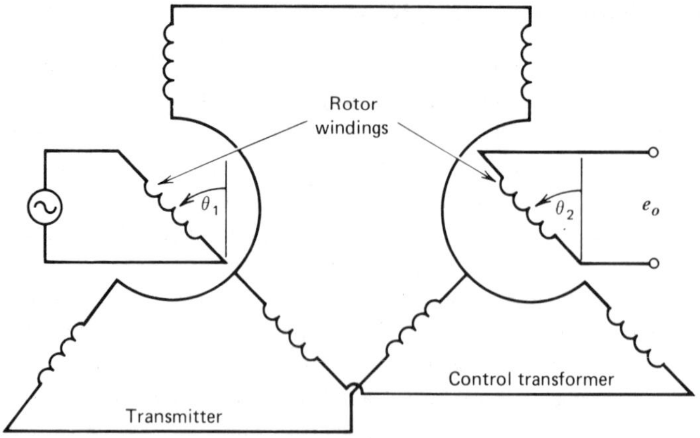


**Figure 6.13 Synchro transmitter-control transformer.**

Velocity measurements in control systems are most commonly obtained with a *tachometer*. This is essentially a dc generator (the reverse of a dc motor). The input is mechanical (a velocity). The output is a generated voltage proportional to the velocity. Translational velocity is usually measured by converting it to angular velocity with gears, for example. Tachometers using ac signals are also available.

Other velocity transducers include a magnetic pickup that generates a pulse every time a gear tooth passes. If the number of gear teeth is known, a pulse counter and timer can be used to compute the angular velocity. A similar principle is employed by an *optical encoder*. A light beam is broken by a rotating slotted disk, and a photoelectric cell converts this information into pulses that can be analyzed as before. The outputs of these devices are especially suitable for digital control purposes.

## Temperature Transducers

When two wires of dissimilar metals are joined together as in Figure 6.14, a voltage is generated if the junctions are at difference temperatures. If the reference junction is kept at a fixed, known temperature, the thermocouple can be calibrated to indicate the temperature at the other junction in terms of the voltage $v$. An ice-water bath can be used for the reference temperature, but many modern systems provide an electronic equivalent.

Electrical resistance changes with temperature. Platinum gives a linear relation between resistance and temperature, while nickel is less expensive and gives a large resistance change for a given temperature change. Semiconductors designed with this property are called *thermistors.*

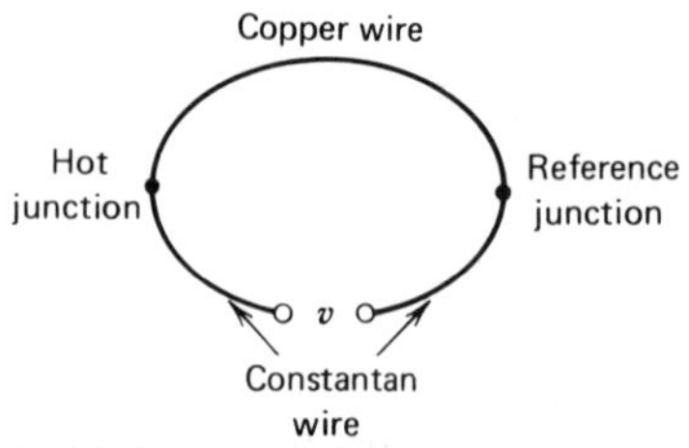


**Figure 6.14 Type-*T* thermocouple.**

Different metals expand at different rates when the temperature is increased. This fact is used in the bimetallic strip transducer found in most home thermostats (Figure 6.15). Two dissimilar metals

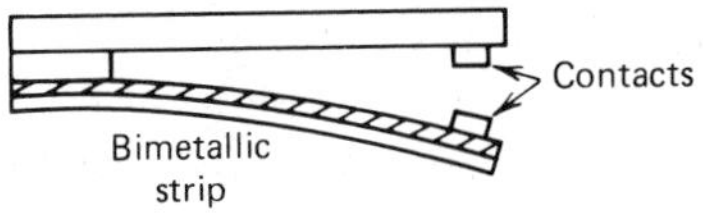


**Figure 6.15 Electrical temperature sensor utilizing a bimetallic strip.**

are bonded together to form the strip. In most thermostats the strip is actually coiled, but the figure illustrates the principle. As the temperature rises the strip curls, breaking contact and shutting off the furnace. The temperature gap can be adjusted by changing the distance between the contacts. The motion also moves a pointer on the temperature scale of the thermostat.

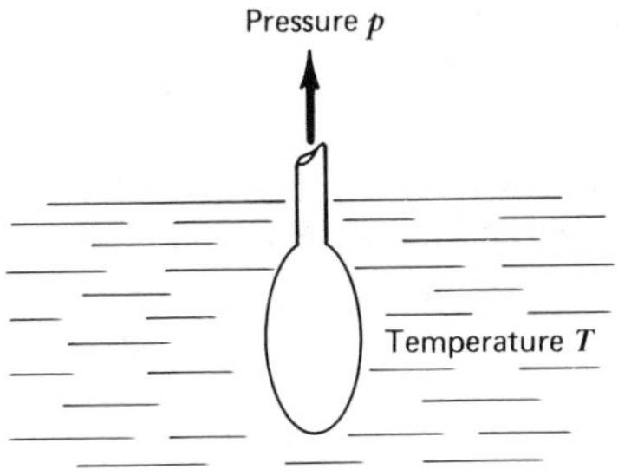


**Figure 6.16 Pneumatic temperature sensor.**

The pressure of a fluid inside a bulb will change as its temperature changes (Figure 6.16). If the bulb fluid is air, the device is suitable for use in pneumatic temperature controllers. If the air mass in the bulb is constant and if the thermal resistance and capacitance of the bulb walls are negligible, the perfect gas law can be applied to obtain

$$p = KT \tag{6.3-1}$$

where the constant $K$ depends on the gas and the bulb volume.

## Flow Transducers

Flow rates can be measured by introducing a flow restriction, such as an orifice plate, and measuring the pressure drop across the restriction. The flow-rate–pressure relation is $\Delta p = Rq^2$, where $R$ can be found from calibration of the device. The pressure drop can be sensed by converting it into the motion of a diaphragm. Figure 6.17 illustrates a related technique. The venturi-type flowmeter measures the static pressures in the constricted and unconstricted flow regions. Bernoulli's principle relates the pressure difference to the flow rate. This pressure difference is converted into displacement by the diaphragm.

## Error Detectors

The error detector is simply a device for finding the difference between two signals. This function is sometimes an integral feature of sensors, such as with the synchro transmitter-transformer combination. A beam on a pivot provides a way of comparing

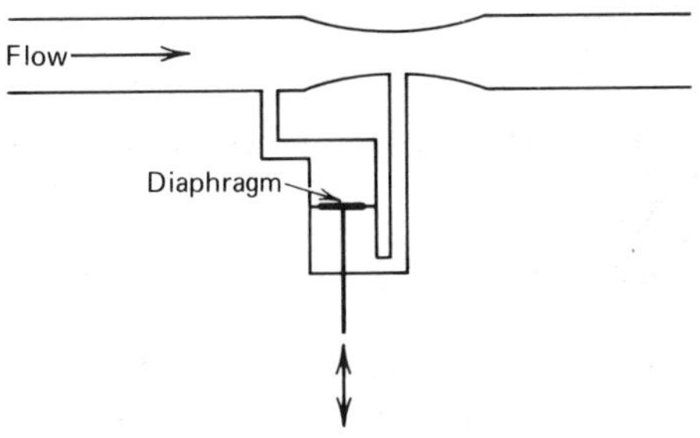


**Figure 6.17 Venturi-type flowmeter. The diaphragm displacement indicates the flow rate.**

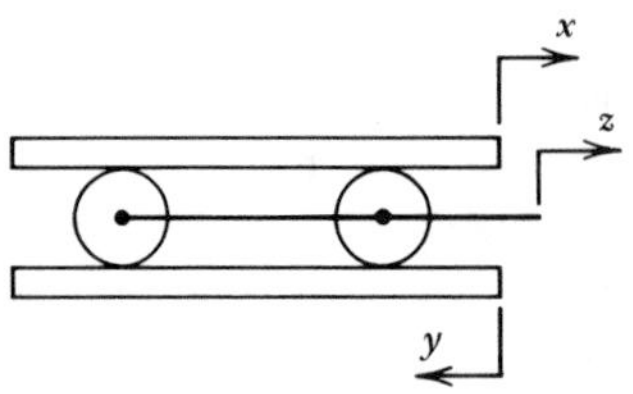


**Figure 6.18** **Rack-and-gear differential for detecting displacement differences $z = \frac{1}{2}(x - y)$.**

displacements, forces, or pressures, as was done with the pneumatic level controller (Figure 6.10). A similar concept is used with the diaphragm element shown in Figure 6.17. A detector for voltage difference can be obtained as with the position control system shown in Figure 6.4. An amplifier intended for this purpose is a *differential amplifier.* Its output is proportional to the difference between the two inputs.

Relative motion can be detected with sufficient accuracy for some applications by the simple device shown in Figure 6.18.

In order to detect differences in other types of signals, such as temperature, they are usually converted to a displacement or pressure. One of the detectors mentioned previously can then be used.

### Dynamic Response

The quantitative transducer and detector models presented earlier are static models, and as such imply that the components respond instantaneously to the variable being sensed. Of course, any real component has a dynamic response of some sort, and this response time must be considered in relation to the controlled process when a sensor is selected. For example, the time constant of a thermocouple in air typically is between 10 and 100 sec. If the thermocouple is in a liquid, the thermal resistance is different, and so will be the time constant (it will be smaller). If the controlled process has a time constant at least 10 times greater, we probably would be justified in using a static sensor model. The effect of this assumption will be considered in a later example.

We will assume in general that the sensors have been selected properly in relation to the plant dynamics, and we will thus use a static model in most cases.

## 6.4 ACTUATORS

An *actuator* is the final control element that operates on the low-level control signal to produce a signal containing enough power to drive the plant for the intended purpose. The armature-controlled dc motor, the hydraulic servomotor, and the pneumatic bellows are common examples of actuators. Dynamic models of their behavior were developed earlier.

### Electromechanical Actuators

From Example 5.18, the transfer function for the armature-controlled dc motor is

$$\frac{\Omega(s)}{V(s)} = \frac{K_T}{LIs^2 + (RI + cL)s + cR + K_e K_T} \qquad (6.4\text{-}1)$$

where $\Omega(s)$ and $V(s)$ are the angular velocity and input voltage transforms. The armature inductance is often negligible. In that case the transfer function becomes a first-order one.

$$\frac{\Omega(s)}{V(s)} = \frac{K_T}{RIs + cR + K_e K_T} \tag{6.4-2a}$$

$$= \frac{K}{\tau s + 1} \tag{6.4-2b}$$

Another motor configuration is the field-controlled dc motor, whose model was developed in Chapter Five (problem 5.6). In this case the armature current is kept constant, and the field voltage $v$ is used to control the motor. The transfer function is function is

$$\frac{\Omega(s)}{V(s)} = \frac{K_T}{(R + Ls)(Is + c)} \tag{6.4-3}$$

where $R$ and $L$ are the resistance and inductance of the field circuit, and $K_T$ is the torque constant. No back emf exists in this motor to act as a self-braking mechanism. The field inductance usually is not negligible, so the model is second order. The field control requires less power than the armature-controlled motor. However, a constant current is more difficult to provide than the constant voltage needed for a constant field (this field can also be provided by a permanent magnet).

Two-phase ac motors can be used to provide a low-power, variable-speed actuator. This motor type can accept the ac signals directly from LVDTs and synchros without demodulation. However, it is difficult to design ac amplifier circuitry to do other than proportional action (see Section 6.5). For this reason the ac motor is not found in control systems as often as dc motors. The transfer function for this type is of the form of (6.4-2*b*).

An actuator especially suitable for digital systems is the *stepper motor*, a special dc motor that takes a train of electrical input pulses and converts each pulse into an angular displacement of a fixed amount. Motors are available with resolutions ranging from about four steps per revolution to as many as 800 steps per revolution. For 36 steps per revolution the motor will rotate by $10^\circ$ for each pulse received. When not being pulsed, the motors lock in place. Thus they are excellent for precise positioning applications such as required with printers and computer tape drives. A disadvantage is that they are low-torque devices. Mathematical models are difficult to develop for them because of the complexity of stepping motor construction and operation. Available models are piecewise-linear at best and are inaccurate when linearized. However, if the input pulse frequency is not near the resonant frequency of the motor, we can take the output rotation to be directly related to the number of input pulses and use that description as the motor model.

## Hydraulic Actuators

The hydraulic servomotor with a load was analyzed in Example 5.6. Its transfer function is

$$\frac{X(s)}{Z(s)} = \frac{C_1 A}{C_2 m s^2 + (cC_2 + A^2)s + C_2 k A} \tag{6.4-4}$$

where $z$ is the input displacement and $x$ is the output displacement. For no damping and no stiffness, and a negligible load, it was shown that (6.4-4) reduces to the first-order model

$$\frac{X(s)}{Z(s)} = \frac{C_1}{As} \tag{6.4-5}$$

Rotational motion can be obtained with a *hydraulic motor*, which is, in principle, a pump acting in reverse (fluid input and mechanical rotation output). Such motors can achieve higher torque levels than electric motors. A hydraulic pump driving a hydraulic motor constitutes a *hydraulic transmission*.

A popular actuator choice is the *electrohydraulic* system, which uses an electric actuator to control a hydraulic servomotor or transmission by moving the pilot valve or the swash-plate angle of the pump. Such systems combine the power of hydraulics with the advantages of electrical systems. For example, modern aircraft control systems use hydraulic actuators to move the aerodynamic control surfaces. These motions are commanded by the pilot's stick motions, which are converted into electrical signals to be sent to the actuators. This avoids the need to run long, damage-prone hydraulic lines from the cockpit to the actuators. The small electrical lines can be run in multiple sets for redundancy.

Figure 6.19 shows a hydraulic motor whose pilot valve motion is caused by an armature-controlled dc motor. The transfer function between the motor voltage and the piston displacement is

$$\frac{X(s)}{V(s)} = \frac{K_1 K_2 C_1}{As^2(\tau s + 1)} \tag{6.4-6}$$

If the rotational inertia of the electric motor is small, then $\tau \approx 0$.

## Pneumatic Actuators

Pneumatic actuators are commonly used because they are simple to maintain and use a readily available working medium. Compressed air supplies with the pressures required are commonly available in factories and laboratories. No flammable fluids or electrical sparks are present, so these devices are considered the safest to use with chemical processes. Their power output is less than that of hydraulic systems, but greater than that of electric motors.

A device for converting pneumatic pressure into displacement is the bellows considered in Example 2.19 and shown in Figure 6.20. The transfer function for a linearized model is of the form

$$\frac{X(s)}{P(s)} = \frac{K}{\tau s + 1} \tag{6.4-7}$$

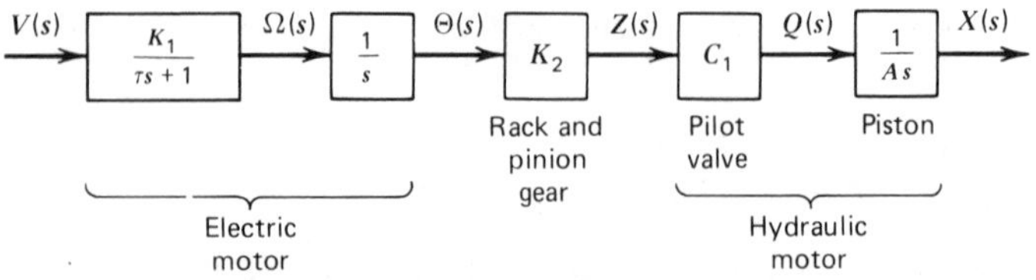


**Figure 6.19 Electrohydraulic system for translation.**

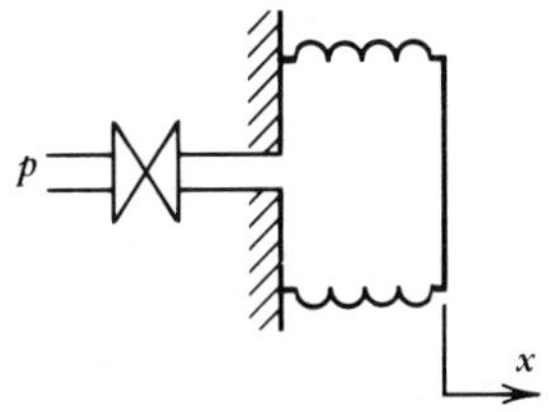


**Figure 6.20 Pneumatic bellows.**

where $x$ and $p$ are deviations of the bellows displacement and input pressure from nominal values. The restriction in the input line acts as a resistance and the bellows volume as a capacitance. The time constant $\tau$ is the product of the resistance and capacitance.

In many control applications a device is needed to convert small displacements into relatively large pressure changes. The *nozzle-flapper* serves this purpose (Figure 6.21*a*). The input displacement $y$ moves the flapper with little effort required. This changes the opening at the nozzle orifice. For a large enough opening the nozzle back pressure is approximately the same as atmospheric pressure $p_a$. At the other extreme position with the flapper completely blocking the orifice, the back pressure equals the supply pressure $p_s$. This variation is shown in Figure 6.21*b*. Typical supply pressures are between 30 and 100 psia. The orifice diameter is approximately 0.01 in. Flapper displacement is usually less than one orifice diameter.

The nozzle-flapper is operated in the linear portion of the back pressure curve. Let $p$ and $x$ denote the deviations in back pressure and flapper displacement from the nominal values. The input displacement $y$ is zero at the nominal condition. Then the linearized back pressure relation is

$$p = -K_f x \tag{6.4-8}$$

where $-K_f$ is the slope of the curve and is a very large number. From the geometry of similar triangles we have

$$x = \frac{a}{a+b}y \tag{6.4-9}$$

Thus

$$p = -\frac{aK_f}{a+b}y \tag{6.4-10}$$

In its operating region the nozzle-flapper's back pressure is well below the

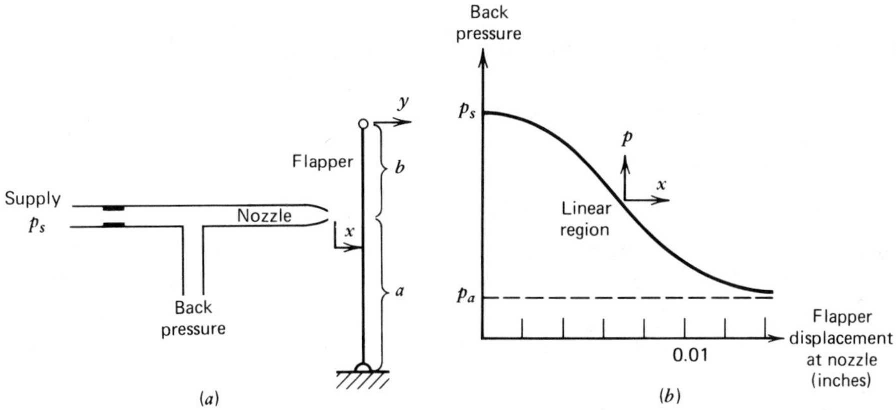


**Figure 6.21 Pneumatic nozzle-flapper amplifier and its characteristic curve.**

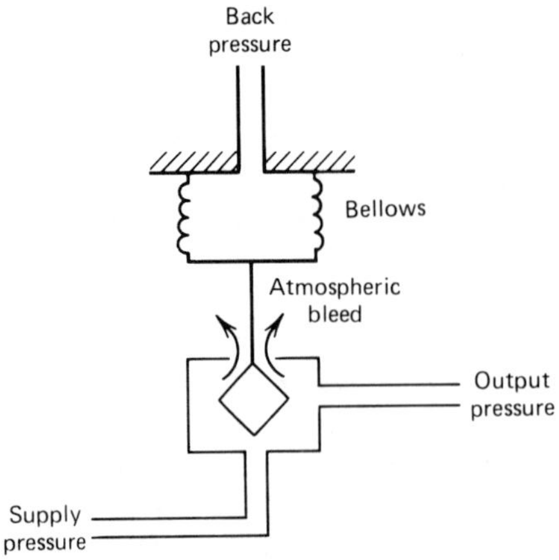


**Figure 6.22 Pneumatic relay.**

supply pressure. If more output pressure is required, a pneumatic relay or amplifier can be used. Figure 6.22 illustrates this concept. As the back pressure increases, the relay closes off the supply line, and the output pressure approaches atmospheric pressure. As the back pressure decreases, the valve shuts off the atmospheric bleed, and the output pressure approaches the supply pressure. The relay is said to be reverse-acting because an increase in back pressure produces a decrease in output.

The output pressure from the relay can be used to drive a final control element like the pneumatic actuating valve shown in Figure 6.23. The pneumatic pressure acts on the upper side of the diaphragm and is opposed by the return spring.

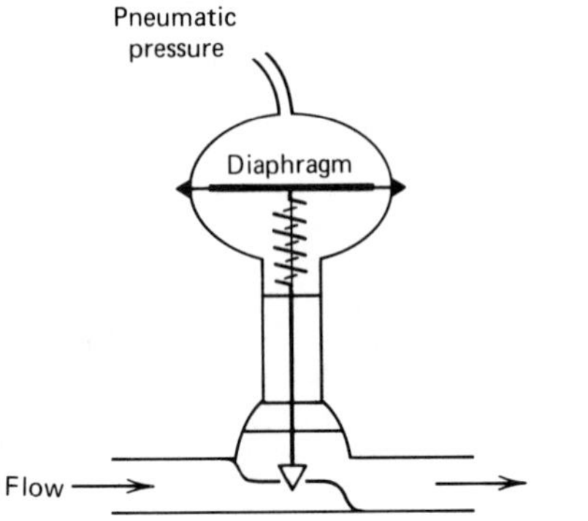


**Figure 6.23 Pneumatic flow-control valve.**

The nozzle-flapper is a *force-distance* type of actuator because its input is a displacement and its output a force (pressure). The other type of pneumatic actuator is the *force-balance* type in which both the input and output are pressures. The pressures are made to act across diaphragms whose resulting motion activates a pneumatic relay. Details can be found in the specialized literature (e.g., Reference 7).

## 6.5 CONTROL LAWS

The control logic elements are designed to act on the actuating (error) signal to produce the control signal. The algorithm that is physically implemented for this purpose is the *control law* or *control action*. A non-zero error signal results from either a change in command or a disturbance. The general function of the controller is to keep the controlled variable near its desired value when these occur. More specifically, the control objectives might be stated as follows.

1. Minimize the steady-state error.

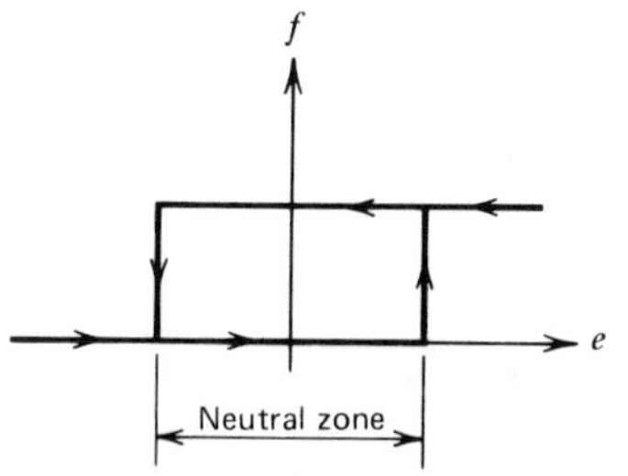


**Figure 6.24 Transfer characteristics of the on-off controller. The actuating error is $e = r - c$, where $r$ = set point, $c$ = controlled variable, and $f$ = control signal.**

2. Minimize the settling time.
3. Achieve other transient specifications, such as minimizing the maximum overshoot.

In practice, the design specifications for a controller are more detailed. For example, the bandwidth might also be specified along with a safety margin for stability. We never know the numerical values of the system's parameters with true certainty, and some controller designs can be more sensitive to such parameter uncertainties than other designs. So a parameter sensitivity specification might also be included. We will return to this topic later, but for now a general understanding of control objectives is sufficient for our purpose.

The following two control laws form the basis of many control systems.

## Two-Position Control

Two-position control is the most familiar type perhaps because of its use in home thermostats. The control output takes on one of two values. With the *on-off* controller, the controller output is either on or off (fully open or fully closed). Such is the case with the thermostat-furnace system. The controller output is determined by the magnitude of the error signal. The switching diagram for the on-off controller with hysteresis is shown in Figure 6.24.

An example of an application of an on-off controller to a liquid-level system is shown in Figure 6.25*a*. The time response shown in Figure 6.25*b* with a solid line is for an ideal system in which the control valve acts instantaneously. The controlled variable cycles with an amplitude that depends on the width of the neutral zone or gap. This zone is provided to prevent frequent on-off switching, or *chattering*, which can shorten the life of the device. The cycling frequency also depends on the time constant of the controlled process and the magnitude of the control signal.

In a real system, as opposed to ideal, the sensor and control valve will not respond instantaneously, but have their own time constants. The valve will not close

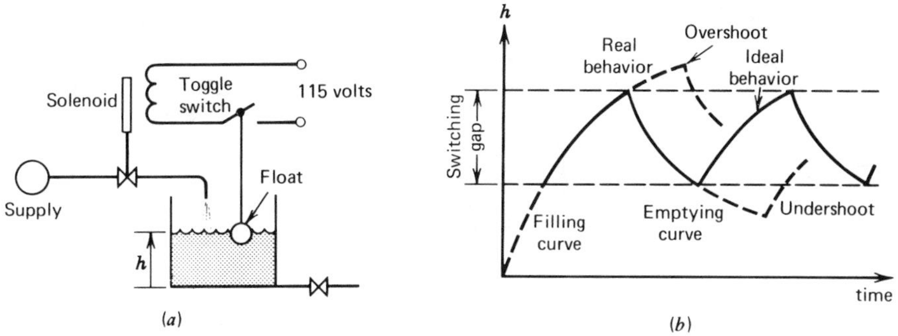


**Figure 6.25 (*a*) Liquid-level control with on-off action. (*b*) Time response.**

at the instant the height reaches the desired level. There will be some delay during which flow continues into the tank. The result is shown by the dotted line in Figure 6.25*b*. The opposite occurs when the valve is turned on. This unwanted effect can be reduced by decreasing the neutral zone, but the cycling frequency increases if this is done.

The overshoot and undershoot in on-off control will be acceptable only when the time constant of the process is large compared to the time lag of the control elements. This lag is related to the time constants of the elements, as well as to their distance from the plant. If the control valve in Figure 6.25*a* is far upstream from the tank, a significant lag can exist between the time of control action and its effect on the plant. Another source of time lag is the capacitance of the controller itself. For example, if the heater capacitance in the temperature controller of Section 6.2 is appreciable, the heater will continue to deliver energy to the oven even after it has been turned off.

An example close to home demonstrates how the capacitance of the plant affects the suitability of on-off control. On-off control of the hot water valve in a shower obviously will be uncomfortable, but it is acceptable for a bath, because the thermal capacitance is greater.

Another type of two-position control is the *bang-bang* controller whose switching diagram is shown in Figure 6.26*a*. This controller is distinguished from on-off control by the fact that the direction or sign of the control signal can have two values. A motor with constant torque that can reverse quickly might be modeled as a bang-bang device. Because such perfect switching is impossible, a more accurate model would include a dead zone (Figure 6.26*b*). When the error is within the zone, the controller output is zero.

Two-position controllers applied to linear systems lead to piecewise-linear models, which can be analyzed as in Section 3.11 or with the phase plane techniques presented in Section 5.10. The bang-bang controller without a dead zone is identical to the Coulomb friction effect analyzed in Section 5.10.

## Proportional Control

Two-position control is acceptable for many applications in which the requirements are not too severe. In the home heating application, the typical 2°F temperature gap is hardly detectable by the occupants. Thus the system is acceptable. However, many situations require finer control.

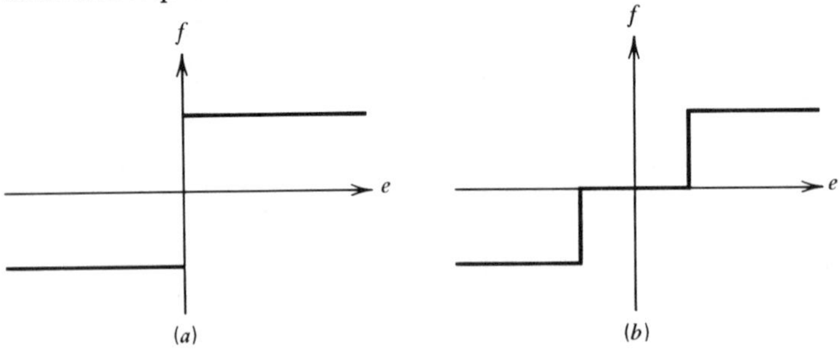


**Figure 6.26** Transfer characteristics. (*a*) Ideal bang-bang control. (*b*) Bang-bang control with a dead zone. The control signal is $f$; the error signal is $e$.

Consider the tank system shown in Figure 6.25*a*. To replace the two-position controller, we might try setting the control valve manually to achieve a flow rate that balances the system at the desired level. We might then add a controller that adjusts this setting in proportion to the deviation of the level from the desired value. This is *proportional control*, the algorithm in which the change in the control signal is proportional to the error. Recall the convention that block diagrams for controllers are drawn in terms of the deviations from a zero-error equilibrium condition. Applying this convention to the general terminology of Figure 6.7, we see that proportional control is described by

$$F(s) = K_p E(s) \tag{6.5-1}$$

where $F(s)$ is the deviation in the control signal and $K_p$ is the *proportional gain*. If the total valve displacement is $y(t)$ and the manually created displacement is $x$, then

$$y(t) = K_p e(t) + x$$

The percent change in error needed to move the valve full scale is the *proportional band*. It is related to the gain as follows.

$$K_p = \frac{100}{\text{band \%}} \tag{6.5-2}$$

The zero-error valve displacement $x$ is the *manual reset*.

## Proportional Control of a First-Order System

To investigate the behavior of proportional control, consider the speed-control system shown in Figure 6.27; it is identical to the position controller shown in Figure 6.4, except that a tachometer replaces the feedback potentiometer. A linear differential amplifier produces an output proportional to the difference between the input voltages. If the power amplifier is also linear, we can combine their gains into one, denoted $K_p$. The system is thus seen to have proportional control in which the motor voltage is proportional to the difference between the command voltage and the feedback voltage from the tachometer.

The equations for a field-controlled dc motor can be developed following the manner of Example 5.3, with no back emf effect. The resulting model is

$$L\frac{di_f}{dt} = v - Ri_f \tag{6.5-3}$$

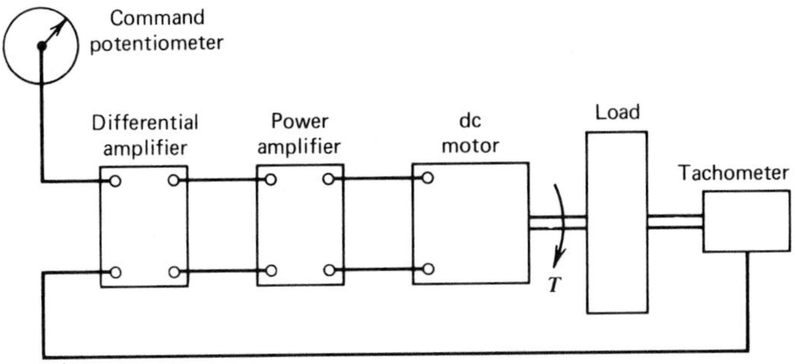


**Figure 6.27** **Velocity-control system using a dc motor.**

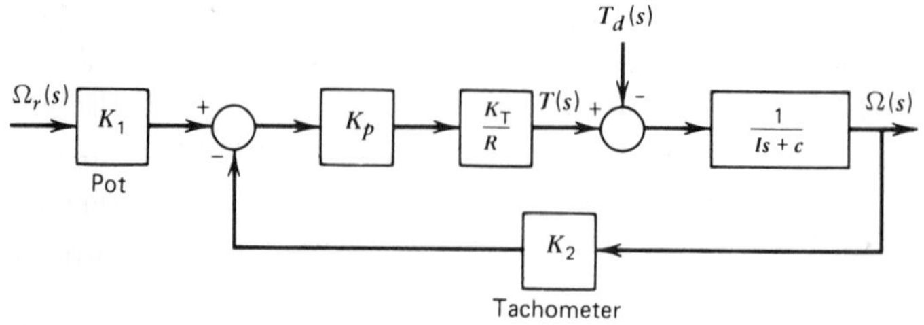


**Figure 6.28 Block diagram of the velocity-control system of Figure 6.27.**

$$I\frac{d\omega}{dt} = T - c\omega \tag{6.5-4}$$

$$T = K_T i_f \tag{6.5-5}$$

where $R$ and $L$ are the parameters of the field circuit. For now we assume that the inductance $L$ is negligible. This gives

$$T = \frac{K_T}{R}v \tag{6.5-6}$$

The disturbance is a torque $T_d$, for example, resulting from friction. Choose the reference equilibrium condition to be $T_d = T = 0$ and $\omega_r = \omega = 0$. The block diagram is shown in Figure 6.28. For a meaningful error signal to be generated, $K_1$ and $K_2$ should be chosen to be equal. With this simplication the diagram becomes that shown in Figure 6.29, where $K = K_1K_pK_T/R$.

The transfer functions are

$$\frac{\Omega(s)}{\Omega_r(s)} = \frac{K}{Is+c+K} \tag{6.5-7}$$

$$\frac{\Omega(s)}{T_d(s)} = \frac{-1}{Is+c+K} \tag{6.5-8}$$

A change in desired speed can be simulated by a step input for $\omega_r$. Linearity allows us to use a unit step and scale the results accordingly. For $\Omega_r(s) = 1/s$,

$$\Omega(s) = \frac{K}{Is+c+K}\frac{1}{s}$$

The response can be computed by partial fraction expansion as before. The velocity approaches the steady-state value

$$\omega_{ss} = \lim_{s\to 0} s\frac{K}{Is+c+K}\frac{1}{s} = \frac{K}{c+K} < 1$$

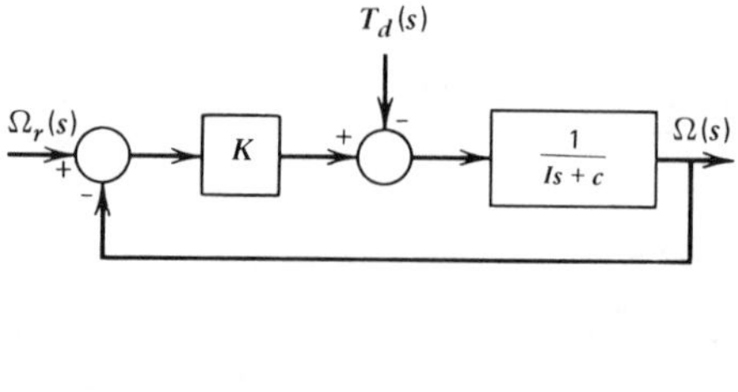


**Figure 6.29 Simplified form of Figure 6.28 for the case $K_1 = K_2$; $K = K_1K_pK_T/R$.**

Thus the final value is less than the desired value of 1, but it might be close enough if the damping $c$ is small. The time required to reach this value is approximately four time constants, or $4\tau = 4I/(c + K)$.

A sudden change in load torque can also be modeled by a unit step function $T_d(s) = 1/s$. The response due solely to the disturbance is found from (6.5-8).

$$\Omega(s) = \frac{-1}{Is + c + K}\frac{1}{s}$$

The steady-state effect of the disturbance is found with the final value theorem to be $-1/(c + K)$. If $(c + K)$ is large, this error will be small.

The performance of the proportional control law thus far can be summarized as follows. For a first-order system whose inputs are step functions,

1. The output never reaches its desired value if resistance is present ($c \neq 0$), although it can be made arbitrarily close by choosing the gain $K$ large enough. This is *offset* error.
2. The output approaches its final value without oscillation. The time to reach this value is inversely proportional to $K$.
3. The output deviation due to the disturbance at steady state is inversely proportional to the gain $K$. This error is present even in the absence of resistance ($c = 0$).

As the gain $K$ is increased, the time constant becomes smaller and the response faster. Thus the chief disadvantage of proportional control is that it results in steady-state errors and can only be used when the gain can be selected large enough to reduce the effect of the largest expected disturbance. Since proportional control gives zero error only for one load condition (the reference equilibrium), the operator must change the manual reset by hand (hence the name).

An advantage to proportional control is that the control signal responds to the error instantaneously (in theory at least). It is used in applications requiring rapid action. Processes with time constants too small for the use of two-position control are likely candidates for proportional control.

The results of this analysis can be applied to any first-order system of the form in Figure 6.29. The temperature controller and the liquid-level controller in Section 6.2 are proportional controllers, and their block diagrams can easily be put into this form.

## Proportional Control of a Second-Order System

Proportional control of a neutrally stable second-order plant is represented by the position controller of Figure 6.6 if the amplifier transfer function is a constant $G_a(s) = K_a$. Let the motor transfer function be $G_m(s) = K_T/R$ as before. The modified block diagram is given in Figure 6.30 with $K = K_1 K_a K_T/R$. The transfer functions are

$$\frac{\Theta(s)}{\Theta_r(s)} = \frac{K}{Is^2 + cs + K} \tag{6.5-9}$$

$$\frac{\Theta(s)}{T_d(s)} = \frac{-1}{Is^2 + cs + K} \tag{6.5-10}$$

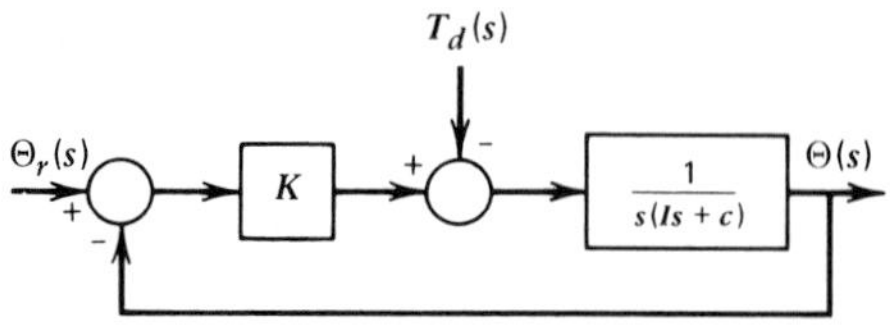


**Figure 6.30 Position servo with proportional control.**

The closed-loop system is stable if $I$, $c$, and $K$ are positive. For no damping ($c = 0$), the closed-loop system is neutrally stable.

With no disturbance and a unit step command, $\Theta_r(s) = 1/s$, the steady-state output is

$$\theta_{ss} = \frac{K}{K} = 1$$

The offset error is thus zero if the system is stable ($c > 0$, $K > 0$). The steady-state output deviation due to a unit step disturbance is $-1/K$. This deviation can be reduced by choosing $K$ large.

The transient behavior is indicated by the damping ratio.

$$\zeta = \frac{c}{2\sqrt{IK}}$$

For slight damping the response to a step input will be very oscillatory and the overshoot large. The situation is aggravated if the gain $K$ is made large to reduce the deviation due to the disturbance. We conclude therefore that proportional control of this type of second-order plant is not a good choice unless the damping constant $c$ is large. We will see shortly how to improve the design.

## 6.6 INTEGRAL CONTROL

The offset error that occurs with proportional control is a result of the system reaching an equilibrium in which the control signal no longer changes. This allows a constant error to exist. If the controller is modified to produce an increasing signal as long as the error is nonzero, the offset might be eliminated. This is the principle of *integral control.* In this mode the change in the control signal is proportional to the *integral* of the error. In the terminology of Figure 6.7, this gives

$$F(s) = \frac{K_I}{s} E(s) \tag{6.6-1}$$

where $F(s)$ is the deviation in the control signal and $K_I$ is the *integral gain*. In the time domain the relation is

$$f(t) = K_I \int_0^t e(t)\, dt \tag{6.6-2}$$

if $f(0) = 0$. In this form it can be seen that the integration cannot continue indefinitely because it would theoretically produce an infinite value of $f(t)$. This implies that special care must be taken to reinitialize the controller. This requirement is considered later.

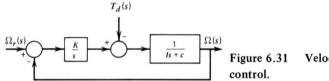

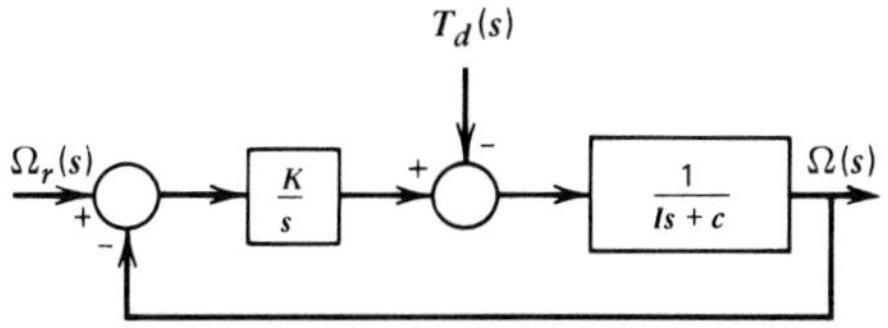


**Figure 6.31** Velocity servo with integral control.

## Integral Control of a First-Order System

Integral control of the velocity in the system of Figure 6.27 has the block diagram shown in Figure 6.31, where $K = K_1 K_I K_T / R$. The integrating action of the amplifier is physically obtained by the techniques to be presented in Section 6.8. The closed-loop transfer functions are

$$\frac{\Omega(s)}{\Omega_r(s)} = \frac{K}{Is^2 + cs + K} \tag{6.6-3}$$

$$\frac{\Omega(s)}{T_d(s)} = \frac{-s}{Is^2 + cs + K} \tag{6.6-4}$$

The control system is stable for $I$, $c$, and $K$ positive. For a unit step command input, $\omega_{ss} = K/K = 1$; so the offset error is zero. For a unit step disturbance, the steady-state deviation is zero if the system is stable. The steady-state performance using integral control is thus excellent, for this plant with step inputs.

The damping ratio is

$$\zeta = \frac{c}{2\sqrt{IK}}$$

For slight damping the response will be oscillatory rather than exponential as with proportional control. Improved steady-state performance has thus been obtained at the expense of degraded transient performance. The conflict between steady-state and transient specifications is a common theme in control system design. As long as the system is underdamped, the time constant is $\tau = 2I/c$ and is not affected by the gain $K$, which only influences the oscillation frequency in this case. It might be physically possible to make $K$ small enough so that $\zeta \geqslant 1$, and the nonoscillatory feature of proportional control recovered, but the response would tend to be sluggish. Transient specifications for fast response generally require $\zeta < 1$. The difficulty with $\zeta < 1$ is that $\tau$ is fixed by $c$ and $I$. If $c$ and $I$ are such that $\zeta < 1$, then $\tau$ is large if $I \gg c$.

## Integral Control of a Second-Order System

Proportional control of the position servomechanism in Figure 6.29 gives a nonzero steady-state deviation due to the disturbance. Integral control applied to this system results in the block diagram of Figure 6.32 and the transfer functions

$$\frac{\Theta(s)}{\Theta_r(s)} = \frac{K}{Is^3 + cs^2 + K} \tag{6.6-5}$$

$$\frac{\Theta(s)}{T_d(s)} = \frac{-s}{Is^3 + cs^2 + K} \tag{6.6-6}$$

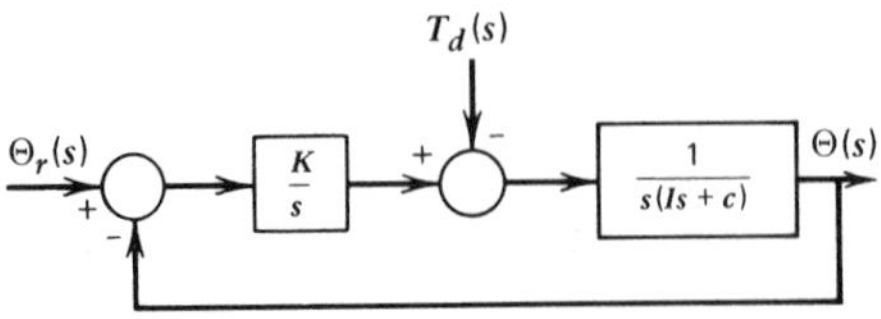


**Figure 6.32** **Position servo with integral control.**

With the Routh criterion we immediately see that the system is not stable because of the missing $s$ term. The final value theorem thus cannot be applied.

Integral control is useful in improving steady-state performance, but in general it does not improve and may even degrade transient performance. Improperly applied, it can produce an unstable control system. It is best used in conjunction with other control modes.

## Proportional-plus-Integral Control

Integral control raised the order of the system by one in the preceding examples, but did not give a characteristic equation with enough flexibility to achieve acceptable transient behavior. The instantaneous response of proportional control action might introduce enough variability into the coefficients of the characteristic equation to allow both steady-state and transient specifications to be satisfied. This is the basis for using *proportional-plus-integral control* (PI control). The algorithm for this two-mode control is

$$F(s) = K_p E(s) + \frac{K_I}{s} E(s) \tag{6.6-7}$$

The integral action provides an automatic, not manual, reset of the controller in the presence of a disturbance. For this reason it is often called *reset action*.

The algorithm is sometimes expressed as

$$F(s) = K_p \left(1 + \frac{1}{T_I s}\right) E(s) \tag{6.6-8}$$

where $T_I$ is the *reset time*. The reset time is the time required for the integral action signal to equal that of the proportional term, if a constant error exists (a hypothetical situation). The reciprocal of reset time is expressed as repeats per minute and is the frequency with which the integral action repeats the proportional correction signal.

The proportional control gain must be reduced when used with integral action. The integral term does not react instantaneously to a zero-error signal but continues to correct, which tends to cause oscillations if the designer does not take this effect into account.

## PI-Control of a First-Order System

PI action applied to the speed controller of Figure 6.27 gives the diagram shown in Figure 6.33. The transfer functions are

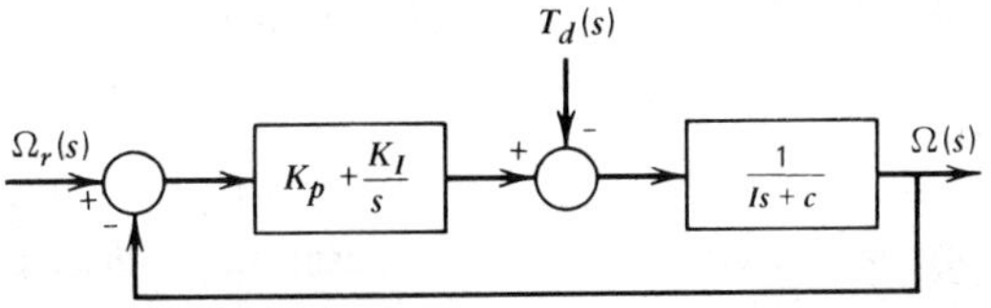


**Figure 6.33** Velocity servo with PI control.

$$\frac{\Omega(s)}{\Omega_r(s)} = \frac{K_p s + K_I}{Is^2 + (c + K_p)s + K_I} \tag{6.6-9}$$

$$\frac{\Omega(s)}{T_d(s)} = \frac{-s}{Is^2 + (c + K_p)s + K_I} \tag{6.6-10}$$

where the gains $K_p$ and $K_I$ are related to the component gains as before. The system is stable for positive values of $K_p$ and $K_I$. For $\Omega_r(s) = 1/s$, $\omega_{ss} = K_I/K_I = 1$, and the offset error is zero as with only integral action. Similarly, the deviation due to a unit step disturbance is zero at steady state. The damping ratio is

$$\zeta = \frac{c + K_p}{2\sqrt{IK_I}} \tag{6.6-11}$$

The presence of $K_p$ allows the damping ratio to be selected without fixing the value of the dominant time constant. For example, if the system is underdamped, the time constant is

$$\tau = \frac{2I}{c + K_p}, \qquad (\zeta < 1) \tag{6.6-12}$$

The gain $K_p$ can be picked to obtain the desired time constant, while $K_I$ is used to set the damping ratio. A similar flexibility exists if $\zeta = 1$. Complete description of the transient response requires that the numerator dynamics present in the transfer functions be accounted for (see Section 5.7).

## PI-Control of a Second-Order System

Integral control for the position servomechanism of Figure 6.29 resulted in a third-order system that is unstable. With a proportional term, the diagram becomes that of Figure 6.34, with the transfer functions

$$\frac{\Theta(s)}{\Theta_r(s)} = \frac{K_p s + K_I}{Is^3 + cs^2 + K_p s + K_I} \tag{6.6-13}$$

$$\frac{\Theta(s)}{T_d(s)} = \frac{-s}{Is^3 + cs^2 + K_p s + K_I} \tag{6.6-14}$$

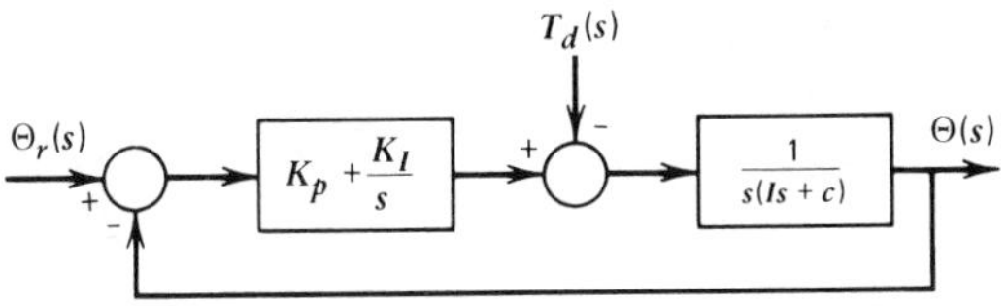


**Figure 6.34** Position servo with PI control.

The steady-state performance is acceptable as before if the system is assumed to be stable. This is true if the Routh criterion is satisfied; that is, if $I$, $c$, $K_p$, and $K_I$ are positive and $cK_p - IK_I > 0$. The difficulty here occurs when the damping is slight. For small $c$, the gain $K_p$ must be large in order to satisfy the last condition, and this can be difficult to implement physically. Such a condition can also result in an unsatisfactory time constant, but a detailed analysis of the cubic equation is beyond our scope now. The root-locus method of Chapter Eight provides the tools for analyzing this design further.

## 6.7 DERIVATIVE CONTROL

Integral action tends to produce a control signal even after the error has vanished, which suggests that the controller be made aware that the error is approaching zero. One way to accomplish this is to design the controller to react to the derivative of the error with the *derivative control* law.

$$F(s) = K_D sE(s) \tag{6.7-1}$$

where $K_D$ is the *derivative gain*. This algorithm is also called *rate action*. It is used to damp out oscillations.

Since it depends only on the error rate, derivative control should never be used alone. When used with proportional action, the following PD-control algorithm results.

$$\begin{aligned} F(s) &= (K_p + K_D s)E(s) \\ &= K_p(1 + T_D s)E(s) \end{aligned} \tag{6.7-2}$$

where $T_D$ is the *rate time* or *derivative time*. With integral action included, the proportional-plus-integral-plus-derivative (PID) control law is obtained.

$$F(s) = \left(K_p + \frac{K_I}{s} + K_D s\right) E(s) \tag{6.7-3}$$

This is a three-mode controller.

### PD-Control of a Second-Order System

Design of a controller with all three modes increases the cost of the system (except perhaps for digital systems, where the only change is a software modification). There are applications of the position servomechanism in which a nonzero deviation resulting from the disturbance can be tolerated, but an improvement in transient response over the proportional control result is desired. Integral action would not be required and rate action can be substituted to improve the transient response. Application of PD control to this system gives the block diagram of Figure 6.35 and the following transfer functions.

$$\frac{\Theta(s)}{\Theta_r(s)} = \frac{K_p + K_D s}{Is^2 + (c + K_D)s + K_p} \tag{6.7-4}$$

$$\frac{\Theta(s)}{T_d(s)} = \frac{-1}{Is^2 + (c + K_D)s + K_p} \tag{6.7-5}$$

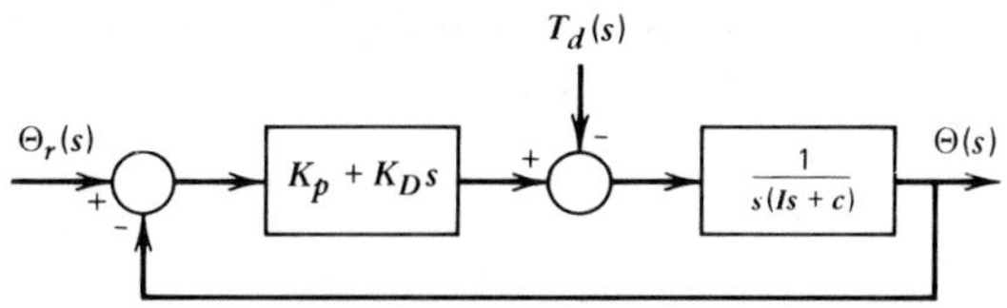


**Figure 6.35 Position servo with PD control.**

The system is stable for positive values of $K_D$ and $K_p$. The presence of rate action does not affect the steady-state response, and the steady-state results are identical to those with P control; namely, zero offset error and a deviation of $-1/K_p$ due to the disturbance. The damping ratio is

$$\zeta = \frac{c + K_D}{2\sqrt{IK_p}}$$

For P control, $\zeta = c/2\sqrt{IK_p}$. Introduction of rate action allows the proportional gain $K_p$ to be selected large to reduce the steady-state deviation, while $K_D$ can be used to achieve an acceptable damping ratio. The rate action also helps to stabilize the system by adding damping (if $c = 0$ the system with P control is not stable).

The feasibility of constructing a differentiating device is contradicted by the principle of integral causality (Chapter Two). However, in the next section techniques are presented for obtaining an approximation of such a device. For now we note that in the present example the equivalent of derivative action can be obtained by using a tachometer to measure the angular velocity of the load. The block diagram is shown in Figure 6.36. The gain of the amplifier-motor-potentiometer combination is $K_1$, and $K_2$ is the tachometer gain. The transfer functions are

$$\frac{\Theta(s)}{\Theta_r(s)} = \frac{K_1}{Is^2 + (c + K_1K_2)s + K_1} \tag{6.7-6}$$

$$\frac{\Theta(s)}{T_d(s)} = \frac{-1}{Is^2 + (c + K_1K_2)s + K_1} \tag{6.7-7}$$

Comparison with (6.7-4) and (6.7-5) shows that the system with tachometer feedback does not possess numerator dynamics. This system will therefore be somewhat more sluggish than the system with pure PD control. Otherwise the tachometer feedback

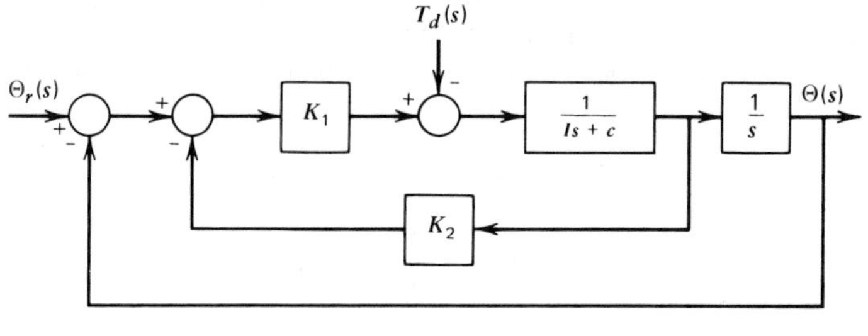


**Figure 6.36 Tachometer feedback arrangement to replace PD control for the position servo.**

arrangement gives a similar characteristic equation. The gains $K_1$ and $K_2$ can be chosen to yield the desired damping ratio and steady-state deviation as was done with $K_p$ and $K_D$.

## PID-Control

The position servomechanism design with PI control is not completely satisfactory because of the difficulties encountered when the damping $c$ is small. This problem can be solved by the use of the full PID-control law. From Figure 6.37 the following transfer functions are derived.

$$\frac{\Theta(s)}{\Theta_r(s)} = \frac{K_D s^2 + K_p s + K_I}{Is^3 + (c + K_D)s^2 + K_p s + K_I} \tag{6.7-8}$$

$$\frac{\Theta(s)}{T_d(s)} = \frac{-s}{Is^3 + (c + K_D)s^2 + K_p s + K_I} \tag{6.7-9}$$

A stable system results if all gains are positive and if

$$(c + K_D)K_p - IK_I > 0 \tag{6.7-10}$$

The presence of $K_D$ relaxes somewhat the requirement that $K_p$ be large to achieve stability. The steady-state errors are zero, and the transient response can be improved because three of the coefficients of the characteristic equation can be selected. To make further statements requires the analysis techniques of later chapters.

Proportional, integral, and derivative actions and their various combinations are not the only control laws possible, but they are the most common. It has been estimated that 90% of all controllers are of the PI type. This percentage will probably decrease as digital control with its great flexibility becomes more widely used. But the PI and PID controllers for some time will remain the standard against which any new designs must compete.

The conclusions reached concerning the performance of the various control laws are strictly true only for the plant model forms considered. These are the first-order model without numerator dynamics and the second-order model with a root at $s = 0$ and no numerator zeros.

The analysis of a control law for any other linear system follows the preceding pattern. The overall system transfer functions are obtained, and all of the linear system analysis techniques can be applied to predict the system's performance. If the performance is unsatisfactory, a new control law is tried and the process repeated. When this process fails to achieve an acceptable design, more systematic methods of altering the system's structure are needed; they appear in Chapters Seven and Eight.

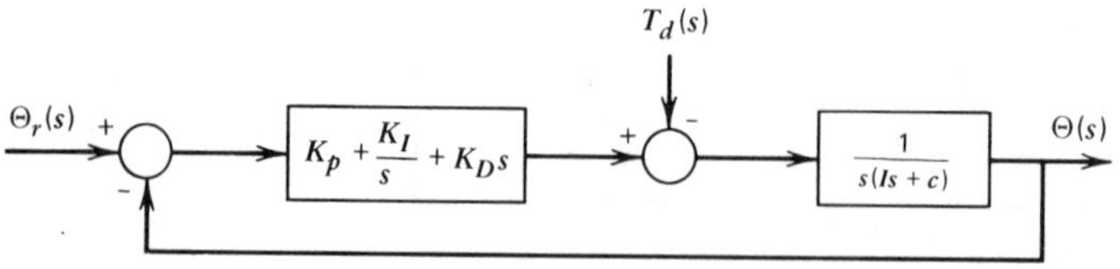


**Figure 6.37 Position servo with PID control.**

Also, the methods of these two chapters, plus those of Chapters Nine and Ten, are needed if the design requires a detailed analysis of a model of third order or higher.

We have used step functions as the test signals because they are the most common and perhaps represent the severest test of system performance. Impulse, ramp, and sinusoidal test signals are also employed. The type to use should be made clear in the design specifications.

## 6.8 ELECTRONIC CONTROLLERS

The control law must be implemented by a physical device before the control engineer's task is complete. The earliest devices were purely kinematic and were mechanical elements such as gears, levers, and diaphragms that usually obtained their power from the controlled variable. Most controllers now are analog electronic, hydraulic, pneumatic, or digital electronic devices. We now consider the analog electronic type. Digital control is taken up at the end of the chapter.

### Feedback Compensation and Controller Design

Most controllers that implement versions of the PID algorithm are based on the following feedback principle. Consider the single loop system shown in Figure 6.38. If the open-loop transfer function is large enough that $|G(s)H(s)| \gg 1$, the closed-loop transfer function is approximately given by

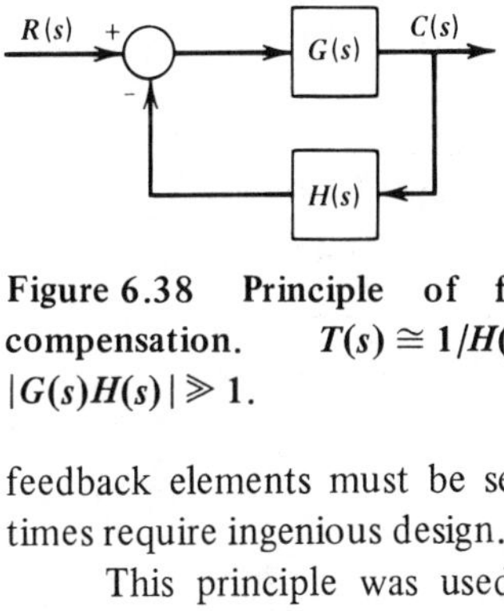


**Figure 6.38 Principle of feedback compensation. $T(s) \cong 1/H(s)$ if $|G(s)H(s)| \gg 1$.**

$$T(s) = \frac{G(s)}{1 + G(s)H(s)} \cong \frac{G(s)}{G(s)H(s)} = \frac{1}{H(s)}$$

The principle states that a power unit $G(s)$ can be used with a feedback element $H(s)$ to create any desired transfer function $T(s)$. The power unit must have a gain high enough that $|G(s)H(s)| \gg 1$, and the feedback elements must be selected so that $H(s) = 1/T(s)$. The latter task can sometimes require ingenious design.

This principle was used in Chapter One to explain the design of a feedback amplifier. The power unit is the uncompensated amplifier with a high gain of $K$. The feedback element is a resistance network (Figure 6.39). Thus $G(s) = K$ and $H(s) =$

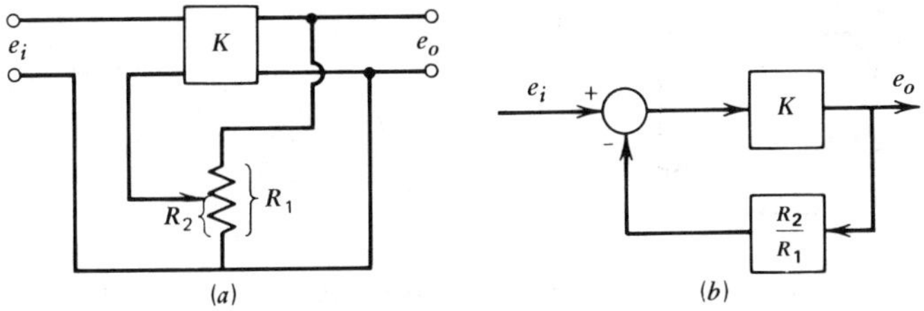


**Figure 6.39 Feedback compensation of an amplifier.**

$R_2/R_1$. If $KR_2/R_1 \gg 1$, then $T(s) \cong R_1/R_2$. This design eliminates the difficulty associated with the age and temperature variability of the gain $K$. The feedback amplifier gain $R_1/R_2$ is more stable but smaller than the original gain. Although the original problem was not a control problem, the design can be used as a proportional controller. The gain $R_1/R_2$ can be adjusted for a particular application if one resistance is a potentiometer.

The feedback amplifier model is a static one in which $G(s)$ and $H(s)$ are constants. For this case it is easy to interpret the meaning of the inequality $|G(s)H(s)| \gg 1$. When dynamic elements are present the inequality obviously depends on the value of the Laplace operator $s$. In practice, the inequality is considered to be satisfied if it is true for the range of frequencies $\omega$ over which the device will be operated, where $s = i\omega$.

## Op-Amp Circuits

The op amp is a high-gain amplifier with a high input impedance. In Section 2.4 an integrating circuit was designed by making use of these properties. The procedure is generalized here for the purpose of designing PID controllers. The uncompensated amplifier is shown in Figure 6.40*a*, in which the sign reversal property is displayed. This will require the use of an inverter to maintain proper voltage signs.

A circuit diagram of the op amp with general feedback and input elements is shown in Figure 6.40*b*. A similar but simplified form is given in Figure 6.40*c*. The impedance $T_i(s)$ of the input elements is defined such that

$$E_i(s) - E_a(s) = T_i(s)I_1(s)$$

For the feedback elements,

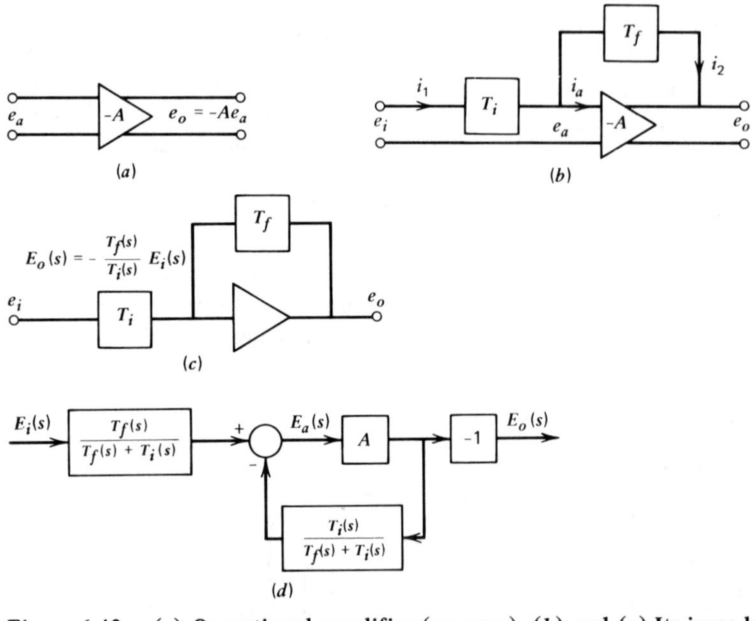


**Figure 6.40** (*a*) Operational amplifier (op amp). (*b*) and (*c*) Its impedance representations. (*d*) Its block diagram.

$$E_a(s) - E_o(s) = T_f(s) I_2(s)$$

(Recall that impedance is the dynamic operator equivalent to resistance in the static case.)

The high internal impedance of the op amp implies that $i_a \cong 0$, and thus $i_1 \cong i_2$. The final relation we need is the amplifier relation

$$e_o = -Ae_a$$

or

$$E_o(s) = -A E_a(s) \tag{6.8-1}$$

When the preceding relations are used to eliminate $I_1(s)$ and $I_2(s)$, the result is

$$E_a(s) = \frac{T_f(s)}{T_f(s) + T_i(s)} E_i(s) + \frac{T_i(s)}{T_f(s) + T_i(s)} E_o(s)$$

With this and (6.8-1), the block diagram of Figure 6.40*d* can be constructed. The transfer function between $E_i(s)$ and $E_o(s)$ is

$$T(s) = \frac{E_o(s)}{E_i(s)} = -\frac{T_f(s)}{T_f(s) + T_i(s)} \frac{A}{1 + A H(s)}$$

where

$$H(s) = \frac{T_i(s)}{T_f(s) + T_i(s)}$$

Since $A$ is large (of the order of $10^5$ to $10^8$), $|A H(s)| \gg 1$ and we obtain

$$T(s) \cong \frac{T_f(s)}{T_i(s)}$$

and

$$\frac{E_o(s)}{E_i(s)} = -\frac{T_f(s)}{T_i(s)} \tag{6.8-2}$$

This is the basic relation for op-amp applications.

## Proportional Control

A proportional controller can be obtained with two resistors, as shown in Figure 6.41*a*. For this circuit, $T_f(s) = R_f$, $T_i(s) = R_i$, and

$$\frac{E_o(s)}{E_i(s)} = -\frac{R_f}{R_i} \tag{6.8-3}$$

The gain can be made adjustable by using a potentiometer for one of the resistances.

An *inverter* is such a circuit with $R_f = R_i$. It simply inverts the sign of the input voltage without changing its magnitude. It can be used in cascade with other elements to maintain proper sign relations, as shown in Figure 6.41*b*.

This multiplier circuit can be modified to act as an adder (Figure 6.42). The output relation is

$$e_o = -\frac{R_3}{R_1} e_1 - \frac{R_3}{R_2} e_2 \tag{6.8-4}$$

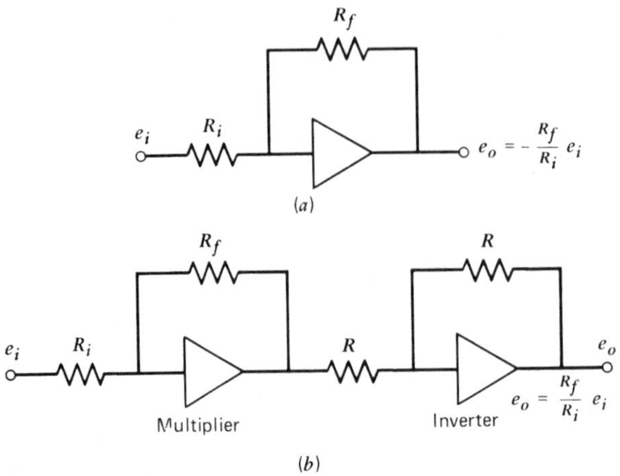


**Figure 6.41** (*a*) **Op-amp implementation of proportional control.** (*b*) **Use of an inverter to maintain voltage polarity.**

When used with an inverter, this circuit implements the proportional control law plus the manual reset term. The voltage $e_1$ corresponds to the error, $R_3/R_1$ is the proportional gain, and $R_3e_2/R_2$ is the manual reset. An application requiring such a circuit is the temperature controller of Figure 6.8.

## PI Controllers

The impedance of a capacitor is found from its voltage-current relation in the Laplace domain

$$E(s) = \frac{1}{Cs} I(s)$$

The impedance is $1/Cs$. An integral controller is obtained by using this element in the feedback circuit, with a resistance as the input element (Figure 6.43). For this circuit,

$$\frac{E_o(s)}{E_i(s)} = -\frac{1}{RCs} \tag{6.8-5}$$

The integral gain is $K_I = 1/RC$. If a larger gain is required, a multiplier can be cascaded with this circuit.

Equation (6.8-5) is true if the initial voltage on the capacitor is zero. Any nonzero initial voltage $e_o(0)$ is added to the output to give

$$e_o(t) = -\frac{1}{RC}\int_0^t e_i(t)\,dt + e_o(0) \tag{6.8-6}$$

Many industrial controllers provide the operator with a choice of control modes, and the operator can switch from one mode to another when the process characteristics or control objectives change. When a switch occurs it is necessary to provide any

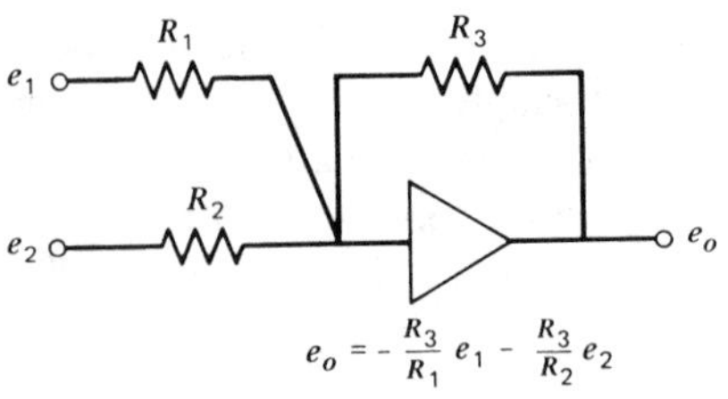


**Figure 6.42 Op-amp adder circuit.**

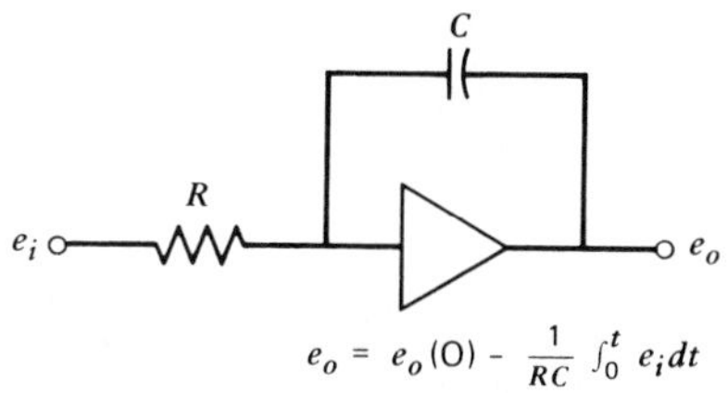


**Figure 6.43 Op-amp integrator circuit.**

integrators with the proper initial voltages, or else undesirable transients will occur when the integrator is switched into the system. Commercially available controllers usually have built-in circuits for this purpose. A diagram of one such design for PI control is given in Appendix C of Reference 9.

PI control can be implemented with the circuit of Figure 6.44. The impedance of the series resistance and capacitance is $T_f(s) = R_f + 1/Cs$. If the capacitance is initially discharged,

$$\frac{E_o(s)}{E_i(s)} = -\frac{R_f Cs + 1}{R_i Cs} = -\frac{R_f}{R_i} - \frac{1}{R_i Cs} \tag{6.8-7}$$

The corresponding gains are

$$K_p = \frac{R_f}{R_i}$$

$$K_I = \frac{1}{R_i C}$$

## Output Limitation

In practice the final control elements are always incapable of delivering energy to the controlled system above a certain rate. Also, we will see that integral control can suffer from a nonlinear effect called *reset windup*, which is in part caused by the finite capacity of the final control elements. This problem can be eliminated by limiting the output of the controller so that it cannot command the final control elements to deliver more power than they can. This topic is discussed further in Section 7.3. For now we consider how such a limitation can be accomplished. (Some power amplifiers will already contain such circuitry; consult the manufacturer's data.)

One way to limit the output of a multiplier is to use diodes and bias voltages $V_1$ and $V_2$, as shown in Figure 6.45*a*. The diodes short circuit the feedback circuit whenever the output $e_o$ lies outside of the desired range. The input-output relation is shown in Figure 6.45*b* for $V_1 < V_2$ ($V_1 > 0$, $V_2 > 0$). Other transfer characteristics can be achieved with slightly different circuits (see, for example, Chapter 3 of Reference 9 and Chapter 10 of Reference 10). For PI control, the feedback resistor $R_f$ is placed in series with a capacitor, and the same limiting circuit design can be used. The concept can be extended to other control circuits.

## PD Control

In theory a differentiator can be created by interchanging the resistance and capaci-

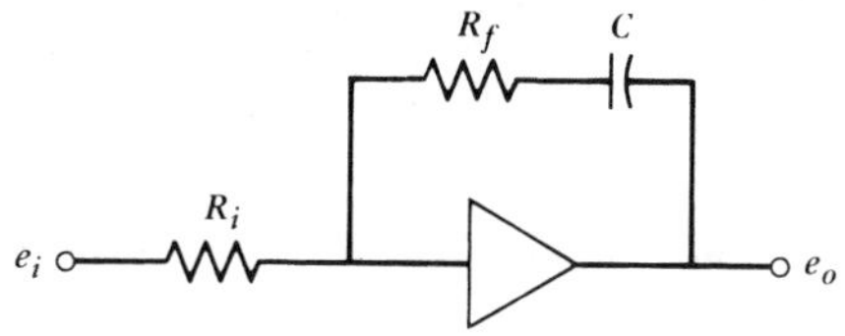


**Figure 6.44 Op-amp implementation of PI control.**

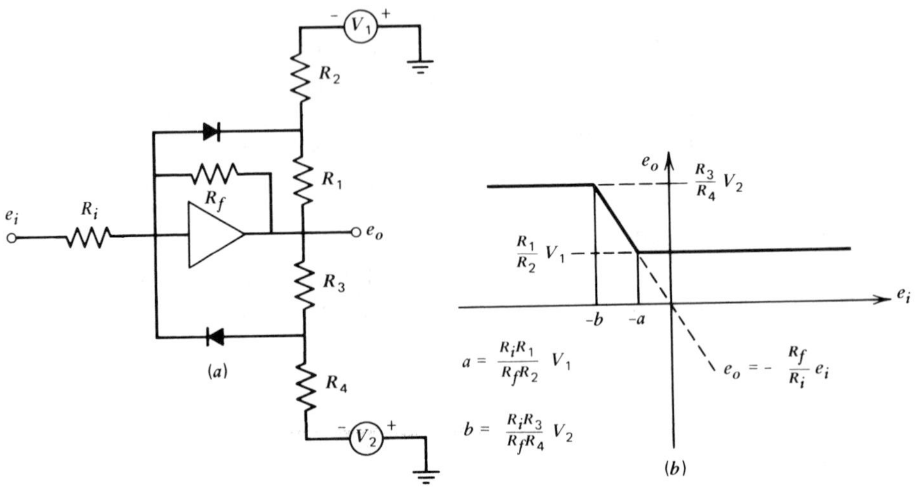


**Figure 6.45** (*a*) **Diode circuit for limiting the output of an op-amp multiplier.** (*b*) **Transfer characteristic.**

tance in the integrator. The result is shown in Figure 6.46*a*. The input-output relation for this ideal differentiator is

$$\frac{E_o(s)}{E_i(s)} = -RCs \tag{6.8-8}$$

The difficulty with this design is that no electrical signal is "pure." Contamination always exists as a result of voltage spikes, ripple, and other transients generally categorized as "noise." These high-frequency signals have large slopes compared with the more slowly varying primary signal, and thus they will dominate the output of the differentiator. In practice this problem is solved by filtering out high-frequency signals either with a low-pass filter inserted in cascade with the differentiator, or by using a redesigned differentiator such as the one shown in Figure 6.46*b*. Its transfer function is

$$\frac{E_o(s)}{E_i(s)} = -\frac{RCs}{R_1Cs + 1} \tag{6.8-9}$$

The frequency-response plot of this transfer function shows that it acts like the ideal differentiator for frequencies up to about $\omega = 1/R_1C$. For higher frequencies, the attenuation curve has zero slope rather than the 20 db/decade slope required for

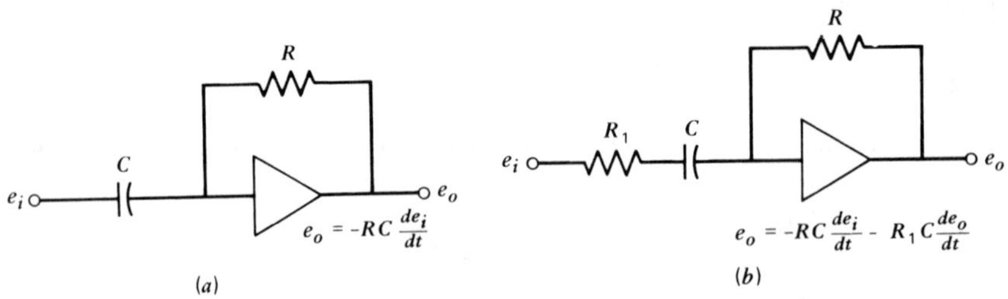


**Figure 6.46** **Op-amp implementations of a differentiator.** (*a*) **Ideal differentiator.** (*b*) **Practical differentiator.**

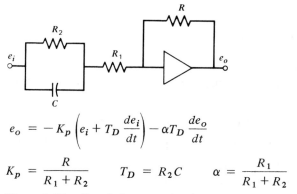

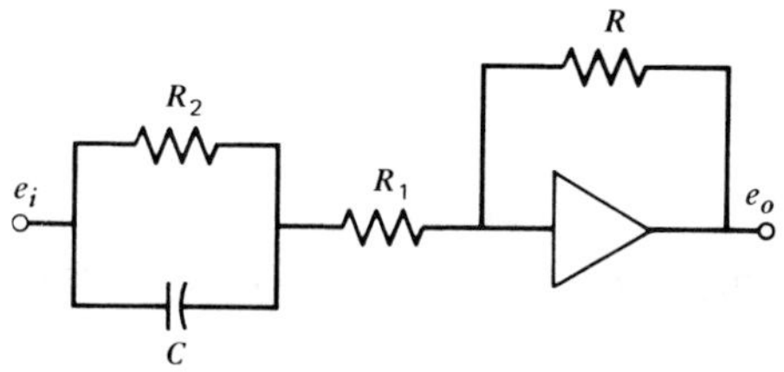


$$e_o = -K_p\left(e_i + T_D\frac{de_i}{dt}\right) - \alpha T_D\frac{de_o}{dt}$$

$$K_p = \frac{R}{R_1+R_2} \qquad T_D = R_2C \qquad \alpha = \frac{R_1}{R_1+R_2}$$

**Figure 6.47 Practical op-amp implementation of PD control.**

differentiation, and thus the circuit does not differentiate the high-frequency signals but merely amplifies them. Since their amplitudes are generally small, this effect is negligible. For $\omega < 1/R_1C$, the derivative gain of (6.8-9) is $K_d = RC$.

A practical PD controller can be constructed in a similar manner (Figure 6.47). The impedance of the input elements is

$$T_i(s) = \frac{R_1 + R_2 + R_1R_2Cs}{R_2Cs + 1}$$

The transfer function reduces to

$$\frac{E_o(s)}{E_i(s)} = -\frac{R(R_2Cs + 1)}{R_1 + R_2 + R_1R_2Cs} \tag{6.8-10}$$

For the ideal PD controller, $R_1 = 0$ and

$$\frac{E_o(s)}{E_i(s)} = -\frac{R(R_2Cs + 1)}{R_2} = -\frac{R}{R_2}(R_2Cs + 1) \tag{6.8-11}$$

The attenuation curve for the ideal controller breaks upward at $\omega = 1/R_2C$ with a slope of 20 db/decade. The curve for the practical controller does the same but then becomes flat for $\omega > (R_1 + R_2)/R_1R_2C$. This provides the same limiting effect at high frequencies as that of the practical differentiator (6.8-9).

The transfer function (6.8-10) can be written as

$$\frac{E_o(s)}{E_i(s)} = -\frac{K_p(1 + T_Ds)}{(1 + \alpha T_Ds)} \tag{6.8-12}$$

where

$$K_p = \frac{R}{R_1 + R_2}$$

$$T_D = R_2C$$

$$\alpha = \frac{R_1}{R_1 + R_2}$$

The circuit limits frequencies above $\omega = 1/\alpha T_D$. For lower frequencies the transfer function is that of PD control (6.8-26). The proportional gain is $K_p$ and the rate constant is $T_D$.

## PID Control

PID control can be implemented by joining the PI and PD controllers in parallel, but this is expensive because of the number of op amps and power supplies required. Instead, the usual implementation is that shown in Figure 6.48. The impedances for the series and parallel $RC$ elements were given previously. From (6.8-2), the transfer function is

$$\frac{E_o(s)}{E_i(s)} = -\frac{(RCs+1)(R_2C_1s+1)}{Cs(R_1+R_2+R_1R_2C_1s)} \tag{6.8-13a}$$

$$= -\left(\frac{RC+R_2C_1}{R_2C}+\frac{1}{R_2Cs}+RC_1s\right)\frac{\beta}{(\beta R_1C_1s+1)}$$

$$= -\left(K_p+\frac{K_I}{s}+K_Ds\right)\frac{1}{(\beta R_1C_1s+1)} \tag{6.8-13b}$$

*Where*

$$\beta = \frac{R_2}{R_1+R_2}$$

$$K_p = \beta\frac{RC+R_2C_1}{R_2C}$$

$$K_I = \frac{\beta}{R_2C}$$

$$K_D = \beta RC_1$$

The denominator term $\beta R_1C_1s+1$ limits the effect of frequencies above $\omega = 1/\beta R_1C_1$. When $R_1 = 0$, ideal PID control results. This is sometimes called the *noninteractive*

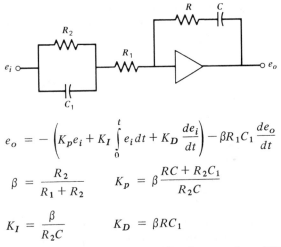

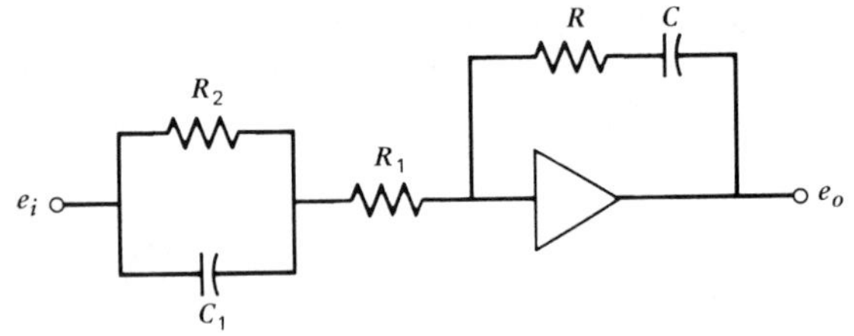


$$e_o = -\left(K_pe_i + K_I\int_0^t e_i\,dt + K_D\frac{de_i}{dt}\right) - \beta R_1C_1\frac{de_o}{dt}$$

$$\beta = \frac{R_2}{R_1+R_2} \qquad K_p = \beta\frac{RC+R_2C_1}{R_2C}$$

$$K_I = \frac{\beta}{R_2C} \qquad K_D = \beta RC_1$$

**Figure 6.48 Practical op-amp implementation of PID control.**

algorithm because the effect of each of the three modes is additive, and they do not interfere with one another. The form given by (6.8-13*a*) for $R_1 \neq 0$ is the *real* or *interactive* algorithm. This name results from the fact that historically it was difficult to implement noninteractive PID control with mechanical or pneumatic devices.

Preliminary calculation of the control gains is made as in Sections 6.5, 6.6, and 6.7, using the transient and steady-state performance specifications. The resistance and capacitance values can then be selected in terms of the gains. Usually there is sufficient freedom to choose feasible values for these electrical components because there are more such elements than control gains. In the PID controller, there are three resistors and two capacitors, but only three gains. If the limiting upper frequency is specified, the problem involves four constraints and five values to be chosen. Four of the values can be related to the fifth one, which can be selected to satisfy another criterion (cost, size, etc.).

## 6.9 PNEUMATIC CONTROLLERS

The nozzle-flapper introduced in Section 6.4 is a high-gain device that is difficult to use without modification. The gain $K_f$ is known only imprecisely and is sensitive to changes induced by temperature and other environmental factors. Also, the linear region over which (6.4-8) applies is very small. However, the device can be made useful by compensating it with feedback elements, with the general principle of Figure 6.38 as a guide. The overall gain is reduced, but the result is a controller with an increased linear range and decreased sensitivity to parameter changes and uncertainties.

### Proportional Control

In order to obtain proportional action, the feedback compensation principle states that the feedback element around the nozzle-flapper should have a constant transfer function. Let the input to the controller be the flapper displacement $y$ and the output be a controlled pressure (see Figures 6.21*a* and 6.23). Then we can use the flapper itself as an error-detecting beam. The feedback element must then be a transducer capable of converting the output pressure into a displacement of the lower end of the flapper. The result is shown in Figure 6.49*a*. The controlled output pressure $p_o$ represents a deviation from the nominal output pressure corresponding to the reference equilibrium operating condition. The total pressure ($p_o$ plus nominal) can be used to operate a pneumatic flow valve, for example.

A pneumatic *relay* or *booster* allows a large airflow and amplifies the back pressure, as in Figure 6.22. As the actuating displacement $y$ moves to the right, the feedback bellows acts as a pivot and the back pressure decreases. This causes the booster bellows to retract and open the booster; the output pressure then increases. This increased pressure causes the feedback bellows to expand to the left and to reduce the nozzle-flapper distance (the upper end of the flapper is held constant at its new position during this motion).

The system is modeled as follows in terms of linearized variables. The geometric relation for the flapper is

$$x = \frac{a}{a+b}y - \frac{b}{a+b}z$$

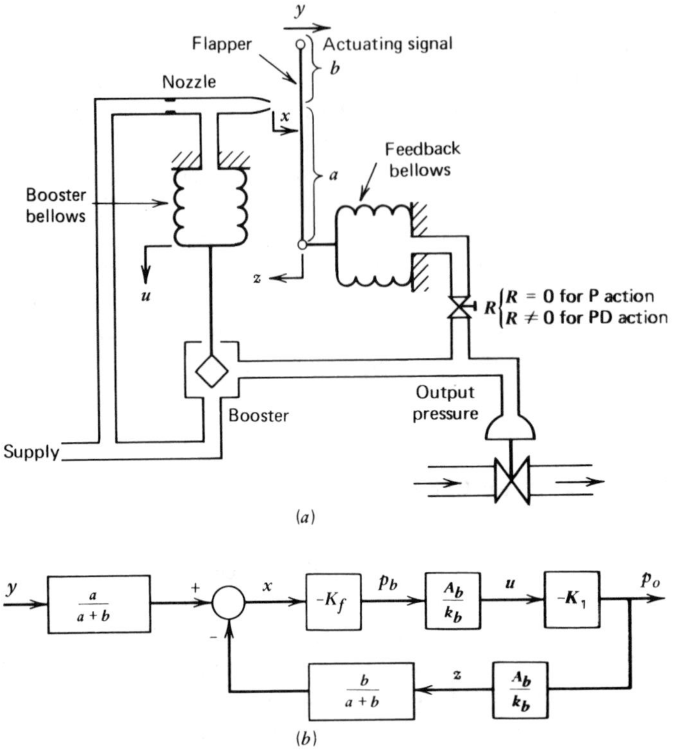


**Figure 6.49** **(*a*) Pneumatic implementation of P and PD control via feedback compensation of the nozzle-flapper element. (*b*) Block diagram for the P-control case ($R = 0$).**

Assume both bellows are identical with areas $A_b$ and equivalent spring constants $k_b$. Newton's law applied to a negligible mass gives for the feedback bellows

$$p_o A_b = k_b z$$

and

$$p_b A_b = k_b u$$

for the booster bellows. The time constants of the bellows are negligible if their inlet resistance is small ($R = 0$). Finally, the linearized relation for the booster operation is

$$p_o = -K_1 u$$

where $K_1 > 0$. The resulting block diagram is shown in Figure 6.49*b*.

The transfer function for the controller is approximately

$$\frac{p_o}{y} = \frac{a k_b}{b A_b} \tag{6.9-1}$$

if $K_f \gg (a+b)k_b^2/bA_b^2K_1$. The action is proportional. Note that we need not be concerned with the precise value of $K_f$ as long as we know that the latter inequality is satisfied. The gain in (6.9-1) is smaller than the uncompensated gain $K_f$. This is the price paid for reduced sensitivity to uncertainties in $K_f$.

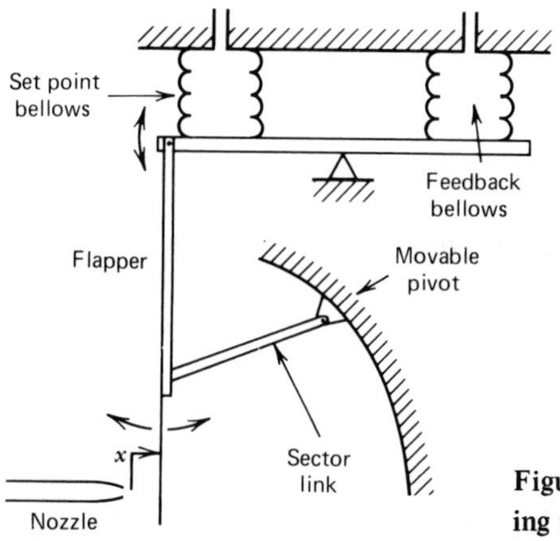


**Figure 6.50 Typical mechanism for adjusting the gain of a nozzle-flapper controller.**

To demonstrate how the feedback improves the linearity, derive the relation between $x$ and $y$. From the block diagram,

$$x = \frac{a}{a+b} \frac{y}{1 + \dfrac{b}{a+b}\left(\dfrac{A_b}{k_b}\right)^2 K_1 K_f}$$

Since $K_f$ is large, $x$ remains small and thus within the linear region for a wider range of $y$ values than with the uncompensated system.

The proportional gain can be adjusted by using a linkage that changes the ratio $a/b$. An example of such a design is shown in Figure 6.50. This can be used with the pivoted-beam, double-bellows arrangement shown in Figure 6.10. The force on the left side of the beam results from the set-point pressure. The force on the right results from the feedback bellows. If the feedback pressure is below the set-point pressure, the flapper moves away from the nozzle and the output pressure is increased. The sensitivity of $x$ to the beam motion is affected by the pivot location of the sector link. Moving the pivot clockwise decreases the sensitivity and thus decreases the proportional gain.

## PD Control

Derivative action can be synthesized pneumatically with an adjustable restriction in the input line of a bellows, as shown in Figure 6.20. The transfer function is given by (6.4-7). PD control is obtained from the system in Figure 6.49 if the inlet resistance $R$ of the feedback bellows is nonzero. In this case the time constant $\tau$ in (6.4-7) is not zero, and the bellows pressure lags behind changes in the output pressure $p_o$. The resulting model is given by Figure 6.51. The bellows constant $K$ is $A_b/k_b$. The system's transfer function is

$$\frac{P_o(s)}{Y(s)} = \frac{a}{a+b} \frac{K_f K_1 K}{1 + \dfrac{b K_f K_1 K^2}{(a+b)(\tau s + 1)}}$$

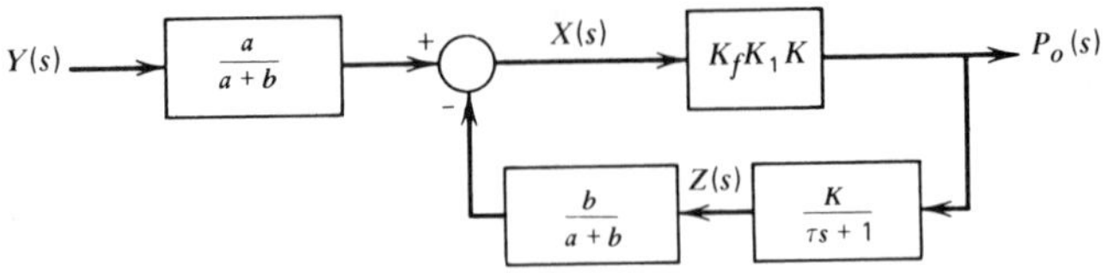


**Figure 6.51 Block diagram of the system shown in Figure 6.49 for PD control ($R \neq 0$).**

If $K_f > |(a+b)(\tau s+1)/bK_1K^2|$,

$$\frac{P_o(s)}{Y(s)} = \frac{a}{bK}(\tau s + 1) \tag{6.9-2}$$

The proportional gain is: $K_p = a/bK$; the derivative gain is $K_D = a\tau/bK$. Note that this result reduces to (6.9-1) when the bellows time constant $\tau$ is zero. This occurs when the adjustable resistance $R$ is fully open ($R = 0$). When the resistance is fully closed ($R = \infty$), the feedback loop does not operate, and the system becomes a two-position controller because the nozzle back pressure will take either one of the two extreme values ($p_a$ or $p_s$), depending on the direction of motion of $y$.

## PID Control

PI and PID control can be obtained with two opposing bellows in the feedback path, as shown in Figure 6.52*a*. For simplicity the booster subsystem is not shown but can be included as before. The plate used as the interface between the bellows is connected to the lower end of the flapper. Assume that the bellows have the same area $A_b$, spring constant $k_b$, and pneumatic capacitance $C$. Usually the resistances are such that $R_I \gg R_D$ so that the time constant $R_IC$ of bellows 2 is larger than that for bellows 1. A force balance at the interface gives, for negligible system mass,

$$k_bZ(s) = \frac{A}{R_DCs+1}P_o(s) - \frac{A}{R_ICs+1}P_o(s)$$

This gives the feedback loop shown in Figure 6.52*b*, with $K = A_b/k_b$. The transfer function of the feedback loop reduces to

$$H(s) = \frac{bK}{a+b}\frac{(R_I - R_D)Cs}{(R_ICs+1)(R_DCs+1)} \tag{6.9-3}$$

The simplified diagram is given in Figure 6.52*c*. The closed-loop transfer function is

$$\frac{P_o(s)}{Y(s)} = \frac{a}{a+b}\frac{-K_f}{1-K_fH(s)}$$

If $K_f \gg |1/H(s)|$,

$$\frac{P_o(s)}{Y(s)} = \frac{a}{(a+b)H(s)} \tag{6.9-4}$$

PI control is obtained if the inlet resistance to bellows 1 is zero ($R_D = 0$). In this case

$$H(s) = \frac{bK}{a+b}\frac{R_ICs}{R_ICs+1}$$

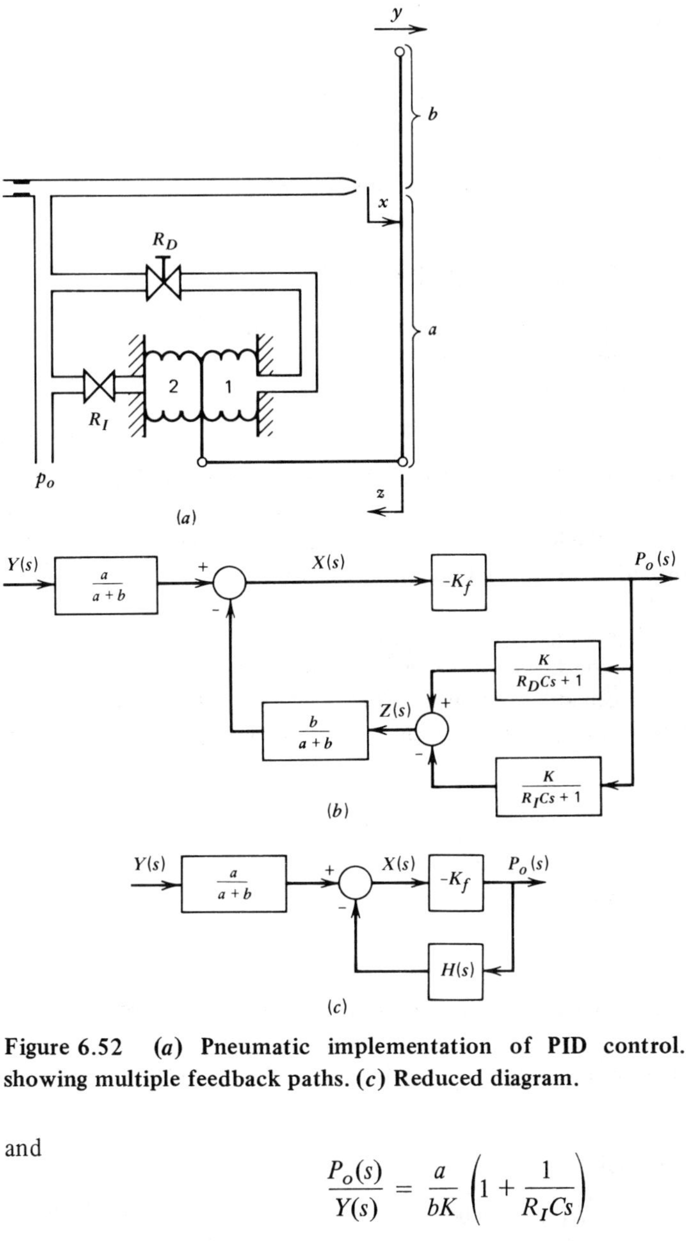


**Figure 6.52** **(*a*) Pneumatic implementation of PID control. (*b*) Block diagram showing multiple feedback paths. (*c*) Reduced diagram.**

and

$$\frac{P_o(s)}{Y(s)} = \frac{a}{bK}\left(1 + \frac{1}{R_I Cs}\right) \tag{6.9-5}$$

The proportional gain is $K_p = a/bK$. The reset time is $T_I = R_I C$, and the integral gain is $K_I = K_p/T_I$. Note that if the inlet to bellows 2 is closed ($R_I = \infty$), the expression (6.9-5) reduces to that of proportional control (6.9-1).

The full PID-control law is obtainable when both resistances are nonzero and $R_I \gg R_D$. With this inequality, $R_I \pm R_D \cong R_I$, and (6.9-3) reduces to

$$H(s) = \frac{bK}{a+b}\,\frac{R_I Cs}{R_I R_D C^2 s^2 + R_I Cs + 1}$$

The system transfer function is

$$\frac{P_o(s)}{Y(s)} = \frac{a}{bK}\left(1 + \frac{1}{R_I C s} + R_D C s\right) \tag{6.9-6}$$

which is the PID-control law. The proportional gain is $K_p = a/bK$. The reset and rate times $T_I$ and $T_D$ are simply the time constants of the bellows.

## 6.10 HYDRAULIC CONTROLLERS

The basic unit for synthesis of hydraulic controllers is the hydraulic servomotor analyzed in Example 5.6. For a negligible load the model is

$$\frac{X(s)}{Z(s)} = \frac{K}{s} \tag{6.10-1}$$

where $K = C_1/A$ and where $z$ and $x$ are the pilot valve and power piston displacements. Use of this model for design purposes presumes that other, more powerful final control elements are used if the load is appreciable. We note in passing that the nozzle-flapper concept is also used in hydraulic controllers. Reference 11 discusses these devices and others in some detail.

In contrast to the high-gain amplifier used as the basis for electronic and pneumatic controllers, the hydraulic servomotor is an integrator. Nevertheless, the principle of feedback compensation can still be applied. With a servomotor in the forward path, the structure of a hydraulic controller is as shown in Figure 6.53. The closed-loop transfer function is

$$\frac{X(s)}{Y(s)} = \frac{K/s}{1 + (K/s)H(s)}$$

where $y$ is a command displacement like that for the pneumatic flapper. If $K \gg |s/H(s)|$, the transfer function becomes

$$\frac{X(s)}{Y(s)} = \frac{1}{H(s)} \tag{6.10-2}$$

The preceding inequality is satisfied if the servovalve constant $C_1$ is large in relation to the power piston area $A$. Physically this means that the valve flow rate must be large compared to the volume rate swept out by the piston; that is, the unit's response must be fast.

### Proportional Control

Equation (6.10-2) shows that $H(s)$ must be constant in order for proportional control to be obtained. Since $H(s)$ relates two displacements, this suggests a pivoted beam. The design is shown in Figure 6.54$a$. The beam is called a *walking beam* because of its motion during the unit's operation. Its function is similar to that of the pneumatic flapper. Assume that the displacements $x$, $y$, and $z$ are measured from reference

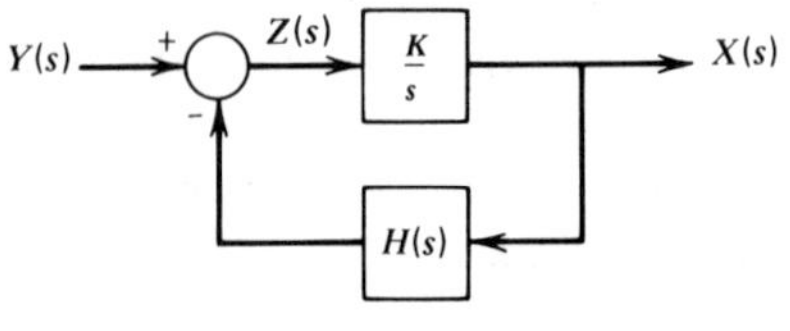


**Figure 6.53 Feedback compensation of a hydraulic integrator.**

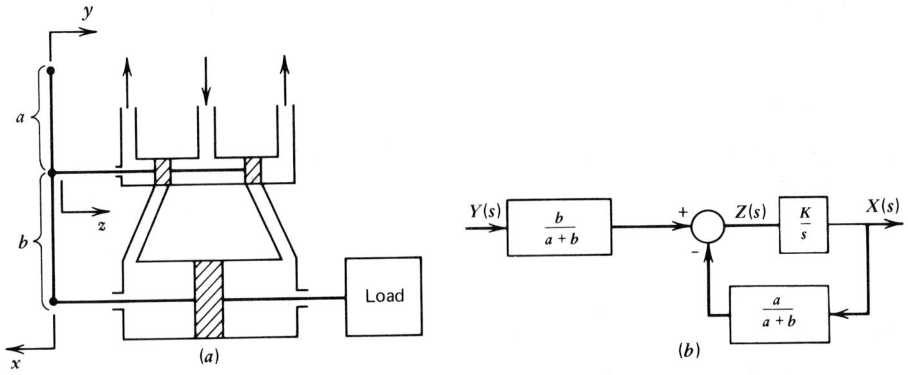


**Figure 6.54 Walking-beam servomotor and its block diagram.**

positions corresponding to the line-on-line configuration of the pilot valve and some rest position of the load. For small deviations from this reference equilibrium, the geometry of the beam is such that

$$z = \frac{b}{a+b}y - \frac{a}{a+b}x \tag{6.10-3}$$

This equation and (6.10-1) complete the model. The block diagram is shown in Figure 6.54*b*. The feedback transfer function is

$$H(s) = \frac{a}{a+b}$$

If $K \gg |(a+b)s/a|$, the closed-loop transfer function is

$$\frac{X(s)}{Y(s)} = \frac{b}{a+b}\,\frac{a+b}{a} = \frac{b}{a} \tag{6.10-4}$$

The proportional gain is $K_p = b/a$. It can be adjusted by varying the ratio $b/a$ with a linkage.

In addition to a proportional controller, the servomotor with a walking beam can also be used as a servomechanism. If we assume that the gain $K$ is not high enough for the preceding approximation to be valid, the following transfer function applies.

$$\frac{X(s)}{Y(s)} = \frac{b/a}{\tau s + 1} \tag{6.10-5}$$

where $\tau = (a+b)/aK$. The operation is as follows. For an actuating displacement $y$ to the right, the lower end of the beam acts as a pivot because the inertia and friction of the power piston are greater than those of the pilot wave. This motion moves the pilot valve to the right and causes the power piston to move to the left. The beam now pivots about the upper end, which is held fixed by the mechanism that produces the actuating displacement. This motion moves the pilot valve to the left until the flow is shut off and the load comes to rest at a new position. If $y$ is a step input, the load reaches its new position in approximately $4\tau$ time units. If the gain $K$ is high, $\tau$ is small and the motion approximates the instantaneous action of proportional control (6.10-4).

## PI Control

To obtain PI control, (6.10-2) indicates that the feedback transfer function $H(s)$ must be of the form

$$H(s) = \frac{a_1 s}{a_2 + a_3 s} = \frac{W(s)}{X(s)}$$

where $w$ is a displacement created by the feedback element in response to $x$. The displacement $x$ and $w$ must be related by

$$a_3 \dot{w} + a_2 w = a_1 \dot{x}$$

With some thought this leads us to try the damper-spring combination shown in Figure 6.55. For a negligible mass, the force balance gives

$$c\dot{w} + kw = c\dot{x}$$

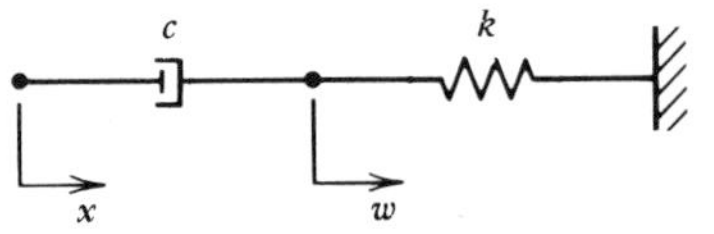


**Figure 6.55 Spring-damper subsystem used to create PI control.**

which is in the required form. The motion $w$ occurs at the lower end of the walking beam and affects the displacement $z$ in the same way that $x$ did in (6.10-3). With this in mind, we can envision the system as shown in Figure 6.56, with its associated block diagram.

The transfer function for the closed-loop system is

$$\frac{X(s)}{Y(s)} = \frac{b}{a}\left(1 + \frac{k}{cs}\right) \tag{6.10-6}$$

if $K \gg |k(a+b)(cs+k)/ac|$. The proportional gain $b/a$ is adjustable with a linkage, while the integral gain is adjusted by changing the damping constant $c$. This cannot be done with the typical dashpot we have seen thus far (for example, see Figure 2.21), but the design shown in Figure 6.57 is suitable for this purpose. The damping is varied by changing the resistance $R$. It can be shown with the approach of Section 2.6 that the damping constant is approximately given by $c = RA^2$, where $R$ relates the volume flow rate to the pressure drop, and $A$ is the area of the damper's piston.

## Derivative Action

In order to obtain PD control, $H(s)$ must be of the form

$$H(s) = \frac{b_1}{b_2 + b_3 s} = \frac{W(s)}{X(s)}$$

Thus

$$b_3 \dot{w} + b_2 w = b_1 x$$

This action can be synthesized by interchanging the locations of the spring and damper elements in Figure 6.56 so that

$$c\dot{w} + kw = kx$$

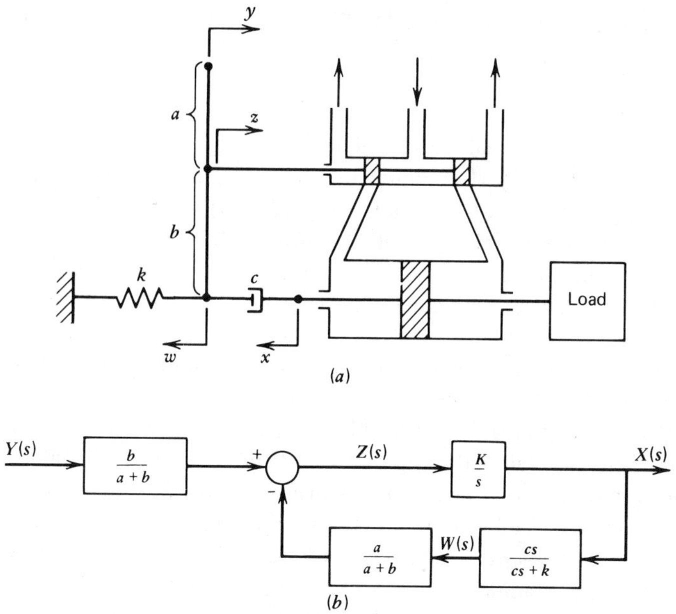


**Figure 6.56 Hydraulic implementation of PI control and its block diagram.**

The resulting overall system transfer function for a high-gain $K$ is

$$\frac{X(s)}{Y(s)} = \frac{b}{a}\left(\frac{c}{k}s + 1\right) \tag{6.10-7}$$

This is the desired PD-control law.

PID control was synthesized pneumatically by using two opposing bellows, one for the integral action and one for derivative action. The same principle can be applied here with the spring-damper pairs playing the role of the bellows, with the time constant of one pair being much greater than the other. The algebra is similar to the previous cases.

We note that derivative action has not seen much use in hydraulic controllers. This action supplies damping to the system, but hydraulic systems are usually highly damped intrinsically because of the viscous working fluid. PI control is the algorithm most commonly implemented with hydraulics.

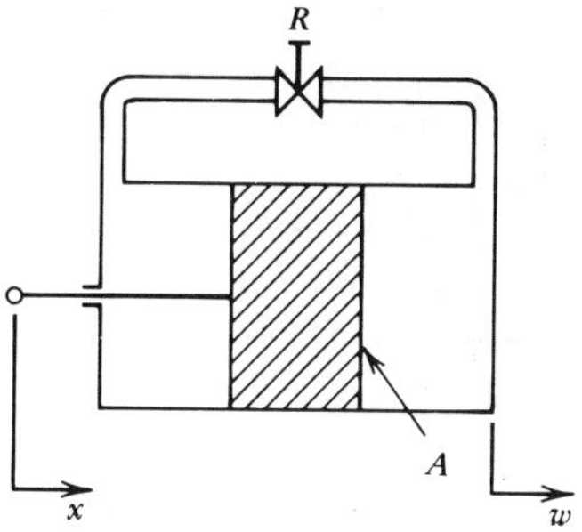


**Figure 6.57 Damper with adjustable resistance.**

## 6.11 DIGITAL IMPLEMENTATION OF CONTROL ALGORITHMS

There are two types of applications of digital computers to control problems. The first is *supervisory control* in which analog controllers are directly involved with the plant to be controlled, while the digital computer provides command signals to the controllers. The second application is *direct digital control* (*DDC*) in which the digital device acts at the lowest levels of the system, in direct control of the plant.

There are several advantages to using digital computers in control systems. Complex control algorithms can be implemented easily because the algorithm is created in software. Thus the design difficulties that exist in electronic, pneumatic, and hydraulic implementation of control laws do not occur. Nonlinear algorithms to prevent saturation of the final control elements and to ease start-up problems can be easily programmed (see Section 7.3). Since the software can be easily created, it can be easily changed, either to alter the algorithm itself or to change gain values. The latter is especially useful when the command signal drives the system far away from the linearization reference point, and new gains are required for the new operating condition. Finally, the computer is valuable for keeping records of energy consumption, downtime, etc., for administrative purposes. The sometimes higher cost of the digital system can often be justified because of these advantages. Development of inexpensive microcomputer systems is making this justification less necessary all the time.

Because digital-control algorithms are not as limited by hardware as analog controllers are, the algorithm is essentially limited only by the designer's imagination. One of the reasons for the widespread use of the PI-control algorithm is that it can be physically implemented relatively easily. While this algorithm is often implemented in DDC systems with good reason, some of the justifications for its use historically are no longer relevant, and the designer should feel free to consider algorithms that previously were out of the question; some of these are considered in Section 7.8. Here we limit ourselves to DDC versions of the PID family of algorithms.

The basic structure of a single-loop DDC controller is shown in Figure 6.58. The computer with its internal clock drives the D/A and A/D converters. It compares the command signals with the feedback signals and generates the control signals to be sent to the final control elements. These control signals are computed from the control algorithm stored in the memory. Slightly different structures exist, but Figure 6.58 shows the important aspects. For example, the comparison between the com-

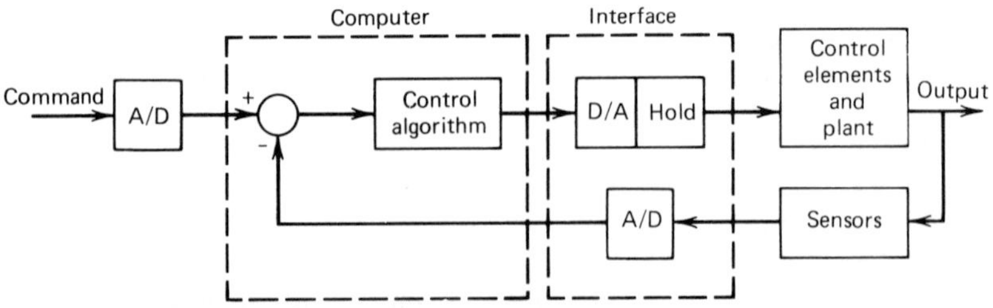


**Figure 6.58 Structure of a digital control system.**

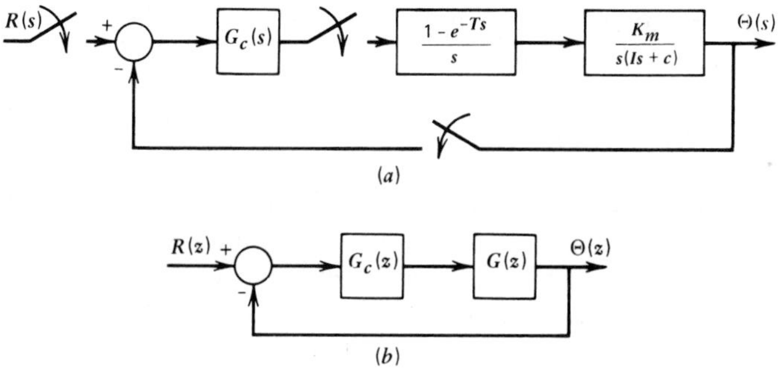


**Figure 6.59 Digital control of a dc motor. (*a*) Sampled data diagram in the *s* domain. (*b*) Diagram in the *z* domain.**

mand and feedback signals can be done with analog elements, and the A/D conversion made on the resulting error signal.

## Digital Control of a Motor

An example of a DDC system is obtained by using a displacement sensor and digital controller with the dc motor system shown in Figure 5.37. The resulting position controller is shown in Figure 6.59*a*. The transform of the control algorithm is $G_c(z)$. The transfer function of the sample-and-hold, motor-and-load combination was obtained in Section 5.11 (5.11-5) and is repeated here.

$$G(z) = \frac{K}{b^2}\frac{(bT-1+a)z+1-a-bTa}{(z-1)(z-a)} \tag{6.11-1}$$

where $K = K_m/I$, $b = c/I$, $a = e^{-bT}$, and $T$ is the sampling period. This can be further simplified to

$$G(z) = K\frac{b_1 z + b_2}{(z-1)(z-a)} \tag{6.11-2}$$

where the identification of $b_1$ and $b_2$ is easily made.

With (6.11-2) we can analyze the system in the $z$ domain entirely without further consideration of the sampling. The system diagram is shown in Figure 6.59*b*. The transfer function is

$$T(z) = \frac{\Theta(z)}{R(z)} = \frac{G_c(z)G(z)}{1+G_c(z)G(z)}$$

or

$$T(z) = \frac{KG_c(z)(b_1 z + b_2)}{(z-1)(z-a)+KG_c(z)(b_1 z+b_2)} \tag{6.11-3}$$

Design of the control algorithm can proceed in a manner similar to that used in Sections 6.5, 6.6, and 6.7 for analog controllers. For example, we can try proportional control $G_c(z) = K_p$. In this case, (6.11-3) gives

$$T(z) = \frac{KK_p(b_1 z + b_2)}{z^2 + (KK_p b_1 - 1 - a)z + a + KK_p b_2} \tag{6.11-4}$$

In the analog case, only one parameter, $K_p$, needs to be selected. However, in the digital case we often must choose more parameters because the sampling period and the machine's word length might be under the designer's control. For the sake of simplicity let us assume that these quantities have been specified for this example, and that only the control gains are to be found.

For given values of $K_m$, $I$, $c$, and $T$, (6.11-4) is completely determined except for the value of $K_p$. The question now becomes *how* to determine $K_p$. The results of Example 5.23 can be used to determine the values of $K_p$ that will give a stable system. The conditions that must be satisfied are (5.12-7) and (5.12-8). For the present system these stability requirements are

$$\frac{1-a-KK_pb_2}{KK_p(b_1+b_2)} > 0 \tag{6.11-5}$$

$$\frac{2(1+a)+KK_p(b_2-b_1)}{KK_p(b_1+b_2)} > 0 \tag{6.11-6}$$

With all parameters except $K_p$ specified, these conditions are easy to use. However, the reader is urged to consider how the situation changes if $T$ must also be selected.

In addition to the stability requirement, $K_p$ must also be selected to yield the desired response characteristics, which are frequently difficult to visualize in the $z$ domain. For example, the damping ratio provides a simple test in the $s$ domain for the existence of oscillatory behavior. No simple method exists in the $z$ domain because such behavior can be caused by negative as well as complex roots. Our conclusion is that it is often preferable to specify the response characteristics in terms of continuous-time features and to convert the root locations in the $s$ domain into corresponding root locations in the $z$ domain. The procedure for doing this was presented in Section 5.12 (see especially Figures 5.38, 5.39, and 5.40).

## Position and Velocity Algorithms

Extending this line of reasoning a bit further, we see that there are two ways to describe the form of the digital control law. One way is to select a controller transfer function $G_c(s)$ in the $s$ domain, such as a PID-type law, and convert it to an equivalent $G_c(z)$ in the $z$ domain. We will present methods for doing this shortly. The other way is to select $G_c(z)$ directly, as was done for the proportional controller in the previous example. With this approach, the question now arises as to what are discrete-time equivalents of the integral and derivative actions. Two commonly used forms are the *proportional-plus-sum* and the *proportional-plus-difference* algorithms. These are

$$f(k) = K_p e(k) + K_I T \sum_{i=0}^{k} e(i) \tag{6.11-7}$$

$$f(k) = K_p e(k) + \frac{K_D}{T}[e(k) - e(k-1)] \tag{6.11-8}$$

where $f(k)$ and $e(k)$ are the control and error signals and $T$ is the sampling period. The analogy to PI and PD control is apparent. In transform notation these become

$$F(z) = \left(K_p + K_I T \frac{z}{z-1}\right) E(z) \tag{6.11-9}$$

$$F(z) = \left[K_p + \frac{K_D}{T}(1 - z^{-1})\right] E(z) \tag{6.11-10}$$

Generalization of PID control is made in the obvious way. These algorithms are the *position* versions of the PI and PD laws.

The *incremental* or *velocity* versions of the algorithms determine the *change* in the control signal $f(k) - f(k-1)$. To obtain these, decrement $k$ by 1 in (6.11-7) and (6.11-8) and subtract the results from (6.11-7) and (6.11-8). This gives

$$f(k) = f(k-1) + K_p\,[e(k) - e(k-1)] + K_I T e(k) \tag{6.11-11}$$

$$f(k) = f(k-1) + K_p\,[e(k) - e(k-1)] + \frac{K_D}{T}\,[e(k) - 2e(k-1) + e(k-2)] \tag{6.11-12}$$

or

$$F(z) = \frac{(K_p + K_I T)z - K_p}{z-1} E(z) \tag{6.11-13}$$

$$F(z) = \frac{\left(K_p + \frac{K_D}{T}\right) z^2 - \left(K_p + \frac{2K_D}{T}\right) z + \frac{K_D}{T}}{z(z-1)} E(z) \tag{6.11-14}$$

Suppose that the control signal $f(k)$ is a valve position. The position version of the algorithm is so named because it specifies the valve position directly as a function of the error signal. The incremental algorithm, on the other hand, specifies the change in valve position. The incremental version has the advantage that the valve will maintain its last position in the event of failure or shutdown of the control computer. Also, the valve will not "saturate" at start-up if the controller is not matched to the current valve position. In addition to having these safety features, the incremental algorithm is also well suited for use with incremental output devices such as stepper motors. These differences between the position and the velocity algorithms do not show up in a transform analysis, for (6.11-9) reduces to (6.11-13), and (6.11-10) reduces to (6.11-14).

## Some Practical Considerations

Some practical aspects to the implementation of these digital algorithms that do not appear in analog design are related to the step changes that occur in the digital signals between the sample times and to the finite word length of the control computer.

Consider the integral term in (6.11-7). The change in this term is given in (6.11-11) and is $K_I T e(k)$. If $T$ and $e(k)$ are small, the finite word length of the machine can result in a zero change in the integral output. For example, assume that the word length of the controller is 12 bits. If the full-scale value is 4095, the smallest nonzero value that can be represented is 1. If $K_I = 0.001$ and $T = 1$ second, any value of the error $e(k)$ less than 1000 (24% of full scale) will result in the term $K_I T e(k)$ being less than 1. The change in the output of the integral mode will be zero. The fact that this nonzero error causes no change in the control signal results in a steady-state offset error, which never occurs with I action in analog systems.

Two remedies are available. The first is to improve the resolution by increasing the word length of the computer, but this can be expensive. An alternative solution is a modification of the software. Before the term $K_I T e(k)$ is computed, any "ineffective" portion of $e(k)$ is removed and saved to be added to the next error sample. In the preceding example, if $e(k) = 1200$, the additional value of 200 will not influence the controller's output [the output will be 1 both for $e(k) = 1000$ and for $e(k) = 1200$]. So the ineffective portion of value 200 is removed and saved. If the next error sample is 1900, it is modified by the remainder term to become 2100. The controller output is then 2, not 1.

The output from D action in analog controllers is constant if the error signal increases at a constant rate. However, D action in digital controllers can produce a fluctuating output for such an error signal. The effect again results from the round-off required by the finite word length of the machine. The derivative term from (6.11-8) is $K_D [e(k) - e(k-1)]/T$. Suppose $K_D = 1000$ and $T = 1$ second, with a 12-bit controller. If the error signal is increasing at a rate of 0.5 parts/second, the sampled error series looks like 0, 1, 1, 2, 2, 3, 3, . . . . The resulting value of the derivative term will be 1000, 0, 1000, 0, . . . . This behavior is referred to as *derivative mode kick*. It can be reduced by improving the resolution or by using an improved approximation to the derivative.

Increasing the sampling period $T$ will not eliminate the preceding problems with the integral and derivative terms. The dynamics of the discrete-time system are heavily influenced by the value of $T$; thus the values of $K_I$ and $K_D$ depend on $T$. Increasing $T$ in order to increase the product $K_I T$ and to decrease the ratio $K_D/T$ might not work because the correct values of $K_I$ and $K_D$ will shift accordingly. Finally, the larger $T$ is, the less accurate will be the discrete approximation.

The approximation to the derivative can be improved by using values of the sampled error signal at more instants. For example, in the velocity algorithm (6.11-12), we can replace the D-action term with one obtained from a four-point central-difference technique. Let $m$ be the mean of the previous four error samples.

$$m = \frac{e(k) + e(k-1) + e(k-2) + e(k-3)}{4} \tag{6.11-15}$$

For $\hat{e}(k) = e(k) - m$, the new D-action term is

$$\begin{aligned} f_D(k) &= \frac{K_D}{T}\left[\frac{\dfrac{\hat{e}(k)}{1.5T} + \dfrac{\hat{e}(k-1)}{0.5T} + \dfrac{\hat{e}(k-2)}{0.5T} + \dfrac{\hat{e}(k-3)}{1.5T}}{4}\right] \\ &= \frac{1}{6T}\left[e(k) + 3e(k-1) - 3e(k-2) - e(k-3)\right] \end{aligned} \tag{6.11-16}$$

This requires slightly more programming and storage for two additional values of the error sample.

The I-action term in (6.11-7) represents the rectangular integration formula, and its accuracy can be improved by substituting a more sophisticated algorithm such as the trapezoidal rule. With this, the I-action term becomes

$$f_I(k) = \sum_{i=0}^{k} \tfrac{1}{2}[e(i) + e(i-1)]K_I T \tag{6.11-17}$$

The improvements given by (6.11-16) and (6.11-17) can also be applied to the velocity algorithm. The steps are similar to those used to derive (6.11-11) and (6.11-12).

Another form of derivative kick occurs when the command input is a step function. The D action is the most sensitive to the resulting rapid change in the error samples. This effect can be eliminated by reformulating the control algorithm as follows (Reference 8, Chapter 11). To do this, I action must be included. The velocity algorithm for PID control is

$$\begin{aligned} f(k) &= f(k-1) + K_p[e(k) - e(k-1)] + K_I T\, e(k) \\ &\quad + \frac{K_D}{T}[e(k) - 2e(k-1) + e(k-2)] \end{aligned} \tag{6.11-18}$$

The error is $e(k) = r(k) - c(k)$, where $r$ and $c$ are the set point and output. The key step is to treat $r$ as a constant temporarily and to write the algorithm in terms of $r$ and $c$. This gives

$$\begin{aligned} f(k) &= f(k-1) + K_p[c(k-1) - c(k)] + K_I T[r - c(k)] \\ &\quad + \frac{K_D}{T}[-c(k) + 2c(k-1) - c(k-2)] \end{aligned} \tag{6.11-19}$$

The set point $r$ can now be replaced with $r(k)$. Note that integral action is required since $r$ now appears only in this term.

## 6.12 DEVELOPMENT OF THE CONTROL LAW FROM THE ANALOG FORM

Equations (6.11-7) through (6.11-14) are digital versions of the PID family of control laws that were obtained by mimicking the attributes of the analog versions. For example, the sum in the $I$-action term in (6.11-7) is supposed to represent the behavior of an analog integrator.

It is instructive to see how a digital control algorithm can be obtained from the analog law by other means. Three common ways of converting a continuous-time model into a discrete-time form are the Euler method, the Tustin method, and the $z$-transform method. Examples of these appear in Chapter Four. The PID algorithm in analog form is

$$f(t) = K_p\, e(t) + K_I \int_0^t e\, dt + K_D \frac{de}{dt} \tag{6.12-1}$$

or its equivalent

$$\frac{F(s)}{E(s)} = K_p + \frac{K_I}{s} + K_D s \tag{6.12-2}$$

Differentiating (6.12-1) gives

$$\frac{df}{dt} = K_p \frac{de}{dt} + K_I e + K_D \frac{d^2 e}{dt^2} \tag{6.12-3}$$

Application of the Euler method with a step size $T$ results in a difference equation that is identical to the velocity algorithm given by (6.11-18). Thus this algorithm suffers from the same inaccuracies as the Euler method *unless* the algorithm is applied in the manner to be indicated. For later reference, the transfer function of (6.11-18) is

$$\frac{F(z)}{E(z)} = \frac{a_1 z^2 + a_2 z + a_3}{z(z-1)} \tag{6.12-4}$$

where

$$a_1 = K_p + K_I T + a_3 \tag{6.12-5}$$

$$a_2 = -(K_p + 2a_3) \tag{6.12-6}$$

$$a_3 = \frac{K_D}{T} \tag{6.12-7}$$

The line of reasoning is easier to follow if we assume that no D action is present. With $K_D = 0$, (6.12-4) becomes

$$\frac{F(z)}{E(z)} = \frac{a_1 z + a_2}{z-1} \tag{6.12-8}$$

with $a_3 = 0$ and

$$a_2 = -K_p = K_I T - a_1 \tag{6.12-9}$$

The Tustin approximation is

$$\mathscr{Z}\left(\frac{d^n y}{dt^n}\right) = [D(z)]^n Y(z) \tag{6.12-10}$$

where

$$D(z) = \frac{2}{T}\frac{z-1}{z+1} \tag{6.12-11}$$

Application of this relation to (6.12-3) with $K_D = 0$ gives a transfer function identical in form to (6.12-8) except that

$$a_1 = K_p + \frac{K_I T}{2} \tag{6.12-12}$$

$$a_2 = -K_p + \frac{K_I T}{2} \tag{6.12-13}$$

Finally, with $K_D = 0$, use Table 4.3 to find the $z$-domain transfer function equivalent to (6.12-2). This is

$$\frac{F(z)}{E(z)} = K_p + \frac{K_I z}{z-1} = \frac{(K_p + K_I)z - K_p}{z-1} \tag{6.12-14}$$

which is of the form (6.12-8) with

$$a_1 = K_p + K_I \tag{6.12-15}$$

$$a_2 = -K_p \tag{6.12-16}$$

Thus for PI control, the Euler, Tustin, and $z$-transform approximations lead to the same *form* of the digital control law transform (6.12-8). The only difference is in the relation of the coefficients $a_1$ and $a_2$ to $T$, $K_I$, and $K_p$. We will show that this difference is irrelevant for many applications and that only the form of (6.12-8) is important.

## *Example 6.1*

Consider PI control of the first-order plant whose transfer function is $1/(2s + 1)$. The block diagram in Figure 6.33 applies with $I = 2$, $c = 1$. For analog control the appropriate transfer function is (6.6-9). Let us suppose that no oscillations are wanted with a step input so that $\zeta = 1$ is specified. Assume that the desired time constant is $\tau = 1$. This gives $K_p = 3$, $K_I = 2$.

Design a digital control law for this system as shown in Figure 6.60*a*. Assume that the sampling rate is one-tenth of the desired time constant, so that $T = 0.1$. Neglect any disturbances.

We demonstrate two approaches to the problem.

(a) Use the procedure of Example 4.14 to convert the plant's transfer function with a zero-order hold into one in the $z$-domain. The result is given in Figure 6.60*b*. $G_c(z)$ is the control law, which is designed by converting the performance specifications in the $s$-domain into equivalent ones in the $z$-domain. Refer to (5.12-11) and note that for $\zeta = 1$ the corresponding $z$-roots are repeated and equal to

$$z = e^{-\omega_n T} \tag{6.12-17}$$

where $\omega_n$ is arbitrary. Its value can be obtained by noting that $\zeta = 1$ and $\tau = 1$ imply that $s = -1, -1$. Thus, $\omega_n = 1$. For $T = 0.1$, the two required roots are both located at $z = \exp[-0.1] = 0.905$.

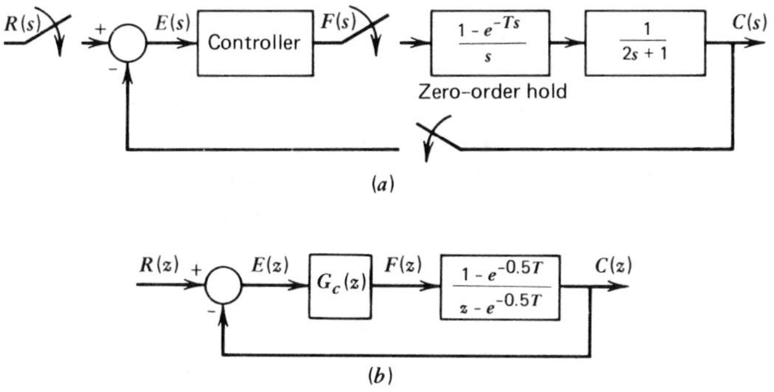


**Figure 6.60 Digital control of a first-order plant. (*a*) *s* domain. (*b*) *z* domain.**

Using the form of PI control by (6.12-8), we obtain

$$G_c(z) = \frac{a_1 z + a_2}{z-1} = \frac{F(z)}{E(z)} \quad \textbf{(6.12-18)}$$

$$\frac{C(z)}{R(z)} = \frac{(1-b)(a_1 z + a_2)}{z^2 + (a_1 - 1 - b - a_1 b)z + b + a_2 - a_2 b} \quad \textbf{(6.12-19)}$$

where $b = e^{-0.5T}$ With $T = 0.1$, the values of $a_1$ and $a_2$ required to place the roots at $z = 0.905$ are $a_1 = 2.8996$ and $a_2 = -2.716$. The difference equation that the control computer must implement is, from (6.12-18),

$$f(k) = f(k-1) + a_1\, e(k) + a_2\, e(k-1) \quad \textbf{(6.12-20)}$$

The response to a unit step input can be computed analytically from (6.12-19). The analog controller gives a response that crosses $c = 1$ at $t = 2$ and has an overshoot of 2.5% at $t = 3$. The digital controller's response crosses $c = 1$ at $t = 2.0$ and has an overshoot of 2.2% at $t = 2.9$. After four time constants, $c(4) = 1.018$ for the analog controller, while the response for the digital controller reaches this value at $t = 3.5$. Thus the digital system responds similarly to the analog system. The overshoot in both cases results from the numerator dynamics. The slight differences in response result from the discrete approximations for the transfer function of the PI-control law and from the effects of the sample-and-hold operation.

(b) The second approach is to convert the acceptable analog control law with $K_p = 3$, $K_I = 2$ into discrete form. Inserting these values into the Euler coefficients (6.12-9) we obtain $a_1 = 3.2$ and $a_2 = -3$. From (6.12-9), the characteristic roots are $z = 0.926, 0.870$. The system is approaching the instability region where $|z| > 1$.

The Tustin coefficients from (6.12-22) and (6.12-23) are $a_1 = 3.1$, $a_2 = -2.9$. These give $z = 0.91, 0.88$. For the equivalent $z$-transform method, (6.12-25) and (6.12-26) give $a_1 = 5$, $a_2 = -3$, and $z = 0.854 \pm i0.276$. The system is stable but oscillatory.

The response for the Euler and Tustin approximations resembles that for the design in part a. However, for the third method, the oscillatory response has a maximum overshoot of 43%.

It is impossible to draw general conclusions about the relative merits of the three approximation methods on the basis of one example. However, the example clearly shows the undesirable response that can be generated when using a discrete approximation with the gain values computed for an analog control law. Decreasing the sample period $T$ improves the approximation, but it is not always possible to do so. For example, it is not unusual to have an A/D conversion time greater than 0.1 sec. The sample period must be large enough to accommodate this time as well as that required for D/A conversion and for processing the control algorithm. At present there are significant application areas, like robotics, in which these times are large enough to limit the performance of the control system in a serious way. A good design requires a balance between the complexity of the algorithm and the speed of the computer.

## Comparison of Methods

We can categorize the design situations that occur as follows:

### *Case 1*

The controller design is done in the $s$ domain. The resulting analog control law $G_c(s)$ must be converted to discrete-time form by an approximation technique. (This was the approach in part b of Example 6.1.)

### *Case 2*

The performance specifications are given in terms of the desired continuous-time response and/or desired root locations in the $s$ plane. From these the corresponding root locations in the $z$ plane are found, and a discrete control law $G_c(z)$ is designed. (This was the method used in part a of Example 6.1.)

### *Case 3*

The performance specifications are given in terms of the desired discrete-time response and/or desired root locations in the $z$ plane. The rest of the procedure follows as in Case 2.

Obviously, Case 3 is the most direct since the $s$ plane is bypassed entirely. However, since most of our applications involve analog plants, it is difficult to state specifications in the $z$ domain. Therefore, the method described in Case 2 is the most practical one to use. It avoids the approximation errors that are inherent in the method of Case 1.

If the sampling period is small, the approximation method of Case 1 can be successfully applied. The technique is widely used for two reasons. When existing analog controllers are converted to digital control, the form of the control law and the values of its associated gains are known to have been satisfactory. Therefore, a digital version is designed. Second, because analog design methods are well established, many engineers prefer to take this route and then convert the design into a discrete-time equivalent. Other methods for developing discrete equivalents from analog transfer functions can be developed with frequency-response techniques (these are treated in advanced works on digital control; see Reference 12, for example).

If D action is used in the algorithm, the three approximation methods do not yield the same discrete-time form. Following the same procedure as before, it can be shown that the Euler method and Table 4.3 both give a control law of the form (6.12-4). Of course, the values of $a_1, a_2,$ and $a_3$ will be different for the two methods. However, the Tustin method gives a different denominator form for PID.

$$\frac{F(z)}{E(z)} = \frac{a_1 z^2 + a_2 z + a_3}{z^2 - 1} \qquad (6.12\text{-}21)$$

The coefficients are derived in the chapter problems. To obtain the discrete equivalent of PID with design Case 2 or 3, one of the forms (6.11-19), (6.12-4), or (6.12-21) can be used.

## 6.13 SUMMARY

This chapter introduced the basic concepts, devices, and algorithms of feedback control. We have emphasized the PID family of control laws since they are satisfactory for many applications. A proposed design is analyzed on paper in light of specifications relating to stability, accuracy, and speed of response. If the control law appears satisfactory, the engineer chooses the control medium: analog electronic, pneumatic, hydraulic, or digital. This choice involves many factors such as the type of plant being controlled, weight, power, reliability, or cost, and it is impossible to make general recommendations in this regard.

Often, however, simple adjustment of the gains in a PID algorithm will not yield satisfactory performance. In this case, a new control law must be found or more controllers added to the system. This brings us to the topic of the next chapter, which treats alternate control system configurations.

## REFERENCES

1. O. Mayr, *The Origins of Feedback Control*, MIT Press, Cambridge, Mass., 1970.
2. A. N. Burstall, *A History of Mechanical Engineering*, MIT Press, Cambridge, Mass., 1969.
3. N. Wiener, *Cybernetics*, John Wiley, New York, 1948.
4. E. Doebelin, *Measurement Systems*, McGraw-Hill, New York, 1975.
5. *Transducer Compendium*, Instrument Society of America, Pittsburg.
6. L. S. Marks, *Mechanical Engineer's Handbook*, McGraw-Hill, New York, 1967.
7. J. Truxal, ed., *Control Engineer's Handbook*, McGraw-Hill, New York, 1958.
8. Y. Takahashi, M. Rabins, and D. Auslander, *Control*, Addison-Wesley, Reading, Mass., 1970.
9. R. Phelan, *Automatic Control Systems*, Cornell Univ. Press, Ithaca, New York, 1977.
10. J. Truxal, *Introductory Systems Engineering*, McGraw-Hill, New York, 1971.
11. D. McCloy and H. Martin, *The Control of Fluid Power*, Longman, London, 1973.
12. G. Franklin and J. Powell, *Digital Control of Dynamic Systems*, Addison-Wesley, Reading, Mass., 1980.

## PROBLEMS

6.1 Consider the dc motor position controller shown in Figure 6.4. Assume that the amplifier has a gain $K_a$. Modify the block diagram in Figure 6.6 and find the transfer function between $\theta$ and $\theta_r$, that between $\theta$ and $T_d$, for the following cases.

(a) An armature-controlled dc motor. Do not neglect the motor's inductance.

(b) A field-controlled dc motor. Do not neglect the motor's inductance.

6.2 Consider the temperature control system shown in Figure 6.8. Modify the block diagram shown in Figure 6.9 to include the effect of the heater's thermal capacitance $C_1$ and thermal resistance $R_1$ (see Figure 5.2). Find the transfer function between $x$ and $u$ and that between $x$ and $v$.

6.3 A common error detector is the Wheatstone bridge, shown in Figure P6.3. A fixed voltage $V$ is applied, while $R_3$ and $R_4$ are potentiometers and represent the reference input and the controlled variable, respectively. Find the relation between the error voltage $e$ and the resistances $R_3$ and $R_4$.

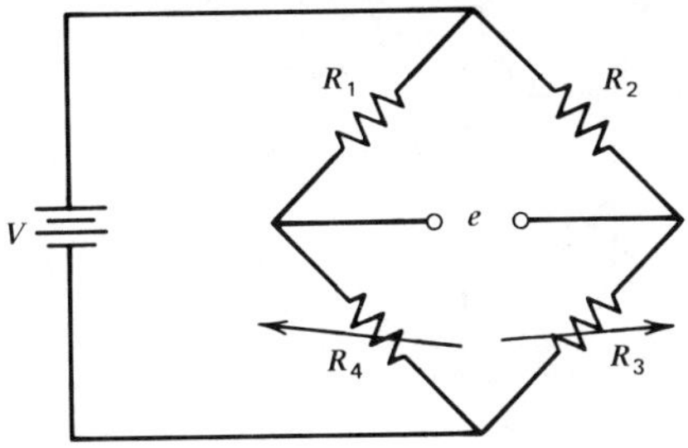


Figure P6.3

6.4 The model for a one-degree-of-freedom gyroscope is

$$I\ddot{\theta} + c\dot{\theta} + k\theta = H\omega_i$$

where $\theta$ is the angular displacement of the gyro's spin axis and is the unit's output (Figure P6.4). The inertia about the output axis is $I$, and $H$ is the angular momentum of the wheel, which spins at a constant velocity.

The gyro can be used as a sensor to measure either the angular displacement (an *integrating* gyro) or angular velocity (a *rate* gyro).

(a) Show that the gyro measures displacement if $k = 0$.

(b) Show that the gyro measures velocity if $k \neq 0$.

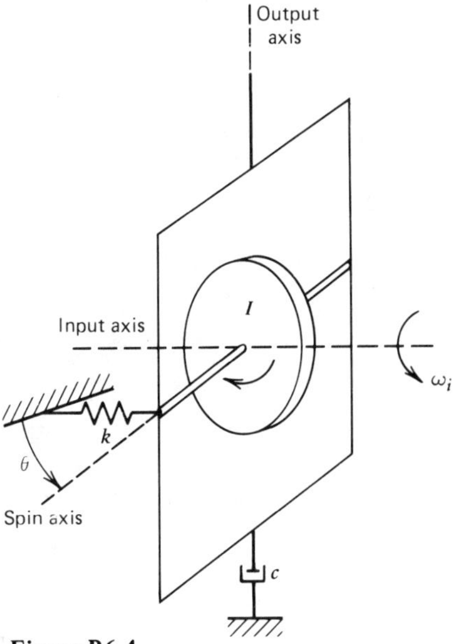


Figure P6.4

6.5 Hydraulic controllers are used to drive cutting tools in machining applications. (see Figure P6.5). The command input $x_d$ is a voltage representing the desired position of the cutting tool as a function of time. It can be generated by a cam drive, a punched paper tape, or a computer program. Develop a block diagram for the system and find the transfer function between the tool displacement $x$ and $x_d$. Neglect the inertia, inductance, and shaft elasticity of the motor. Treat the reactive force $F$ of the cutting tool as a disturbance.

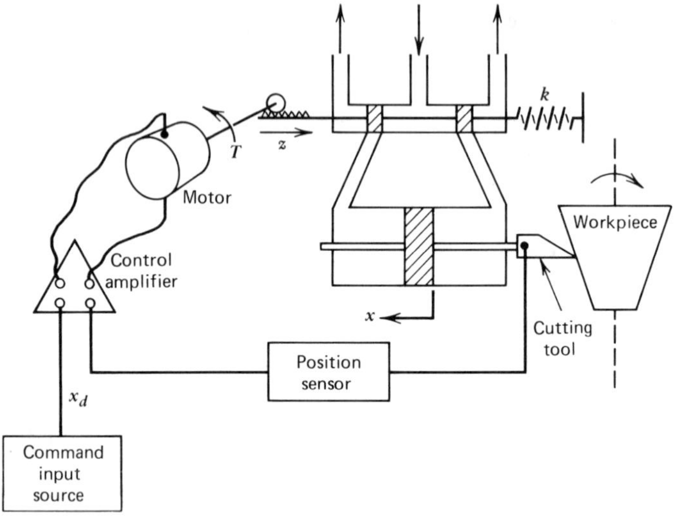


**Figure P6.5**

6.6 Figure P6.6 shows a hydraulic system for controlling the flow rate $q$. The adjustment screw is used to select the desired flow rate $q_d$. It adjusts the opposing force on the pilot valve by adjusting the compression in the spring. Explain the operation of the system and draw the block diagram with $q_d$ as the input and $q$ as the output. Find the transfer function. Name two possible disturbances to the system.

6.7 The motion of an oscillatory system is to be controlled by a relay controller. The relay has a maximum available force magnitude of $f_{max} = 50$. Plot the phase plane trajectories of the system for the following.

(a) No dead zone in the relay.

(b) A dead zone of total width 2 (see Figure P6.7).

6.8 For the liquid-level system shown in Figure 6.25, the outlet resistance is $R = 5/6\ \text{min/ft}^2$, the tank area is $C = 12\ \text{ft}^2$, and the inlet flow rate when switched on is a constant $12\ \text{ft}^3/\text{min}$. The desired height is 5 ft, and the switching gap is 2 ft, so that $h = 5 \pm 1$ ft in the ideal case.

(a) Assume that $h(0) = 4$ ft (the start of the filling curve). Plot the height as

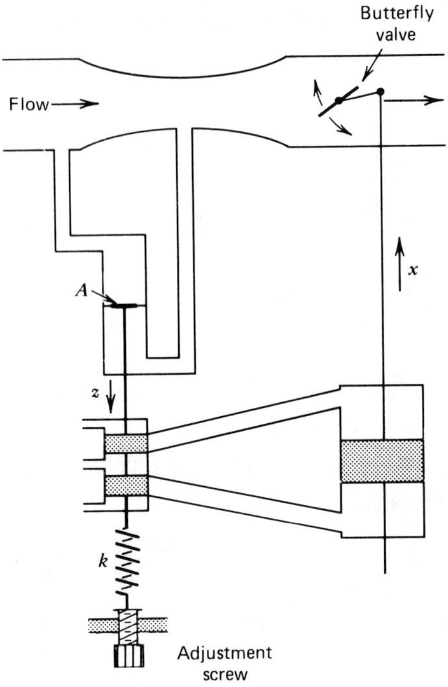


Figure P6.6

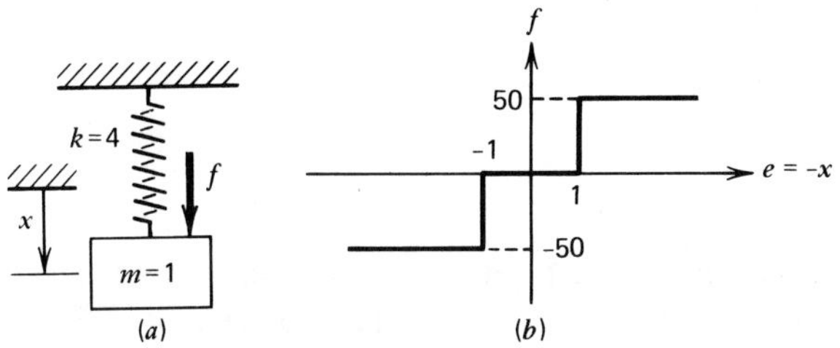


Figure P6.7

a function of time for one cycle. What is the cycle time? Assume that the switch and the float act instantaneously.

**(b)** Now assume that the float is a first-order system with a time constant $\tau$ so that the transfer function between the height measurement $h_m$ and the actual height $h$ is

$$\frac{H_m(s)}{H(s)} = \frac{1}{\tau s + 1}$$

Assume that $h(0) = 4\,\text{ft}$ and plot two cycles of the response $h(t)$ for the two following cases. Compare with the ideal case analyzed in part a.

**(1)** $\tau = 2$ min.

**(2)** $\tau = 0.5$ min.

**6.9** Figure P6.9 shows a position control system with PD control. Find the values of the gains $K_p$ and $K_D$ to meet all of the following specifications.

1. No steady-state error with a step input.
2. A damping ratio of 0.9.
3. A dominant time constant of 1.

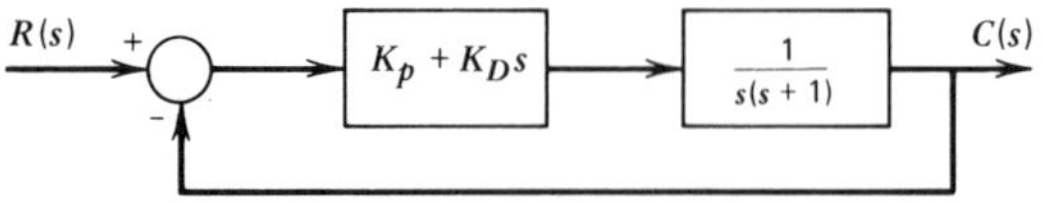


**Figure P6.9**

**6.10** Figure P6.10 shows a PID-control system with the same plant as in Problem 6.9. For the values of $K_p$ and $K_D$ found previously,

(a) Find the maximum value $K_I$ can have without causing the system to be unstable.

(b) If we relax the requirement that $\zeta = 0.9$, but still require that $\tau = 1$, then $s = -1$ is one characteristic root. Is it possible to select $K_I$ so that $s = -1$ is the dominant root? If so, find the value of $K_I$ required to do this.

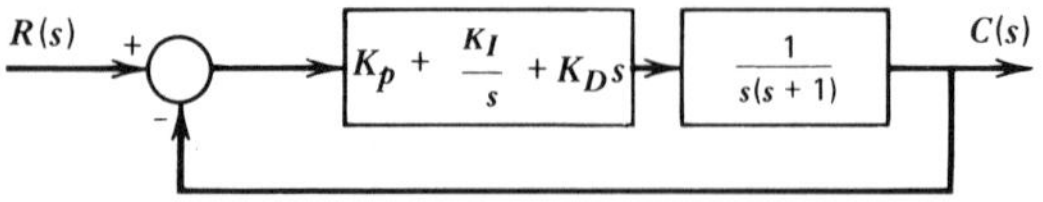


**Figure P6.10**

**6.11** In the absence of a disturbance, the original system shown in Figure P6.11*a* gave a satisfactory steady state output of $c_{ss} = 1$ when the input was a unit step, and it had a satisfactory time response (the output took 4/5 sec to reach 98% of its final value). However, with a step disturbance, the performance is not satisfactory, so the controller $G(s)$ and the feedback loop have been added to compensate for the disturbance without changing the satisfactory aspects of the original system. This new system is shown in Figure P6.11*b*.

Design the controller. The choice is limited to *P*, *I*, or PI control. With unit step command and disturbance inputs, the specifications are

1. No steady-state offset error in the absence of a disturbance.
2. No steady-state deviation resulting from the disturbance.
3. The output must respond as quickly as possible to the command input, but without oscillation.
4. The output must take no longer then 4/5 sec to reach and stay within approximately 98% of the final value, in the absence of a disturbance. (The dominant-root approximation is acceptable here.)

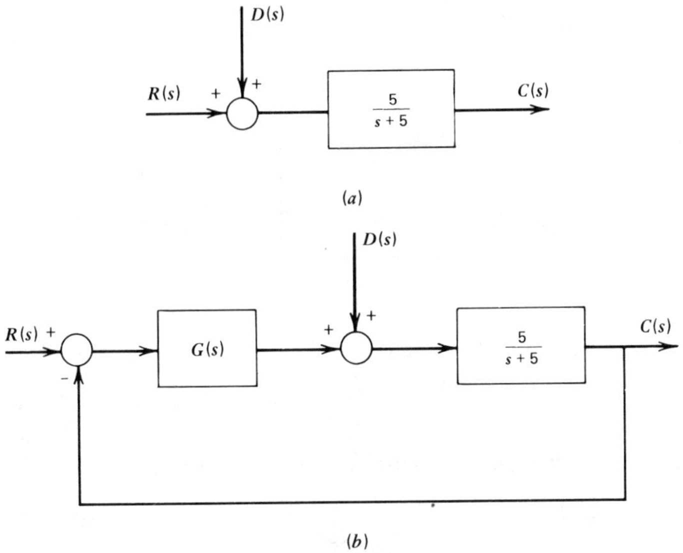


**Figure P6.11** (*a*) Original system. (*b*) New system.

**6.12** A speed-control system with proportional control is shown in Figure P6.12. Find the minimum gain $K_p$ such that the speed deviation due to a 500 lb-in. step disturbance will not exceed 2 rad/sec. The plant's time constant is 2 sec and $I = 200$ lb-in.-sec$^2$. The gain $K_p$ must be positive for physical reasons.

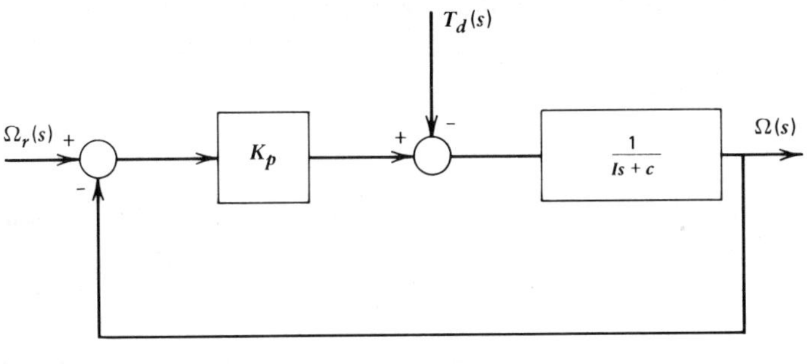


**Figure P6.12**

**6.13** Integral control of the plant

$$G_p(s) = \frac{K}{\tau s + 1}$$

results in a system that is too oscillatory. Will D action improve this situation?

**6.14** Consider the first-order system shown in Figure P6.14 and suppose that $I = 10$ and $c = 0.1$. Consider the following performance specifications.

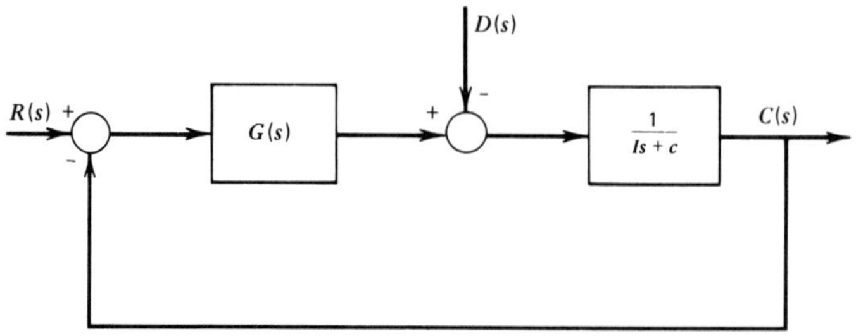


Figure P6.14

1. The magnitude of the steady-state error must be no more than 0.01 when $r(t)$ is a unit ramp and $d(t) = 0$.
2. The damping ratio must be unity.
3. The dominant time constant must be no greater than 0.1.

(a) Design a controller to meet the first specification. What is the damping ratio that results? What is the time constant?

(b) Design a controller $G(s)$ to meet the first two specifications. Evaluate the resulting time constant.

(c) Plot the response in part b when $r(t)$ is a unit step. How can we modify the design in part b to eliminate or at least reduce the overshoot? Why does an overshoot occur even with $\zeta = 1$?

(d) For the design obtained in part b, what is the deviation caused by a unit step disturbance? By a unit ramp disturbance?

(e) Design a controller $G(s)$ to meet all three specifications. Plot the response for $r(t)$ a unit step and $d(t) = 0$. Discuss the results in light of the second and third specifications.

**6.15** Consider the system shown in Figure P6.15.

(a) Design a controller $G(s)$ that minimizes the steady-state error when $r(t)$ is a unit ramp and $d(t) = 0$.

(b) Evaluate the steady-state deviation for the design in part a when $d(t)$ is a unit ramp.

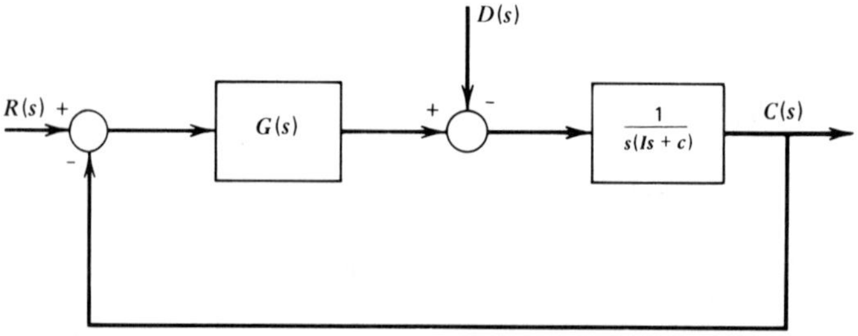


Figure P6.15

**6.16** Consider the system shown in Figure P6.16. Suppose that $I = 10$, $c = 0.1$, and $a = 1$.

**(a)** Design a controller $G(s)$ to give a minimum steady-state error when $r(t)$ is a unit step and $d(t) = 0$. The dominant time constant should be no greater than 0.1 sec and the damping ratio $\zeta$ should be unity.

**(b)** Compute the step response of the design in part a. What is the effect of the numerator dynamics $(s + a)$ on the design and its performance?

**(c)** Compute the steady-state error of the design in part a if $r(t)$ is a unit ramp and $d(t) = 0$.

**(d)** Evaluate the steady-state deviation resulting from a unit step disturbance.

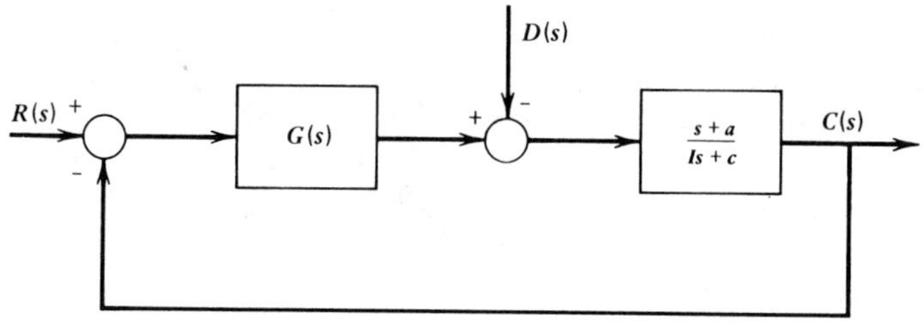


**Figure P6.16**

**6.17** The plant shown in Figure P6.17 is representative of a vibratory system. Suppose that $m = 1$, $c = 1$, and $k = 4$. Suppose also that $r(t)$ is a unit step and $d(t) = 0$.

**(a)** Design a controller $G(s)$ so that the magnitude of the steady-state error is no greater than 0.1. Evaluate the resulting time constant and damping ratio, and compare with those of the plant.

**(b)** Suppose we wish the dominant time constant to be no greater than 1 sec. Design the controller. Evaluate the damping ratio.

**(c)** Now suppose that instead of specifying the time constant, we specify that the damping ratio should be no less than 0.707. Design the controller and evaluate the time constant.

**(d)** What type of controller might be used to satisfy all three specifications ($e_{ss}$, $\tau$ and $\zeta$)? Do not compute the gains.

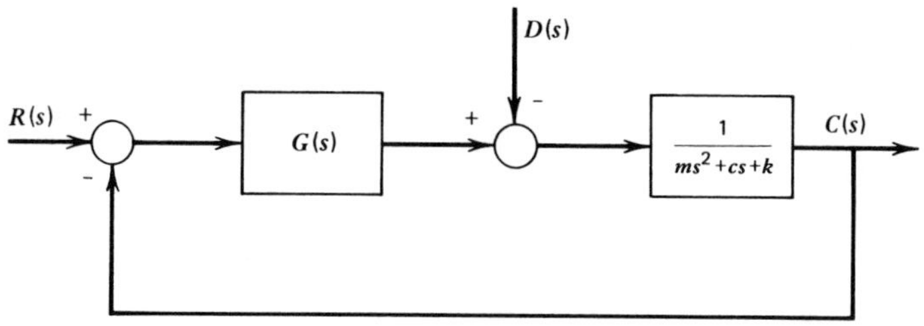


**Figure P6.17**

**6.18** The attitude of a vehicle moving through a fluid represents an inherently unstable plant because the fluid forces do not act through the vehicle's center of mass, but through a point called the center of pressure. They thus create a net torque that acts to rotate the vehicle. The attitude can be controlled by control surfaces such as elevators and ailerons on missiles and aircraft, by thrusters on rockets, and by diving planes and fins on submarines and other underwater vehicles.

An example of such a system is shown in Figure P6.18. A linearized model of the missile's dynamics is

$$I\ddot{\theta} = T + C_n L \theta$$

where $\theta$ is the missile's attitude angle, $I$ is the missile's inertia about the indicated axis of rotation, $C_n$ is the normal-force coefficient, $L$ is the distance between the center of mass and the center of pressure, and $T$ is the applied control torque.

The plant is unstable, and the primary control objective is to stabilize the missile's attitude so that $\theta \cong 0$. Develop a control law for this purpose.

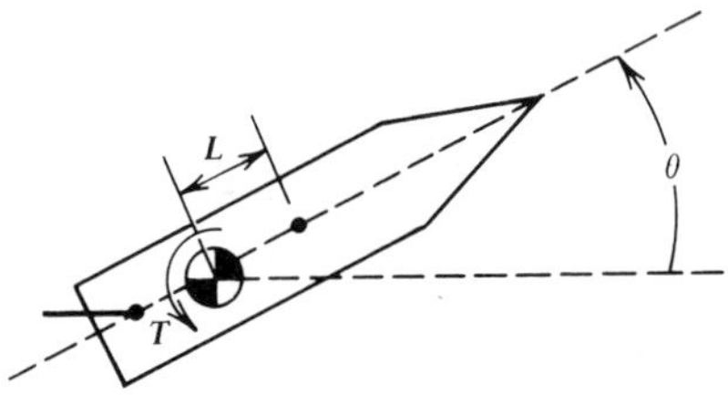


**Figure P6.18**

**6.19** The control variable for the tank system shown in Figure P6.19 is $q_1$, the flow rate into tank 1. It is desired to control the height $h_2$ with a measurement of $h_2$.

Develop a control law that will give a zero steady-state error for a step command $h_{2d}$ and a zero steady-state deviation for a step disturbance $q_d$.

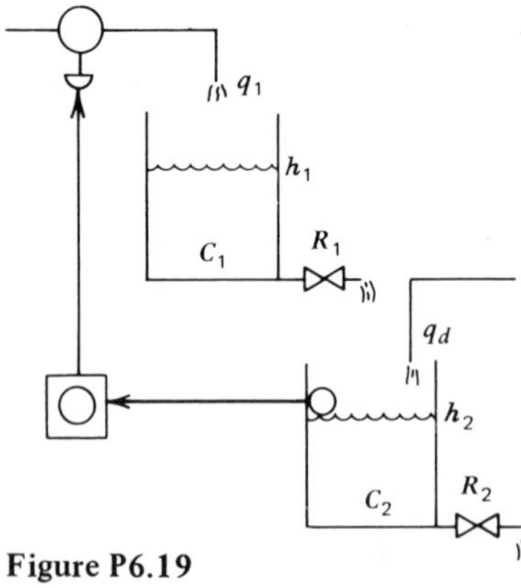


**Figure P6.19**

**6.20** The system shown in Figure P6.20 can be thought of as a representation of the problem of stabilizing the attitude of a rocket during takeoff. The applied force

$f$ represents that from the side thrusters of the rocket. The linearized form of Newton's law for the system reduces to

$$ML\ddot{\theta} - (M + m)g\theta = f$$

where $f$ is the control variable.

Design a control law to maintain $\theta$ near zero. The specifications are $\zeta = 0.707$ and a 2% settling time of 10 sec. The parameter values are $M = 50$ slugs, $m = 10$ slugs, $L = 30$ ft, and $g = 32.2$ ft/sec$^2$.

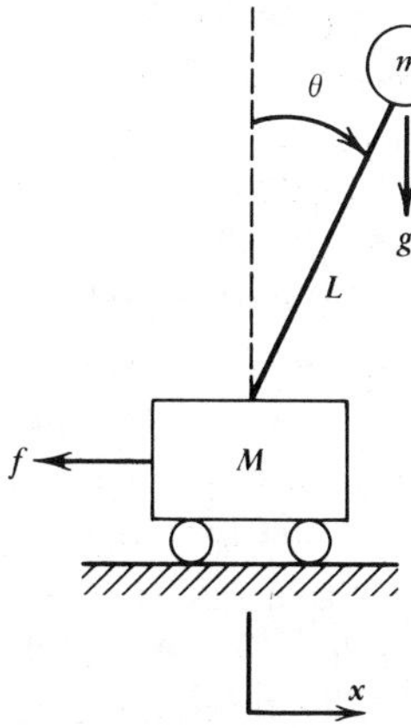

**Figure P6.20**

**6.21** In many applications, such as tape drives or cutting tools, the controlled object must be accelerated to a certain speed, held at that speed, and then decelerated to zero speed at prescribed times. One such speed profile is shown in Figure P6.21. Suppose that the plant is $1/(s + 1)$ and the motor-amplifier to be used has a gain of $K_m = 10$.

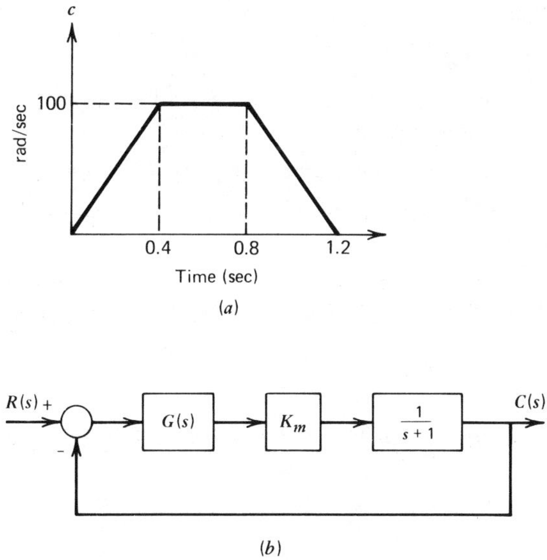

**Figure P6.21**

(a) Design a controller $G(s)$ such that the speed $c(t)$ at $t = 0.4$ is within 2 rad/sec of the desired value of 100 rad/sec at that time, assuming that $c(0) = \dot{c}(0) = 0$.

(b) Plot the response for one cycle and compare it to the desired profile. What is the maximum deviation from the desired speed on each part of the profile?

**6.22** The problem is to design a controller to position an unbalanced load of mass $m$. We can represent the unbalance as a pendulum with a point mass $m$ located at the center of mass, a distance $L$ from the center of rotation (see Figure P6.22). The control variable is the torque $T$ to be supplied by a motor whose dynamics we will ignore for now. We will also assume for now that the motor can produce as much torque as needed. Take $mg = 1$ lb, $g = 32.2\text{ ft/sec}^2$, and $L = 0.25$ ft.

(a) It is desired to position the mass $m$ at $\theta = \pi$ rad. Obtain the governing dynamic equations and linearize them. Use the linearized model to design a controller that will position $m$ at $\theta = \pi$ rad exactly at steady state. The transient specifications call for a damping ratio $\zeta \geqslant 0.707$ and a 2% settling time of no greater than 2 sec.

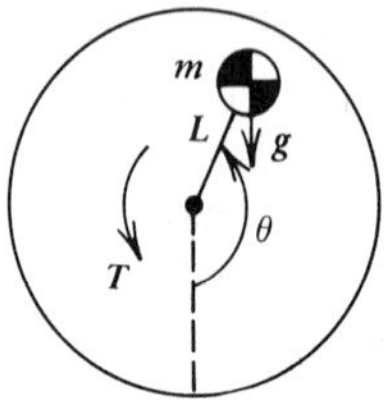


**Figure P6.22**

(b) Compute the response for the design obtained in part a, using the linearized model with the initial conditions

(1) $\theta(0) = 2.9$ rad, $\dot{\theta}(0) = 0$.

(2) $\theta(0) = 1.6$ rad, $\dot{\theta}(0) = 0$.

Does the response satisfy that demanded by the specifications? (i.e., does the response settle quickly without much overshoot?).

(c) Simulate the response of the design obtained in part a using the *nonlinear* dynamic model with the initial conditions given in part b. Evaluate the performance of the controller.

(d) What is the maximum torque that the motor must be capable of supplying in order for the load to be positioned at $\theta = \pi$ rad regardless of the initial conditions?

**6.23** Repeat Problem 6.22 except position the load at $\theta = 0$. For the simulations use

(1) $\theta(0) = 0.2, \dot{\theta}(0) = 0$

(2) $\theta(0) = 1.5, \dot{\theta}(0) = 0$

Discuss how the performance of this design compares with that found in Problem 6.22.

**6.24** The following problem is typical of those found in the textile and wire industries. A reel containing yarn, wire, or cable, for example, must be unwound while maintaining a constant linear cable velocity. The tachometer voltage indicates this velocity $v$ and is compared to a voltage from the potentiometer that represents the desired velocity $v_d$ (see Figure P6.24). As the cable unwinds due

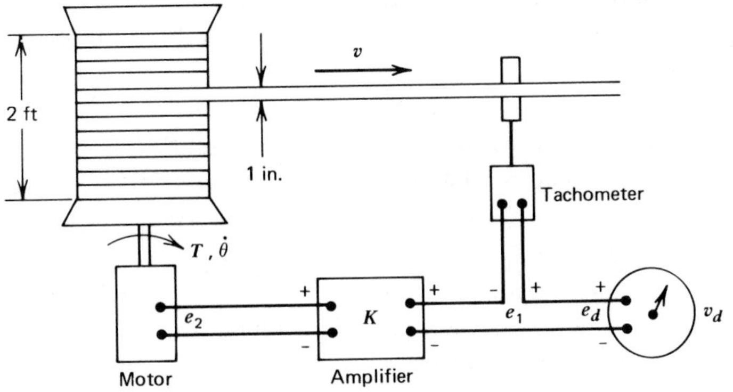


**Figure P6.24**

to its own weight, the effective reel radius decreases and thus the reel's angular velocity $\dot{\theta}$ must increase.

Suppose that the reel's effective radius is 2 ft when empty and 3 ft when full. The reel's shape resembles that of a hollow cylinder, and its inertia is taken to be

$$I = 16.5R^4 - 150 \text{ ft-lb-sec}^2$$

where $R$ is the effective radius of the reel, which will be a function of time. The reel width is 2 ft, and the cable diameter is 1 in. Assume that the motor's inertia is negligible and take its transfer function to be

$$\frac{T(s)}{E_2(s)} = \frac{5}{2s + 1}$$

where $T(s)$ and $E_2(s)$ are the motor torque and voltage. Let the tachometer time constant be 1 sec so that

$$\frac{E_1(s)}{V(s)} = \frac{2}{s + 1}$$

where $E_1(s)$ is the output voltage of the tachometer and $V(s)$ is the cable velocity. The potentiometer constant is $K_1 = 2$ volts/ft/sec. The amplifier gain $K$ implements proportional control.

(a) For a given layer of cable, we assume that $R$ and $I$ are constant to obtain an invariant model. Draw the block diagram and obtain the transfer function $V(s)/V_d(s)$ in terms of $R$ and $I$.

(b) Use the Routh-Hurwitz criterion to find the range of $K$ over which the system will be stable. Compare the ranges at the two limits represented by an empty and a full reel.

**6.25** Determine the resistance values required to obtain an electronic PI controller with $K_p = 2$ and $K_I = 0.04$. Use a 1 $\mu f$ capacitor.

6.26 (a) Determine the resistance values to obtain an electronic PD controller with $K_p = 1$, $T_D = 1$ sec. The circuit should limit frequencies above 10 rad/sec. Use a 1 $\mu f$ capacitor.

(b) Plot the frequency response of the circuit. Compare with the plot of the PD controller without frequency limiting.

6.27 (a) Determine the resistance values to obtain an electronic PID controller with $K_p = 5$, $K_I = 0.7$, and $K_D = 2$. The circuit should limit frequencies above 250 rad/sec. Take one capacitance to be 1 $\mu f$.

(b) Plot the frequency response and compare with that of the PID controller without frequency limiting.

6.28 A pneumatic temperature controller using a bulb transducer is shown in Figure P6.28. Explain its operation. Draw the block diagram and find the transfer function. The input is the desired temperature $T_d$, and the output is the actual temperature $T$.

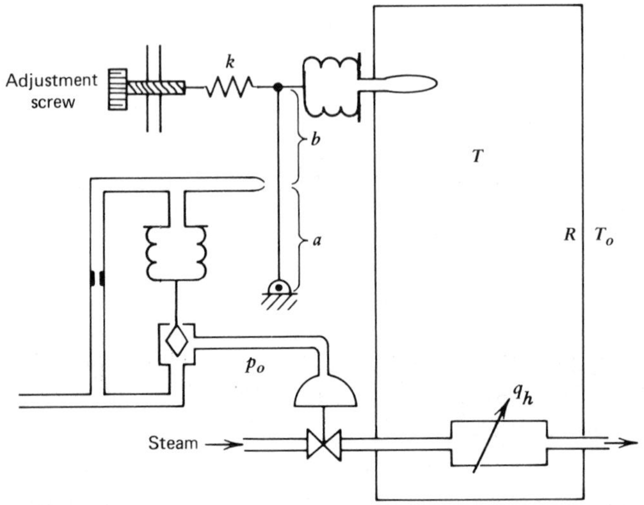


Figure P6.28

6.29 The pneumatic stack controller shown in Figure P6.29 can be operated with the upper outlet restriction either fully closed or partly open. Obtain the transfer function between $p_o$ and $(p_r - p_c)$ for each mode of operation. Identify the control action in each case.

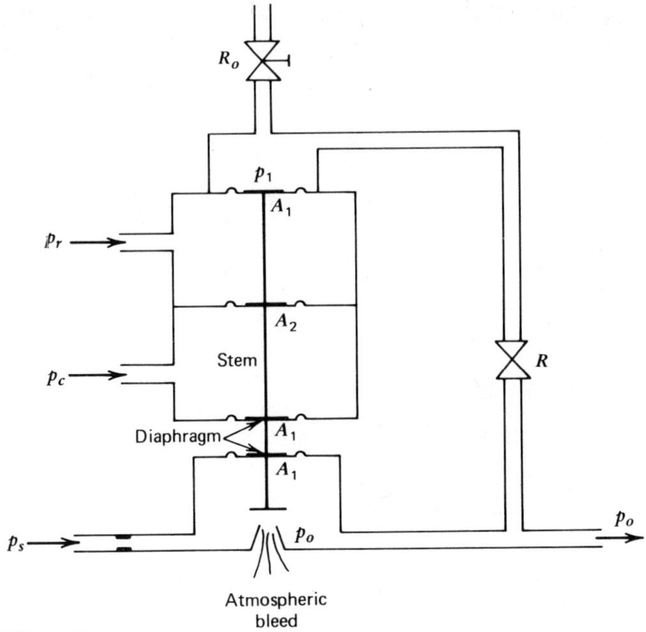


**Figure P6.29**

**6.30** Final control elements often have their own feedback loops to improve their performance. An example of this is the pneumatic valve positioner shown in Figure P6.30. If the position of the actuating valve does not correspond to that demanded by the control pressure, the positioner will act until the valve position is correct. Explain the operation of the system and draw a schematic block diagram.

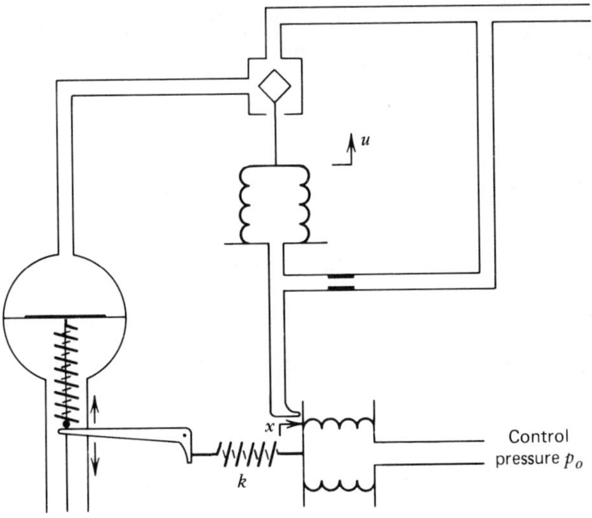


**Figure P6.30**

**6.31** The hydraulic jet pipe controller shown in Figure P6.31*a* is an alternative to the spool valve design. Instead of moving the spool valve, the input displacement $z$ moves the jet pipe, or nozzle, which ejects fluid at high pressure. When the pipe is in the neutral position in the center, no fluid enters either side of the piston chamber.

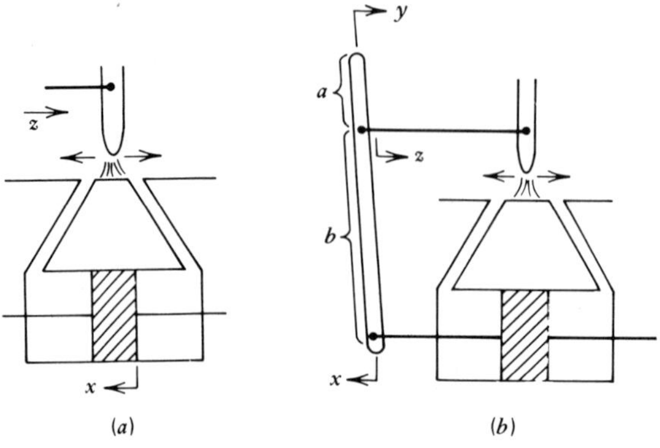


**Figure P6.31**

(a) Derive the transfer function between the output displacement $x$ and the input displacement $z$. What type of control action results?

(b) The system in part a has been modified with a linkage. Derive the transfer function between $y$ and $x$. What type of control action results?

(c) How could the jet pipe unit be used to produce PI action?

**6.32** Hydraulic controllers are used extensively in aircraft and ships for controlling the motion of rudders, stabilizers, flaps, and other control surfaces. The system shown in Figure P6.32 is a unit for controlling an elevator. Draw the block diagram and find the transfer function between $\theta$ and $y$.

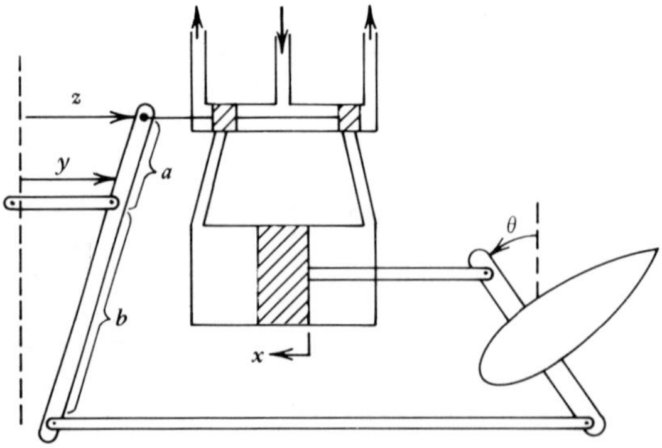


**Figure P6.32**

**6.33** Each of the following problems refers to a previous problem in which an analog controller was developed. For each problem assume that a sampler and zero-order hold is inserted between the controller $G(s)$ nd the plant $G_p(s)$ as shown in Figure P6.33. Assume also that the command input and any feedback measurements are sampled. The disturbance input is not sampled.

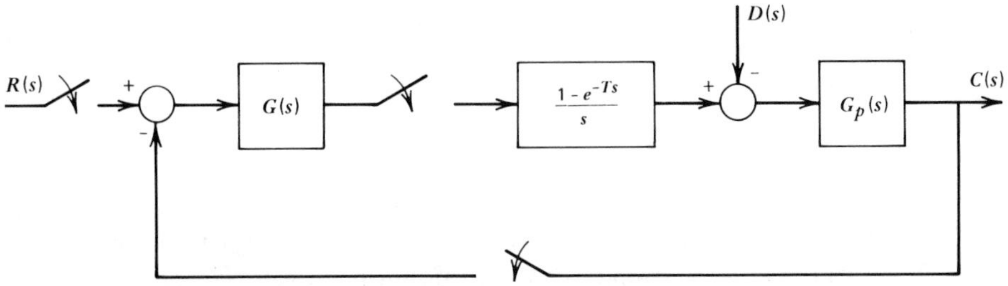


**Figure P6.33**

For each problem, design a digital controller $G(z)$ to meet the specifications given for the analog system in the problem cited. Evaluate the resulting design to see if it meets the continuous-time specifications. Use the sampling period $T$ given for each problem. This value corresponds to 1/10 or less of the smallest relevant time constant of the plant or as required by the specifications. If a satisfactory design cannot be achieved with the given sampling period, find a suitable value for $T$.

**(a)** Problem 6.9, $T = 0.1$.

**(b)** Problem 6.10, $T = 0.1$.

**(c)** Problem 6.11, $T = 0.08$.

**(d)** Problem 6.12, $T = 0.2$.

**(e)** Problem 6.14, $T = 0.01$.

**(f)** Problem 6.16, $T = 0.01$.

**(g)** Problem 6.17, $T = 0.1$.

**(h)** Problem 6.20, $T = 0.01$.

**6.34** The diagram shown in Figure P6.34 represents digital control of a dc motor system with a disturbance torque included. Obtain the relation for the output

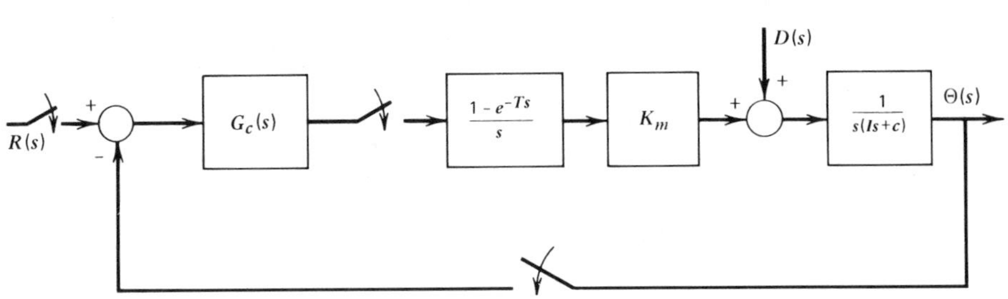


**Figure P6.34**

$\Theta(z)$ in terms of $D(z)$ for

**(a)** Proportional control $G_c(s) = K_p$.

**(b)** Proportional-plus-integral control $G_c(s) = K_p + \dfrac{K_I}{s}$.

**6.35** In Figure P6.35*a* the controller is a PI type with

$$G(s) = 3 + \frac{2}{s}$$

and the plant is

$$G_p(s) = \frac{1}{2s + 1}$$

The gains $K_p = 3$, $K_I = 2$ were selected to give $\zeta = 1$, $\tau = 1$, and zero steady-state error for a step input. Use these gain values with $T = 0.1$ to obtain a digital controller $G(z)$ for the sampled data system shown in Figure P6.35*b*.

**(a)** Use the rectangular integration formula (6.11-7) for the I action.

**(b)** Use the trapezoidal integration form in (6.11-17) for the I action.

Find the unit step response of each design. Compare them with each other and with the analog controller's response.

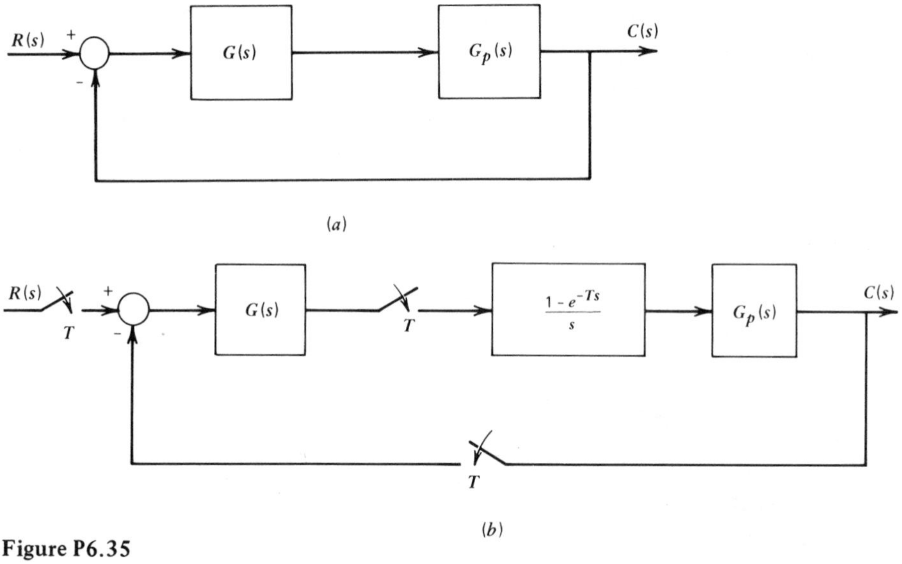


**Figure P6.35**

**6.36** Refer to Figure P6.35. The controller is now a PD type with

$$G(s) = 202 + 20s$$

and the plant is

$$G_p(s) = \frac{1}{s^2 - 2}$$

The gains $K_p = 202$, $K_D = 20$ were selected to give $\zeta = 0.707$, $\tau = 0.1$, and a 1%

steady-state error with a step input. Use these gain values with $T = 0.01$ to obtain a digital controller $G(z)$ for the sampled data system shown in Figure P6.35*b*.

**(a)** Use the proportional-plus-difference law (6.11-8).

**(b)** Use the four-point central difference formula (6.11-16) for the D action.

Find the unit step response of each design. Compare them with each other and with that of the analog controller.

**6.37** Refer to Figure P6.35. The controller is a PID type with

$$G(s) = 120 + \frac{100}{s} + 20s$$

and the plant is

$$G_p(s) = \frac{1}{s(s+1)}$$

The gains $K_p = 120$, $K_I = 100$, $K_D = 20$ were selected to give a dominant root at $s = -1$, with the remaining roots at $s = -10$, and a zero steady-state error for a step input. Use these gain values with $T = 0.1$ to obtain a digital controller $G(z)$ for the sampled data system shown in Figure P6.35*b*.

**(a)** Use (6.11-18).

**(b)** Use (6.11-19).

Find the unit step response of each design. Compare the responses with each other and with that of the analog controller.

**6.38** Consider the Tustin approximation for the PID law (6.12-21). Derive the coefficients $a_1$, $a_2$, and $a_3$ in terms of the analog gains $K_p$, $K_I$, and $K_D$.

**6.39** Repeat Example 6.1 with (a) $T = 0.2$ and (b) $T = 0.05$. Discuss the results.

# CHAPTER SEVEN

# Control System Design: Modeling Considerations, Compensation, and Alternate Control Structures

The models used to describe the plant, sensors, and actuators of a control system greatly influence the design and performance of a control law. Once the form of the law is selected, its gains must be computed from the system model and the given performance specifications. Often a transfer function model of the plant is not available, and the gains must be computed from experimentally determined open-loop response data. Also, systematic procedures for calculating the control gains to achieve the desired stability, accuracy, and speed of response characteristics can be difficult to develop because such specifications are often loosely stated. For example, the results of choosing the gains to give a dominant root with a damping ratio of $\zeta = 0.7$ can be unclear because the other roots might have a significant effect on the response. Thus more direct procedures for computing the gains are sometimes required. Methods for treating such cases are developed in Section 7.1.

Assumptions made to reduce the order of the model and to obtain a linear model can have a significant effect on the controller's performance. In particular, when performance specifications are stringent and large control gains must be used, behavior not predicted by a reduced-order model can become significant. In addition, selection of high gain values tends to drive the control elements to such an extent that they overload or "saturate" and thus exhibit nonlinear behavior. Design procedures that avoid these unwanted effects are covered in Sections 7.2 and 7.3.

When a PID-type control law fails to yield the desired performance, it must be either modified or replaced by an entirely different control scheme. Insertion of additional control elements to modify the control law is called *compensation*, and Section 7.4 discusses several types of compensation schemes. Compensation often leads to a satisfactory design if the PID law is capable of satisfying most of the performance specifications.

When compensation of the PID law still does not produce an acceptable design, a different control scheme must be developed. Multiple feedback loops have been found to be effective, either by using all the state variables to form the actuating signal (Section 7.5), or by using an inner loop to modify the controller's output. Section 7.6 presents one such scheme that has recently been developed, called *pseudo-derivative feedback*. It has advantages over the traditional PID algorithms and it is often selected first, instead of trying to achieve satisfactory performance by compensating the PID law.

Unacceptable performance often results when two different control loops affect each other's behavior. If this interaction is significant, the design must be approached from the viewpoint of the total system, using a multiple-input, multiple-output control scheme (Section 7.7).

The chapter's concepts are mostly developed with respect to analog implementation, but the techniques are easily applied in digital form as well. Therefore, Section 7.8 concludes the chapter with a discussion of control schemes that are unique to digital systems.

## 7.1 SELECTION OF CONTROLLER GAINS

Once the form of the control law has been selected, the gains must be computed in light of the performance specifications. In the examples of the PID family of control laws in Sections 6.5, 6.6, and 6.7, the damping ratio, dominant time constant, and steady-state error were taken to be the primary indicators of system performance in the interests of simplicity. In practice, the criteria are usually more detailed. For example, the rise time and maximum overshoot, as well as the other transient response specifications of Section 5.7, may be encountered. Requirements can also be stated in terms of frequency response characteristics, such as bandwidth, resonant frequency, and peak amplitude. Whatever specific form they take, a complete set of specifications for control system performance generally should include the following considerations, for given forms of the command and disturbance inputs.

1. Equilibrium specifications.
   - (a) Stability.
   - (b) Steady-state error.
2. Transient specifications.
   - (a) Speed of response.
   - (b) Form of response (degree of damping).
3. Sensitivity specifications.
   - (a) Sensitivity to parameter variations.
   - (b) Sensitivity to model inaccuracies.
   - (c) Noise rejection (bandwidth, etc.).
4. Nonlinear effects.
   - (a) Stability.
   - (b) Final control element capabilities.

In addition to these performance stipulations, the usual engineering considerations of initial cost, weight, maintainability, etc., must be taken into account. The considerations are highly specific to the chosen hardware, and it is difficult to deal with such issues in a general way. Here we limit ourselves to the performance considerations listed.

Two approaches exist for designing the controller. The proper one depends on the quality of the analytical description of the plant to be controlled. If an accurate model of the plant is easily developed, the approach is to design a specialized controller for the particular application. The range of adjustment of controller gains in this case can usually be made small because the accurate plant model allows the gains to be precomputed with confidence. This technique reduces the cost of the controller and can usually be applied to electromechanical systems.

The second approach is used when the plant is relatively difficult to model, which is often the case in process control. A standard controller with several control modes and wide ranges of gains is used, and the proper mode and gain settings are obtained by testing the controller on the process in the field. Some guidelines for such tuning are given in this section. This approach should be considered when the cost of developing an accurate plant model might exceed the cost of controller tuning in the field. Of course, the plant must be available for testing for this approach to be feasible.

## Performance Indices

The performance criteria encountered thus far require a set of conditions to be specified – for example, one for steady-state error, one for damping ratio, and one for the dominant time constant. If there are many such conditions, and if the system is of high order with several gains to be selected, the design process can get quite complicated because transient and steady-state criteria tend to drive the design in different directions. An alternative approach is to specify the system's desired performance by means of one analytical expression called a *performance index*. Powerful analytical and numerical methods are available that allow the gains to be systematically computed by minimizing (or maximizing) this index. This approach forms the basis for many advanced design methods. Here we introduce the concept.

To be useful a performance index must be selective. The index must have a sharply defined extremum in the vicinity of the gain values that give the desired performance. If the numerical value of the index does not change very much for large changes in the gains from their optimal values, the index will not be selective. We will only discuss minimizing the index. Applications requiring maximization of an index can be converted to minimization problems by changing the sign of the index.

Any practical choice of a performance index must be easily computed, either analytically, numerically, or experimentally. Four common choices for an index are the following.

$$J = \int_0^\infty |e(t)|\, dt \qquad \text{(IAE)} \tag{7.1-1}$$

$$J = \int_0^\infty t|e(t)|\, dt \qquad \text{(ITAE)} \tag{7.1-2}$$

$$J = \int_0^\infty e^2(t)\, dt \qquad \text{(ISE)} \tag{7.1-3}$$

$$J = \int_0^\infty te^2(t)\, dt \qquad \text{(ITSE)} \tag{7.1-4}$$

where $e(t)$ is the system error. This error usually is the difference between the desired and the actual values of the output. However, if $e(t)$ does not approach zero as $t \to \infty$, the preceding indices will not have finite values. In this case, $e(t)$ can be defined as $e(t) = c(\infty) - c(t)$, where $c(t)$ is the output variable. If the index is to be computed numerically or experimentally, the infinite upper limit can be replaced by the limit $t_f$, where $t_f$ is large enough that $e(t)$ is negligible for $t > t_f$.

The *integral absolute-error* (IAE) criterion (7.1-1) expresses mathematically that the designer is not concerned with the sign of the error, only its magnitude. In some applications the IAE criterion describes the fuel consumption of the system. The index says nothing about the relative importance of an error occurring late in the response versus an error occurring early. Because of this, the index is not as selective as the *integral-of-time-multiplied absolute-error* criterion (ITAE) (7.1-2). Since the multiplier $t$ is small in the early stages of the response, this index weights early errors less heavily than later errors. This makes sense physically. No system can respond instantaneously, and the index is lenient accordingly, while punishing any system that allows a nonzero error to remain for a long time. Neither criterion allows highly underdamped or highly overdamped systems to be optimum. The ITAE criterion usually results in a system whose step response has a slight overshoot and well-damped oscillations.

The *integral squared-error* (ISE) and *integral-of-time-multiplied squared-error* (ITSE) criteria are analogous to the IAE and ITAE criteria, except that the square of the error is employed, for three reasons: (1) in some applications, the squared error represents the system's power consumption; (2) squaring the error weights large errors much more heavily than small errors; (3) the squared error is much easier to handle analytically. The derivative of a squared term is easier to compute than that of an absolute value and does not have a discontinuity at $e = 0$. These differences are important when the system is of high order with multiple error terms. This case is treated at an advanced level (see, for example, Chapter Ten and Chapter 10 of Reference 1).

The ISE and ITSE criteria are similar to the least-squares technique for curve fitting. We can think of the process as one of finding the response curve that minimizes the total squared error. The ITSE index is analogous to weighted least-squares, with the weighting factor being proportional to time. Systems designed with the ISE criterion tend to be fast and oscillatory. The ITSE index provides better selectivity than the ISE. The following example illustrates the procedure.

## *Example 7.1*

It is desired to use proportional control to keep the state $x$ and the rate $\dot{x}$ near zero, for some arbitrary initial condition $x(0)$. The state equation is $\dot{x} + x = u$, where the input $u$ is given by the proportional law $u = -Kx$. (Here the desired output is $x = 0$; thus the error is $0 - x = -x$.) Find the gain $K$.

Since we want to keep both $x$ and $\dot{x}$ near zero, the performance index should reflect this. Choosing a weighted sum of the squares of the deviations, we obtain

$$J = \int_0^\infty [w_1 x^2(t) + w_2 \dot{x}^2(t)]\, dt \qquad \textbf{(7.1-5)}$$

where $w_1$ and $w_2$ are the constant weighting factors. With the control law $u = -Kx$, the system equation is $\dot{x} + x = -Kx$. Its solution is $x(t) = x(0) \exp[-(1 + K)t]$. Substituting this expression into (7.1-5) and integrating gives

$$J = x^2(0)[w_1 + w_2(1 + K)^2] \frac{1}{2(1 + K)} \tag{7.1-6}$$

To minimize $J$ with respect to $K$, we require that

$$\frac{\partial J}{\partial K} = 0, \qquad \frac{\partial^2 J}{\partial K^2} \geqslant 0$$

For a stable system, $1 + K > 0$, and the minimizing value of $K$ is

$$K = -1 + \sqrt{\frac{w_1}{w_2}} \tag{7.1-7}$$

Some insight is gained by examining the extreme values of the weights. For $w_1 = 0$, no weight is given to minimizing $x$, and the solution $K = -1$ results in $x(t) = x(0)$. The velocity is thus minimized. For $w_2 = 0$, $K \to \infty$, which indicates that the controller is trying to drive the state to zero in an infinitesimal amount of time. The general solution for $K$ lies between these two extremes. If we weight $x$ more heavily than $\dot{x}$, $w_1 > w_2$ and $K$ is large. We might argue that the problem has not been solved, but merely has been shifted from one of finding $K$ to one of finding $w_1$ and $w_2$. In a sense this is true, and we rarely use the performance index concept for such simple systems. However, when many gains must be computed, this approach offers a systematic way of doing so.

The general second-order system with zero steady-state error for a unit step input has the form

$$\frac{C(s)}{R(s)} = \frac{\omega_n^2}{s^2 + 2\zeta\omega_n s + \omega_n^2}$$

If we introduce a new time scale $T$ such that $t = \omega_n T$, the transfer function becomes

$$\frac{C(s)}{R(s)} = \frac{1}{s^2 + 2\zeta s + 1} \tag{7.1-8}$$

The step response can easily be found in terms of the parameter $\zeta$ and the various performance indices can be computed, but the algebra required is enough to make us think twice about attempting an analytical approach with a third-order system. For $e(t) = 1 - c(t)$, the results show that the ITAE index and the index

$$J = \int_0^\infty t(|e| + |\dot{e}|)\, dt$$

have the most sharply defined minima as functions of $\zeta$. For these two indices the optimum value of $\zeta$ occurs at approximately $\zeta = 0.7$. The other indices might be more selective when applied to a different system and should not be discarded on the basis of this example.

For the third-order system

$$\frac{C(s)}{R(s)} = \frac{a_0}{s^3 + a_2 s^2 + a_1 s + a_0} \tag{7.1-9}$$

subjected to a unit step input, the values of $a_2$ and $a_1$ that minimize the ITAE criterion are $a_2 = 1.75a_0$, $a_1 = 2.15a_0$. The step response has an overshoot of about 5%, with quickly damped oscillations.

In conclusion, we note that the closed-form solution for the response is not required to evaluate a performance index. For a given set of parameter values, the response and the resulting index value can be computed numerically – for example, with the Runge-Kutta method of Chapter 3 and Appendix C. Using systematic search procedures such as those given in Reference 2, the optimum solution can be obtained; this makes the procedure suitable for use with nonlinear systems.

## The Ziegler-Nichols Rules

The difficulty of obtaining accurate transfer function models for some processes has led to the development of empirically based rules of thumb for computing the optimum gain values for a controller. Commonly used guidelines are the Ziegler-Nichols rules, which have proved so helpful that they are still in use 40 years after their development (References 3 and 4). The rules actually consist of two separate methods. The first method requires the open-loop step response of the plant, while the second utilizes the results of experiments performed with the controller already installed. While primarily intended for use with systems for which no analytical model is available, the rules are also helpful even when a model can be developed.

Ziegler and Nichols developed their rules from experiments and analysis of various industrial processes. Using the IAE criterion with a unit step response, they found that controllers adjusted according to the following rules usually had a step response that was oscillatory but with enough damping so that the second overshoot was less than 25% of the first (peak) overshoot. This is the *quarter-decay* criterion and is sometimes used as a specification.

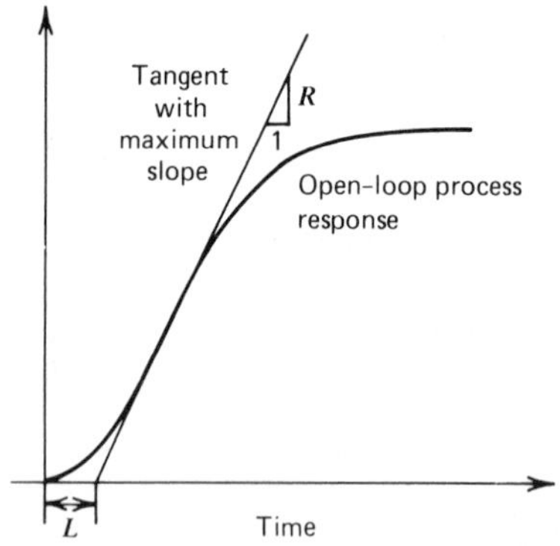


**Figure 7.1 Process signature for a unit step input. The parameters $R$ and $L$ are determined for use with the process-reaction method of Ziegler and Nichols.**

The first method is the *process-reaction method* and relies on the fact that many processes have an open-loop step response like that shown in Figure 7.1. This is the process *signature* and is characterized by two parameters, $R$ and $L$. $R$ is the slope of a line tangent to the steepest part of the response curve, and $L$ is the time at which this line intersects the time axis. First- and second-order linear systems do not yield positive values for $L$, and so the method cannot be applied to such systems. However, third-order and higher linear systems with sufficient damping do yield such a response. If so, the Ziegler-Nichols rules recommend the controller settings given in Table 7.1.

TABLE 7.1 The Ziegler-Nichols Rules

Controller transfer function $G(s) = K_p \left(1 + \frac{1}{T_I s} + T_D s\right)$

| Control Mode | Process-Reaction Method | Ultimate-Cycle Method |
|---|---|---|
| P control | $K_p = \frac{1}{RL}$ | $K_p = 0.5\,K_{pu}$ |
| PI control | $K_p = \frac{0.9}{RL}$ | $K_p = 0.45\,K_{pu}$ |
| | $T_I = 3.3L$ | $T_I = 0.83\,P_u$ |
| PID control | $K_p = \frac{1.2}{RL}$ | $K_p = 0.6\,K_{pu}$ |
| | $T_I = 2L$ | $T_I = 0.5\,P_u$ |
| | $T_D = 0.5L$ | $T_D = 0.125\,P_u$ |

The *ultimate-cycle method* utilizes experiments with the controller in place. All control modes except proportional are turned off, and the process is started with the proportional gain $K_p$ set at a low value. The gain is slowly increased until the process begins to exhibit sustained oscillations. Denote the period of this oscillation by $P_u$ and the corresponding ultimate gain by $K_{pu}$. The Ziegler-Nichols recommendations are given in Table 7.1 in terms of these parameters. The proportional gain is lower for PI control that for P control and is higher for PID control because I action increases the order of the system and thus tends to destabilize it; thus a lower gain is needed. On the other hand, D action tends to stabilize the system; hence the proportional gain can be increased without degrading the stability characteristics. Because the rules were developed for a typical case out of many types of processes, final tuning of the gains in the field is usually necessary. The following example shows how the ultimate cycle method can be applied to an analytical model.

### *Example 7.2*

Proportional control of a third-order plant without numerator dynamics leads to the following characteristic equation.

$$a_3 s^3 + a_2 s^2 + a_1 s + a_0 + K_p = 0 \tag{7.1-10}$$

where $K_p$ is the proportional gain. Find the ultimate period $P_u$ and the associated gain $K_{pu}$.

Sustained oscillation occurs when a pair of roots is purely imaginary, and the rest of the roots have negative real parts. To find when this occurs, set $s = i\omega_u$, where $\omega_u$ is the as-yet-unknown frequency of oscillation. Then $s^2 = -\omega_u^2$ and $s^3 = -i\omega_u^3$.

Substituting these into (7.1-10) and collecting real and imaginary parts, we obtain

$$(a_0 + K_{pu} - a_2\omega_u^2) + i(a_1\omega_u - a_3\omega_u^3) = 0$$

Both the real and imaginary terms must be zero, so we obtain two equations to solve for the two unknowns, $\omega_u$ and $K_{pu}$.

$$a_0 + K_{pu} - a_2\omega_u^2 = 0$$

$$\omega_u(a_1 - a_3\omega_u^2) = 0$$

The latter equation yields the nonoscillatory solution $\omega_u = 0$, as well as the one of interest – namely,

$$\omega_u = \pm\sqrt{\frac{a_1}{a_3}} \tag{7.1-11}$$

Substituting this into the first equation we see that the ultimate gain is given by

$$K_{pu} = a_2\omega_u^2 - a_0 = \frac{a_2 a_1}{a_3} - a_0 \tag{7.1-12}$$

The ultimate period is

$$P_u = \frac{2\pi}{\omega_u} \tag{7.1-13}$$

In order for sustained oscillations to occur, the third root must be negative. The first two roots are $s_1 = i\omega_u$ and $s_2 = -i\omega_u$. The third root can be determined by multiplying the three factors and comparing with the polynomial (7.1-10) with the cubic coefficient normalized to unity. This gives

$$(s - i\omega_u)(s + i\omega_u)(s - s_3) = 0$$

or

$$s^3 - s_3 s^2 + \omega_u^2 s - s_3\omega_u^2 = 0$$

From (7.1-10) we immediately see that $s_3 = -a_2/a_3$. If $s_3 < 0$, the values given by (7.1-12) and (7.1-13) can be used with Table 7.1 to compute the optimum gains. Note that this analysis utilizes only proportional control, but the gains for other control laws can be computed with these results. The power of the method is shown by the fact that we did not have to solve for the roots of the cubic (7.1-10). Note also that I action applied to this system would result in a fourth-order characteristic equation to solve were it not for the Ziegler-Nichols rules.

For most systems the value of $K_{pu}$ represents the largest gain that can be used without producing an unstable system. This is a convenient result, as is the result for the third root $s_3$. In general, for the polynomial

$$s^n + a_{n-1}s^{n-1} + \ldots + a_1 s + a_0 = 0 \tag{7.1-14}$$

the sum of the roots equals $-a_{n-1}$. This fact is frequently very useful.

## 7.2 DESIGN WITH LOW-ORDER MODELS

The development of any model is influenced by our limited knowledge of the system under study as well as by our desire to keep the model simple enough to be useful.

A real system always has more energy-storage modes than can be described by a model, given the time and cost constraints that inevitably exist for the analysis. Thus the designer is always working with a model of an order lower than is necessary to describe the system dynamics completely. The system responses that are omitted from the model are termed the *parasitic modes*, and the parasitic roots are the additional roots that these modes would add to the characteristic equation.

The art of engineering includes the ability to decide which modes are important and to estimate the effects of any parasitic modes on the performance of the system. Here we investigate this process as it pertains to the design of a feedback-control system.

## Design of a DC Servo System

Consider PI control of velocity using the armature-controlled dc motor shown in Figure 5.4. The block diagram of this control system is given in Figure 7.2. The salient points of the example are more easily understood if we use specific and realistic numerical values for the parameters. We take those given in Example 5.18.

| | |
|---|---|
| $K_e = 0.199$ volts/rad/sec | $R = 0.43$ ohms |
| $K_T = 26.9$ oz-in./amp | $c = 0.07$ in.-oz/rad/sec |
| $I = 0.1$ oz-in.-sec$^2$ | $L = 0.0021$ henrys |
| $K_1 = 0.027$ volt/rad/sec | $K_2 = 5$ volt/volt |

where the gain $K_1$ is that of the command pot and the tachometer, and $K_2$ is the gain of the power amplifier.

The time constants of the armature and the mechanical subsystems are

$$\tau_a = \frac{L}{R}$$

$$\tau_m = \frac{I}{c}$$

With the values described, the time constants are $\tau_a = 0.0049$ sec and $\tau_m = 1.429$ sec. Since $\tau_m$ is 290 times larger than $\tau_a$, we conjecture that the armature inductance might be neglected to reduce the order of the system model by one. With this assumption we can attempt the PI controller design with a second-order model. With the inductance retained in the model, a third-order system must be analyzed, and the analytical expressions for the roots in terms of $K_p$ and $K_I$ are difficult to use to select the gains.

## Design Assuming Negligible Inductance

Neglect the inductance temporarily and set $L = 0$. The resulting transfer function for the reference input velocity $\omega_R$ is obtained from Figure 7.2.

$$T_1(s) = \frac{\Omega(s)}{\Omega_R(s)} = \frac{K_1K_2K_T(K_ps + K_I)}{RIs^2 + (cR + K_eK_T + K_1K_2K_TK_p)s + K_1K_2K_TK_I} \tag{7.2-1}$$

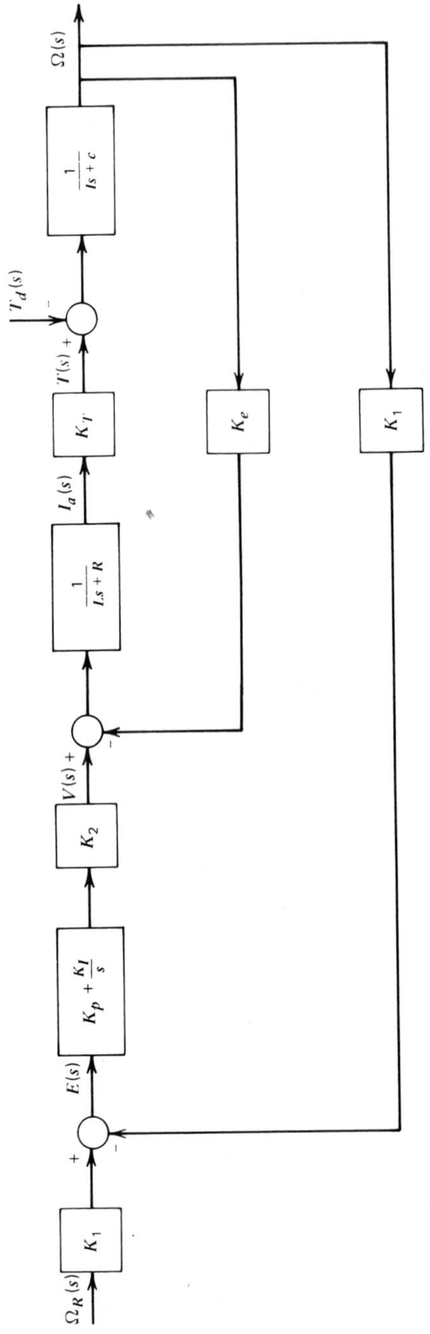


**Figure 7.2** PI control of velocity with an armature-controlled dc motor.

For the given values,

$$T_1(s) = \frac{3.6315(K_p s + K_I)}{0.043s^2 + (5.3832 + 3.6315K_p)s + 3.6315K_I} \tag{7.2-2}$$

The disturbance transfer function is

$$T_2(s) = \frac{\Omega(s)}{T_d(s)} = \frac{-Rs}{RIs^2 + (cR + K_T K_e + K_T K_1 K_2 K_p)s + K_T K_1 K_2 K_I} \tag{7.2-3}$$

or

$$T_2(s) = \frac{-0.43s}{0.043s^2 + (5.3832 + 3.6315K_p)s + 3.6315K_I} \tag{7.2-4}$$

For now assume that the performance specifications require the system to deal with step changes in the reference velocity and in the disturbance torque. For P action alone ($K_I = 0$), (7.2-2) shows that the system is first order and that the steady-state response to a set-point change has the value

$$\omega_{ss} = \frac{3.6315K_p}{5.3832 + 3.6315K_p}$$

For $\omega_{ss}$ to be close to unity, as desired, $K_p$ must be large. Equation (7.2-4) with $K_I = 0$ shows that P action alone cannot give a zero steady-state offset due to the disturbance. Thus the usual disadvantages of proportional control are encountered here.

If we attempt to use I action alone ($K_p = 0$), (7.2-4) shows that the step disturbance effects will be eliminated, but that the damping in the characteristic equation is now beyond our control. The equation is

$$0.043s^2 + 5.3832s + 3.6315K_I = 0$$

For stability $K_I$ must be positive. We can select $K_I$ to achieve the degree of damping desired, but the time constant is then beyond our control. For example, for $\zeta \leqslant 1$, the time constant is 0.016 sec regardless of the $K_I$ value as long as $K_I \geqslant 46.4$. Note that when $K_p = 0$, most of the damping is caused by the back emf. If this damping is not great enough to achieve the desired time constant, proportional control must be added.

PI control will give a zero offset at steady state for both the set-point and disturbance inputs for any stable combination of $K_p$ and $K_I$. The gains can then be used to select the damping ratio and dominant time constant. Let us assume that we can accept a 5% overshoot in the response to a step set-point change. From Figure 5.20*a*, a 5% overshoot specification is seen to correspond to $\zeta \leqslant 0.707$, approximately.

From (7.2-2), the roots of the denominator can be seen to give the following time constant, if $\zeta \leqslant 1$.

$$\tau = \frac{0.086}{5.3832 + 3.6315K_p} \tag{7.2-5}$$

Suppose that a time constant no larger than 0.01 sec is desired. For $\tau = 0.01$, the proportional gain must be $K_p = 0.8856$, and therefore the integral gain must be $K_I = 237$ for $\zeta = 0.707$. The characteristic roots will be $s = -100 \pm 100i$.

Now suppose that we want $\tau \leqslant 0.001$. Then (7.2-5) gives $K_p = 22.2$. For $\zeta = 0.707$, $K_I = 23{,}700$. The roots are $s = -1000 \pm 1000i$. These two cases are tabulated as Cases 1 and 2 in the first column of Table 7.2.

TABLE 7.2 The Effects of Parasitic Armature Inductance $L$

Armature time constant = 0.0049 sec

| Second-Order Design ($L = 0$) | Third-Order Behavior ($L = 0.0021$) |
|---|---|
| 1. $\tau \leqslant 0.01$, $\zeta = 0.707$<br>$K_p = 0.8856$<br>$K_I = 237$<br>$s = -100 \pm 100i$ | $\tau = 0.027$, $\zeta = 0.21$<br>$s = -131$, $-37 \pm 172i$<br>Dominant time constant too large<br>Damping ratio too small |
| 2. $\tau \leqslant 0.001$, $\zeta = 0.707$<br>$K_p = 22.2$<br>$K_I = 23{,}708$<br>$s = -1000 \pm 1000i$ | $s = -167$, $206 \pm 788i$<br>An unstable system |
| 3. $\tau_1 = 0.05$, $\tau_2 = 0.005$<br>$\zeta > 1$<br>$K_p = 1.123$<br>$K_I = 47.37$<br>$s = -20, -200$ | $\tau = 0.0505$, $\zeta > 1$<br>$s = -19.8$, $-92.8 \pm 181i$<br>Dominant root is close to desired value |

## The Effects of Armature Inductance

Let us check the performance of each design by predicting its behavior more accurately with the armature inductance included in the model. From Figure 7.2, the two transfer functions are

$$T_1(s) = \frac{\Omega(s)}{\Omega_R(s)} = \frac{K_T K_1 K_2 (K_p s + K_I)}{\Delta(s)} \tag{7.2-6}$$

$$T_2(s) = \frac{\Omega(s)}{T_d(s)} = \frac{-s(Ls + R)}{\Delta(s)} \tag{7.2-7}$$

where the characteristic polynomial is

$$\Delta(s) = LIs^3 + (RI + cL)s^2 + (cR + K_e K_T + K_T K_1 K_2 K_p)s + K_T K_1 K_2 K_I \tag{7.2-8}$$

For $L = 0.0021$ and the other parameter values as given before, we obtain

$$\Delta(s) = 0.00021s^3 + 0.04315s^2 + (5.3832 + 3.6315K_p)s + 3.6315K_I \tag{7.2-9}$$

Examination of the transfer functions shows that the presence of a nonzero inductance $L$ does not change the steady-state performance as predicted by the second-order model. Let us see if the same is true for the transient performance. For Case 1, substitute the values $K_p = 0.8856$ and $K_I = 237$ into (7.2-9) and solve for the roots. These are $s = -131$, $-37 \pm 172i$. The dominant behavior is given by the conjugate pair. The dominant time constant is $\tau = 1/37 = 0.027$, which is almost three times larger than can be tolerated according to the given specifications. The damping ratio for the dominant pair is $\zeta = 0.21$ and indicates an overshoot of approximately 50%, ten times larger than desired. Obviously the design is poor because the roots of the quadratic model do not approximate the dominant roots of the cubic model.

For Case 2, with $K_p = 22.2$ and $K_I = 23{,}700$, the roots of (7.2-9) are $s = -617$, $206 \pm 788i$. The third-order model is thus unstable, whereas the second-order model predicted the dominant roots would be $s = -1000 \pm 1000i$. These results are summarized in Table 7.2.

Case 2 shows that the quadratic approximation is degraded as the desired time constant is made smaller. This gives a clue as to what the underlying difficulty is. The armature time constant $\tau_a$ is neglected in the quadratic approximation. For Case 1 the desired time constant $\tau$ is about twice the value of $\tau_a$ (0.01 versus 0.0049), whereas in Case 2 $\tau$ is about five times *smaller* than $\tau_a$ (0.001 vs 0.0049). Thus we have attempted to obtain a system whose time constant is barely dominant relative to the neglected time constant $\tau_a$ in Case 1 and not dominant in Case 2.

The solution to the difficulty lies in specifying a time constant that is much larger than $\tau_a$. For $\zeta < 1$, (7.2-5) shows that the largest time constant obtainable in the quadratic model is $\tau = 0.016$, if $K_p$ must be positive for physical reasons. This value is only three times larger than $\tau_a$, so we must look at the overdamped case. For a dominant time constant $\tau = 0.05 = 10\tau_a$, and a secondary time constant much smaller, say 0.005, the required gains from the quadratic model are $K_p = 1.123$ and $K_I = 47.37$. The predicted roots are $s = -20, -200$. With these gain values substituted into (7.2-9), the cubic roots are $s = -19.8, -92.8 \pm 181i$. Thus the dominant root of the third-order model is very close to that of the second-order model. The response of the third-order model will show oscillations that are not predicted by the second-order one, but these will decay five times more quickly than the transients produced by the dominant root. This is Case 3 in Table 7.2.

The imaginary designer in this example sought to avoid using a third-order model to select the gains $K_p$ and $K_I$. In fact, such a model is often not too difficult to work with, if the techniques of the following chapters are employed. However, the example was constructed to illustrate the effect of parasitic roots. In practice, for this particular application, the designer has the third-order model available, but might not have an accurate value for the armature inductance.

In other applications, however, it might be difficult or expensive to develop the higher-order model necessary to predict the effects of parasitic modes. Such modes commonly result from the neglected dynamics of sensors and final control elements. For example, the interaction of the water surface and the float in Figure 6.25*a* can produce a significant parasitic mode if proportional control is used with a high gain (small time constant). Another example is the neglected elasticity of the shaft connecting a load to a motor. If the system is a high-performance one, the motor torques will be large and easily reversible, and the shaft twist might be significant. In general, parasitic modes will be important in systems whose specifications call for a time constant close to that of any neglected mode. The model to be used in such a design should therefore include such modes.

## 7.3 NONLINEARITIES AND CONTROLLER PERFORMANCE

All physical systems have nonlinear characteristics of some sort, although they can often be modeled as linear systems provided the deviations from the linearization reference condition are not too great. Under certain conditions, however, the nonlinearities have significant effects on the system's performance. Here we consider some of these effects and how they can be avoided or accounted for in control system design.

One such situation can occur during the start-up of a controller if the initial conditions are much different from the reference condition for linearization. The linearized model is then not accurate, and nonlinearities govern the behavior. If the nonlinearities are mild, there might not be much of a problem. Where the nonlinearities are severe, such as in process control, special consideration must be given to start-up. Usually in such cases the control signal sent to the final control elements is manually adjusted until the system variables are within the linear range of the controller. Then the system is switched into automatic mode. Digital computers are often used to replace the manual adjustment process because they can be readily coded to produce complicated functions for the start-up signals. Care must also be taken when switching from manual to automatic. For example, the integrators in electronic controllers must be provided with the proper initial conditions. An example of a circuit with this feature is given in Reference 5, pp. 237–239.

## Limit Cycles

A *limit cycle* is a self-sustaining oscillation whose characteristics are determined by the *system itself* and not by the initial conditions or the inputs. Limit cycles are found only in nonlinear systems. The sustained oscillations found in neutrally stable linear systems, such as an undamped mass-spring system, are not limit cycles because their amplitude depends on the initial conditions. The oscillation produced at steady state by periodic forcing of a linear system is also not a limit cycle, because the frequency of oscillation is determined by the input frequency.

The existence of a limit cycle is characterized by a closed curve on the phase plane plot. If all the state trajectories approach the limit-cycle curve, it is said to be a stable limit cycle. An example is given by the liquid-level controller shown in Figure 6.25 with on-off control. The familiar system model is

$$\dot{h} = -\frac{1}{RC}h + \frac{1}{C}q_m \tag{7.3-1}$$

for the filling curve and

$$\dot{h} = -\frac{1}{RC}h \tag{7.3-2}$$

for the emptying curve. The flow rate $q_m$ is the maximum flow rate through the supply valve. If we wish to keep the height at $h_d$, a constant flow rate of $q_d = h_d/R$ is required. If the controller allows a fluctuation $\Delta$ in the level above and below $h_d$, the controller switches from the filling to the emptying curve when $h = h_d + \Delta$. When $h = h_d - \Delta$, the controller switches back to the filling curve.

These characteristics can be easily summarized with a phase plot of $\dot{h}$ vs $h$. On this plot, (7.3-1) is a straight line with a slope of $-1/RC$ and an intercept of $q_m/C$. The emptying curve (7.3-2) has the same slope but a zero intercept. Figure 7.3 shows the limit cycle for a specific case with the following parameter values.

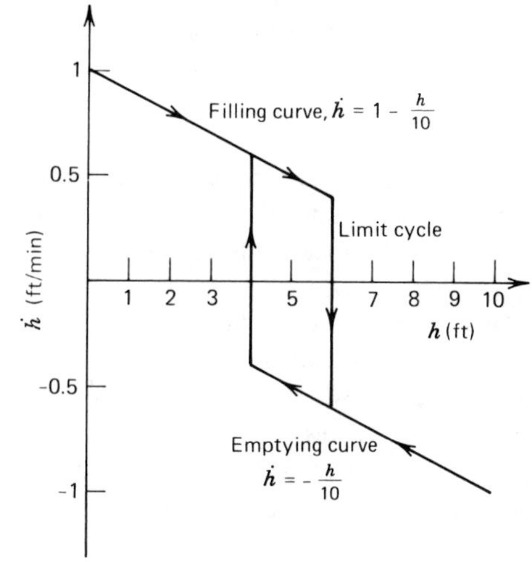


**Figure 7.3 Limit cycle produced by on-off control of the liquid-level system of Figure 6.25. The tank's time constant is $RC = 10$ min, and the switching gap is 2 ft.**

$$R = \tfrac{5}{6}\ \text{min/ft}^2 \qquad C = 12\ \text{ft}^2$$

$$h_d = 5\ \text{ft} \qquad \Delta = 1\ \text{ft}$$

$$q_d = 6\ \text{ft}^3/\text{min} \qquad q_m = 12\ \text{ft}^3/\text{min}$$

The only trajectories possible are those shown in the figure. For any initial height greater than 4 ft, the system follows the emptying curve until the switching at $h = 4$ ft. The switching is assumed to be instantaneous; thus $\dot{h}$ is discontinuous at that point. Similarly, for $h(0) < 6$ ft, the filling curve is followed until $h = 6$ ft. Once the height is within the range $4 \leqslant h \leqslant 6$, the system follows the limit cycle, which is the closed curve. The solutions of (7.3-1) and (7.3-2) can be used to show that 8.1 min are required for one cycle. Because all the trajectories lead to the limit cycle, it is stable.

When the system model is of first or second order and is piecewise-linear, it is not too demanding to use the phase plot to determine the existence of a limit cycle. Otherwise it is generally difficult to do so, and more advanced methods are required (see for example Chapter 12 of Reference 1). Digital simulation can be used to detect such cycles, but the step size must be chosen carefully. If not, round-off and truncation errors can either mask the presence of a cycle or induce oscillations that might be falsely interpreted as a cycle.

Control systems designed to be stable on the basis of a linearized model can be unstable if any nonlinearities present are strong enough to cause a limit cycle. This can occur in servomechanisms in which Coulomb friction or gear backlash is present. Another cause of this behavior is saturation of the final control elements when overdriven by the controller.

## Reset Windup

The saturation nonlinearity shown in Figure 7.4 represents the practical behavior of many actuators and final control elements. For example, a motor-amplifier combination can produce a torque proportional to the input voltage over a limited range. However, no amplifier can supply an infinite current; there is a maximum current and

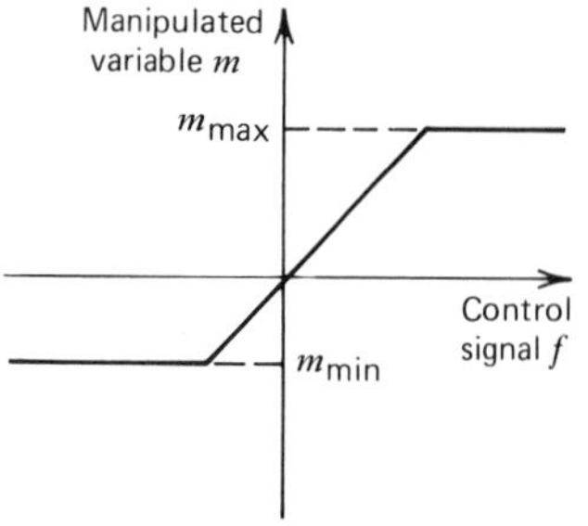


Figure 7.4 **Saturation nonlinearity.**

thus a maximum torque that the system can produce in either direction (clockwise or counterclockwise). Another example is a flow-control valve. Once it is fully open, the flow rate cannot be made greater without increasing the supply pressure. This shows that the saturation nonlinearity is not always symmetric; in the case $m$ is flow rate, $f$ is valve position, and $m_{min} = 0$.

The final control elements are said to be overdriven when they are commanded by the controller to do something they cannot do. Since the limitations of the final control elements are ultimately due to the limited rate at which they can supply energy, it is important that all system performance specifications and controller designs be consistent with the energy-delivery capabilities of the elements to be used.

Any controller with integral action can exhibit the phenomenon called *reset windup* or *integrator buildup* when overdriven, if it is not properly designed. Consider the PI-control law.

$$m(t) = K_p e(t) + K_I \int_0^t e(t)\,dt \tag{7.3-3}$$

For a step change in set point, the proportional term responds instantly and saturates immediately if the set-point change is large enough. On the other hand, the integral term does not respond as fast. It integrates the error signal and saturates some time later if the error remains large for a long enough time. As the error decreases, the proportional term is no longer saturated. However, the integral term continues to increase as long as the error has not changed sign, and thus the manipulated variable remains saturated. Even though the output is very near its desired value, the manipulated variable remains saturated until after the error has reversed sign. The result can be a large overshoot in the response of the controlled variable.

Consider PI control of a plant that is a pure integrator, such as a unit inertia with torque input, velocity output, and no damping (Figure 7.5). Assume that the torque is limited to a maximum magnitude $M$ and that the slope of the linear region of the saturation function is unity, for simplicity. For system operation within the linear region, we can utilize the transfer function

$$\frac{\Omega(s)}{\Omega_R(s)} = \frac{K_p s + K_I}{s^2 + K_p s + K_I} \tag{7.3-4}$$

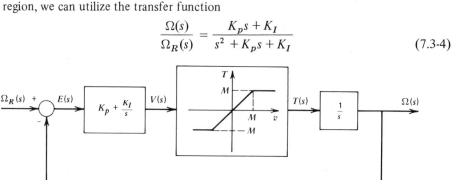

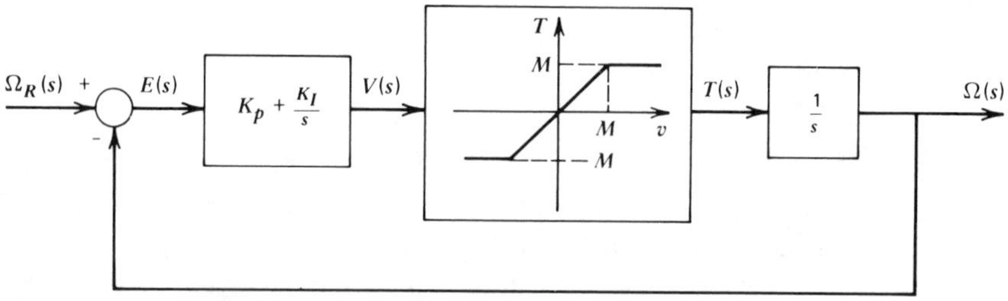


Figure 7.5 **PI control of an inertia load with a saturating-torque motor.**

where $T(s) = V(s)$ for $|v(t)| < M$. The system will be critically damped if $K_p^2 = 4K_I$. In this case the time constant will be $\tau = 2/K_p$.

Take as an example the case where $M = 1$ and the input is a unit step. For the specifications $\zeta = 1$ and $\tau = 0.4$, the gains must be $K_p = 5, K_I = 6.25$. For zero initial velocity, the error is $e(0) = 1 - 0 = 1$. The system immediately saturates because the proportional term is $K_p e = 5$. In the saturated mode, $T(t) = 1$ and $\dot{\omega} = 1$. Thus at start-up, $\omega(t) = t$ and $e(t) = 1 - t$, until $v = 1$. The voltage function during start-up is found from (7.3-3), with $m = v$, to be

$$v(t) = 5 + 1.25t - 3.12t^2$$

The system enters the linear region when $v(t) = 1$ at $t = 1.35$.

For $t > 1.35$ the response can be obtained from (7.3-4) with the two solutions matched at $t = 1.35$. The governing equation is

$$\ddot{\omega} + 5\dot{\omega} + 6.25\omega = 5\dot{\omega}_R + 6.25\omega_R$$

where $\omega_R = 1$ and $\dot{\omega}_R = 0$ for $t > 1.35$. The solution is

$$\omega(t) = 54.58te^{-2.5t} - 63.46e^{-2.5t} + 1$$

with the matching conditions $\omega(1.35) = 1.35$, $\dot{\omega}(1.35) = 1$. The remaining variables

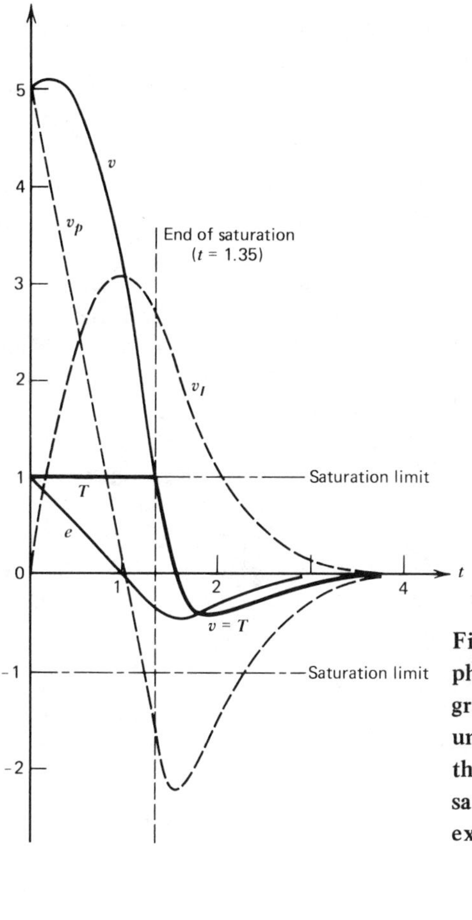


**Figure 7.6 Example of the reset windup phenomenon in a PI controller. The integral term $v_I$ does not begin to decrease until the error $e$ changes sign. This extends the time during which the controller is saturated ($|v| \geq 1$). The lower limit is not exceeded here since $v = v_p + v_I > -1$.**

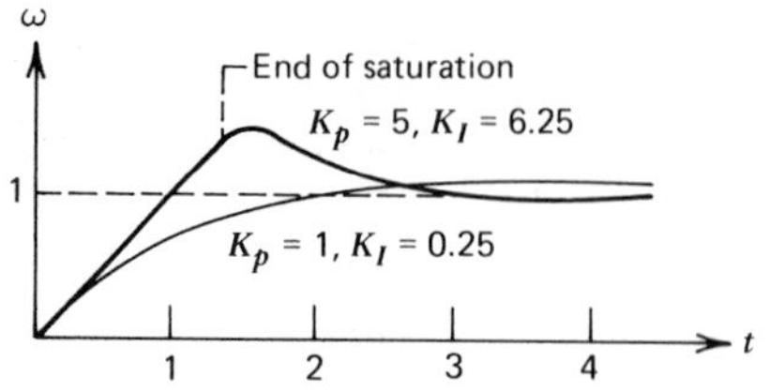


**Figure 7.7 Unit step response of a PI controller with a saturation nonlinearity. The overshoot for the case $K_p = 5, K_I = 6.25$ results from reset windup. The overshoot is decreased when the gains are lowered to prevent saturation.**

are obtained from $\omega(t)$ through their defining relations. The results are plotted in Figure 7.6. The proportional action's contribution to the voltage $v(t)$ is denoted $v_p(t)$, while that due to the integral action is $v_I(t)$. Note that the error changes sign at $t = 1$, but the integral term $v_I$ keeps the system saturated until $t = 1.35$. The velocity output is plotted in Figure 7.7 and displays the overshoot that is often caused by the reset windup phenomenon.

## Design with Power-Limited Elements

It has been stated that control system performance can be improved if the limited energy-delivering capacity of the final control elements is taken into account. The previous example of the reset windup phenomenon provides a convenient illustration of how this improvement can be accomplished.

Reset windup can be avoided if we make sure the controller never allows the final control elements to saturate. One way of accomplishing this limitation for electronic controllers was discussed in Section 6.5 (see Figure 6.45). If such limiters were installed on the PI controller of the previous example, the voltages $v_p(t)$ and $v_I(t)$ would not be allowed to exceed the value required to induce saturation of the torque, and the overshoot in the velocity response would be reduced. This technique is useful as a precaution in applications where the magnitudes of the inputs are difficult to predict in advance.

Another method is to select the gains so that saturation will never occur. This requires knowledge of the maximum input magnitude that the system will encounter. For example, suppose the error signal is $e(t) = r(t) - c(t)$, where $r$ and $c$ are the input and output signals. The maximum error in a well-designed system will occur at $t = 0$ for a step input and zero initial output. Thus in a PI controller described by (7.3-3), saturation will first occur because of the proportional term, since it responds instantly to the error. The integral term might cause saturation also, but at a later time, because its output requires time to accumulate.

Denote the maximum expected value of $r(t)$ by $r_{\max}$ and the value of the controller output at saturation by $m_{\max}$. Then under the stated conditions, the maximum error is $r_{\max}$. Since the integral contribution is zero at $t = 0$, the maximum gain for linear operation is

$$K_p = \frac{m_{\max}}{r_{\max}} \tag{7.3-5}$$

The time constant for the model (7.3-4) is $\tau = 2/K_p$ if $\zeta \leqslant 1$. Thus $K_p$ is usually set to the value given above to minimize $\tau$. The required value of $K_I$ for (7.3-4) can be expressed in terms of $\zeta$ as

$$K_I = \frac{K_p^2}{4\zeta} \tag{7.3-6}$$

For the previous example, $m_{max} = M = 1$, and we assume that the system's step inputs will have a magnitude no greater than unity, so that $r_{max} = 1$. From (7.3-5), $K_p = 1$. With $\zeta = 1$ as before, (7.3-6) gives $K_I = 0.25$. Using these values in (7.3-4) with a unit step input, we obtain the response

$$\omega(t) = 0.5te^{-0.5t} - e^{-0.5t} + 1$$

The other variables are found from $\omega(t)$.

$$e(t) = 1 - \omega(t)$$

$$v_p(t) = e(t)$$

$$v_I(t) = 0.25te^{-0.5t}$$

$$v(t) = v_p(t) + v_I(t)$$

From these expressions it can be seen that $v_p$, $v_I$, and $v$ do not exceed the saturation limit of 1. The system is always within the linear region. Its response is shown in Figure 7.7, along with that of the previous example. With the present choice of gains, no reset windup occurs and the overshoot is much less than before. However, the response now takes longer to settle down to its steady-state value. The new time constant is 2, as opposed to 0.4 for the previous design.

In order to decide which response is optimal, the designer must look more closely at the requirements for the particular application. In many systems the occurrence of saturation and response overshoot indicates that the system is using more energy than is necessary to accomplish the task. If so, the preceding design method is preferred. The technique was applied to PI control of a pure integrator, but it can be generalized as follows.

1. For a model whose only nonlinearity results from saturation, assume the system is operating in the linear region, and compute the response for the most severe inputs the system is expected to deal with.
2. Compute the maximum gain values that can be allowed without inducing saturation. Steps 1 and 2 can be done analytically for first- and second-order systems. For higher-order systems, computer simulation can be used.
3. Use other criteria to decide whether the gains should be set to these maximum values or to lower values. Such criteria might be the desired values of time constants or damping ratio, or the performance indices (IAE, ISE, etc.).

## 7.4 COMPENSATION

It is sometimes impossible to adjust the gains in a PID-type control law in order to satisfy all of the performance specifications. If a control law can be found that satisfies most of the specifications, a common design technique is to insert a *compensator* into the system. This is a device that alters the response of the controller so that the overall system will have satisfactory performance.

Compensation alters the structure of the control system, and compensation

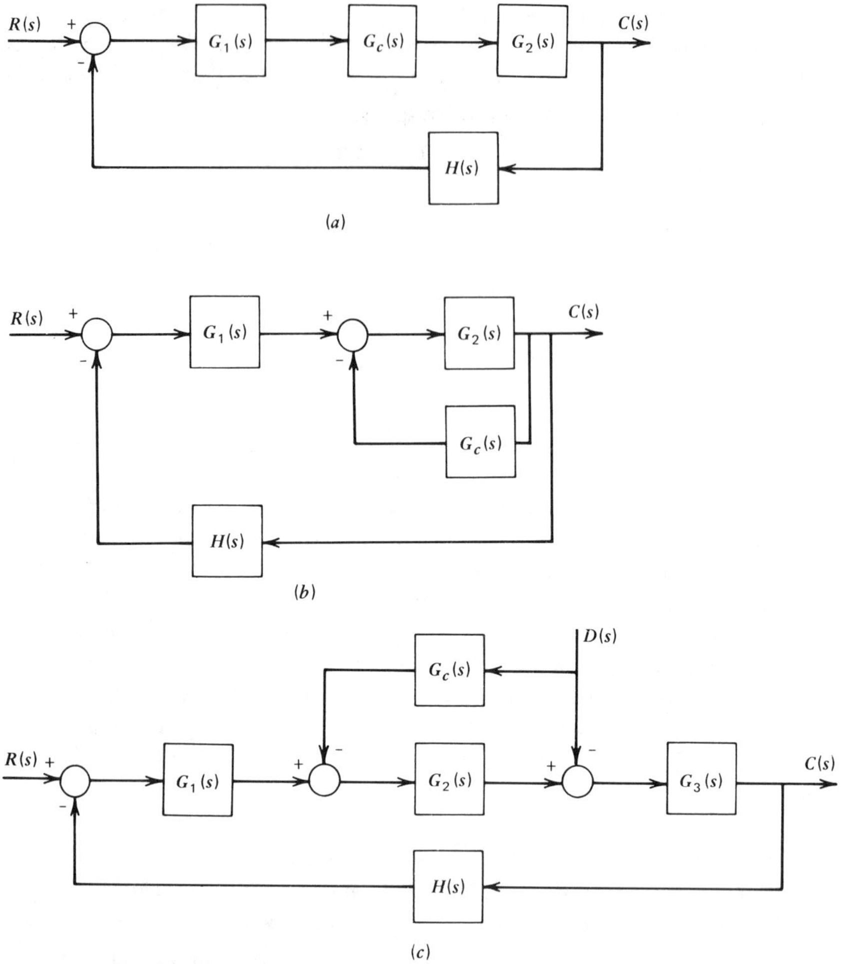


**Figure 7.8 General structures of the three compensation types. (*a*) Series. (*b*) Parallel (or feedback). (*c*) Feed-forward. The compensator transfer function is $G_c(s)$.**

techniques are categorized accordingly. The three categories generally recognized are *series compensation*, *parallel* (or *feedback*) *compensation*, and *feed-forward compensation*. The three structures are loosely illustrated in Figure 7.8, where we assume the final control elements have a unity transfer function. The transfer function of the controller is $G_1(s)$. The feedback elements are represented by $H(s)$ and the compensator by $G_c(s)$. We assume that the plant is unalterable, as is usually the case in control system design.

The choice of compensation structure depends on what type of specifications must be satisfied. The following examples are intended to provide some insight into this selection. More examples of compensation are given in the following chapter.

The physical devices used as compensators are similar to the pneumatic, hydraulic, and electrical devices treated previously. Once the transfer function is determined, a device can be selected with criteria similar to those used to select controller elements.

## Feed-Forward Compensation

The control algorithms considered thus far have counteracted disturbances by using measurements of the output. One difficulty with this approach is that the effects of the disturbance must show up in the output of the plant before the controller can begin to take action. On the other hand, if we can measure the disturbance, the response of the controller can be improved by using the measurement to augment the control signal sent from the controller to the final control elements. This is the essence of feed-forward compensation.

Consider the two-tank system shown in Figure 7.9. The height $h_2$ in the second tank is to be controlled by adjusting the flow rate $q_1$ into the first tank. A measurement of the height $h_2$ is used to do this. The system disturbance is the flow rate $q_d$ that enters the second tank at unpredictable times and might result from catalysts being recycled into the process, for example. The block diagram is given in Figure 7.10. The feed-forward compensator is a simple-gain $K_f$ in this case; dynamic elements can be used if necessary. PI action could be used for the controller transfer function $G_1(s)$ to cancel the effects of the disturbance, but the integral term increases the order of the system. This makes it more oscillatory, less stable, and more difficult to analyze.

The performance of the feed-forward compensator is revealed by the disturbance transfer function.

$$\frac{H_2(s)}{Q_d(s)} = \frac{R_2(R_1C_1s + 1 - K_fK_v)}{(R_1C_1s + 1)(R_2C_2s + 1) + R_2K_vK_p} \tag{7.4-1}$$

where P action is used for the controller $[G_1(s) = K_p]$. If $K_p$ is positive, the system is stable. The steady-state deviation resulting from a unit-step disturbance is

$$\Delta h_{2ss} = \frac{R_2(1 - K_fK_v)}{1 + R_2K_vK_p} \tag{7.4-2}$$

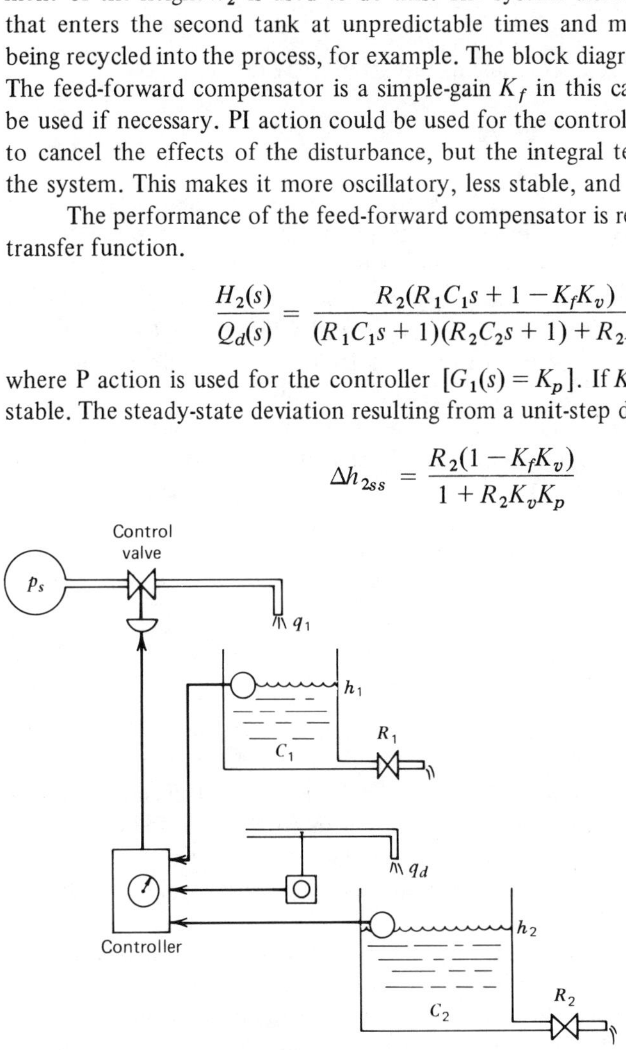


**Figure 7.9 Liquid-level system used for feed-forward compensation and cascade-control examples.**

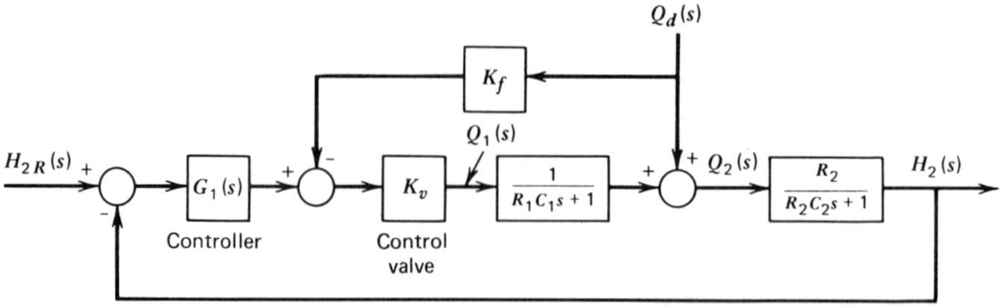


**Figure 7.10 Feed-forward compensation applied to the liquid-level system of Figure 7.9.**

The deviation can be made zero for a step disturbance of any magnitude if

$$K_f = \frac{1}{K_v} \tag{7.4-3}$$

The feed-forward compensation does not affect the response of $h_2$ to the command $h_{2R}$. In particular, $K_f$ does not affect the characteristic roots of the system. Its effect on the disturbance response is shown by the step response of a second-order system with numerator dynamics analyzed in Section 5.7, Example 5.19. The numerator zero of (7.4-1) is $s = 0$ if (7.4-3) is satisfied, and is $s = -1/R_1C_1$ if $K_f = 0$ (no compensation). From the results of Example 5.19, we can conclude that the feed-forward compensator improves the speed of response to a disturbance.

A difficulty with feed-forward compensation is that it is an open-loop technique in that it contains no self-correcting action. If the value of $K_v$ is not accurately known, the gain $K_f$ selected from (7.4-3) will not cancel the disturbance completely. The usefulness of the method must be determined in light of the specific performance requirements and the uncertainty in $K_v$. If the uncertainty is large, self-correcting action can be obtained by using a PI control law for $G_1(s)$ in conjunction with the feed-forward compensation.

## Feedback Compensation and Cascade Control

We have already encountered several examples of feedback compensation. The use of a tachometer to obtain velocity feedback, as in Figure 6.36, is such a case. The feedback-compensation principle of Figure 6.38 is another. Here we consider another form, called *cascade control*, in which another controller is inserted within the loop of the original control system. The new controller can be used to achieve better control of variables within the forward path of the system. Its set point is manipulated by the first controller.

Cascade control can be applied to the system shown in Figure 7.9 to reduce the effects of a change in the supply pressure $p_s$. The block diagram is shown in Figure 7.11 with the cascade controller inserted. For now we neglect the disturbance $q_d$ and the feed-forward compensator. The constant $K_s$ relates the change in flow rate through the control valve to the change in supply pressure. The idea of cascade control as applied here is to measure the internal variable $h_1$ and to use this information to adjust the

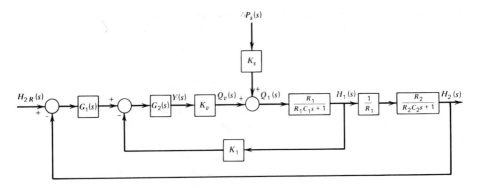

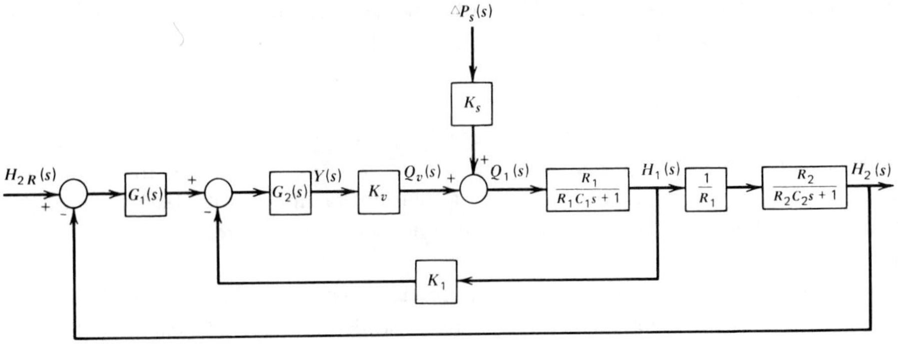


**Figure 7.11** **Cascade control applied to the liquid-level system of Figure 7.9.**

position $y$ of the control valve. The controller $G_2(s)$ and the feedback gain $K_1$ have been inserted for this purpose. When a change in supply pressure occurs, the effect is seen in the height $h_1$ before $h_2$. The use of cascade control should increase the speed of response of the control system in reacting to the disturbance.

The effect of the cascade controller is seen more easily with specific values for the parameters. To this end, let

$$R_1 = R_2 = C_1 = 1, \qquad C_2 = 2$$

$$K_1 = K_v = K_s = 1$$

and assume that P action is used for both controllers, so that $G_1(s) = K_{p1}$, $G_2(s) = K_{p2}$. The resulting transfer function for the disturbance is

$$\frac{H_2(s)}{\Delta P_s(s)} = \frac{1}{2s^2 + (3 + 2K_{p2})s + 1 + K_{p1}K_{p2} + K_{p2}} \tag{7.4-4}$$

If no compensation is used $[K_1 = 0, G_2(s) = 1]$,

$$\frac{H_2(s)}{\Delta P_s(s)} = \frac{1}{2s^2 + 3s + 1 + K_{p1}} \tag{7.4-5}$$

For a unit step disturbance, the compensated system gives a steady-state deviation of

$$\Delta h_{2ss} = \frac{1}{K_{p2}(K_{p1} + 1)} \tag{7.4-6}$$

With no compensation,

$$\Delta h_{2ss} = \frac{1}{1 + K_{p1}} \tag{7.4-7}$$

We immediately see that the damping coefficient of the uncompensated system is not affected by the control gain $K_{p1}$. This gain can be selected to achieve a desired time constant, damping ratio, or steady-state deviation, but it is highly unlikely that a single value will satisfy all three criteria. On the other hand, the compensated system has the damping ratio

$$\zeta = \frac{3 + 2K_{p2}}{2\sqrt{2(1 + K_{p1}K_{p2} + K_{p2})}} \tag{7.4-8}$$

The inner loop gain $K_{p2}$ can be adjusted to increase the speed of response, while the outer loop gain $K_{p1}$ can be used to minimize the steady-state deviation.

To see this, assume that the maximum gain value available is 9, due to physical constraints. The minimum deviation for the uncompensated system is $\Delta h_{2ss} = 0.1$ with $K_{p1} = 9$; the time constant is 0.67, and the damping ratio is 0.34; this indicates a large overshoot. For the compensated system with $K_{p1} = K_{p2} = 9$, the results are $\Delta h_{2ss} = 0.01$, $\tau = 0.095$, and $\zeta = 0.78$. The improvement is substantial, but must be weighed against the cost of an extra controller and an extra sensor to measure the height $h_1$.

If the feed-forward compensator is reintroduced to handle the disturbance $q_d$, the gain $K_f$ must be recomputed because the cascade controller has changed the gain in the forward path. For previous parameter values, $K_f$ should be unity without cascade control, but $K_f = 1/9$ is the proper choice otherwise.

Cascade control exists in disguise in many systems because the final control elements usually include feedback loops to improve their performance. For example, a power amplifier has internal circuits to regulate voltage level while supplying current to the load. We usually assume that such subsystems act as perfect open-loop devices. If not, cascade control must be used to regulate their performance. For example, the control valve in Figure 7.9 will have improved performance in the presence of a supply-pressure disturbance, if the flow rate $q_1$ is measured and used with a controller to adjust the valve position $y$.

Cascade control is frequently used when the plant cannot be satisfactorily approximated with a model of second order or lower. This is because the difficulty of analysis and control increases rapidly with system order. The characteristic roots of a second-order system can easily be expressed in analytical form. This is not so for third-order or higher, and few general design rules are available. When faced with the problem of controlling a high-order system, the designer should first see if the performance requirements can be relaxed so that the system can be approximated with a low-order model. The discussion of parasitic roots in Section 7.2 should serve as a guide in this regard. If this is not possible, the designer should attempt to divide the plant into subsystems, each of which is second order or lower. A controller is then designed for each subsystem. Finally, if the expense of extra hardware is prohibitive, or if it is impossible to insert controllers at internal points in the system, the design techniques of the following chapters can be applied.

## 7.5 STATE-VARIABLE FEEDBACK

There are techniques for improving system performance that do not fall entirely into one of the three compensation categories considered in the previous section. In some forms these techniques can be viewed as a type of feedback compensation, while in other forms they constitute a modification of the control law. We now consider two such techniques in this and the following section.

State-variable feedback (SVFB) is a technique that uses information about all the system's state variables to modify either the control signal or the actuating signal. These two forms are illustrated in Figure 7.12. The former is a version of internal feedback compensation relative to the outer control loop. An example of this

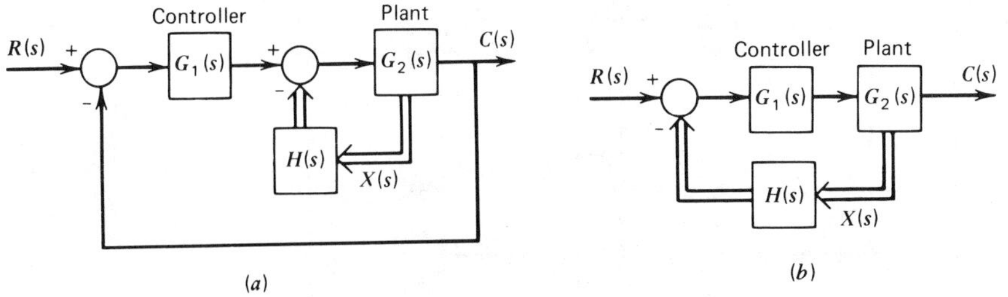


**Figure 7.12 Two forms of state-variable feedback. (*a*) Internal compensation of the control signal. (*b*) Modification of the actuating signal.**

approach is the cascade-control scheme shown in Figure 7.8, where the state variables are the heights $h_1$ and $h_2$. The tachometer feedback scheme shown in Figure 6.36 illustrates the latter form. Both forms require that the state vector **x** be measurable or at least derivable from other information. For example, a position signal can theoretically be passed through a differentiator to obtain the velocity, if a tachometer cannot be used. Devices or algorithms used to obtain state variable information other than directly from measurements are variously termed *state reconstructors, estimators, observers*, or *filters* in the literature.

The advantages of SVFB can be summarized as follows. In some applications instrumentation costs might be less than the cost of implementing a complex control algorithm or adding extra actuators. If so, it would be more economical to measure all the state variables and use this information to reduce the controller's complexity. For example, P control can sometimes be substituted for PD control if the state variable corresponding to rate is measured. In addition to cost consideration, SVFB offers significant improvement in performance because the state variables contain a complete description of the plant's dynamics. With SVFB, we thus have a better chance of placing the characteristic roots of the closed-loop system in locations that will give desirable performance. This concept is explored in more detail for high-order systems in Chapter Ten.

## Avoiding D-Action in the Controller

We have seen that D action can be used to improve the speed of response of a system. However, it is advantageous to avoid using D action in the controller if it is possible to obtain an equivalent effect with a compensator elsewhere in the system. The reason for this is simple. When a change in command input occurs, the error signal and therefore the controller are instantaneously affected. If the change is sudden, as with a step input, the physical limitations of the differentiating device mean that an accurate derivative is not computed, and the actual performance of the system will be degraded relative to the ideal performance predicted by the mathematical model. If D action is required, a good design practice is to place the differentiator at a point in the loop where the signals are more slowly varying. Then the physical behavior of the differentiator will more closely correspond to its behavior as predicted by the mathematical model. A frequent choice for a differentiator location is in the outer feedback loop of the system. The output is usually the result of several integrations and therefore will

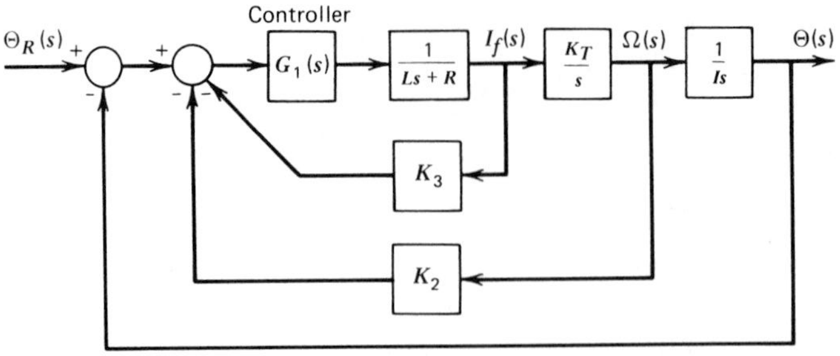


**Figure 7.13 Position servo with a field-controlled dc motor and state-variable feedback.**

exhibit behavior that is smoothed (filtered) and slowly varying with respect to the other signals in the system. Compare the time history of an inertia's displacement with the time history of its driving torque to see this point. With SVFB compensation, we can often dispense with D action in the controller.

## SVFB in a Position Servo

Consider SVFB applied to a position control system using a field-controlled dc motor (Figure 7.13). The state variables are $\theta$, $\omega$, and the field current $i_f$. The transfer function $G_1(s)$ is assumed to contain the controller algorithm, the amplifier gain, and the feedback gain for $\theta$. The feedback gains for $\omega$ and $i_f$ are $K_2$ and $K_3$. We neglect the system damping $c$.

A similar system was treated in Section 6.7, where field inductance was neglected. It was found that D action could be avoided in the controller by the use of tachometer feedback, but the resulting P-control system is not as fast as the one with PD control. Here we indicate how SVFB can be used with P control to improve the response.

Consider the PD-control case first. Let $G_1(s) = K_p + K_D s$, and $K_2 = K_3 = 0$ in Figure 7.13. This represents PD control with only position feedback. The transfer function is

$$\frac{\Theta(s)}{\Theta_R(s)} = \frac{K_T(K_p + K_D s)}{LIs^3 + RIs^2 + K_T K_D s + K_T K_p} \tag{7.5-1}$$

The system is stable if $K_p/K_D < R/L$. The numerator dynamics will speed up the response, but the coefficient of $s^2$ in the characteristic polynomial is unaffected by the control gains. Therefore it might not be possible to select these gains to achieve the desired transient response.

Now consider P control with SVFB. The gains $K_2$ and $K_3$ are nonzero, and $G_1(s) = K_1$. The transfer function is

$$\frac{\Theta(s)}{\Theta_R(s)} = \frac{K_1 K_T}{LIs^3 + (RI + IK_1K_3)s^2 + IK_1K_2K_T s + K_1 K_T} \tag{7.5-2}$$

The system is stable if $(R + K_1K_3)K_2 > L/I$. The coefficient of $s^2$ is now affected by the gains, as are the coefficient of $s$ and the constant term. We therefore have more influence over the transient behavior with this design – achieved by feeding back all the state variables. If the field current is not fed back, $K_3 = 0$ and the $s^2$ coefficient is again beyond our influence. The system is unstable if no velocity feedback is used ($K_2 = 0$).

## 7.6 PSEUDO-DERIVATIVE FEEDBACK

By now it should be clear that a great variety of control system structures and algorithms can be applied to the solution of a given problem. The final choice is intimately connected to the accuracy of the plant model and to the performance requirements. The field has seen many developments within the last 20 years, not only in terms of new hardware like microprocessors, but also in terms of new algorithms for control schemes as well as for analysis of such schemes. The state vector-matrix methods are an example of this. Another, more recent development is the concept of *pseudo-derivative feedback* (PDF), a new control structure that captures the advantages of D action without the attendant difficulties caused by a differentiator located in the forward path of the controller.

The concept of PDF was developed by R. C. Phelan of Cornell University. His monograph (Reference 5) is a thorough presentation of the properties of PDF controllers and a comparison of their performance with that of traditional control systems. Here we present an introduction to PDF and some examples to illustrate its advantages.

### The Principle of One Master

Criticism can be leveled at the PID family of control laws because they ask the controller to respond simultaneously to signals that can be conflicting. If $f(t)$ is the control signal, the actuating signal $e(t)$ produces $f(t)$ according to the differential equation

$$\frac{df}{dt} = K_I e + K_P \frac{de}{dt} + K_D \frac{d^2e}{dt^2}$$

For example, if $e(t)$ is sinusoidal, the terms $\dot{e}$ and $\ddot{e}$ are 90° and 180° out of phase with $e(t)$, and the controller is forced to reconcile these differing signals to produce $f(t)$.

The result of this observation is the *principle of one master*, in the terminology of Reference 5. It states that the control algorithm should contain no more than one operation in its forward path. From the discussion in Section 7.5, it is obvious that this single operation cannot be D action. Under some circumstances where the performance requirements are not severe, the action can be on-off control or proportional control. The latter can be used with PDF when the plant is first order with negligible damping (a pure integrator), no disturbances occur, and fast response is not required. However, a criticism of P action in the forward path is that it unrealistically says that the control element can respond instantaneously to a step input. Thus Phelan recommends I action for the control algorithm because of its physical realizability and its ability to counteract disturbances.

### PI vs. PDF-Control of a Second-Order Plant

To illustrate the concept of PDF-control, consider the control of the displacement of a pure inertia plant, as shown in Figure 7.14. For PID control, $H(s) = 1$ and

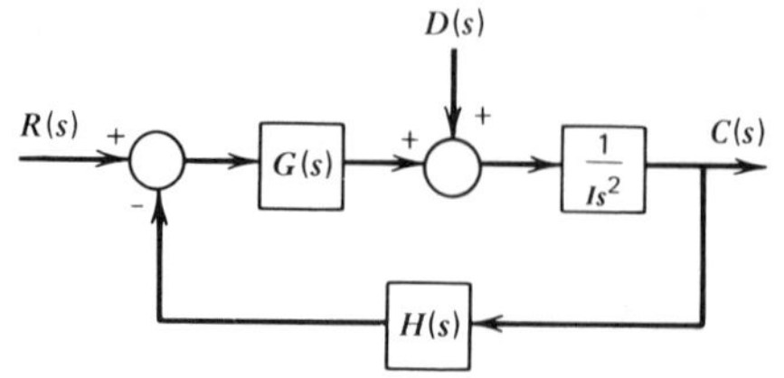


**Figure 7.14** Control of a second-order plant.

$$G(s) = \left(K_p + \frac{K_I}{s} + K_D s\right)$$

The transfer function for the command input is

$$\frac{C(s)}{R(s)} = \frac{K_D s^2 + K_p s + K_I}{Is^3 + K_D s^2 + K_p s + K_I} \tag{7.6-1}$$

Each control gain corresponds to one coefficient in the characteristic polynomial. This eases the task of selecting the gains to achieve the desired transient performance. The disturbance transfer function has an $s$ in the numerator, and the steady-state deviation for a step disturbance will be zero. The design has promise except for the physically unrealistic use of a differentiator in the forward path.

Suppose the D action is removed from the forward path and used in the feedback loop. The resulting design is a PI controller with proportional-plus-derivative feedback. In terms of Figure 7.14,

$$G(s) = K_p + \frac{K_I}{s}$$

$$H(s) = 1 + K_D s$$

The disturbance rejection properties are unchanged, but the command transfer function becomes

$$\frac{C(s)}{R(s)} = \frac{K_p s + K_I}{Is^3 + K_p K_D s^2 + (K_p + K_I K_D)s + K_I} \tag{7.6-2}$$

The controller might be more difficult to tune because the coefficients of $s$ and $s^2$ in the characteristic polynomial depend on more than one gain. Adjustment of one control gain will change two coefficients in the polynomial. One approach is to take a cue from the Ziegler-Nichols ultimate-cycle method and set $K_I = 0$. Select the gains $K_p$ and $K_D$ to provide critical damping with a large value for $K_p$. It can be shown with the Routh criterion that this procedure gives a system that cannot be made unstable when $K_I$ is reintroduced, in theory at least. In practice, Phelan notes, the fact that a physical differentiator cannot produce a pure derivative means that the system will be unstable if $K_I$ is large enough. Therefore, $K_I$ should be made just large enough to provide acceptable response.

The prediction of the response of (7.6-2) is made more difficult by the numerator dynamics. For such systems the response is not entirely described by the characteristic polynomial, and both numerators and denominators must be analyzed when comparing systems. The occurrence of numerator dynamics is a consequence of the

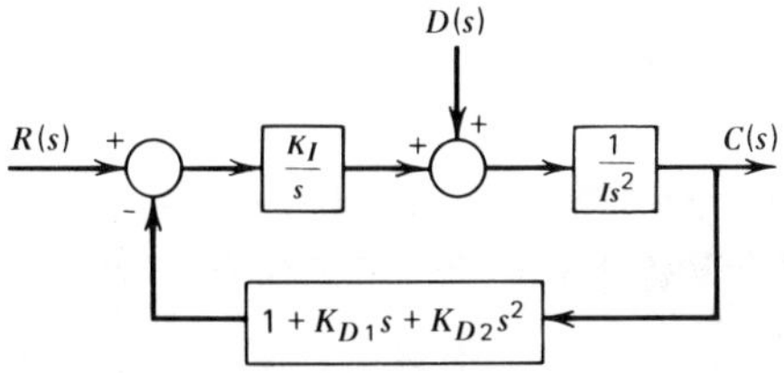


**Figure 7.15 Hypothetical system used to motivate the development of PDF control logic.**

violation of the principle of one master. If the principle is followed, only a constant will appear in the numerator of the command transfer function. Thus, only the command variable, and not its integrals or derivatives, will appear as a forcing function in the system's differential equations – except for disturbances, of course.

The difficulties with tuning and with numerator dynamics disappear if PDF control is used. Its logic can be derived as follows. The numerator dynamics in (7.6-2) are caused by the $K_p$ term. Therefore, we reserve P action for the feedback loop. With only I action in the forward path, the control signal will depend on the output, its derivative, and its integral only if the feedback path has the transfer function

$$H(s) = 1 + K_{D1}s + K_{D2}s^2$$

This is shown in Figure 7.15. The command transfer function is

$$\frac{C(s)}{R(s)} = \frac{K_I}{Is^3 + K_I K_{D2}s^2 + K_I K_{D1}s + K_I} \tag{7.6-3}$$

The numerator dynamics are gone and the polynomial coefficients are under the designer's influence, but the physical system will require devices to compute not only the first but also the second derivative of the output. Reliable second-order derivatives of signals are impossible to obtain except in a few applications, and we therefore seek a better way.

The way out of the difficulty is to note that the system in Figure 7.15 is integrating the derivatives of the output, an operation that makes little sense physically. Removal of one $s$-operator from the $K_{D1}$ and $K_{D2}$ terms is equivalent to integration. If this is done and the result inserted directly into the control signal as it leaves the integrator, the same effect will be achieved except for multiplication by $K_I$. The result is the PDF controller shown in Figure 7.16. Its command transfer function is

$$\frac{C(s)}{R(s)} = \frac{K_I}{Is^3 + K_{D2}s^2 + K_{D1}s + K_I} \tag{7.6-4}$$

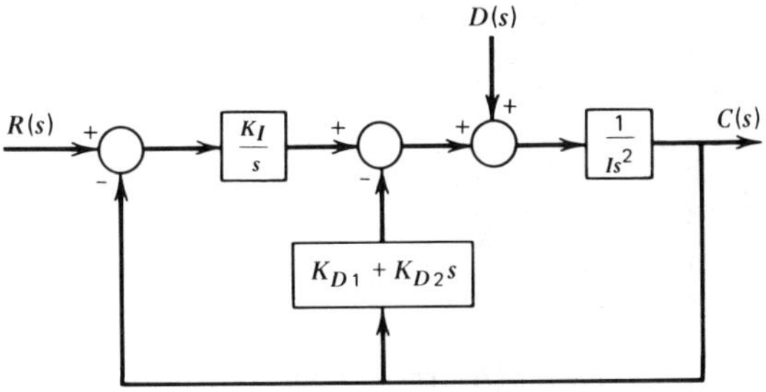


**Figure 7.16 PDF control of a second-order plant.**

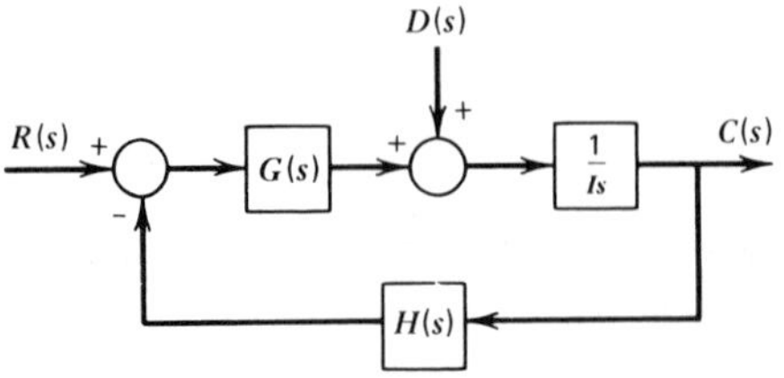


**Figure 7.17 Traditional control system structure for a first-order plant with no damping.**

The disturbance transfer function has a simple $s$-term in the numerator. A single gain is associated with each polynomial coefficient, thus simplifying tuning of the controller. Another advantage is that the system can be treated as one with an inner and an outer loop, which simplifies the procedure for selecting the control gain. We will return to this after developing some insights from PDF control applied to a first-order plant.

## PDF-Control of a First-Order Plant

Consider the problem of controlling a first-order plant with no damping term, as shown in Figure 7.17. One application is a velocity servo. For PI control and unity feedback, $G(s) = K_p + K_I/s$ and $H(s) = 1$. The command transfer function is

$$\frac{C(s)}{R(s)} = \frac{K_p s + K_I}{Is^2 + K_p s + K_I} \tag{7.6-5}$$

In contrast to the second-order plant, no derivative action is required for stability, and thus no differentiator is required. Changing $K_p$ or $K_I$ only changes one coefficient in the characteristic polynomial. Finally, the disturbance transfer function has only an $s$ in the numerator, so steady-state deviations are canceled. Except for violation of the one-master principle, an admittedly somewhat vague criterion at this point, the design seems to be satisfactory.

Following the one-master principle, we remove the proportional term from the controller. (The integral term must remain to cancel the disturbance.) However, with $K_p = 0$, the system is not stable, as can be seen from the denominator of (7.6-5). Therefore, we must change the structure to retain influence over the $s$-term in the characteristic polynomial. This can be done with I control and proportional-plus-derivative feedback, but, as before, it makes no sense physically to differentiate a signal only to integrate it immediately afterward. Therefore, we insert a term proportional to the output immediately after the I action, which gives the PDF controller shown in Figure 7.18. The command transfer function is

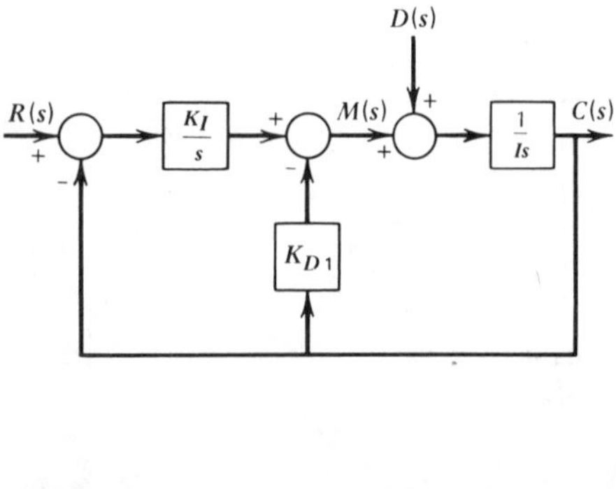


**Figure 7.18 PDF control of a first-order plant with no damping.**

$$\frac{C(s)}{R(s)} = \frac{K_I}{Is^2 + K_{D1}s + K_I} \tag{7.6-6}$$

Each polynomial coefficient depends on only one gain, while the numerator dynamics have been eliminated. The step disturbance is rejected at steady state. The system damping ratio is

$$\zeta = \frac{K_{D1}}{2\sqrt{IK_I}} \tag{7.6-7}$$

and the undamped natural frequency is

$$\omega_n = \sqrt{\frac{K_I}{I}} \tag{7.6-8}$$

No overshoot will occur in the step response for $\zeta \geqslant 1$ because of the absence of numerator dynamics. Simulation studies by Phelan and others have shown that the critically damped system provides the best performance for most cases, but these results are for specific values of the parameter $I$ and the controller saturation limits. Therefore, values of $\zeta$ other than unity should not necessarily be ignored in the design.

The step-response curve can be normalized with the parameters $\zeta$ and $\omega_n$. This is shown in Figure 5.18, in which the nondimensional response is plotted versus the nondimensional time $\omega_n t$. Such general plots are useful but must be carefully interpreted. Figure 5.18 shows the overshoot greatly increases as $\zeta$ decreases, but this conclusion is valid only if we are comparing systems that have the same value of $\omega_n$. The same is true for the measures of response time (rise time, settling time, etc.). When the comparison is for different values of $\omega_n$, the response should be presented in dimensional form as a function of true time, without any normalizing constants.

When this was done for the specific parameter values used in Reference 5, it was seen that the step-response curves were close to one another for values of $\zeta$ from about 0.6 to 1.0. A traditional design precept advises that $\zeta$ should be less than unity to achieve a response that reaches the vicinity of its final value rather quickly; a commonly recommended value is $\zeta = 0.7$. However, if the response curves for various $\zeta$ values are bunched together, there is little to be gained by designing $\zeta$ to be less than 1.0, especially if the existence of overshoot indicates excessive energy consumption.

For the step response of (7.6-6), the IAE and ITAE performance indices are minimized for $\zeta \approx 1$ and show relatively slight increase for $\zeta > 1$ for the parameter values considered. By plotting the response versus true time, the system in these cases was not observed to be very sluggish even for large values of $\zeta$. In fact, the response was always faster than that achieved with P control with proportional feedback. Thus highly damped systems should not be ruled out as possible designs. As will be seen shortly, a large damping ratio will allow the use of a large integral gain $K_I$ to counteract disturbances more effectively.

## Selecting the PDF Gains

The presence of integration in the PDF algorithm means that reset windup can occur if the gains are not chosen properly. This can be accomplished in a manner similar to that used in Section 7.3 for PI control. For a step input, it was easy to see from (7.3-3) that the maximum error, and therefore the maximum control signal, occurs at $t = 0$

for PI control. However, this is not true for most control algorithms in general and for PDF control in particular. The maximum control signal will usually occur at some time after the application of the input. This time must be determined by an extremum-seeking technique. For relatively simple systems this can be done analytically. Here we illustrate the method for the PDF controller shown in Figure 7.18 for the critically damped case.

As before, let $r_{max}$ be the largest expected magnitude of the step input $r(t)$ and $m_{max}$ be the value of the manipulated variable at which saturation occurs. From Figure 7.18,

$$M(s) = \frac{K_I}{s} E(s) - K_{D1} C(s) \tag{7.6-9}$$

or

$$M(s) = Is\, C(s) \tag{7.6-10}$$

For $\zeta = 1$, the step response with zero initial conditions is found from (7.6-6).

$$c(t) = r_{max} \left[1 - \left(1 + \frac{K_{D1}}{2I} t\right) e^{-(K_{D1}/2I)t}\right] \tag{7.6-11}$$

From (7.6-10),

$$m(t) = I \frac{dc}{dt} = r_{max} \frac{K_{D1}^2}{4I} t\, e^{-(K_{D1}/2I)t} \tag{7.6-12}$$

The maximum value of $m(t)$ is attained when $dm/dt$ is zero. Applying this condition to (7.6-12), we find the maximum occurs at

$$t_{max} = \frac{2I}{K_{D1}} \tag{7.6-13}$$

and has the value

$$m(t_{max}) = r_{max} \frac{K_{D1}}{5.44} \tag{7.6-14}$$

Note that the latter result is independent of $I$.

In order for the system response to remain within the linear region,

$$m(t_{max}) \leqslant m_{max} \tag{7.6-15}$$

From (7.6-14) we obtain the maximum value allowed for $K_{D1}$.

$$K_{D1,max} = 5.44 \frac{m_{max}}{r_{max}} \tag{7.6-16}$$

Since $\zeta = 1$, (7.6-7) gives the corresponding value of $K_I$.

$$K_{I,max} = \frac{K_{D1,max}^2}{4I} \tag{7.6-17}$$

## PDF versus PI Control

The analysis of PI control following (7.6-5) indicated that the design might be acceptable. Let us compare its behavior with that of PDF control (7.6-6). A valid comparison can be made between competing designs only when the performance of each is optimized for identical constraints. In particular, the test inputs and the available energy-delivery capability of the final control elements should be the same for each system, and each system should have its gain adjusted to utilize this maximum capability. Therefore, we assume that the gains have been chosen with (7.6-16) and (7.6-17) for PDF control and with (7.3-5) and (7.3-6) for PI control.

Assume that oscillations are not desirable and let $\zeta = 1$ for the PI case. Then the integral gain $K_I$ for PDF is 29.6 times the value of $K_I$ for the PI case. We would therefore expect the effects of a disturbance to be reduced faster with the PDF design. This will be shown to be true.

Consider the unit step response first, and assume that $m_{\max} > 1$ so that no saturation occurs. For PDF control, the response is given by (7.6-11) with $r_{\max} = 1$. For PI control, the critically damped response is

$$c(t) = 1 - \left(1 - \frac{K_p}{2I}t\right) e^{-(K_p/2I)t} \tag{7.6-18}$$

This relation can be used to show that an overshoot of 13.5% occurs at $t = 4I/K_p$. No overshoot exists for the PDF system.

One's immediate reaction might be to increase the damping ratio of the PI design to eliminate the overshoot. For $\zeta > 1$ the analytical solution for the unit step response of (7.6-5) shows that the overshoot occurs at

$$t_p = \frac{I\zeta}{K_p\sqrt{\zeta^2 - 1}} \ln\left(\frac{x - y}{x + y}\right) \tag{7.6-19}$$

where $x = 1 - 2\zeta^2$ and $y = 2\zeta\sqrt{\zeta^2 - 1}$. An overshoot always exists for $\zeta > 1$; its magnitude becomes smaller and the time of occurrence is later as $\zeta$ is increased. For example, if $\zeta = 2$, the overshoot is 4.8% at $t = 6.08\ I/K_p$. The overshoot cannot be eliminated because the numerator coefficients of (7.6-5) are coupled to those of the denominator, and the latter are partly constrained by the need to avoid saturation of the final control elements.

A similar effect occurs when a step disturbance is applied to the system. Both disturbance transfer functions are of the form

$$\frac{C(s)}{D(s)} = \frac{s}{as^2 + bs + c}$$

where the denominator is given by that of either (7.6-5) or (7.6-6). Thus both systems eliminate the disturbance effects at steady state, but the transient behavior is quite different. For $\zeta = 1$ and a unit-step input disturbance, the response of the PI design is

$$c(t) = \frac{t}{I} e^{-(K_p/2I)t} \tag{7.6-20}$$

The peak is

$$c_{\max} = \frac{0.735}{K_p} \tag{7.6-21}$$

and occurs at

$$t_p = \frac{2I}{K_p} \tag{7.6-22}$$

For the critically damped PDF design, the results are identical to the preceding equations, with $K_p$ replaced by $K_{D1}$. Comparing (7.3-5) with (7.6-16), we see that

$K_{D1} = 5.44K_p$. Thus $c_{max}$ and the time to recover, $t_p$, are 5.44 times greater for the PI controller. In addition, the IAE index can be shown to be 29.6 times larger for the PI controller.

Increasing the damping ratio of the PI controller does not improve the situation. For $\zeta > 1$, the unit-step disturbance response is

$$c(t) = \frac{I\zeta}{K_p\sqrt{\zeta^2-1}}\left(e^{s_1 t} - e^{s_2 t}\right) \tag{7.6-23}$$

where the characteristic roots are

$$s_{1,2} = \frac{K_p}{2I}\left(-1 \pm \frac{\sqrt{\zeta^2-1}}{\zeta}\right) \tag{7.6-24}$$

The peak is

$$c_{max} = \frac{-1}{s_2}\left(\frac{s_2}{s_1}\right)^{s_1/(s_1-s_2)} \tag{7.6-25}$$

and occurs at

$$t_p = \frac{1}{s_1 - s_2}\ln\frac{s_2}{s_1} \tag{7.6-26}$$

As $\zeta$ is increased, $c_{max}$ and $t_p$ increase. For example, with $\zeta = 2$, $c_{max} = 0.874/K_p$ and $t_p = 3.04\,I/K_p$.

To see why this counterintuitive behavior occurs, look at the characteristic equation of the PI design.

$$s^2 + as + b = 0$$

where $a = K_p/I$ and $b = K_p^2/4I^2\zeta^2$. With $K_p$ fixed from (7.3-5) in terms of the saturation constraints, we can think of the system as a mass-spring-damper combination with a fixed mass and a fixed damping constant. As the damping ratio is increased, the equivalent spring constant $b$ is decreased. Thus the restoring force of the system is smaller, and the peak caused by a disturbance is larger. This is another example in which nondimensional response plots in terms of $\omega_n$ and $\zeta$ can be misleading. As we change the damping ratio of this system, the undamped natural frequency $\omega_n$ is also changed.

The relative performance of the PI controller is not entirely negative, however. The step input is the most severe input to be encountered because of its suddenness; therefore, we will continue to use the gains computed previously for a step input, while investigating the response to a unit-ramp command input. The PI design for $\zeta = 1$ gives

$$c(t) = t(1 - e^{-(K_p/2I)t}) \tag{7.6-27}$$

while for PDF control,

$$c(t) = t - A + (t + A)e^{-2t/A} \tag{7.6-28}$$

where $A = 4I/K_{D1}$. The steady-state error is zero for PI control. For PDF control, $e_{ss} = A$. If the design specifications place heavy weight on the ramp response, the previous disadvantages of the PI system might be ignored.

The main points of this comparison can be summarized as follows. PDF control generally gives performance superior to that obtained by PI control, at least for the first-order plant with no damping term. Its performance with other types of plants and inputs must be investigated on an individual basis. When making comparisons, use nondimensional results with care and allow all competing systems to have final control elements with the same rate of energy delivery. Select the gains to utilize this delivery capability for a fair comparison.

## Selection of PDF Gains for a Second-Order Plant

In theory, the gains of the PDF controller shown in Figure 7.16 can be selected in the same way as was done for the first-order plant. However, the analytical expressions would be extremely cumbersome. For given plant parameters and saturation limits, numerical simulation might be used, but three gains must be selected, and thus the number of possible combinations can be large. Here we present some guidelines developed by Phelan. The approach is also useful for cascade control problems that involve inner-loop and outer-loop controllers.

As shown in Figure 7.19, the PDF controller of Figure 7.16 can be broken down into two loops. With no disturbance, the inner loop has the transfer function

$$\frac{C(s)}{R_i(s)} = \frac{1}{Is^2 + K_{D2}s + K_{D1}} \tag{7.6-29}$$

with the damping ratio

$$\zeta_i = \frac{K_{D2}}{2\sqrt{IK_{D1}}} \tag{7.6-30}$$

Certainly no overshoot will occur if $\zeta_i \geqslant 1$, but values of $\zeta_i < 1$ can also be used sometimes since the input $r_i$ can never be as severe as a step function. This is because $r_i$ is the output of an integrator. In particular, the results of many numerical simulations for a wide range of gain values have shown that the smoothest, quickest step response without overshoot for the total system occurs when $\zeta_i \cong 0.7$ and

$$K_{D1} \cong 8\frac{m_{max}}{r_{max}} \tag{7.6-31}$$

For the specific case, $I = 1$, $m_{max} = 0.25$ and $r_{max} = 1$, it has been shown that for a step input, the ITAE index is minimized for $\zeta \cong 0.7$, and that the index increases only slightly for $\zeta \gg 1$ (References 5 and 6).

A design procedure is as follows. For $\zeta_i = 0.7$, (7.6-30) and (7.6-31) yield values for $K_{D1}$ and $K_{D2}$. With these values, analyze the command input step response for increasingly larger values of $K_I$ until saturation or overshoot occurs. A larger value of $K_I$ is desired to compensate for disturbances. This procedure should in most cases result in a ballpark estimate of the required gains. Further simulations would be used to fine-tune the results.

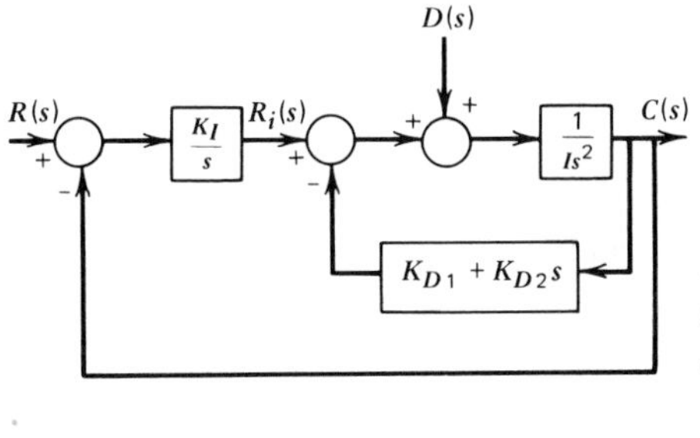


**Figure 7.19** Inner-outer loop configuration of the PDF controller for a second-order plant.

The controller output is

$$m(t) = K_I \int e\,dt - K_{D1}c - K_{D2}\dot{c} \tag{7.6-32}$$

The only increase in $m$ comes from the integral term, and the speed of response and possibility of overshoot both increase with increasing values of $K_I$. With the response relatively insensitive to $\zeta_i$, we see that the choice of $K_I$ is the most critical. The $K_{D1}c$ term is most important near steady state when $c$ is large. Thus $K_{D1}$ can be adjusted to eliminate overshoot. In the early stages of the response, $c$ and the error integral are small, and the controller output is heavily influenced by the $K_{D2}\dot{c}$ term. Increasing $K_{D2}$ will therefore tend to decrease the maximum value of $m$, prevent saturation, and reduce oscillations.

Modified versions of PDF control are also useful. These are explored in the problems at the end of the chapter.

## 7.7 INTERACTING CONTROL LOOPS

We have seen that it is important for the control system designer to be aware of the effects of the parasitic modes of the system even though they are not explicitly modeled. A related effect occurs when simplified models are used to design controllers for complex systems in which several variables are to be controlled. If these variables affect each other, the control loops are said to be interacting. In some cases, the interaction is so slight that we can neglect it and design the controller for each loop as if it were a separate system. The stringency of the performance specifications will also determine whether or not this approach can be used.

As the demands on technology become more severe for safety, environmental, and efficiency considerations, many traditional systems with separate-loop control designs must now be redesigned to include the interactive effects. For example, a steam-turbine plant for generating electricity requires control of the turbine speed, steam flow rate, steam temperature, and steam pressure, among others. Historically, separate controllers were designed for each of these variables, but the system's performance could not be improved beyond a certain point without taking the interactions into account. Modern control theory has led to improvements in this respect, as it has also done with internal combustion engine and jet engine control.

As a simple example, consider the liquid-level system shown in Figure 7.20. A chemical reaction requires that the concentration as well as the temperature of the reactants be closely controlled. This requires that the level $h$ and the liquid temperature $T$ be controlled. We do this here by changing the flow rate and the hot-cold ratio of the incoming liquid streams. A simple model can be derived as follows. Conservation of liquid mass gives

$$A\dot{h} = -\frac{1}{R}h + Q_i \tag{7.7-1}$$

where $Q_i$ is the input volume flow rate. Let $c$ and $w$ represent the specific heat and density of the liquid and assume these are constants. The heat energy in the tank liquid is $cwAhT$, where $T$ is the temperature of the tank liquid and is assumed to be

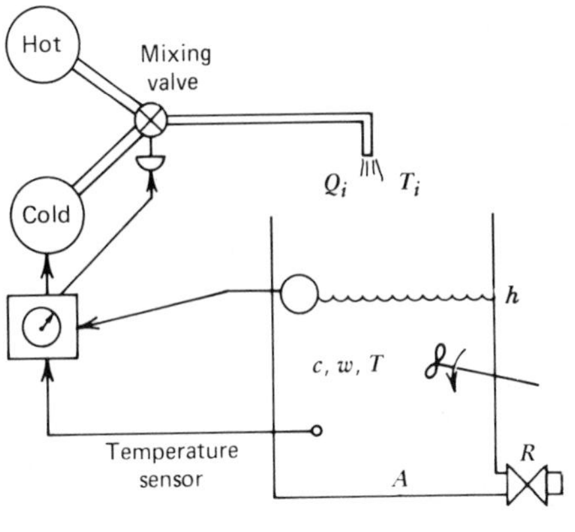


**Figure 7.20 System in which the level-control and temperature-control loops interact.**

uniform from stirring. The rate of change of this energy equals the incoming rate minus the outflow rate, or

$$\frac{d}{dt}(cwAhT) = cwQ_iT_i - \frac{1}{R}hcwT \tag{7.7-2}$$

Carrying out the differentiation gives

$$AT\dot{h} + Ah\dot{T} = Q_iT_i - \frac{1}{R}hT$$

Substitution of (7.7-1) for $\dot{h}$ results in

$$Ah\dot{T} = Q_i(T_i - T) \tag{7.7-3}$$

Equations (7.7-1) and (7.7-3) form the model of the system. We note that (7.7-1) is uncoupled from the temperature model, but the derivative coefficient in (7.7-3) is a function of $h$. Therefore (7.7-3) is nonlinear unless the height $h$ is constant. Also, the input variable $Q_i$ for the height subsystem appears in the temperature equation. Thus the temperature subsystem is coupled to the height subsystem in two ways.

## Possible Design Methods

The system is to be controlled by changing the manipulated variables $Q_i$ and $T_i$. Three approaches to designing a control system to do this are the following.

1. Treat the coupling variables ($h$ and $Q_i$ here) as disturbances to the temperature-control loop, and take the temperature and height systems to be separate. Design a controller for each. This requires that the temperature model be linearized about nominal values for $h$ and $Q_i$.
2. Linearize the coupled model set (7.7-1) and (7.7-3) and treat the problem as a linear multivariable system. The two outputs would be $h$ and $T$, and the two command inputs would be $h_R$ and $T_R$, the desired height and temperature. This results in a higher-order model than the first approach, but it provides a better description of the interaction between the subsystems. Linear control theory can be used with both approaches.
3. Treat the problem as a coupled, nonlinear system. No general control theory is available to support this approach, although many ad hoc procedures exist. Their success depends on the nature of the nonlinearity.

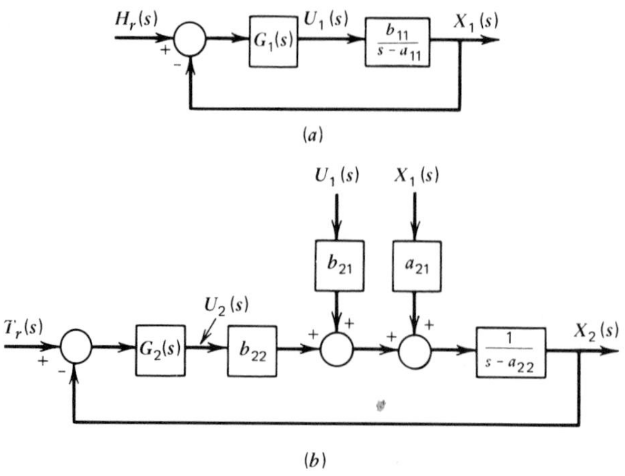


**Figure 7.21 Separate control loops for the system (7.7-4) and (7.7-5). The coupling variables are treated as disturbances to the second loop (*b*).**

As an example, consider the linearized system model.

$$\dot{x}_1 = a_{11} x_1 + b_{11} u_1 \tag{7.7-4}$$

$$\dot{x}_2 = a_{21} x_1 + a_{22} x_2 + b_{21} u_1 + b_{22} u_2 \tag{7.7-5}$$

where we have assumed that the $x_2$ subsystem and the input $u_2$ do not affect the $x_1$ subsystem. Hence $a_{12} = b_{12} = 0$.

The first design approach would use a model like that shown in Figure 7.21. The control laws $G_1(s)$ and $G_2(s)$ are designed separately. The latter's design would treat $u_1$ and $x_1$ as disturbances, modeled perhaps as step functions. If the response of the resulting control system in Figure 7.21*b* is fast relative to the changes in $u_1$ and $x_1$, the design might be satisfactory.

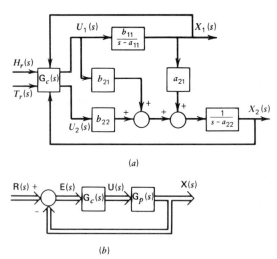

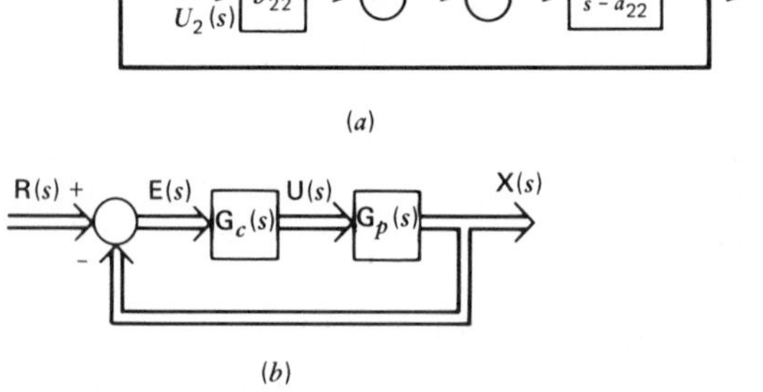


**Figure 7.22 Control system with interacting loops. (*a*) Specific diagram for (7.7-4) and (7.7-5). (*b*) General scheme for the vector case.**

The second approach can be described by a block diagram like that of Figure 7.22*a*. In this case the control would be expressed as a transfer function matrix $\mathbf{G}_c(s)$. The general form is shown in Figure 7.22*b*, where $\mathbf{G}_p(s)$ is the transfer function matrix for the plant. The design of control laws for multivariable systems will be considered in more detail in Chapter Ten. For now we assume that the interaction between the two subsystems is undesirable, and we consider how a controller might be designed to reduce the interaction.

## Noninteraction in Control Systems

The absence of interaction between the $x_1$ and $x_2$ subsystems is indicated by a diagonal transfer function matrix $\mathbf{T}(s)$ between $\mathbf{X}(s)$ and $\mathbf{R}(s)$. The transfer function matrix for the system in Figure 7.22*b* is obtained by writing the equations for each component and solving for $\mathbf{X}(s)$ in terms of $\mathbf{R}(s)$.

$$\mathbf{X}(s) = \mathbf{G}_p(s)\mathbf{U}(s)$$

$$\mathbf{U}(s) = \mathbf{G}_c(s)\mathbf{E}(s)$$

$$\mathbf{E}(s) = \mathbf{R}(s) - \mathbf{X}(s)$$

Thus

$$\mathbf{X}(s) = \mathbf{G}_p(s)\mathbf{G}_c(s)[\mathbf{R}(s) - \mathbf{X}(s)]$$

$$[\mathbf{I} + \mathbf{G}_p(s)\mathbf{G}_c(s)]\mathbf{X}(s) = \mathbf{G}_p(s)\mathbf{G}_c(s)\mathbf{R}(s)$$

or

$$\mathbf{X}(s) = \mathbf{T}(s)\mathbf{R}(s) \tag{7.7-6}$$

where

$$\mathbf{T}(s) = [\mathbf{I} + \mathbf{G}_p(s)\mathbf{G}_c(s)]^{-1}\mathbf{G}_p(s)\mathbf{G}_c(s) \tag{7.7-7}$$

The matrix $\mathbf{I}$ is the identity matrix.

If $\mathbf{x}$ and $\mathbf{r}$ are two-dimensional, $\mathbf{T}(s)$ is a $(2 \times 2)$ matrix. The system is uncoupled if the command $r_1$ has no effect on $x_2$ and if $r_2$ has no effect on $x_1$. This is true if $\mathbf{T}(s)$ has the form

$$\mathbf{T}(s) = \begin{bmatrix} T_{11}(s) & 0 \\ 0 & T_{22}(s) \end{bmatrix} \tag{7.7-8}$$

This condition provides the basis for designing the control law $\mathbf{G}_c(s)$.

The desired forms of $T_{11}(s)$ and $T_{22}(s)$ can be found with the performance specifications for the design. Thus $\mathbf{T}(s)$ can be taken as a given function, and (7.7-7) can be solved for $\mathbf{G}_c(s)$. To do this, multiply both sides from the left to clear the inverse matrix. This gives

$$[\mathbf{I} + \mathbf{G}_p(s)\mathbf{G}_c(s)]\mathbf{T}(s) = \mathbf{G}_p(s)\mathbf{G}_c(s)$$

Now solve for $\mathbf{G}_c(s)$

$$\mathbf{G}_c(s) = \mathbf{G}_p^{-1}(s)\mathbf{T}(s)[\mathbf{I} - \mathbf{T}(s)]^{-1} \tag{7.7-9}$$

We have assumed that the inverse of the plant matrix $\mathbf{G}_p(s)$ exists.

The plant transfer function matrix $\mathbf{G}_p(s)$ relates $\mathbf{X}(s)$ to $\mathbf{U}(s)$, and can be obtained either by reducing the block diagram shown in Figure 7.22*a*, or by using the plant's resolvent matrix and the result (5.2-30), repeated here.

$$\mathbf{G}_p(s) = [s\mathbf{I} - \mathbf{A}]^{-1}\mathbf{B} \tag{7.7-10}$$

The matrices **A** and **B** are found from the linear state equations, of which (7.7-4) and (7.7-5) are examples. Here these matrices have the form

$$A = \begin{bmatrix} a_{11} & 0 \\ a_{21} & a_{22} \end{bmatrix}, \qquad B = \begin{bmatrix} b_{11} & 0 \\ b_{21} & b_{22} \end{bmatrix} \tag{7.7-11}$$

The design procedure is best illustrated with a specific example.

*Example 7.3*

Compute the controller transfer function required to achieve noninteraction in the system described by (7.7-11). Assume that

$$\mathbf{A} = \begin{bmatrix} -3 & 0 \\ 0 & -4 \end{bmatrix}, \qquad \mathbf{B} = \begin{bmatrix} 1 & 0 \\ -1 & 2 \end{bmatrix}$$

The desired closed-loop time constants for the control loops are $\tau_1 = 1, \tau_2 = 2$.

The last condition indicates that the desired overall transfer function is

$$\mathbf{T}(s) = \begin{bmatrix} \dfrac{1}{s+1} & 0 \\ 0 & \dfrac{1}{2s+1} \end{bmatrix} \tag{7.7-12}$$

and the term in (7.7-9) becomes

$$\mathbf{T}(s)[\mathbf{I} - \mathbf{T}(s)]^{-1} = \begin{bmatrix} \dfrac{1}{s} & 0 \\ 0 & \dfrac{1}{2s} \end{bmatrix} \tag{7.7-13}$$

Using (7.7-10) to compute $G_p(s)$, we obtain

$$\mathbf{G}_p(s) = \begin{bmatrix} \dfrac{1}{s+3} & 0 \\ \dfrac{1}{s+4} & \dfrac{2}{s+4} \end{bmatrix} \tag{7.7-14}$$

From (7.7-9), the required controller matrix is

$$\mathbf{G}_c(s) = \begin{bmatrix} s+3 & 0 \\ \dfrac{s+3}{2} & \dfrac{s+4}{2} \end{bmatrix} \begin{bmatrix} \dfrac{1}{s} & 0 \\ 0 & \dfrac{1}{2s} \end{bmatrix}$$

$$= \begin{bmatrix} \dfrac{s+3}{s} & 0 \\ \dfrac{s+3}{2s} & \dfrac{s+4}{4s} \end{bmatrix}$$

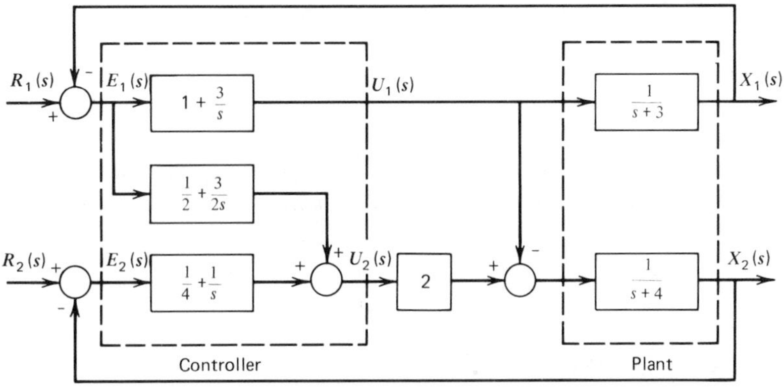


**Figure 7.23 Noninteracting control system design of Example 7.3.**

The controller elements are seen to be PI algorithms. For example, the manipulated variable $u_2$ is produced by

$$U_2(s) = \left(\frac{1}{2} + \frac{3}{2s}\right) E_1(s) + \left(\frac{1}{4} + \frac{1}{s}\right) E_2(s)$$

where $e_1$ and $e_2$ are the errors in $x_1$ and $x_2$; that is,

$$e_1 = r_1 - x_1$$

$$e_2 = r_2 - x_2$$

The control system's block diagram is shown in Figure 7.23.

If the diagram in Figure 7.23 is reduced to find the transfer functions relating $x_1$ and $x_2$ to $r_1$ and $r_2$, the matrix given by (7.7-12) would be obtained. In so doing, some numerator terms would be canceled by denominator terms. These canceled terms represent patterns of the system response that have been isolated from the inputs $r_1$ and $r_2$ by the design of the controller. A problem occurs when the system is acted on by disturbances that have not been accounted for in the model. These disturbances could excite these hidden response patterns and produce uncontrolled behavior.

Obviously, these more advanced control schemes are getting us into deeper water, and to analyze them properly we need additional methods based on state-space vector-matrix concepts. In particular, the concepts of controllability and observability are useful here. These are covered in Chapter Nine.

## 7.8 SOME ADVANTAGES OF DIGITAL CONTROL

The approach to the design of analog controllers has been governed by a great extent by the limitations on the physical realization of the control algorithm. The PID family of algorithms is realizable with electronic, hydraulic, and pneumatic devices because of some very inventive designs. However, one should try to envision the difficulties that would be encountered if a more complicated control law were to be implemented, especially if the number of active elements like op amps is limited for reliability purposes.

When implementing the control law with a computer, such difficulties vanish. Limiting the controller output to avoid saturation of the control elements and the reset windup phenomenon is easily accomplished with digital control. Also, the objection that true derivative action is not obtainable physically is not relevant with digital control in which the corresponding difference action is easily implemented. Because almost any algorithm can be implemented digitally, we can specify the desired response and work backward to find the required control algorithm. This is the *direct design* method.

## Direct Design of Digital Controllers

Consider the control system shown in Figure 7.24*a*. The pulse transfer function diagram is shown in Figure 7.24*b*. The transfer function is

$$T(z) = \frac{C(z)}{R(z)} = \frac{G(z)D(z)}{1 + G(z)D(z)} \tag{7.8-1}$$

With the direct design method, $T(z)$ is specified, the plant-hold transfer function $G(z)$ is given, and the control algorithm $D(z)$ is found from (7.8-1). This gives

$$D(z) = \frac{T(z)}{G(z)[1 - T(z)]} \tag{7.8-2}$$

Alternately, the input $R(z)$ and desired output $C(z)$ can be specified. These, of course, determine $T(z)$, and (7.8-2) can still be used.

### *Example 7.4*

Find the control algorithm $D(z)$ so that the response to a unit step function will be $c(t) = 1 - e^{-t}$. The plant transfer function is $G_p(s) = 1/(10s + 1)$. Assume that the sampling time is $T = 2$ sec.

From the results of Example 4.14 (4.6-18), we obtain the plant-hold transfer function.

$$G(z) = \frac{1 - e^{-2/10}}{z - e^{-2/10}} = \frac{0.18}{z - 0.82} \tag{7.8-3}$$

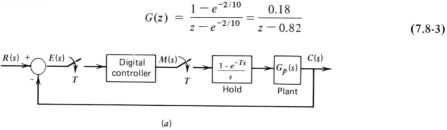

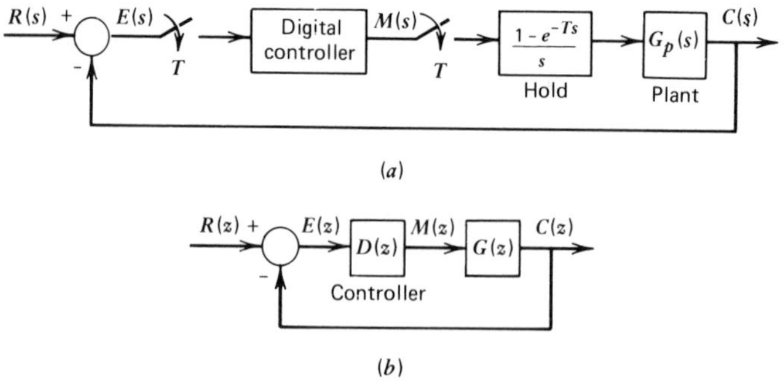


**Figure 7.24 Digital controller configurations in the *s* domain and the *z* domain.**

From the input information, $R(s) = 1/s$, so $R(z) = z/(z-1)$. Also, the specified output form implies that

$$C(s) = \frac{1}{s} - \frac{1}{s+1}$$

Therefore, with $T = 2$,

$$C(z) = \frac{z}{z-1} - \frac{z}{z-e^{-2}} = \frac{0.86z}{(z-1)(z-0.14)} \tag{7.8-4}$$

and

$$T(z) = \frac{C(z)}{R(z)} = \frac{z-1}{z} \frac{0.86z}{(z-1)(z-0.14)}$$

$$= \frac{0.86}{z-0.14} \tag{7.8-5}$$

Substitution of this into (7.8-2) gives

$$D(z) = \frac{z-0.82}{0.18} \frac{0.86}{z-1}$$

$$= \frac{4.8 - 3.9z^{-1}}{1-z^{-1}} = \frac{M(z)}{E(z)} \tag{7.8-6}$$

The difference equation that must be implemented in the software of the computer control system is

$$m(k) = m(k-1) + 4.8e(k) - 3.9e(k-1) \tag{7.8-7}$$

Several considerations are important when specifying the transfer function $T(z)$. If $T(z)$ has the form

$$T(z) = \frac{C(z)}{R(z)} = \frac{b_0 + b_1 z^{-1} + b_2 z^{-2} + \ldots}{a_0 + a_1 z^{-1} + a_2 z^{-2} + \ldots} \tag{7.8-8}$$

causality requires that $b_0 = 0$. This can be seen by writing the relation as a difference equation.

$$a_0 c(k) = -a_1 c(k-1) - a_2 c(k-2) - \ldots$$
$$+ b_0 r(k) + b_1 r(k-1) + b_2 r(k-2) + \ldots \tag{7.8-9}$$

The output of a physical system will require at least one sampling period to react to the input. Thus $r(k)$ cannot affect $c(k)$ in a real system, and $b_0$ must be zero. Given the denominator of (7.8-8), the same argument shows that the numerator cannot contain terms such as $z^i$, $i \geqslant 0$.

The controller transfer function $D(z)$ resulting from (7.8-2) will produce the desired output $C(z)$ with the specified input $R(z)$, but how will it perform with other input functions or disturbances? These questions will be important for some applications and should be answered before accepting the design as final.

The algorithm designed in Example 7.4 might be useful for starting up the process according to the prescribed function $c(t)$. The control software can be programmed to switch, when the process reaches $c = 1$, to another control algorithm that is more suitable for regulation in the presence of disturbances.

## 7.9 SUMMARY

It is apparent that some of the issues and control schemes discussed in this chapter will often require analytical techniques more sophisticated than those presented thus far. The full PID law has three gains to be selected. This selection can be difficult to do when the plant model must be of high order to include important parasitic modes. The inclusion of compensation devices can also raise the order of the system, and state variable feedback requires that several feedback gains be computed. Thus practical design problems can easily involve high-order models and multiple gains.

Fortunately several techniques exist to aid the designer in this matter. Chapter Eight presents some useful graphical techniques, and Chapters Nine and Ten develop matrix methods. Both are helpful for high-order systems and allow the designer to take advantage of the computational power of the digital computer.

## REFERENCES

1. Y. Takahashi, M. Rabins, and D. Auslander, *Control*, Addison-Wesley, Reading, Mass., 1970.
2. L. Hasdorff, *Gradient Optimization and Nonlinear Control*, John Wiley, New York, 1976.
3. J. G. Ziegler and N. B. Nichols, "Optimum Settings for Automatic Controls." *ASME Transactions*, Vol. 64, No. 8, 1942, p. 759.
4. J. G. Ziegler and N. B. Nichols, "Process Lags in Automatic Control Circuits," *ASME Transactions*, Vol. 65, No. 5, 1943, p. 433.
5. R. Phelan, *Automatic Control Systems*, Cornell Univ. Press, Ithaca, New York, 1977.
6. A. G. Ulsoy, "Optimal Pseudo-Derivative Feedback Control," M.S. Thesis, Cornell University, Ithaca, New York, 1975.

## PROBLEMS

7.1 (a) The following table gives the measured open-loop response of a system to a unit step input. Use the process-reaction method to find the controller gains for P, PI, and PID control.

| Time (min) | Response |
|---|---|
| 0 | 0 |
| 0.5 | 0.02 |
| 1.0 | 0.05 |
| 1.5 | 0.16 |
| 2.0 | 0.28 |
| 2.5 | 0.42 |
| 3.0 | 0.58 |
| 3.5 | 0.70 |
| 4.0 | 0.80 |
| 4.5 | 0.86 |
| 5.0 | 0.92 |
| 5.5 | 0.95 |
| 6.0 | 0.97 |
| 7.0 | 0.98 |

(b) A certain plant has the open-loop transfer function

$$G_p(s) = \frac{0.5}{(s^2 + s + 1)(s + 0.5)}$$

Use the ultimate-cycle method to compute the controller gains for P, PI, and PID control.

(c) The response in part a was actually generated from the plant in part b for the purpose of this example. Compare the controller gains computed with each method.

7.2 Integral control applied to the plant

$$G_p(s) = \frac{k}{ms^2 + cs + k}$$

results in a closed-loop transfer function of the form

$$T(s) = \frac{K_I k}{ms^3 + cs^2 + ks + K_I k}$$

Assume that $m = 2$, $c = 1$, and that $k$ can be selected by the designer along with the integral gain $K_I$. Design a system that will minimize the ITAE criterion for a unit step input. Plot the resulting response.

7.3 (a) Consider part b of Problem 6.8, where the float dynamics are significant. Plot the phase plane trajectories. Does this model have a limit cycle? Compare with the results for the ideal case shown in Figure 7.3.

(b) Consider part b of Problem 5.44. Plot the results on the phase plot. Does a limit cycle exist?

7.4 Use analytical methods or numerical simulation to determine the value of $\zeta$ that minimizes the IAE index for the step response of the following system.

$$\frac{C(s)}{R(s)} = \frac{1}{s^2 + 2\zeta s + 1}$$

Plot the unit step response for the optimal value of $\zeta$, and find the maximum percent overshoot.

7.5 Consider the PI-control system shown in Figure 7.5, except that now the plant is such that

$$\Omega(s) = \frac{1}{s+2} T(s)$$

It is desired to obtain a closed-loop system having $\zeta = 1$ and $\tau = 0.1$.

(a) Find the required values of $K_p$ and $K_I$, neglecting any saturation of the control elements.

(b) Let $m_{\max} = 1 = r_{\max}$. Find $K_p$ and $K_I$. Compare the unit step response of the two designs.

7.6 Many of the causes of damping are nonlinear in nature, such as Coulomb friction. For this reason, we try to model their effects as disturbances to the system so that a linear system model can be used. Consider a mass with a control force $f$ applied to it, along with the Coulomb friction force. The equations of motion are

$$m\ddot{x} = f - \mu mg, \qquad \dot{x} > 0$$

$$m\ddot{x} = f + \mu mg, \qquad \dot{x} < 0$$

(a) Draw the block diagram of a system to control the position $x$ with the friction modeled as a disturbance.

(b) What type of disturbance input could the designer use to test the response of the controller?

(c) Design a controller to position the mass at $x = 0$.

7 7 Given the plant

$$G_p(s) = \frac{1}{s(\tau_p s + 1)}$$

where $\tau_p \ll 1$.

(a) Neglect $\tau_p$ and use proportional control to design a system whose closed-loop time constant is 1.

(b) Same as part a but obtain a time constant of 0.1.

(c) Suppose that $\tau_p = 0.06$. Evaluate the performance of the designs in parts a and b in terms of meeting their specifications.

7.8 The system shown in Figure P7.8*a* is a representation of a common industrial situation in which the displacement or velocity of a mass $m_2$ must be controlled.

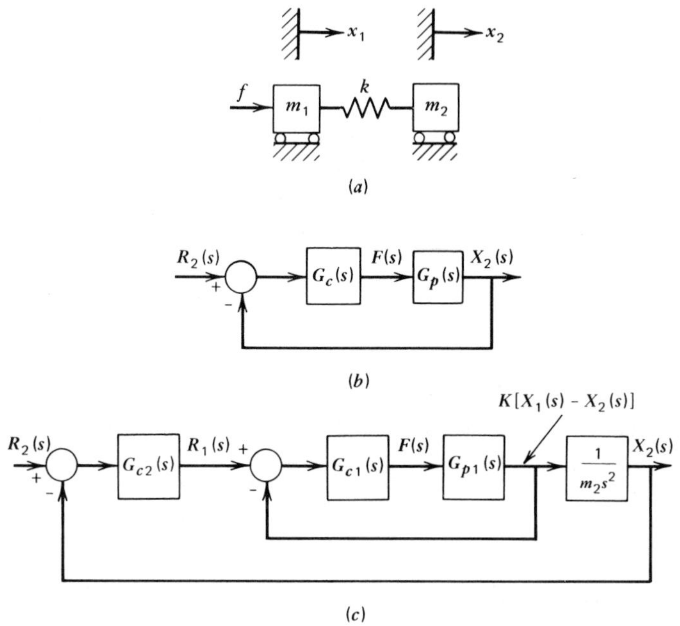


Figure P7.8

The force (or torque) that must do the controlling is constrained to act indirectly on $m_2$ through the mass $m_1$ and the elastic element $k$. For example, $m_1$ might represent the mass of the motor's armature and the gear reducer, $f$ the applied motor torque, and $k$ the elasticity of the coupling shaft.

(a) Find the transfer function $G_p(s)$ between the displacement $x_2$ and the force $f$ as input.

(b) What difficulties will be encountered if we attempt to design a single-loop controller, as shown in Figure P7.8*b*, to control the displacement $x_2$?

(c) Formulate a cascade-control system to control $x_2$. Make the inner-loop control the force $k(x_1 - x_2)$ acting on the mass $m_2$. Assume that $k(x_1 - x_2)$ can be measured by a force transducer, and use this measurement in an inner loop. Use the outer loop to control the set point $r_1$ of the inner loop. The set point $r_1$ is the desired value of the force $k(x_1 - x_2)$. Find $G_{p1}(s)$ and suggest forms for the control laws $G_{c1}(s)$ and $G_{c2}(s)$, as shown in Figure P7.8*c*.

7.9 A system with feed-forward compensation is shown in Figure P7.9.

(a) With no compensation ($K_f = 0$), find the value of the proportional gain $K_p$ to give $\zeta = 1$.

(b) Find the value of $K_f$ required to give zero steady-state error to a unit step disturbance. Use the value of $K_p$ from part a.

(c) Compare the steady-state and transient responses of the designs for the following.

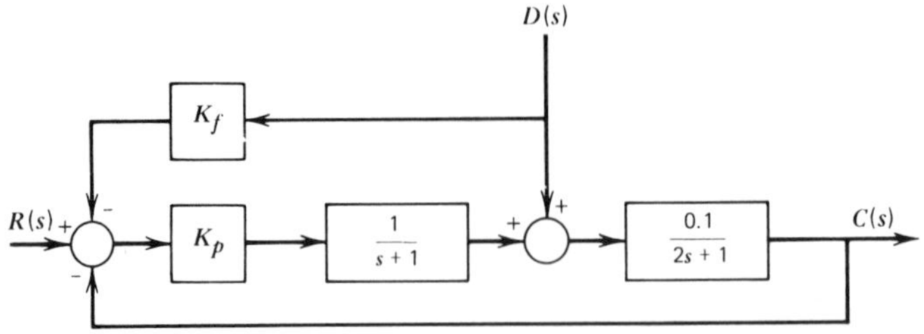


**Figure P7.9**

(1) A unit step disturbance.

(2) A unit impulse disturbance.

(3) A unit ramp disturbance.

**7.10** The control system shown in Figure P7.10 has a PD type compensator inserted in the command path. Find the value of $K_D$ required to give a zero steady-state error for a ramp input of slope $m$.

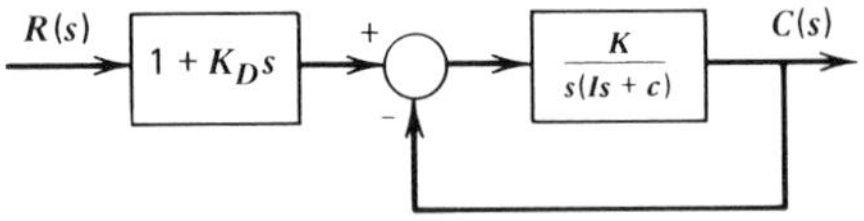


**Figure P7.10**

**7.11** The problem of balancing a pencil or broomstick in one's hand was discussed in Section 6.1 as an example of an inherently unstable system. Variations of this problem constitute some of the classical and most interesting problems in control theory. A similar engineering application is the control of the angular pitch of a rocket during takeoff. Here we consider a simple version of the problem. A more detailed version is treated in Chapter Ten.

The situation is shown in Figure P7.11. The input $u$ is an acceleration provided by the control system and is applied in the horizontal direction to the lower end of the rod. The horizontal displacement of the lower end is $y$. The linearized form of Newton's law for small angles gives

$$mL\ddot{\theta} = mg\theta - mu$$

Put this into state variable form by letting $x_1 = \theta$ and $x_2 = \dot{\theta}$. Construct a state variable feedback controller by letting $u = k_1 x_1 + k_2 x_2$. Over what ranges of values of $k_1$ and $k_2$ will the controller stabilize the system (keep $\theta$ near 0)? What does this formulation imply about the displacement $y$?

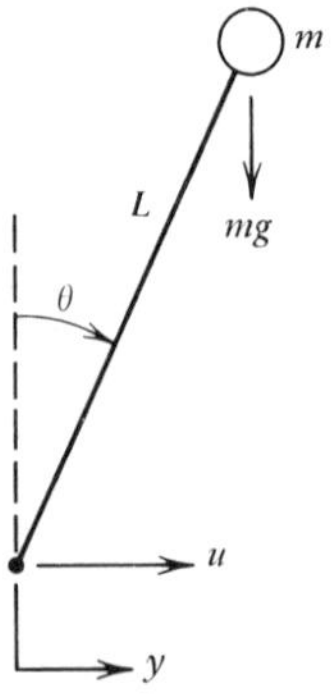


**Figure P7.11**

**7.12** It is desired to stabilize the plant $1/(s^2 - 2)$ with a feedback controller. The closed-loop system should also have a damping ratio of $\zeta = 0.707$ and a dominant time constant of $\tau = 0.1$.

(a) Use the arrangement shown in Figure P7.12*a*. Find the values of $K_p$ and $K_D$.

(b) Use the arrangement shown in Figure P7.12*b*. Find the values of $K_p$ and $K_1$.

(c) Compare the two designs, in light of their unit step response.

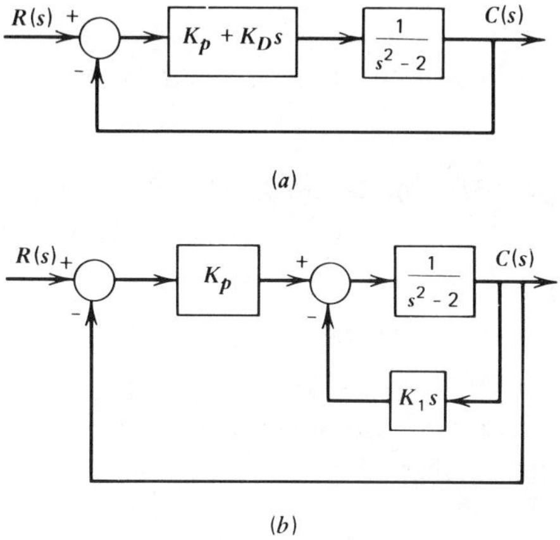


Figure P7.12

**7.13** Compare P control with PDF control (Figure 7.18) for the case $I = 2, m_{max} = 4$, and $r_{max} = 8$, with step inputs of the following magnitudes.

(a) 4.

(b) 8.

(c) 12.

**7.14** Compare the performance of the critically damped PI and PDF controllers for the plant $1/Is$ with the following.

(a) A unit ramp disturbance.

(b) A sinusoidal disturbance.

(c) A sinusoidal command input.

**7.15** Consider PDF control as in Figure 7.18 except that the plant's transfer function is $1/(Is + c)$.

(a) Find the transfer functions $C(s)/R(s)$ and $C(s)/D(s)$.

(b) Discuss the effect of $c$ on the choice of $K_{D1}$.

(c) Discuss the effect of $c$ on the choice of $K_I$.

(d) Obtain $m(t)$ for a unit step reference input when $\zeta = 1$. Find $t_{max}$, the time when $m(t)$ reaches its maximum.

(e) Let $I = c = 1$, and $r_{max} = m_{max} = 1$. Use numerical simulation to design a PDF controller and compare its performance with that of a PI controller. (*Hint.* Start with the gain values for the case $c = 0$, and use the results of parts b and c to choose values of $K_I$ and $K_{D1}$ for the simulation.)

**7.16** Consider the PDF controller shown in Figure 7.19 with $I = 1$, $r_{max} = 1$ and $m_{max} = 0.25$. Select the gains $K_{D1}$, $K_{D2}$, and $K_I$ to eliminate overshoot.

**7.17** Repeat Problem 7.16 using the minimum IAE results (7.1-9). Compare the step response of each design.

**7.18** A modification of PDF control called *PDF plus* is obtained by including a derivative operation in the inner feedback loop. Figure P7.18 shows PDF-plus control applied to a first-order plant.

(a) Obtain the transfer function $C(s)/R(s)$ and $C(s)/D(s)$.

(b) Show that if $K_{D2}$ can be made arbitrarily large, the response to the disturbance $D(s)$ can be made arbitrarily small without changing the speed of response of the basic PDF controller (with $K_{D2} = 0$).

(c) What limits the magnitude of $K_{D2}$ in practice?

(d) How can the concepts of PDF-plus control be extended to handle a second-order plant $1/Is^2$?

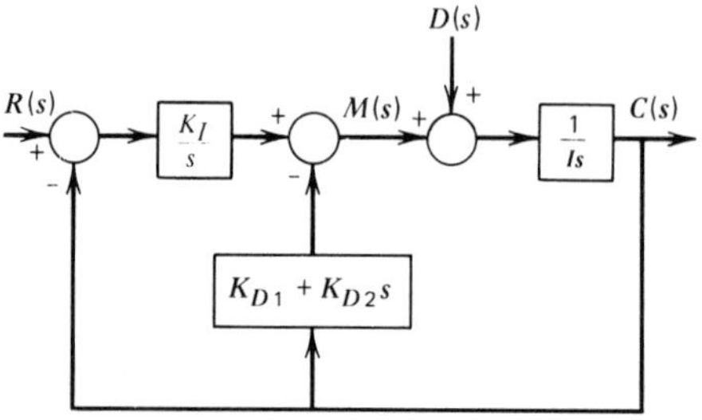


**Figure P7.18**

**7.19** When the coupled nonlinear equations of the temperature–liquid-level system (7.7-1) and (7.7-3) are linearized about the equilibrium solution with $T_i$ and $Q_i$ constant, the linearized equations are uncoupled. Explain this effect. (*Hint.* Consider the linearization of the equation $\dot{y} = y^2$.)

**7.20** In Example 7.3 suppose that because of aging effects, the system matrix changes to

$$\mathbf{A} = \begin{bmatrix} -3.3 & 0 \\ 0 & -4 \end{bmatrix}$$

Analyze the effect of this change on the performance of the controller designed in that example.

**7.21** Repeat Example 7.4 using a sampling period of $T = 1$ sec. Compare the responses for the two values of $T$.

**7.22** Suppose that the plant in the control system shown in Figure 7.24 is

$$G_p(s) = \frac{1}{2s + 1}$$

with $T = 1$.

(a) Find the control algorithm $D(z)$ so that the closed-loop transfer function will be

$$T(z) = z^{-1}$$

(b) Determine $c(k)$, $e(k)$, and $m(k)$ for a unit step input.

**7.23** It is desired that the open-loop system shown in Figure P7.23 will exhibit the following response to a unit step input.

$$c(0) = 0, \qquad c(1) = 0.5, \qquad c(2) = c(3) = \ldots = 1$$

(a) Find the required transfer function $G(z)$.

(b) The given unit step response suggests a first-order system. Is this correct? Explain.

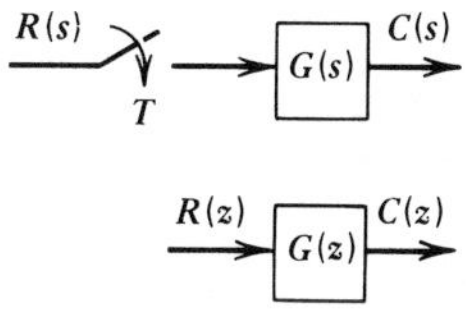


**Figure P7.23**

# CHAPTER EIGHT

# Graphical Design Methods: The Root-Locus, Nyquist, and Bode Plots

It is not difficult to analyze a second-order linear system because its characteristic equation can be solved for the roots in closed form. Rooting formulas exist for third- and fourth-order polynomials, but their complexity is such that they do not generate much insight into the system's behavior. The Routh-Hurwitz criterion is useful for stability analysis, but it cannot give a complete picture of the transient response. Thus we are in need of design methods for systems of third order and higher.

Here we develop graphical methods for representing the system's behavior. The *root-locus plot* is a plot of the location of the characteristic roots in terms of some system parameter, such as the proportional gain. The theory of polynomial equations is well developed enough to allow us to sketch the general behavior of the roots without actually solving for them in many cases. This information is often sufficient to make design decisions. This method is developed in Sections 8.1 through 8.6.

The frequency-response plots of the system's open-loop transfer function contain much information about the behavior of the closed-loop system. These plots are easily generated even for high-order systems, and they allow the proper control gain to be selected simply by adjusting the scale factor on the plot. The technique is especially useful because it does not require the values of the characteristic roots. This is helpful for the analysis of high-order systems and systems with dead time. The latter are especially difficult to treat with rooting methods since they possess an infinite number of roots. The technique is covered in Section 8.7.

When a gain adjustment must be supplemented with a series compensator, the open-loop frequency-response plots enable the designer to select the appropriate compensator parameters. In this context we introduce two widely used compensators, the *series lead* and the *series lag*, and develop graphical design methods for them (Sections 8.8 and 8.9).

We continue our usual format in this chapter by introducing the concepts in terms of analog systems and later discussing any required modifications for digital implementation. The root-locus technique is just as useful for digital design, while the frequency response methods are less so. In Section 8.10 we illustrate how continuous-time performance specifications given in terms of $s$-plane root locations can be used with the $z$-plane root locus to design digital controllers. Finally, we introduce the finite settling time algorithm – an example of a control algorithm that is difficult to implement with analog controllers, but is often easily accomplished digitally.

Although the methods of this chapter were developed before the widespread use of digital computers and calculators for computation, the methods' usefulness has been enhanced by such machines. The methods require frequent evaluation of polynomial, logarithmic, and trigonometric functions, and these are easily done with machine aids. The plots can be sketched by hand or generated with a computer-based plotter, if available.

## 8.1 THE ROOT-LOCUS CONCEPT

The root locus is a plot of the locations of the roots of an equation as some real-valued parameter varies, possibly from $-\infty$ to $+\infty$. To see this more clearly, consider the polynomial equation

$$ms^2 + cs + k = 0 \tag{8.1-1}$$

Suppose that $m = c = 1$ and we wish to display the root locations as $k$ varies. For now let $k \geqslant 0$. The roots are

$$s = \frac{-1 \pm \sqrt{1-4k}}{2} \tag{8.1-2}$$

It is easily seen that if $k < 0.25$, the roots are real and distinct. They are repeated if $k = 0.25$ and are complex conjugates if $k > 0.25$. In the latter case, the time constant is always 2. The root locations for $0 \leqslant k < \infty$ can thus be plotted in the $s$ plane as in Figure 8.1*a*. The root locations when $k = 0$ are denoted by $\times$ and are at $s = 0, -1$. The arrows on the plot indicate the direction of root movement as $k$ increases. As $k$ approaches $\infty$, the roots approach $s = -0.5 \pm i\infty$.

Equation (8.1-1) is the characteristic equation of the commonly found transfer function

$$T(s) = \frac{1}{ms^2 + cs + k} \tag{8.1-3}$$

Suppose the desired damping ratio is $\zeta = 0.5$, with $m = c = 1$. The corresponding characteristic roots are $s = -0.5 \pm i0.866$ and are marked in Figure 8.1*a*. These roots can be obtained from the plot by marking off the line corresponding to $\cos\beta = \zeta = 0.5$,

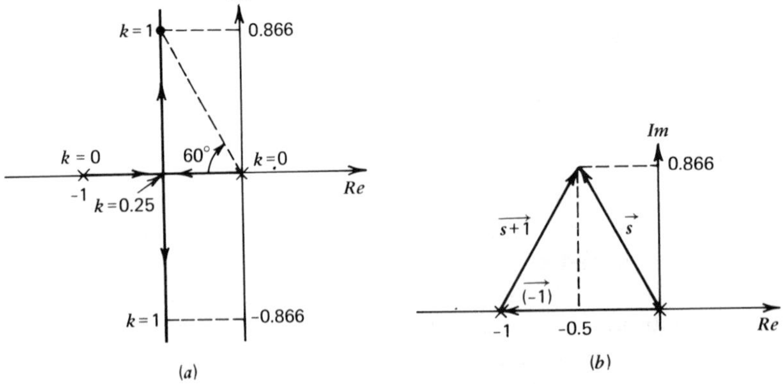


**Figure 8.1** (*a*) Root-locus plot for $s^2 + s + k = 0, k \geqslant 0$. (*b*) Vector diagram showing the calculation of $k$ at the point $s = -0.95 + i0.866$.

or $\beta = 60°$. The intersection of this line with the root-locus plot gives the upper root of the conjugate pair. At this point the reader might wish to review the graphical interpretations of the parameters $\tau$, $\omega_n$, and $\zeta$ in terms of the root locations. See Section 5.4 (Figures 5.14 and 5.15).

The root-locus plot thus can be used to determine characteristic root locations that give specified behavior, such as $\zeta = 0.5$, because the plot represents the solution of the characteristic equation (8.1-1). Once the root locations have been determined, the plot can be used to compute the value of $k$ required to give these characteristic roots. To see how this is done, rewrite (8.1-1) with $m = c = 1$.

$$k = -s(s+1) \tag{8.1-4}$$

Since $k$ is a real number, the right-hand side must also be real. At the desired root locations $s = -0.5 \pm i0.866$, (8.1-4) gives

$$\begin{aligned} k &= -(-0.5 + i0.866)(-0.5 + 1 + i0.866) \\ &= -(-\tfrac{1}{4} - \tfrac{3}{4}) = 1 \end{aligned} \tag{8.1-5}$$

Thus if $k = 1$, the damping ratio will be 0.5. This result could have been obtained directly from (8.1-1), but we are preparing the way for more difficult problems.

The vector properties of complex numbers can be used to simplify the calculation of $k$. (A review of these properties is given in Appendix A, Section A.4.) With reference to (8.1-4) and Figure 8.1$b$, we note that the vector $\vec{s} = -0.5 + i0.866$ has its head at this point and its tail at the origin. From the figure we see that the property of vector addition gives

$$\vec{s} = \overrightarrow{(-1)} + \overrightarrow{(s+1)} \tag{8.1-6}$$

From this we deduce that $\overrightarrow{(s+1)}$ has its head at $s = -0.5 + i0.866$ and its tail at $-1$. In general, the number $(s + a)$ is represented by a vector with its head at the point $s$ and its tail at $-a$.

The magnitude of a product of complex numbers is the product of the magnitudes. Therefore, for $k \geqslant 0$, (8.1-4) can be expressed as

$$|k| = k = |-s(s+1)| = |-s||s+1| = |s||s+1| \tag{8.1-7}$$

Since the magnitude of a complex number is the length of its corresponding vector, we can evaluate $k$ by measuring on the plot the lengths of the vectors $\vec{s}$ and $\overrightarrow{(s+1)}$. This gives

$$k = 1(1) = 1$$

When many such factors are involved, it is easier to measure vector lengths on the plot, rather than to multiply several complex numbers as was required in (8.1-5). The resulting accuracy depends on the size of the graph, but acceptable accuracy is usually obtained with a graph of moderate size.

## Sensitivity to Parameter Variation

With the design selected to be $k = 1$, we now wish to see what is the effect of the parameter $c$ not having its nominal value of $c = 1$. This is a common problem because we

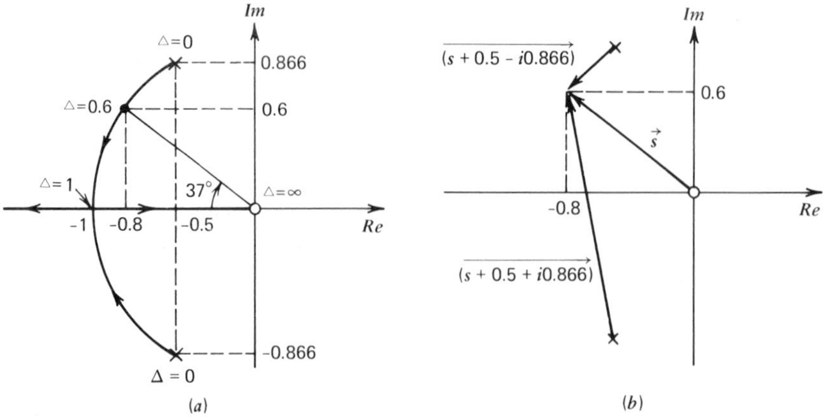


**Figure 8.2** **(*a*) Root-locus plot for $s^2 + (1 + \Delta)s + 1 = 0$, $\Delta \geqslant 0$. (*b*) Vector diagram showing the calculation of $\Delta$ at $s = -0.8 + i0.6$.**

never know parameter values exactly. To this end, we write (8.1-1) with $m = k = 1$ and $c = 1 + \Delta$.

$$s^2 + (1 + \Delta)s + 1 = 0 \tag{8.1-8}$$

Here $\Delta$ represents the deviation of $c$ from its nominal value. To simplify the discussion, we consider only $\Delta \geqslant 0$. The roots of (8.1-8) are

$$s = \frac{-(1 + \Delta) \pm \sqrt{2\Delta + \Delta^2 - 3}}{2} \tag{8.1-9}$$

We can plot the roots as shown in Figure 8.2. Even though we normally would consider only small deviations from the nominal, we show the roots for $0 \leqslant \Delta < \infty$ for general interest. The root locations when $\Delta = 0$ are marked by × and coincide with the roots when $c = k = 1$. As $\Delta \to \infty$, one root approaches $s = -\infty$ while the other approaches $s = 0$. A finite termination point is denoted by a circle ○. The arrows here represent the motion of the roots as $\Delta$ increases.

This plot can be used to determine how much of a change in $c$ is required to change the damping ratio by a certain amount from its desired value of $\zeta = 0.5$. The method is similar to that used earlier. Rewrite (8.1-8) as

$$\Delta = \frac{-(s^2 + s + 1)}{s} = \frac{-(s + 0.5 - i0.866)(s + 0.5 + i0.866)}{s} \tag{8.1-10}$$

To see what value of $\Delta$ will increase $\zeta$ from 0.5 to 0.8, we use a protractor to draw the line corresponding to $\cos\beta = 0.8$, or $\beta = 37°$. It intersects the locus at $s = -0.8 + i0.6$, as shown in Figure 8.2*a*. To evaluate $\Delta$ at this point we note that (8.1-10) involves vectors whose heads are at this point and whose tails are at the points denoted by ○ and ×. We measure the lengths of these vectors to be

$$|s| = 1$$

$$|(s + 0.5 - i0.866)| = 0.4$$

$$|(s + 0.5 + i0.866)| = 1.5$$

Equation (8.1-10) gives

$$\Delta = |\Delta| = \frac{|s + 0.5 - i0.866|\,|s + 0.5 + i0.866|}{|s|}$$

$$= \frac{0.4(1.5)}{1} = 0.6$$

If $\Delta = 0.6$ (or $c = 1.6$), the damping ratio will be 0.8.

## Formulation of the General Problem

The general form of an equation whose roots can be studied with the root locus method is

$$D(s) + KN(s) = 0 \tag{8.1-11}$$

where $K$ is the parameter to be varied. For now we take the functions $D(s)$ and $N(s)$ to be polynomials in $s$ with constant coefficients. Later we will allow them to include the dead-time element $e^{-Ds}$. At first we consider the case: $K \geqslant 0$, and later extend the results to $K < 0$.

The guides to be developed for plotting the root locus require that the coefficients of the highest powers of $s$ in both $N(s)$ and $D(s)$ are normalized to unity. The multipliers required to do this are absorbed into the parameter to be varied. The result is $K$. For example, consider the variation of the parameter $b$ in the equation

$$3s^2 + 5bs + 2 = 0$$

This can be written as

$$s^2 + \frac{2}{3} + \frac{5b}{3}s = 0$$

From (8.1-11) we see that $D(s) = s^2 + \frac{2}{3}$, $N(s) = s$ and $K = 5b/3$. The root-locus plot would be made in terms of the parameter $K$, and the values for $b$ would be recovered from $K$.

Comparison of the general form (8.1-11) with the two previous examples shows that for the first example with $m = c = 1$ and $k$ the parameter, $K = k$, $N(s) = 1$ and $D(s) = s^2 + s$. For the second example, (8.1-8) shows that $K = \Delta$, $N(s) = s$, and $D(s) = s^2 + s + 1$.

Another standard form of the problem is obtained by rewriting (8.1-11) as

$$1 + KP(s) = 0 \tag{8.1-12}$$

where

$$P(s) = \frac{N(s)}{D(s)} \tag{8.1-13}$$

## Terminology

The standard terminology is as follows. The roots of $N(s) = 0$ are called the *zeros* of the problem. The name refers to the fact that these are the finite values of $s$ that make $P(s)$ zero. The roots of $D(s) = 0$ are the *poles* of the problem. These are the finite values of $s$ that make $P(s)$ become infinite.

Referring to (8.1-11) we see that when $K = 0$, the roots of the equation are the poles. As $K \to \infty$, the roots of (8.1-11) approach the zeros. These facts will be useful.

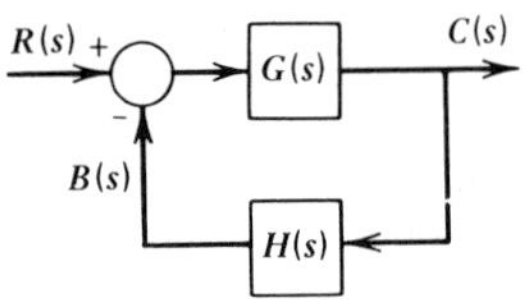


**Figure 8.3 Terminology for the single-loop system.**

For most of our applications, the root-locus method will be applied to the characteristic equation of a feedback system. Therefore it is helpful to relate the preceding terminology to such a system. A general single-loop system is shown in Figure 8.3 without a disturbance input. The transfer function is

$$T(s) = \frac{C(s)}{R(s)} = \frac{G(s)}{1 + G(s)H(s)} \tag{8.1-14}$$

In general $G(s)$ and $H(s)$ will consist of numerator and denominator polynomials, denoted by

$$G(s) = \frac{N_G(s)}{D_G(s)} \tag{8.1-15}$$

$$H(s) = \frac{N_H(s)}{D_H(s)} \tag{8.1-16}$$

The transfer function can be written as

$$T(s) = \frac{N_G(s)D_H(s)}{D_G(s)D_H(s) + N_G(s)N_H(s)} \tag{8.1-17}$$

The characteristic equation is

$$D_G(s)D_H(s) + N_G(s)N_H(s) = 0 \tag{8.1-18}$$

The *open-loop transfer function* of this system is $G(s)H(s)$ and is so named because it is the transfer function relating the feedback signal $B(s)$ to the input $R(s)$ if the loop is opened or broken at $B(s)$. That is,

$$B(s) = G(s)H(s)R(s) \tag{8.1-19}$$

The *closed-loop transfer function* is $T(s)$ given by (8.1-14). Comparing (8.1-18) with (8.1-11), we see that the poles of the characteristic equation are also the poles of $G(s)H(s)$ if the parameter $K$ is a multiplicative factor in $N_G(s)N_H(s)$, as is often the case. Similarly, the zeros of the problem are also the zeros of $G(s)H(s)$. Thus the terms *open-loop poles* and *open-loop zeros* are often used as terms for the poles and zeros on the root-locus plot. The *closed-loop poles* are the characteristic roots since they are the finite values of $s$ that make the closed-loop transfer function become infinite.

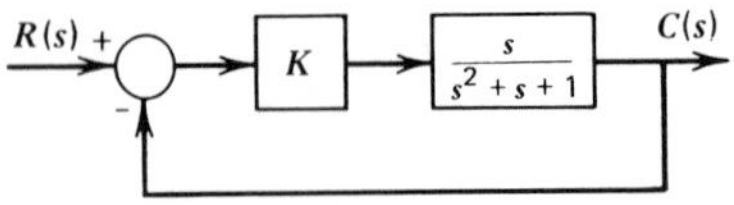


**Figure 8.4 Feedback system with open-loop poles at $s = -0.5 \pm i0.866$ and an open-loop zero at $s = 0$. The root locus for this system for $K \geqslant 0$ is given by Figure 8.2$a$, with $\Delta = K$.**

For example, consider proportional control of a second-order plant, as shown in Figure 8.4. The open- and closed-loop transfer functions are

$$G(s)H(s) = \frac{Ks}{s^2 + s + 1} \tag{8.1-20}$$

$$T(s) = \frac{Ks}{s^2 + (1 + K)s + 1} \tag{8.1-21}$$

The open-loop poles are found from $s^2 + s + 1 = 0$ to be $s = -0.5 \pm i0.866$. The open-loop zero is $s = 0$. With $K = \Delta$, the characteristic equation for this system is the same as that given by (8.1-8). Therefore, the root-locus plot is given by Figure 8.2*a*. Note that the open-loop poles are the characteristic root locations when $K = 0$ and represent the starting points of the locus. The open-loop zero is one of the characteristic roots when $K = \infty$ and represents the termination point of one of the root-locus paths. Finally, we observe that the open-loop poles and zeros can be found without having a value for $K$. The same is not true for the closed-loop poles.

## Angle and Magnitude Criteria

The general problem in the form (8.1-12) can be written as

$$KP(s) = -1 \tag{8.1-22}$$

From this we see that two requirements must be met if $s$ is to be a root of this equation. Equation (8.1-22) is an equation between two complex numbers. As such the equality requires that the magnitudes of each side be equal and that the angles be the same. These are the magnitude and angle criteria, respectively.

$$|KP(s)| = 1 \tag{8.1-23}$$

$$\measuredangle KP(s) = \measuredangle(-1) = n180°, \quad n = \pm 1, \pm 3, \pm 5, \ldots \tag{8.1-24}$$

For now we consider only $K \geqslant 0$. Thus $|K| = K$ and $\measuredangle K = 0°$, so that (8.1-23) becomes

$$K = \frac{1}{|P(s)|} \tag{8.1-25}$$

Also, since the angle of a product is the sum of the angles (see Section A.4 in Appendix A),

$$\measuredangle KP(s) = \measuredangle K + \measuredangle P(s) = \measuredangle P(s)$$

Equation (8.1-24) implies that

$$\measuredangle P(s) = n180°, \quad n = \pm 1, \pm 3, \pm 5, \ldots \tag{8.1-26}$$

Equations (8.1-25) and (8.1-26) are the only two conditions that every point on the root locus must satisfy. It is interesting to note that the angle criterion (8.1-26) does not contain $K$. The *shape* of the root-locus plot is determined *entirely* by the angle criterion. All of the plotting guides to follow, except two, are the result of this single condition. The magnitude criterion (8.1-25) is used only to assign the associated value of $K$ to a designated point $s$ on the root locus. This process is referred to as *scaling* the locus.

# 8.2 PLOTTING GUIDES

Let us collect the insights generated thus far and formalize them as guides for plotting the root locus. Our primary reference will be (8.1-12) and its associated magnitude and angle criteria, (8.1-25) and (8.1-26), for $K \geqslant 0$. The poles are the finite $s$ values that make $P(s)$ infinite. The zeros are the finite values of $s$ that make $P(s)$ zero. We assume that the order of the numerator of $P(s)$ is less than or equal to the order of the denominator. It is required that the real and imaginary axes have the same scale. Each path on the plot corresponds to one root of (8.1-12). These paths are referred to as the loci or branches.

### Guide 1

*The root-locus plot is symmetric about the real axis.* This is because complex roots occur in conjugate pairs. Thus we need only deal with the upper half plane of the plot.

### Guide 2

*The number of loci equals the number of poles of* P(s). This guide tells us how many paths we must account for. The proof is taken from (8.1-11). If the order of $D(s)$ is greater than that of $N(s)$, then the order of the equation, and thus the number of roots, is determined by $D(s)$. But the order of $D(s)$ is the number of poles of $P(s)$.

### Guide 3

*The loci start at the poles of* P(s) *with* K = 0*, and terminate with* K = ∞ *either at the zeros of* P(s) *or at infinity.* When $K = 0$, (8.1-11) shows that the roots are given by $D(s) = 0$; in other words, by the poles. When $K \to \infty$, there are only two ways that (8.1-11) can be satisfied. The first way requires that $N(s) \to 0$; that is, that $s$ approach a zero of $P(s)$. If $N(s)$ does not approach zero as $K \to \infty$, then $D(s)$ must approach infinity, which can occur only if $s \to \infty$. This is the basis for the guide.

The assumption concerning the relative order of $D(s)$ and $N(s)$ is true in control system applications, as well as in most other types of problems. When the assumption is not satisfied, it is often because of an improper formulation of the problem. For example, consider again (8.1-1), but this time with $c = k = 1$ and $m$ as the parameter $K$. Rearranging in terms (8.1-12), we obtain

$$1 + \frac{Ks^2}{s+1} = 0$$

Thoughtless application of Guides 2 and 3 results in a contradiction. Guides 2 and 3 say that there is one path and that it starts at $s = -1$. Guide 3 says that two termination points exist at $s = 0$ because there are two zeros at that point. The problem is improperly formulated because when $m = K = 0$ the order of the equation drops from two to one. When $K \neq 0$ there are two roots; when $K = 0$ there is only one.

The variation of the leading coefficient can be studied if the order of the equation remains the same throughout the variation. For example, if $c = k = 1$ in (8.1-1) and the nominal value of $m$ is 1, let $m = 1 + K$, and write (8.1-1) as

$$1 + \frac{Ks^2}{s^2 + s + 1} = 0$$

The guides for $K \geqslant 0$ can now be applied without any difficulty. Both paths terminate at $s = 0$.

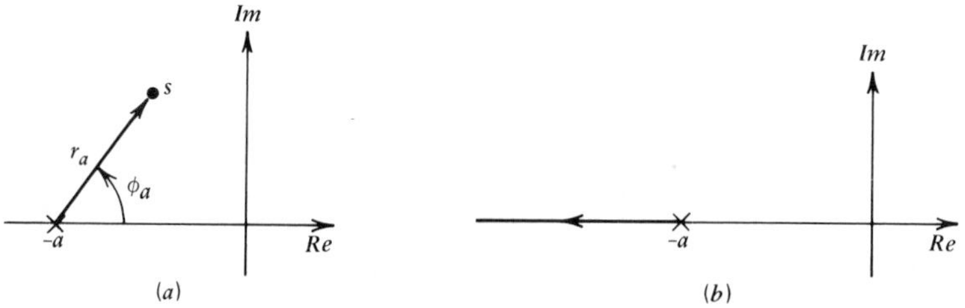


**Figure 8.5** **Single-pole system $s + a + K = 0$. (*a*) Complex exponential representation. (*b*) Root-locus plot for $K \geq 0$.**

## Behavior of the Loci on the Real Axis

In order to develop more plotting guides, we now apply the angle criterion to some typical problems.

Consider proportional control of a first-order plant $1/(s + a)$ with a proportional gain $K$. The characteristic equation is

$$K = -(s + a)$$

If we express the number $(s + a)$ in complex exponential form (Section A.4 in Appendix A) we have

$$K = -r_a e^{i\phi_a}$$

See Figure 8.5*a*. By definition, $r_a > 0$. Thus the preceding equation implies that $K = r_a$ and $\exp(i\phi_a) = -1$. This requires that $\phi_a = n180°$, $n = \pm 1, \pm 3, \ldots$. It says that the root locus lies entirely on the real axis and entirely on the left of the pole at $s = -a$, as shown in Figure 8.5*b*. Of course, we could have deduced this without the angle criterion, but the approach is useful.

Now suppose the proportional control is applied to a second-order plant $1/(s + a)(s + b)$. The characteristic equation is

$$K = -(s + a)(s + b)$$

or

$$K = -r_a r_b e^{i(\phi_a + \phi_b)}$$

See Figure 8.6*a*. The requirements are $K = r_a r_b$ and

$$\phi_a + \phi_b = n180°, \quad n = \pm 1, \pm 3, \ldots \tag{8.2-1}$$

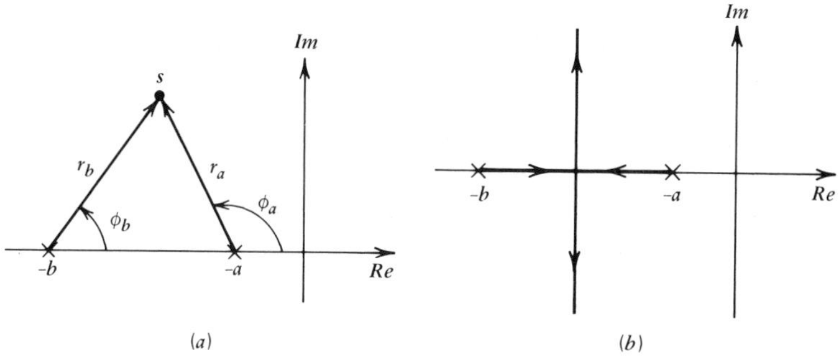


**Figure 8.6** **System with two real poles $(s + a)(s + b) + K = 0$. (*a*) Complex representation. (*b*) Root-locus plot for $K \geq 0$.**

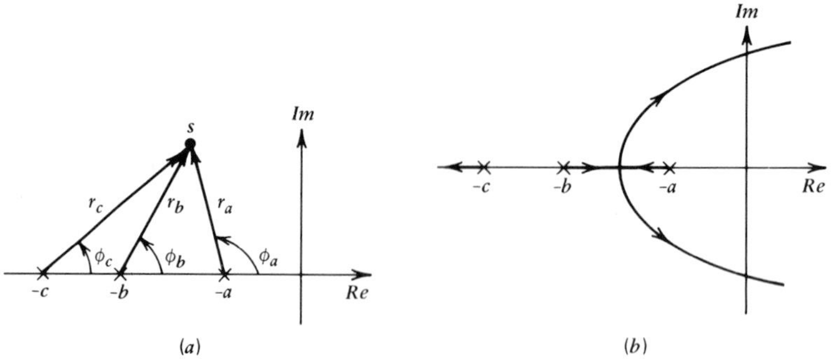


**Figure 8.7 System with three real poles $(s + a)(s + b)(s + c) + K = 0$. (*a*) Complex representation. (*b*) Root-locus plot for $K \geqslant 0$.**

The latter condition implies that the root locus *must* exist on the real axis between the poles and *cannot* exist anywhere else on the real axis. This can be seen by noting that the tips of the vectors must touch, and by trying different values of $\phi_a$ and $\phi_b$ that satisfy this constraint and the relation (8.2-1). The only values that satisfy both on the real axis are: $\phi_a = n180°$, $n = \pm 1, \pm 3, \ldots$, and $\phi_b = m360°$, $m = 0, \pm 1, \pm 2, \ldots$. The root locus is shown in Figure 8.6*b*, but discussion of how the behavior off the real axis was determined is deferred for now.

A third-order plant $1/(s + a)(s + b)(s + c)$ with proportional control gives the characteristic equation

$$K = -(s + a)(s + b)(s + c)$$

or

$$K = -r_a r_b r_c e^{i(\phi_a + \phi_b + \phi_c)}$$

The two conditions for the root locus are $K = r_a r_b r_c$ and

$$\phi_a + \phi_b + \phi_c = n180°, \quad n = \pm 1, \pm 3, \ldots \tag{8.2-2}$$

Analysis of Figure 8.7*a* in the same manner as the previous case shows that the only possible root locations on the real axis correspond to

$$\phi_a = n180°, \quad n = \pm 1, \pm 3, \ldots$$
$$\phi_b = \phi_c = m360°, \quad m = 0, \pm 1, \pm 2, \ldots$$

or

$$\phi_a = \phi_b = n180°, \quad n = \pm 1, \pm 3, \ldots$$
$$\phi_c = m360°, \quad m = 0, \pm 1, \pm 2, \ldots$$

In other words, the root locus exists on the real axis to the left of the pole $s = -c$ and between the poles at $s = -a$ and $s = -b$. The root locus is shown in Figure 8.7*b*.

We can generalize these results to obtain Guide 4. For a rigorous proof of this and the other guides, see Reference 1.

### *Guide 4*

*The root locus can exist on the real axis only to the left of an odd number of real poles and/or zeros; furthermore, it must exist there.* Our numbering system is as

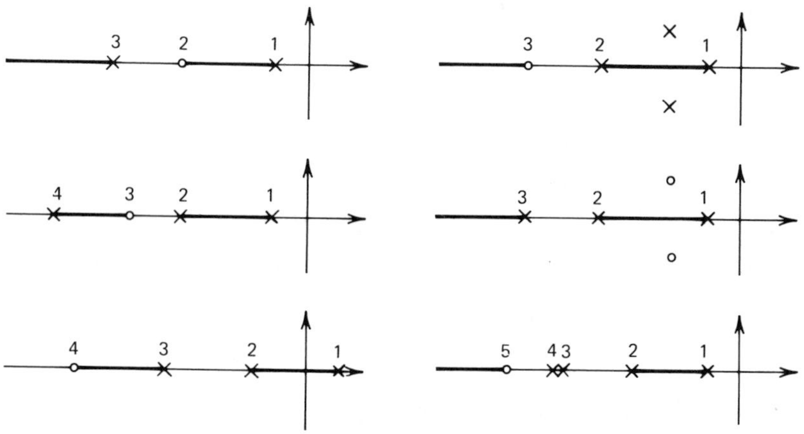


**Figure 8.8 Location of the root locus on the real axis for various pole-zero configurations. The locus off the real axis is not shown.**

follows. The real pole or zero that lies the farthest to the right is number 1; the next real pole or zero is number 2; etc. If a pole or zero falls on top of another, number each one. The phrase *only to the left* is interpreted to mean that the locus *cannot* exist on the real axis to the *right* of an odd-numbered real pole or zero. Some examples are shown in Figure 8.8. In this figure, no implication is made concerning the location of the locus off the real axis.

If we refer to either Figure 8.1*a* or 8.6*b*, we can see that the two loci depart from the poles and head toward each other. It is impossible for two root paths to coincide over any finite length (but they can cross). So the two loci must leave the real axis at some point (the *breakaway point*). From Guide 1, they must do this in a symmetric way.

A method for determining the breakaway point can be developed by noting that the loci approach each other as $K$ increases. Thus at the breakaway point, $K$ attains the *maximum* value it has on the real axis in the vicinity of the breakaway point. The value of $s$ corresponding to the breakaway point can be found by computing $dK/ds$ from (8.1-25), setting $dK/ds = 0$ and solving for the value of $s$. In general, multiple solutions will occur. The extraneous ones can be discarded with a knowledge of the location of the locus on the real axis from Guide 4. Usually the second derivative need not be computed to distinguish between a minimum and a maximum.

Figure 8.2*a* shows that the locus can also enter the real axis. The point at which this occurs is the *break-in* point. After the locus has entered the real axis, $K$ continues to increase. Thus at a break-in point, $K$ attains the *minimum* value it has on the real axis in the vicinity of the break-in point. The location of the break-in point is determined from $dK/ds = 0$ in exactly the same manner as with a breakaway point.

Although the same method is used to find the breakaway and break-in points, no difficulty is encountered in identifying the type of point since this is usually obvious once the first four guides have been applied. It is possible that there are multiple breakaway or break-in points. Hence we speak of $K$ attaining a maximum or minimum value only in the vicinity of each such point.

### *Guide 5*

*The locations of breakaway and break-in points are found by determining where the parameter* K *attains a local maximum or minimum on the real axis.*

You should use this guide to convince yourself of the locations of the breakaway and break-in points shown in Figures 8.1*a* and 8.2*a*. We now illustrate Guide 5 with an example.

### *Example 8.1*

Determine the root-locus plot for the equation

$$1 + \frac{K(s+6)}{s(s+4)} = 0 \tag{8.2-3}$$

The poles are $s = 0, -4$; the zero is $s = -6$. Thus there are two loci. One starts at $s = 0$ and the other at $s = -4$, with $K = 0$. One path terminates at $s = -6$, while the other must terminate at $s = \infty$. This information is given by Guides 2 and 3.

Guide 4 shows that the locus exists on the real axis between $s = 0$ and $s = -4$ and to the left of $s = -6$. Therefore, the two paths must break away from the real axis between $s = 0$ and $s = -4$. From the location of the termination point at $s = -6$, we know that the locus must return to the real axis. From Guide 1 we see that both paths must break in at the same point.

These points are found from Guide 5 as follows. Solve (8.2-3) for $K$ and compute $dK/ds$.

$$K = -\frac{s(s+4)}{s+6}$$

$$\frac{dK}{ds} = -\frac{(s+6)(2s+4) - s(s+4)}{(s+6)^2} = 0$$

This is satisfied for finite $s$ if the numerator is zero.

$$s^2 + 12s + 24 = 0$$

The candidates are

$$s = -6 \pm 2\sqrt{3} = -2.54, -9.46$$

There is no need to check for a minimum or a maximum, because we know from Guide 4 that the breakaway point must be at $s = -2.54$ and the break-in point at $s = -9.46$. This leaves only the shape of the locus off the real axis to be determined. We note that the breakaway and break-in points are symmetrically placed with a distance of 3.46 from the zero at $s = -6$. The simplest way for this to occur is with a circle of radius 3.46 centered at the zero. We draw this circle, as shown in Figure 8.9*a*. To confirm this, we can select trial points along the circle and apply the angle criterion. For example, the point $s = -6 + 3.46i$ lies on the locus because

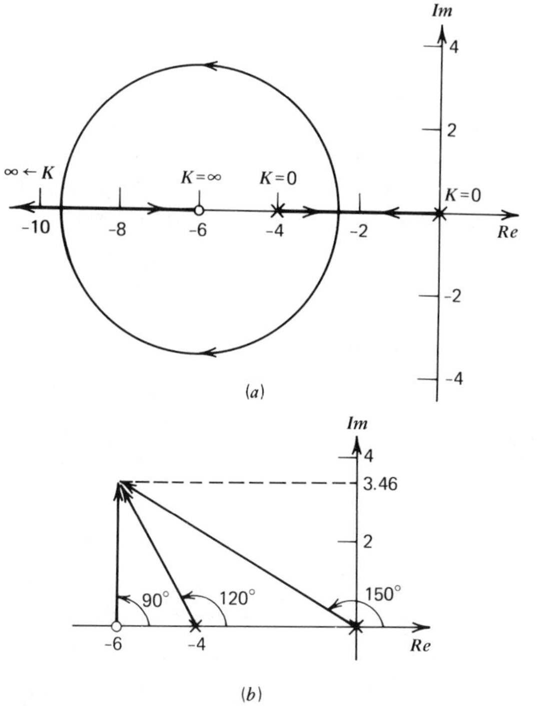


**Figure 8.9** **(*a*) Root-locus plot for $s(s+4)+K(s+6)=0$, for $K \geqslant 0$. (*b*) Angle criterion applied to the test point $s=-6+i3.46$.**

$$\measuredangle \frac{s(s+4)}{s+6} = \measuredangle s + \measuredangle(s+4) - \measuredangle(s+6)$$

$$= 150^\circ + 120^\circ - 90^\circ$$

$$= 180^\circ$$

Several points can be quickly checked with a protractor. This would show that the locus is indeed a circle.

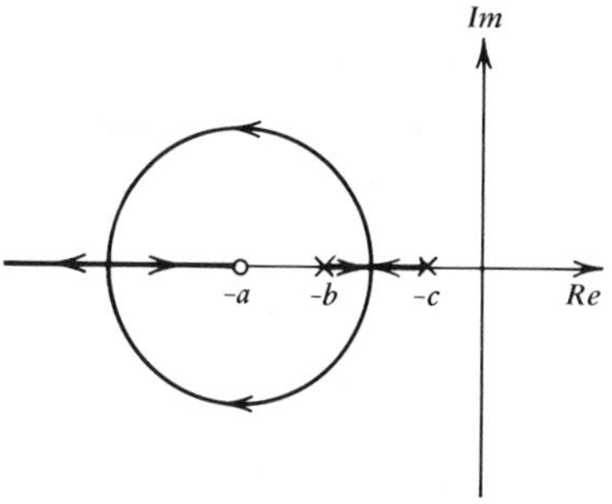


**Figure 8.10** **Root-locus plot for $(s+b)(s+c)+K(s+a)=0$ for $K \geqslant 0$, and $a > b > c > 0$.**

High-school geometry can be used to show that the root locus of the equation

$$(s+b)(s+c)+K(s+a) = 0 \tag{8.2-4}$$

is a circle centered on the zero at $s=-a$, if $a, b, c > 0$ and $a > b > c$. This case is shown in Figure 8.10. The radius of the circle can be determined once the break-away and break-in points are found.

## Behavior of the Locus for Large $K$

As $K \to \infty$ the loci approach either a zero or $s = \infty$. We now investigate how to determine the shape of the loci in the latter case. The vectors to a point on the locus at $s = \infty$ from the poles and zeros are parallel and make an angle $\theta$ with the real axis. This is shown in Figure 8.11 for a particular pole-zero pattern. If we let $P$ and $Z$ denote the number of poles and zeros, respectively, the angle criterion states that

$$\measuredangle P(s) = Z\theta - P\theta = n180°, \quad n = \pm 1, \pm 3, \ldots$$

Solving for $\theta$, we obtain

$$\theta = \frac{n180°}{Z - P}, \quad n = \pm 1, \pm 3, \ldots \tag{8.2-5}$$

This leads to Guide 6.

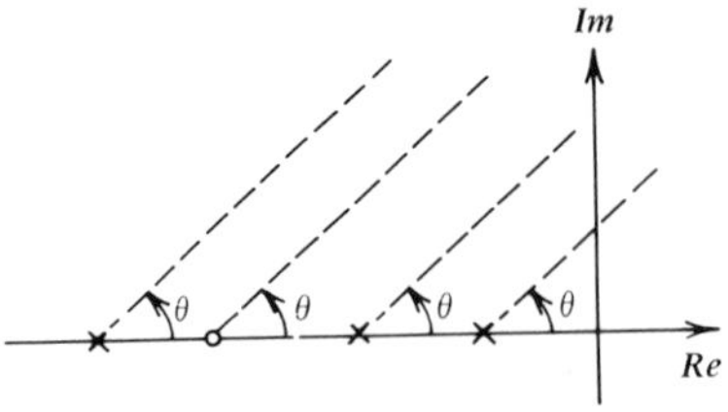


**Figure 8.11** Asymptotic angle $\theta$ for a particular pole-zero configuration.

### *Guide 6*

*The loci that do not terminate at a zero approach infinity along asymptotes. The angles that the asymptotes make with the real axis are found from (8.2-5), where* n *is chosen successively as* n $= +1, -1, +3, -3, \ldots$, *until enough angles have been found.* Figure 8.12 shows the most commonly found patterns.

Unless $\theta$ is 0° or 180°, the asymptotes cannot be drawn until we know where they intersect the real axis. This is given by Guide 7. The proof follows from the angle criterion, but is detailed (see Reference 1).

### *Guide 7*

*The asymptotes intersect the real axis at the common point*

$$\sigma = \frac{\Sigma s_p - \Sigma s_z}{P - Z} \tag{8.2-6}$$

*where* $\Sigma s_p$ *and* $\Sigma s_z$ *are the algebraic sums of the values of the poles and zeros.*

Note that this guide states that the asymptotes all intersect at the same point. Referring to (8.1-1) with $m = c = 1$ and $K = k$, the poles are $s = 0, -1$, and

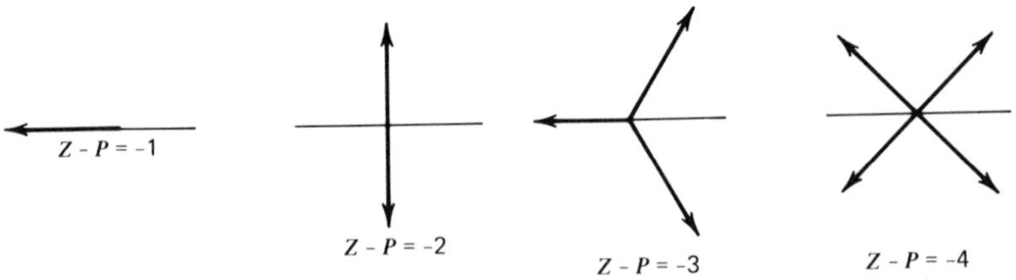


**Figure 8.12** Asymptotic angles for commonly occurring cases.

there are no zeros. Thus $P = 2, Z = 0$, and

$$\Sigma s_p = 0 + (-1) = -1$$

$$\Sigma s_z = 0$$

We need two asymptotes. Therefore, from (8.2-5)

$$\theta_1 = \frac{180°}{0-2} = -90°$$

$$\theta_2 = \frac{-180°}{0-2} = +90°$$

From (8.2-6)

$$\sigma = \frac{-1-0}{2-0} = -\frac{1}{2}$$

As shown in Figure 8.1*a*, the locus lies on the asymptotes. The asymptote in the upper-half plane has the angle + 90°. Both asymptotes intersect at $s = -\frac{1}{2}$.

An example with complex poles is given by (8.1-8) with $K = \Delta$. The poles are $s = -0.5 \pm i0.866$. The zero is $s = 0$. Thus $P = 2, Z = 1$, and

$$\Sigma s_p = (-0.5 + i0.866) + (-0.5 - i0.866) = -1$$

$$\Sigma s_z = 0$$

Therefore, the one asymptote required has the angle

$$\theta = \frac{180°}{1-2} = -180°$$

This is shown in Figure 8.2*a*. The asymptote lies on the real axis, and we do not need to calculate an intersection point.

Some systems become unstable for certain values of the parameter $K$. This occurs if any path crosses the imaginary axis into the right-half plane. The value of $K$ at which this happens can be determined from the Routh-Hurwitz criterion (Section 5.5). Often the substitution $s = i\omega$ into the polynomial equation of interest is quicker and gives the crossing location as well as the value of $K$. This procedure was applied to the cubic polynomial in Section 7.1, Example 7.2. This gives Guide 8.

### *Guide 8*

*The points at which the loci cross the imaginary axis and the associated values of* K *can be found with the Routh-Hurwitz criterion or by substituting* s = iω *into the equation of interest. The frequency ω is the* crossover frequency.

### *Example 8.2*

Plot the root locus for the equation

$$s^3 + 3s^2 + 2s + K = 0 \tag{8.2-7}$$

for $K \geqslant 0$.

We can factor the equation as follows.

$$1+\frac{K}{s(s+1)(s+2)}=0$$

The poles are $s=0, -1, -2$, and there are no zeros. All three paths approach asymptotes as $K\to\infty$. The locus exists on the real axis between $s=0$ and $-1$ and to the left of $s=-2$.

To find the breakaway point, compute $dK/ds$.

$$K = -(s^3+3s^2+2s)$$

$$\frac{dK}{ds} = -(3s^2+6s+2)=0$$

The candidates are $s=-0.423, -1.58$. The breakaway point obviously must be $s=-0.423$ because the locus cannot exist between $s=-1$ and $s=-2$. With this value of $s$, (8.2-7) gives $K=0.385$. The significance of the second solution will be discussed later.

The three asymptotic angles are found from (8.2-5).

$$\theta = \frac{n180^\circ}{0-3}=-n60^\circ, \quad n=\pm 1, \pm 3, \ldots$$

Thus

$$\theta = -60^\circ, +60^\circ, \pm 180^\circ$$

Note that for the last angle it does not matter whether we use $n=+3$ or $n=-3$. The intersection is found from (8.2-6).

$$\sigma = \frac{0+(-1)+(-2)-0}{3-0}=-1$$

The two paths that start at $s=0$ and $-1$ approach the $\pm 60^\circ$ asymptotes. The path starting at $s=-2$ follows the $180^\circ$ asymptote and thus lies entirely on the real axis.

The $60^\circ$ asymptotes indicate that two paths will cross the imaginary axis and generate two unstable roots. Substitution of $s=i\omega$ into (8.2-7) gives

$$-i\omega^3-3\omega^2+2i\omega+K = 0$$

or

$$i\omega(2-\omega^2) = 0$$

$$K = 3\omega^2$$

The solution $\omega=0$, $K=0$ corresponds to the pole at $s=0$. The solution of interest is $\omega=\pm\sqrt{2}, K=6$. The locus can now be sketched. This is shown in Figure 8.13*a*. The system has three roots. All are real and negative for $0\leqslant K\leqslant 0.385$. For $0.385<K<6$ the system is stable with two complex roots, and one real root. For $K>6$ the system is unstable due to two complex roots with positive real parts.

In many applications it is not necessary to determine the precise location of the locus off the real axis, and the first eight guides are often quite sufficient to obtain ample information about the system's behavior. When more accuracy is required the angle criterion can be employed as follows (see Figure 8.13*b*). We know the position

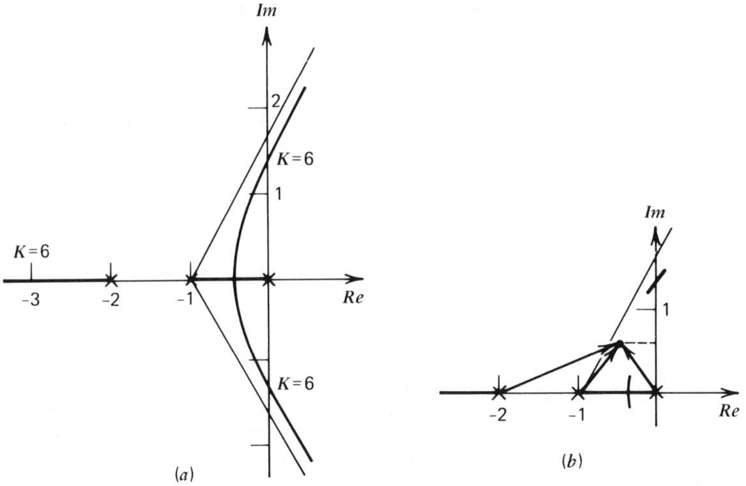


**Figure 8.13** **(*a*) Root-locus plot for $s(s+1)(s+2)+K=0$, for $K \geqslant 0$. (*b*) Use of test points with the angle criterion to plot the locus off-axis.**

of the locus at the breakaway point and at the crossover point. We select a series of test points along a line parallel to the real axis, starting to the left of where the locus could be, for example. At the leftmost point we draw the vectors from the poles and zeros and apply the angle criterion. Unless we are extremely lucky, $\measuredangle P(s)$ will not equal $n180°$, $n = \pm 1, \pm 3, \ldots$. In the previous example,

$$\measuredangle P(s) = \measuredangle \frac{1}{s(s+1)(s+2)} = -\measuredangle s - \measuredangle(s+1) - \measuredangle(s+2)$$

Suppose the first test point gives $\measuredangle P(s) = 200°$. We select another point to the right of the first. Suppose this gives $175°$. We have passed the point that would give $180°$ and therefore know that the locus lies between the two test points. This iteration procedure is repeated until the desired accuracy is reached. Then a new line of test points is selected above the first, and the plot is extended in this way.

The root-locus method is very useful in the design of aircraft control systems. Thirty years ago a significant part of the time of many engineers employed in that industry was taken up with constructing root-locus points with this trial-and-error method, which is still useful for sketching the locus. A device that helps the process is the Spirule (available from The Spirule Co., 9728 El Venado Ave., Whittier, CA 90603).

Today we are fortunate to have computer programs to generate root-locus plots. The author has found the program listed in Reference 2 to be useful. You must have a working knowledge of the plotting guides when using such a program in order to interpret the plots and to apply them. For those not wishing to support a general-purpose root-locus program, note that the analytical solutions are available for the roots of polynomials up to order four. Since the majority of applications give models of this order or less, you can develop a very useful root-locus program by coding these solutions.

## Locus Behavior Near Complex Poles and Zeroes

In order to apply the preceding iterative procedure, it is helpful to know the angle at which the locus leaves a complex pole (the *angle of departure*) and the angle at which it terminates at a complex zero (the *angle of arrival*). To determine these angles, we again call on the angle criterion.

### Guide 9

*Angles of departure and angles of arrival are determined by choosing an arbitrary point infinitesimally close to the pole or zero in question and applying the angle criterion.*

The guide is easily explained with an example. Equation (8.1-8) with $K = \Delta$ gives the root locus shown in Figure 8.2$a$. Assume that we do not know the location of the locus. The pole-zero pattern is shown in Figure 8.14. To determine the departure angle for the upper pole, we select an arbitrary test point and draw the vectors to this point from the poles and zeros. The test point is infinitesimally close to the upper pole, but the figure exaggerates the distance from the pole to the test point so that we may see the vector properly. The equation in question can be written as (8.1-10), repeated here as

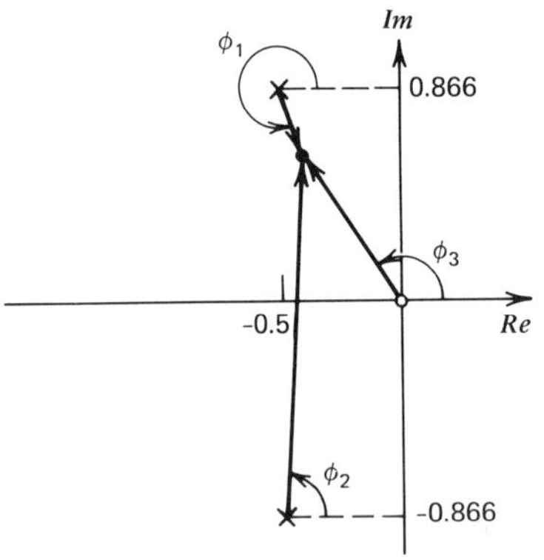


**Figure 8.14 Application of the angle criterion to determine the angle of departure $\phi_1$.**

$$K = -\frac{(s + 0.5 - i0.866)(s + 0.5 + i0.866)}{s}$$

The angle criterion requires that

$$\measuredangle K = 0° = \measuredangle(-1) + \phi_1 + \phi_2 - \phi_3$$

or

$$\phi_1 + \phi_2 - \phi_3 = n180°, \quad n = \pm 1, \pm 3, \ldots \tag{8.2-8}$$

The key step is to assume that the point $s$ is so close to the upper pole that the angles $\phi_2$ and $\phi_3$ are given approximately by the angles to the upper pole. These are $\phi_2 = 90°$, $\phi_3 = 90° + \tan^{-1}(0.5/0.866) = 120°$. From (8.2-8), $\phi_1$ must be

$$\phi_1 = 120° - 90° + n180°$$

With $n = 1$, $\phi_1 = 210°$. Any other choice of odd-numbered $n$ would given an equivalent result. A glance at Figure 8.2$a$ will show that this departure angle is correct. From symmetry we can easily obtain the departure angle for the lower pole.

## Additional Guides

Two more guides can be obtained from the theory of equations and the magnitude criterion.

*Guide 10*

*For the polynomial equation*

$$s^n + a_{n-1}s^{n-1} + \ldots + a_1 s + a_0 = 0 \tag{8.2-9}$$

*the sum of the roots* $r_1, r_2, \ldots, r_n$ *is*

$$r_1 + r_2 \ldots + r_n = -a_{n-1} \tag{8.2-10}$$

Note the unity coefficient of $s^n$. This guide is useful for determining the location of the remaining roots, once some roots have been found. Consider the problem of Example 8.2 (8.2-7). When $K = 6$, two of the roots were found to be $s = \pm i1.41$. The third root can be found with $r_1 = +i1.41, r_2 = -i1.41$ in (8.2-10).

$$i1.41 - i1.41 + r_3 = -3$$

Thus $r_3 = -3$ when $K = 6$.

Also, given a root or root pair that is suspected of being dominant, Guide 10 sometimes provides a quick way of telling whether or not the remaining roots are close to the candidate for dominance.

The final guide formally establishes the usefulness of the magnitude criterion.

*Guide 11*

*Once the root locus is drawn, it is scaled with the magnitude criterion (8.1-25).*

For example, the circular locus of Example 8.1 corresponds to (8.2-3). The magnitude criterion states that

$$|K| = K = \frac{|s||s+4|}{|s+6|}$$

Given any point $s$ on the locus, $K$ can be computed from this equation either by direct substitution of values of $s$ or by measuring the lengths of the vectors $s$, $(s + 4)$, $(s + 6)$, and combining the lengths according to the preceding equation.

Figure 8.15 shows some other common root-locus plots for third-order systems. They can be used to test your knowledge of the plotting guides.

## 8.3 SOME NUMERICAL AIDS

Sometimes the root-locus guides require the solution for the roots of a higher-order equation. For example, the condition $dK/ds = 0$ used to solve for breakaway or break-in points can easily result in a cubic equation. While an analytical result is known, it is often easier to use *Newton iteration* (Appendix A, Section A.2). This is because from Guide 4 we know the general location of the breakaway and break-in points and thus have a good starting guess for the iteration. In addition, we only need to find as many roots as there are breakaway and break-in points.

The second case requiring numerical methods is in the solution for the crossover frequency and ultimate gain (Guide 8). When $s = i\omega$ is substituted into the equation of interest, two equations result: one for the real part and one for the imaginary part.

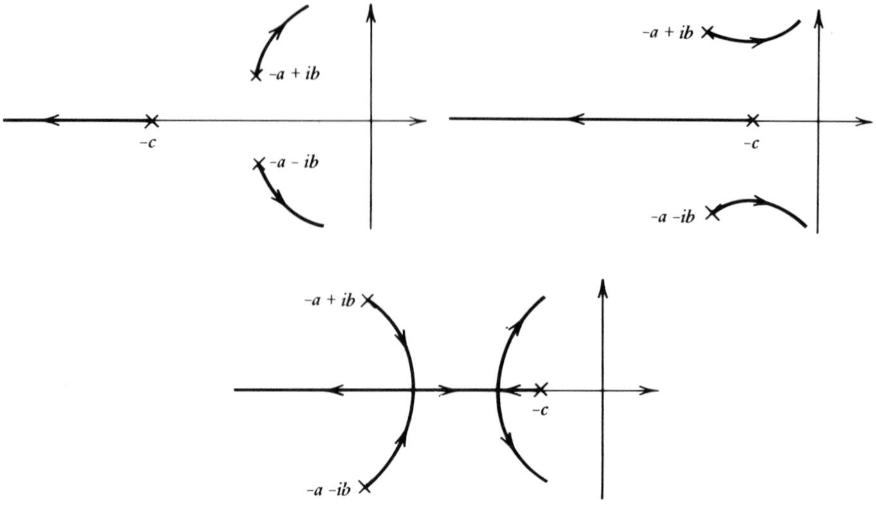


(a)

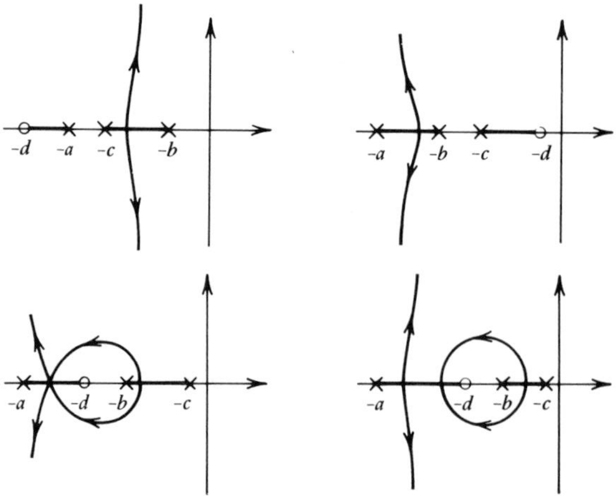


(b)

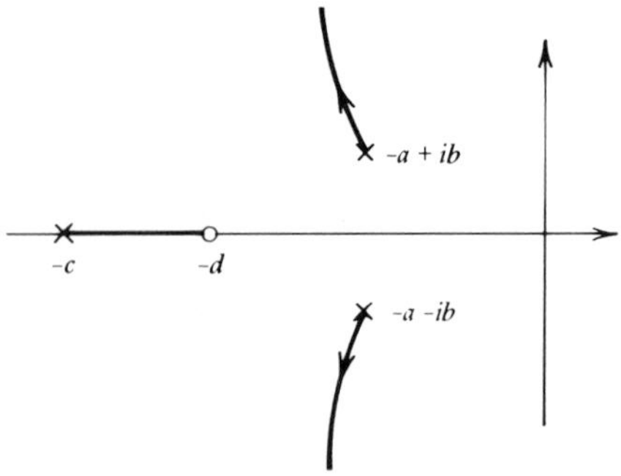


(c)

Figure 8.15 Some common root-locus forms not seen in the chapter examples. (a) $(s + c)(s + a + ib)(s + a - ib) + K = 0$, $K \geqslant 0$. (b) $(s+a)(s+b)(s+c)+K(s+d) = 0$, $K \geqslant 0$ (see also Figures 8.16 and 8.38c). (c) $(s + c)(s + a + ib)(s + a - ib) + K(s + d) = 0$, $K \geqslant 0$.

The equation for $\omega$ is of lower order than the original. When the original equation is of seventh order, a cubic equation must be solved for $\omega$. In this case and those of higher order, Newton iteration can be used to advantage. Again, a good starting guess is usually available from the value of $s$ where the asymptote crosses the imaginary axis.

As shown in Section A.4, Newton iteration is based on the truncated Taylor series expansion of a function $f(x)$ about a point $x_o$ (see Appendix A, Section A.1 for a discussion of the Taylor series). When terms of second order and higher are neglected, the series becomes

$$f(x) \cong f(x_o) + \left(\frac{df}{dx}\right)_{x=x_o}(x - x_o) \tag{8.3-1}$$

If $x$ is the root of the equation $f(x) = 0$, and $x_o$ is an initial guess for $x$, we can set (8.3-1) equal to zero and solve for the estimate of the root $x$. This gives

$$x \cong x_o - \frac{f(x_o)}{f'(x_o)} \tag{8.3-2}$$

where

$$f'(x_o) = \left(\frac{df}{dx}\right)_{x=x_o} \tag{8.3-3}$$

We can repeat the process until it converges to an answer with the desired accuracy. Thus the algorithm can be expressed as

$$x_k \cong x_{k-1} - \frac{f(x_{k-1})}{f'(x_{k-1})}, \quad k = 1, 2, \ldots \tag{8.3-4}$$

*Example 8.3*

Sketch the root locus for the equation

$$1 + \frac{K(s + 1.5)}{s(s + 1)(s + 2)} \tag{8.3-5}$$

The poles are $s = 0, -1, -2$, and the zero is $s = -1.5$ (see Figure 8.16). A breakaway point exists between $s = 0$ and $s = -1$. It is found as follows. From (8.3-5)

$$K = -\frac{s(s + 1)(s + 2)}{s + 1.5} = -\frac{s^3 + 3s^2 + 2s}{s + 1.5}$$

$$\frac{dK}{ds} = -\frac{(s + 1.5)(3s^2 + 6s + 2) - (s^3 + 3s^2 + 2s)}{(s + 1.5)^2} = 0$$

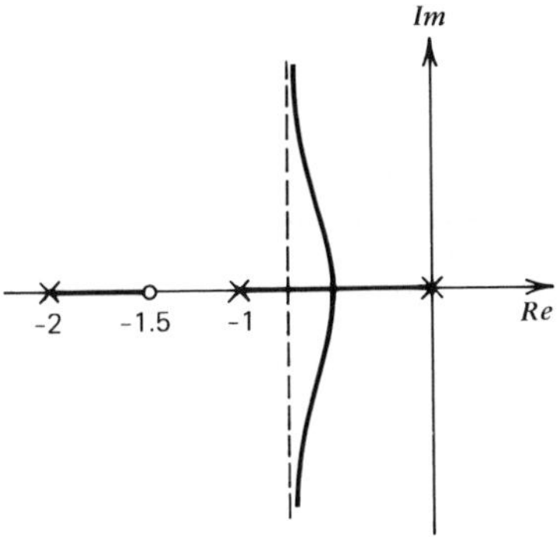


**Figure 8.16** **Root-locus plot for $s(s + 1)(s + 2) + K(s + 1.5) = 0, K \geqslant 0$.**

Setting the numerator to zero gives

$$f(s) = 2s^3 + 7.5s^2 + 9s + 3 = 0$$

from which

$$f'(s) = 6s^2 + 15s + 9$$

The Newton iteration formula is

$$s_k \cong s_{k-1} - \frac{f(s_{k-1})}{f'(s_{k-1})} \tag{8.3-6}$$

The logical trial point is the center of the region $(-1, 0)$, or $s_0 = -0.5$. Instead, to show the details of the Newton iteration, we choose to start at $s_0 = 0$. The steps are

$$s_1 = s_0 - \frac{f(s_0)}{f'(s_0)} = 0 - \frac{f(0)}{f'(0)} = -0.333$$

$$s_2 = -0.333 - \frac{f(-0.333)}{f'(-0.333)} = -0.333 - 0.163$$

$$= -0.496$$

$$s_3 = -0.496 - 0.045 = -0.541$$

$$s_4 = -0.541 - 0.003 = -0.544$$

The process has stabilized in the second decimal place, so we stop because no more decimal places can be shown on our plot. The breakaway point is approximately at $s = -0.54$.

If we had chosen the logical trial point $s_0 = -0.5$, the iteration would have been

$$s_1 = -0.5 - 0.042 = -0.542$$

The answer is obtained with one step. The information provided by Guide 4 concerning the location of the locus on the real axis is invaluable in choosing a starting point for the Newton iteration. If a little common sense or insight is included in making the choice, the iteration's convergence can be greatly improved.

The rest of the locus can be sketched with the remaining guides. The asymptotic angles are

$$\theta = \frac{n180^\circ}{1-3} = \pm 90^\circ$$

Their intersection occurs at

$$\sigma = \frac{0 - 1 - 2 - (-1.5)}{3 - 1} = \frac{-1.5}{2} = -0.75$$

The results of this example can be used to compare the performance of two control systems (Figure 8.17). The plant is $1/(s+1)(s+2)$. Figure 8.17*a* shows an integral controller with this plant. The root locus of this sytem is shown in Figure 8.13*a*. From this it was seen that the system is unstable for $K \geqslant 6$. Figure 8.17*b* shows the same system with a series compensator inserted after the controller. The compensator's transfer function $(s + 1.5)$ produces the zero in the root-locus plot shown in Figure 8.16. The effect of the zero is to shift the locus away from the right-half plane. The new system is always stable, and the good response characteristics of the

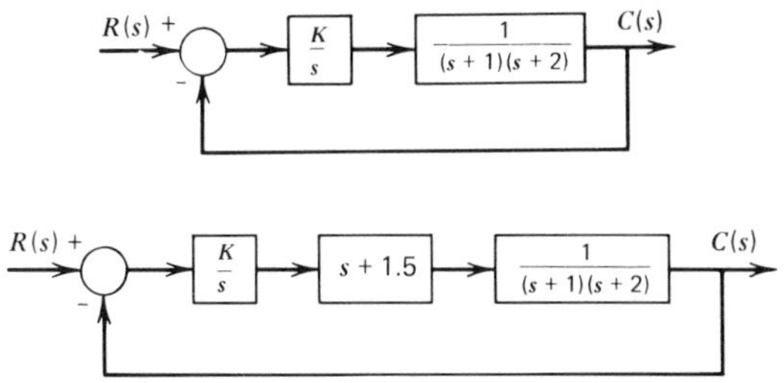


**Figure 8.17 Effect of an open-loop zero on system performance. (*a*) Integral control of a second-order plant. (*b*) Series compensation of the controller in part *a*. The corresponding root loci are shown in Figures 8.13*a* and 8.16.**

original controller for $\zeta = 1$ are preserved. We will return to the subject of series compensators later in the chapter.

## Sensitivity

The truncated Taylor series (8.3-1) can be written as

$$\Delta f \cong \left(\frac{df}{dx}\right)_{x=x_o} \Delta x \tag{8.3-7}$$

where $\Delta f = f(x) - f(x_o)$ is the change in $f$ resulting from a change in $x$ of $\Delta x = x - x_o$. This formula is useful in quickly obtaining estimates of the effects of a slight change in one of the systems' parameters. In control system design there are two common applications of this method. The first concerns the variation in the characteristic roots due to a parameter change. This deals with the *root sensitivity*. The second deals with the change in a transfer function and is called the *transmittance sensitivity* or *system sensitivity*.

*Example 8.4*

As an example of root sensitivity analysis, consider the equation

$$s^3 + 9s^2 + 23s + K = 0 \tag{8.3-8}$$

Suppose we know that the roots are $s = -1, -3, -5$ when $K = 15$, and we wish to estimate the new roots that will occur if $K$ changes to 16. From (8.3-7), the change $\Delta s_i$ in the root $s_i$ is related to the change $\Delta K$ by

$$\Delta s_i \cong \left(\frac{ds}{dK}\right)_{\substack{K=15\\ s=s_i}} \Delta K \tag{8.3-9}$$

To compute $ds/dK$, differentiate (8.3-8) with respect to $K$.

$$3s^2 \frac{ds}{dK} + 18s \frac{ds}{dK} + 23 \frac{ds}{dK} + 1 = 0$$

Solve for $ds/dK$.

$$\frac{ds}{dK} = \frac{-1}{3s^2 + 18s + 23} \tag{8.3-10}$$

Evaluate this derivative for each root. For $s_1 = -1$,

$$\frac{ds}{dK} = \frac{-1}{8}$$

and

$$\Delta s_1 = -\tfrac{1}{8}\Delta K = -\tfrac{1}{8}(16 - 15) = -0.125$$

Similarly for $s_2 = -3$ and $s_3 = -5$,

$$\Delta s_2 = \tfrac{1}{4}\Delta K = 0.25$$

$$\Delta s_2 = -\tfrac{1}{8}\Delta K = -0.125$$

The estimates of the roots when $K = 16$ are

$$s_1 = -1 + \Delta s_1 = -1 - 0.125 = -1.125$$

$$s_2 = -3 + \Delta s_2 = -3 + 0.25 = -2.75$$

$$s_3 = -5 + \Delta s_3 = -5 - 0.125 = -5.125$$

For comparison we present the true root values when $K = 16$. To four figures, these are $s = -1.139, -2.746$, and $-5.115$.

One of the principal reasons for using feedback is to reduce the effect of parameter variations. A measure of the effectiveness of the system in this regard is given by the system sensitivity $S_K^T$, defined as

$$S_K^T = \frac{\partial T}{\partial K}\frac{K}{T} \tag{8.3-11}$$

where $T(s)$ is the system's transfer function and $K$ is the parameter. The sensitivity is seen to be the ratio of the relative change in $T$ to the relative change in $K$. $S_K^T$ is a function of $s$ and is analyzed by letting $s = 0$ to yield the steady-state sensitivity or by letting $s = i\omega$ and applying frequency-response methods.

## 8.4 RESPONSE CALCULATIONS FROM THE ROOT LOCUS

Once the system's characteristic roots have been determined with the root-locus plot, the plot can be used to simplify the calculations required to obtain the transient and frequency responses. For a given input function, the transform of the system's output is expressed in a partial fraction expansion (see Appendix B). The coefficients of the expansion are easily calculated with the vector interpretation of complex numbers, and the transient response is obtained from this expansion. Frequency response is computed similarly, with the factors in the transfer function taken to be vector functions of the frequency $\omega$. Two examples are sufficient to demonstrate the techniques.

*Example 8.5*

Proportional control of a second-order system is shown in Figure 8.18. This system

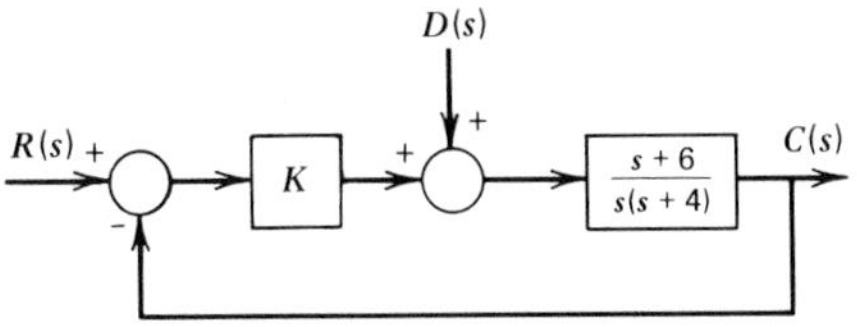


**Figure 8.18 Proportional control of a second-order system.**

has the circular locus shown in Figure 8.9*a*. Find the gain $K$ required to give a damping ratio of $\zeta = 0.9$. With this gain, compute the response to a unit step disturbance.

The desired damping ratio corresponds to a line making an angle of $\cos^{-1} 0.9 = 26°$ with the negative real axis. This line is shown in Figure 8.19*a*. It intersects the locus at two places. We choose the intersection at $s = -6.95 + i3.4$ because this gives the smallest time constant. The required gain is found from the locus to be

$$K = \frac{|s||s+4|}{|s+6|} = \frac{7.74(4.50)}{3.53} = 9.87$$

See Figure 8.19*a*.

The disturbance transfer function is

$$\frac{C(s)}{D(s)} = \frac{s+6}{s(s+4) + K(s+6)}$$

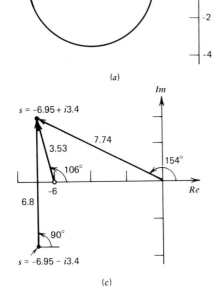

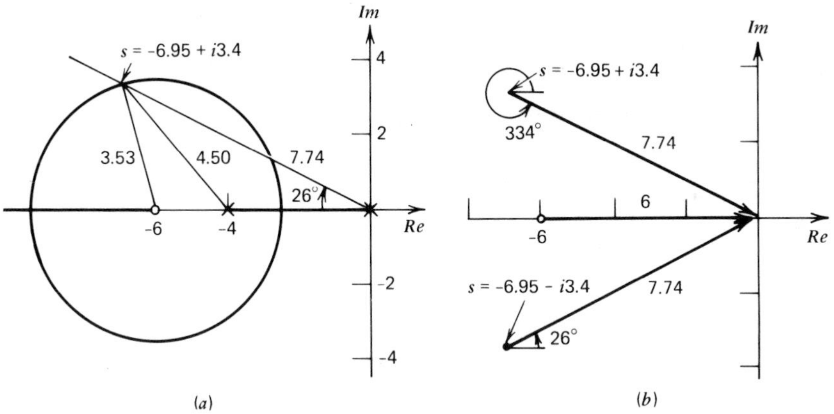


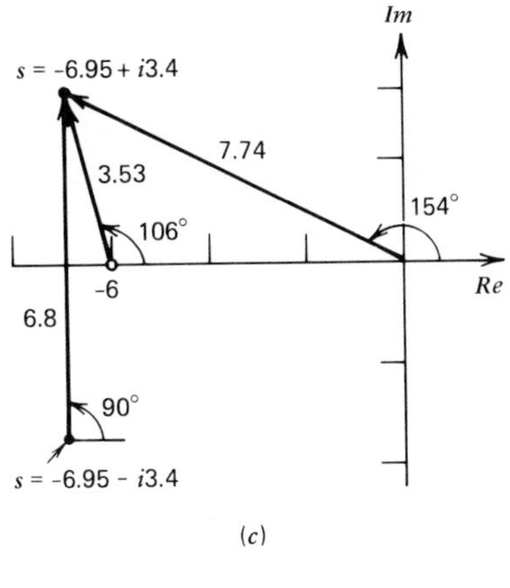


**Figure 8.19 Response calculations with the locus of the system in Figure 8.18. (*a*) Calculation of the gain at the point $s = -6.95 + i3.4$. (*b*) Calculation of the constant $C_1$ in (8.4-3). (*c*) Calculation of the constant $C_2$ in (8.4-4).**

With $K = 9.87$ and $D(s) = 1/s$, we obtain

$$C(s) = \frac{s+6}{s(s+6.95-i3.4)(s+6.95+i3.4)} \tag{8.4-1}$$

The partial fraction expansion is

$$C(s) = \frac{C_1}{s} + \frac{C_2}{s+6.95-i3.4} + \frac{C_3}{s+6.95+i3.4} \tag{8.4-2}$$

*Where*

$$C_1 = \lim_{s\to 0}\left[\frac{s+6}{(s+6.95-i3.4)(s+6.95+i3.4)}\right] \tag{8.4-3}$$

$$C_2 = \lim_{s\to -6.95+i3.4}\left[\frac{s+6}{s(s+6.95+i3.4)}\right] \tag{8.4-4}$$

$$C_3 = \text{complex conjugate of } C_2 \tag{8.4-5}$$

The constant $C_1$ can be evaluated using the vector representation of each factor with the tips of the vectors at $s = 0$. This is shown in Figure 8.19*b*. Thus if a complex number is represented by $r\measuredangle\phi$, where $r$ is the magnitude and $\phi$ the phase angle,

$$C_1 = \frac{6\measuredangle 0^\circ}{(7.74\measuredangle 334^\circ)(7.74\measuredangle 26^\circ)} = 0.100\measuredangle 0^\circ$$

where we have used the multiplication/division properties of complex numbers [see (A.5-15) and (A.5-16)].

Similarly, for $C_2$ the tips of the vectors are at $s = -6.95 + i3.4$ and we obtain

$$C_2 = \frac{3.53\measuredangle 106^\circ}{(7.74\measuredangle 154^\circ)(6.8\measuredangle 90^\circ)} = 0.067\measuredangle 222^\circ$$

See Figure 8.19*c*. Therefore $C_3 = 0.067\measuredangle 138^\circ$.

With these values the response is given by the inverse transform of (8.4-2).

$$c(t) = 0.1 + 0.067e^{i222^\circ}e^{(-6.95+i3.4)t} + 0.067e^{i138^\circ}e^{(-6.95-i3.4t)} \tag{8.4-6}$$

This can be reduced to the form of a sines and cosines by the techniques of Chapter Five. However, a simpler method is to replace the calculation of $C_2$ with the complex factor $R(-a+ib)$ developed in Section 5.7. See (5.7-23), (5.7-28), and (5.7-29). Comparing (8.4-1) with (5.7-23), we see that

$$C(s) = \frac{s+6}{s[(s+6.95)^2+(3.4)^2]} \tag{8.4-7}$$

and

$$P(s) = \frac{s+6}{s}$$

From (5.7-28),

$$R(-6.95+i3.4) = \lim_{s\to -6.95+i3.4}\left(\frac{s+6}{s}\right)$$

$$= \frac{3.53\measuredangle 106^\circ}{7.74\measuredangle 154^\circ} = 0.456\measuredangle 312^\circ$$

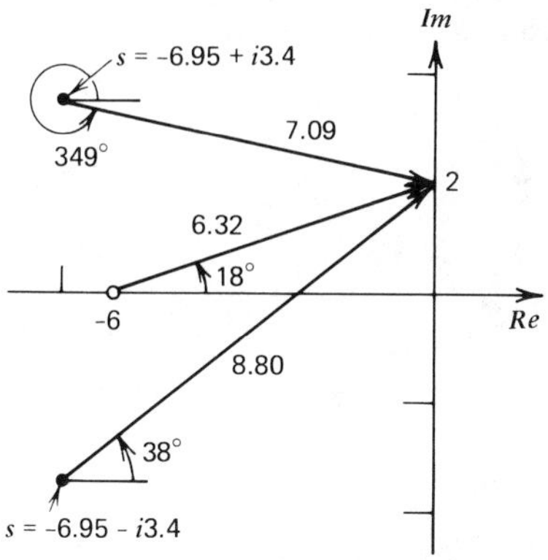


**Figure 8.20 Calculation of the frequency response from the root locus.**

where $a = 6.95$ and $b = 3.4$. From (5.7-29), the contribution to the response from the complex roots is

$$0.134e^{-6.95t}\sin(3.4t + 312°)$$

This is added to the response terms due to the other roots. Here this is $C_1/s$, and the response is

$$c(t) = 0.1 + 0.134e^{-6.95t}\sin(3.4t + 312°)$$

*Example 8.6*

Demonstrate how the closed-loop frequency response to the command $R(s)$ can be determined from the root-locus plot. Use the system shown in Figure 8.18 with the gain $K = 9.87$ as computed in the previous example.

The desired transfer function with $s = i\omega$ is

$$T(i\omega) = \frac{C(i\omega)}{R(i\omega)} = \frac{9.87(i\omega + 6)}{(i\omega + 6.95 - i3.4)(i\omega + 6.95 + i3.4)} \tag{8.4-8}$$

The vector for each factor has its tip at $s = i\omega$. We select a value for $\omega$ and compute the magnitude and phase angle for $T(i\omega)$ at this frequency. For example, at $\omega = 0$, the vectors are the same as those shown in Figure 8.19*b*. Thus

$$T(0) = \frac{9.87(6\measuredangle 0°)}{(7.74\measuredangle 334°)(7.74\measuredangle 26°)} = 0.989\measuredangle 0°$$

For $\omega = 2$, Figure 8.20 shows that

$$T(i2) = \frac{9.87(6.32\measuredangle 18°)}{(7.09\measuredangle 349°)(8.80\measuredangle 38°)} = 0.999\measuredangle -9°$$

The process is repeated until the desired range of frequencies is covered. The magnitude ratio and phase angle curves can then be plotted.

## 8.5 THE COMPLEMENTARY ROOT LOCUS

The development of the plotting guides was based on the angle and magnitude criteria, (8.1-23) and (8.1-24), with the assumption that the parameter $K$ is positive or zero.

We now consider the construction of the locus when $K \leqslant 0$ for the equation

$$1 + KP(s) = 0 \tag{8.5-1}$$

This is the *complementary root locus.* Its plotting guides are found from the same criteria, repeated here

$$|KP(s)| = 1 \tag{8.5-2}$$

$$\measuredangle KP(s) = \measuredangle -1 \tag{8.5-3}$$

If $K \leqslant 0$, (8.5-2) becomes

$$-K\,|P(s)| = 1$$

or

$$K = \frac{-1}{|P(s)|} \tag{8-5.4}$$

Thus the same scaling procedure is used, with a sign reversal included. Equation (8.5-3) becomes

$$\measuredangle K + \measuredangle P(s) = \measuredangle -1$$

or

$$180^\circ + \measuredangle P(s) = 180^\circ$$

The angle criterion reduces to

$$\measuredangle P(s) = n360^\circ, \quad n = 0, \pm 1, \pm 2, \ldots \tag{8.5-5}$$

For this reason the complementary locus is sometimes referred to as the 0° locus, while the primary locus ($K \geqslant 0$) is called the 180° locus.

The guides for plotting the complementary locus can be obtained from (8.5-4) and (8.5-5) in the same way as those for the primary locus. We limit ourselves to their statement and denote them as 1a, 2a, etc., to distinguish them from the guides for the primary locus. Any differences are noted in the list of guides. We continue to assume that $P(s)$ is a ratio of two finite polynomials with constant coefficients.

### *Guide 1a*

*The locus is symmetric about the real axis.*

### *Guide 2a*

*The number of loci equals the number of poles of* P(s).

### *Guide 3a*

*The loci start at the poles of* P(s) *with* K = *0, and terminate with* K = − ∞ *either at the zeros of* P(s) *or at infinity.*

### *Guide 4a*

*The locus exists in a section on the real axis only if the number of real poles and/or zeros to the right of the section is even; furthermore, it must exist there.* Thus the complementary locus exists on the real axis wherever the primary locus is absent. For the purpose of this guide, the number zero is even.

## *Guide 5a*

*The locations of breakaway and break-in points are found by determining where the parameter* K *attains a local minimum or maximum on the real axis.* This is the same as Guide 5 for the primary locus. The extraneous roots that were discarded previously when applying Guide 5 are now seen to be the breakaway or break-in points of the complementary locus. This occurred in Example 8.2 with the root $s = -1.58$.

## *Guide 6a*

*The loci that do not terminate at a zero approach infinity along asymptotes. The asymptotic angles relative to the real axis are found from*

$$\theta = \frac{n360°}{Z-P}, \quad n = 0, \pm 1, \pm 2, \ldots \tag{8.5-6}$$

*where* Z *and* P *are the number of zeros and poles, and* n *is increased until enough angles have been found.* Some of these possibilities are shown in Figure 8.21.

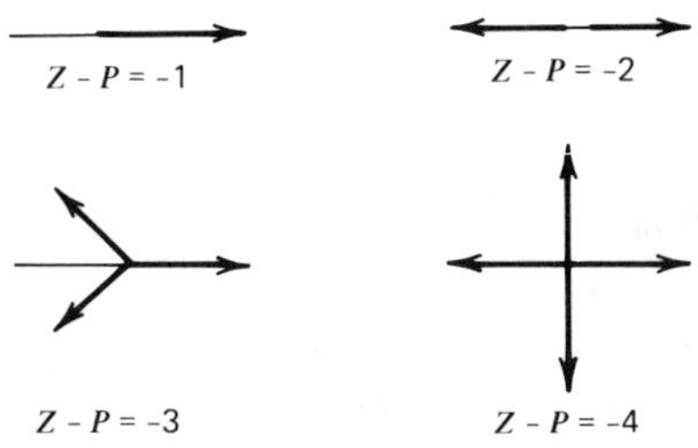


**Figure 8.21 Common asymptote configurations for $K \leqslant 0$.**

## *Guide 7a*

*All the asymptotes for both the primary and complementary loci intersect at a common point on the real axis. This point is given by (8.2-6), repeated here.*

$$\sigma = \frac{\Sigma s_p - \Sigma s_z}{P-Z} \tag{8.5-7}$$

*where the summations are for the values of the poles and zeros.*

## *Guide 8a*

*The points at which the loci cross the imaginary axis and the associated values of* K *can be found as with Guide 8, namely with the Routh-Hurwitz criterion or by substituting* $s = i\omega$ *into (8.5-1).*

## *Guide 9a*

*The angles of departure from complex poles, and the angles of arrival at complex zeros, can be found by applying the angle criterion (8.5-5) to a trial point infinitesimally close to the pole or zero in question.* Note that the angle criterion for the complementary locus must be used in this case.

*Guide 10a*

*This is the same as Guide 10. See* (*8.2-9*) *and* (*8.2-10*). This guide does not result from the angle and magnitude criteria.

*Guide 11a*

*Once the complementary locus is drawn, it is scaled with the magnitude criterion* (*8.5-4*).

## Examples

The primary root locus of the equation

$$1 + \frac{K}{s(s+1)} \tag{8.5-8}$$

is shown in Figure 8.1$a$. Its complementary locus is easily obtained from the complementary guides and is shown in Figure 8.22.

For the equation

$$1 + \frac{Ks}{s^2 + s + 1} = 0 \tag{8.5-9}$$

the primary locus is given by Figure 8.2$a$. The solutions for the breakaway or break-in points are $s = \pm 1$. The break-in point at $s = -1$ is for the primary locus, while that at $s = +1$ is for the complementary locus. At $s = 1$, $K = -3$. For $K < 0$ Guide 9a gives the departure angle as $\phi = 120° - 90° - n360°$, or $\phi_1 = 30°$. From Guide 8a, the crossover frequency is $\omega = 1$ at $K = -1$. The entire locus off the real axis is seen to be a unit circle centered at $s = 0$, as shown in Figure 8.23.

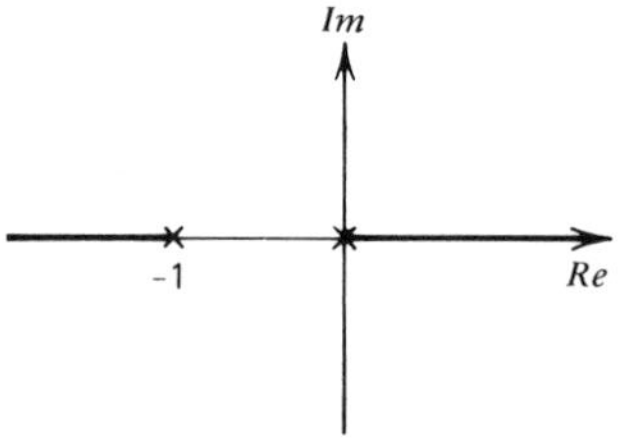


**Figure 8.22** **Root locus for $s(s+1) + K = 0, K \leq 0$.**

The latter example illustrates the usual situation in which the complementary locus is useful. Control system gains are usually positive for physical reasons, and the

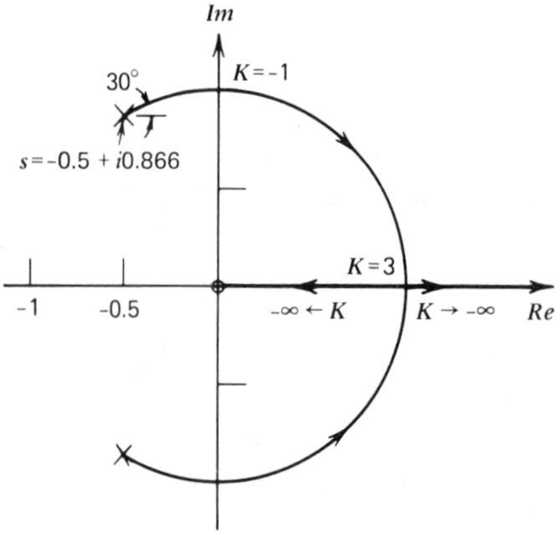


**Figure 8.23** **Root locus for $s(s+1) + Ks = 0, K \leq 0$.**

primary locus can be used to determine the proper gain values. Then a sensitivity study is often performed to ascertain the effects of the variation or uncertainty in another parameter. For example, the primary locus in Figure 8.1*a* was used to set the gain $k$ to obtain the desired response. With $k$ set equal to 1, the coefficient $c$ in (8.1-1) was allowed to vary about its nominal value of 1. Figure 8.2*a* is the locus with the variation $\Delta$ as the parameter. It is valid only for $\Delta \geqslant 0$. The complementary locus given by Figure 8.23 for $\Delta \leqslant 0$ is required to obtain the effects of variations in $c$ below the nominal value of 1. Normally, we are interested only in relatively small variations about the nominal, so the plot need not be constructed for the entire range $-\infty \leqslant \Delta \leqslant +\infty$.

Another application of the complementary root locus is in the analysis of *non-minimum phase* systems, which possess poles and zeros in the right-half plane. The name is descriptive only of the frequency-response characteristics (see Reference 3). A positive feedback system can be nonminimum phase. Another example is the *reverse-reaction process.* The response of this process to a positive step input exhibits a negative initial slope, for zero initial conditions. This is a model of some boiling processes in which the liquid level temporarily drops when bubbles collapse as the result of an input of cold liquid.

### *Example 8.7*

Proportional control with a gain $K_p$ is to be applied to a reverse-reaction process described by

$$\frac{1-s}{(s+1)(s+2)}$$

Plot the root locus for $K_p \geqslant 0$.

The characteristic equation for the system with P control is

$$1+\frac{-K_p(s-1)}{(s+1)(s+2)}=0 \qquad \textbf{(8.5-10)}$$

The root-locus parameter is $K=-K_p$. Since $K \leqslant 0$, we use the plotting guides for the complementary root locus. The breakaway point is at $s=-1.45$, and the break-in

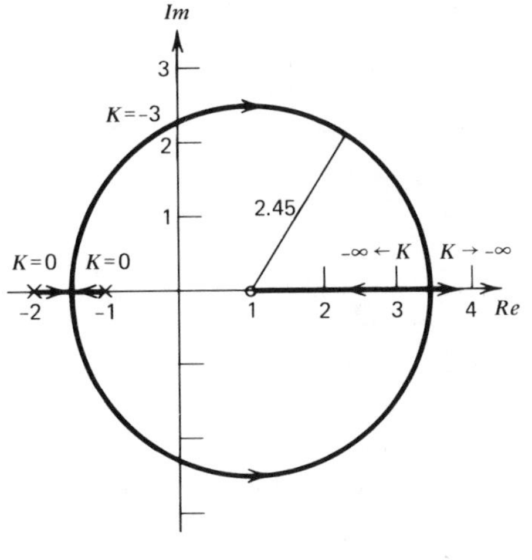


**Figure 8.24** Root locus for the reverse reaction process $(s+1)(s+2)+K(s-1)=0, K \geqslant 0$.

point is at $s = 3.45$. The locus is shown in Figure 8.24. The control system is unstable if $K < -3$; that is, if $K_p > 3$.

## 8.6 ROOT LOCUS FOR SYSTEMS WITH DEAD-TIME ELEMENTS

Some systems have an unavoidable time delay in the signal flow between components. The delay usually results from the physical separation of the components and typically occurs as a delay between a change in the manipulated variable and its effect on the plant, or as a delay in the measurement of the output. Examples of each as shown in Figures 8.25 and 8.26.

The first shows a temperature-control system in which hot air is delivered to the space to be controlled. The heating element is located a distance $L$ from the plant. This distance can be long, as in large buildings heated by a central source. The velocity of the air in the duct is $v$, so that a delay $D = L/v$ exists between a change in the temperature of the air leaving the heating element and its effect on the temperature of the space. If the plant is modeled as a first-order system, a proportional controller would have the block diagram shown. The heat flow rate from the heater is $q_1(t)$, and that affecting the plant is $q_2(t)$. Thus $q_2(t) = q_1(t - D)$, so that $Q_2(s) = e^{-sD}\,Q_1(s)$.

The second example involves a measurement delay (Figure 8.26). Ribbons of hot metal are drawn through rollers at high speed to produce sheet metal of a desired thickness. The upper roller's position is controlled to obtain this thickness. For obvious reasons the thickness sensor cannot be placed at the rollers; therefore it is placed downstream. The delay between the measured thickness $c(t - D)$ and the

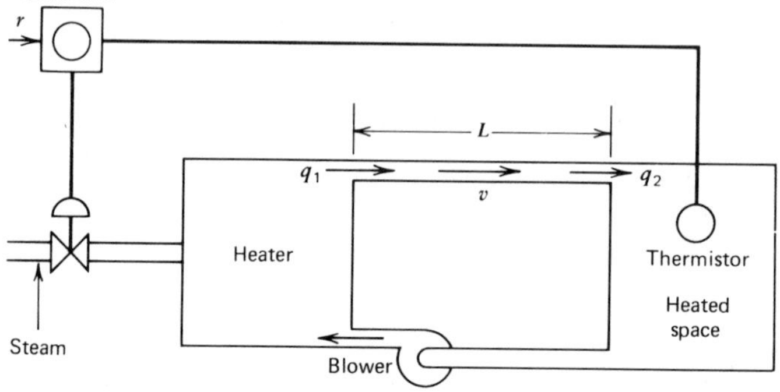


(a)

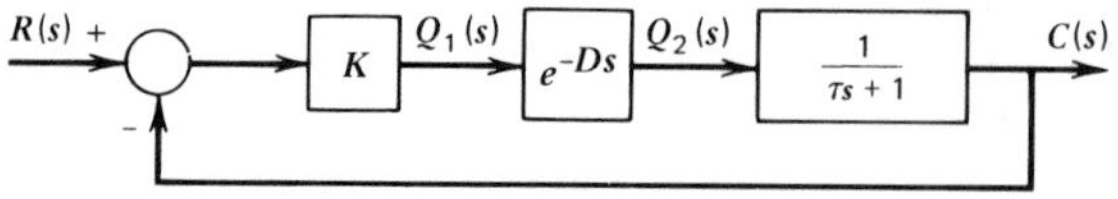


(b)

**Figure 8.25** **(a) Temperature controller with dead time in the forward path. (b) Block diagram for proportional control and a first-order plant.**

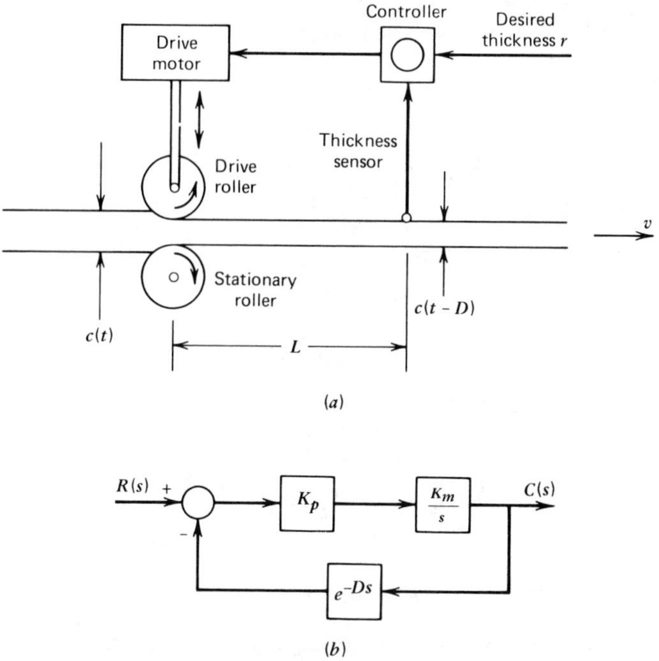


**Figure 8.26 (*a*) System for metal rolling. (*b*) Thickness-control system with proportional action and dead time in the feedback loop.**

thickness $c(t)$ at the rollers is $D = L/v$ and can be significant because the high ribbon velocity requires a fast control system. The drive motor and roller displacement can be modeled as an integrator. The system diagram for proportional control is given in Figure 8.26*b*. The delay element now appears in the feedback path.

The time delay described in the examples mentioned is variously referred to as *dead time*, *transportation lag*, or *pure time delay* to distinguish it from the delay associated with the time constant of a first-order element. We will see that the presence of such delays tends to cause oscillations and to make the system less stable. They should be avoided or reduced if at all possible. This cannot always be done, so we must know how to analyze systems with dead time. The root locus provides one means of doing so.

## Plotting Guides for Dead-Time Elements

The basic equation studied with the root locus previously is (8.1-12). Here we generalize its form to include the effect of dead time. Our basic form is

$$1 + KP(s)e^{-Ds} = 0 \tag{8.6-1}$$

where $P(s)$ is the ratio of two polynomials with constant coefficients. Write $s$ as $s = \sigma + i\omega$ to show that

$$|e^{-Ds}| = |e^{-D\sigma}e^{-i\omega D}| = e^{-D\sigma} \tag{8.6-2}$$

since

$$|e^{-i\omega D}| = |\cos \omega D - i \sin \omega D| = 1 \tag{8.6-3}$$

Therefore the magnitude criterion for (8.6-1) is

$$|K| = \frac{1}{e^{-D\sigma}|P(s)|} \tag{8.6-4}$$

The angle criterion of (8.6-1) becomes

$$\measuredangle K + \measuredangle P(s) + \measuredangle e^{-Ds} = \measuredangle -1 \tag{8.6-5}$$

Note that

$$\begin{aligned} \measuredangle e^{-Ds} &= \measuredangle e^{-D\sigma} + \measuredangle e^{-i\omega D} \\ &= 0^\circ + \measuredangle(\cos \omega D - i \sin \omega D) \\ &= -\omega D \end{aligned} \tag{8.6-6}$$

Thus for $K \geqslant 0$, (8.6-5) gives

$$\measuredangle P(s) = n180^\circ + \omega D, \quad n = \pm 1, \pm 3, \ldots \tag{8.6-7}$$

For $K \leqslant 0$ we get

$$\measuredangle P(s) = n360^\circ + \omega D, \quad n = 0, \pm 1, \pm 2, \ldots \tag{8.6-8}$$

The presence of the imaginary part $\omega$ means that the angle criterion depends on the root location. If $D = 0$, there are $n$ points in the $s$-plane that satisfy the angle criterion, where $n$ is the order of the denominator of $P(s)$. However, if $D \neq 0$, the presence of $\omega$ in the angle criterion means that there will be an infinite number of roots for any given $K$ value.

The plotting guides can be developed as before from the magnitude and angle criterion. We do not state all of the guides explicitly here, because some of them are identical to the case where $D = 0$. For example, the symmetry property of Guide 1 is unchanged. The guides for finding the locus on the real axis (Guide 4) are unchanged because $\omega = 0$ on this axis, and thus the angle criterion is identical to the case with no dead time. Similarly, the procedure for determining the locations of breakaway and break-in points remains the same (Guide 5). Also, the crossover points are found with the substitution $s = i\omega$, but the Routh-Hurwitz criterion does not apply (Guide 8). Finally, Guides 9 and 11 apply without changes.

The remaining guides require some modifications. To distinguish them, we denote them as Guide 2b, 3b, etc., and simultaneously consider the case for $K \geqslant 0$ and $K \leqslant 0$.

### *Guide 2b*

*The number of loci is infinite for a dead-time system.* This follows from the discussion of the angle criterion.

### *Guide 3b*

*The loci start at the poles of* P(s) *and at* $\sigma = -\infty$, *with* $K = 0$, *and terminate at the zeroes of* P(s) *or at* $\sigma = +\infty$, *with* $K = \pm\infty$. The proof of this guide is similar to that of Guide 3, except that the term $e^{-D\sigma}$ now appears in the magnitude criterion (8.6-4).

From this equation we see that $K = 0$ when $\sigma = -\infty$. Similarly, when $\sigma = +\infty, K = \pm\infty$. Note that we must determine the asymptotes that describe the starting location at $\sigma = -\infty$. These are found with Guide 6b.

### *Guide 6b*

*The loci that do not terminate at a zero approach infinity along asymptotes. There is an infinite number of asymptotes, and they are all parallel to the real axis. The intersections of these asymptotes and the* $K = 0$ *asymptotes with the imaginary axis are given by*

$$\omega = \frac{N\pi}{D} \tag{8.6-9}$$

*Let* P *be the number of finite poles and* Z *the number of finite zeros. Then choose* N *as follows. Let* n *take the values* $n = \pm 1, \pm 3, \ldots,$ *and* m *the values* $m = 0, \pm 2, \pm 4, \ldots$. *For the termination asymptotes* ($K = \pm\infty$),

1. $K \geqslant 0$

$$N = n \tag{8.6-10}$$

2. $K \leqslant 0$

$$N = m \tag{8.6-11}$$

*For the starting asymptotes* ($K = 0$),

3. $K = 0+$

$$N = \begin{cases} n & \text{if} \quad P - Z \text{ even} \\ m & \text{if} \quad P - Z \text{ odd} \end{cases} \tag{8.6-12}$$

4. $K = 0-$

$$N = \begin{cases} m & \text{if} \quad P - Z \text{ even} \\ n & \text{if} \quad P - Z \text{ odd} \end{cases} \tag{8.6-13}$$

Note that Guide 7 does not apply because the asymptotes do not intersect the real axis. With respect to Guide 8, we note that there will be an infinite number of intersections of the locus with the imaginary axis. In general, we need not find all of these because the dominant roots of the system usually lie on the first path to cross the imaginary axis. This is shown in Example 8.8. Guide 10 applies only to polynomials with a finite number of roots and cannot be applied to a transcendental equation like (8.6-1).

### *Guide 11b*

*Once the root locus is drawn, it is scaled with the magnitude criterion* (8.6-4).

### *Example 8.8*

Plot the root locus for the equation

$$1 + \frac{Ke^{-Ds}}{s} = 0 \tag{8.6-14}$$

The finite pole is $s = 0$, and there are no finite zeros. The locus for $K > 0$ exists on the real axis to the left of $s = 0$, and that for $K < 0$ exists to the right. On the real axis, with $K \geqslant 0$, the locus starts at $s = 0$ *and also* at $s = -\infty$. These two paths move toward each other, so there must be a breakaway point. This is found from

$$\frac{dK}{ds} = \frac{d}{ds}(-se^{Ds}) = 0$$

or

$$-Ds\,e^{Ds} - e^{Ds} = 0$$

This gives

$$s = -\frac{1}{D}$$

This path will approach one of the asymptotes. To find these, we use Guide 6b. For this problem, $P - Z = 1$. Thus the termination asymptotes intersect the imaginary axis at $\omega = N\pi/D$ where

$$N = 0, \pm 2, \pm 4, \ldots \qquad \text{for } K \leqslant 0$$

$$N = \pm 1, \pm 3, \ldots \qquad \text{for } K \geqslant 0$$

The starting asymptotes intersect the axis at the points given by

$$N = \pm 1, \pm 3, \ldots \qquad \text{for } K \leqslant 0$$

$$N = 0, \pm 2, \pm 4, \ldots \qquad \text{for } K \geqslant 0$$

These are shown by dotted lines in Figure 8.27.

To find the crossover point, let $s = i\omega$ to obtain from (8.6-14),

$$i\omega + K e^{-i\omega D} = 0$$

or

$$i(\omega - K \sin \omega D) + K \cos \omega D = 0$$

The crossover points are infinite in number and are given by

$$\cos \omega D = 0$$

$$K = \frac{\omega}{\sin \omega D}$$

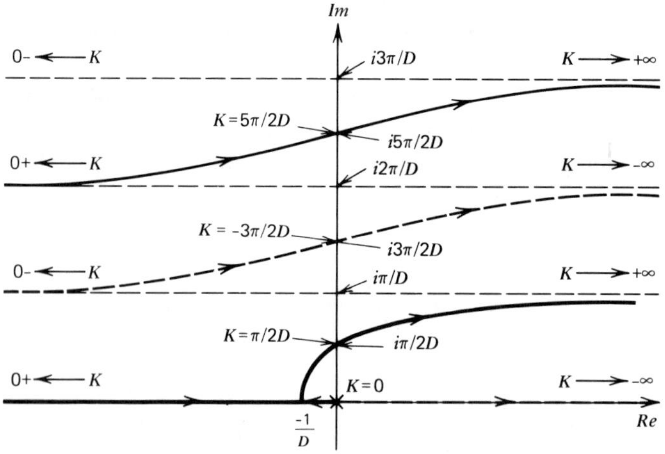


**Figure 8.27 Root-locus plot for $s + Ke^{-sD} = 0, -\infty \leqslant K \leqslant \infty$.**

Thus if $K > 0$, the crossover points for $\omega > 0$ are

$$\omega D = \frac{\pi}{2}, \frac{5\pi}{2}, \ldots$$

$$K = \omega$$

For $K < 0$, they are

$$\omega D = \frac{3\pi}{2}, \frac{7\pi}{2}, \ldots$$

$$K = -\omega$$

for $\omega > 0$. For $\omega < 0$, the crossing points are the negative of the preceding values.

The complete root locus can now be sketched. It is shown in Figure 8.27, where the heavy dotted lines are the loci for $K < 0$. The system is stable if $0 < K < \pi/2D$, the value for the gain at the first crossover point. The plot shows that the roots on the branches that lie outside the region $-\pi \leqslant \omega D \leqslant \pi$ are far to the left of the roots on the primary branch for stable values of $K$. Thus we usually consider only the primary branch when choosing a value for $K$. The selection of $K$ is made in the same way as before.

The root locus shows that the system will always exhibit oscillatory behavior because of the infinite number of paths with nonzero imaginary parts. However, if the roots on the primary branch are selected to be real, the oscillations will not dominate the response.

## 8.7 SYSTEM DESIGN WITH OPEN-LOOP FREQUENCY-RESPONSE PLOTS

Although we already have some powerful methods for designing linear control systems, there are several advantages to performing the design with the system's open-loop frequency-response characteristics. First, frequency-response data are often easier to obtain experimentally and are especially useful when it is difficult to develop a transfer-function model for the plant and actuators from basic principles. Second, the method to be developed is easier to use for higher-order systems with dead-time elements than is the root-locus method. Third, a compensator to improve the system's response is sometimes more easily designed with this approach. Finally, this technique is sometimes useful for determining the existence of limit cycles and instability in nonlinear systems. We will pursue the first three topics; the last consists of a group of specialized methods whose applications are limited to specific problem types (see Chapter 12 of Reference 3).

### The Nyquist Stability Theorem

Recall that a plot of the frequency-transfer function $T(i\omega)$ in vector form is the polar plot. On it we plot the location of the tip of the vector as $\omega$ varies from 0 to $\infty$. The axes of the plot are the real and imaginary parts of $T(i\omega)$. Thus the angle and magnitude of $T(i\omega)$ are represented on the same plot.

The *Nyquist stability theorem* is a powerful tool for linear system analysis. Its

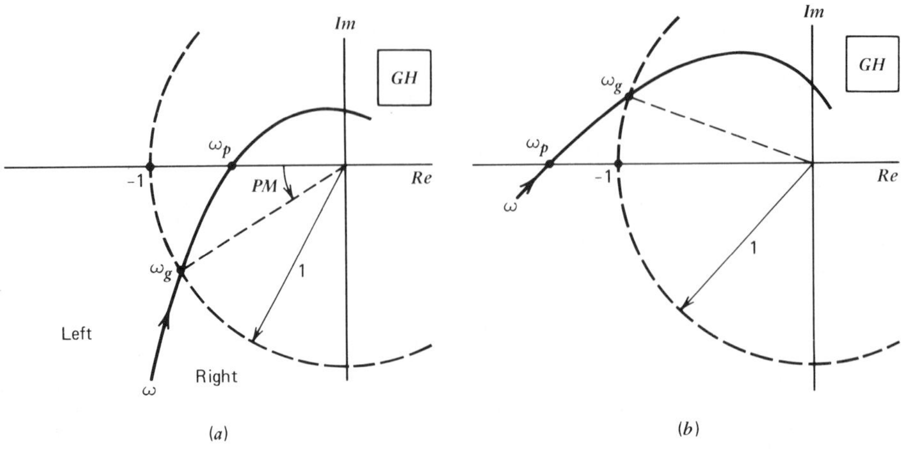


**Figure 8.28 Nyquist plots. (*a*) Stable system. (*b*) Unstable system.**

proof requires considerable development and will not be attempted here (see Reference 4). Instead, we will concentrate on the aspect of the theorem that is useful for control systems design. In many control applications, the plant has no poles with positive real parts, and typical controllers do not introduce poles of this type. If the open-loop system has no poles with positive real parts, we can concentrate our attention on the region around the point $-1 + i0$ on the polar plot of the *open-loop* transfer function. Because of its relation to the theorem, the polar plot is also known as the *Nyquist plot.*

Figure 8.28 shows the polar plot of the open-loop transfer function of two arbitrary systems, both of which are assumed to be open-loop stable. The Nyquist stability theorem is stated as follows.

The system is closed-loop stable if and only if the point $-1 + i0$ lies to the left of the open-loop Nyquist plot relative to an observer traveling along the plot in the direction of increasing frequency $\omega$.

Therefore the system described by Figure 8.28*a* is closed-loop stable. A system exhibits sustained oscillations (neutral stability) if the plot passes through the $-1 + i0$ point. It is unstable if the point lies to the right of the plot, as in Figure 8.28*b*.

The basis for the Nyquist theorem is easily understood. Let $e(t)$ and $b(t)$ be the actuating and feedback signals, respectively (see Figure 8.29). Suppose that the input $r(t)$ is sinusoidal with frequency $\omega_f$, and assume that the phase angle of the open-loop transfer function is $-180°$. This means that at steady state, the signals $e(t)$ and $b(t)$ will be sinusoidal with the frequency $\omega_f$ and will be 180° out of phase with each other [when $e(t)$ reaches a peak, $b(t)$ is at a minimum, and vice versa]. When the signal $b(t)$ has its sign changed at the comparator, the result, $-b(t)$, will be in phase with the signal $e(t)$.

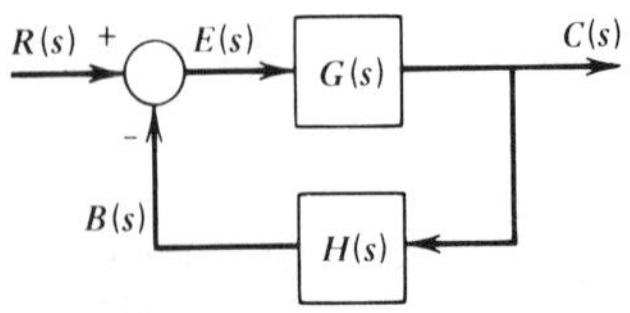


**Figure 8.29 Single-loop system used in the development of the Nyquist stability theorem.**

Imagine that the input is now switched off. When the open-loop plot passes through the $-1 + i0$ point at the frequency $\omega_f$, the gain relating $b(t)$ to $e(t)$ is unity at this frequency. Therefore, $e(t) = -b(t)$ and the original oscillation will be sustained at a constant amplitude. If the $-1 + i0$ point lies to the left of the plot, the gain between $e(t)$ and $b(t)$ will be less than unity. In this case the amplitude of $b(t)$ will be less than that of $e(t)$, and the oscillations will gradually diminish. The system is then stable. Finally, if the $-1 + i0$ point lies to the right of the plot, the amplitude of $b(t)$ will be larger than that of $e(t)$, and the oscillations will grow in magnitude as the signal travels around the loop. The system is then unstable.

## Phase and Gain Margins

The Nyquist theorem provides a convenient measure of the relative stability of a system. A measure of the proximity of the plot to the $-1 + i0$ point is given by the angle between the negative real axis and a line from the origin to the point where the plot crosses the unit circle (see Figure 8.28*a*). The frequency corresponding to this intersection is denoted $\omega_g$. This angle is the *phase margin* (PM) and is positive when measured down from the negative real axis. The absence of a positive or zero phase margin thus indicates an unstable system (Figure 8.28*b*). The phase margin is the phase at the frequency $\omega_g$ where the magnitude ratio or "gain" of $G(i\omega)H(i\omega)$ is unity (0 db).

The frequency $\omega_p$, the phase crossover frequency, is the frequency at which the phase angle is $-180°$. The *gain margin* (GM) is the difference in decibels between the unity gain condition (0 db) and the value of $|GH|$ db at the phase crossover frequency. Thus

$$\text{gain margin} = -|G(i\omega_p)H(i\omega_p)| \text{ db} \tag{8.7-1}$$

A system is stable only if the phase and gain margins are both positive.

H. W. Bode contributed greatly to the development of design methods based on the open-loop frequency response. For this reason the rectangular frequency-response plots are sometimes called *Bode plots.* The phase and gain margins can be illustrated on the Bode plots shown in Figure 8.30*a*. The situation shown in Figure 8.30*b* represents an unstable system that lacks positive gain and phase margins.

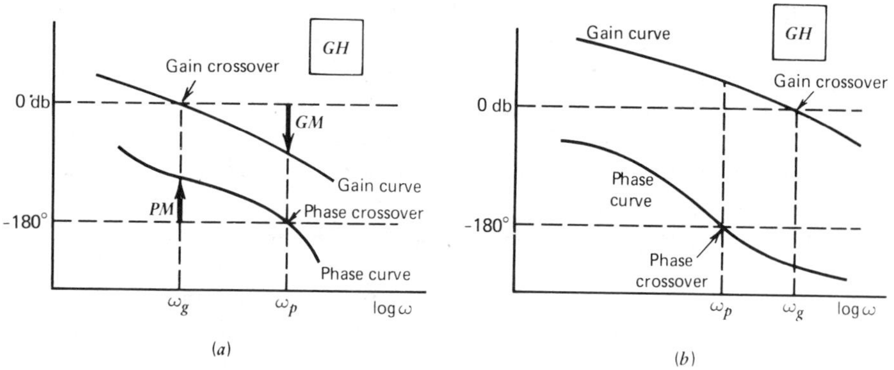


Figure 8.30 **Bode plots. (*a*) Stable system. (*b*) Unstable system.**

The phase and gain margins can be stated as safety margins in the design specifications. A typical set of such specifications is as follows.

$$\text{gain margin} \geqslant 8 \text{ db} \quad \text{and} \quad \text{phase margin} \geqslant 30^\circ \tag{8.7-2}$$

In common design situations, only one of these equalities can be met, and the other margin is allowed to be greater than its minimum value. It is not desirable to make the margins too large since this results in a low gain. The design will be sluggish and might have a large steady-state error. We note that another common set of specifications is

$$\text{gain margin} \geqslant 6 \text{ db} \quad \text{and} \quad \text{phase margin} \geqslant 40^\circ \tag{8.7-3}$$

The 6 db limit corresponds to the quarter amplitude decay response obtained with the gain settings given by the Ziegler-Nichols ultimate-cycle method (Table 7.1).

## Open-Loop Design for PID-Control

Some general comments can be made about the effects of proportional, integral, and derivative control actions on the phase and gain margins. P action does not affect the phase curve at all and thus can be used to raise or lower the open-loop gain curve until the specifications for the gain and phase margins are satisfied. If I action or D action is included, the proportional gain is selected last. Therefore, when using this approach to the design, it is best to write the PID algorithm with the proportional gain factored out, as

$$F(s) = K_p\left(1 + \frac{1}{T_I s} + T_D s\right)E(s) \tag{8.7-4}$$

D action affects both the phase and gain curves. Therefore, the selection of the derivative gain is more difficult than the proportional gain. The increase in phase margin due to the positive phase angle introduced by D action is partly negated by the derivative gain, which reduces the gain margin. Increasing the derivative gain increases the speed of response, makes the system more stable, and allows a larger proportional gain to be used to improve the system's accuracy. However, if the phase curve is too steep near $-180^\circ$, it is difficult to use D action to improve the performance.

I action also affects both the gain and phase curves. It can be used to increase the open-loop gain at low frequencies. However, it lowers the phase crossover frequency $\omega_p$ and thus reduces some of the benefits provided by D action. If required, the D-action term is usually designed first, followed by I action and P action, resepectively.

The classical design methods based on the Bode plots obviously have a large component of trial and error because usually both the phase and gain curves must be manipulated to achieve an acceptable design. Given the same set of specifications, two designers can use these methods and arrive at substantially different designs. Many rules of thumb and ad hoc procedures have been developed, but a general foolproof procedure does not exist. Many such procedures and examples can be found in older references and in the technical literature (see References 5 through 12). An experienced designer can often obtain a good design quickly with these techniques. The use of a computer plotting routine greatly speeds up the design process.

Some applications of this method are given in Section 8.8. We now consider how it can be used with dead-time systems.

## Systems With Dead-Time Elements

The Nyquist theorem is particularly useful for systems with dead-time elements, especially when the plant is of an order high enough to make the root-locus method cumbersome. A delay in either the manipulated variable or the measurement will result in an open-loop transfer function of the form

$$G(s)H(s) = e^{-Ds}P(s) \tag{8.7-5}$$

For this case, (8.6-3) and (8.6-6) show that

$$|G(i\omega)H(i\omega)| = |P(i\omega)||e^{-i\omega D}| = |P(i\omega)| \tag{8.7-6}$$

and

$$\measuredangle G(i\omega)H(i\omega) = \measuredangle P(i\omega) - \measuredangle e^{-i\omega D}$$

$$= \measuredangle P(i\omega) - \omega D \tag{8.7-7}$$

Thus the dead time decreases the phase proportionally to the frequency $\omega$, but it does not change the gain curve. This makes the analysis of its effects easier to accomplish with the open-loop frequency response plot.

The following example shows how these methods can be applied to a system whose transfer function cannot be obtained analytically.

### *Example 8.9*

The frequency response for a particular plant was determined experimentally, and the results are shown in Figure 8.31*a*. It is intended to use proportional control for this plant with a unity feedback loop as shown in Figure 8.31*b*. It is known that there will be dead time between the controller action and its effect on the plant.

(a) Design the controller to achieve the specifications given by (8.7-2), namely, $GM \geqslant 8$ db and $PM \geqslant 30^\circ$.

(b) How large can the dead time be before the system becomes unstable?

(a) Neglect the dead-time for now and design the controller. The plot in Figure 8.31*a* shows that if the gain $K$ is unity, the phase crossover frequency $\omega_p$ is 0.01 rad/sec, and the gain crossover occurs at $\omega_g = 0.003$ rad/sec. The phase margin will be $70^\circ$ and the gain margin 10.5 db.

Increasing the value of $K$ above unity affects only the gain curve. This curve can be raised by $10.5 - 8 = 2.5$ db without violating the specifications given for $PM$ and $GM$. Thus, for $GM = 8$ db, $K$ must be such that $20 \log K = 2.5$, or $K = 1.334$. With this value of $K$, the phase margin becomes $60^\circ$, and the new gain crossover frequency is $\omega_g = 0.0045$ rad/sec. This can be seen by translating the gain curve upward by 2.5 db.

(b) From (8.7-6) and (8.7-7) we see that the dead time affects only the phase margin. The system is stable only if the phase margin is positive. This occurs only if the dead time's contribution to the phase curve at the *new* gain crossover frequency is greater than $-60^\circ$. The stability requirement thus is

$$0.0045D \leqslant 60^\circ = 1.047 \text{ rad}$$

or

$$D \leqslant 232.7 \text{ sec}$$

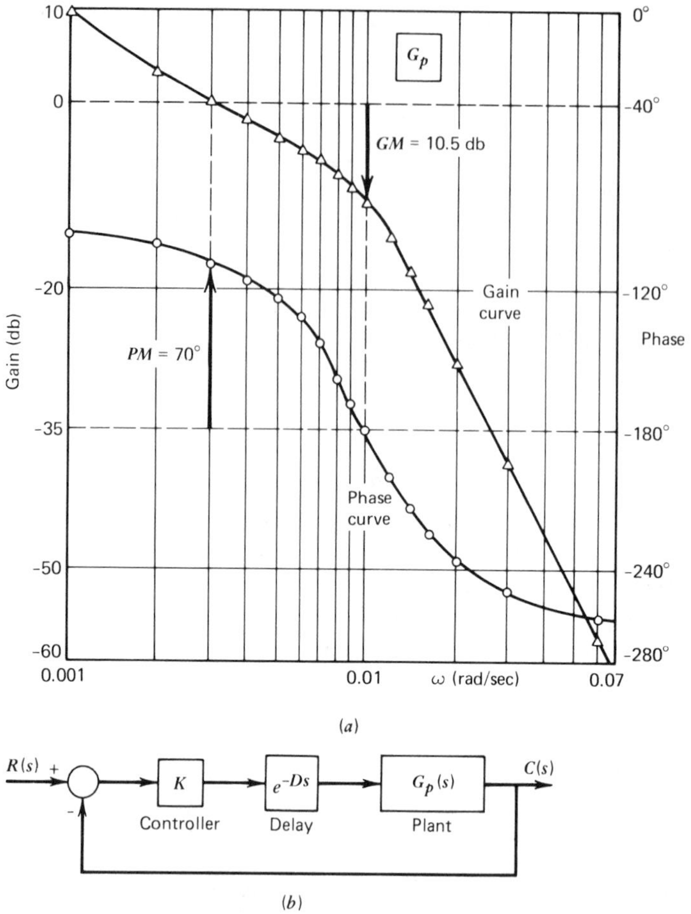


**Figure 8.31** **Design information for Example 8.9. (*a*) Experimentally determined open-loop frequency response for the plant. (*b*) Proportional controller with unity feedback and dead time.**

## Closed-Loop Response and the Nichols Chart

We have seen how the open-loop Bode diagrams can be used to set the parameters of the controller. Once this is done, computation of the closed-loop response can be made with the same diagrams. The system shown in Figure 8.29 has the closed-loop frequency transfer function

$$T_c(i\omega) = \frac{G(i\omega)}{1 + G(i\omega)H(i\omega)} \tag{8.7-8}$$

Define the magnitude and phase of this transfer function as

$$m_c = |T_c(i\omega)| \text{ db} \tag{8.7-9}$$

$$\phi_c = \measuredangle T_c(i\omega) \tag{8.7-10}$$

From (8.7-8),

$$T_c(i\omega) = \frac{G(i\omega)H(i\omega)}{1 + G(i\omega)H(i\omega)} \frac{1}{H(i\omega)} \tag{8.7-11}$$

and thus

$$m_c = \frac{G(i\omega)H(i\omega)}{1 + G(i\omega)H(i\omega)} \text{ db} - |H(i\omega)| \text{ db} \tag{8.7-12}$$

$$\phi_c = \measuredangle \frac{G(i\omega)H(i\omega)}{1 + G(i\omega)H(i\omega)} - \measuredangle H(i\omega) \tag{8.7-13}$$

If a computer is used to calculate the complex numbers $G(i\omega)$ and $H(i\omega)$ to generate the open-loop plots, these numbers can also be used in (8.7-12) and (8.7-13) to generate the closed-loop plots simultaneously. If the system has unity feedback $[H(s) = 1]$, (8.7-12) and (8.7-13) are simplified considerably.

The *Nichols chart* provides a manual technique for obtaining closed-loop response information from the open-loop data (Reference 13). The use of the chart is usually stated in terms of a unity feedback system. If nonunity feedback exists, the block diagram can be rearranged to give an equivalent unity feedback system. With $H(s) = 1$, define $m_o$ and $\phi_o$ to be the magnitude and phase of the open-loop transfer function.

$$m_o = |G(i\omega)| \text{ db} \tag{8.7-14}$$

$$\phi_o = \measuredangle G(i\omega) \tag{8.7-15}$$

The Nichols chart is a plot of $m_o$ vs. $\phi_o$. Superimposed on the chart are contours of constant $m_c$ and $\phi_c$ values. Figure 8.32 shows a Nichols chart with a few contours displayed. Most design work requires many contours. These can be obtained with commercially available templates (for example, Berol's Rapidesign #331) or by computer-generated plots.

The procedure for obtaining the closed-loop response from the Nichols chart is as follows.

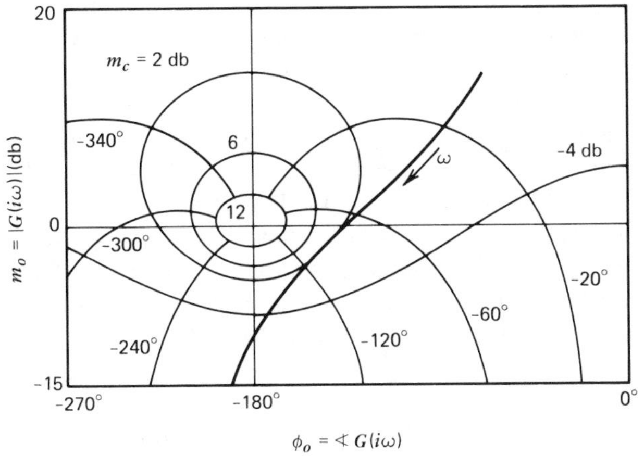


**Figure 8.32 Nichols chart. The heavy line is a plot of the open-loop response for a particular system.**

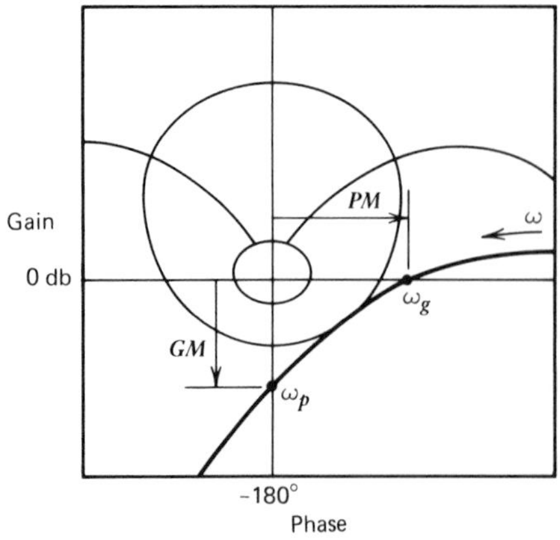


**Figure 8.33 Interpretation of phase and gain margins on the Nichols chart.**

1. From the open-loop Bode plots, the values of $m_o$ and $\phi_o$ can be read off for a particular frequency. These are used to plot a point on the Nichols chart. This step is repeated for a number of frequency values. The result is a single curve on the Nichols chart, such as is shown by the heavy line in Figure 8.32.
2. From the intersection of this curve with the contours at particular frequencies, values of $m_c$ and $\phi_c$ can be read off the plot. Use these values to construct the closed-loop Bode plots.

This procedure can be modified to handle systems with nonunity feedback. Equations (8.7-12) and (8.7-13) provide the basis for this modification.

The intersections of the closed-loop response curve with the magnitude and phase axes are given by the phase and gain margins, respectively (see Figure 8.33). Thus the curve for a stable system must pass below the origin of the magnitude-phase axes. Interpretations for unstable systems are meaningless because the frequency-response plots are not defined for such systems, since a steady-state does not exist.

Construction of the closed-loop response plots from the Nichols chart can be tedious. However, the chart is useful for determining the gain required to achieve a specified maximum (resonance) peak value $m_{cp}$ in the closed-loop response. It can be shown with a Fourier series analysis that the frequency components of an input signal near the resonance peak dominate the system's response (Reference 3, Section 9-4). Thus the specification of this peak value is an important design criterion. The control gain required to achieve the desired $m_{cp}$ is found by plotting the closed-loop response line on the chart for a unity gain. The amount of gain that must be added (in decibels) is then found from the number of decibels by which the line must be translated to become tangent to the $m_{cp}$ contour. A specification in common use is that $M_{cp}$ should be less than 2 ($m_{cp}$ less than 6 db). A typical value is $M_{cp} = 1.3$; that is, $m_{cp} = 2.28$ db. For many systems, this specification is approximately equivalent to the gain and phase margins specified by (8.7-2) – namely, $GM \geqslant 8$ db and $PM \geqslant 30°$.

## 8.8 SERIES COMPENSATION AND PID CONTROL

Of the types of compensation discussed in Section 7.4, series (or cascade) compensation is particularly well suited for design with the open-loop Bode plots. Usually the system is analyzed with proportional control action included (often with a unity gain). The series compensator is then included to meet the performance specifications. The effect of the compensator is readily seen on the Bode plot because the compensator's gain and phase have an additive effect on the plot of the uncompensated open-loop system.

## PD Control as Series Compensation

P control with either I action or D action, or both, may be considered as a form of series compensation. The PD law is

$$F(s) = K_p(1 + T_D s)E(s) \tag{8.8-1}$$

The term $(1 + T_D s)$ can be considered as a series compensator to the proportional controller. The D action adds an open-loop zero at $s = -1/T_D$. Its Bode plots are shown in Figure 8.34*a*. From these it can be seen that the usefulness of D action is that it adds phase shift at higher frequencies. It is thus said to give phase "lead." However, it also increases the gain at these frequencies, since the derivative term gives more response for rapidly changing signals.

The effect of D action as a series compensator can also be seen with the root locus. For example, a plant with the transfer function $1/s(s+1)(s+2)$, when subjected to proportional control, has the root locus shown in Figure 8.13*a*. If the proportional gain is high enough, the system will be unstable. If D action is used to put an open-loop zero at $s = -1.5$, the resulting root locus is given by Figure 8.16. The D action prevents the system from becoming unstable.

## PI Control as Series Compensation

PI control is described by

$$F(s) = K_p\left(1 + \frac{1}{T_I s}\right)E(s) = \frac{K_p}{s}\left(s + \frac{1}{T_I}\right)E(s) \tag{8.8-2}$$

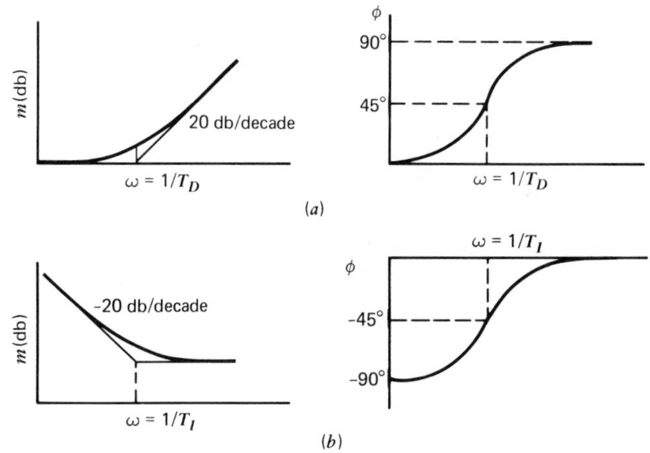


**Figure 8.34 Bode plots for series compensators. (*a*) PD action $1 + T_D s$. (*b*) PI action $(s + 1/T_I)/s$.**

The integral action can thus be considered to add an open-loop pole at $s = 0$, and a zero at $s = -1/T_I$. The Bode plots for the term $(T_I s + 1)/T_I s$ are shown in Figure 8.34*b*. The compensation adds gain at the lower frequencies but decreases the phase. It can be considered as a phase-lag compensator.

Proportional control of the plant $1/(s+1)(s+2)$ gives a root locus like that shown in Figure 8.6*b*, with $a = 1$ and $b = 2$. A steady-state error will exist for a step input. With the PI compensator applied to this plant, the root locus is given by Figure 8.16, with $T_I = 2/3$. The steady-state error is eliminated, but the response of the system has been slowed because the dominant paths of the root locus of the compensated system lie closer to the imaginary axis than for the uncompensated system.

## PI Control of a Second-Order Plant

PI control of a second-order plant leads to a third-order model. The methods of previous chapters are not always sufficient to deal with such a problem. However, the root-locus and Bode plots provide the necessary design tools. For example, let the plant-transfer function be

$$G_p(s) = \frac{1}{s^2 + a_2 s + a_1} \tag{8.8-3}$$

where $a_2 > 0$ and $a_1 > 0$. PI control applied to this plant gives the closed-loop command-transfer function

$$T_1(s) = \frac{K_p s + K_I}{s^3 + a_2 s^2 + a_1 s + K_p s + K_I} \tag{8.8-4}$$

This is the same form as that of the dc motor problem of Section 7.2, with the armature inductance included in the model [see (7.2-6)].

Note that the Ziegler-Nichols rules cannot be used to set the gains $K_p$ and $K_I$. The second-order plant (8.8-3) does not have the S-shaped signature of Figure 7.1, so the process-reaction method does not apply. The ultimate-cycle method requires $K_I$ to be set to zero and the ultimate gain $K_{pu}$ determined. With $K_I = 0$ in (8.8-4), the resulting system is stable for all $K_p > 0$, and thus a positive ultimate gain does not exist.

Take the form of the PI-control law given by (8.8-2) and assume that the characteristic roots of the plant (8.8-3) are real values $-r_1$ and $-r_2$ such that $-r_2 < -r_1$. In this case the open-loop transfer function of the control system is

$$G(s)H(s) = \frac{K_p(s + 1/T_I)}{s(s + r_1)(s + r_2)} \tag{8.8-5}$$

One design approach is to select $T_I$ and plot the locus with $K_p$ as the parameter. If the zero at $s = -1/T_I$ is located to the right of $s = -r_1$, the dominant time constant cannot be made as small as is possible with the zero located between the poles at $s = -r_1$ and $s = -r_2$ (Figure 8.35). A large integral gain (small $T_I$ and large $K_p$) is desirable for reducing the overshoot due to a disturbance, but the zero should not be placed to the left of $s = -r_2$ because the dominant time constant will be larger than that obtainable with the placement shown in Figure 8.35 for large values of $K_p$. Sketch the root-locus plots to see this. A similar situation exists if the poles of the plant are complex.

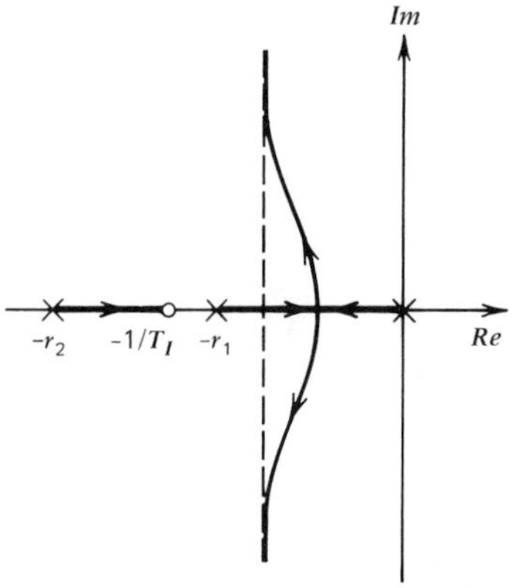


**Figure 8.35** **Root-locus plot for PI control of a second-order plant. The characteristic equation is $s(s+r_1)(s+r_2)+K_p(s+1/T_I)=0$.**

The steady-state error for a step input will be zero. The values of $K_p$ and $T_I$ can then be selected from the root locus to achieve specifications on the transient performance.

### Design Considerations with the Bode Plot

It is difficult to state a rigid set of rules to follow for design of a series compensator because of the variety of specifications and plant types that occur. However, the following considerations should be kept in mind.

1. In order to minimize the steady-state error, the open-loop gain should be kept as high as possible in the low-frequency range of the Bode plot.
2. At intermediate frequencies (near the gain crossover frequency), a slope of $-20$ db/decade in the gain curve will help to provide an adequate phase margin.
3. At high frequencies, small gain is desirable in order to attenuate high-frequency disturbances such as electronic noise, or mechanical vibrations induced by gear teeth, shaft elasticity, or hydraulic and pneumatic pressure fluctuations, etc.

In addition to shaping the gain curve, the phase curve must often be shaped by means of a series compensator in order to achieve phase and gain margin specifications, for example. The PID-type compensators do not always allow sufficient flexibility to do this. We now consider more general types of compensators, called lead and lag compensators, that provide this flexibility.

## 8.9 LEAD AND LAG COMPENSATION

We now consider three commonly used compensators, the lead, lag, and lag-lead compensators. These can be easily realized electrically with $RC$ networks and perhaps an amplifier. Mechanical equivalents are developed in the chapter problems. Specifications used to design these compensators include the types seen thus far, as well as the static error coefficients, now to be presented.

## Static Error Coefficients

Consider the single-loop system shown in Figure 8.36. The error signal is related to the input as follows.

$$E(s) = \frac{1}{1 + G(s)H(s)} R(s) \tag{8.9-1}$$

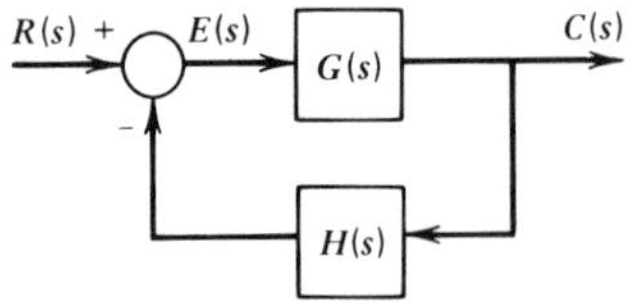


**Figure 8.36 System configuration for definition of static error coefficients.**

Assuming that the final value theorem can be applied, it gives

$$e_{ss} = \lim_{s \to 0} \frac{sR(s)}{1 + G(s)H(s)} \tag{8.9-2}$$

If the input is a unit step, we obtain the steady-state error

$$e_{ss} = \lim_{s \to 0} \frac{1}{1 + G(s)H(s)} = \frac{1}{1 + C_p} \tag{8.9-3}$$

where

$$C_p = \lim_{s \to 0} G(s)H(s) \tag{8.9-4}$$

The constant $C_p$ is the *static position error coefficient.* The name derives from servomechanism applications in which the output is a position. The error between the desired and the actual positions at steady state is given by (8.9-3). The larger $C_p$ is, the smaller will be the error. $C_p$ is finite and $e_{ss}$ nonzero when P control is applied to a first-order plant. When I control is applied to such a plant, the resulting $C_p$ will be infinite, and the error $e_{ss}$ will be zero. In general, it can be shown that a unity feedback system will have a nonzero steady-state error if no integration occurs in the forward path. Systems for which $C_p$ is finite are called *type 0 systems*; for systems of *type 1* and higher, $C_p$ is infinite.

The *static velocity error coefficient* is defined for a unit ramp input as follows.

$$e_{ss} = \lim_{s \to 0} \frac{s}{1 + G(s)H(s)} \frac{1}{s^2}$$

$$= \frac{1}{C_v} \tag{8.9-5}$$

where the velocity coefficient is

$$C_v = \lim_{s \to 0} sG(s)H(s) \tag{8.9-6}$$

Note that the error here is not an error in velocity, but a position error that results when a unit ramp input is applied. For type 0 systems, $C_v = 0$; for type 1 systems, $C_v$ is finite but nonzero; for type 2 and higher, $C_v = \infty$. Thus to eliminate the steady-state error in a unity feedback system with a ramp input, at least two integrations are required in the forward loop. For a type 1 system with unity feedback, the output velocity at steady state equals that of the input (the slope of the ramp), but an error exists between the desired and the actual positions.

Other error coefficients can be defined in a similar manner. For example, the *static acceleration error coefficient* $C_a$ is defined for a unit acceleration input $R(s) =$

$1/s^3$, as

$$C_a = \lim_{s \to 0} s^2 G(s)H(s) \tag{8.9-7}$$

Thus

$$e_{ss} = \frac{1}{C_a} \tag{8.9-8}$$

Reduced steady-state error can thus be seen to result from large values for the error coefficients (high gain) and from increasing the number of integrations in the open-loop transfer-function.* Both remedies can decrease the relative stability of the system, however. A good design must achieve a balance between these objectives.

## Lead and Lag Compensator Circuits

A variety of compensator circuits have been developed, but the lead and lag compensators shown in Figures 8.37 and 8.39 are the simplest ways of adjusting gains and phases in a variety of frequency ranges. In addition to the PID compensators, they are the most commonly used.

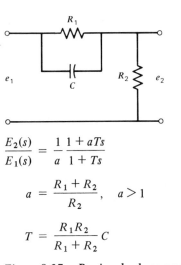

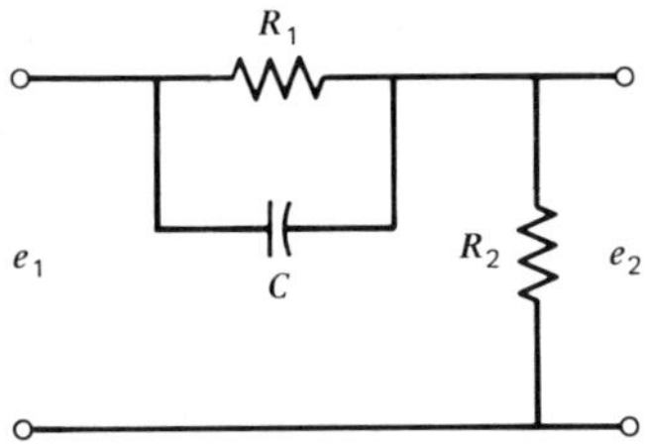


$$\frac{E_2(s)}{E_1(s)} = \frac{1}{a}\frac{1+aTs}{1+Ts}$$

$$a = \frac{R_1+R_2}{R_2}, \quad a > 1$$

$$T = \frac{R_1R_2}{R_1+R_2}C$$

**Figure 8.37 Passive lead compensator.**

When used as series compensators, the circuits shown must see a small impedance at the source ($e_1$) and a large impedance at the load ($e_2$). Sometimes an isolating amplifier is inserted to ensure the validity of this assumption. Taking these impedances to be zero and infinity, respectively, we can derive the circuits' transfer functions with the methods of Chapter Two. For the lead compensator,

$$\frac{E_2(s)}{E_1(s)} = \frac{R_2 + R_1R_2Cs}{R_1+R_2+R_1R_2Cs} = \frac{1}{a}\frac{1+aTs}{1+Ts} = \frac{s+\dfrac{1}{aT}}{s+\dfrac{1}{T}} \tag{8.9-9}$$

where

$$a = \frac{R_1+R_2}{R_2}, \quad a > 1 \tag{8.9-10}$$

$$T = \frac{R_1R_2}{R_1+R_2}C \tag{8.9-11}$$

Note that the circuit would be useless as a series compensator without the resistance $R_1$, since it would not pass dc signals.

* The number of these integrations is thus seen to be equal to the system type number.

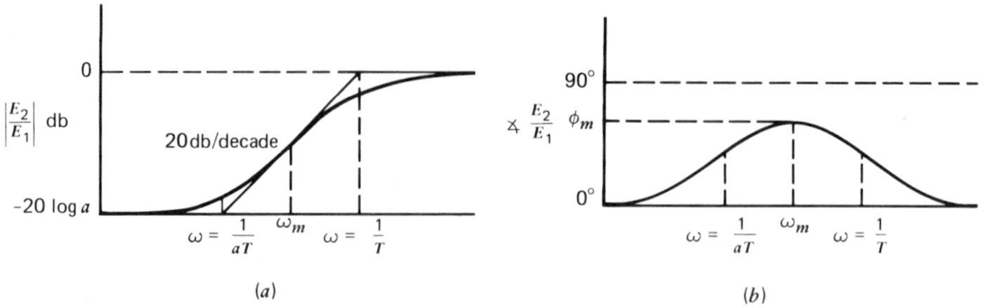


**Figure 8.38 Bode plots for the lead compensator shown in Figure 8.37.**

For the lag compensator,

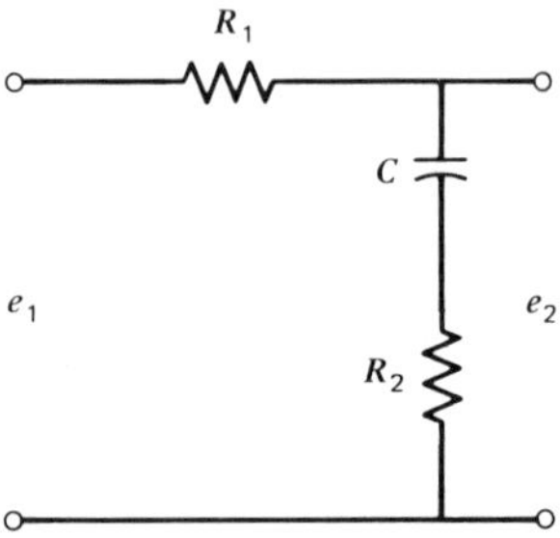


**Figure 8.39 Passive lag compensator.**

$$\frac{E_2(s)}{E_1(s)} = \frac{1 + R_2Cs}{1 + (R_1 + R_2)Cs} = \frac{1 + aTs}{1 + Ts}$$

$$= a \frac{s + \dfrac{1}{aT}}{s + \dfrac{1}{T}} \tag{8.9-12}$$

where

$$a = \frac{R_2}{R_1 + R_2}, \quad a < 1 \tag{8.9-13}$$

$$T = C(R_1 + R_2) \tag{8.9-14}$$

The Bode plots of the lead and lag compensators are shown in Figures 8.38 and 8.40. The compensators' usefulness can be briefly described as follows. When used in series with a proportional gain $K_p$, the lead compensator increases the gain margin. This allows the gain $K_p/a$ to be made larger than is possible without the compensator. The result is a decrease in the closed-loop bandwidth and an increase in the speed of response.

On the other hand, the lag compensator is used when the speed of response and damping of the closed-loop system are satisfactory, but the steady-state error is too large. The lag compensator allows the gain to be increased without substantially changing the resonance frequency $\omega_r$ and the resonance peak $m_p$ of the closed-loop system.

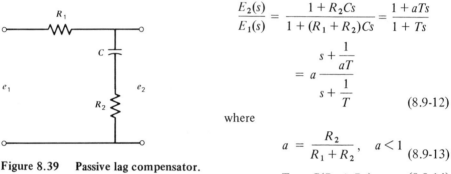

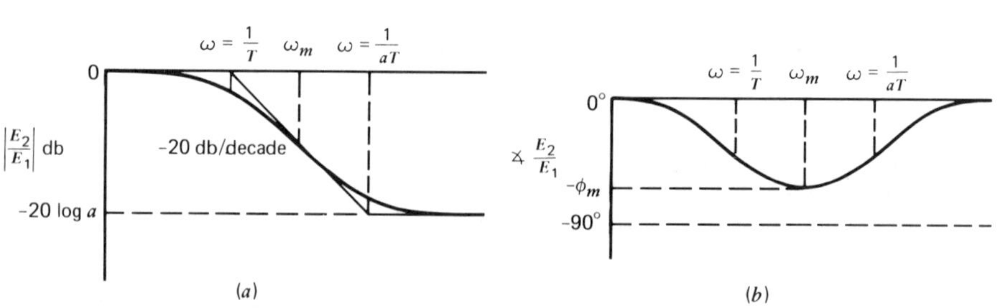


**Figure 8.40 Bode plots for the lag compensator shown in Figure 8.39.**

## Root-Locus Design of Lead Compensators

The effects of the compensator in terms of time-domain specifications (characteristic roots) can be shown with the root-locus plot. Consider the second-order plant with the real distinct roots $s = -\alpha, -\beta$. The root locus for this system with proportional control is shown in Figure 8.41*a*. The smallest dominant time constant obtainable is $\tau_1$, marked in the figure. With lead compensation, the root locus becomes that shown in Figure 8.41*b*. The pole and zero introduced by the compensator reshapes the locus so that a smaller dominant time constant can be obtained. This is done by choosing the proportional gain high enough to place the roots close to the asymptotes.

Design of a lead compensator with the root locus is done as follows. We assume that the specifications do not include required values for the static error coefficients. If they do, the design should be done with the Bode plot. The same is true if phase and gain margins are specified.

1. From the time-domain specifications (time constant, damping ratio, etc.), the required locations of the dominant closed-loop poles are determined.
2. From the root-locus plot of the uncompensated system, determine whether or not the desired closed-loop poles can be obtained by adjusting the open-loop gain. If not, determine the net angle associated with the desired closed-loop pole by drawing vectors to this pole from the open-loop poles and zeros. The difference between this angle and $-180°$ is the *angle deficiency.*
3. Locate the pole and zero of the compensator so that they will contribute the angle required to eliminate the deficiency. A method for doing this is presented in the example to follow.
4. Compute the required value of the open-loop gain from the root-locus plot.
5. Check the design to see if the specifications are met. If not, adjust the locations of the compensator's pole and zero.

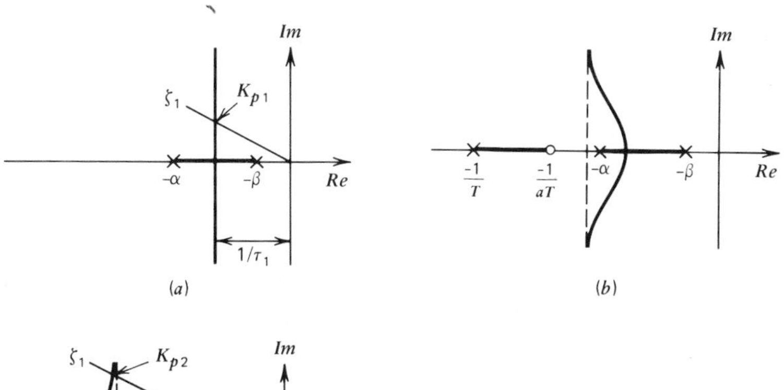


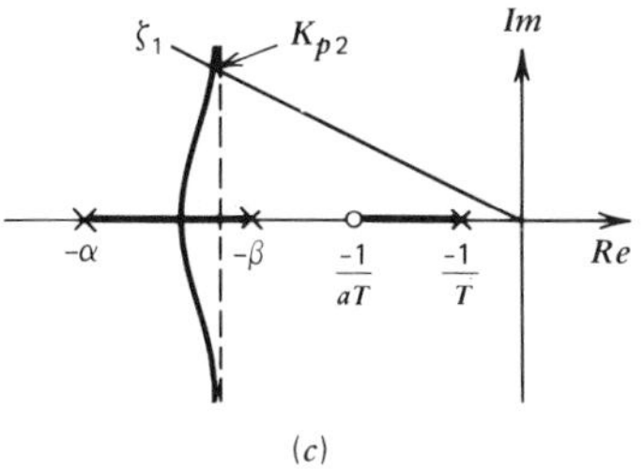


**Figure 8.41** Effects of series lead and lag compensators. (*a*) Uncompensated system's root locus. (*b*) Root locus with lead compensation. (*c*) Root locus with lag compensation.

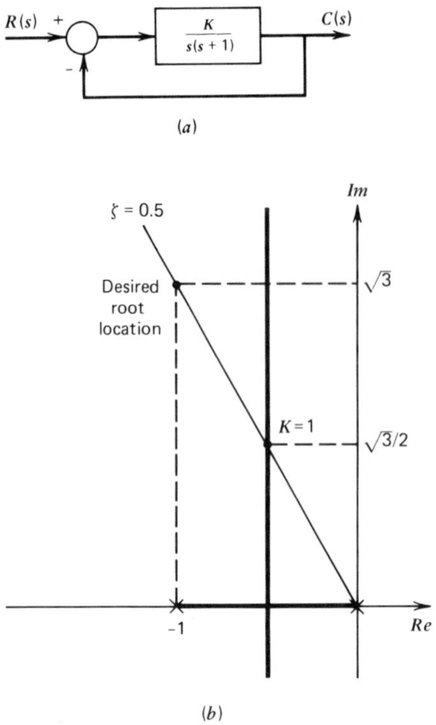


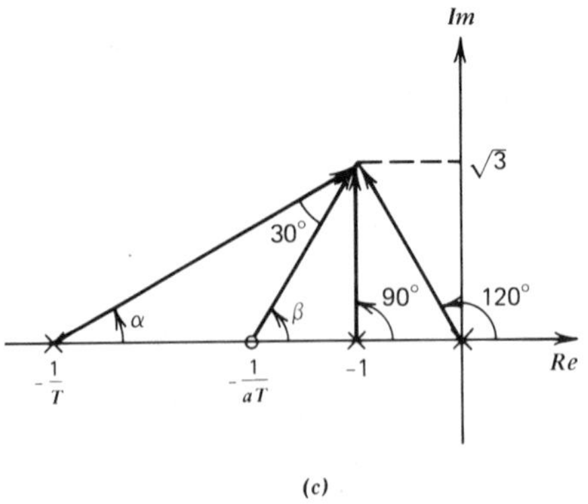


**Figure 8.42** (*a*) **System for Example 8.11.** (*b*) **Root-locus plot for the system shown in part** *a* **showing the actual and desired root locations.** (*c*) **Vector diagram showing the contributions of the poles and zeros at the desired root location.**

*Example 8.10*

Consider the system shown in Figure 8.42*a*. Its root locus plot is given in Figure 8.42*b*. In Section 8.1 (Figure 8.1) it was shown that a value of $K = 1$ gives $\zeta = 0.5$, and closed-loop poles at $s = -\frac{1}{2} \pm i\frac{1}{2}\sqrt{3}$. The resulting time constant is 2. Assume that the transient specifications require that $\zeta = 0.5$ with a time constant of $\tau = 1$ and that no steady-state error specifications are given. This information implies that the closed-loop poles should be at $s = -1 \pm i\sqrt{3}$. This performance cannot be obtained with a gain change in the present system. Design a compensator to meet the specifications.

Since the specifications are given in terms of time-domain characteristics, we use the root-locus method. From Figure 8.42*c*, the angle of the uncompensated system at the desired root location is

$$\measuredangle \left. \frac{1}{s(s+1)} \right|_{s=-1+i3} = -90^\circ - 120^\circ = -210^\circ$$

Thus the angle deficiency is $-210^\circ + 180^\circ = -30^\circ$. We therefore need a lead compensator to increase the phase angle by $30^\circ$.

The choice of the poles and zeros of the compensator is somewhat arbitrary. We must pull the original locus to the left, so we place the pole and zero to the left of the pole at $s = -1$. The difference between the angle contributions of the pole and zero of the compensator must be $30^\circ$. This specifies the size of the apex angle of the triangle shown in Figure 8.42*c*. Since the parameter $a$ appears in the denominator of the transfer function (8.9-9), we should select its value to be as small as possible in order to minimize the additional amount of gain required to cancel its effect. Keeping physical realizability in mind, we choose $a = 10$. Since $\beta = \alpha + 30°$, simple trigonometry applied to Figure 8.42*c* gives the required value for $T$. The tangents of $\alpha$ and $\beta$ are

$$\tan\alpha = \frac{T\sqrt{3}}{1-T}$$

$$\tan\beta = \frac{aT\sqrt{3}}{1-aT}$$

Also

$$\tan\beta = \tan(\alpha + 30°) = \frac{\tan\alpha + \tan 30°}{1 - \tan\alpha \tan 30°}$$

Elimination of $\alpha$ and $\beta$ yields

$$4aT^2 + 2(1-2a)T + 1 = 0$$

For $a = 10$, $T = 0.027$, $0.923$. The second solution results in the pole lying to the left of $s = -1$ and the zero lying to the right of $s = -1$, so we use $T = 0.027$. From (8.9-10) and (8.9-11), the resistances required with a $1\,\mu f$ capacitance are $R_1 = 270\,k\Omega$ and $R_2 = 30\,k\Omega$. The compensator pole is at $s = -37.0$, and the zero is at $s = -3.7$.

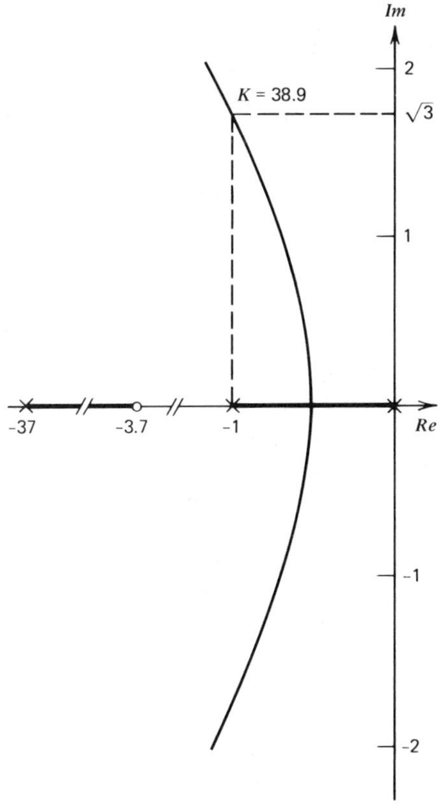


**Figure 8.43 Root-locus plot for the compensated system of Example 8.10.**

The open-loop transfer function of the compensated system is

$$G_c(s)G(s)H(s) = \frac{K(s+3.7)}{s(s+37)(s+1)}$$

Its root locus is shown in Figure 8.43. From this the gain $K$ required to place the roots at $s = -1 \pm i\sqrt{3}$ is $K = 38.9$. The third root is far to the left at $s = -36$ for this value of $K$. Its effect on the transient behavior is probably slight. This can be checked by analysis or simulation before the design is made final.

## Root-Locus Design of Lag Compensators

With reference to the system shown in Figure 8.41*a*, suppose that the desired damping ratio $\zeta_1$ and desired time constant $\tau_1$ are obtainable with a proportional gain of $K_{p1}$, but the resulting steady-state error $\alpha\beta/(\alpha\beta + K_{p1})$ to a step input is too large. We need to increase the gain while preserving the desired damping ratio and time constant. With the lag compensator, the root locus is as shown in Figure 8.41*c*. By considering specific numerical values, one can show that for the compensated system, roots with a damping ratio $\zeta_1$ correspond to a high value of the proportional gain. Call this value $K_{p2}$. Thus $K_{p2} > K_{p1}$, and the steady-state error will be reduced.

The effect of the lag compensator on the time constant can be seen as follows. The open-loop transfer function is

$$G(s)H(s) = \frac{aK_p\left(s + \dfrac{1}{aT}\right)}{(s+\alpha)(s+\beta)\left(s + \dfrac{1}{T}\right)} \tag{8.9-15}$$

If the value of $T$ is chosen large enough the pole at $s = -1/T$ in (8.9-15) is approximately canceled by the zero at $s = -1/aT$, and the open-loop transfer function is given approximately by

$$G(s)H(s) = \frac{aK_p}{(s+\alpha)(s+\beta)} \tag{8.9-16}$$

Thus the system's response is governed approximately by the complex roots corresponding to the gain value $K_{p2}$. By comparing Figure 8.41*a* with 8.41*c*, we see that the compensation leaves the time constant relatively unchanged.

From (8.9-16) it can be seen that since $a < 1$, $K_p$ can be selected as the larger value $K_{p2}$. The ratio of $K_{p1}$ to $K_{p2}$ is approximately given by the parameter $a$.

Design by pole-zero cancellation can be difficult to accomplish because a response pattern of the system is essentially ignored. The pattern corresponds to the behavior generated by the canceled pole and zero, and this response can be shown to be beyond the influence of the controller (see Chapter Nine). In this example, the canceled pole gives a stable response because it lies in the left-hand plane. However, another input not modeled here, such as a disturbance, might excite the response and cause unexpected behavior. The designer should therefore proceed with caution. None of the physical parameters of the system are known exactly, so exact pole-zero cancellation is not possible. A root-locus study of the effects of parameter uncertainty and a simulation study of the response are often advised before the design is accepted as final.

Based on this example, we can outline an approach to the design of a lag compensator with the root locus as follows.

1. From the root locus of the uncompensated system, the gain $K_{p1}$ is found that will place the roots at the locations required to give the desired relative stability and transient response.
2. Let $K_{p2}$ denote the value of the gain required to achieve the desired steady-state performance. The parameter $a$ is the ratio of these two gain values $a = K_{p1}/K_{p2} < 1$.
3. The value of $T$ is then chosen large so that the compensator's pole and zero are close to the imaginary axis. This placement should be made so that the compensated locus is relatively unchanged in the vicinity of the desired closed-loop poles. This will be true if the angle contribution of the lag compensator is close to zero.
4. Locate the desired closed-loop poles on the compensated locus and set the open-loop gain so that the dominant roots are at this location (neglecting the existence of the compensator's pole and zero).
5. Check the design to see if the specifications are met. If not, adjust the locations of the compensator's pole and zero.

*Example 8.11*

Consider the system shown in Figure 8.42*a*. If $K = 1$, then $\zeta = 0.5$ and $s = -\frac{1}{2} \pm i\frac{1}{2}\sqrt{3}$. The static velocity error coefficient is $C_v = 1$. Suppose that the transient response thus obtained is satisfactory, but that the error must be decreased by increasing $C_v$ to 10. Design a compensator to do this.

A lag compensator is indicated since the steady-state error is too large. The gain $K_{p1}$ required to achieve the desired transient performance has already been established as $K_{p1} = K = 1$. The second step is to determine the value of the parameter $a$. For this system, the coefficient $C_v$ is

$$C_v = \lim_{s\to 0} s \frac{K_p}{s(s+1)} = K_p$$

and $K_{p2}$ is the value of $K_p$ that gives $C_v = 10$. Thus $K_{p2} = 10$, and the parameter $a = K_{p1}/K_{p2} = 1/10$. The compensator's pole and zero must be placed close to the imaginary axis, with the ratio of their distances being $1/10$. Noting that the plant has a pole at $s = -1$, we select locations well to the right of this pole – say, at $s = -0.01$ and $s = -0.1$ for the pole and zero, respectively. This gives $T = 100$.

The open-loop transfer function of the compensated system is thus

$$G_c(s)G(s)H(s) = \frac{0.1K_c(s+0.1)}{s(s+1)(s+0.01)}$$

The root locus is shown in Figure 8.44. For the desired damping ratio of $\zeta = 0.5$, the locus shows that the dominant roots are at $s = -0.449 \pm i0.778$ with $K_c = 9$. The third root is at $s = -0.111$.

The error coefficient is thus $C_v = 9$ and is less than the desired value of 10. Also, the dominant roots differ somewhat from the desired locations at $s = -0.5 \pm i0.866$.

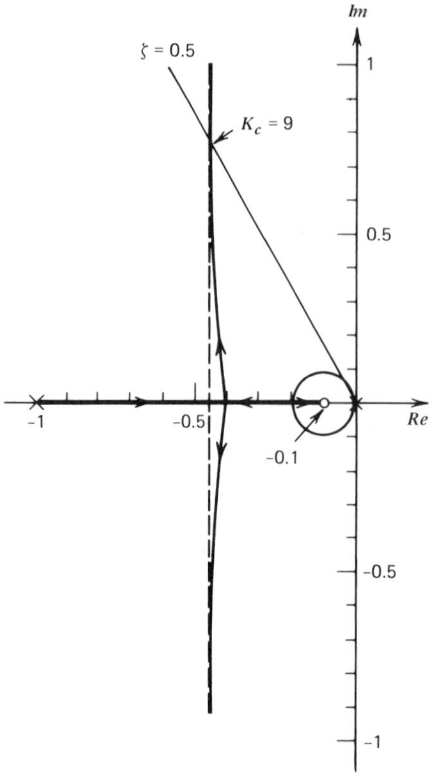


**Figure 8.44** Root-locus plot for $G(s)H(s) = 1/s(s+1)$ with the lag compensator $G_c(s) = 0.1K_c(s+0.1)/(s+0.01)$.

If these differences are too large, the compensator's pole and zero can be placed closer to the imaginary axis – say, at $s = -0.01$ and $s = -0.001$, respectively, with $T = 1000$. This will decrease the compensator's influence on the locus near the desired root locations.

With $T = 100$ and the capacitance $C = 1\ \mu f$, $R_1 = 90\ M\Omega$ and $R_2 = 10\ M\Omega$. The improved performance obtained with $T = 1000$ requires that $R_1 = 900\ M\Omega$ and $R_2 = 100\ M\Omega$.

## The Root-Contour Method

The root-contour method can be used with the root-locus plots to select the values of more than one parameter. For lead or lag compensation the pertinent parameters are $a$ and $T$. To use the method, one parameter is fixed at a reasonable value, and the other is varied to generate a root-locus plot. From this plot, reasonable values of the second parameter are chosen, and the locus is constructed for the variation in the first parameter. For example, lag compensation requires that $0 < a < 1$, so we can choose $a = 0.5$, say, and plot the locus for $0 < T < \infty$. Each point on this $T$-locus is a potential pole for a locus in terms of the deviation $\Delta$, where $a = 0.5 + \Delta$. The $\Delta$-loci can be constructed for $-0.5 \leqslant \Delta \leqslant 0.5$ at strategic points along the $T$-locus. The values of $a$ and $T$ are then selected with the aid of these plots.

For lead compensation, $a > 1$; so a starting value of $a = 1$ might be used. In this case, $a = 1 + \Delta$ and $0 \leqslant \Delta \leqslant \infty$. The number of loci generated with this approach can be quite large, and a computer plotting routine is advised.

## Bode Design of Lead Compensators

For any particular value of the parameter $a$, the lead compensator can provide a maximum phase lead $\phi_m$ (see Figure 8.38). This value, and the frequency $\omega_m$ at which it occurs, can be found as a function of $a$ and $T$ from the Bode plot. The frequency $\omega_m$ is the geometric mean of the two corner frequencies of the compensator. Thus

$$\omega_m = \frac{1}{T\sqrt{a}} \tag{8.9-17}$$

By evaluating the phase angle of the compensator at this frequency, we can show that

$$\sin \phi_m = \frac{a-1}{a+1} \tag{8.9-18}$$

or

$$a = \frac{1 + \sin \phi_m}{1 - \sin \phi_m} \tag{8.9-19}$$

These relations are useful in designing a lead compensator with the Bode plot.

System design in the frequency domain usually is done with phase and gain margin requirements. We assume that the specifications include these as well as one on the allowable steady-state error. The purpose of the lead compensator is to utilize the maximum phase lead of the compensator to increase the phase of the open-loop system near the gain crossover frequency, while not changing the gain curve near that frequency. This usually is not entirely possible because the gain crossover frequency is increased in the process, and a compromise must be sought between the resulting increase in bandwidth and the desired values of the phase and gain margins.

A suggested design method is as follows.

1. Set the gain $K_p$ of the uncompensated system to meet the steady-state error requirement.
2. Determine the phase and gain margins of the uncompensated system from the Bode plot, and estimate the amount of phase lead $\phi$ required to achieve the margin specifications. The extra phase lead required can be used to estimate the value for $\phi_m$ to be provided by the compensator. Thus $a$ can be found from (8.9-19).
3. Choose $T$ so that $\omega_m$ from (8.9-17) is located at the gain crossover frequency of the compensated system. One way to do this is to find the frequency at which the gain of the uncompensated system equals $-20 \log \sqrt{a}$. Choose this frequency to be the new gain crossover frequency. This frequency corresponds to the frequency $\omega_m$ at which $\phi_m$ occurs.
4. Construct the Bode plot of the compensated system to see if the specifications have been met. If not, the choice for $\phi_m$ needs to be evaluated and the process repeated. It is possible that a solution does not exist.

The procedure can be quickly accomplished in the hands of an experienced designer, because intuition based on past designs can play a large role in the process. It is difficult to state rules that are foolproof.

The attempt to design a lead compensator can be unsuccessful if the required value of $a$ is too large. This can occur with plants that either are not stable, or have a low relative stability with a rapidly decreasing phase curve near the gain crossover frequency. In the former case the extra phase lead required can be too large. In the latter case the phase angle at the compensated gain crossover frequency is much less than at the uncompensated gain crossover. Thus the extra phase required can be excessive. The difficulty with a large value of $a$ is that the resistance and capacitance values that result might be incompatible or not possible to obtain physically. The usual range for $a$ is $1 < a < 20$. Additional phase lead can sometimes be obtained by cascading more than one lead compensator.

In spite of these potential difficulties, the lead compensator has a record of many successful applications.

## *Example 8.12*

The system shown in Figure 8.42$a$ can yield a static-velocity error coefficient of $C_v = 10/\text{sec}$ if $K$ is set to 10. However, if this is done the transient performance is unsatisfactory. Design a compensator to give a gain margin of at least 6 db and a phase margin of at least 40°.

The Bode plot of the open-loop uncompensated system is shown in Figure 8.45. The phase margin is 17°, and the gain margin is infinite because the phase curve never falls below $-180°$. Thus $40° - 17° = 23°$ must be added to the phase curve in order to meet the specifications, so we choose to try a lead compensator

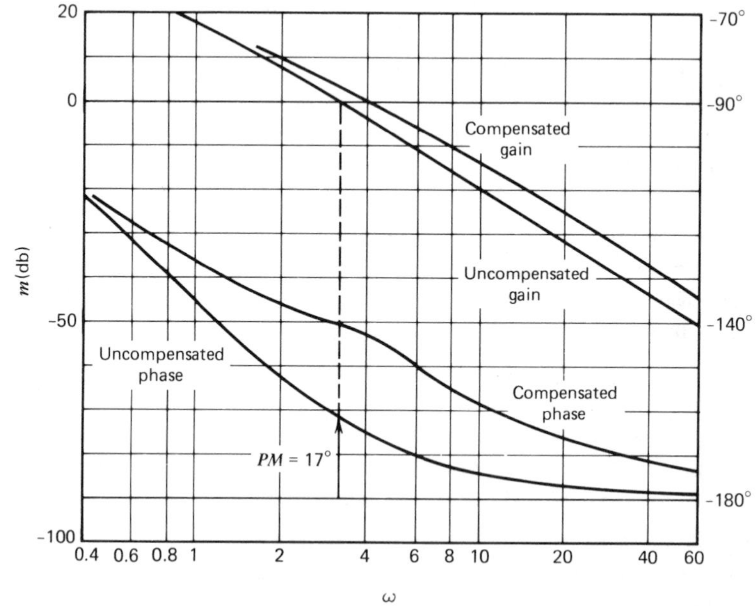


**Figure 8.45 Bode plots for the uncompensated and compensated systems of Example 8.12.**

with $\phi_m = 23^\circ$. From (8.9-19) we have

$$a = \frac{1 + \sin 23^\circ}{1 - \sin 23^\circ} = 2.283$$

From step 3, we compute

$$-20 \log \sqrt{a} = -3.585 \text{ db}$$

The open-loop gain of the uncompensated system is $-3.585$ db at approximately $\omega = 4$ rad/sec. Take this frequency to be $\omega_m$ and solve for $T$ from (8.9-17).

$$T = \frac{1}{4\sqrt{2.283}} = 0.1655$$

From (8.9-9) these values of $a$ and $T$ give a compensator with the transfer function

$$G_c(s) = \frac{1}{2.283} \frac{0.3778s + 1}{0.1655s + 1}$$

The open-loop transfer function of the compensated system is

$$G_c(s)G(s)H(s) = \frac{K(0.3778s + 1)}{2.283s(s + 1)(0.1655s + 1)} \tag{8.9-20}$$

The original choice of $K = 10$ no longer gives $C_v = 10$ because the compensator introduces the attenuation factor $1/2.283$. Thus we must choose $K = 10(2.283) = 22.83$ to achieve $C_v = 10$. With $K$ equal to 22.83, the Bode plot of the compensated system is shown in Figure 8.45. The phase margin is $37^\circ$, and the gain margin is still infinite, so the specifications have not been met exactly. The phase margin is $3^\circ$ less than the required value. This illustrates the trial-and-error nature of this method. The guides listed in the design steps are based on approximations to the real phase and gain curves of the compensator. The difficulty could have been avoided if we had aimed for a phase margin higher than that specified – say, $45^\circ$. Then $\phi_m$ would have been used as $45^\circ - 17^\circ = 28^\circ$, and $a$ and $T$ would differ accordingly.

## Bode Design of Lag Compensators

Lag compensation uses the high-frequency attenuation of the network to keep the phase curve unchanged near the gain crossover frequency while this frequency is lowered. A suggested design procedure is as follows.

1. Set the open-loop gain $K_p$ of the uncompensated system to meet the steady-state error requirements.
2. Construct the Bode plots for the uncompensated system and determine the frequency at which the phase curve has the desired phase margin. Determine the number of decibels required at this frequency to lower the gain curve to 0 db. Let this amount be $m' > 0$ db, and this frequency be $\omega_g'$. Then $a$ is found from

$$a = 10^{-m'/20} \tag{8.9-21}$$

3. The first two steps alter the phase curve. However, this curve will not be changed appreciably near $\omega_g'$ if $T$ is chosen so that $\omega_g' \gg 1/aT$. A good choice is to place the frequency $1/aT$ one decade below $\omega_g'$. Any larger separation might result in a system with a slow response.
4. Construct the Bode plots of the compensated system to see if all the specifications are met. If not, choose another value for $T$ and repeat the process.

As with lead compensation, the procedure is one of trial and error and might not work. The physical elements must be realizable, so a common range for $a$ is $0.05 < a < 1$. The compensator introduces a lag angle that is not accounted for in the preceding procedure. To account for this effect, the designer might add 5° to 10° to the specified phase margin before starting the design process.

### *Example 8.13*

Consider the uncompensated system of Example 8.12 shown in Figure 8.42*a*. The characteristic roots when $K = 1$ are $s = -0.5 \pm i0.866$. Suppose that the transient response given by these roots is acceptable, but we wish $C_v = 10$. Setting $K = 10$ will accomplish this but will alter the transient performance. Design a compensator to improve the system so that $C_v = 10$, $PM \geq 40°$, and $GM \geq 6$ db.

Since the transient response is satisfactory, a lag compensator is indicated. Set $K = 10$ to achieve the desired value of $C_v$, and construct the Bode plots for the uncompensated open-loop system. These are the same as those shown in Figure 8.45 and are repeated in Figure 8.46. Keeping in mind the lag introduced by the compen-

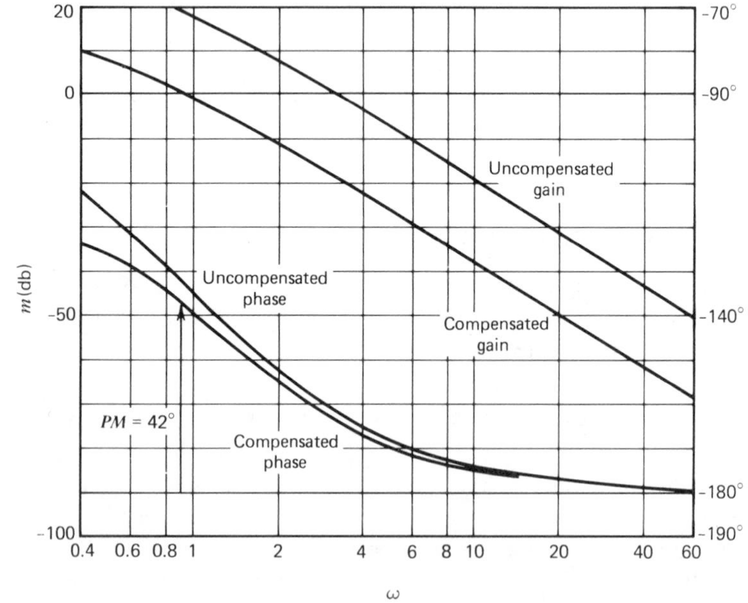


**Figure 8.46** Bode plots for the uncompensated and compensated systems of Example 8.13.

sator, we attempt to achieve a phase margin of $40^\circ + 5^\circ = 45^\circ$. The uncompensated system would have a phase margin of $45^\circ$ if the gain crossover occurs at $\omega = 1$. Thus $\omega_g' = 1$. The gain at this frequency is 18 db, so $m' = 18$. From (8.9-21),

$$a = 10^{-18/20} = 0.126$$

For step 3, we choose $T$ to place $\omega = 1/aT$ one decade below $\omega = 1$. Thus, $1/aT = 0.1$, and $T = 79.4$. The lag compensator that results is

$$G_c(s) = \frac{10s + 1}{79.4s + 1}$$

The open-loop transfer function of the compensated system is

$$G_c(s)G(s)H(s) = \frac{K(10s + 1)}{s(s + 1)(79.4s + 1)}$$

With $K = 10$, $C_v = 10$, and the Bode plots are given in Figure 8.46. The phase margin is $42^\circ$ and the gain margin is infinite, so the specifications have been met.

On the surface it appears that Examples 8.12 and 8.13 are concerned with solving the same problem; namely, compensate the system of Figure 8.42*a* so that $C_v = 10$, $PM \geqslant 40^\circ$, and $GM \geqslant 6$ db. However, this is not the case. The difference lies in what is taken as satisfactory in the performance of the uncompensated system and what must be improved. In Example 8.12, the transient response that results when $K = 10$ was judged to be poor, and the lead compensator was used to improve the transient performance. In Example 8.13 acceptable transient response was obtained when $K = 1$, but the steady-state error was then unacceptable, and a lag compensator was therefore used. The compensation in Example 8.12 was obtained by shifting the phase curve, whereas that in Example 8.13 was obtained by shifting the gain curve.

When $K = 1$, the roots of the uncompensated system are $s = -0.5 \pm i0.866$. With the lag compensator of Example 8.13, the new characteristic roots are $s = -0.107, -0.453 \pm i0.985$ and are thus very close to the desired roots if the effect of the root at $s = -0.107$ is negligible because of the pole-zero cancellation of the compensator. This is the same process that occurred in Example 8.11.

On the other hand, the characteristic roots of the lead-compensated system in Example 8.13 are $s = -3.66, -1.68 \pm i3.69$. These represent an improvement in the transient response since the uncompensated roots lie at $s = -0.5 \pm i3.12$ when $K = 10$.

## Lag-Lead Compensation

The lead and lag compensators are complementary to each other in that one improves the transient performance while the other improves steady-state performance. In situations where one of these fails to produce a satisfactory design, a cascade combination of the two can be used. This is lag-lead compensation. A lag and a lead cascaded together give the transfer function

$$\frac{E_2(s)}{E_1(s)} = \frac{1}{a}\frac{1 + aT_1 s}{1 + T_1 s}\frac{1 + bT_2 s}{1 + T_2 s} \tag{8.9-22}$$

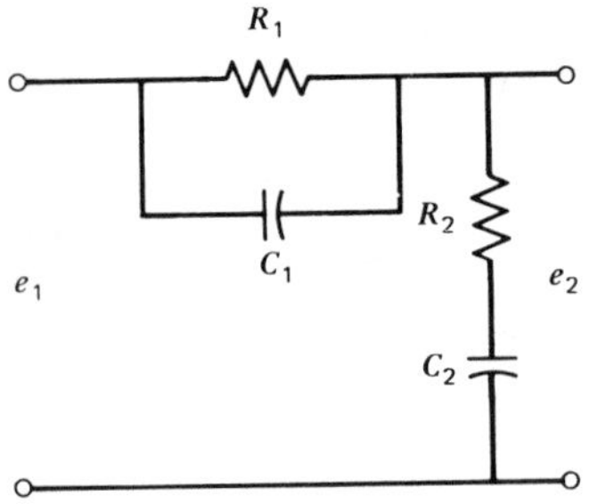


$$\frac{E_2(s)}{E_1(s)} = \frac{1 + aT_1 s}{1 + T_1 s}\,\frac{1 + bT_2 s}{1 + T_2 s}$$

$$aT_1 = R_1C_1, \quad a > 1$$

$$bT_2 = R_2C_2$$

$$T_1 + T_2 = R_1C_1 + R_1C_2 + R_2C_2$$

$$b = \frac{1}{a}$$

**Figure 8.47 Passive lag-lead compensator.**

A simpler approach is to use the single network shown in Figure 8.47. With the usual impedance assumptions, we obtain

$$\frac{E_2(s)}{E_1(s)} = \frac{1 + (R_1C_1 + R_2C_2)s + R_1C_1R_2C_2s^2}{1 + (R_1C_1 + R_1C_2 + R_2C_2)s + R_1C_1R_2C_2s^2}$$

$$= \frac{1 + aT_1 s}{1 + T_1 s}\,\frac{1 + bT_2 s}{1 + T_2 s} \tag{8.9-23}$$

where

$$aT_1 = R_1C_1, \quad a > 1 \tag{8.9-24}$$

$$bT_2 = R_2C_2 \tag{8.9-25}$$

$$T_1 + T_2 = R_1C_1 + R_1C_2 + R_2C_2 \tag{8.9-26}$$

$$b = \frac{1}{a} \tag{8.9-27}$$

The Bode plots for this compensator are shown in Figure 8.48 for $T_2 > bT_2 > aT_1 > T_1$. The maximum phase shift $\phi_m$ occurs at $\omega_{m1} = 1/T_1\sqrt{a}$, and the attenuation at this frequency is equal in magnitude but opposite in sign to that of the lead compensator alone. Thus the lag-lead compensator can be more effective than the lead compensator. The plots also show that the compensator affects the gain and phase only in the intermediate frequency range from $\omega = 1/T_2$ to $1/T_1$.

Let the compensator transfer function in (8.9-23) be written as

$$G_c(s) = G_1(s)G_2(s) \tag{8.9-28}$$

where $G_1(s)$ represents the lead compensation and $G_2(s)$ the lag compensation.

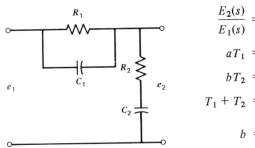

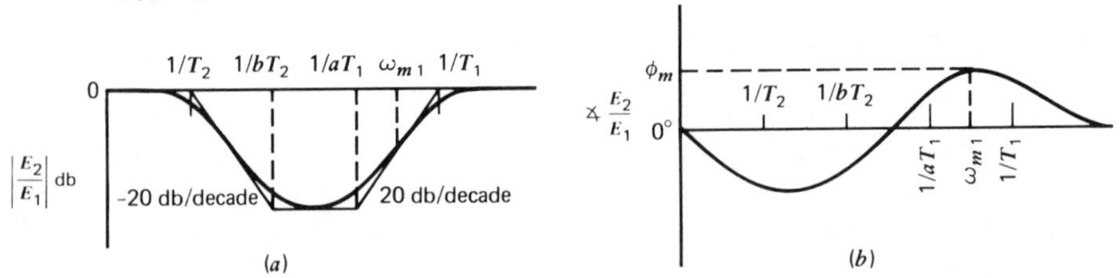


**Figure 8.48 Bode plots for the lag-lead compensator shown in Figure 8.47. The frequency $\omega_{m1}$ is the geometric mean frequency for the lead compensator alone: $\omega_{m1} = 1/T_1\sqrt{a}$.**

$$G_1(s) = \frac{1 + aT_1 s}{1 + T_1 s} \tag{8.9-29}$$

$$G_2(s) = \frac{1 + \dfrac{T_2}{a} s}{1 + T_2 s} \tag{8.9-30}$$

The uncompensated open-loop transfer function is $G(s)H(s)$, so that of the compensated system is $G_c(s)G(s)H(s)$.

The root-locus approach to designing the lag-lead compensator is a combination of the approaches used for the lead and lag compensators.

1. Determine the desired locations of the dominant closed-loop poles from the transient performance specifications and calculate the phase-angle deficiency $\phi$. This deficiency must be supplied by $G_1(s)$.

2. Use the steady-state error specifications to determine the required open-loop gain for $G_c(s)G(s)H(s)$.

3. Let $s_d$ be the desired location of the dominant closed-loop poles. Assume the lag coefficient $T_2$ is large enough to achieve pole-zero cancellation when $s$ is near $s_d$; that is,

$$|G_2(s_d)| \cong 1 \tag{8.9-31}$$

With (8.9-31) satisfied, $T_1$ and $a$ can be found from the magnitude criterion for the locus and the phase lead requirement from step 1; that is,

$$|G_1(s_d)||G(s_d)H(s_d)| = 1 \tag{8.9-32}$$

$$\measuredangle G_1(s_d) = \phi \tag{8.9-33}$$

4. With the parameters $a$ and $T_1$ now selected, choose $T_2$ so that condition (8.9-31) is satisfied. As before, check the design to see if the specifications are satisfied and if the circuit elements are realizable.

Design of a lag-lead compensator with the Bode plot follows the procedures for the lead and lag compensators discussed previously. Presumably the lag-lead is to be designed because of a deficiency in both the transient and steady-state performance of the uncompensated system. A designer who thinks that the transient response constitutes the most serious deficiency can choose to apply the procedures for the lead compensator first. When this part of the design is completed, the lag compensation can be designed. The opposite procedure could be used if the steady-state performance were worse than the transient.

## 8.10 APPLICATIONS TO DIGITAL CONTROL

The root-locus, frequency-response, and compensation techniques developed in this chapter can be applied to the design of digital controllers. Here we outline how this is done.

## Root Locus for Discrete-Time Systems

The root locus is simply a presentation of the locations of the roots of an equation in terms of some parameter. Thus the root-locus plotting guides apply to the characteristic equation of a discrete-time system without any modifications. The variable $s$ is simply replaced with the variable $z$. However, the interpretation of the resulting root-locus plot is different for discrete-time systems because of the different relationships between the $s$ and $z$ variables and the time response. These relationships are summarized in Section 5.12 (Figure 5.39). The following example will illustrate the method.

*Example 8.14*

Design a digital version of integral control to be used with a first-order plant whose transfer function is

$$G_p(s) = \frac{1}{\tau s + 1} \tag{8.10-1}$$

where $\tau = 2$ sec. When a unit step input is applied, the minimum overshoot should be less than 20%, the 100% rise time should be less than 5 sec, and the 2% settling time less than 20 sec.

With a sample-and-hold device, the pulse transfer function for it and the plant together can be found with the results of (4.6-18). It is

$$G_p(z) = \frac{1-a}{z-a} \tag{8.10-2}$$

$$a = e^{-T/\tau} \tag{8.10-3}$$

The discrete form of the analog $I$-action $K_I/s$ is $K_1 z/(z-1)$. See Section 6.12, (6.12-8). The relation between $K_1$ and $K_I$ is irrelevant for our purposes, since we will determine $K_1$ directly from the specifications. Thus the digital controller transfer function is

$$D(z) = \frac{M(z)}{E(z)} = \frac{K_1 z}{z-1} \tag{8.10-4}$$

The characteristic equation is thus

$$1 + \frac{K_1(1-a)z}{(z-1)(z-a)} = 0 \tag{8.10-5}$$

The block diagram and root locus of this system are shown in Figure 8.49. The root-locus parameter is $K = K_1(1-a)$. The plot is obtained with the guides of Section 8.2 with $z$ replacing $s$. Off the real axis the locus is a circle of radius $\sqrt{a}$. From (8.10-3) this radius is always less than unity. The system becomes unstable if the gain $K$ is large enough to make the locus in the left-hand plane cross the unit circle. This will occur for a combination of large $K_1$ and small $a$ (that is, for a sampling time $T$ large relative to $\tau$).

Select the sampling period $T$ to be a fraction of the plant's time constant. Thus, try $T = \frac{1}{4}(2) = 0.5$ sec. In this case, $a = \exp(-0.5/2) = 0.78$. The radius of the locus is $\sqrt{a} = 0.88$.

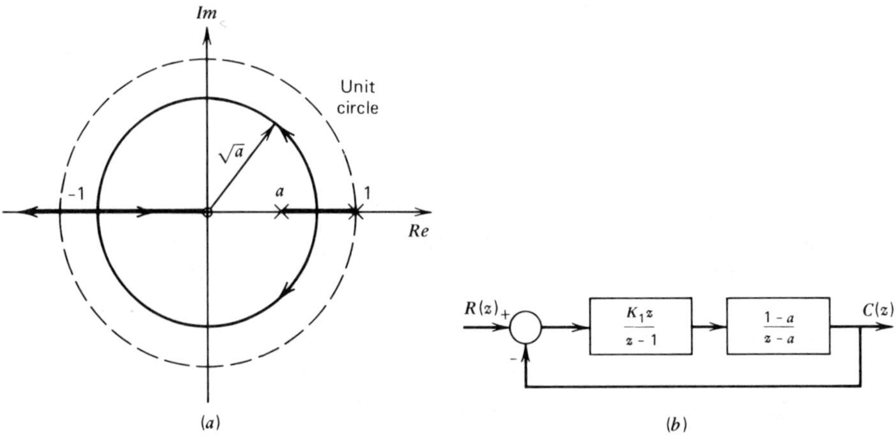


**Figure 8.49 Root locus and block diagram for the system of Example 8.14.**

Referring to Figure 5.20$a$, we see that the overshoot specification requires that $\zeta \geqslant 0.45$. From Figure 5.20$c$, this implies that $\omega_n t_r \geqslant 2.3$. Therefore the rise-time specification calls for $\omega_n \geqslant 2.3/5 = 0.46$. Finally, the settling time from (5.7-22) requires that $\zeta\omega_n \geqslant 4/20 = 0.2$. Note that if $\zeta \geqslant 0.45$ and $\omega_n \geqslant 0.46$, the last requirement is always satisfied.

The $z$-plane relation (5.12-11) is repeated here.

$$z = e^{-\zeta\omega_n T} e^{i\omega_n T\sqrt{1-\zeta^2}} \tag{8.10-6}$$

For $\omega_n = 0.46$ and $T = 0.5$, (8.10-6) gives

$$z = e^{-0.23\zeta} e^{i0.23\sqrt{1-\zeta^2}} \tag{8.10-7}$$

This specifies two roots. The upper root is plotted in Figure 8.50 for $\zeta \geqslant 0.45$.

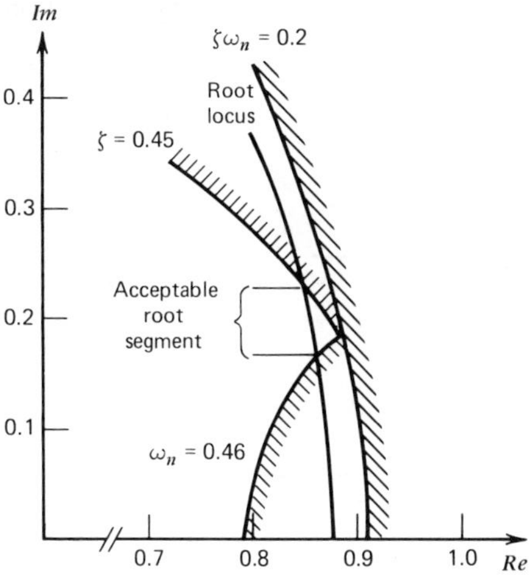


**Figure 8.50 Root-locus segment and boundaries corresponding to the specifications of Example 8.14.**

For $\zeta = 0.45$, (8.10-6) gives

$$z = e^{-0.225\,\omega_n} e^{i0.4465\,\omega_n} \tag{8.10-8}$$

This curve is plotted in Figure 8.50 for $\omega_n \geqslant 0.46$. Also shown is the line corresponding to $\zeta\omega_n = 0.2$, or

$$|z| = e^{-0.1} = 0.905$$

The characteristic roots of the system must not lie on the hatched side of these three curves. The appropriate segment of the root locus is also displayed. From this we can see that the only acceptable upper roots lie in the region indicated. The value of $K_1$ can be obtained in the usual manner with the magnitude criterion. For example, with $z = 0.86 + i0.2$,

$$K = K_1(1-a) = \frac{|z-1||z-a|}{|z|} = \frac{0.24(0.22)}{0.88} = 0.06$$

or

$$K_1 = \frac{0.06}{0.22} = 0.27$$

With the value for $K_1$, the closed-loop transfer function is

$$T(z) = \frac{C(z)}{R(z)} = \frac{0.06z}{z^2 - 1.72z + 0.78} \tag{8.10-9}$$

The corresponding difference equation is

$$c(k) = 1.72c(k-1) - 0.78c(k-2) + 0.06r(k-1) \tag{8.10-10}$$

For a unit step input, $r(i) = 0$, $i < 0$ and $r(i) = 1$, $i \geqslant 0$. The system (8.10-10) was simulated with this input and the zero initial conditions $c(-1) = c(0) = 0$. The results are shown in Figure 8.51. The maximum overshoot is 18%, the rise time is 4.3 sec, and the settling time is approximately 15 sec.

Our choice of $T = 0.5$ thus appears to be acceptable. The difference equation to be implemented by the control computer is found from (8.10-4) to be

$$m(k) = m(k-1) + 0.27e(k) \tag{8.10-11}$$

The next step in the design process would be to check the sensitivity of the design to variations in the parameter $a$, due to uncertainty in the value of the plant's time constant.

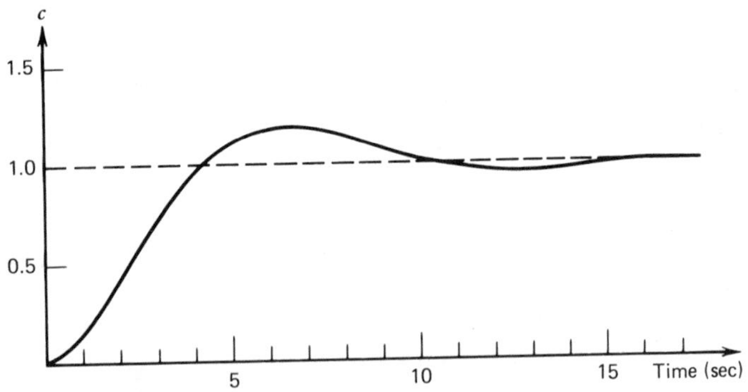


**Figure 8.51 Response of (8.10-10) for a unit step input.**

## Frequency Response and Digital Compensation

As we have seen, analog system performance can be improved by the introduction of new poles and zeros with a compensator. The same is true for digital systems, and the design can be made with either the root-locus or the frequency-response plots. The Nyquist stability criterion remains essentially the same for discrete-time systems, and phase and gain margins can be used as before. Bode showed that for a minimum phase system with a rational transfer function, the amplitude curve must have a slope of no more than about $-20$ db/decade at the gain crossover in order for the phase to stay above $-180°$ (the stability boundary).

The difficulty with discrete-time systems is that the frequency-transfer functions are obtained by substituting $z = \exp(i\omega T)$. Thus the transfer functions are not rational, and Bode's technique does not apply. An alternative is to transform the variable to obtain rational transfer functions in the new plane. Franklin and Powell (Reference 14) have proposed the transformation

$$w = \frac{2}{T}\frac{z-1}{z+1} \tag{8.10-12}$$

or

$$w = \frac{2}{T}\frac{e^{sT}-1}{e^{sT}+1} = \frac{2}{T}\tanh\frac{sT}{2} \tag{8.10-13}$$

From (8.9-6) we see that if $s = i\omega$ (the stability boundary), then $z = \exp(i\omega T)$. The latter expression specifies the unit circle as $\omega$ varies from 0 to infinity. Equation (8.10-13) in this case gives

$$w = i\frac{2}{T}\tan\frac{\omega T}{2} = iv \tag{8.10-14}$$

Thus as $z$ goes around the unit circle, $v$, the frequency in the $w$-plane, goes from 0 to $\infty$.

Given the open-loop transfer function $G(z)H(z)$, we substitute

$$z = \frac{1 + wT/2}{1 - wT/2} \tag{8.10-15}$$

The open-loop transfer function is now $G(w)H(w)$, and the Bode design procedure is the same as before. However, the introduction of the sample-and-hold operation can introduce additional zeros whose locations depend on the sampling period. As the sampling period is decreased, these additional zeros become less important in the design analysis. If they occur in the right-half plane, they make the system nonminimum phase (see Section 8.5), and Bode's relationship between gain and phase no longer applies. A comprehensive design example with this effect is given in Reference 14 (pp. 114–117).

## Compensation with Finite-Time Settling Algorithms

The lead and lag compensation methods of Section 8.9 can be applied to digital systems either by performing the design in the $s$-domain and converting it to difference form, or by converting the specifications into root locations in the $z$-domain. In the latter case, lead and lag compensation simply consists of introducing additional poles

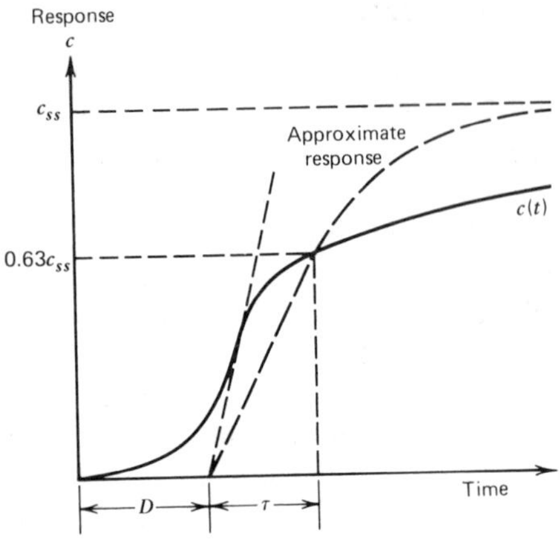


**Figure 8.52 Approximation of process response with a first-order lag $\tau$ and a dead-time element $D$. The time $D$ is found from the maximum slope of the response curve.**

and zeros to the open-loop transfer function in terms of $z$. The techniques are similar to those used with analog systems.

We now indicate how a digital controller can be designed to compensate for the effects of dead time. With the direct design method of Section 7.8, the required controller transfer function $D(z)$ can be found in terms of that of the plant-and-hold $G(z)$ and the desired closed-loop transfer function $T(z)$ (see Figure 7.24). The result (7.8-2) is repeated here.

$$D(z) = \frac{T(z)}{G(z)[1 - T(z)]} \tag{8.10-16}$$

Suppose that the plant has a step response like the $S$-shaped signature shown in Figure 8.52. This is similar to that shown in Figure 7.1 for the Ziegler-Nichols process reaction method. Thus a simple model of the plant consists of a first-order lag with a dead-time element; that is,

$$G_p(s) = \frac{e^{-Ds}}{\tau s + 1} \tag{8.10-17}$$

The pulse-transfer function of the first-order lag with a sample-and-hold is given by (8.10-2). The dead time results in a term $z^{-n}$, where $n$ relates the dead time to the sampling period, as

$$D = nT \tag{8.10-18}$$

We assume that the sampling period is chosen so that $n$ is an integer, for simplicity. Thus

$$G_p(z) = z^{-n}\frac{1-a}{z-a} \tag{8.10-19}$$

where $a = \exp(-T/\tau)$.

Consider the closed-loop transfer function

$$T(z) = \frac{C(z)}{R(z)} = z^{-(n+1)} \tag{8.10-20}$$

Then

$$C(z) = z^{-(n+1)}R(z) \tag{8.10-21}$$

From Tables 4.3 and 4.5, (8.10-21) implies that

$$c(k) = r(k-n-1) \qquad (8.10\text{-}22)$$

If the input $r(k)$ is a unit step, $r(k) = 1, k \geqslant 0; r(k) = 0, k < 0$, then

$$c(k) = 0, \quad k \leqslant n$$

$$c(k) = 1, \quad k \geqslant n+1$$

That is, the output lags the input signal by $(n + 1)$ sampling periods, one more sampling period than is in the dead time $D$. We assume that the extra sample period is required because no system can respond instantaneously to an input.

The preceding response is desirable for start-up operations, and the controller transfer function required to obtain it is given by substituting (8.10-20) and (8.10-19) into (8.10-16). This yields

$$\frac{M(z)}{E(z)} = D(z) = \frac{1}{1-a}\frac{1-az^{-1}}{1-z^{-(n+1)}} \qquad (8.10\text{-}23)$$

Such a fast response might require a value for the manipulated variable beyond the capability of the final control element. To check this, obtain $M(z)$ in terms of $R(z)$. From block-diagram manipulation, the required relation is found to be

$$M(z) = \frac{T(z)}{G_p(z)}R(z) = \frac{1-az^{-1}}{1-a}R(z) \qquad (8.10\text{-}24)$$

If $R(z)$ represents a unit step input,

$$M(z) = \frac{1-az^{-1}}{1-a}\frac{z}{z-1} = \frac{a}{1-a} + \frac{z}{z-1}$$

and

$$m(k) = \frac{a}{1-a}\delta(k) + 1$$

where $\delta(k) = 1$ when $k = 0$, and $\delta(k) = 0$ otherwise. The maximum value of the manipulated variable occurs at $k = 0$ and is $1 + a/(1-a) = 1/(1-a)$. For a fixed dead time $D$, an increase in the sampling period $T$ causes an increase in this maximum value.

The algorithm for $D(z)$, (8.10-23), is called a *finite-settling time* algorithm because the response reaches the desired value in a finite, prescribed time. Such algorithms produce a closed-loop transfer function that is a finite polynomial in $z^{-1}$, such as (8.10-20). The principle can be applied to other plant types. It should be noted, however, that the controller transfer function cancels the plant pole at $z = a$, and the design is therefore subject to the criticism given in Section 8.9. A further analysis of the point is presented in Section 9.4.

## 8.11 SUMMARY

The methods of this chapter are very powerful and constitute the most commonly used design tools for control systems. Nevertheless, some situations require more advanced methods, such as when more than one control loop or control variable exists. Although a given system will have only one characteristic equation, the selection of multiple gains is difficult to do with the root-locus or frequency-response plots. Also, the methods can be difficult to apply if the performance specifications are not expressed in terms of either dominant root locations, bandwidth, or phase and gain margins.

For this purpose, in the next two chapters we consider a generalization of the performance index method of Section 7.1, along with other matrix methods that can deal with multiple-loop, multiple-gain systems in a systematic way.

## REFERENCES

1. R. H. Cannon, *Dynamics of Physical Systems*, McGraw-Hill, New York, 1967.
2. J. L. Melsa and S. K. Jones, *Computer Programs for Computational Assistance in the Study of Linear Control Theory*, McGraw-Hill, New York, 1973.
3. Y. Takahashi, M. Rabins, and D. Auslander, *Control*, Addison-Wesley, Reading, Mass., 1970.
4. S. H. Lehnigk, *Stability Theorems for Linear Motions*, Prentice-Hall, Englewood Cliffs, N.J., 1966.
5. R. Oldenburger (ed.), *Frequency Response*, Macmillan, New York, 1956.
6. J. G. Truxal (ed.), *Control Engineers Handbook*, McGraw-Hill, New York, 1958.
7. H. Chestnut and R. W. Mayer, *Servomechanisms and Regulating System Design*, Vol. I, 2nd ed., John Wiley, New York, 1960.
8. *ASME Quarterly Transactions on Dynamic Systems, Measurement and Control* (formerly the *Journal of Basic Engineering*), ASME, New York.
9. *IEEE Monthly Transactions on Automatic Control*, IEEE, New York.
10. *Control Engineering* (monthly), Penton Press, Cleveland.
11. *Instruments and Control Systems* (monthly), Chilton Publishing Co., Philadelphia.
12. *Automatica* (monthly), Pergamon Press, Elmsford, N.Y.
13. H. M. James, N. B. Nichols, and R. S. Philips, *Theory of Servomechanisms*, McGraw-Hill, New York, 1947.
14. G. F. Franklin and J. D. Powell, *Digital Control of Dynamic Systems*, Addison-Wesley, Reading, Mass., 1980.

## PROBLEMS

8.1 In the following equations, identify the root-locus plotting parameter $K$ and its range in terms of the parameter $b$ and its indicated range.

(a) $s^2 + 4s + 3b = 0, \quad b \geqslant 0$

(b) $s^2 + (2 + b)s + 3 + 2b = 0, \quad b \geqslant 0$

(c) $s^2 + (2 + b)s + 2 + 2b = 0, \quad b \leqslant 0$

(d) $s^3 + 4bs^2 + 3s + b = 0, \quad b \geqslant 0$

8.2 In parts a through i, sketch the root-locus plot for the given characteristic equation for $K \geqslant 0$. Use all the guides that are applicable.

(a) $s(s + 4) + K = 0$

(b) $s(s + 4)(s + 6) + K = 0$

(c) $s^2 + 2s + 2 + K(s + 2) = 0$

(d) $s(s + 2) + K(s + 3) = 0$

(e) $s(s^2 + 2s + 2) + K = 0$

(f) $s(s + 1)(s + 8) + K(s + 2) = 0$

(g) $s + 1 + Ke^{-Ds} = 0$

(h) $s(s + 1) + Ke^{-Ds} = 0$

(i) $s(s + 2) + K(3 - s) = 0$

8.3 In parts a through g, sketch the root-locus plot for $K \leqslant 0$ (the complementary locus), for the given characteristic equation. Use all applicable guides.

(a) $s(s + 4) + K = 0$

(b) $s^2 + 2s + 2 + K(s + 2) = 0$

(c) $s(s^2 + 2s + 2) + K = 0$

(d) $s(s + 4)(s + 6) + K = 0$

(e) $s(s + 2) + K(s + 3) = 0$

(f) $s + 1 + Ke^{-Ds} = 0$

(g) $s(s + 1) + Ke^{-Ds} = 0$

8.4 The root-locus plot obtained in part d of Problem 8.2 is for a unity-feedback system with the open-loop transfer function:

$$G(s) = \frac{K(s + 3)}{s(s + 2)}$$

(a) Set the gain $K$ so that the damping ratio is $\zeta = 0.95$ with the smallest possible dominant time constant.

(b) For this value of $K$, compute the response to a unit step input. Use the root-locus plot to evaluate the coefficients in the partial fraction expansion.

8.5 A common situation in position-control systems is that the inertia of the object to be positioned changes during the control process. The control of the angular position of a satellite (to aim a telescope, for example) must account for the change in the satellite's inertia due to the fuel consumption of the control jets. The inertia of a tape reel or a paper roll changes as it unwinds, and the controller must be designed to handle this variation.

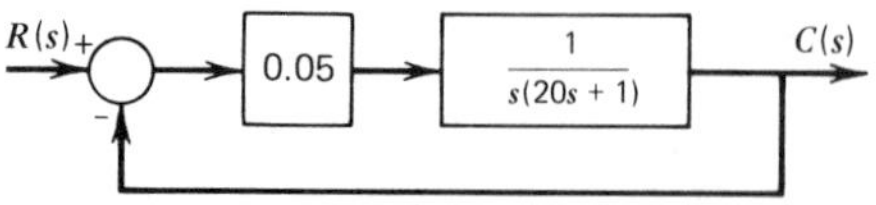


**Figure P8.5**

Figure P8.5 shows P control of a second-order plant with an inertia $I = 20$. Use the root locus to investigate the change in the characteristic roots of the closed-loop system if $I$ can vary by $\pm$ 10%.

8.6 Consider Problem 6.20. Use the root locus to evaluate the performance of the resulting controller if the smaller mass $m$ has a range of uncertainty of $5 \leqslant m \leqslant 15$. The relevant specifications are $\zeta = 0.707$ and $t_s = 10$ sec.

8.7 Consider Problem 6.24. Plot the root locus for the system in terms of the amplifier gain $K$ for the two limits represented by (a) an empty reel and (b) a full reel.

8.8 Consider part e of Problem 6.14. Use the root locus to evaluate the performance of the resulting control system in light of the given specifications.

**(a)** When the final control elements have a transfer function

$$G_c(s) = \frac{1}{\tau s + 1} \qquad 0 \leqslant \tau \leqslant 1$$

**(b)** When the feedback sensor has a transfer function of the form in part a and the dynamics of the final control elements are negligible.

8.9 Consider Problem 6.20. Use the root locus to evaluate the performance of the resulting controller in light of the given specifications.

**(a)** When the final control elements have a transfer function

$$G_c(s) = \frac{1}{\tau s + 1} \qquad 0 \leqslant \tau \leqslant 1$$

**(b)** When the feedback sensor has a transfer function of the form in part a and the dynamics of the final control elements are negligible.

8.10 Consider part a of Problem 6.16. Use the root locus to evaluate the performance of the resulting controller relative to the specifications $\tau = 0.1$, $\zeta = 1$ when the following parameters vary about their nominal values.

**(a)** $0.05 \leqslant c \leqslant 0.15$

**(b)** $0.5 \leqslant a \leqslant 1.5$

**8.11** Consider part a of Problem 7.12. Use the root locus to evaluate the performance of the resulting controller in light of the specifications $\zeta = 0.707, \tau = 0.1$ if the plant transfer function $G_p(s)$ has an uncertainty $\Delta$, where

$$G_p(s) = \frac{1}{s^2 - 2 - \Delta} \qquad -1 \leqslant \Delta \leqslant 1$$

**8.12** A certain plant has the transfer function

$$G_p(s) = \frac{4p}{(s^2 + 4\zeta s + 4)(s + p)}$$

where the nominal values of $\zeta$ and $p$ are $\zeta = 0.5, p = 1$.

**(a)** Use Ziegler-Nichols tuning to compute the PID gains. Find the resulting closed-loop characteristic roots.

**(b)** Use the root locus to determine the effect of a variation in the parameter $\zeta$ over the range $0.4 \leqslant \zeta \leqslant 0.6$.

**(c)** Use the root locus to determine the effect of a variation in the parameter $p$ over the range $0.5 \leqslant p \leqslant 1.5$.

**8.13** The open-loop transfer function of a certain plant is

$$G_p(s) = \frac{1}{s(s + 1)(s + 8)}$$

It is not possible with proportional control to achieve a dominant time constant of less than 2.08 sec for this plant. Use PD control to improve the response, so that the dominant time constant is 0.5 sec or less, and the damping ratio is 0.5 or greater. To do this, first select a suitable value for $T_D$ and plot the locus with $K_p$ as the variable.

**8.14** Proportional control of a liquid level system gives the open-loop transfer function

$$G(s) = \frac{K}{\tau s + 1}$$

where unity feedback is used $[H(s) = 1]$, and $\tau$ is the time constant of the plant. This transfer function assumes that the float used to measure the liquid height responds instantaneously to changes in the liquid level. However, the float possesses dynamics of its own that resemble those of a mechanical oscillator. These dynamics introduce parasitic roots into the system.

Assume that $\tau = 1$ and that the float dynamics are such that the feedback transfer function is

$$H(s) = s^2 + 4s + 8$$

Sketch the root-locus plots for the unity feedback and nonunity feedback cases. For both cases select $K$ to give the smallest dominant time constant possible. Discuss how the parasitic roots affect this choice. What is the effect of using a large value of $K$ in each case?

**8.15** Consider the velocity control of a load driven by an armature-controlled dc motor, as discussed in Section 7.2 (see Figure 7.2).

(a) If we neglect the armature inductance $L$, the plant is first order. With the parameter values given in Section 7.2, the open-loop transfer function for PI control is

$$G(s)H(s) = \frac{3.6315K_p(s + 1/T_I)}{s(0.043s + 5.3832)}$$

It is desired that $\zeta = 0.707$ and that the dominant time constant be less than 0.01 sec. Two roots will satisfy these conditions if the zero at $s = -1/T_I$ is selected to the left of the plant's pole. Choose $T_I$ and plot the locus for $K_p$. Find the values of $K_p$ that correspond to both root locations. Which one gives the smallest dominant time constant?

(b) If the armature inductance is included, the open-loop transfer function with PI control is

$$G(s)H(s) = \frac{3.6315K_p(s + 1/T_I)}{s[0.1Ls^2 + (0.07L + 0.043)s + 5.3832]}$$

For each of the two designs generated in part a, plot the root locus in terms of $L$ as the locus variable for $0 \leq L \leq 0.003$. Which design is more sensitive to a nonzero inductance $L$?

**8.16** Design a PI controller for the system described in part b of Problem 8.15. Take $L = 0.0021$. The specifications state that the dominant roots must have $\zeta \geq 0.5$ and a time constant of no more than 0.02 sec. Choose a trial value for $T_I$ and plot the locus for $K_p$. If a solution is not obtained, try a new $T_I$ and repeat the process.

**8.17** Two mechanical compensators are shown in Figure P8.17. Derive the transfer functions for each. The input is the displacement $x_i$, and the output is the displacement $x_o$. Show that the system in part a is a lead compensator and that the one in part b is a lag compensator.

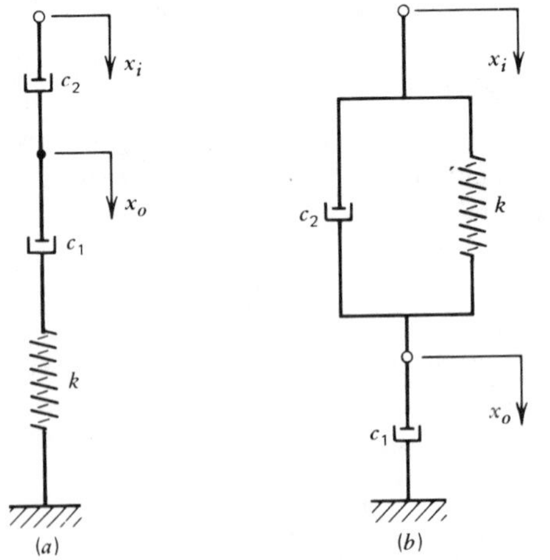


**Figure P8.17**

**8.18** Consider a plant whose open-loop transfer function is

$$G(s)H(s) = \frac{1}{s(s^2 + 4s + 13)}$$

The complex poles near the origin give only slightly damped oscillations that are considered undesirable. Insert a gain $K_c$ and a compensator $G_c(s)$ in series to speed up the closed-loop response of the system. Consider the following for $G_c(s)$.

**(a)** The lead compensator (8.9-9).

**(b)** The lag compensator (8.9-12).

**(c)** The so-called reverse-action compensator

$$G_c(s) = \frac{1 - T_1 s}{T_2 s + 1}$$

Sketch the root-locus plots for the compensated system using each compensator. Use $K_c$ as the locus parameter. Without computing specific values for the compensator parameters, determine which compensator gives the best response.

**8.19** A control system with a dead time of $D$ sec has the open-loop transfer function

$$G(s)H(s) = \frac{10e^{-sD}}{0.1s + 1}$$

How large can the dead time be if the phase margin must be at least 40°?

**8.20** The Pade approximation to the dead-time element is

$$e^{-sD} = \frac{1}{[1 + (Ds/n)]^n}$$

The approximation improves as $n$ becomes larger. For relatively high-order systems with dead time, the techniques presented in this chapter can be difficult to apply, and the Pade approximation for low values of $n$ is often used to model the effects of dead time.

Use the Pade approximation to construct the root locus for the system

$$G(s)H(s) = \frac{Ke^{-s}}{s}$$

for $n = 1$ and $n = 2$. Compare with the exact plot given in Figure 8.27.

**8.21** The open-loop transfer function of a certain system is

$$G(s)H(s) = \frac{0.5K}{s(s + p)}$$

where $H(s) = 1$. The nominal value of $p$ is $p = 1$. The desired root locations are $s = -0.5 \pm i0.5$, so that $K$ is set to $K = 1$. Use the root sensitivity method to determine what parameter, $p$ or $K$, must have the most stringent tolerance on its nominal value.

8.22 The transfer functions for the most basic lead and lag compensators are

(a) $T(s) = K(\tau s + 1)$ (lead)

(b) $T(s) = \dfrac{K}{\tau s + 1}$ (lag)

Plot the Bode diagrams for these and compare with those of the compensator circuits shown in Figures 8.37 through 8.40. Discuss the relative performance of each.

8.23 A particular control system with unity feedback has the open-loop transfer function

$$G(s) = \frac{K}{s(s+2)}$$

With $K = 4$, the damping ratio is $\zeta = 0.5$, the natural frequency is $\omega_n = 2$ rad/sec, and $C_v = 2$/sec.

(a) Design a compensator to obtain $\omega_n = 4$ while keeping $\zeta = 0.5$. Find the compensator's resistances if $C = 1\ \mu f$.

(b) Design a compensator that will give a static velocity error coefficient of $C_v = 20$/sec, a phase margin of at least 40°, and a gain margin of at least 6 db. How do the open-loop poles and zeros of the resulting compensated system compare with those of the system designed in part a?

(c) Suppose that with $K = 4$, the original system gives a satisfactory transient response, but $C_v$ must be increased to $C_v = 20$/sec. Design a compensator to do this.

8.24 The bandwidth of a closed-loop system is usually given roughly by the gain crossover frequency of the open-loop system. This is a useful design guide. Design a lead compensator for a system whose open-loop transfer function is

$$G(s)H(s) = \frac{72}{(s+1)(s+3)^2}$$

The compensated system should have a phase margin of 40° or more, and a closed-loop bandwidth approximately the same as the gain crossover frequency of the open-loop uncompensated system.

8.25 Consider the system whose open-loop transfer function is given in the previous problem. Design a lag-lead compensator to give a phase margin of no less than 40°, $C_p \geqslant 10$ and the same gain crossover frequency as with the uncompensated system.

8.26 A unity feedback system has the open-loop transfer function

$$G(s) = \frac{0.625}{s(0.5s + 1)(0.125s + 1)}$$

Design a compensator to give $s = -2 \pm i2\sqrt{3}$ and $C_v = 80/\text{sec}$.

8.27 The system shown in Figure P8.27*a* represents the digital equivalent of I action, while in Figure P8.27*b* the controller is a delayed I action. Plot the root locus for each system and determine the stability limit.

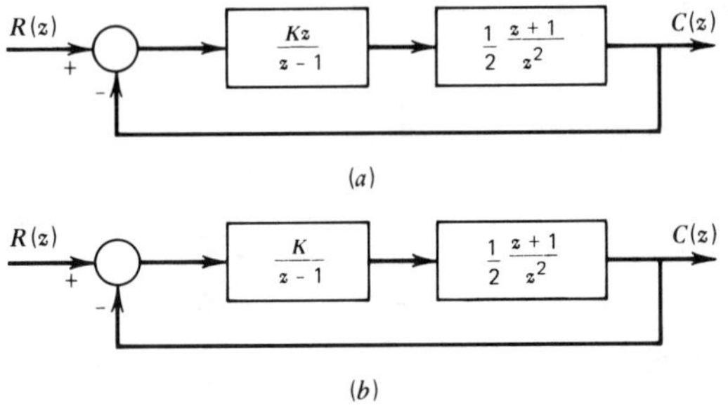


**Figure P8.27**

8.28 Consider the proportional controller shown in Figure P8.28, where

$$G(s) = \frac{1}{s(s + 1)}$$

Let $T = 1$.

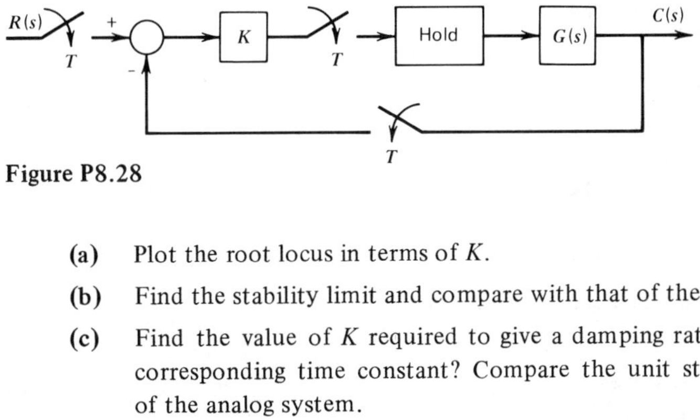


**Figure P8.28**

(a) Plot the root locus in terms of $K$.

(b) Find the stability limit and compare with that of the analog system.

(c) Find the value of $K$ required to give a damping ratio of 0.5. What is the corresponding time constant? Compare the unit step response with that of the analog system.

(d) Is it possible to find a value of $K$ to give a damping ratio of 1 with a time constant of 2? Explain.

8.29 Consider the system shown in Figure P8.28 where

$$G(s) = \frac{1}{s(s + 1)(s + 3)}$$

and $T = 0.1$. Plot the root locus in terms of $K$.

8.30 P control of a pure integrator with dead time is considered. Suppose that the dead time is $D = 1$.

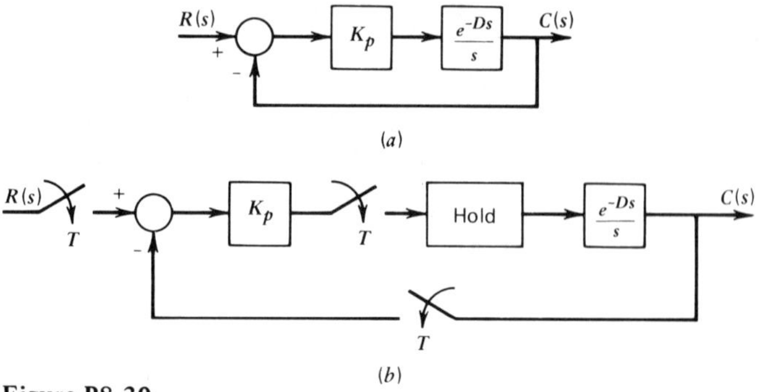


**Figure P8.30**

(a) For the analog controller shown in Figure P8.30*a*, compute the proportional gain $K_p$ in the following ways.

(1) By the Ziegler-Nichols response method.

(2) By the Ziegler-Nichols ultimate-cycle method.

(b) For the digital controller shown in Figure P8.30*b*, compute the stability limit for $K_p$ and compare with that found for the analog system.

(c) Determine the characteristic root locations using the two values of $K_p$ found in part a, assuming the following.

(1) $T = D = 1$

(2) $T = 0.5D = 0.5$

Compare the results.

8.31 Consider the system shown in Figure P8.28, where

$$G(s) = \frac{e^{-Ds}}{4s + 1}$$

Suppose that the dead time is $D = 1$.

(a) Plot the root locus in terms of $K$ for the following.

(1) $T = D = 1$

(2) $T = 0.5D = 0.5$

(3) $T = 0.25D = 0.25$

(b) What is the effect on the root locus of the choice of $n = D/T$?

# CHAPTER NINE
# Advanced Matrix Methods for Dynamic Systems Analysis

In Chapter Five we saw that a linear continuous-time model can be expressed in state variable form as $\dot{\mathbf{x}} = \mathbf{A}\mathbf{x} + \mathbf{B}\mathbf{u}$, with the output equation: $\mathbf{y} = \mathbf{C}\mathbf{x} + \mathbf{D}\mathbf{u}$. The power of this notation is that it allows relations and algorithms to be developed regardless of the model's order, which is reflected only in the dimensions of the matrices and vectors. In this chapter we exploit this advantage to develop vector-matrix methods for analysis. These methods are particularly well suited for computer use and can be applied to obtain results for high-order or multiinput systems that are difficult to handle with the methods of previous chapters.

We have emphasized the importance of keeping the model's order as low as possible for analytical convenience. Sometimes, however, the important features of the system's behavior can be determined only with a high-order model. In Section 9.1 we discuss typical applications of such models. A vector interpretation of the free response of a high-order system is given in Section 9.2. This leads to better understanding of the system dynamics and allows the development of efficient algorithms for computing the response. These algorithms are the topics of Section 9.3. Any model is an abstraction of reality, and sometimes this abstraction can produce a model in which either the input is incapable of completely influencing the model's response, or the measurements are insufficient for determining the current values of the state variables. This concerns the properties of *controllability* and *observability* and these are treated in Section 9.4. Matrix methods for discrete-time systems are very similar to those for continuous time. These similarities are summarized in Section 9.5.

## 9.1 EXAMPLES OF MATRIX MODELS

Matrix models provide a convenient representation of the system dynamics when the number of energy-storage elements is large. This commonly occurs, for example, when there are many separate masses in a mechanical system, or when the motion consists of rotations or translations in several directions. When the mass is continuously distributed throughout the system, the order is infinite, but the system can often be approximated by a finite-order, lumped-parameter model. We now present examples of such models to illustrate some applications of the methods of this chapter.

### Multimass Systems

In many of our previous motor examples we have lumped the mass of the motor, gears, shafts, and load into one mass. Such a model is useful for estimating the domi-

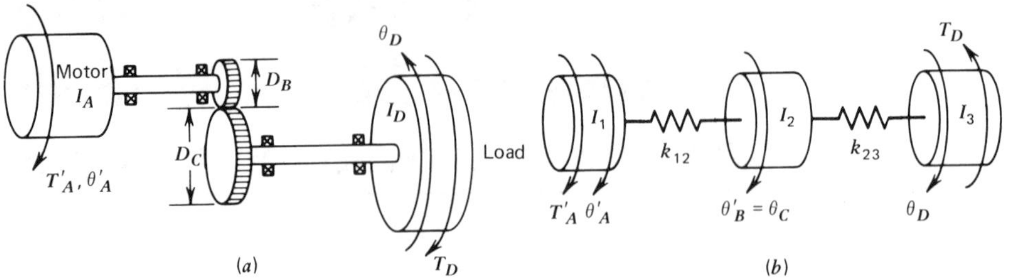


**Figure 9.1 Motor-gear-load system. (*a*) Geometry. (*b*) Three-mass model with the gears and shafts lumped.**

nant time constant but cannot predict the dynamic behavior of the stresses in the shafts, for example. A typical configuration is shown in Figure 9.1*a*. If we are not interested in the gear-tooth stresses, we may lump the two gear inertias together, along with the shaft inertias. The resulting three-mass model is shown in Figure 9.1*b*, in which the motions, torques, stiffnesses, and inertias have been "reflected" to the load shaft using the formulas of Table 2.12. Thus $\theta'_A$ is the displacement $\theta_A$ reflected through the gears such that $\theta'_A = \theta_A/R$ where the gear ratio $R = D_C/D_B$. Similarly $\theta'_B, T'_A, k_{12}, I_1$, and $I_2$ are also reflected values. The friction torques resulting from the bearings and gears are assumed to be negligible relative to the motor torque $T'_A$ and the load torque $T_D$.

Applying Newton's law to each mass gives

$$\begin{aligned} I_1\ddot{\theta}'_A &= T'_A - k_{12}(\theta'_A - \theta_C) \\ I_2\ddot{\theta}_C &= k_{12}(\theta'_A - \theta_C) - k_{23}(\theta_C - \theta_D) \\ I_3\ddot{\theta}_D &= k_{23}(\theta_C - \theta_D) - T_D \end{aligned} \tag{9.1-1}$$

If the load torque results from viscous friction, such as with a pump impeller, a linear model would be

$$T_D = c\dot{\theta}_D \tag{9.1-2}$$

The choice of the relative twists $(\theta'_A - \theta_C)$ and $(\theta_C - \theta_D)$ as state variables is convenient for analyzing the shaft stresses. Thus the state vector can be taken as

$$\left.\begin{aligned} x_1 &= \dot{\theta}'_A - \dot{\theta}_C \qquad & x_2 &= \theta'_A - \theta_C \\ x_3 &= \dot{\theta}_C - \dot{\theta}_D \qquad & x_4 &= \theta_C - \theta_D \\ x_5 &= \dot{\theta}'_A \end{aligned}\right\} \tag{9.1-3}$$

A sixth state variable, say $\theta'_A$ or $\theta_D$, can be chosen if the angular position of the motor or load is important. With $T_D$ specified by (9.1-2), the input is the scalar $u = T'_A$, and (9.1-1) can be written in the standard form: $\dot{\mathbf{x}} = \mathbf{Ax} + \mathbf{Bu}$.

## Motion in Multiple Directions

In general the motion of a body in one direction is coupled to motions in other directions. Summation of the torques relative to the center of mass results in a second-order equation for each axis. Similarly, summation of the forces acting through the center of mass gives a second-order equation for each of the $x, y, z$ directions. Thus if there is significant energy storage associated with every rectilinear and rotational displacement and velocity, the model can be twelfth order. Often, however, the mass distribution of the body or constraints on the motion are such that negligible coupling exists between some coordinates. In addition, we can frequently separate the force-summation and torque-summation equations into two sets.

### *Example 9.1*

Develop a model to investigate the ride-quality effects of road-surface variations in the direction of travel. Assume small angles and use the representation shown in Figure 9.2, where $m$ and $I$ are the car's mass and inertia about the axis indicated by the rotation $\theta$. All displacements are relative to the equilibrium position of the vehicle. The model lumps the elasticity and damping of the tires and suspension, and neglects their masses.

For small angles

$$y_5 = y_3 + \frac{L_1}{L}(y_4 - y_3) \tag{9.1-4}$$

$$L\theta = y_4 - y_3 \tag{9.1-5}$$

Summation of forces in the vertical direction gives

$$m\ddot{y}_5 = k_1(y_1 - y_3) + k_2(y_2 - y_4) + c_1(\dot{y}_1 - \dot{y}_3) + c_2(\dot{y}_2 - \dot{y}_4) \tag{9.1-6}$$

Summation of torques about the center of mass gives

$$\begin{aligned} I\ddot{\theta} &= -L_1k_1(y_1 - y_3) + (L - L_1)k_2(y_2 - y_4) \\ &\quad - L_1c_1(\dot{y}_1 - \dot{y}_3) + (L - L_1)c_2(\dot{y}_2 - \dot{y}_4) \end{aligned} \tag{9.1-7}$$

Choosing the state and input vectors as

$$\mathbf{x} = \begin{bmatrix} y_5 \\ \dot{y}_5 \\ \theta \\ \dot{\theta} \end{bmatrix}, \qquad \mathbf{u} = \begin{bmatrix} y_1 \\ y_2 \end{bmatrix}$$

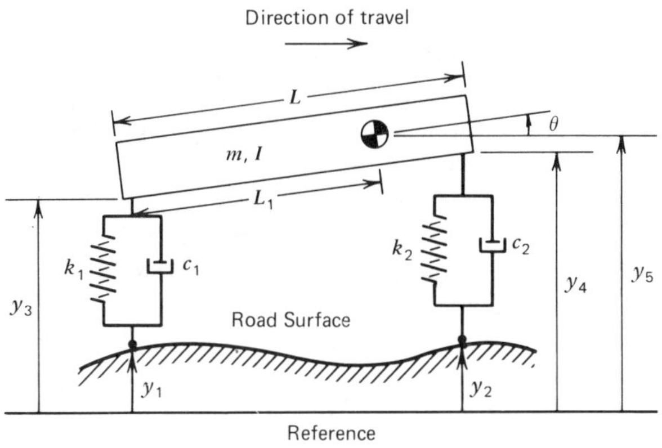


**Figure 9.2 Ride-quality model.**

we can use (9.1-4) and (9.1-5) to eliminate $y_3$ and $y_4$ from (9.1-6) and (9.1-7). The model is linear because of the small angle approximation. It is fourth order because the vertical displacement $y_5$ is coupled to the angular displacement $\theta$.

A more detailed ride-quality model might include the mass of the passengers and the elasticity of the seats, as well as separate elements to model the tires and suspension.

In general, the coupling between coordinates results in a nonlinear model. The general equations for the rotation of the rigid body are the *Euler equations*.

$$I_1\dot{\omega}_1 + (I_3 - I_2)\omega_2\omega_3 = T_1 \tag{9.1-8}$$

$$I_2\dot{\omega}_2 + (I_1 - I_3)\omega_3\omega_1 = T_2 \tag{9.1-9}$$

$$I_3\dot{\omega}_3 + (I_2 - I_1)\omega_1\omega_2 = T_3 \tag{9.1-10}$$

where $I_i$, $\omega_i$, and $T_i$ are the inertia, angular velocity, and torque about the $i$th principal axis. A *principal axis* of a rigid body has the property that if the body rotates about it, then the direction of the angular momentum is the same as that of the angular velocity. Any axis of symmetry is a principal axis if the body's mass is uniformly distributed. The Euler equations can be expressed for an axis set other than principal axes, but the resulting expressions are not as simple as (9.1-8) to (9.1-10).

The Euler equations are nonlinear and coupled as a result of the cross-products of the angular velocities. These terms disappear for some shapes. For example, if $I_3 = I_2$ in (9.1-8), the motion $\omega_1$ is independent of $\omega_2$ and $\omega_3$. Also, if $T_2 = T_3 = 0$, and if $\omega_2(0) = \omega_3(0) = 0$, then $\omega_2\omega_3 = 0$ for all time, and the motion $\omega_1$ is again uncoupled. However, in practice these conditions are rarely met exactly. It is difficult to ensure that the initial conditions for $\omega_2$ and $\omega_3$ are zero or to align the driving force exactly so that its only nonzero torque component appears about the $\omega_1$ axis. An example of this is shown in Figure 9.3, in which the main thrust $F$ generates torques about the three axes. Smaller jets with a thrust $f$ are installed around the rim of the vehicle and are operated by feedback controllers to maintain the desired vehicle orientation and velocities.

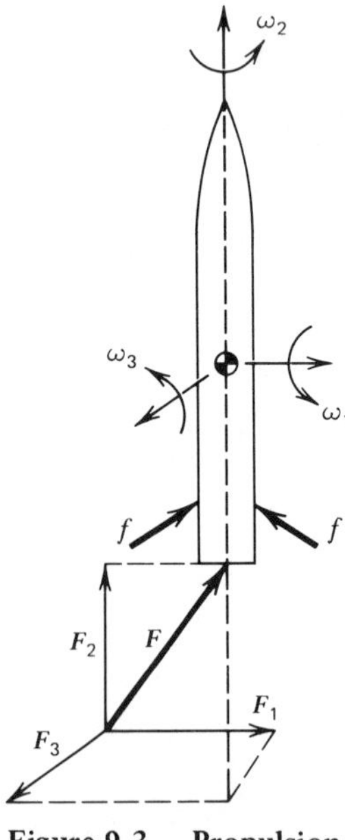


**Figure 9.3 Propulsion and control forces acting on a rocket.**

The Euler equations only describe rotation and must be supplemented with the force summation equations to obtain information on the rectilinear motion. This is the case, for example, when one or more of the torque components depends on the body's rectilinear velocity, such as with aerodynamic control surfaces. In such applications it is often more convenient to uncouple the dynamics, not on the basis of rotational versus translational motions, but on the basis of lateral and longitudinal motions. This type of decoupling is used to design control systems for the aileron,

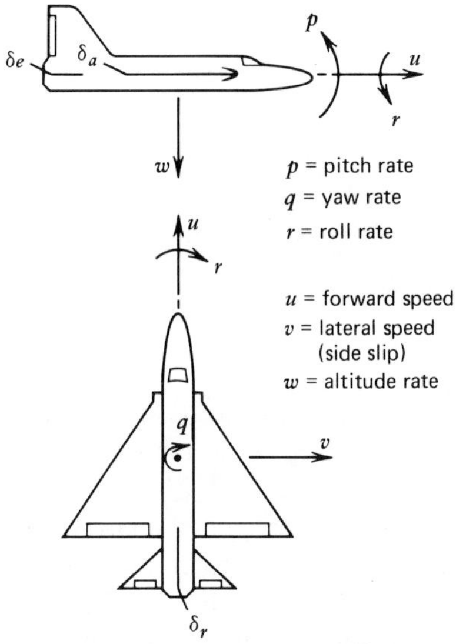


**Figure 9.4 Coordinates and control inputs for an aircraft-stabilization system.**

elevator, and rudder deflections $\delta_a, \delta_e, \delta_r$ in order to improve the stability properties of an aircraft (Figure 9.4). The coupling between the lateral and longitudinal motions is often small, and a model for the lateral motion is then taken to be a linearized fifth-order model with the state and input vectors

$$\mathbf{x} = \begin{bmatrix} \theta \\ r \\ v \\ \phi \\ q \end{bmatrix}, \qquad \mathbf{u} = \begin{bmatrix} \delta_r \\ \delta_a \end{bmatrix} \tag{9.1-11}$$

where $\dot{\theta} = r$ and $\dot{\phi} = q$. A controller for the longitudinal motion might use a fifth-order model also, but with the vectors

$$\mathbf{x} = \begin{bmatrix} z \\ w \\ u \\ \psi \\ p \end{bmatrix}, \qquad \mathbf{u} = \delta_e \tag{9.1-12}$$

where $\dot{z} = w$ and $\dot{\psi} = p$.

Many fascinating problems in dynamic systems and control are currently found in the rapidly developing field of robotics. Mechanical manipulators are needed for industrial automation and for space and undersea exploration. A typical manipulator like that shown in Figure 9.5 has six joints (five for rotation and one for translation), seven links, and a gripper. Each joint is driven by its own motor and can have position and velocity sensors. The dynamics are quite complicated because the system's inertias and the gravity torques depend on the manipulator's instantaneous configuration. The model has the form

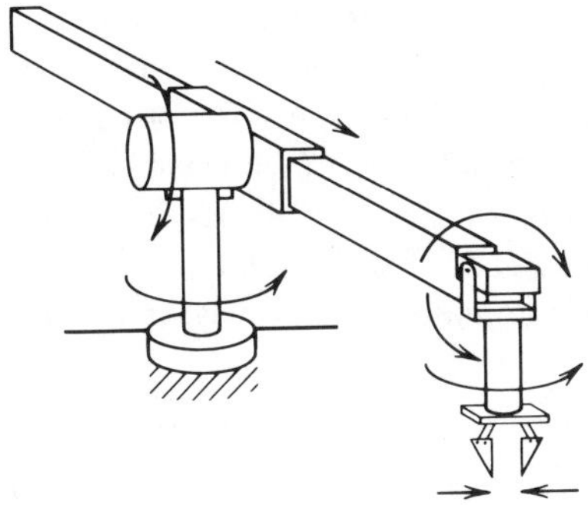

**Figure 9.5 Mechanical manipulator with an arm, wrist, and parallel-jaws gripper.**

$$\mathbf{I}(\mathbf{y})\dot{\mathbf{v}} = -\mathbf{d}(\mathbf{v}) - \mathbf{g}(\mathbf{y}) - \mathbf{c}(\mathbf{y}, \mathbf{v}) + \mathbf{f} \tag{9.1-13}$$

$$\dot{\mathbf{y}} = \mathbf{v} \tag{9.1-14}$$

*Where*

$\mathbf{y}$ = vector of joint displacements [(6 × 1)]

$\mathbf{v}$ = vector of joint velocities [(6 × 1)]

$\mathbf{I}(\mathbf{y})$ = inertia matrix [(6 × 6)]

$\mathbf{d}(\mathbf{y})$ = vector of viscous damping forces

$\mathbf{g}(\mathbf{y})$ = vector of gravitational forces

$\mathbf{c}(\mathbf{y}, \mathbf{v})$ = vector of Coriolis and centrifugal forces

$\mathbf{f}$ = vector of input forces and torques

The vector $\mathbf{c}(\mathbf{y}, \mathbf{v})$ contains cross-products and is thus a nonlinear function. Equations (9.1-13) and (9.1-14) can be put into state variable form by using the state vector $\mathbf{x} = [\mathbf{y}, \mathbf{v}]^T$. The resulting model is twelfth order.

The actuators on each joint are under coordinated computer control. The computer must repeatedly evaluate (9.1-13) and (9.1-14) to find the control torques and forces required to produce the desired motion of the object being held at the endpoint, given the appropriate feedback signals about the endpoint's actual motion. Such computations are difficult but must still be done quickly for accurate motion control. In addition, the desired motions and feedback information from external sources such as television systems are frequently specified in a cartesian coordinate system and not the natural joint coordinates of the manipulator. The resulting coordinate transformations place another burden on the control computer. The proper model formulation, control schemes, and numerical techniques for this application are an active area of research.

## Lumped-Parameter Models of Distributed-Parameter Processes

Distributed-parameter models take the form of partial differential equations. These are more difficult than ordinary differential equations to solve, both analytically and numerically. The solution of such equations is beyond the scope of this work, but we can still deal with distributed-parameter processes by using a lumped-parameter approximation. The state-variable approach is particularly useful for such applications.

Consider a circular rod perfectly insulated all around except at one end (Figure 9.6*a*). The temperature $T$ at any distance $x$ from the left end is a function of time $t$. Thus $T = T(x, t)$. The most accurate model for $T(x, t)$ would involve both spatial and time derivatives and would be a partial differential equation. Instead of formulating this model, we now approximate the rod by a series of "lumps," each having its own representative temperature, which we take to be constant throughout the lump. Assume that the only thermal resistance occurs at the interfaces of the lump, and that each lump is a pure capacitance. An energy balance on the lump $i$ gives the system model

$$C_i \frac{dT_i}{dt} = \frac{1}{R_i}(T_{i-1} - T_i) - \frac{1}{R_{i+1}}(T_i - T_{i+1}) \qquad i = 1, 2, \ldots, n-1 \tag{9.1-15}$$

$$C_n \frac{dT_n}{dt} = \frac{1}{R_n}(T_{n-1} - T_n) \tag{9.1-16}$$

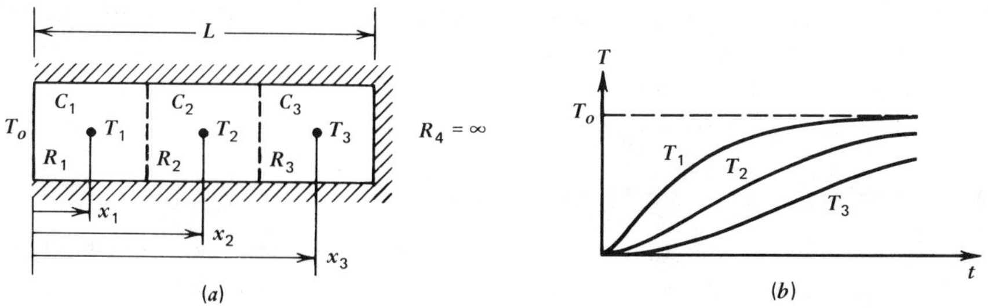


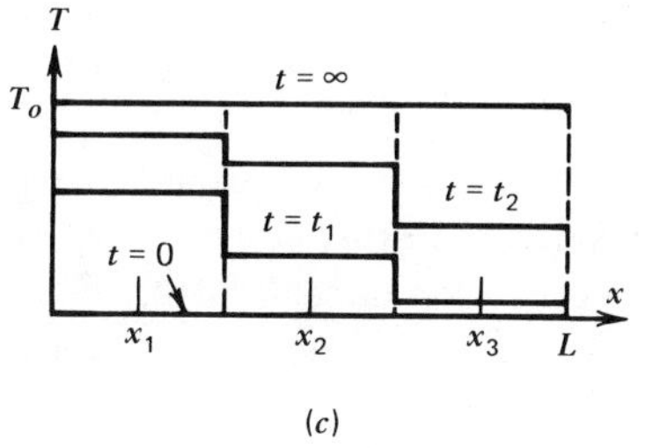


**Figure 9.6 Lumped-parameter model of temperature dynamics in an insulated rod. (*a*) Three-lump model. (*b* and *c*) Response to a step change in $T_o$.**

where $n$ is the number of lumps. The state vector and the input can be taken to be

$$\mathbf{x} = \begin{bmatrix} T_1 \\ T_2 \\ \cdot \\ \cdot \\ \cdot \\ T_n \end{bmatrix}, \qquad u = T_o$$

The matrix **A** will have a "banded" structure with zero entries everywhere except on the main diagonal and the first upper and lower subdiagonals.

Figure 9.6*a* shows a three-lump approximation. The step response of a three-lump model of the rod would look like Figures 9.6*b* and 9.6*c*. The actual response, of course, will not be discontinuous but smooth. The response of the lumped-parameter model will approach the actual behavior as the number of lumps is increased. If great accuracy is required, a state model of very high order can easily result. The model can be used as a control model, for example, when $T_o$ is a temperature input used to control the rod temperature. Then the model's utility will be lessened if it is of high order. A compromise between utility and accuracy must be sought.

There are several considerations relative to developing lumped-parameter approximations.

1. Exact solutions of partial differential equations are available for simple geometries for some heat-transfer, fluid-flow, and mechanical-vibration problems; they can be found in the appropriate reference works for these subjects. The solutions often give insight for analyzing the behavior of systems with similar but more complex geometry.

2. Physical insight can be used to improve the choice of lump size. For example, if the rod had a diameter that varied along its length, lumps of uneven length might be chosen in order to obtain lumps of equal capacitance.

3. The required number of lumps can often be found relative to the desired accuracy by increasing the number of lumps, simulating the process, and comparing the results. As the number is increased, the results should converge to the same values.

### Multivariable Control Systems

Recent improvements in computer and sensor technology have resulted in the development of control systems for a wider range of applications. They usually have more than one control signal and are termed *multivariable control systems*. One of the most exciting of these areas is engine control, for both automotive and jet engines. These applications rely heavily on matrix methods.

The control variables that can be used to improve the fuel economy and emissions characteristics of an automotive engine are the air-fuel ratio, the spark advance, and the exhaust gas recirculation. The driver is in direct control of the throttle position and the times of gear shifting. These are additional input variables. A possible model is

$$\dot{\mathbf{x}} = \mathbf{f}(\mathbf{x}, \mathbf{u})$$

*Where*

$x_1$ = manifold pressure

$x_2$ = engine angular velocity

$x_3$ = vehicle velocity

and $\mathbf{u}$ consists of the five input and control variables just mentioned. The vector function $\mathbf{f}(\mathbf{x}, \mathbf{u})$ is quite complicated but can be developed by using curve-fitting techniques with test data. Examples of such models are given in Reference 1 for two engine types, along with a control analysis.

## 9.2 VECTOR REPRESENTATIONS OF THE FREE RESPONSE: SYSTEM MODES

At this point it is convenient to reformulate the results of Section 5.3 in terms of state vector notation. This formulation will provide insight into the model's dynamic behavior and will allow our results to be extended to higher-order models more easily.

### Exponential Substitution: The Vector Case

Insight into the free response of the vector state equation

$$\dot{\mathbf{x}} = \mathbf{A}\mathbf{x} + \mathbf{B}\mathbf{u} \tag{9.2-1}$$

can be obtained by setting $\mathbf{u} = \mathbf{0}$ and substituting an exponential vector form $\mathbf{x}(t) = \mathbf{p}e^{st}$ into (9.2-1). For the second-order case this form reduces to

$$x_1(t) = p_1 e^{st}, \qquad x_2(t) = p_2 e^{st} \tag{9.2-2}$$

where $s$, $p_1$, and $p_2$ are unknown constants. Making the substitution and cancelling

$e^{st}$ from each equation, we obtain

$$(s-a_{11})p_1 - a_{12}p_2 = 0 \tag{9.2-3}$$

$$-a_{21}p_1 + (s-a_{22})p_2 = 0 \tag{9.2-4}$$

This is a set of two equations in three unknowns, $s$, $p_1$, and $p_2$. The theory of linear algebraic equations tells us that if all the right-hand sides are zero, a nonzero solution for $p_1$ and $p_2$ exists only if the determinant of the left-hand side is zero. Consider a numerical example where the equations are

$$p_1 - p_2 = 0$$

$$2p_1 - 2p_2 = 0$$

The determinant is

$$\begin{vmatrix} 1 & -1 \\ 2 & -2 \end{vmatrix} = -2+2=0$$

The only information the set gives is that $p_1 = p_2$. Obviously the second equation is just the first multiplied by two. This is the implication of a determinant of value zero; both equations are identical (one is *redundant*), and the solution is *indeterminate*. If, however, the equations are

$$p_1 - p_2 = 0$$

$$p_1 - 2p_2 = 0$$

the determinant has the value $-1$, and the only solution is the trivial one $p_1 = p_2 = 0$.

In order for our assumed solution forms (9.2-2) to be general, $p_1$ and $p_2$ cannot be zero. Thus we require the determinant of the set (9.2-3) and (9.2-4) to be zero. This determinant is identical to $\Delta(s)$ as given by (5.2-32).

$$\Delta(s) = \begin{vmatrix} (s-a_{11}) & -a_{12} \\ -a_{21} & (s-a_{22}) \end{vmatrix} = 0 \tag{9.2-5}$$

This requirement results in the characteristic equation for the model and produces the two values for $s$, namely $s_1$ and $s_2$. The assumed solution forms (9.2-2) are valid for both $s_1$ and $s_2$, and the general solution is a linear combination of the two.

$$x_1(t) = p_{11}e^{s_1 t} + p_{12}e^{s_2 t} \tag{9.2-6}$$

$$x_2(t) = p_{21}e^{s_1 t} + p_{22}e^{s_2 t} \tag{9.2-7}$$

where we have introduced a second subscript to emphasize that two values are required for both $p_1$ and $p_2$ (one for each root).

More information can be extracted from (9.2-3) and (9.2-4) by letting $s = s_1$ and solving for the values of $p_1$ and $p_2$ associated with the root $s_1$. These are denoted $p_{11}$ and $p_{21}$. From (9.2-3),

$$(s-a_{11})p_{11} - a_{12}p_{21} = 0$$

or, if $a_{12} \neq 0$,

$$p_{21} = \frac{s_1 - a_{11}}{a_{12}} p_{11} = m_1 p_{11} \tag{9.2-8}$$

where the constant $m_1$ is introduced to simplify notation. Since the second equation (9.2-4) is redundant, it will give the same answer.

Repeat this procedure for $s = s_2$ to obtain

$$p_{22} = \frac{s_2 - a_{11}}{a_{12}} p_{12} = m_2 p_{12} \tag{9.2-9}$$

We now have two unknowns left, $p_{11}$ and $p_{12}$. These will be determined from the initial conditions on $x_1$ and $x_2$. The solution now takes the form

$$x_1(t) = p_{11} e^{s_1 t} + p_{12} e^{s_2 t} \tag{9.2-10}$$

$$x_2(t) = m_1 p_{11} e^{s_1 t} + m_2 p_{12} e^{s_2 t} \tag{9.2-11}$$

These can be expressed in vector form as follows.

$$\begin{bmatrix} x_1(t) \\ x_2(t) \end{bmatrix} = p_{11} \begin{bmatrix} 1 \\ m_1 \end{bmatrix} e^{s_1 t} + p_{12} \begin{bmatrix} 1 \\ m_2 \end{bmatrix} e^{s_2 t} \tag{9.2-12}$$

This form shows that the free response consists of a linear combination of two vectors $\mathbf{v}_1$ and $\mathbf{v}_2$, called *eigenvectors*, where

$$\mathbf{v}_1 = \begin{bmatrix} 1 \\ m_1 \end{bmatrix} \tag{9.2-13}$$

$$\mathbf{v}_2 = \begin{bmatrix} 1 \\ m_2 \end{bmatrix} \tag{9.2-14}$$

and

$$\mathbf{x}(t) = p_{11} \mathbf{v}_1 e^{s_1 t} + p_{12} \mathbf{v}_2 e^{s_2 t} \tag{9.2-15}$$

The eigenvectors are weighted in the solution by the initial condition factors $p_{11}$ and $p_{12}$ and by the exponential time functions. The vectors $\mathbf{v}_1 e^{s_1 t}$ and $\mathbf{v}_2 e^{s_2 t}$ are the *eigenmodes* of the system. The following example illustrates their physical significance.

*Example 9.2*

Discuss the physical significance of the eigenmodes of the system shown in Figure 9.7.

Assume that $\omega_1 > \omega_2$. Summation of torques on each inertia equals the product of its inertia times its angular velocity. The resulting model is

$$I_1 \dot{\omega}_1 = -c_1(\omega_1 - \omega_2) \tag{9.2-16}$$

$$I_2 \dot{\omega}_2 = c_1(\omega_1 - \omega_2) - c_2 \omega_2 \tag{9.2-17}$$

For the sake of discussion, assume that $I_1 = I_2 = 1$.

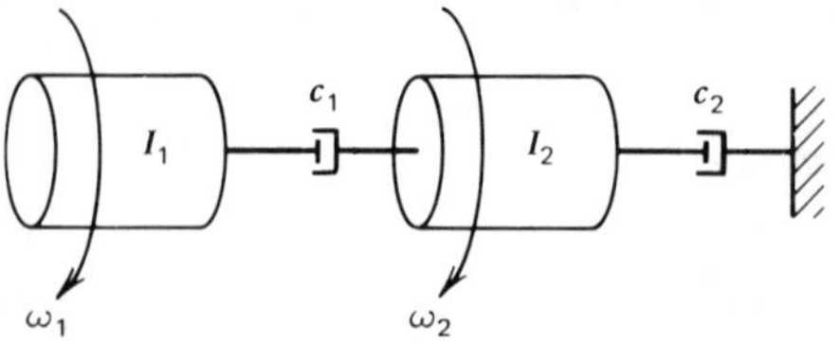


**Figure 9.7** Two masses in rotation with damping. The modes of this system are discussed in Example 9.2.

(a) First consider the case in which the second damper does not exist ($c_2 = 0$), and let $c_1 = 1$. Then

$$\dot{\omega}_1 = -\omega_1 + \omega_2 \tag{9.2-18}$$

$$\dot{\omega}_2 = \omega_1 - \omega_2 \tag{9.2-19}$$

and $a_{11} = a_{22} = -1$, $a_{12} = a_{21} = 1$ in the general state variable form ($x_1 = \omega_1$, $x_2 = \omega_2$). The characteristic equation is

$$s^2 + 2s = 0$$

and the roots are $s_1 = 0$, $s_2 = -2$. From (9.2-8) and (9.2-9), $m_1 = 1$ and $m_2 = -1$. The free response is

$$\begin{bmatrix} \omega_1(t) \\ \omega_2(t) \end{bmatrix} = p_{11} \begin{bmatrix} 1 \\ 1 \end{bmatrix} e^0 + p_{12} \begin{bmatrix} 1 \\ -1 \end{bmatrix} e^{-2t} \tag{9.2-20}$$

where the eigenvectors are

$$\mathbf{v}_1 = \begin{bmatrix} 1 \\ 1 \end{bmatrix}, \qquad \mathbf{v}_2 = \begin{bmatrix} 1 \\ -1 \end{bmatrix}$$

This says that the general motion of the system consists of a linear combination of two motions. The first consists of both inertias rotating in the same direction at the same speed. This corresponds to the first eigenvector $\mathbf{v}_1$. In the second motion both inertias move with equal speeds, but in the opposite direction.

The eigenmodes indicate the relative importance of these motions as a function of time. The characteristic root $s_1 = 0$ indicates that the first motion does not disappear with time, while the root $s_2 = -2$ shows that the second motion will disappear exponentially with a time constant of 1/2. This makes sense physically. If we start the inertias rotating together with the same speed, they will continue to do so without any change in speed because no net velocity, and thus no net torque, exists between them (remember that $c_2 = 0$ here).

If the initial velocities of the inertias are not the same, the action of the damper will eventually reduce the velocity *difference* to zero. This is the meaning of the second eigenmode. The velocity difference decays exponentially with a time constant of 1/2.

The preceding observation suggests that it is possible to describe the motion in terms of the sum and the difference of the two speeds. Let $z_1 = \omega_1 + \omega_2$, and $z_2 = \omega_1 - \omega_2$. It is easy to show by substitution in (9.2-18) and (9.2-19) that the model becomes

$$\dot{z}_1 = 0 \tag{9.2-21}$$

$$\dot{z}_2 = -2z_2 \tag{9.2-22}$$

The solution of these uncoupled equations is

$$z_1(t) = z_1(0) = \omega_1(0) + \omega_2(0) \tag{9.2-23}$$

$$z_2(t) = z_2(0)e^{-2t} = [\omega_1(0) - \omega_2(0)]e^{-2t} \tag{9.2-24}$$

This says that the total speed ($z_1$) remains constant while the relative speed ($z_2$)

decays, as we have seen. The eigenvectors have suggested a new set of variables in which the system's differential equations are simplified (uncoupled). We will see this to be the case in general.

(b) The simple sum and difference of speeds usually do not represent the eigenmodes. To see this, now insert the second damper and let $c_1 = c_2 = 1$. The model is

$$\dot{\omega}_1 = -\omega_1 + \omega_2 \tag{9.2-25}$$

$$\dot{\omega}_2 = \omega_1 - 2\omega_2 \tag{9.2-26}$$

and the roots are

$$s_1 = -0.382, \qquad s_2 = -2.62 \tag{9.2-27}$$

The resulting eigenmodes are shown in the following solution.

$$\begin{bmatrix} \omega_1(t) \\ \omega_2(t) \end{bmatrix} = p_{11} \begin{bmatrix} 1 \\ 0.62 \end{bmatrix} e^{-0.382t} + p_{12} \begin{bmatrix} 1 \\ -1.62 \end{bmatrix} e^{-2.62t} \tag{9.2-28}$$

The first eigenmode now consists of a motion in which the inertia $I_2$ is moving in the same direction as $I_1$, but with 62% of the speed of $I_1$. This motion decays exponentially with a time constant of $1/0.382 = 2.62$. In the second eigenmode, $I_2$ rotates oppositely to $I_1$, with 162% of the speed of $I_1$. Because of the opposing directions, this motion decays much faster, with a time constant of $1/2.62 = 0.382$.

The initial conditions determine the magnitude of each eigenmode ($p_{11}$ and $p_{12}$). If we start both inertias rotating in the same direction, with $\omega_2(0) = 0.62\,\omega_1(0)$, they will continue to rotate with their speeds maintaining this 62% ratio, with both slowing down exponentially with a time constant of 0.382. The second eigenmode will not appear in the motion ($p_{12} = 0$). If, on the other hand, $\omega_2(0) = -1.62\,\omega_1(0)$, only the second eigenmode will appear; here the opposite speeds will maintain the 162% ratio while decaying to zero in approximately $4(2.62) = 10.5$ time units. Any set of initial conditions different from these will excite both eigenmodes and the motion will be a combination of the two ($p_{11} \neq 0, p_{12} \neq 0$).

## The General Eigenvalue Problem

We now summarize the preceding results in terms of general vector notation. The exponential substitution is

$$\begin{aligned} \mathbf{x}(t) &= \mathbf{p}e^{st} \\ \dot{\mathbf{x}}(t) &= s\mathbf{p}e^{st} \end{aligned} \tag{9.2-29}$$

When substituted into the homogeneous state equation, $\dot{\mathbf{x}} = \mathbf{A}\mathbf{x}$, this gives

$$s\mathbf{p}e^{st} = \mathbf{A}\mathbf{p}e^{st}$$

or

$$s\mathbf{p} = \mathbf{A}\mathbf{p} \tag{9.2-30}$$

The problem of finding a vector $\mathbf{p}$ and a constant $s$ that satisfy (9.2-30) for a given matrix $\mathbf{A}$ is an *eigenvalue problem*. The resulting solutions for $s$ are the eigenvalues of the matrix $\mathbf{A}$ and are the same as the characteristic roots of the state equation, $\dot{\mathbf{x}} = \mathbf{A}\mathbf{x}$. (*Eigen* means "characteristic" in German.) The solutions for $\mathbf{p}$ are the eigenvectors of $\mathbf{A}$. If the matrix $\mathbf{A}$ is $(n \times n)$, it will have $n$ eigenvalues and $n$ associated eigenvectors of dimension $(n \times 1)$. If the eigenvalues are distinct, the eigenvectors are linearly independent. Here we treat only the distinct eigenvalue case; if any eigenvalues are repeated, the analysis is somewhat different.

Rewrite (9.2-30) as

$$s\mathbf{I}\mathbf{p} - \mathbf{A}\mathbf{p} = \mathbf{0}$$

or

$$(s\mathbf{I} - \mathbf{A})\mathbf{p} = \mathbf{0} \tag{9.2-31}$$

where $\mathbf{I}$ is the identity matrix. This equation is the vector equivalent of the linear algebraic equation set (9.2-3) and (9.2-4). To obtain a nonzero solution for $\mathbf{p}$, we require that

$$|s\mathbf{I} - \mathbf{A}| = 0 \tag{9.2-32}$$

This is the characteristic equation from which the $n$ eigenvalues are obtained. It is an $n$th-order polynomial in $s$.

For each eigenvalue $s_i$ there corresponds an $(n \times 1)$ eigenvector $\mathbf{p}_i$ such that (9.2-31) is satisfied; that is,

$$(s_i\mathbf{I} - \mathbf{A})\mathbf{p}_i = \mathbf{0}, \qquad i = 1, 2, \ldots, n \tag{9.2-33}$$

For a given $i$, (9.2-33) represents $n$ linear equations. One of these equations is redundant if the eigenvalues are distinct. These can be solved for $(n - 1)$ elements of the vector $\mathbf{p}_i$, in terms of the remaining element. This solution is given by (9.2-8) and (9.2-9) for the two-dimensional case. The free response, as before, is the linear combination of the eigenmodes.

$$\mathbf{x}(t) = \sum_{i=1}^{n} \mathbf{p}_i e^{s_i t} \tag{9.2-34}$$

Each $\mathbf{p}_i$ will have one undetermined constant, for a total of $n$ constants in the solution. These can be found with the initial values of the $n$ state variables.

## Invariance of the Eigenvalues

We would expect from physical considerations that the eigenvalues of a model remain the same regardless of the choice of state variables used to represent the model. This fact can be proved mathematically (Reference 2).

## Diagonalization of the State Equations

The coordinate system in which the coordinates are the state variables is called the *state space*. The eigenvectors $\mathbf{v}_1$ and $\mathbf{v}_2$ can be represented by arrows in this coordinate system (Figure 9.8). The system's behavior is due to the vector combination of the motion of each eigenmode along its eigenvector. These motions take place at different rates, with each rate determined by the time constant of the eigenmode. The projection of the initial condition vector $\mathbf{x}(0)$ onto each eigenvector determines the initial magnitude associated with the eigenmode. If the vector $\mathbf{x}(0)$ lies exactly on one of the eigenvectors, only that eigenmode will appear in the motion.

If the eigenvectors are linearly independent they can be used to form a new set of coordinate axes.* Let the new coordinates be $z_1$ and $z_2$, and let the unit-length

* See Appendix A, Section A.4, for the definition of linear independence.

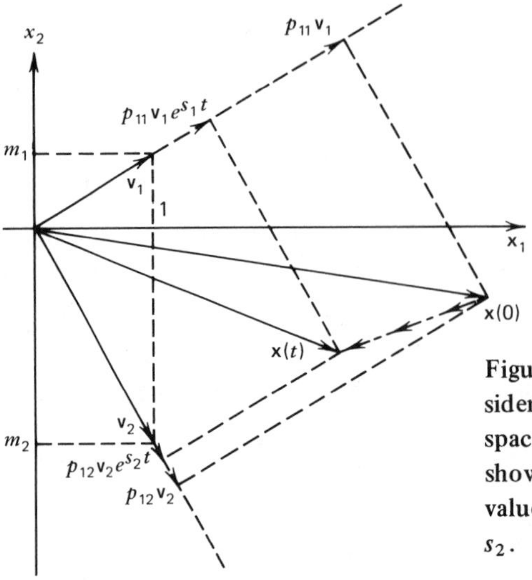


**Figure 9.8** Eigenvectors $\mathbf{v}_1$ and $\mathbf{v}_2$ considered as a coordinate basis in the state space. The motion of the state vector **x** is shown from its initial value $\mathbf{x}(0)$ to a new value $\mathbf{x}(t)$ for negative eigenvalues $s_1$ and $s_2$.

basis vectors in this sytem be denoted $\hat{\mathbf{e}}_1$ and $\hat{\mathbf{e}}_2$. If the basis vectors lie along the eigenvectors $\mathbf{v}_1$ and $\mathbf{v}_2$, they are also eigenvectors. They are found by normalizing the vectors $\mathbf{v}_1$ and $\mathbf{v}_2$ by dividing by their lengths; that is,

$$\hat{\mathbf{e}}_1 = \frac{1}{\sqrt{m_1^2+1}}\begin{bmatrix}1\\ m_1\end{bmatrix}$$

$$\hat{\mathbf{e}}_2 = \frac{1}{\sqrt{m_2^2+1}}\begin{bmatrix}1\\ m_2\end{bmatrix}$$

The general vector **x** expressed in these coordinates is

$$\mathbf{x} = z_1\hat{\mathbf{e}}_1 + z_2\hat{\mathbf{e}}_2$$

and its derivative is

$$\dot{\mathbf{x}} = \dot{z}_1\hat{\mathbf{e}}_1 + \dot{z}_2\hat{\mathbf{e}}_2$$

The free response satisfies $\dot{\mathbf{x}} = \mathbf{A}\mathbf{x}$; therefore,

$$\begin{aligned}\dot{z}_1\hat{\mathbf{e}}_1 + \dot{z}_2\hat{\mathbf{e}}_2 &= \mathbf{A}(z_1\hat{\mathbf{e}}_1 + z_2\hat{\mathbf{e}}_2)\\ &= z_1\mathbf{A}\hat{\mathbf{e}}_1 + z_2\mathbf{A}\hat{\mathbf{e}}_2 \qquad (9.2\text{-}35)\end{aligned}$$

because $z_1$ and $z_2$ are scalars. From the definition of the eigenvectors, $\hat{\mathbf{e}}_1$ and $\hat{\mathbf{e}}_2$ must satisfy

$$\mathbf{A}\hat{\mathbf{e}}_1 = s_1\hat{\mathbf{e}}_1$$

$$\mathbf{A}\hat{\mathbf{e}}_2 = s_2\hat{\mathbf{e}}_2$$

Thus

$$\dot{z}_1\hat{\mathbf{e}}_1 + \dot{z}_2\hat{\mathbf{e}}_2 = s_1z_1\hat{\mathbf{e}}_1 + s_2z_2\hat{\mathbf{e}}_2$$

Comparing components we see that

$$\dot{z}_1 = s_1z_1 \qquad (9.2\text{-}36)$$

$$\dot{z}_2 = s_2z_2 \qquad (9.2\text{-}37)$$

This shows that the state equations become uncoupled when expressed in coordinate axes along the eigenvector directions.

It can be shown by analytic geometry in the two-dimensional case, or by linear algebra in general, that the transformation matrix between the **x** and the **z** coordinate systems is the *modal matrix* **M** whose columns are the eigenvectors (Reference 2). For the two-dimensional case,

$$\mathbf{M} = [\mathbf{v}_1, \mathbf{v}_2] \tag{9.2-38}$$

where $\mathbf{x} = \mathbf{Mz}$. This implies that $\dot{\mathbf{x}} = \mathbf{M\dot{z}}$, and from the state equation, $\dot{\mathbf{x}} = \mathbf{Ax} + \mathbf{Bu}$, we get

$$\mathbf{M\dot{z}} = \mathbf{AMz} + \mathbf{Bu}$$

It can be shown that $\mathbf{M}^{-1}$ exists for the distinct eigenvalues case (Reference 2). Thus,

$$\dot{\mathbf{z}} = \mathbf{M}^{-1}\mathbf{AMz} + \mathbf{M}^{-1}\mathbf{Bu} \tag{9.2-39}$$

Comparison with (9.2-36) and (9.2-37) for $\mathbf{u} = \mathbf{0}$ shows that

$$\mathbf{M}^{-1}\mathbf{AM} = \begin{bmatrix} s_1 & 0 \\ 0 & s_2 \end{bmatrix}$$

This property also holds for the general vector case, if the eigenvalues of **A** are distinct; that is,

$$\mathbf{M}^{-1}\mathbf{AM} = \mathbf{\Lambda} \tag{9.2-40}$$

where $\mathbf{\Lambda}$ is a diagonal matrix whose elements are the system's eigenvalues. The state equation in terms of the vector **z** is then

$$\dot{\mathbf{z}} = \mathbf{\Lambda z} + \mathbf{M}^{-1}\mathbf{Bu} \tag{9.2-41}$$

The output equation $\mathbf{y} = \mathbf{Cx} + \mathbf{Du}$ becomes

$$\mathbf{y} = \mathbf{CMz} + \mathbf{Du} \tag{9.2-42}$$

## Eigenmodes from the Transfer Function

The technique used in Section 5.2 for deriving state models from the transfer functions can also be used to find the eigenmodes. If the roots of $|s\mathbf{I} - \mathbf{A}| = 0$ are known, the denominator polynomial of the transfer functions can be expressed in factored form and a partial fraction expansion performed. If the roots are distinct, the expansion for the transfer function $T(s)$ is

$$T(s) = \frac{Y(s)}{U(s)} = \frac{K_1}{s - s_1} + \frac{K_2}{s - s_2} + \ldots + \frac{K_n}{s - s_n} \tag{9.2-43}$$

where the roots are $s_1, s_2, \ldots, s_n$, and

$$K_i = \lim_{s \to s_i} T(s)(s - s_i) \tag{9.2-44}$$

Figure 9.9*a* shows the diagram for a transfer function of the form

$$\frac{Z_i(s)}{U(s)} = \frac{K_i}{s - s_i} \tag{9.2-45}$$

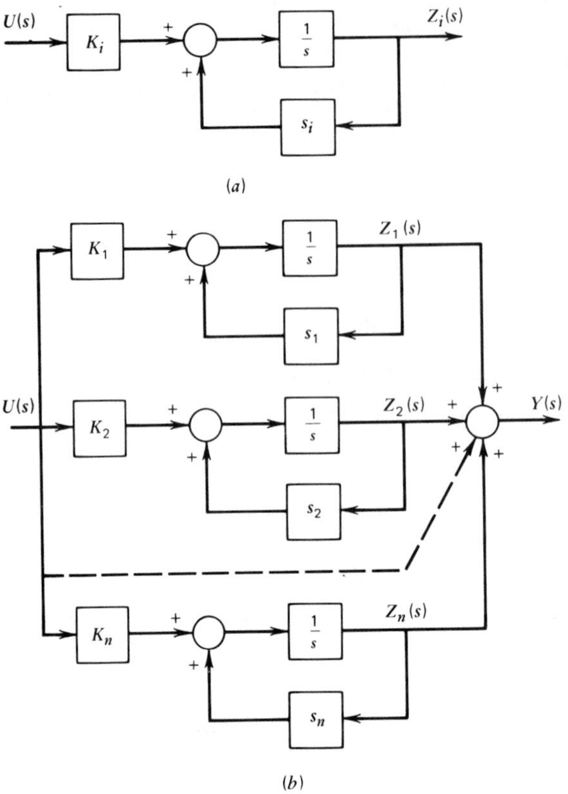


**Figure 9.9 Modal diagrams for a system with distinct roots. (*a*) Diagram for the *i*th mode. (*b*) Diagram for the complete system.**

The corresponding differential equation is

$$\dot{z}_i = s_i z_i + K_i u \tag{9.2-46}$$

Thus by defining the state variables $z_i$ with this relation, (9.2-43) can be written as

$$Y(s) = Z_1(s) + Z_2(s) + \ldots + Z_n(s)$$

Figure 9.9*b* results from this relation. The state diagram in this form consists of parallel first-order feedback loops in which the feedback gains are the eigenvalues, and the feed-forward gains are the residues $K_i$ from the partial fraction expansion. The state equations in vector form are

$$\dot{\mathbf{z}} = \begin{bmatrix} s_1 & 0 & 0 & \ldots & 0 & 0 \\ 0 & s_2 & 0 & \ldots & 0 & 0 \\ \cdot & \cdot & \cdot & \ldots & \cdot & \cdot \\ 0 & \cdot & \cdot & \ldots & 0 & s_n \end{bmatrix} \mathbf{z} + \begin{bmatrix} K_1 \\ K_2 \\ \cdot \\ \cdot \\ \cdot \\ K_n \end{bmatrix} u \tag{9.2-47}$$

$$y = [1 \quad 1 \ldots 1]\,\mathbf{z} \tag{9.2-48}$$

We immediately see that the resulting matrix $\mathbf{A}$ is diagonal and equal to $\mathbf{\Lambda}$.

In order to solve (9.2-46) we need the initial values $\mathbf{z}(0)$. These are obtained from the initial conditions on the output $y$ and its derivatives. (Remember that we have no idea of the physical meaning of the components of $\mathbf{z}$ when they are derived in this manner.) The free response of the model (9.2-46) can be used with the output equation (9.2-48) to show that

$$y(t) = \sum_{i=1}^{n} z_i(0)e^{s_i t}$$

Thus

$$\dot{y}(t) = \sum_{i=1}^{n} z_i(0)s_i e^{s_i t}$$

and so forth. Then

$$\left.\frac{d^k y}{dt^k}\right|_{t=0} = \sum_{i=1}^{n} z_i(0)s_i^k \triangleq g_{k+1}(0) \tag{9.2-49}$$

where $k = 0, 1, 2, \ldots, n-1$. The defined term $g(0)$ represents the given initial conditions. Then (9.2-49) can be written in matrix form as

$$\begin{bmatrix} 1 & 1 & \cdots & 1 \\ s_1 & s_2 & \cdots & s_n \\ s_1^2 & s_2^2 & \cdots & s_n^2 \\ \vdots & \vdots & \cdots & \vdots \\ s_1^{n-1} & s_2^{n-1} & \cdots & s_n^{n-1} \end{bmatrix} \begin{bmatrix} z_1(0) \\ z_2(0) \\ z_3(0) \\ \vdots \\ z_n(0) \end{bmatrix} = \begin{bmatrix} g_1(0) \\ g_2(0) \\ g_3(0) \\ \vdots \\ g_n(0) \end{bmatrix} \tag{9.2-50}$$

or

$$\mathbf{Q}\mathbf{z}(0) = \mathbf{g}(0) \tag{9.2-51}$$

where $\mathbf{Q}$ is the transformation matrix defined in (9.2-50). It can be shown that $Q^{-1}$ exists if the eigenvalues are distinct, and thus

$$\mathbf{z}(0) = Q^{-1}\,\mathbf{g}(0)$$

This relation gives the required initial conditions for $\mathbf{z}$.

The preceding result suggests a useful transformation matrix for use in diagonalizing a model. The transfer function relation (9.2-43) can be written as

$$D(s)Y(s) = N(s)U(s) \tag{9.2-52}$$

where $N(s)$ and $D(s)$ are the numerator and denominator polynomials of $T(s)$, and $D(s)$ has the form

$$D(s) = s^n + a_1 s^{n-1} + \ldots + a_n \tag{9.2-53}$$

If we revert (9.2-52) back to the time domain, an $n$th-order differential equation is obtained. From our discussion of state models in Sections 5.1 and 5.2, we see that a suitable set of state variables for this system would be

$$\left.\begin{aligned} x_1 &= y \\ x_2 &= \dot{y} \\ x_3 &= \ddot{y} \\ &\vdots \\ x_n &= \frac{d^{n-1}y}{dt^{n-1}} \end{aligned}\right\} \tag{9.2-54}$$

Such a choice of the state variables as successive derivatives of the output $y$ is called the *phase variable form*. With this choice we see that $\mathbf{x}(t) = \mathbf{g}(t)$, with the definition of $\mathbf{g}(t)$ an obvious extension of that given in (9.2-49). The matrix $\mathbf{A}$ for the phase variable model is

$$A = \begin{bmatrix} 0 & 1 & 0 & 0 & \cdots & 0 \\ 0 & 0 & 1 & 0 & \cdots & 0 \\ \vdots & \vdots & \vdots & \vdots & \cdots & \vdots \\ 0 & 0 & \cdot & \cdot & \cdots & 1 \\ -a_n & -a_{n-1} & \cdot & \cdot & \cdots & -a_1 \end{bmatrix} \tag{9.2-55}$$

If we now attempt a transformation of the form of (9.2-51) – that is,

$$\mathbf{Q}\mathbf{z}(t) = \mathbf{g}(t) = \mathbf{x}(t) \tag{9.2-56}$$

where $\mathbf{Q}$ is given in (9.2-50) – we find that this transformation will diagonalize the matrix $\mathbf{A}$ given by (9.2-55).

## Eigenmodes for Repeated Roots

If the denominator of the transfer function has repeated roots, the form of the partial fraction expansion differs from (9.2-43). Consider a second-order system with the repeated root $s_1$. The expansion is

$$T(s) = \frac{Y(s)}{U(s)} = \frac{N(s)}{D(s)} = \frac{K_1}{(s-s_1)^2} + \frac{K_2}{s-s_1}$$

where $D(s) = (s-s_1)^2$ and

$$K_1 = \lim_{s\to s_1} T(s)(s-s_1)^2$$

$$K_2 = \lim_{s\to s_1} \frac{d}{ds}[T(s)(s-s_1)^2]$$

Let

$$Z_1(s) = \frac{K_1}{(s-s_1)^2} U(s)$$

$$Z_2(s) = \frac{K_2}{s-s_1} U(s)$$

By considering $Z_1(s)$ to result from two cascaded transfer functions of the form $1/(s-s_1)$, we can draw the diagram shown in Figure 9.10. From this diagram we

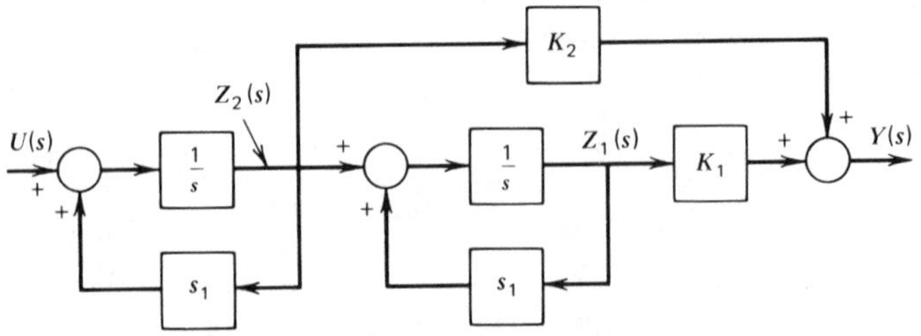


**Figure 9.10 Modal diagram for a second-order system with repeated roots.**

can write the state equations

$$\dot{\mathbf{z}} = \mathbf{\Lambda}_R \mathbf{z} + \begin{bmatrix} 0 \\ 1 \end{bmatrix} u$$

$$y = [K_1 \quad K_2]\mathbf{z} \tag{9.2-57}$$

where

$$\mathbf{\Lambda}_R = \begin{bmatrix} s_1 & 1 \\ 0 & s_1 \end{bmatrix} \tag{9.2-58}$$

The new system matrix is not diagonal as it was for distinct roots, and the subscript $R$ is used to distinguish $\mathbf{\Lambda}_R$ from $\mathbf{\Lambda}$ given by (9.2-47). The form (9.2-58) is the usual form for repeated roots, although for certain transfer functions a diagonal matrix can occur (Reference 2). When the new system matrix is not diagonal, $n$ independent eigenvectors do not exist.

A third-order system with two repeated roots $s_1$ and a distinct root $s_2$ has the expansion

$$T(s) = \frac{Y(s)}{U(s)} = \frac{K_1}{(s-s_1)^2} + \frac{K_2}{s-s_1} + \frac{K_3}{s-s_2}$$

Direct extension of the preceding procedure gives the state model

$$\dot{\mathbf{z}} = \begin{bmatrix} s_1 & 1 & 0 \\ 0 & s_1 & 0 \\ 0 & 0 & s_2 \end{bmatrix} \mathbf{z} + \begin{bmatrix} 0 \\ 1 \\ 1 \end{bmatrix} u \tag{9.2-59}$$

$$y = [K_1 \quad K_2 \quad K_3]\mathbf{z}$$

The meaning of this result will be discussed shortly.

## Eigenmodes for Complex Roots

The occurrence of complex characteristic roots is included in the distinct-roots case. However, the eigenvectors and the diagonal elements of the matrix $\mathbf{\Lambda}$ will be complex. In this case the physical interpretation of the eigenmodes is difficult. For example, the eigenvectors are no longer coordinate axes in the real plane containing the state variables. Also, for some computational purposes, it is desirable to avoid complex quantities. Therefore, we now discuss the complex-roots case from a slightly different viewpoint.

Consider the second-order case where the matrix $\mathbf{A}$ has the eigenvalues $s = -a \pm ib$ $(b > 0)$. It is easily shown that the eigenvectors $\mathbf{v}_1$ and $\mathbf{v}_2$ are complex conjugates such that

$$\mathbf{v}_1 = \mathbf{q} + i\mathbf{r}$$

$$\mathbf{v}_2 = \mathbf{q} - i\mathbf{r}$$

In place of the modal matrix $\mathbf{M}$, defined by (9.2-38), use instead the matrix

$$\mathbf{M}_C = [\mathbf{q}, \mathbf{r}] \tag{9.2-60}$$

Note that $\mathbf{M}_C$ has real elements. To derive the properties of $\mathbf{M}_C$, we use the definition

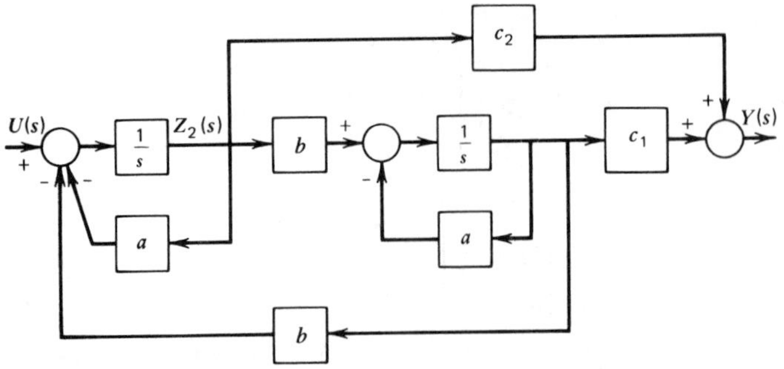


**Figure 9.11** **Modified modal diagram for a system with complex roots.**

of the eigenvector $\mathbf{v}_1$. It must satisfy

$$\mathbf{A}\mathbf{v}_1 = (-a + ib)\mathbf{v}_1$$

or

$$\mathbf{A}(\mathbf{q} + i\mathbf{r}) = (-a + ib)(\mathbf{q} + i\mathbf{r})$$

$$= (-a\mathbf{q} - b\mathbf{r}) + i(b\mathbf{q} - a\mathbf{r})$$

Thus

$$\mathbf{A}\mathbf{q} = -a\mathbf{q} - b\mathbf{r}$$

$$\mathbf{A}\mathbf{r} = b\mathbf{q} - a\mathbf{r}$$

Rearranging into matrix form we get

$$\mathbf{A}[\mathbf{q}, \mathbf{r}] = [\mathbf{q}, \mathbf{r}]\begin{bmatrix} -a & b \\ -b & -a \end{bmatrix}$$

or

$$\mathbf{A}\mathbf{M}_C = \mathbf{M}_C \mathbf{\Lambda}_C$$

where

$$\mathbf{\Lambda}_C = \begin{bmatrix} -a & b \\ -b & -a \end{bmatrix} \qquad (9.2\text{-}61)$$

With the transformation $\mathbf{x} = \mathbf{M}_C\mathbf{z}$, the state equation $\dot{\mathbf{x}} = \mathbf{A}\mathbf{x} + \mathbf{B}\mathbf{u}$ becomes (for the second-order case)

$$\dot{\mathbf{z}} = \mathbf{\Lambda}_C\mathbf{z} + \mathbf{M}_C^{-1}\mathbf{B}\mathbf{u} \qquad (9.2\text{-}62)$$

Figure 9.11 shows the diagram for this system for the case of a single input $u$ and a single output $y = [c_1, c_2]\mathbf{z}$. This system has the transfer function

$$T(s) = \frac{Y(s)}{U(s)} = \frac{N(s)}{(s + a)^2 + b^2} \qquad (9.2\text{-}63)$$

*Example 9.3*

For a particular mass-spring-damper system, $m = 1$, $c = 2$, and $k = 5$ (Figure 9.12). Obtain the modified modal representation for this system.

With the given parameter values,

$$\mathbf{A} = \begin{bmatrix} 0 & 1 \\ -5 & -2 \end{bmatrix}, \qquad \mathbf{B} = \begin{bmatrix} 0 \\ 1 \end{bmatrix}$$

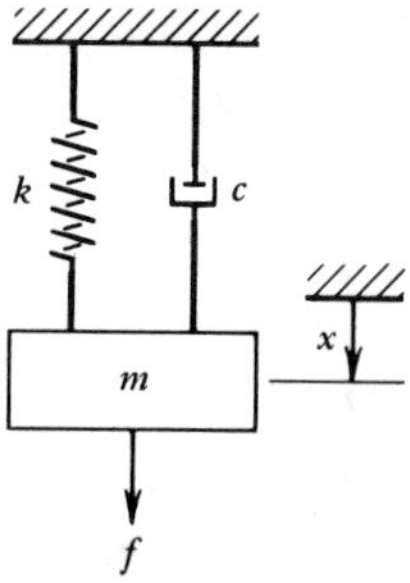


$$\dot{\mathbf{x}} = \begin{bmatrix} 0 & 1 \\ -\dfrac{k}{m} & -\dfrac{c}{m} \end{bmatrix} \mathbf{x} + \begin{bmatrix} 0 \\ \dfrac{1}{m} \end{bmatrix} u$$

$$x_1 = x$$

$$x_2 = \dot{x}$$

$$u = f$$

**Figure 9.12 Vibratory system and its state-variable model.**

The eigenvalues are obtained from $|s\mathbf{I} - \mathbf{A}| = 0$, or $s^2 + 2s + 5 = 0$. This gives $s = -1 \pm 2i$, and $a = 1$, $b = 2$. The eigenvectors are found from

$$\begin{bmatrix} s & -1 \\ 5 & (s+2) \end{bmatrix} \mathbf{v} = \mathbf{0}$$

For $\mathbf{v}_1$, this gives

$$(-1 + 2i)v_{11} - v_{12} = 0$$

and

$$\mathbf{v}_1 = v_{11} \begin{bmatrix} 1 \\ -1 + 2i \end{bmatrix}$$

Similarly for $\mathbf{v}_2$,

$$\mathbf{v}_2 = v_{21} \begin{bmatrix} 1 \\ -1 - 2i \end{bmatrix}$$

We can normalize by choosing $v_{11} = v_{21} = 1$. Thus,

$$\mathbf{r} = \begin{bmatrix} 1 \\ -1 \end{bmatrix}, \qquad \mathbf{q} = \begin{bmatrix} 0 \\ 2 \end{bmatrix}$$

and

$$\mathbf{M}_C = \begin{bmatrix} 1 & 0 \\ -1 & 2 \end{bmatrix}$$

$$\mathbf{M}_C^{-1}\mathbf{B} = \frac{1}{2} \begin{bmatrix} 2 & 0 \\ 1 & 1 \end{bmatrix} \begin{bmatrix} 0 \\ 1 \end{bmatrix} = \frac{1}{2} \begin{bmatrix} 0 \\ 1 \end{bmatrix}$$

$$\mathbf{\Lambda}_C = \begin{bmatrix} -1 & 2 \\ -2 & -1 \end{bmatrix}$$

The modified modal form is

$$\dot{\mathbf{z}} = \begin{bmatrix} -1 & 2 \\ -2 & -1 \end{bmatrix} \mathbf{z} + \frac{1}{2} \begin{bmatrix} 0 \\ 1 \end{bmatrix} u$$

where $\mathbf{x} = \mathbf{M}_C \mathbf{z}$, and $u$ is the applied force $f$.

## Canonical Forms

We have seen that for distinct eigenvalues it is always possible to transform the state model to obtain a diagonal matrix in the form of (9.2-47). If any of the eigenvalues are repeated, this diagonalization is usually not possible, and the form (9.2-58) results.

These forms are the *canonical forms* of the state equation. Their usefulness will become apparent shortly. For a higher-order system we can state that it is always possible to find a transformation matrix **M** to obtain the *Jordan canonical form*.

$$\mathbf{\Lambda} = \mathbf{M}^{-1}\mathbf{A}\mathbf{M} = \begin{bmatrix} s_1 & \alpha_1 & 0 & 0 & \cdots & & 0 \\ 0 & s_2 & \alpha_2 & 0 & \cdots & & 0 \\ \cdot & \cdot & \cdot & \cdot & \cdots & & \cdot \\ 0 & \cdot & \cdot & \cdot & 0 & s_{n-1} & \alpha_{n-1} \\ 0 & \cdot & \cdot & \cdot & 0 & 0 & s_n \end{bmatrix} \tag{9.2-64}$$

where the $\alpha_i$, $i = 1, \ldots, n-1$ can be either zero or one. For example, we have seen that a third-order system with two repeated roots $s_1$ and one distinct root $s_2$ gives the form

$$\mathbf{\Lambda} = \left[\begin{array}{cc|c} \multicolumn{2}{c|}{\multirow{2}{*}{$\mathbf{\Lambda}_R$}} & 0 \\ & & 0 \\ \hline 0 & 0 & \mathbf{\Lambda}_D \end{array}\right]$$

where $\mathbf{\Lambda}_D = s_2$ and

$$\mathbf{\Lambda}_R = \begin{bmatrix} s_1 & 1 \\ 0 & s_1 \end{bmatrix}$$

In general, $\mathbf{\Lambda}$ will consist of blocks of submatrices corresponding to the canonical matrix $\mathbf{\Lambda}_i$ for each system mode set. Two repeated roots correspond to one set of two modes that are usually coupled, and $\mathbf{\Lambda}_i$ is a $(2 \times 2)$ matrix. For three repeated roots, $\mathbf{\Lambda}_i$ is $(3 \times 3)$, and so forth. For each distinct root, $\mathbf{\Lambda}_i = s_i$.

When complex roots occur, the Jordan form will have complex entries, and therefore it is sometimes rewritten using the form $\mathbf{\Lambda}_C$ given by (9.2-61). For example, consider a 6th-order system model with two repeated roots $s_1$, two complex roots $-a \pm ib$, and two distinct roots $s_5$ and $s_6$. If the eigenvectors corresponding to $s_1$ are dependent, the modified Jordan form is

$$\mathbf{\Lambda} = \left[\begin{array}{cc|cc|c|c} s_1 & 1 & 0 & 0 & 0 & 0 \\ 0 & s_1 & 0 & 0 & 0 & 0 \\ \hline 0 & 0 & -a & b & 0 & 0 \\ 0 & 0 & -b & -a & 0 & 0 \\ \hline 0 & 0 & 0 & 0 & s_5 & 0 \\ \hline 0 & 0 & 0 & 0 & 0 & s_6 \end{array}\right]$$

## 9.3 THE TRANSITION MATRIX

In Section 5.3 we obtained the free response of the second-order linear system in reduced form $m\ddot{x} + c\dot{x} + kx = f(t)$. For many applications this is perhaps the most convenient form. However, in some situations the required computations are more easily handled if the free response is cast in terms of the state model. This form also allows us to generalize our results to higher-order systems. The reduced form is rarely used for systems of order greater than three.

## Solution from the Eigenmodes

Consider the diagonal form (9.2-47) of the state equation. The equation for each modal coordinate $z_i$ is

$$\dot{z}_i = s_i z_i + K_i u \tag{9.3-1}$$

The free response is easily shown to be

$$z_i(t) = z_i(0)e^{s_i t}$$

or in matrix form

$$\mathbf{z}(t) = \begin{bmatrix} e^{s_1 t} & 0 & 0 & \cdots & . & 0 \\ 0 & e^{s_2 t} & 0 & \cdots & . & 0 \\ . & . & . & \cdots & . & . \\ 0 & . & . & \cdots & 0 & e^{s_n t} \end{bmatrix} \mathbf{z}(0) \tag{9.3-2}$$

Let the matrix of exponentials in (9.3-2) be denoted by $\boldsymbol{\varphi}_M(t)$. Then

$$\mathbf{z}(t) = \boldsymbol{\varphi}_M(t)\mathbf{z}(0) \tag{9.3-3}$$

The matrix $\boldsymbol{\varphi}_M(t)$ is the *modal transition matrix*. It describes the transition of each modal coordinate from one point in time to another, just as the scalar transition function does for the first-order system (Chapter Three).

The diagonal form (9.2-47) was obtained by the transformation $\mathbf{x}(t) = \mathbf{M}\mathbf{z}(t)$. Thus, $\mathbf{z}(0) = \mathbf{M}^{-1}\mathbf{x}(0)$, and from (9.3-3),

$$\mathbf{M}\mathbf{z}(t) = \mathbf{M}\boldsymbol{\varphi}_M(t)\mathbf{z}(0)$$

or

$$\mathbf{x}(t) = \mathbf{M}\boldsymbol{\varphi}_M(t)\mathbf{M}^{-1}\mathbf{x}(0) \tag{9.3-4}$$

We have now recovered the free response of the equation $\dot{\mathbf{x}} = \mathbf{A}\mathbf{x} + \mathbf{B}\mathbf{u}$ in terms of the original state variables. In analogy with the first-order case, we write (9.3-4) as

$$\mathbf{x}(t) = \boldsymbol{\varphi}(t)\mathbf{x}(0) \tag{9.3-5}$$

$$\boldsymbol{\varphi}(t) = \mathbf{M}\boldsymbol{\varphi}_M(t)\mathbf{M}^{-1} \tag{9.3-6}$$

where $\boldsymbol{\varphi}(t)$ is the *state transition matrix*. It is analogous to the transition function for first-order systems. If $\mathbf{A}$ is an $(n \times n)$ matrix, so is $\boldsymbol{\varphi}$.

## Properties of the Transition Matrix

Some useful properties of the state transition matrix can be derived from (9.3-5) and (9.3-6). Set $t = 0$ in (9.3-5) to show that

$$\boldsymbol{\varphi}(0) = \mathbf{I} \tag{9.3-7}$$

For $\mathbf{u} = \mathbf{0}$, $\dot{\mathbf{x}} = \mathbf{A}\mathbf{x}$. Use this with the derivative of (9.3-5) to obtain

$$\dot{\mathbf{x}} = \dot{\boldsymbol{\varphi}}(t)\mathbf{x}(0) = \mathbf{A}\mathbf{x}(t)$$

$$= \mathbf{A}\boldsymbol{\varphi}(t)\mathbf{x}(0)$$

or

$$\dot{\boldsymbol{\varphi}}(t) = \mathbf{A}\boldsymbol{\varphi}(t) \tag{9.3-8}$$

Thus $\boldsymbol{\varphi}(t)$ is the solution of the differential equation (9.3-8) with the initial condition (9.3-7). Solve (9.3-6) for $\boldsymbol{\varphi}_M(t)$.

$$\boldsymbol{\varphi}_M(t) = \mathbf{M}^{-1}\boldsymbol{\varphi}(t)\mathbf{M}$$

The inverse of a diagonal matrix is found by simply inverting the diagonal elements. Thus $\boldsymbol{\varphi}_M^{-1}(t) = \boldsymbol{\varphi}_M(-t)$. It follows that

$$\boldsymbol{\varphi}^{-1}(t) = \boldsymbol{\varphi}(-t) \tag{9.3-9}$$

This property avoids the need to perform the usually tedious calculations in finding the matrix inverse.

Equation (9.3-5) relates $x(t)$ to the initial state at time zero. However, because we are dealing with the free response of a constant coefficient equation, we could also write the solution in terms of the state at any time, say $t_1$, as

$$\mathbf{x}(t + t_1) = \boldsymbol{\varphi}(t)\mathbf{x}(t_1) \tag{9.3-10}$$

But $\mathbf{x}(t_1) = \boldsymbol{\varphi}(t_1)\mathbf{x}(0)$, and thus

$$\mathbf{x}(t + t_1) = \boldsymbol{\varphi}(t)\boldsymbol{\varphi}(t_1)\mathbf{x}(0)$$

If we had used (9.3-5) immediately, we would have obtained

$$\mathbf{x}(t + t_1) = \boldsymbol{\varphi}(t + t_1)\mathbf{x}(0)$$

Comparison of the last two equations shows that

$$\boldsymbol{\varphi}(t + t_1) = \boldsymbol{\varphi}(t)\boldsymbol{\varphi}(t_1) \tag{9.3-11}$$

for any $t$ and $t_1$. Also, since $\boldsymbol{\varphi}(t + t_1) = \boldsymbol{\varphi}(t_1 + t)$, the commutative property holds.

$$\boldsymbol{\varphi}(t)\boldsymbol{\varphi}(t_1) = \boldsymbol{\varphi}(t_1)\boldsymbol{\varphi}(t) \tag{9.3-12}$$

Finally, if we let $t = t_1 = \Delta$ in (9.3-11), we see that

$$\boldsymbol{\varphi}(2\Delta) = \boldsymbol{\varphi}(\Delta)\boldsymbol{\varphi}(\Delta)$$

or in general,

$$\boldsymbol{\varphi}(n\Delta) = \boldsymbol{\varphi}^n(\Delta) \tag{9.3-13}$$

This property will be especially useful in numerical solutions.

*Example 9.4*

(a) Find the transition matrix for a mass-spring-damper system with $m = 1$, $c = 7$, and $k = 10$ (Figure 9.12).

(b) Find the displacement $x_1$ and the velocity $x_2$ of the mass at the times $t = 0.1$ and $t = 0.2$ if the initial displacement is $x_1(0) = 3$ and the initial velocity is $x_2(0) = 0$.

(c) If the displacement and velocity at $t = 0.1$ are measured to be $x_1(0.1) = 0.5013$ and $x_2(0.1) = -0.2141$, what initial conditions produced this motion?

(a) For this model the eigenvalues are $s_1 = -2$ and $s_2 = -5$. The eigenvectors can be normalized as

$$\mathbf{v}_1 = \begin{bmatrix} 1 \\ -2 \end{bmatrix}, \quad \mathbf{v}_2 = \begin{bmatrix} 1 \\ -5 \end{bmatrix}$$

and the modal transformation matrix with its inverse are

$$\mathbf{M} = \begin{bmatrix} 1 & 1 \\ -2 & -5 \end{bmatrix}, \quad \mathbf{M}^{-1} = \frac{1}{3}\begin{bmatrix} 5 & 1 \\ -2 & -1 \end{bmatrix}$$

The modal transition matrix is

$$\boldsymbol{\varphi}_M(t) = \begin{bmatrix} e^{-2t} & 0 \\ 0 & e^{-5t} \end{bmatrix}$$

The transition matrix is found from (9.3-6).

$$\begin{aligned}\boldsymbol{\varphi}(t) &= \mathbf{M}\boldsymbol{\varphi}_M(t)\mathbf{M}^{-1} \\ &= \frac{1}{3}\begin{bmatrix} (5e^{-2t} - 2e^{-5t}) & (e^{-2t} - e^{-5t}) \\ (10e^{-2t} + 10e^{-5t}) & (-2e^{-2t} + 5e^{-5t}) \end{bmatrix}\end{aligned}$$

(b) Here the evaluation times $t = 0.1, 0.2$ are separated by the same interval, and we can evaluate $\boldsymbol{\varphi}(t)$ once for $t = 0.1$ and use (9.3-10).

$$\boldsymbol{\varphi}(0.1) = \frac{1}{3}\begin{bmatrix} 2.8806 & 0.21220 \\ -2.1220 & 1.3952 \end{bmatrix}$$

Thus

$$\begin{aligned}\mathbf{x}(0.1) &= \boldsymbol{\varphi}(0.1)\mathbf{x}(0) = \boldsymbol{\varphi}(0.1)\begin{bmatrix} 3 \\ 0 \end{bmatrix} \\ &= \begin{bmatrix} 2.8806 \\ -2.1220 \end{bmatrix}\end{aligned}$$

Also,

$$\mathbf{x}(0.2) = \boldsymbol{\varphi}(0.1)\mathbf{x}(0.1) = \begin{bmatrix} 2.6159 \\ -3.0244 \end{bmatrix}$$

(c) Given $\mathbf{x}(0.1)$, to find $\mathbf{x}(0)$ we use property (9.3-9).

$$\mathbf{x}(0) = \boldsymbol{\varphi}^{-1}(0.1)\mathbf{x}(0.1) = \boldsymbol{\varphi}(-0.1)\mathbf{x}(0.1)$$

or

$$\mathbf{x}(0) = \frac{1}{3}\begin{bmatrix} 2.8095 & -0.4273 \\ 4.273 & 5.8008 \end{bmatrix}\begin{bmatrix} 0.5013 \\ -0.2141 \end{bmatrix} = \begin{bmatrix} 0.5 \\ 0.3 \end{bmatrix}$$

## Solution from the Laplace Transform

The Laplace transform can also be used to find the matrix $\boldsymbol{\varphi}(t)$. The transform of the state equation $\dot{\mathbf{x}} = \mathbf{A}\mathbf{x} + \mathbf{B}\mathbf{u}$ resulted in (5.2-24).

$$(s\mathbf{I} - \mathbf{A})\mathbf{X}(s) = \mathbf{x}(0) + \mathbf{B}\mathbf{U}(s) \tag{5.2-24}$$

The transform solution is

$$\mathbf{X}(s) = (s\mathbf{I} - \mathbf{A})^{-1}\mathbf{x}(0) + (s\mathbf{I} - \mathbf{A})^{-1}\mathbf{B}\mathbf{U}(s)$$

To consider the free response, set $\mathbf{U}(s) = \mathbf{0}$, and perform the inverse Laplace transformation to obtain

$$\mathbf{x}(t) = \mathscr{L}^{-1}[(s\mathbf{I} - \mathbf{A})^{-1}]\mathbf{x}(0) \tag{9.3-14}$$

Comparison with (9.3-5) shows that

$$\boldsymbol{\varphi}(t) = \mathscr{L}^{-1}[(s\mathbf{I} - \mathbf{A})^{-1}] \tag{9.3-15}$$

Thus the transition matrix is the inverse transform of the resolvent matrix inverse $\mathbf{R}^{-1}(s)$.

### *Example 9.5*

Use the Laplace transform to find $\boldsymbol{\varphi}(t)$ for the vibration model given in Example 9.4.
The matrix $\mathbf{A}$ is

$$\mathbf{A} = \begin{bmatrix} 0 & 1 \\ -10 & -7 \end{bmatrix}$$

and the resolvent matrix is

$$\mathbf{R}(s) = s\mathbf{I} - \mathbf{A} = \begin{bmatrix} s & -1 \\ 10 & (s+7) \end{bmatrix}$$

Thus

$$\mathbf{R}^{-1}(s) = \frac{1}{s(s+7)+10} \begin{bmatrix} (s+7) & 1 \\ -10 & s \end{bmatrix}$$

Perform the partial fraction expansion of each of the four elements of $\mathbf{R}^{-1}(s)$. For example, the upper-left element gives

$$\frac{s+7}{s(s+7)+10} = \frac{s+7}{(s+2)(s+5)} = \frac{5}{3}\frac{1}{s+2} - \frac{2}{3}\frac{1}{s+5}$$

The inverse transform gives $\varphi_{11}(t)$.

$$\varphi_{11}(t) = \tfrac{5}{3}e^{-2t} - \tfrac{2}{3}e^{-5t}$$

Repeating this process for the remaining three elements gives the same $\boldsymbol{\varphi}(t)$ as given in Example 9.4.

Comparing the results of the last two examples, we see that both the modal-matrix method and the Laplace transform method for finding $\boldsymbol{\varphi}(t)$ require us to solve the characteristic polynomial $|s\mathbf{I} - \mathbf{A}| = 0$. Both also require a matrix inversion to find $\mathbf{M}^{-1}$ and $\mathbf{R}^{-1}(s)$, respectively. However, the modal matrix $\mathbf{M}$ consists only of numbers and thus is suitable for inversion by a computer. On the other hand, the resolvent matrix cannot be inverted numerically because it is a function of the variable $s$. This difference gives the modal-matrix method an advantage for high-order models. For example, a fourth-order model would have a $(4 \times 4)$ resolvent matrix to be inverted. After the inversion, 16 partial fraction expansions must be performed. Thus the Laplace transform method is usually not employed for models of fourth order or higher. Calculation of $\mathbf{R}^{-1}(s)$ can be avoided by a technique that finds the response to an impulse input at each integrator in the state diagram. The difficulty with this method is that it can result in a reduction of a cumbersome state diagram. This and other methods based on more advanced matrix theory are given in References 2 and 3.

When the eigenvalues are complex, the modal transformation matrix will contain imaginary terms, and the method becomes more difficult to use. In such cases, the modified modal form given by (9.2-61) is useful, when employed together with the Laplace transform method.

*Example 9.6*

Find $\boldsymbol{\varphi}_C(t)$, the transition matrix for the following system corresponding to the modified canonical form (9.2-61).

$$\dot{z} = \begin{bmatrix} -a & b \\ -b & -a \end{bmatrix} \mathbf{z} \tag{9.3-16}$$

Recall that the eigenvalues are $s = -a \pm ib$. Thus

$$\mathbf{R}(s) = \begin{bmatrix} (s+a) & -b \\ b & (s+a) \end{bmatrix}$$

$$\mathbf{R}^{-1}(s) = \frac{1}{(s+a)^2 + b^2} \begin{bmatrix} (s+a) & b \\ -b & (s+a) \end{bmatrix}$$

We can revert to the time domain directly by using the transforms given in Table B.1 in Appendix B.

$$\mathscr{L}(e^{-at} \sin bt) = \frac{b}{(s+a)^2 + b^2} \tag{9.3-17}$$

$$\mathscr{L}(e^{-at} \cos bt) = \frac{s+a}{(s+a)^2 + b^2} \tag{9.3-18}$$

Therefore

$$\boldsymbol{\varphi}_C(t) = e^{-at} \begin{bmatrix} \cos bt & \sin bt \\ -\sin bt & \cos bt \end{bmatrix} \tag{9.3-19}$$

The modified modal transformation $\mathbf{x} = \mathbf{M}_C \mathbf{z}$ can be applied to the transition relation $\mathbf{x}(t) = \boldsymbol{\varphi}(t)\mathbf{x}(0)$ to obtain

$$\mathbf{M}_C \mathbf{z}(t) = \boldsymbol{\varphi}(t)\mathbf{M}_C \mathbf{z}(0)$$

This implies that

$$\mathbf{z}(t) = \mathbf{M}_C^{-1} \boldsymbol{\varphi}(t)\mathbf{M}_C \mathbf{z}(0)$$

This says that the transition matrix $\boldsymbol{\varphi}_C(t)$ for the modified system is

$$\boldsymbol{\varphi}_C(t) = \mathbf{M}_C^{-1} \boldsymbol{\varphi}(t)\mathbf{M}_C \tag{9.3-20}$$

If we know $\boldsymbol{\varphi}_C(t)$, we can obtain $\boldsymbol{\varphi}(t)$ from this relation, as

$$\boldsymbol{\varphi}(t) = \mathbf{M}_C \boldsymbol{\varphi}_C(t)\mathbf{M}_C^{-1} \tag{9.3-21}$$

*Example 9.7*

Find $\boldsymbol{\varphi}(t)$ for the vibration model of Example 9.3, with $m = 1$, $c = 2$, and $k = 5$.

From the results of Examples 9.3 and 9.6, $a = 1$, $b = 2$, and

$$\mathbf{M}_C = \begin{bmatrix} 1 & 0 \\ -1 & 2 \end{bmatrix} \qquad \mathbf{M}_C^{-1} = \tfrac{1}{2} \begin{bmatrix} 2 & 0 \\ 1 & 1 \end{bmatrix}$$

$$\boldsymbol{\varphi}_C(t) = e^{-t} \begin{bmatrix} \cos 2t & \sin 2t \\ -\sin 2t & \cos 2t \end{bmatrix}$$

From (9.3-21), the answer is

$$\boldsymbol{\varphi}(t) = \begin{bmatrix} 1 & 0 \\ -1 & 2 \end{bmatrix} \begin{bmatrix} \cos 2t & \sin 2t \\ -\sin 2t & \cos 2t \end{bmatrix} \begin{bmatrix} 2 & 0 \\ 1 & 1 \end{bmatrix} \frac{e^{-t}}{2}$$

$$= e^{-t} \begin{bmatrix} (\cos 2t + \frac{1}{2} \sin 2t) & \frac{1}{2} \sin 2t \\ -\frac{5}{2} \sin 2t & (\cos 2t - \frac{1}{2} \sin 2t) \end{bmatrix}$$

Table 5.1 gives the transition matrix for the general second-order constant-coefficient model for the distinct root cases.

## Series Solution for $\phi(t)$

Sometimes we do not require the solution for $x(t)$ in functional form, but only at discrete points in time. If we wish to investigate a variety of initial conditions, the transition matrix is particularly well suited. In such cases, we can use a numerical method for computing $\boldsymbol{\varphi}(t)$, especially for high-order models where the preceding methods become cumbersome.

The development of the series method to be presented now can be motivated by noting that for the first-order model $\dot{x} = ax$,

$$\phi(t) = e^{at} = 1 + at + \frac{(at)^2}{2!} + \ldots + \frac{(at)^k}{k!} + \ldots$$

This suggests a solution for the vector case of the form

$$\boldsymbol{\varphi}(t) = \mathbf{I} + \mathbf{A}t + \frac{\mathbf{A}^2 t^2}{2!} + \ldots + \frac{\mathbf{A}^k t^k}{k!} + \ldots \tag{9.3-22}$$

That this is the solution can be verified by noticing that it satisfies property (9.3-7), $\boldsymbol{\varphi}(0) = \mathbf{I}$, and that it satisfies the differential equation (9.3-8), $\dot{\boldsymbol{\varphi}}(t) = \mathbf{A}\boldsymbol{\varphi}(t)$. To show this, differentiate (9.3-22).

$$\dot{\boldsymbol{\varphi}}(t) = \mathbf{A} + \mathbf{A}^2 t + \ldots + \frac{A^k t^{k-1}}{(k-1)!} + \ldots$$

$$= \mathbf{A} \left( \mathbf{I} + \mathbf{A}t + \ldots + \frac{\mathbf{A}^{k-1} t^{k-1}}{(k-1)!} + \ldots \right)$$

$$= \mathbf{A}\boldsymbol{\varphi}(t)$$

This completes the proof of (9.3-22). Because of the similarity of this series expression with the series for $e^{at}$, the transition matrix $\boldsymbol{\varphi}(t)$ is sometimes termed the *matrix exponential* and written as $e^{\mathbf{A}t}$.

The series expression for $\boldsymbol{\varphi}(t)$ is particularly well suited for computer application because it merely requires repeated multiplication of the matrix $\mathbf{A}$. Suppose we desire the free response at a series of instants a time $\Delta$ apart. First, compute $\boldsymbol{\varphi}(\Delta)$ from (9.3-22) with $t = \Delta$. The number of terms in the series obviously determines its accuracy. H. M. Paynter of MIT has suggested the following approach (Reference 4).

1. Define $p$ to be the element of the matrix $\mathbf{A}\Delta$ that has the maximum abso-

lute value; that is, $p = \max |a_{ij}\Delta|$, where $a_{ij}$ represents the elements of the $(n \times n)$ system matrix $\mathbf{A}$.

2. Determine a number $q$ such that the following relation is satisfied.

$$\frac{1}{q!}(np)^q e^{np} \leqslant 0.001 \tag{9.3-23}$$

3. Compute $\boldsymbol{\varphi}(\Delta)$ by truncating the series (9.3-22) after the term involving $\mathbf{A}^q$.

The criterion (0.001) in step 2 was determined from simulations of somewhat typical cases, and therefore is only an approximate guide.

With $\boldsymbol{\varphi}(\Delta)$ thus computed, properties (9.3-10) and (9.3-13) can be used to determine $\mathbf{x}(t)$ at the times $\Delta, 2\Delta, 3\Delta, \ldots$. For a given system matrix $\mathbf{A}$, $\boldsymbol{\varphi}(\Delta)$ need be computed only once. If the response is desired for several sets of initial conditions $\mathbf{x}(0)$, this method has advantages over the methods discussed in Chapter Three. Note that the larger $\Delta$ is chosen, the more terms must be carried in the series for comparable accuracy.*

### *Example 9.8*

Use the series method to find $\boldsymbol{\varphi}(0.1)$ for the model given in Example 9.5.

For this model

$$\mathbf{A}\Delta = 0.1\begin{bmatrix} 0 & 1 \\ -10 & -7 \end{bmatrix}$$

Thus $p = 0.1(10) = 1$. From the criterion for $q$,

$$\frac{1}{q!}2^q e^2 \leqslant 0.001$$

The value of $q$ is most easily found by incrementing $q$ upward until the left-hand side exceeds 0.001 (This is easily done with a loop on $q$ in a computer program.) The result here is $q = 11$. Thus to achieve the accuracy suggested by the criterion, all terms up to and including the $\mathbf{A}^{11}$ term should be retained in the series.

Let us investigate the accuracy if only the terms through $\mathbf{A}^2$ are retained; that is,

$$\boldsymbol{\varphi}(t) \cong \mathbf{I} + \mathbf{A}t + \tfrac{1}{2}\mathbf{A}^2 t^2$$

This gives

$$\boldsymbol{\varphi}(t) = \begin{bmatrix} 1 & 0 \\ 0 & 1 \end{bmatrix} + \begin{bmatrix} 0 & 1 \\ -10 & -7 \end{bmatrix} t + \tfrac{1}{2}\begin{bmatrix} -10 & -7 \\ 70 & 39 \end{bmatrix} t^2$$

$$= \begin{bmatrix} (1 - 5t^2) & (t - 3.5t^2) \\ (-10t + 35t^2) & (1 - 7t + 19.5t^2) \end{bmatrix}$$

For $t = \Delta = 0.1$,

$$\boldsymbol{\varphi}(0.1) = \begin{bmatrix} 0.950 & 0.065 \\ -0.650 & 0.495 \end{bmatrix}$$

---

* The Cayley-Hamilton theorem (Problem 9.11) can also be used to compute $\boldsymbol{\varphi}(t)$. This requires the solution of simultaneous linear equations and thus is similar to the eigenvalue method (see References 2 and 3).

The analytical solution obtained in Example 9.4 gives

$$\boldsymbol{\varphi}(0.1) = \begin{bmatrix} 0.960 & 0.071 \\ -0.707 & 0.465 \end{bmatrix}$$

The agreement is close considering that only a few terms were used in the series. However, if we were to use $\boldsymbol{\varphi}(0.1)$ in repeated matrix multiplications to obtain the response, as in

$$\mathbf{x}((k+1)\Delta) = \boldsymbol{\varphi}(\Delta)\mathbf{x}(k\Delta)$$

we would need a higher degree of accuracy in $\boldsymbol{\varphi}(\Delta)$ in order to minimize the error propagation into the later response calculations.

## The Transition Matrix for Time-Varying Systems

So far in this section we have treated only the time-invariant case in which the system matrix **A** consists of constant elements. However, the transition matrix is a consequence of the linearity property of a model. Therefore, even if **A** is time-varying, a transition matrix exists, but it is more difficult to obtain than in the time-invariant case.

The basic relation for the free response is

$$\mathbf{x}(t) = \boldsymbol{\varphi}(t, t_0)\mathbf{x}(t_0) \tag{9.3-24}$$

where $t_0$ is a fixed but arbitrary time. Thus, the transition matrix in general depends on two parameters, $t$ and $t_0$. For the time-invariant case it is easily shown that the transition matrix depends only on the difference between $t$ and $t_0$; that is,

$$\boldsymbol{\varphi}(t, t_0) = \boldsymbol{\varphi}(t - t_0) \tag{9.3-25}$$

if **A** is constant. This is not true in the time-varying case, and thus properties (9.3-11) and (9.3-13) can no longer be employed. However, the other properties still hold. In terms of the new notation with the parameters $t$ and $t_0$, these properties are the transition property

$$\boldsymbol{\varphi}(t_2, t_0) = \boldsymbol{\varphi}(t_2, t_1)\boldsymbol{\varphi}(t_1, t_0) \tag{9.3-26}$$

the inversion property

$$\boldsymbol{\varphi}(t_0, t_1) = \boldsymbol{\varphi}^{-1}(t_1, t_0) \tag{9.3-27}$$

and the defining differential equation

$$\frac{\partial \boldsymbol{\varphi}(t, t_0)}{\partial t} = \dot{\boldsymbol{\varphi}}(t, t_0) = \mathbf{A}(t)\boldsymbol{\varphi}(t, t_0) \tag{9.3-28}$$

with the initial condition

$$\boldsymbol{\varphi}(t_0, t_0) = \mathbf{I} \tag{9.3-29}$$

Some analytical methods exist for determining $\boldsymbol{\varphi}(t, t_0)$, but they are restricted to special functional forms of $\mathbf{A}(t)$ (Reference 2). A numerical procedure can be developed from properties (9.3-28) and (9.3-29), as follows. For a given $t_0$, let the columns of $\boldsymbol{\varphi}(t, t_0)$ be denoted by the column vector $\boldsymbol{\Psi}_i(t)$. Then

$$\boldsymbol{\varphi}(t, t_0) = [\boldsymbol{\Psi}_1(t), \boldsymbol{\Psi}_2(t), \ldots, \boldsymbol{\Psi}_n(t)] \tag{9.3-30}$$

Then from (9.3-29),

$$\boldsymbol{\psi}_1(t_0) = \begin{bmatrix} 1 \\ 0 \\ \cdot \\ \cdot \\ \cdot \\ 0 \end{bmatrix}, \qquad \boldsymbol{\psi}_2(t_0) = \begin{bmatrix} 0 \\ 1 \\ 0 \\ \cdot \\ \cdot \\ \cdot \\ 0 \end{bmatrix} \tag{9.3-31}$$

and so forth. From (9.3-28), we get $n$ differential equations, one for each column vector of $\boldsymbol{\varphi}(t, t_0)$; that is,

$$\dot{\boldsymbol{\psi}}_i(t) = \mathbf{A}(t)\boldsymbol{\psi}_i(t) \qquad i = 1, 2, \ldots, n \tag{9.3-32}$$

whose initial condition is given by (9.3-31). Therefore, if we solve (9.3-32), with (9.3-31) as the initial conditions, using a numerical integration scheme such as Runge-Kutta or predictor-corrector, we will obtain $\boldsymbol{\varphi}(t, t_0)$ numerically at whatever times we desire. The solution for $\mathbf{x}(t)$ is then obtained from (9.3-24). This technique is powerful if the same system is to be analyzed for a variety of initial conditions $\mathbf{x}(t_0)$.

## Convolution and Forced Response

Our discussion of the transition matrix thus far has focused on the free response of the state model $\dot{\mathbf{x}} = \mathbf{Ax} + \mathbf{Bu}$. However, the transition matrix is also useful in computing the forced response. Consider the time-invariant case in which $\mathbf{A}$ and $\mathbf{B}$ are constant. Write the state model as

$$\dot{\mathbf{x}}(t) - \mathbf{Ax}(t) = \mathbf{Bu}(t)$$

Premultiply by $\boldsymbol{\varphi}(-t)$.

$$\boldsymbol{\varphi}(-t)[\dot{\mathbf{x}}(t) - \mathbf{Ax}(t)] = \boldsymbol{\varphi}(-t)\mathbf{Bu}(t)$$

The series solution for $\boldsymbol{\varphi}(t)$ can be used to show that $\boldsymbol{\varphi}(t)\mathbf{A} = \mathbf{A}\boldsymbol{\varphi}(t)$. This result and property (9.3-8) imply that

$$\begin{aligned} \frac{d}{dt}[\boldsymbol{\varphi}(-t)\mathbf{x}(t)] &= \frac{d\boldsymbol{\varphi}(-t)}{dt}\mathbf{x}(t) + \boldsymbol{\varphi}(-t)\frac{d\mathbf{x}(t)}{dt} \\ &= \boldsymbol{\varphi}(-t)[\dot{\mathbf{x}}(t) - \mathbf{Ax}(t)] \end{aligned}$$

Therefore,

$$\frac{d}{dt}[\boldsymbol{\varphi}(-t)\mathbf{x}(t)] = \boldsymbol{\varphi}(-t)\mathbf{Bu}(t)$$

Integrate both sides from 0 to $t$, and use $\boldsymbol{\varphi}(0) = \mathbf{I}$ to obtain

$$\boldsymbol{\varphi}(-t)\mathbf{x}(t) - \mathbf{x}(0) = \int_0^t \boldsymbol{\varphi}(-\tau)\mathbf{Bu}(\tau)\, d\tau$$

or

$$\mathbf{x}(t) = \boldsymbol{\varphi}(t)\mathbf{x}(0) + \int_0^t \boldsymbol{\varphi}(t-\tau)\mathbf{Bu}(\tau)\, d\tau \tag{9.3-33}$$

This form clearly shows the solution to be the sum of the free response and the forced response. This result can also be derived via the convolution theorem for Laplace transforms. The integral in (9.3-33) is the convolution integral.

The solution for time-varying systems is similar and is given by

$$\mathbf{x}(t) = \boldsymbol{\varphi}(t, t_0)\,\mathbf{x}(t_0) + \int_0^t \boldsymbol{\varphi}(t, \tau)\mathbf{B}(\tau)\,\mathbf{u}(\tau)\,d\tau \tag{9.3-34}$$

where the integral is the superposition integral. A derivation is given in Reference 2.

### *Example 9.9*

Determine the response of the time-invariant system $\dot{\mathbf{x}} = \mathbf{Ax} + \mathbf{Bu}$ to an input vector $\mathbf{u}$ consisting of (a) step functions starting at $t = 0$, and (b) ramp functions starting at $t = 0$.

(a) Let the vector $\mathbf{u}$ be $(m \times 1)$. Then $\mathbf{B}$ is $(n \times m)$. Assume the step inputs have different magnitudes and denote the $(m \times 1)$ vector of the magnitudes by $\mathbf{p}$. The solution from (9.3-33) is

$$\mathbf{x}(t) = \boldsymbol{\varphi}(t)\,\mathbf{x}(0) + \int_0^t \boldsymbol{\varphi}(t - \tau)\mathbf{Bp}\,d\tau$$

From property (9.3-11), $\boldsymbol{\varphi}(t - \tau) = \boldsymbol{\varphi}(t)\,\boldsymbol{\varphi}(-\tau)$. Thus

$$\mathbf{x}(t) = \boldsymbol{\varphi}(t)\,\mathbf{x}(0) + \boldsymbol{\varphi}(t)\int_0^t \boldsymbol{\varphi}(-\tau)d\tau\,\mathbf{Bp}$$

From the series expression for $\boldsymbol{\varphi}(t)$

$$\boldsymbol{\varphi}(-\tau) = \mathbf{I} - \mathbf{A}\tau + \frac{\mathbf{A}^2\tau^2}{2!} - \ldots$$

and

$$\begin{aligned}\int_0^t \boldsymbol{\varphi}(-\tau)\,d\tau &= \mathbf{I}t - \frac{\mathbf{A}t^2}{2!} + \frac{\mathbf{A}^2t^3}{3!} - \ldots \\ &= \mathbf{A}^{-1}[\mathbf{I} - \boldsymbol{\varphi}(-t)] \\ &= \mathbf{A}^{-1}[\mathbf{I} - \boldsymbol{\varphi}^{-1}(t)]\end{aligned}$$

if $\mathbf{A}^{-1}$ exists. Since $\mathbf{A}\boldsymbol{\varphi} = \boldsymbol{\varphi}\mathbf{A}$, $\boldsymbol{\varphi}\mathbf{A}^{-1} = \mathbf{A}^{-1}\boldsymbol{\varphi}$, and we get

$$\mathbf{x}(t) = \boldsymbol{\varphi}(t)\,\mathbf{x}(0) + \mathbf{A}^{-1}[\boldsymbol{\varphi}(t) - \mathbf{I}]\,\mathbf{Bp} \tag{9.3-35}$$

(b) A vector of ramp functions starting at $t = 0$ can be written as

$$\mathbf{u}(t) = t\mathbf{q}$$

where $\mathbf{q}$ is the $(m \times 1)$ vector of the slopes of the ramps. The response is

$$\begin{aligned}\mathbf{x}(t) &= \boldsymbol{\varphi}(t)\,\mathbf{x}(0) + \int_0^t \boldsymbol{\varphi}(t - \tau)\mathbf{B}\tau\mathbf{q}\,d\tau \\ &= \boldsymbol{\varphi}(t)\,\mathbf{x}(0) + \boldsymbol{\varphi}(t)\int_0^t \boldsymbol{\varphi}(-\tau)\tau\,d\tau\,\mathbf{Bq}\end{aligned}$$

$$= \boldsymbol{\varphi}(t)\,\mathbf{x}(0) + \boldsymbol{\varphi}(t)\left(\frac{\mathbf{I}t^2}{2} - \frac{2\mathbf{A}t^3}{3!} + \frac{3\mathbf{A}^2}{4!}t^4 - \ldots\right)\mathbf{B}\mathbf{q}$$

Inserting the series expression for $\boldsymbol{\varphi}(t)$ and multiplying the two series, we can show that

$$\mathbf{x}(t) = \boldsymbol{\varphi}(t)\mathbf{x}(0) + \mathbf{A}^{-2}[\boldsymbol{\varphi}(t) - \mathbf{I} - \mathbf{A}t]\mathbf{B}\mathbf{q} \tag{9.3-36}$$

The preceding results are especially convenient to use if the transition matrix is already available from an analysis of the free response.

## 9.4 CONTROLLABILITY AND OBSERVABILITY

In control-system design we must develop a controller to drive the system so that the output is a specified function of time. Thus it is desirable to be able to predict whether or not such a goal is attainable. This requires an understanding of the *controllability* of the system. A *state* $\mathbf{x}_1$ of a system is controllable if it is possible for the input vector to transfer any state $\mathbf{x}_0$ at any previous time $t_0$ to the state $\mathbf{x}_1$ in a finite amount of time. If all states $\mathbf{x}_1$ of the system are controllable, the *system* is controllable (this is sometimes called *complete controllability*).

On the other hand, in order to control a system we must be able to monitor its performance with feedback. The question of whether our measured variables are adequate for the task is related to the *observability* of the system. If our knowledge of the input $\mathbf{u}(\tau)$ and output $\mathbf{y}(\tau)$ on a finite time interval $t_0 < \tau \leqslant t$ allows us to determine the state $\mathbf{x}(t)$ for some given $t$, that *state* is said to be observable. The *system* is observable if all its states are observable (system observability is sometimes termed *complete observability*).

Any mathematical model is a simplification of the real physical system. Thus, while most physical systems are controllable and observable, their models might not be, and it is important to know when this occurs.

The first-order model $\dot{x} = ax + bu$ is controllable if $b \neq 0$. To show this, use the convolution result

$$x(t) = e^{at}x(t_0) + e^{at}\int_{t_0}^{t} e^{-a\tau}bu(\tau)\,d\tau$$

Controllability requires that this equation is true for arbitrary values of $x(t_0)$, $x(t)$, $t_0$, and $t$. A constant input $u$ can accomplish this task if no constraints are put on its magnitude. To see this carry out the integration for arbitrary but constant $u$, and solve for $u$.

If the output of the above system is $y = cx$, the system is observable if $c \neq 0$. This is easily shown by noting that the single-state variable $x$ constitutes a complete description of the system's behavior. If $c \neq 0$, we can recover $x$ from the measurement $y$.

### Controllability with Distinct Eigenvalues

When considering higher-order systems, controllability and observability properties are not so apparent. For example, consider the second-order system

$$\dot{x}_1 = -2x_1 + u \tag{9.4-1}$$

$$\dot{x}_2 = x_1 - x_2 - u \tag{9.4-2}$$

Obviously, the state $x_1$ is controllable, but the situation for $x_2$ is not as clear. The answer is found by recalling that the modes of a model form the structure of its dynamic behavior. In order to control the system's behavior, we must be able to influence each of its modes separately. Therefore, let us examine the modal composition of the model. The eigenvalues are $s = -1, -2$, and the eigenvectors are

$$\mathbf{v}_1 = \begin{bmatrix} 0 \\ 1 \end{bmatrix}, \qquad \mathbf{v}_2 = \begin{bmatrix} 1 \\ -1 \end{bmatrix}$$

[For this model, $a_{12} = 0$, so (9.2-8) and (9.2-9) cannot be applied.] The modal transformation matrix $\mathbf{M}$ and its inverse are

$$\mathbf{M} = \begin{bmatrix} 0 & 1 \\ 1 & -1 \end{bmatrix}, \qquad \mathbf{M}^{-1} = \begin{bmatrix} 1 & 1 \\ 1 & 0 \end{bmatrix}$$

The modes are $\mathbf{z} = \mathbf{M}^{-1}\mathbf{x}$, or

$$z_1 = x_1 + x_2$$

$$z_2 = x_1$$

The modal equations are

$$\dot{\mathbf{z}} = \mathbf{\Lambda z} + \mathbf{M}^{-1}\mathbf{B}u$$

$$= \begin{bmatrix} -1 & 0 \\ 0 & -2 \end{bmatrix} \mathbf{z} + \begin{bmatrix} 0 \\ 1 \end{bmatrix} u \tag{9.4-3}$$

or

$$\dot{z}_1 = -z_1$$

$$\dot{z}_2 = -2z_2 + u$$

Since neither $u$ nor $z_2$ appears in the first modal equation, the mode $z_1$ is uncontrollable, and therefore the system is uncontrollable.

This example has revealed a convenient test for controllability. If every mode is to be affected by the input vector $\mathbf{u}$, then at least one element of $\mathbf{u}$ must appear in each mode equation. For distinct eigenvalues this means that the system is controllable if and only if every row of $\mathbf{M}^{-1}\mathbf{B}$ has at least one nonzero element.

## Controllability with Repeated Eigenvalues

To investigate the controllability of a system with repeated eigenvalues, consider the second-order case with the repeated root $s_1$ and a scalar input $u$. The modal form is

$$\begin{bmatrix} \dot{z}_1 \\ \dot{z}_2 \end{bmatrix} = \begin{bmatrix} s_1 & 1 \\ 0 & s_1 \end{bmatrix} \begin{bmatrix} z_1 \\ z_2 \end{bmatrix} + \mathbf{M}^{-1}\mathbf{B}u \tag{9.4-4}$$

or

$$\dot{z}_1 = s_1 z_1 + z_2 + \alpha_1 u$$

$$\dot{z}_2 = s_1 z_2 + \alpha_2 u$$

where

$$\mathbf{M}^{-1}\mathbf{B} = \begin{bmatrix} \alpha_1 \\ \alpha_2 \end{bmatrix}$$

The constants $\alpha_1$ and $\alpha_2$ depend on **B** and $s_1$. We see that the mode $z_2$ is independent of the mode $z_1$, but $z_1$ is dependent on $z_2$. Thus if $\alpha_2 = 0$ the input $u$ cannot affect $z_2$, even indirectly. Thus we require $\alpha_2 \neq 0$ for controllability of the mode $z_2$. On the other hand, if $\alpha_1 = 0$, the mode $z_1$ can still be affected by the input through $z_2$. This can be shown by setting $\alpha_1 = 0$ and obtaining a single second-order equation for $z_1$. Using an argument similar to that for the first-order system, it can be shown that only $\alpha_2$ must be nonzero for controllability of both modes.

We can generalize this result to higher-order systems. A root that is repeated $q$ times will have a Jordan block of dimension $(q \times q)$. If all the eigenvectors corresponding to this root are not independent, the Jordan block will have 1s on the upper subdiagonal. Our earlier reasoning applies to this case, and all the modes corresponding to this repeated root will be controllable if any row of $\mathbf{M}^{-1}\mathbf{B}$ that corresponds to the last row of each such Jordan block has at least one nonzero element.

When the eigenvectors corresponding to a repeated root are independent, some zero elements will appear above the diagonal of the $\mathbf{\Lambda}$ matrix. When this happens, the purely diagonal blocks form Jordan blocks themselves. For example, suppose a fourth-order system has the root $s_1$ repeated three times and a distinct root $s_4$. If one of the three eigenvectors corresponding to $s_1$ is independent of the other two, the matrix will have the form

$$\mathbf{\Lambda} = \begin{bmatrix} s_1 & 1 & 0 & 0 \\ 0 & s_1 & 0 & 0 \\ 0 & 0 & s_1 & 0 \\ 0 & 0 & 0 & s_4 \end{bmatrix} \tag{9.4-5}$$

If this system has two input variables $u_1$ and $u_2$, the modal equations are of the form

$$\dot{\mathbf{z}} = \begin{bmatrix} s_1 & 1 & 0 & 0 \\ 0 & s_1 & 0 & 0 \\ 0 & 0 & s_1 & 0 \\ 0 & 0 & 0 & s_4 \end{bmatrix} \mathbf{z} + \begin{bmatrix} \alpha_1 & \alpha_2 \\ \alpha_3 & \alpha_4 \\ \alpha_5 & \alpha_6 \\ \alpha_7 & \alpha_8 \end{bmatrix} \begin{bmatrix} u_1 \\ u_2 \end{bmatrix} \tag{9.4-6}$$

This system will be controllable if and only if both of the following conditions are satisfied.

1. The vectors $[\alpha_3, \alpha_4]^T$ and $[\alpha_5, \alpha_6]^T$ are linearly independent.
2. Either $\alpha_7$ or $\alpha_8$ is nonzero.

The values of $\alpha_1$ and $\alpha_2$ are irrelevant to the controllability of the system.

We can now state the controllability conditions in general, as follows. Note that by definition, each Jordan block is associated with one and only one linearly independent eigenvector, and vice versa. Also, each distinct eigenvalue corresponds to a

$(1 \times 1)$ Jordan block. For the system: $\dot{\mathbf{x}} = \mathbf{Ax} + \mathbf{Bu}$, $\mathbf{\Lambda} = \mathbf{M}^{-1}\mathbf{AM}$, where $\mathbf{M}$ is the modal transformation matrix. Let $\mathbf{G} = [\mathbf{G}_1, \mathbf{G}_2, \ldots, \mathbf{G}_p]^T = \mathbf{M}^{-1}\mathbf{B}$, where $\mathbf{G}$ is partitioned according to the $p$ Jordan blocks in $\mathbf{\Lambda}$. Let $\mathbf{g}_i^T$ denote the last row in $\mathbf{G}_i$. The system is controllable if and only if both of the following conditions are satisfied.

1. $\mathbf{g}_i, \mathbf{g}_j, \ldots, \mathbf{g}_k$ is a linearly independent set if $J_i, J_j, \ldots, J_k$ are the Jordan blocks with the *same* eigenvalue $s_i$.
2. $\mathbf{g}_p \neq \mathbf{0}$ if $J_p$ is the *only* Jordan block with eigenvalue $s_p$.

For distinct eigenvalues, the "last" row of the Jordan block is the only row. The proof of these results is given in Reference 3.

## Determination of Controllability Directly from $A$ and $B$

The controllability criterion requires that the modal transformation matrix $\mathbf{M}$ and the Jordan matrix $\mathbf{\Lambda} = \mathbf{M}^{-1}\mathbf{AM}$ be known. This criterion tells us exactly which modes are uncontrollable, if any. Such information can be valuable for determining whether or not any approximations made in obtaining the system model have misrepresented the system's dynamics. On the other hand, if we need only a yes or no answer to the controllability question, an alternate criterion is available that requires only the matrices $\mathbf{A}$ and $\mathbf{B}$. This criterion states that the time invariant system $\dot{\mathbf{x}} = \mathbf{Ax} + \mathbf{Bu}$ is controllable if and only if the $(n \times nm)$ matrix

$$\mathbf{Q} = [\mathbf{B} | \mathbf{AB} | \ldots | \mathbf{A}^{n-1}\mathbf{B}] \tag{9.4-7}$$

is of rank $n$. [The matrix $\mathbf{A}$ is $(n \times n)$ and $\mathbf{B}$ is $(n \times m)$.]

The rank of a matrix $\mathbf{Q}$ is $n$ if there exists an $(n \times n)$ submatrix of $\mathbf{Q}$ such that the determinant of the submatrix is nonzero, and the determinant of every $(q \times q)$ submatrix of $\mathbf{Q}$ is zero for $q > n$. Determination of the rank of $\mathbf{Q}$ by this definition can be tedious. An alternate method is to check the determinant of $\mathbf{QQ}^T$. If this is nonzero, then the rank of $\mathbf{Q}$ is $n$.

*Example 9.10*

The second-order system model (9.4-1) and (9.4-2) has the matrices

$$\mathbf{A} = \begin{bmatrix} -2 & 0 \\ 1 & -1 \end{bmatrix}, \qquad \mathbf{B} = \begin{bmatrix} 1 \\ -1 \end{bmatrix}, \qquad \mathbf{AB} = \begin{bmatrix} -2 \\ 2 \end{bmatrix}$$

where $n = 2$, $m = 1$. Investigate its controllability.

We have from (9.4-7),

$$\mathbf{Q} = \begin{bmatrix} 1 & -2 \\ -1 & 2 \end{bmatrix}$$

The only $(2 \times 2)$ submatrix in $\mathbf{Q}$ is the determinant of $\mathbf{Q}$ itself, and this determinant is zero. The rank of $\mathbf{Q}$ is therefore less than two, and the system is not controllable.

This criterion can be proved by using the solution (9.4-33) and the series representation of $\boldsymbol{\varphi}(t)$ to show that any initial state $\mathbf{x}(0)$ can be transferred to the zero state $(\mathbf{x} = \mathbf{0})$ if the condition on the rank of $\mathbf{Q}$ is satisfied.

## Observability

The observability of a system indicates whether or not measurements denoted by the output vector $\mathbf{y}$ will reveal all of the dynamics of the system. The modal transformation $\mathbf{x} = \mathbf{Mz}$, when applied to the output equation $\mathbf{y} = \mathbf{Cx} + \mathbf{Du}$, gives

$$\mathbf{y} = \mathbf{CMz} + \mathbf{Du} \tag{9.4-8}$$

Thus if $\mathbf{y}$ contains information on all the modes (all the elements of $\mathbf{z}$), then each of the columns of the matrix $\mathbf{CM}$ must have at least one nonzero element. If the $i$th column of $\mathbf{CM}$ is identically zero, the mode $z_i$ will not appear in the vector $\mathbf{y}$. This is the observability criterion for the distinct eigenvalues case.

When some eigenvalues are repeated, the modes represented by their associated Jordan block are coupled. Thus only one of these modes (the first within the block) need be measured to determine the behavior of the remaining modes in the block. For example, in (9.4-4) we saw that the mode $z_1$ depends on $z_2$, but not vice versa. Mode $z_1$ contains information on $z_2$, but $z_2$ is ignorant of $z_1$. With this effect noted, we can now state the general observability criterion for time-invariant systems ($\mathbf{A}, \mathbf{B}, \mathbf{C}$ constant). Let the matrix $\mathbf{C}$ be partitioned according to the Jordan block structure of $\Lambda$, as $[\mathbf{C}_1, \mathbf{C}_2, \ldots, \mathbf{C}_p]$ where $p$ is the number of Jordan blocks. Let $\mathbf{c}_i$ denote the first column of $\mathbf{C}_i$. As shown in Reference 3, the system is observable if and only if both of the following conditions are satisfied.

1. $\mathbf{c}_i, \mathbf{c}_j, \ldots, \mathbf{c}_k$ is a linearly independent set if $J_i, J_j, \ldots, J_k$ are the Jordan blocks with the *same* eigenvalue $s_i$.
2. $\mathbf{c}_p \neq \mathbf{0}$ if $J_p$ is the *only* Jordan block with eigenvalue $s_p$.

For example, suppose the fourth-order system described by (9.4-6) has the output relation

$$\mathbf{y} = \left[\begin{array}{cc|c|c} \beta_1 & \beta_3 & \beta_5 & \beta_7 \\ \beta_2 & \beta_4 & \beta_6 & \beta_8 \end{array}\right] \begin{bmatrix} z_1 \\ z_2 \end{bmatrix} + \mathbf{Du}$$

The elements $\beta_1$, $\beta_2$, $\beta_3$, and $\beta_4$ correspond to the first Jordan block, $\beta_5$ and $\beta_6$ to the second block, and $\beta_7$ and $\beta_8$ to the third block. Thus the system is observable if and only if both of the following conditions are satisfied.

1. The vectors $[\beta_1, \beta_2]^T$ and $[\beta_5, \beta_6]^T$ are linearly independent.
2. Either $\beta_7$ or $\beta_8$ is nonzero.

The elements $\beta_3$ and $\beta_4$ are irrelevant to the system's observability.

As with controllability, a criterion is available for the determination of observability directly from the matrices $\mathbf{A}$ and $\mathbf{C}$. However, it does not indicate what states are unobservable. The criterion states that the time invariant system $\dot{\mathbf{x}} = \mathbf{Ax} + \mathbf{Bu}$, $\mathbf{y} = \mathbf{Cx} + \mathbf{Du}$ is observable if and only if the $(qn \times n)$ matrix

$$\mathbf{P} = \begin{bmatrix} \mathbf{C} \\ \hline \mathbf{CA} \\ \hline \vdots \\ \hline \mathbf{CA}^{n-1} \end{bmatrix} \tag{9.4-9}$$

is of rank $n$. [The matrix $\mathbf{A}$ is $(n \times n)$ and $\mathbf{C}$ is $(q \times n)$.] The rank of $\mathbf{P}$ can be checked by computing the determinant of $\mathbf{PP}^T$. If this determinant is nonzero, $\mathbf{P}$ is rank $n$.

For example, if the model given by (9.4-1) and (9.4-2) has an output relation

$$y = [c_1 \quad c_2]\mathbf{x}$$

then

$$\mathbf{CA} = [c_1 \quad c_2]\begin{bmatrix} -2 & 0 \\ 1 & -1 \end{bmatrix} = [(1 - 2c_2) \quad -c_2]$$

$$\mathbf{P} = \begin{bmatrix} c_1 & c_2 \\ (1 - 2c_2) & -c_2 \end{bmatrix}$$

If $c_1 = 0$, $\mathbf{P}$ is of rank 2 for nonzero $c_2$. If $c_2 = 0$, $\mathbf{P}$ is of rank 1 and the system is unobservable because the first mode is independent of the second mode. Measuring the first mode gives no information about the second.

Remembering that the transpose of a product reverses the order of multiplication, we see that (9.4-9) can be written as

$$\mathbf{P}^T = [\mathbf{C}^T \mid \mathbf{A}^T\mathbf{C}^T \mid \cdots \mid (\mathbf{A}^{n-1})^T\mathbf{C}^T] \tag{9.4-10}$$

If $\mathbf{P}^T$ is of rank $n$, so is $\mathbf{P}$.

The direct criterion for observability can be proved by using the solution (9.4-33) and the series representation of $\boldsymbol{\varphi}(t)$ to show that the state $\mathbf{x}(t_1)$ at any previous time is uniquely determined from the output measurements $\mathbf{y}(t)$ over an interval $t_1 \leqslant t \leqslant t_2$ if the rank of $\mathbf{P}$ is $n$.

## System Composition

Consider the system (9.4-1) and (9.4-2) represented in modal form (9.4-3).

$$\dot{z}_1 = -z_1$$

$$\dot{z}_2 = -2z_2 + u$$

We have seen that $z_2$ is controllable but $z_1$ is not. The observability of this system cannot be determined until we specify an output relation. Suppose the scalar output is $y = z_1$. Then mode $z_1$ is observable but not controllable, whereas the mode $z_2$ is controllable but not observable. On the other hand, if the output is $y = z_2$, the mode $z_1$ is no longer observable, but $z_2$ is. We conclude from this that in general a system can be considered to be composed of four types of subsystems: $S_1$, $S_2$, $S_3$, and $S_4$, defined as follows.

Subsystem $S_1$. Controllable and observable.

Subsystem $S_2$. Controllable but unobservable.

Subsystem $S_3$. Uncontrollable but observable.

Subsystem $S_4$. Uncontrollable and unobservable.

We can illustrate this concept by a fourth-order example. Suppose the following modal form appears.

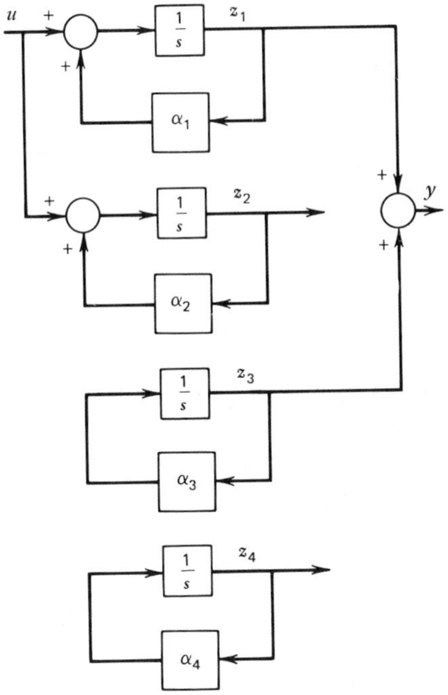


**Figure 9.13** Fourth-order example of system composition.

$$\dot{\mathbf{z}} = \begin{bmatrix} \alpha_1 & 0 & 0 & 0 \\ 0 & \alpha_2 & 0 & 0 \\ 0 & 0 & \alpha_3 & 0 \\ 0 & 0 & 0 & \alpha_4 \end{bmatrix} \mathbf{z} + \begin{bmatrix} 1 \\ 1 \\ 0 \\ 0 \end{bmatrix} u \tag{9.4-11}$$

$$y = [1 \quad 0 \quad 1 \quad 0]\mathbf{z} \tag{9.4-12}$$

From the criteria for controllability and observability, we see that mode $z_1$ is a subsystem of type $S_1$, mode $z_2$ of type $S_2$, mode $z_3$ of type $S_3$, and mode $z_4$ of type $S_4$. The block diagram for this system is shown in Figure 9.13. It clearly shows that the input $u$ affects only the controllable modes $z_1$ and $z_2$, and that the observable modes $z_1$ and $z_3$ are the only modes appearing in the output.

We can represent the general case as shown in Figure 9.14. Dynamics occur in the general case that are not present in the fourth-order example (9.4-11) and (9.4-12). Consider the subsystem $S_2$ in the general case. It can be affected by all of the other subsystems as well as by the input. But because $S_2$ is unobservable, no information can flow from it to the output, even indirectly. Because $S_2$ is controllable, no information can flow from it to any uncontrollable subsystem, such as $S_4$. Thus no signal flows are directed out of $S_2$. Similar interpretations apply to the other subsystems in the diagram.

## Pole-Zero Cancellation in a Transfer Function

Examination of the diagram in Figure 9.14 shows that the only path from the input **u** to the output **y** is through the subsystem $S_1$, the controllable and observable part of

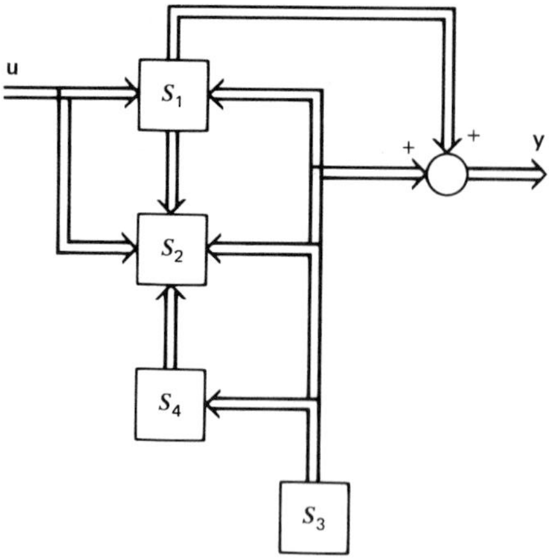


**Figure 9.14 System composition in terms of observability and controllability.**

the system. This means that the transfer functions relating **y** to **u** are a description only of the controllable and observable subsystem. The transfer function description can hide some of the dynamics of the system, and therefore we must exercise care in using it.

Consider the fourth-order model (9.4-11) and (9.4-12). The single transfer function between the scalar input $u$ and the scalar output $y$ is

$$\frac{Y(s)}{U(s)} = T(s) = [1 \quad 0 \quad 1 \quad 0]\begin{bmatrix} (s-\alpha_1)^{-1} & 0 & 0 & 0 \\ 0 & (s-\alpha_2)^{-1} & 0 & 0 \\ 0 & 0 & (s-\alpha_3)^{-1} & 0 \\ 0 & 0 & 0 & (s-\alpha_4)^{-1} \end{bmatrix}\begin{bmatrix} 1 \\ 1 \\ 0 \\ 0 \end{bmatrix}$$

$$= \frac{1}{s-\alpha_1}$$

Only the eigenvalue $\alpha_1$ corresponding to the observable and controllable mode $z_1$ appears in the transfer function.

The transfer function, as a relation only between the input and output, is independent of the choice of the state variables. The mechanism by which the uncontrollable and/or unobservable dynamics are disguised by the transfer function is not apparent in the model (9.4-11) because it has already been expressed in modal form. To show what occurs, return to our first example (9.4-1) and (9.4-2). The transformed equations are

$$sX_1(s) = -2X_1(s) + U(s)$$

$$sX_2(s) = X_1(s) - X_2(s) - U(s)$$

Thus

$$X_1(s) = \frac{1}{s+2}U(s)$$

$$(s+1)X_2(s) = X_1(s) - U(s)$$

Eliminating $X_1(s)$ we get

$$X_2(s) = -\frac{s+1}{(s+1)(s+2)}U(s) = -\frac{1}{s+2}U(s)$$

The root $s = -1$ corresponding to the uncontrollable mode $z_1$ has been cancelled by the pole $s = -1$ in the numerator. If the transfer function had been given to us without the pole-zero cancellation being evident, a clue to the existing difficulty is that the denominator is a first-order polynomial. If it is known that the system is of second order, we would then realize that a cancellation has occurred, and therefore that an uncontrollable and/or unobservable mode exists.

For systems with distinct eigenvalues, the transfer function matrix represents only the controllable and observable part of the system. It might not even provide this information for systems with repeated eigenvalues. For example, if more than one linearly independent eigenvector corresponds to the same eigenvalue, the resulting $\mathbf{\Lambda}$ matrix will be diagonal, and the coupling between that eigenvalue's modes will be hidden in a transfer function description. The danger posed by hidden modes is that they might be unstable. Even though the input cannot affect them, they can be excited by initial conditions and cause unforeseen behavior.

The transfer function is an extremely convenient tool because it reduces a problem in differential equations to one of algebra. However, when using transfer functions, do not forget the physics of the system under study. This is the best way to avoid problems with uncontrollable or unobservable subsystems.

*Example 9.11*

(a) Discuss the observability and controllability of the second-order system:

$$m\ddot{x} + c\dot{x} + kx = b\,u(t)$$

where $m \neq 0$.

(b) Apply the results to the rotational systems shown in Figure 9.15.

(a) Convert the model to state variable form by choosing $x_1 = x$ and $x_2 = \dot{x}$. This gives $\dot{\mathbf{x}} = \mathbf{A}\mathbf{x} + \mathbf{B}u$ where

$$\mathbf{A} = \begin{bmatrix} 0 & 1 \\ -\frac{k}{m} & -\frac{c}{m} \end{bmatrix}, \qquad \mathbf{B} = \begin{bmatrix} 0 \\ \frac{b}{m} \end{bmatrix}$$

The controllability criterion requires that $[\mathbf{B} \,\vdots\, \mathbf{AB}]$ be of rank 2.

$$[\mathbf{B} \,\vdots\, \mathbf{AB}] = \begin{bmatrix} 0 & \frac{b}{m} \\ \frac{b}{m} & -\frac{bc}{m^2} \end{bmatrix} = -\left(\frac{b}{m}\right)^2$$

The system is controllable if $b \neq 0$.

Assume the output relation is $y = c_1x_1 + c_2x_2$. Then

$$\mathbf{C} = [c_1 \quad c_2]$$

and the observability criterion requires that the rank of $[\mathbf{C}^T \,\vdots\, \mathbf{A}^T\mathbf{C}^T]$ be two, where

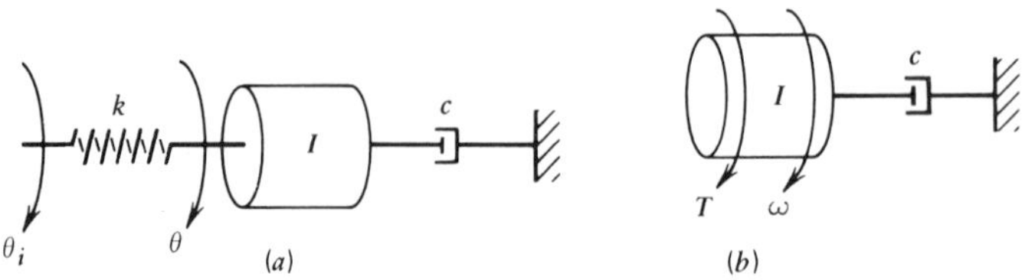


**Figure 9.15** **Rotational systems illustrating controllability and observability. (*a*) Displacement input. (*b*) Torque input.**

$$|\mathbf{C}^T \,\vdots\, \mathbf{A}^T\mathbf{C}^T| = \begin{vmatrix} c_1 & \dfrac{-kc_2}{m} \\ c_2 & \left(c_1 - \dfrac{cc_2}{m}\right) \end{vmatrix} = \Delta$$

Consider the case in which only the displacement $x_1$ is measured ($c_2 = 0$, $c_1 \neq 0$). Then $\Delta = c_1^2 \neq 0$, and the system is observable. Given a displacement as a function of time, we can differentiate it (in theory) and recover the velocity. However, the implementation of such a differentiation scheme, whether by analog or digital hardware, could result in so much error as to make the system unobservable in practice.

For only a velocity measurement, $c_1 = 0$, $c_2 \neq 0$, and $\Delta = kc_2^2/m$. In this case, we have an observable system only if $k \neq 0$. We cannot integrate the velocity to obtain displacement, because the integration introduces a constant that corresponds to the unmeasurable initial displacement. However, we can differentiate $\dot{x}$ (in theory) to obtain $\ddot{x}$, and if $k \neq 0$, we can solve the original differential equation model for $x$ in terms of $\dot{x}$, $\ddot{x}$, and the input $u$.

(b) The model for the system shown in Figure 9.15*a* is

$$I\ddot{\theta} + c\dot{\theta} + k\theta = k\theta_i$$

Using the results from part a, we conclude that the system is controllable with the input displacement $\theta_i$ if $k \neq 0$; it is observable if $\theta$ is measured or if $\dot{\theta}$ is measured and $k \neq 0$. For the system shown in Figure 9.15*b* the model is

$$I\ddot{\theta} + c\dot{\theta} = T.$$

Thus $m = I$, $b = 1$, and $k = 0$ in the general model form. The system is therefore always controllable but is unobservable for only a velocity measurement. However, we must be careful in interpreting this result. No potential energy storage occurs in this system, and therefore the velocity $\dot{\theta}$ is the only state variable required. Letting $\omega = \dot{\theta}$ the model becomes

$$I\dot{\omega} + c\omega = T$$

In this form we see that the system is observable with a velocity measurement. This makes sense physically if we are interested only in the speed of the load $I$ and not in its angular position.

## 9.5 MATRIX ANALYSIS OF DISCRETE-TIME SYSTEMS

The methods of this chapter apply with very little modification to discrete-time systems. State-variable models can be obtained foı such systems either from the pulse-transfer function of the sampled-data system or by direct discretization of the differential equation model. We have seen examples of both methods in Section 5.11.

## Discrete Models from the Transition Matrix $\phi(t)$

The transition matrix for continuous-time systems provides another way of obtaining a discrete-time model. Equation (9.3-35) of Example 9.9 gives the step response of the system $\dot{\mathbf{x}} = \mathbf{A}\mathbf{x} + \mathbf{B}\mathbf{u}$, for $\mathbf{u} = \mathbf{u}(0) = \mathbf{p} = $ constant. It is

$$\mathbf{x}(t) = \boldsymbol{\varphi}(t)\mathbf{x}(0) + \mathbf{A}^{-1}[\boldsymbol{\varphi}(t) - \mathbf{I}]\mathbf{B}\mathbf{p} \qquad (9.5\text{-}1)$$

Recall that the argument $t$ in $\boldsymbol{\varphi}(t)$ actually represents the difference between the time $t$ and the time 0. If we redefine the time variable so that $t = 0$ corresponds to $t_k$, and $t$ corresponds to $t_{k+1}$, then (9.5-1) can be written in difference equation form as

$$\mathbf{x}(t_{k+1}) = \boldsymbol{\varphi}(t_{k+1} - t_k)\mathbf{x}(t_k) + \mathbf{A}^{-1}[\boldsymbol{\varphi}(t_{k+1} - t_k) - \mathbf{I}]\mathbf{B}\mathbf{p}(t_k) \qquad (9.5\text{-}2)$$

where $\mathbf{p}(t_k)$ represents an input that is constant over the interval $(t_k, t_{k+1})$. Let $\mathbf{x}(t_k) = \mathbf{x}(k)$ and $\mathbf{p}(t_k) = \mathbf{u}(k)$, and define the sample period as

$$T = t_{k+1} - t_k \qquad (9.5\text{-}3)$$

then (9.5-2) can be written as

$$\mathbf{x}(k+1) = \mathbf{P}\mathbf{x}(k) + \mathbf{Q}\mathbf{u}(k) \qquad (9.5\text{-}4)$$

where we have defined

$$\mathbf{P} = \boldsymbol{\varphi}(t_{k+1} - t_k) = \boldsymbol{\varphi}(T) \qquad (9.5\text{-}5)$$

$$\mathbf{Q} = \mathbf{A}^{-1}[\boldsymbol{\varphi}(T) - \mathbf{I}]\mathbf{B} \qquad (9.5\text{-}6)$$

If $\mathbf{A}^{-1}$ does not exist, the results of Example 9.9 can still be applied, but they cannot be expressed in as convenient a form.

A sampler with a zero-order hold applied to the input $\mathbf{u}(t)$ will generate a piecewise-constant input $\mathbf{u}(k)$. Since (9.5-4) through (9.5-6) represent the exact solution for such a situation, the resulting difference equation model is equivalent to that obtained from the pulse transfer function of the system with the sample-and-hold. The latter approach resulted in (5.11-9), which is the same as (9.5-4).

Derivation of the state model from the transfer function is the easiest approach when the order of the sampled-data system is relatively low. For higher-order systems the algebra involved in the partial fraction expansion of the transfer function $T(s)$ to obtain the pulse transfer function $T(z)$ is very tedious. In such cases, if $\mathbf{A}$ and $\mathbf{B}$ have known values, we can use a numerical solution for $\boldsymbol{\varphi}(T)$ in (9.5-5) and (9.5-6). This can be obtained with the series method given by (9.2-22).

## Modes of Discrete-Time Response

The modal response methods for discrete-time systems are exactly parallel to those presented in Section 9.2. The free response of (9.5-4) is of the form

$$\mathbf{x}(k) = \mathbf{p}z^k \qquad (9.5\text{-}7)$$

Substituting this into (9.5-4) with $\mathbf{u} = \mathbf{0}$ gives the following eigenvalue problem.

$$(z_i\mathbf{I} - \mathbf{P})\mathbf{p}_i = \mathbf{0} \qquad (9.5\text{-}8)$$

where the $z_i$ are the roots of the characteristic equation

$$|z\mathbf{I} - \mathbf{P}| = 0 \tag{9.5-9}$$

and the $\mathbf{p}_i$ are the eigenvectors. Equation (9.5-8) and (9.5-9) are identical to those of the continuous-time case, with $\mathbf{P}$ and $z$ replacing $\mathbf{A}$ and $s$. The only difference occurs in the form of the resulting response, which is

$$\mathbf{x}(k) = \sum_{i=1}^{n} p_i z_i^k \tag{9.5-10}$$

See (9.2-34) for comparison.

Since the eigenvalue problems are identical, the related properties of invariance and modal transformation also apply here. Thus the model (9.5-4) can be diagonalized by the modal matrix

$$\mathbf{M} = [\mathbf{p}_1, \mathbf{p}_2, \ldots, \mathbf{p}_n] \tag{9.5-11}$$

The modal transformation is $\mathbf{x} = \mathbf{M}\mathbf{w}$, where $\mathbf{w}$ represents the modal coordinates. Thus for a second-order system with two distinct eigenvalues $z_1$ and $z_2$, the diagonalized state matrix is

$$\mathbf{\Lambda} = \mathbf{M}^{-1}\mathbf{P}\mathbf{M} = \begin{bmatrix} z_1 & 0 \\ 0 & z_2 \end{bmatrix} \tag{9.5-12}$$

and (9.5-4) becomes

$$\mathbf{w}(k+1) = \mathbf{\Lambda}\mathbf{w}(k) + \mathbf{M}^{-1}\mathbf{Q}\mathbf{u}(k) \tag{9.5-13}$$

The extension to higher-order systems is obvious. When repeated or complex eigenvalues occur, modified forms like (9.2-58) and (9.2-61) can be used for $\mathbf{\Lambda}$.

## The Transition Matrix

Given the initial condition $\mathbf{x}(0)$, repeated step-by-step evaluation of (9.5-4) shows that

$$\mathbf{x}(k) = \mathbf{P}^k\mathbf{x}(0) + \sum_{j=0}^{k-1} \mathbf{P}^{k-1-j}\mathbf{Q}\mathbf{u}(j) \tag{9.5-14}$$

This constitutes the convolution theorem for discrete systems. From it the value of $\mathbf{x}(k)$ can be obtained for arbitrary $\mathbf{x}(0)$ by brute force evaluation of $\mathbf{P}^k$ with successive matrix multiplication. However, some more efficient means are available.

Equation (9.5-14) shows that the transition matrix for (9.5-4) is

$$\boldsymbol{\varphi}(k) = \mathbf{P}^k \tag{9.5-15}$$

The $z$-transformation of (9.5-4) can be used to obtain another expression for $\boldsymbol{\varphi}(k)$. This gives

$$z\mathbf{X}(z) - z\mathbf{x}(0) = \mathbf{P}\mathbf{X}(z) + \mathbf{Q}\mathbf{U}(z)$$

or

$$\mathbf{X}(z) = (z\mathbf{I} - \mathbf{P})^{-1}[z\mathbf{x}(0) + \mathbf{U}(z)] \tag{9.5-16}$$

Therefore,

$$\boldsymbol{\varphi}(k) = \mathcal{Z}^{-1}[(z\mathbf{I} - \mathbf{P})^{-1}z] \tag{9.5-17}$$

and $\boldsymbol{\varphi}(k)$ can be evaluated with the inverse $z$-transform. Compare this result with (9.3-17). As before, the inverse transformation can be applied either to the original equation (9.5-4) or to the modal form (9.5-13). If $\boldsymbol{\varphi}_M(k)$ denotes the transition matrix in terms of the modal coordinates, then as in (9.3-6), we have

$$\boldsymbol{\varphi}(k) = \mathbf{M}\boldsymbol{\varphi}_M(k)\mathbf{M}^{-1} \tag{9.5-18}$$

For real distinct eigenvalues, the partial-fraction expansion of the elements of $(z\mathbf{I} - \mathbf{P})^{-1}z$ are of the form

$$\frac{C_1 z}{z - z_1} + \ldots + \frac{C_n z}{z - z_n}$$

These have the inverse transforms

$$\mathscr{Z}^{-1}\left(\frac{C_i z}{z - z_i}\right) = C_i z_i^k$$

Thus for $\boldsymbol{\Lambda}$ given by (9.5-12),

$$\boldsymbol{\varphi}_M(k) = \begin{bmatrix} z_1^k & 0 \\ 0 & z_2^k \end{bmatrix} \tag{9.5-19}$$

The results for repeated and complex eigenvalues can be obtained in a similar way.

## Controllability and Observability

Since no dynamic operation is involved with the continuous-time output equation $\mathbf{y} = \mathbf{Cx} + \mathbf{Du}$, we can continue to use the matrices $\mathbf{C}$ and $\mathbf{D}$ for the discrete-time case. Thus the discrete-time output equation is

$$\mathbf{y}(k) = \mathbf{Cx}(k) + \mathbf{Du}(k) \tag{9.5-20}$$

The concepts of observability and controllability also apply to the discrete-time case, and the tests can be developed using the modal transformation just described. We limit ourselves to the statement of the direct criteria. These are identical to the ones for continuous time, with $\mathbf{A}$ and $\mathbf{B}$ replaced by $\mathbf{P}$ and $\mathbf{Q}$. The system described by (9.5-4) and (9.5-20) is controllable if and only if the $(m \times nm)$ matrix

$$[\mathbf{Q} \,|\, \mathbf{PQ} \,|\, \ldots \,|\, \mathbf{P}^{n-1}\mathbf{Q}] \tag{9.5-21}$$

is of rank $n$. Here $\mathbf{P}$ is $(n \times n)$ and $\mathbf{Q}$ is $(n \times m)$.

The system is observable if and only if the rank of the $(qn \times n)$ matrix

$$\begin{bmatrix} \mathbf{C} \\ \hline \mathbf{CP} \\ \hline \vdots \\ \hline \mathbf{CP}^{n-1} \end{bmatrix} \tag{9.5-22}$$

is of rank $n$. Here $\mathbf{C}$ is $(q \times n)$.

## 9.6 SUMMARY

The state-variable matrix methods presented in this chapter are very useful when the design model is of high order. Their advantages can be categorized as providing (1) improved insight into the system response, (2) more general design procedures for control systems, and (3) increased computational power and efficiency.

The response character of a system can be represented most conveniently with the modal variables. Simplification of the state matrix by modal transformation allows the analyst to see more clearly the effect of the system's input and parameters on the response. These results lead directly to the concepts of controllability and observability. These concepts in turn have produced a better understanding of the effects of modeling approximations and classical compensation by pole-zero cancellation.

The algorithm (9.3-35) for computing the step response of a high-order linear system is powerful because it already contains much of the response information in the form of the transition matrix $\boldsymbol{\varphi}(t)$. It is efficient because the effects of the inputs and of the initial conditions each appear as multiplicative factors. When combined with a computer program to implement the series solution for $\boldsymbol{\varphi}(t)$, this algorithm is superior to other models, such as Runge-Kutta integration, when the response must be investigated for many values of input magnitude and initial conditions.

In the next chapter we will apply these methods to obtain new control system structures and design methods.

## REFERENCES

1. R. Prabhakar, S. J. Citron, and R. E. Goodson, "Optimization of Automotive Engine Fuel Economy and Emissions," *ASME J. of Dynamic Systems, Measurement and Control*, June 1977, pp. 109–117.
2. D. M. Wiberg, *State Space and Linear Systems*, McGraw-Hill, New York, 1971.
3. W. L. Brogan, *Modern Control Theory*, Quantum Publishers, New York, 1974.
4. Y. Takahashi, M. J. Rabins, and D. M. Auslander, *Control*, Addison-Wesley, Reading, Mass., 1970.

## PROBLEMS

9.1 Develop a state variable model for the system shown in Figure P9.1. The displacements $y_1$ and $y_2$ represent deviations from equilibrium positions when $f_1 = f_2 = 0$. Let the state and input vectors be $\mathbf{x} = [y_1, \dot{y}_1, y_2, \dot{y}_2]^T$ and $u = [f_1 \;\; f_2]^T$.

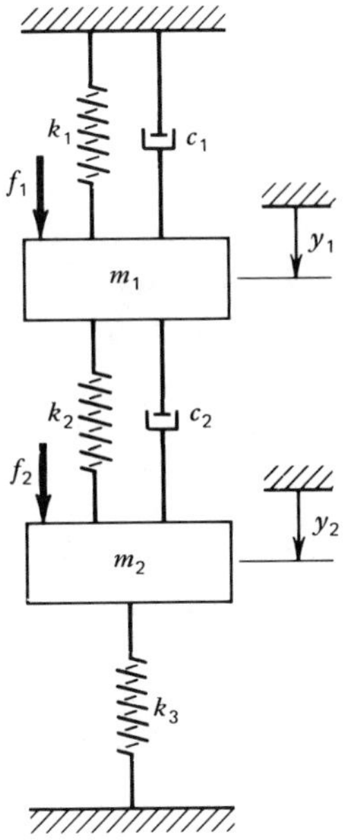


Figure P9.1

9.2 Put the model (9.1-1) and (9.1-2) into state variable form using the state variables defined in (9.1-3).

9.3 Figure P9.3 shows the addition of the mass $m_1$ of the passengers and seats and seat elasticity $k_3$ to the ride quality model of Figure 9.2. Develop the equations of motion and put them into state variable form. Include the damping and elasticities $c_1$, $c_2$, $k_1$, and $k_2$ as shown in Figure 9.2.

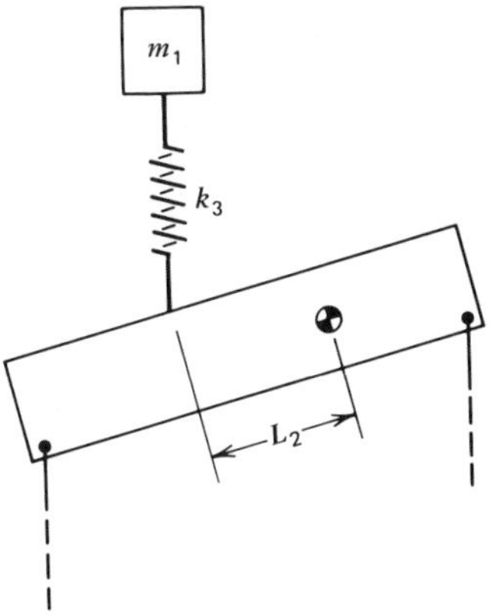


Figure P9.3

9.4 The Euler equations (9.1-8) to (9.1-10) describe the rotational dynamics of a rigid body.

(a) Obtain the state and input matrices **A** and **B** for the linearized form of the Euler equations. The state vector **x** is the deviation of the velocity vector $\boldsymbol{\omega} = [\omega_1, \omega_2, \omega_3]^T$ from its equilibrium value produced when the applied torques are held constant at some reference values. The input vector **u** is the deviation of $\mathbf{T} = [T_1, T_2, T_3]^T$ from these values.

(b) Suppose that the body is rotationally symmetric about the $\omega_1$ axis; that is, $I_2 = I_3$. What effect does this have on the linearized model? Comment on the effects on the accuracy of the linearization of the $\dot{\omega}_1$ equation.

(c) Consider the general case in which $I_1$, $I_2$, and $I_3$ are unequal. Suppose that the reference equilibrium is $\boldsymbol{\omega} = [\omega_1, 0, 0]^T$; that is, no rotation about the second and third axes. What effect does this have on the linearized model?

9.5 The electrohydraulic positioning system shown in Figure P9.5*a* can be represented by the block diagram in Figure P9.5*b*. Typical values for the parameters might be as follows.

$K_1 = 100\ \text{in.}^3\text{sec}^{-1}\text{amp}^{-1}$ $\quad K_2 = 300\ \text{in.}^{-2}\text{sec}^{-2}$

$K_3 = 0.1$ in./in. $\quad K_4 = 200$ volt/in.

$\zeta = 0.8$ $\quad \omega_n = 100$ rad/sec $\quad \tau = 0.01$ sec

(a) Develop a state variable model of the plant with the controller current $i_c$ as input and the displacement $y$ as output.

(b) Assume that proportional control is used so that $G_c(s) = K$. Develop a state model of the system with $y_d$ as input and $y$ as output ($y_d$ is the desired value of $y$).

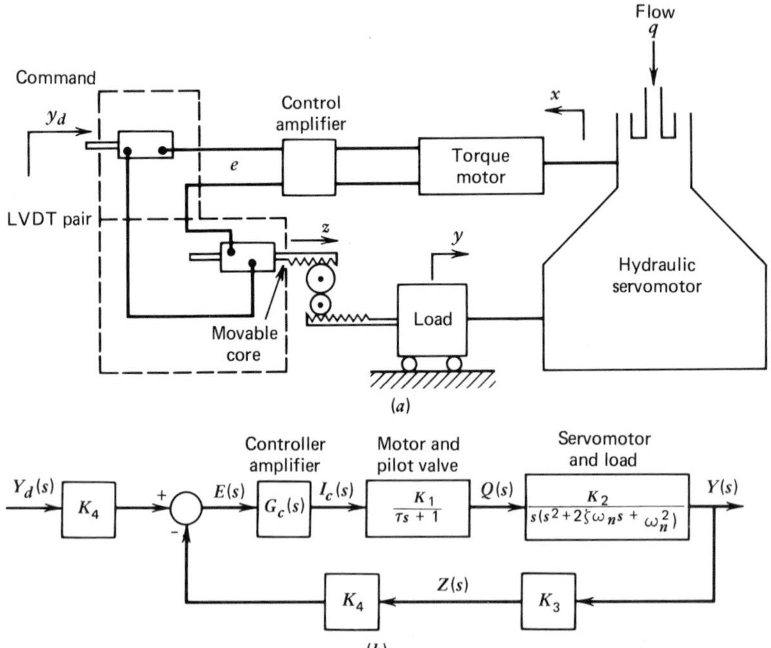


Figure P9.5

9.6 Control systems are used to maintain proper tension in many winding applications in the paper, wire, plastic, and tape industries. For example, it is important to control the tension of paper as it is wound from a master roll and rewound on smaller rolls for distribution. The pinch rollers shown in Figure P9.6 are driven at a speed required to produce the paper speed $v_p$ at the rollers. The paper tension varies as the radius of the rewind roll changes or as the pinch roller speed varies. If the tension is not held constant, the rewound paper will develop internal stresses that will damage the paper.

The paper has an elastic constant $k$ so that the rate of change of tension is

$$\frac{dT}{dt} = k(v_r - v_p)$$

If the paper thickness is $d$, then

$$\frac{dR}{dt} = \frac{d}{2}W$$

The inertia of the rewind roll is $I = \frac{1}{2}\rho\pi WR^4$, where $\rho$ is the paper density and $W$ is the width of the roll. Thus in general, $R$ and $I$ are functions of time.

The figure shows an armature-controlled dc motor for driving the rewind roll. The motor's resistance, inductance, torque, and back emf constants are $R_a$, $L_a$, $K_t$, and $K_e$, respectively. Neglect the motor's viscous damping and its inertia. The viscous damping constant for the roll is $c$.

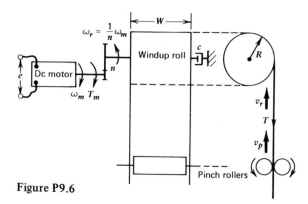

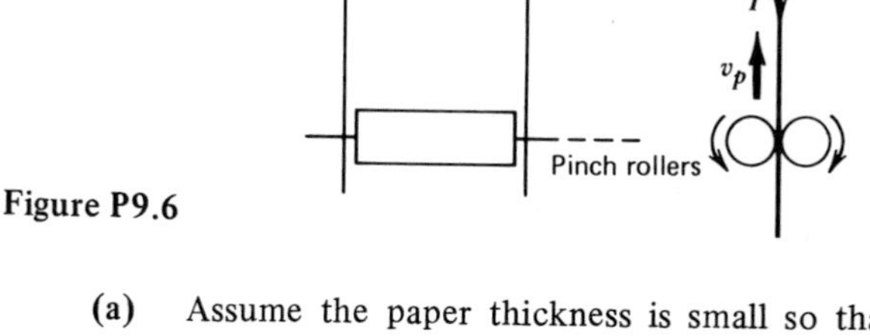


Figure P9.6

(a) Assume the paper thickness is small so that $\dot{R} \cong 0$ over a short enough time. Obtain the state variable equations for the system with $e(t)$, the motor voltage, and $v_p(t)$ as inputs.

(b) Modify the results in part a to include $R$ and $I$ as functions of time.

9.7 Suppose that a copper rod is insulated entirely except for one end, as in Figure 9.6. The rod is 12 in. long and 1 in. in diameter. It is exposed to air at the left end, and the convection coefficient is taken to be $h = 10$ Btu/hr-ft$^2$-°F.

(a) Obtain the state model for the system using the following.

(1) One lump.

(2) Three lumps of equal length.

(3) Five lumps of equal length.

(b) Suppose that the system is in equilibrium at 50°F when the air temperature has a step increase from 50°F to 150°F. Compare the predictions of the center temperature of the rod (halfway down the rod length) using the one-, three-, and five-lump models. If an accuracy to within 10°F is required, is the five-lump model sufficient?

9.8 Find the eigenvectors, the modal matrix, and the Jordan canonical form for the continuous-time systems whose state matrix **A** is given in the following. If complex eigenvalues occur, obtain the modified canonical form instead. Determine whether or not the eigenvectors are linearly independent.

(a) $\frac{1}{3}\begin{bmatrix} -9 & 9 \\ 1 & -9 \end{bmatrix}$  (b) $\begin{bmatrix} 1 & 1 \\ 0 & 1 \end{bmatrix}$

(c) $\begin{bmatrix} 3 & 0 \\ 0 & 3 \end{bmatrix}$  (d) $\begin{bmatrix} 0 & 1 \\ -10 & -2 \end{bmatrix}$

(e) $\begin{bmatrix} 2 & -2 & 3 \\ 1 & 1 & 1 \\ 1 & 3 & -1 \end{bmatrix}$  (f) $\begin{bmatrix} 0 & 6 & -5 \\ 1 & 0 & 2 \\ 3 & 2 & 4 \end{bmatrix}$

(g) $\frac{1}{5}\begin{bmatrix} -9 & 2 & -2 \\ 0 & -5 & 0 \\ -2 & 1 & -6 \end{bmatrix}$  (h) $\begin{bmatrix} 0 & 1 & 0 \\ -1 & -2 & 1 \\ -2 & 0 & 0 \end{bmatrix}$

(i) $\begin{bmatrix} 0 & 1 & 0 \\ 0 & 0 & 1 \\ -6 & -11 & -6 \end{bmatrix}$

9.9 Use two methods to find two modal matrices for the following state matrix **A**. Compare the eigenvectors resulting from each method. (*Hint*. The **A** matrix is in phase-variable form.)

$$\mathbf{A} = \begin{bmatrix} 0 & 1 \\ -12 & -7 \end{bmatrix}$$

9.10 Consider the case in which all the eigenvalues are repeated and only one independent eigenvector exists. Use the fact that the Jordan canonical form is

$$\mathbf{\Lambda} = \mathbf{M}^{-1}\mathbf{A}\mathbf{M} = \begin{bmatrix} p & 1 & 0 & 0 & \cdots \\ 0 & p & 1 & 0 & \cdots \\ \cdot & \cdot & \cdot & \cdot & \cdots \end{bmatrix}$$

to develop an algorithm for generating the dependent eigenvectors from the independent eigenvector. (*Hint.* Use the fact that the columns of **M** are the eigenvectors, and use the equation in this problem to develop a set of recursion relations for the vectors.)

**9.11** The *Cayley-Hamilton theorem* states that every matrix satisfies its own characteristic equation. The characteristic equation for the matrix **A** is $\Delta(s) = |s\mathbf{I} - \mathbf{A}| = 0$. If

$$\mathbf{A} = \begin{bmatrix} 3 & 2 \\ 2 & 3 \end{bmatrix}$$

then $|s\mathbf{I} - \mathbf{A}| = s^2 - 6s + 5$. The theorem states that $\Delta(\mathbf{A}) = \mathbf{0}$. For this example, we have

$$\Delta(\mathbf{A}) = \mathbf{A}^2 - 6\mathbf{A} + 5\mathbf{I} = \mathbf{0}$$

(a) The theorem provides a way of expressing any power of **A** in terms of a linear combination of lower powers of **A**. Use this fact to find $\mathbf{A}^3$, where **A** is given earlier.

(b) Use the theorem to find $\mathbf{A}^{-1}$ in terms of **A**, for **A** as given earlier.

**9.12** Viscous damping limits the amplitude of vibration near the resonant frequencies and often changes the resonant frequency only slightly from its undamped value.
Consider the system shown in Figure P9.1. Let $k_1 = k_2 = k_3 = 1$, and $m_1 = m_2 = 1$.

(a) Let $c_1 = c_2 = 0$. Find the system's modes and eigenvalues.

(b) Let $c_1 = 0.1 = c_2$. Find the eigenvectors and eigenvalues, and compare with those found in part a. How does the damping affect the resonant frequencies?

**9.13** The chief goal of vibration analysis is to estimate the resonant frequencies. Near resonance the driving forces are usually much larger than the damping forces, and the damping is thus often omitted from the model. This results in a considerable reduction in computational complexity. To see this let **z** be the vector of mass displacements. Then a linear vibration model can be expressed as

$$\mathbf{M}\ddot{\mathbf{z}} + \mathbf{C}\dot{\mathbf{z}} + \mathbf{K}\mathbf{z} = \mathbf{f}(t)$$

where **M**, **C**, and **K** are the mass, damping, and stiffness matrices, and $\mathbf{f}(t)$ is the vector of input forces. For no damping, $\mathbf{C} = \mathbf{0}$. The free response at steady state is $\mathbf{z} = \mathbf{A} \sin \omega t$, $\dot{\mathbf{z}} = -\omega \mathbf{A} \cos \omega t$, and $\ddot{\mathbf{z}} = -\omega^2 \mathbf{A} \sin \omega t$. Substituting these into the preceding model with $\mathbf{f} = \mathbf{0}$ and $\mathbf{C} = \mathbf{0}$ gives

$$-\omega^2 \mathbf{M}\mathbf{A} + \mathbf{K}\mathbf{A} = \mathbf{0}$$

(a) Formulate the solution for the amplitude **A** and the resonant frequencies $\omega$ as an eigenvalue problem. What advantage is gained from this formulation as compared to the state variable formulation?

(b) Apply the method in part a to the solution of part a of Problem 9.12.

9.14 For any orthogonal set of body coordinates, the general relation between the angular momentum vector **M** of a rigid body and its angular velocity $\boldsymbol{\omega}$ is $\mathbf{M} = \mathbf{I}\boldsymbol{\omega}$, where **I** is the moment of inertia matrix.

$$\mathbf{I} = \begin{bmatrix} I_{xx} & I_{xy} & I_{xz} \\ I_{yx} & I_{yy} & I_{yz} \\ I_{zx} & I_{zy} & I_{zz} \end{bmatrix}$$

The element $I_{ii}$ is the moment of inertia about the $i$-axis, and $I_{ij}$ is the product of inertia about the $ij$-axes. Note that $I_{ij} = I_{ji}$, so **I** is symmetric.

The principal axes of a rigid body are such that **M** is parallel to $\boldsymbol{\omega}$. Thus $\mathbf{M} = I_p\boldsymbol{\omega}$ where $I_p$ is a scalar, the principal moment of inertia. The principal axes and moments can be found from the eigenvalue problem

$$\mathbf{I}\boldsymbol{\omega} = I_p\boldsymbol{\omega}$$

They are the eigenvectors and eigenvalues of the matrix **I**.

Find the principal moments and axes of a uniform square plate of side length $b$. For the axes shown in Figure P9.14, the matrix **I** is

$$\mathbf{I} = mb^2 \begin{bmatrix} 1/3 & -1/4 & 0 \\ -1/4 & 1/3 & 0 \\ 0 & 0 & 2/3 \end{bmatrix}$$

where $m$ is the plate mass.

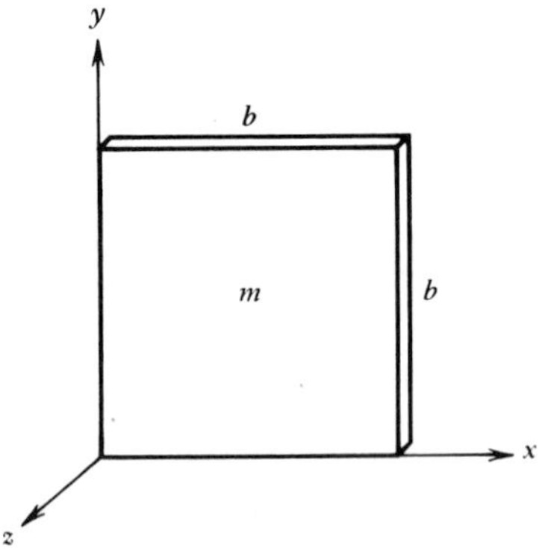

**Figure P9.14**

9.15 Use the modal matrix method and the Laplace transform method to find the transition matrix for the system $\dot{\mathbf{x}} = \mathbf{A}\mathbf{x}$, where **A** is given by the indicated problem number.

(a) Part a of Problem 9.8.

(b) Part b of Problem 9.8.

(c) Part d of Problem 9.8.

9.16 (a) Use a numerical solution method with (9.3-30) and (9.3-32) to find $\boldsymbol{\varphi}(t)$ at $t = 0.1$, where

$$\mathbf{A} = \begin{bmatrix} 0 & 1 \\ 0 & t \end{bmatrix}$$

(b) Use the results of part a to find $\mathbf{x}(0.1)$ if

(1) $\mathbf{x}(0) = [1 \quad 3]^T$

(2) $\mathbf{x}(0) = [2 \quad 1]^T$

9.17 The general Jordan form can be written as

$$\mathbf{\Lambda} = \mathbf{M}^{-1}\mathbf{A}\mathbf{M} = \left[\begin{array}{c|c|c|c} \mathbf{\Lambda}_1 & \mathbf{0} & \mathbf{0} & \cdots \\ \hline \mathbf{0} & \mathbf{\Lambda}_2 & \mathbf{0} & \cdots \\ \hline & & & \end{array}\right]$$

where the $\mathbf{\Lambda}_i$ submatrices are the Jordan blocks corresponding to the various eigenvalues. For complex roots, assume that $\mathbf{\Lambda}_i$ is in the modified form.

(a) Find the expression for $\boldsymbol{\varphi}(t) = \mathscr{L}^{-1}[(s\mathbf{I} - \mathbf{\Lambda})^{-1}]$ where $\mathbf{\Lambda}$ is given earlier, in terms of $\mathbf{\Lambda}_1, \mathbf{\Lambda}_2, \ldots$

(b) Let $\boldsymbol{\varphi}_i(t)$ denote $\mathscr{L}^{-1}[(s\mathbf{I} - \mathbf{\Lambda}_i)^{-1}]$ where $\mathbf{\Lambda}_i$ is an element of $\mathbf{\Lambda}$ (a Jordan block). In the chapter we developed expressions for $\boldsymbol{\varphi}_i(t)$ when $\mathbf{\Lambda}_i$ represents a real distinct root and a pair of complex roots. Find $\boldsymbol{\varphi}_i(t)$ when $\mathbf{\Lambda}_i$ represents a set of repeated roots. Do this by first finding $\boldsymbol{\varphi}_i(t)$ for

$$\mathbf{\Lambda}_i = \begin{bmatrix} p & 1 \\ 0 & p \end{bmatrix}$$

and then for

$$\mathbf{\Lambda}_i = \begin{bmatrix} p & 1 & 0 \\ 0 & p & 1 \\ 0 & 0 & p \end{bmatrix}$$

where $p$ is the root. Generalize the results to the case where the root is repeated $n$ times.

9.18 Apply the results of Problem 9.17 to find $\boldsymbol{\varphi}(t)$ for the following.

(a) $$\mathbf{A} = \begin{bmatrix} 0 & 1 & 0 \\ -1 & -2 & 1 \\ -2 & 0 & 0 \end{bmatrix}$$

(b) The $\mathbf{A}$ matrix for the system shown in Figure P9.1, with $m_1 = m_2 = k_1 = k_2 = 1$, and $k_3 = 0 = c_1 = c_2 = 0$.

9.19 Write a computer program to implement the series solution for the transition matrix (9.3-22) using the recommendation (9.3-23). The inputs should be the matrix $\mathbf{A}$, its dimension $n$, and the desired time interval $\Delta$.

9.20 Use the program developed in Problem 9.19 as a subroutine for a program to implement the step-response solution (9.3-35). The inputs should be the matrices $\mathbf{A}$ and $\mathbf{B}$, their dimensions, $\mathbf{A}^{-1}$ (or this can be computed optionally with another subroutine), $\mathbf{x}(0)$, the step magnitude vector $\mathbf{p}$, and the time step $t$.

9.21 Use the modal method to determine the controllability of the system with $\mathbf{A}$ given by part a of Problem 9.8 and $\mathbf{B}$ given by

(a) $\mathbf{B} = \mathbf{I}$

(b) $\mathbf{B} = \begin{bmatrix} -3 & 3 \\ 1 & -1 \end{bmatrix}$

(c) $\mathbf{B} = \begin{bmatrix} 0 & 0 \\ 0 & 1 \end{bmatrix}$

**9.22** A particular third-order system has the Jordan form

$$\mathbf{\Lambda} = \begin{bmatrix} -1 & 1 & 0 \\ 0 & -1 & 0 \\ 0 & 0 & -2 \end{bmatrix}$$

For each of the following expressions for $\mathbf{M}^{-1}\mathbf{B}$, determine whether or not the system is controllable and identify any uncontrollable modes. The input vector $\mathbf{u}$ is $(2 \times 1)$.

(a) $\mathbf{M}^{-1}\mathbf{B} = \begin{bmatrix} 5 & 1 \\ 0 & 0 \\ 1 & 0 \end{bmatrix}$ (b) $\mathbf{M}^{-1}\mathbf{B} = \begin{bmatrix} 0 & 0 \\ 1 & 0 \\ 0 & 0 \end{bmatrix}$

**9.23** Repeat Problems 9.21 and 9.22 using the direct criterion for controllability.

**9.24** Use the modal method to determine the observability of the system with $\mathbf{A}$ given by part a of Problem 9.8 and $\mathbf{C}$ given by

(a) $\mathbf{C} = \begin{bmatrix} -1 & 3 \\ 1 & -3 \end{bmatrix}$ (b) $\mathbf{C} = \begin{bmatrix} 0 & 0 \\ 1 & 0 \end{bmatrix}$

**9.25** For the third-order system of Problem 9.22, the product $\mathbf{CM}$ is given here for several cases. $\mathbf{M}$ is the modal matrix. Determine which states are unobservable.

(a) $\mathbf{CM} = \begin{bmatrix} 0 & 1 & 2 \\ 0 & 1 & 3 \end{bmatrix}$ (b) $\mathbf{CM} = \begin{bmatrix} 1 & 0 & 1 \\ 0 & 0 & 1 \end{bmatrix}$

**9.26** Use the direct observability criterion for the systems given in Problems 9.24 and 9.25.

**9.27** Determine which modes are uncontrollable for the system whose state matrix $\mathbf{A}$ is given in part c of Problem 9.8 if

(a) $\mathbf{B} = \begin{bmatrix} 0 & 0 \\ 1 & 1 \end{bmatrix}$ (b) $\mathbf{B} = \begin{bmatrix} 1 & 0 \\ 0 & 0 \end{bmatrix}$ (c) $\mathbf{B} = \begin{bmatrix} 1 & 0 \\ 1 & 0 \end{bmatrix}$

**9.28** Determine which modes are unobservable for the system with **A** given in part c of Problem 9.8 if

(a) $\mathbf{C} = \begin{bmatrix} 0 & 1 \\ 0 & 1 \end{bmatrix}$ (b) $\mathbf{C} = \begin{bmatrix} 0 & 0 \\ 1 & 1 \end{bmatrix}$

**9.29** Three tanks are connected as shown in Figure P9.29. The resistances and bottom areas are $R = C = 1$. The input $u$ is the flow rate into the first tank. The output measurement is $h_1$.

(a) Find the state equations.

(b) Find the transfer function and comment on its representation of the system's behavior.

(c) Find the modal matrix and Jordan canonical form.

(d) Investigate the controllability and observability with the direct criteria.

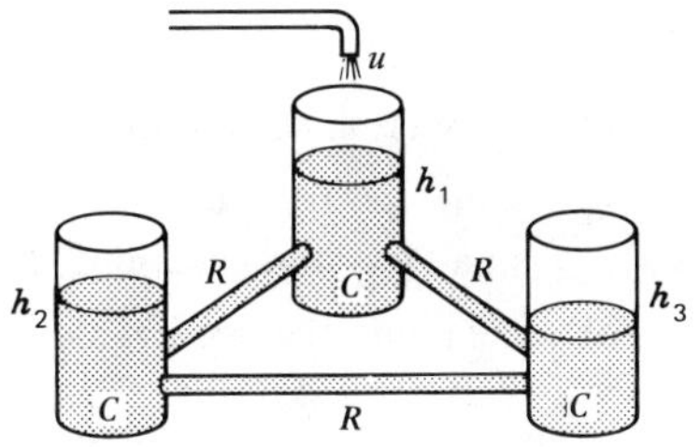

**Figure P9.29**

**9.30** Investigate the controllability and observability of the system shown in Figure P9.30.

(a) Let $m_1 = m_2 = 0$. Use the modal methods.

(b) Let $m_1 = m_2 = 1$ and $k_1 = c_2 = 1$. Use the direct methods.

(c) Comment on the effect of neglecting the system's masses.

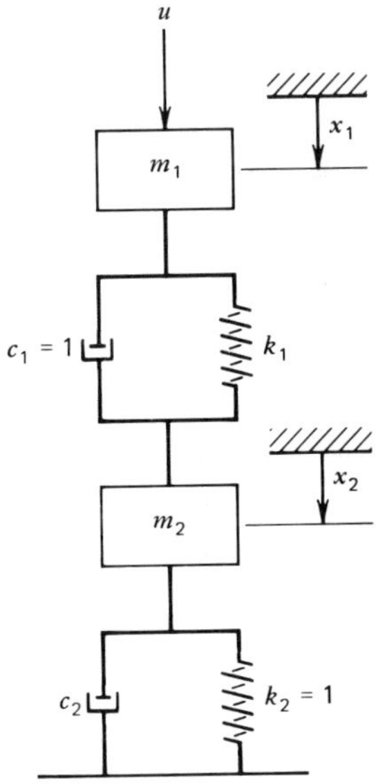

**Figure P9.30**

**9.31** Derive a set of discrete-time state equations in terms of $x_1$ and $x_2$ for the sampled-data system shown in Figure P9.31. Assume that $T = 1$.

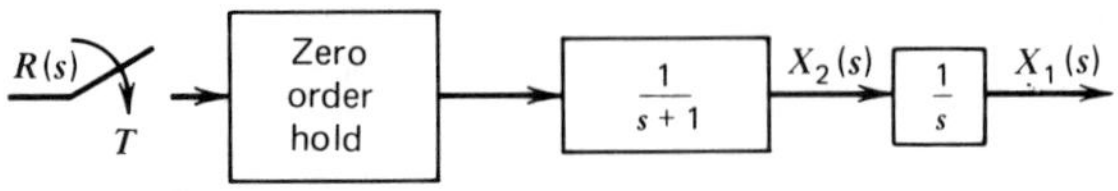


**Figure P9.31**

**9.32** Consider the following system and measurement.

$$\ddot{x} + x = 0, \qquad y = x$$

(a) Show that the continuous-time system is observable.

(b) Find the transition matrix $\phi(T)$ and develop a discrete-time model of the analog system assuming that the sampling time is $T$.

(c) Examine the observability of the sampled system. How does the value of $T$ affect the result?

**9.33** For the system $\mathbf{x}(k+1) = \mathbf{P}\mathbf{x}(k)$, find the transition matrix using (a) the modal method, and (b) equation (9.5-17). Consider two cases.

$$1.\quad \mathbf{P} = \tfrac{1}{3}\begin{bmatrix} -9 & 9 \\ 1 & -9 \end{bmatrix} \qquad 2.\quad \mathbf{P} = \begin{bmatrix} 1 & 1 \\ 0 & 1 \end{bmatrix}$$

(*Hint*. See the solutions to parts a and b of Problem 9.8 for the appropriate modal transformation matrices.)

# CHAPTER TEN
# Matrix Methods for Control System Design

In Section 9.1 there are several examples of systems that require high-order models for proper control system design. We will now see how the matrix methods of Chapter Nine can be applied for this purpose.

Section 10.1 develops the vector formulation of the state variable feedback method for achieving the desired closed-loop poles and response patterns. This method requires that all of the state variables be directly available in the measurements. Section 10.2 presents an algorithm called an *observer* for estimating the states from whatever measurements are available. A control technique that has been successfully applied to multiinput systems is the *linear-quadratic regulator* method, introduced in Section 10.3. It uses an extension of the ISE index to provide a systematic way of computing the many feedback gains that occur in such problems. Finally, Section 10.4 applies the chapter's results to the control of discrete-time systems.

## 10.1 VECTOR FORMULATION OF STATE-VARIABLE FEEDBACK

The vector-matrix methods presented in this chapter provide the basis for modern control theory. These methods have more flexibility than those of classical control theory, and they are more suitable for digital computer solution. The modern methods do not replace the classical methods entirely, however. For relatively low-order systems with one manipulated variable, the classical approach is still very useful.

A foolproof method for the design of linear control systems with multiple inputs and multiple outputs is not available, and research is active in this area. In this and the following sections, we present some techniques that apply to specific problem types. These types illustrate the issues that often arise in practice. More general treatments of these problems require extensive development and are discussed in the more advanced references to be cited.

A general form of a linear multivariable control system is shown in Figure 10.1. The situations that can arise may be categorized as follows. The design problem is considerably more difficult if there is more than one manipulated variable or control signal. The input is a vector $\mathbf{u}$ in this case, and we have separate approaches for the scalar $u$ case and the vector $\mathbf{u}$ case. The scalar case is treated in this section; the vector case is handled in Section 10.3 with a generalization of the ISE index.

If we can measure all the state variables, then $\mathbf{C} = \mathbf{I}$ in Figure 10.1, and the task of control is considerably simplified. If only some of the state variables are measur-

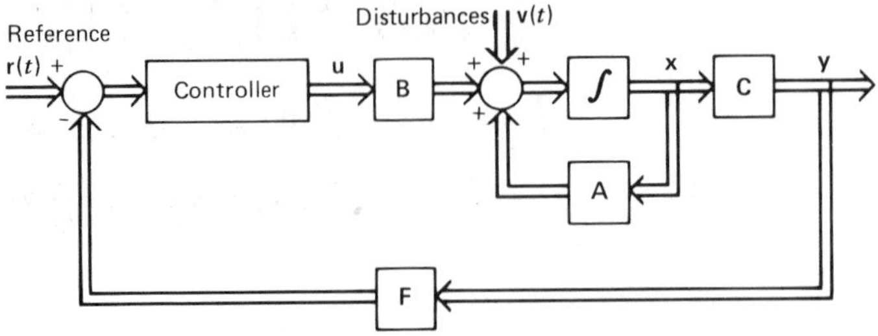


**Figure 10.1** **Multivariable control system.**

able, we can either design a controller that utilizes only the available measurements or construct a state estimator to supply the missing feedback information. An example of the latter approach is given in Section 10.2.

Control structures other than those shown in Figure 10.1 are possible. For example, the techniques of decoupling control and feed-forward compensation covered in Chapter Seven can be generalized with state-vector methods (see, for example, Chapter 10 of Reference 1).

## Design to Achieve a Specified State Matrix

Consider the system shown in Figure 10.1. The state and output equations are

$$\dot{\mathbf{x}} = \mathbf{Ax} + \mathbf{Bu} \tag{10.1-1}$$

$$\mathbf{y} = \mathbf{Cx} \tag{10.1-2}$$

where we assume for simplicity that no direct coupling exists between the output $\mathbf{y}$ and the input $\mathbf{u}$; hence $\mathbf{D} = 0$. Assume also that $\mathbf{B}^{-1}$ and $\mathbf{C}^{-1}$ exist. This implies that the output $\mathbf{y}$ and the input $\mathbf{u}$ have the same dimension as the state vector $\mathbf{x}$.

The model (10.1-1) and (10.2-2) often is the result of a linearization about a reference point $\mathbf{x} = \mathbf{u} = \mathbf{y} = \mathbf{0}$. In this case the desired value of $\mathbf{y}$ is $\mathbf{y} = \mathbf{0}$, and thus the reference input $\mathbf{r}$ is $\mathbf{0}$. We assume for now that this is the case. The performance requirements might be such as to specify the matrix $\mathbf{A}_d$, which represents the desired behavior of the closed-loop control system; that is, the desired response is given by

$$\dot{\mathbf{x}} = \mathbf{A}_d\mathbf{x} \tag{10.1-3}$$

To obtain this behavior, we try a linear feedback law using all the available measurements; that is, the output vector $\mathbf{y}$. Thus from Figure 10.1 we take $\mathbf{F} = \mathbf{I}$, and

$$\mathbf{u}(t) = \mathbf{K}[\mathbf{r}(t) - \mathbf{y}(t)] = -\mathbf{Ky}(t) \tag{10.1-4}$$

where $\mathbf{K}$ is the matrix of proportional control gains we must compute. To do this, substitute (10.1-2) into (10.1-4) to obtain

$$\mathbf{u} = -\mathbf{KCx} \tag{10.1-5}$$

Substitute this into (10.1-1) and compare the result with (10.1-3).

$$\dot{\mathbf{x}} = \mathbf{Ax} - \mathbf{BKCx} = (\mathbf{A} - \mathbf{BKC})\mathbf{x} \tag{10.1-6}$$

Thus

$$\mathbf{A}_d = \mathbf{A} - \mathbf{BKC} \tag{10.1-7}$$

Since $\mathbf{B}^{-1}$ and $\mathbf{C}^{-1}$ are assumed to exist, we can solve for $\mathbf{K}$.

$$\mathbf{K} = \mathbf{B}^{-1}(\mathbf{A} - \mathbf{A}_d)\mathbf{C}^{-1} \tag{10.1-8}$$

The design is thus completed, in theory at least. In practice, however, two problems arise. First, not all applications are such that the dimensions of $\mathbf{u}$, $\mathbf{y}$, and $\mathbf{x}$ are identical. Some techniques are available to treat this case if $\mathbf{C}^{-1}$ exists and if (10.1-1) and (10.1-2) represent a controllable system (see Chapter 10 of Reference 1). Second, it is often difficult to determine what $\mathbf{A}_d$ should be, given performance specifications in terms of dominant time constant, damping ratio, etc. We now present a way around this difficulty for the scalar $u$ case.

## Eigenvalue Placement and Feedback Control

With $\mathbf{r}$ still taken as zero, (10.1-5) and (10.1-6) apply where now $\mathbf{x}$ and $\mathbf{y}$ are $(n \times 1)$, $u$ is $(1 \times 1)$, $\mathbf{A}$ and $\mathbf{C}$ are $(n \times n)$, $\mathbf{B}$ is $(n \times 1)$, and $\mathbf{K}$ is $(1 \times n)$.

If the system is controllable it can be represented in the phase-variable form (9.2-55). This form is

$$\mathbf{A} = \begin{bmatrix} 0 & 1 & 0 & 0 & \cdots & 0 \\ 0 & 0 & 1 & 0 & \cdots & 0 \\ \cdot & \cdot & \cdot & \cdot & \cdots & \cdot \\ 0 & 0 & 0 & 0 & \cdots & 1 \\ -a_n & -a_{n-1} & \cdot & \cdot & \cdots & -a_1 \end{bmatrix}, \qquad \mathbf{B} = \begin{bmatrix} 0 \\ 0 \\ \cdot \\ \cdot \\ \cdot \\ 0 \\ 1 \end{bmatrix} \tag{10.1-9}$$

The reverse is also true. If $\mathbf{A}$ and $\mathbf{B}$ are in the form of (10.1-9), the system is controllable. This can be shown with the direct test for controllability (9.4-7). Note that no restriction is placed on the stability of the open-loop system represented by $\mathbf{A}$. If it is controllable, it can be stabilized by a feedback controller designed according to the following method.

Assuming that $\mathbf{A}$ and $\mathbf{B}$ are in the form (10.1-9), we express the $(1 \times n)$ gain matrix $\mathbf{K}$ as

$$\mathbf{K} = [k_1 \quad k_2 \quad \cdots \quad k_n] \tag{10.1-10}$$

The closed-loop system matrix in (10.1-6) becomes

$$\mathbf{A} - \mathbf{BKC} = \begin{bmatrix} 0 & 1 & 0 & \cdots & 0 \\ 0 & 0 & 1 & \cdots & 0 \\ \cdot & \cdot & \cdot & \cdots & \cdot \\ 0 & 0 & 0 & \cdots & 1 \\ (-a_n - \beta_1) & (-a_{n-1} - \beta_2) & (-a_{n-2} - \beta_3) & \cdots & (-a_1 - \beta_n) \end{bmatrix} \tag{10.1-11}$$

where

$$\beta_j = \sum_{i=1}^{n} k_i c_{ij} \tag{10.1-12}$$

The numbers $c_{ij}$ are the elements of the known output matrix $\mathbf{C}$. The eigenvalues of the matrix $(\mathbf{A} - \mathbf{BKC})$ are given by the characteristic equation

$$|s\mathbf{I} - (\mathbf{A} - \mathbf{BKC})| = 0 \tag{10.1-13}$$

With the form (10.1-11), the characteristic equation is

$$s^n + (a_1 + \beta_n)s^{n-1} + (a_2 + \beta_{n-1})s^{n-2} + \ldots + (a_n + \beta_1) = 0 \tag{10.1-14}$$

If the eigenvalues $s_1, s_2, \ldots, s_n$ of the closed-loop system are specified, the coefficients of the characteristic equation can be obtained and the values $\beta_1, \beta_2, \ldots, \beta_n$ chosen to produce these coefficients. The gain values $k_i$ can then be determined from the $\beta_i$ values. To do this, write (10.1-12) in vector-matrix form as

$$\mathbf{C}^T\mathbf{k} = \boldsymbol{\beta} \tag{10.1-15}$$

where

$$\mathbf{k} = \mathbf{K}^T = \begin{bmatrix} k_1 \\ k_2 \\ \cdot \\ \cdot \\ \cdot \\ k_n \end{bmatrix} \tag{10.1-16}$$

$$\boldsymbol{\beta} = \begin{bmatrix} \beta_1 \\ \beta_2 \\ \cdot \\ \cdot \\ \cdot \\ \beta_n \end{bmatrix} \tag{10.1-17}$$

Equation (10.1-15) represents a set of linear algebraic equations in terms of the unknowns $k_1, k_2, \ldots, k_n$. Its solution is

$$\mathbf{k} = (\mathbf{C}^T)^{-1}\boldsymbol{\beta} \tag{10.1-18}$$

This gives the values for the gains required to place the eigenvalues at the desired locations $s_1, s_2, \ldots, s_n$.

The assumption that $\mathbf{C}^{-1}$ exists means that the output $\mathbf{y}$ is an invertible linear transformation of the state vector $\mathbf{x}$; that is, the state can be found simply from the output as $\mathbf{x} = \mathbf{C}^{-1}\mathbf{y}$. It is therefore tempting to assume that the state is directly measured and to assume that $\mathbf{C} = \mathbf{I}$. In this case, $k_i = \beta_i$, and the solution of (10.1-18) is not required. However, one must keep the physical application in mind. For example, if the plant represents the dynamics of the liquid levels in two tanks, the measured outputs would probably be the two levels. One state vector choice would consist of these levels, and thus $\mathbf{C} = \mathbf{I}$. However, such a state-variable choice would not result in the assumed phase-variable form (10.1-9). To obtain this form, a linear transformation of the original state variables must be made. Now the measurements of the two levels do not correspond to the new state variables, and thus the new $\mathbf{C}$ does not equal $\mathbf{I}$. Implementation of the control law with $\mathbf{C} = \mathbf{I}$ implies that we have sensors available that can measure linear combinations of the two liquid levels, which is not true in general.

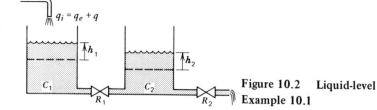

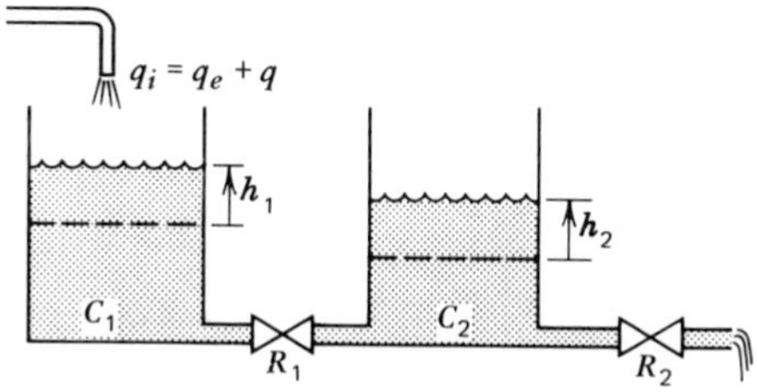


**Figure 10.2** Liquid-level system of Example 10.1

*Example 10.1*

Consider a two-tank system with the controlled inflow rate applied to the first tank (Figure 10.2). Assume that the resistances and areas are unity, so that the state-variable model is

$$\begin{bmatrix} \dot{h}_1 \\ \dot{h}_2 \end{bmatrix} = \begin{bmatrix} -1 & 1 \\ 1 & -2 \end{bmatrix} \begin{bmatrix} h_1 \\ h_2 \end{bmatrix} + \begin{bmatrix} 1 \\ 0 \end{bmatrix} q \tag{10.1-19}$$

where the state variables $h_1$ and $h_2$ represent deviations in the liquid levels from some desired reference equilibrium values, and the control signal $q$ is the deviation of the input flow rate from its corresponding equilibrium value. The required closed-loop time constants are $\tau_1 = 0.2$ and $\tau_2 = 0.05$. Assume that $h_1$ and $h_2$ can be measured.

The model (10.1-19) is not in phase-variable form. To put it in the required form, we can use the methods of Section 5.2. The transfer function between $h_2$ and $q$ is, from Example 5.10,

$$\frac{H_2(s)}{Q(s)} = \frac{1}{s^2 + 3s + 1} = \frac{1/s^2}{1 + 3/s + 1/s^2}$$

or

$$H_2(s) = -\frac{3}{s} H_2(s) - \frac{1}{s^2} H_2(s) + \frac{1}{s^2} Q(s)$$

$$= \frac{1}{s} \left\{ -3H_2(s) - \frac{1}{s} [H_2(s) - Q(s)] \right\}$$

Choose the state variables and control signal to be

$$X_1(s) = H_2(s) \tag{10.1-20}$$

$$X_2(s) = sX_1(s) = sH_2(s) \tag{10.1-21}$$

$$U(s) = Q(s)$$

Thus

$$X_2(s) = \frac{1}{s} [U(s) - 3X_2(s) - X_1(s)]$$

and

$$\begin{bmatrix} \dot{x}_1 \\ \dot{x}_2 \end{bmatrix} = \begin{bmatrix} 0 & 1 \\ -1 & -3 \end{bmatrix} \begin{bmatrix} x_1 \\ x_2 \end{bmatrix} + \begin{bmatrix} 0 \\ 1 \end{bmatrix} u$$

This is the required form, and $a_1 = 3, a_2 = 1$.

The required time constants correspond to the eigenvalue locations $s_1 = -5$, $s_2 = -20$. The corresponding characteristic equation is

$$(s + 5)(s + 20) = s^2 + 25s + 100 = 0$$

Comparison of this with (10.1-14) gives

$$a_1 + \beta_2 = 25$$

$$a_2 + \beta_1 = 100$$

or $\beta_1 = 99, \beta_2 = 22$.

To solve for the gains $k_1$ and $k_2$, we need the output matrix **C**. From (10.1-19) to (10.1-21), the relation between **x** and **h** is

$$x_1 = h_2$$

$$x_2 = \dot{x}_1 = \dot{h}_2 = h_1 - 2h_2$$

or

$$\begin{bmatrix} h_1 \\ h_2 \end{bmatrix} = \begin{bmatrix} 2 & 1 \\ 1 & 0 \end{bmatrix} \begin{bmatrix} x_1 \\ x_2 \end{bmatrix}$$

Given that the measured variables are $h_1$ and $h_2$, the matrix **C** is

$$\mathbf{C} = \begin{bmatrix} 2 & 1 \\ 1 & 0 \end{bmatrix} \tag{10.1-22}$$

From (10.1-15),

$$\begin{bmatrix} 2 & 1 \\ 1 & 0 \end{bmatrix} \begin{bmatrix} k_1 \\ k_2 \end{bmatrix} = \begin{bmatrix} \beta_1 \\ \beta_2 \end{bmatrix} = \begin{bmatrix} 99 \\ 22 \end{bmatrix}$$

or $k_1 = 22, k_2 = 55$. The required feedback control law is

$$u = -\mathbf{KCx} = -\mathbf{Ky} = -[22 \quad 55] \begin{bmatrix} h_1 \\ h_2 \end{bmatrix}$$

or

$$u = -22h_1 - 55h_2$$

Such a matrix technique would normally not be applied to a second-order system, but the example illustrates the features of the method. As the order of the system increases, conversion of the model to phase-variable form becomes more difficult. However, the state-variable formulation and the modal-transformation technique allow a general algorithm to be written for doing the conversion. See for example, Reference 2 of Chapter 9. Such an algorithm can be implemented on a computer (Reference 2).

One disadvantage with design by eigenvalue placement is that the closed-loop system can sometimes be sensitive to parameter variations (Reference 2 of Chapter Nine). It is wise to test the sensitivity of the resulting design to such variations.

## Modal Control

The response properties of a linear dynamic system are determined by its modes and the features of any inputs acting on the system. Therefore, the most direct way of influencing the system's response is to design a control scheme that directly affects each mode of the system. This is the principle of modal control.

A typical application of modal control is in the design of an inner feedback

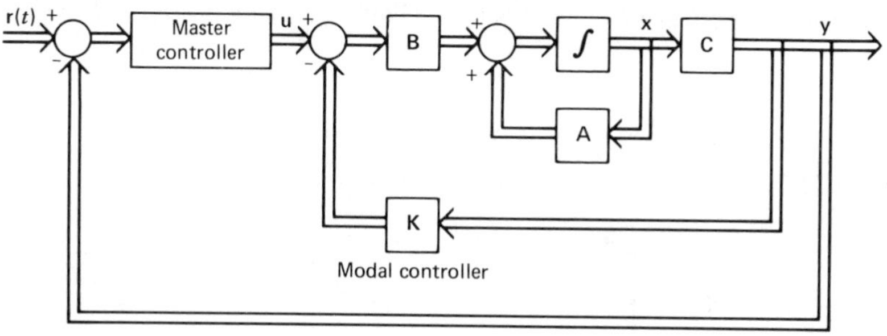


**Figure 10.3** Modal controller as part of a cascaded system. The modal controller improves the transient response; the master controller eliminates steady-state error and allows changes in the reference inputs.

compensation loop to improve the response of the plant. An outer loop with a master controller would be used to make the system respond to changes in reference inputs. The master controller might incorporate integral action in order to have zero steady-state error. A diagram of such a system is shown in Figure 10.3. If disturbances are present, feed-forward compensation might be used.

Assume that the plant's dynamics are described by (10.1-1) and that the desired state is $\mathbf{x} = \mathbf{0}$, as before. Thus $-\mathbf{x}$ represents the error. We limit our attention to proportional control for the modal controller. This means that

$$\mathbf{u} = -\mathbf{K}\mathbf{x} \tag{10.1-23}$$

To take the simplest case, assume that all the states are measurable ($\mathbf{C} = \mathbf{I}$), and that $\mathbf{B}^{-1}$ exists.

The effects of the modal controller on the plant dynamics are described by

$$\dot{\mathbf{x}} = \mathbf{A}\mathbf{x} + \mathbf{B}\mathbf{u} \tag{10.1-24}$$

When the design is done in the original state space, it is difficult to see how the elements of $\mathbf{K}$ affect the response of the system. However, if we transform the state equation to modal form, the controller's influence on the modes is more apparent. Use the modal transformation: $\mathbf{x} = \mathbf{M}\mathbf{z}$ developed in Section 9.2 to transform (10.1-24) into

$$\dot{\mathbf{z}} = \mathbf{\Lambda}\mathbf{z} + \mathbf{M}^{-1}\mathbf{B}\mathbf{u} \tag{10.1-25}$$

where $\mathbf{\Lambda} = \mathbf{M}^{-1}\mathbf{A}\mathbf{M}$ and $\mathbf{M}$ is the modal matrix for (10.1-24). If the plant's eigenvalues are distinct and real, $\mathbf{\Lambda}$ will be diagonal with the eigenvalues appearing on the diagonal. For complex or repeated eigenvalues, the modified transformation matrices for these types should be used (see Example 9.3).

Let the control vector in the modal state space be defined as

$$\mathbf{v} = \mathbf{M}^{-1}\mathbf{B}\mathbf{u} \tag{10.1-26}$$

so that

$$\dot{\mathbf{z}} = \mathbf{\Lambda}\mathbf{z} + \mathbf{v} \tag{10.1-27}$$

Proportional feedback of the modal state vector $\mathbf{z}$ gives

$$\mathbf{v} = -\mathbf{H}\mathbf{z} \tag{10.1-28}$$

where **H** is a constant matrix to be determined. Substitution of this into (10.1-27) results in

$$\dot{\mathbf{z}} = (\mathbf{\Lambda} - \mathbf{H})\mathbf{z} \tag{10.1-29}$$

Suppose that the closed-loop system consisting of the plant and the modal controller is to have the eigenvalues $s_{1d}, s_{2d}, \ldots, s_{nd}$. Then the desired modal state matrix is

$$\mathbf{\Lambda}_d = \begin{bmatrix} s_{1d} & 0 & 0 & \ldots & 0 \\ 0 & s_{2d} & 0 & \ldots & 0 \\ \cdot & \cdot & \cdot & \cdots & \cdot \\ 0 & 0 & 0 & \ldots & s_{nd} \end{bmatrix} \tag{10.1-30}$$

where $\dot{\mathbf{z}} = \mathbf{\Lambda}_d \mathbf{z}$ represents the desired closed-loop behavior. From (10.1-29)

$$\mathbf{\Lambda} - \mathbf{H} = \mathbf{\Lambda}_d$$

or

$$\mathbf{H} = \mathbf{\Lambda} - \mathbf{\Lambda}_d \tag{10.1-31}$$

The control vector **u** and control gains **K** in the original state space can be obtained by combining the previous equations to give

$$\mathbf{M}^{-1}\mathbf{B}\mathbf{u} = \mathbf{v} = -\mathbf{H}\mathbf{z} = -\mathbf{H}\mathbf{M}^{-1}\mathbf{x}$$

or

$$\mathbf{u} = -\mathbf{B}^{-1}\mathbf{M}\mathbf{H}\mathbf{M}^{-1}\mathbf{x} = -\mathbf{K}\mathbf{x} \tag{10.1-32}$$

where

$$\mathbf{K} = \mathbf{B}^{-1}\mathbf{M}\mathbf{H}\mathbf{M}^{-1} \tag{10.1-33}$$

The design procedure for the modal controller is as follows. Given **A** and **B**, find the modal matrix **M** and the resulting matrix **Λ**. From the specified eigenvalue locations, determine $\mathbf{\Lambda}_d$ and thus **H** from (10.1-31). Finally, compute **K** from (10.1-33). The resulting modal controller is as shown in Figure 10.3.

*Example 10.2*

Consider the problem presented in Example 10.1, but now assume that the input flow rate into each tank is under control. The input flow rate deviations are $q_1$ and $q_2$. Design a modal controller to achieve the desired eigenvalues: $s_{1d} = -5, s_{2d} = -20$.

The original state and control vectors are

$$\mathbf{x} = \begin{bmatrix} h_1 \\ h_2 \end{bmatrix} \qquad \mathbf{u} = \begin{bmatrix} q_1 \\ q_2 \end{bmatrix}$$

and

$$\mathbf{A} = \begin{bmatrix} -1 & 1 \\ 1 & -2 \end{bmatrix} \qquad \mathbf{B} = \begin{bmatrix} 1 & 0 \\ 0 & 1 \end{bmatrix}$$

The plant's eigenvalues are found from $|s\mathbf{I} - \mathbf{A}| = 0$ to be $s = -0.382, -2.62$. Using the methods of Section 9.2, we find that

$$\mathbf{M} = \begin{bmatrix} 1 & 1 \\ 0.62 & -1.62 \end{bmatrix} \qquad \mathbf{\Lambda} = \mathbf{M}^{-1}\mathbf{A}\mathbf{M} = \begin{bmatrix} -0.382 & 0 \\ 0 & -2.62 \end{bmatrix}$$

$$\mathbf{M}^{-1} = \begin{bmatrix} 0.723 & 0.446 \\ 0.277 & -0.446 \end{bmatrix}$$

From the eigenvalue specifications,

$$\mathbf{\Lambda}_d = \begin{bmatrix} -5 & 0 \\ 0 & -20 \end{bmatrix}$$

The matrix **H** is found from (10.1-31) to be

$$\mathbf{H} = \begin{bmatrix} 4.618 & 0 \\ 0 & 17.38 \end{bmatrix}$$

The control gain matrix **K** is

$$\mathbf{K} = \mathbf{B}^{-1}\mathbf{M}\mathbf{H}\mathbf{M}^{-1} = \begin{bmatrix} 8.15 & -5.69 \\ -5.73 & 13.8 \end{bmatrix}$$

The feedback-control law in terms of the original state and control variables is found from (10.1-23) with $\mathbf{x} = \mathbf{h}$ and $\mathbf{u} = \mathbf{q}$.

$$q_1 = -8.15h_1 + 5.69h_2$$

$$q_2 = 5.73h_1 - 13.8h_2$$

Note that the scalar control variable $q_1$ was used in Example 10.1 to obtain closed-loop eigenvalues at $s = -5$ and $s = -20$. The difference between this scalar controller and the modal controller using $q_1$ and $q_2$ is that two manipulated variables carry the control load in the latter case. This is preferable since the possibility of saturation of the final control elements is less than when the entire load is carried by one final control element, but the equipment cost is greater.

We have introduced the principle of modal control in its most simple form. When the numbers of states, measurements, and control variables are not equal, design of a modal controller requires considerably more theoretical development (see Section 10-3 of Reference 1).

## 10.2 STATE VECTOR OBSERVERS

The eigenvalue placement and modal control methods of the previous section both require knowledge of the current value of the state variables in order to implement the feedback-control law $\mathbf{u} = -\mathbf{K}\mathbf{x}$. To guarantee that this is possible, we stipulated that the output relation $\mathbf{y} = \mathbf{C}\mathbf{x}$ is such that $\mathbf{C}^{-1}$ exists. Thus given the measured variables $\mathbf{y}$, we can solve for the state variables $\mathbf{x}$ by inverting the matrix $\mathbf{C}$. However, in practice the number of measured variables is often less than the number of state variables. Thus $\mathbf{C}^{-1}$ does not exist in such cases because $\mathbf{C}$ is not square. When this occurs, a unique solution for $\mathbf{x}$ cannot be found with the output relation $\mathbf{y} = \mathbf{C}\mathbf{x}$.

However, algorithms exist for using the available measurements to construct an estimate of the state vector. This estimate can then be used in the feedback-control scheme in place of $\mathbf{x}$. Several types of such algorithms are available for this purpose, and they differ partly according to their treatment of the measurements as deterministic or random signals. The algorithms and the systems implemented with them are generically referred to as state reconstructors, estimators, or filters. Here we consider a specific algorithm called an *observer* that can be used when the measurements contain unwanted high-frequency noise signals.

## State Reconstruction by Output Differentiation

Consider a system with two state variables and no input ($\mathbf{u} = \mathbf{0}$). If the output is a scalar, we cannot obtain values for the two state variables from the single measured variable if we use only the output relation $y = \mathbf{Cx}$. However, suppose that we differentiate this relation, and use the zero-input state equation $\dot{\mathbf{x}} = \mathbf{Ax}$ to obtain

$$\dot{y} = \mathbf{C}\dot{\mathbf{x}} = \mathbf{CAx}$$

Now form a $(2 \times 1)$ vector from the scalars $y$ and $\dot{y}$, and write this vector in terms of $\mathbf{x}$.

$$\begin{bmatrix} y \\ \dot{y} \end{bmatrix} = \begin{bmatrix} \mathbf{Cx} \\ \mathbf{CAx} \end{bmatrix} = \begin{bmatrix} \mathbf{C} \\ \mathbf{CA} \end{bmatrix} \mathbf{x}$$

We can solve this equation for $\mathbf{x}$ if the coefficient matrix is invertible; that is,

$$\mathbf{x} = \begin{bmatrix} \mathbf{C} \\ \mathbf{CA} \end{bmatrix}^{-1} \begin{bmatrix} y \\ \dot{y} \end{bmatrix} \tag{10.2-1}$$

Here, $\mathbf{A}$ is $(2 \times 2)$ and $\mathbf{C}$ is $(1 \times 2)$. Thus, the coefficient matrix in (10.2-1) is $(2 \times 2)$ and its inverse exists if the rank is 2. This condition is also that for observability of the two-dimensional system specified by the matrices $\mathbf{A}$ and $\mathbf{C}$ (see Section 9.4).

We can extend this analysis to a system with $n$ state variables and nonzero inputs to show that the state $\mathbf{x}$ can be reconstructed from the output $\mathbf{y}$ if the matrices $\mathbf{A}$ and $\mathbf{C}$ describe an observable system. If so, $\mathbf{A}$ and $\mathbf{C}$ are said to form an observable "pair."

Practical difficulties can be encountered when the scheme of (10.2-1) is implemented to compute $\mathbf{x}$, because it requires that the measured signal $\mathbf{y}$ be differentiated. We have seen that accurate differentiation with a physical device can be difficult to obtain. Its success depends heavily on the amount of noise in the measured signal.

Having established at least the theoretical possibility of computing $\mathbf{x}$ from $\mathbf{y}$ for an observable system, we now consider how such a scheme can be implemented in practice in light of the differentiation difficulty.

## Design of An Identity Observer

The concept of an observer was developed by Luenberger (References 3 and 4). Here we consider a particular observer algorithm that will give estimates of all the state variables. This is an *identity observer.* The design of a *reduced observer* for estimating only some of the states is discussed in Reference 4.

As usual we assume that the plant and output equations are

$$\dot{\mathbf{x}} = \mathbf{Ax} + \mathbf{Bu} \tag{10.2-2}$$

$$\mathbf{y} = \mathbf{Cx} \tag{10.2-3}$$

with $\mathbf{D}$ taken to be zero for simplicity. Here $\mathbf{x}$ is $(n \times 1)$, $\mathbf{u}$ is $(m \times 1)$, and $\mathbf{y}$ is $(q \times 1)$.

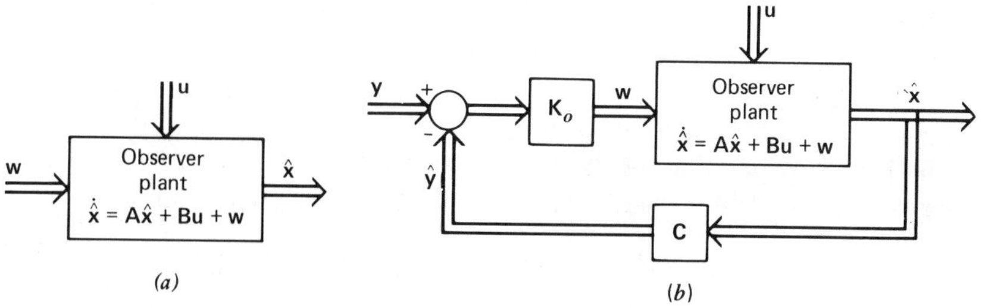


**Figure 10.4 Observer viewed as a feedback-control system. (*a*) Observer plant. (*b*) Closed-loop observer.**

The values of **A**, **B**, and **C** are known. We can formulate the problem as one of designing a feedback-control system (Figure 10.4*a*). That is, let the estimate of x provided by the observer algorithm be denoted $\hat{x}$. The estimate $\hat{x}$ is the output of the observer, whose inputs are the system input **u** and another input w, the observer's control input, which is yet to be determined. The input w is $(n \times 1)$. We consider the observer as a plant that is to be controlled so that its output $\hat{x}$ approaches x.

A logical choice for the observer plant is one whose describing equations are identical to those of the system plant whose states we are trying to estimate. Thus we build an observer so that its state vector $\hat{x}$ has the same state equation as the system plant. To match the system's dynamics further, we include the system's input **u** in the observer. This reasoning leads to the observer model

$$\dot{\hat{x}} = A\hat{x} + Bu + w \tag{10.2-4}$$

To design the observer control input w, note that the error in the state estimate is

$$e = x - \hat{x} \tag{10.2-5}$$

Since we do not know x but we do know y, multiply (10.2-5) by **C** to obtain

$$Ce = Cx - C\hat{x} = y - \hat{y} \tag{10.2-6}$$

where $\hat{y}$ is the estimate of the output, and $y - \hat{y}$ is the error in the output estimate. We can compute $y - \hat{y}$ and use it as an actuating signal to create w. A proportional control scheme for this purpose is shown in Figure 10.4*b*, where

$$w = K_o(y - \hat{y}) = K_o(y - C\hat{x}) \tag{10.2-7}$$

The $(n \times q)$ matrix $K_o$ is the matrix of proportional gains that we will select to obtain an observer with the desired response.

Thus the observer algorithm is

$$\begin{aligned} \dot{\hat{x}} &= A\hat{x} + Bu + K_o(y - C\hat{x}) \\ &= (A - K_oC)\hat{x} + Bu + K_oy \end{aligned} \tag{10-2-8}$$

Note that if $\hat{y} = y$ and $\hat{x}(0) = x(0)$, the observer's response $\hat{x}(t)$ will be identical to $x(t)$.

If the observer is to be made of electronic components such as operational amplifiers, the circuit equations must be identical to (10.2-8), and **u** and y represent input voltages to the circuit. The output voltages represent $\hat{\mathbf{x}}$. For digital implementation, the observer consists of a computer program whose input numbers represent $\mathbf{u}$ and $\mathbf{y}$, with $\hat{\mathbf{x}}$ being the result of the program's computation. To do this, (10.2-8) must be converted to a difference equation.

From (10.2-8) we see that the dynamics of the observer's estimate $\hat{\mathbf{x}}$ are affected by the observer inputs $\mathbf{u}$ and $\mathbf{y}$ and by the observer's eigenvalues. These are given by

$$|s\mathbf{I} - (\mathbf{A} - \mathbf{K}_o\mathbf{C})| = 0 \qquad (10.2\text{-}9)$$

Although we cannot compute the error $\mathbf{e}$, we can design the observer with a knowledge of the error dynamics. These can be obtained as follows. Differentiate $\mathbf{e} = \mathbf{x} - \hat{\mathbf{x}}$, and use (10.2-2) and (10.2-8). This gives

$$\begin{aligned}\dot{\mathbf{e}} &= \dot{\mathbf{x}} - \dot{\hat{\mathbf{x}}} \\ &= \mathbf{Ax} + \mathbf{Bu} - \mathbf{A}\hat{\mathbf{x}} - \mathbf{Bu} - \mathbf{K}_o(\mathbf{y} - \mathbf{C}\hat{\mathbf{x}})\end{aligned}$$

Add and subtract $\mathbf{K}_o\mathbf{Cx}$.

$$\begin{aligned}\dot{\mathbf{e}} &= \mathbf{A}(\mathbf{x} - \hat{\mathbf{x}}) - \mathbf{K}_o\mathbf{y} + \mathbf{K}_o\mathbf{C}\hat{\mathbf{x}} + \mathbf{K}_o\mathbf{Cx} - \mathbf{K}_o\mathbf{Cx} \\ &= \mathbf{A}(\mathbf{x} - \hat{\mathbf{x}}) - \mathbf{K}_o\mathbf{C}(\mathbf{x} - \hat{\mathbf{x}}) + \mathbf{K}_o(\mathbf{Cx} - \mathbf{y})\end{aligned}$$

Because $\mathbf{y} = \mathbf{Cx}$,

$$\dot{\mathbf{e}} = (\mathbf{A} - \mathbf{K}_o\mathbf{C})\mathbf{e} + \mathbf{K}_o(\mathbf{Cx} - \mathbf{y}) = (\mathbf{A} - \mathbf{K}_o\mathbf{C})\mathbf{e} \qquad (10.2\text{-}10)$$

Thus the error dynamics are governed by the eigenvalues of the matrix $(\mathbf{A} - \mathbf{K}_o\mathbf{C})$, and we wish to select $\mathbf{K}_o$ to place these eigenvalues to achieve acceptable performance. We note that the eigenvalues of the matrix $(\mathbf{A} - \mathbf{K}_o\mathbf{C})$ are also those of its transpose $(\mathbf{A} - \mathbf{K}_o\mathbf{C})^T = (\mathbf{A}^T - \mathbf{C}^T\mathbf{K}_o^T)$. In Section 10.1 (10.2-6), we saw that eigenvalue placement is possible with the gain matrix $\mathbf{K}$ if $(\mathbf{A}, \mathbf{B})$ represent a controllable pair. Therefore, eigenvalue placement for the observer is possible with the gain matrix $\mathbf{K}_o$ if $(\mathbf{A}^T, \mathbf{C}^T)$ represent a controllable pair; that is, if the rank of $\mathbf{Q}$ is $n$, where

$$\mathbf{Q} = [\mathbf{C}^T | \mathbf{A}^T\mathbf{C}^T | \ \ldots \ | (\mathbf{A}^T)^{n-1}\mathbf{C}^T] \qquad (10.2\text{-}11)$$

But

$$\mathbf{Q}^T = \begin{bmatrix} \mathbf{C} \\ \hline \mathbf{CA} \\ \hline \cdot \\ \cdot \\ \cdot \\ \hline \mathbf{CA}^{n-1} \end{bmatrix} \qquad (10.2\text{-}12)$$

Therefore, if rank $\mathbf{Q} = n$, then rank $\mathbf{Q}^T = n$. But (10.2-12) is the condition for the observability of the pair $(\mathbf{A}, \mathbf{C})$. Thus if $(\mathbf{A}, \mathbf{C})$ is an observable pair, then $(\mathbf{A}^T, \mathbf{C}^T)$ is a controllable pair. The observability of the system guarantees not only that we can construct an observer, but also that we can use $\mathbf{K}_o$ to place the eigenvalues of the observer.

The transform of the observer estimate for $\hat{\mathbf{x}}(0) = \mathbf{0}$ is, from (10.2-8)

$$\hat{\mathbf{X}}(s) = [s\mathbf{I} - (\mathbf{A} - \mathbf{K}_o\mathbf{C})]^{-1} [\mathbf{BU}(s) + \mathbf{K}_o\mathbf{Y}(s)] \qquad (10.2\text{-}13)$$

If numerator dynamics are present in the relations between either $\mathbf{x}$ and $\mathbf{u}$ or $\mathbf{x}$ and $\mathbf{y}$, the eigenvalues of $(\mathbf{A} - \mathbf{K}_o\mathbf{C})$ do not entirely describe the dynamics of the observer. Therefore, although we can place the eigenvalues at desired locations, the numerator terms might alter the response significantly. This effect can also be seen in the error equation (10.2-10). The elements of $\mathbf{K}_o$ affect all the components of $\mathbf{e}$. In adjusting $\mathbf{K}_o$ to reduce one component of $\mathbf{e}$, we might adversely affect the response of other components of $\mathbf{e}$. The design should be checked by simulating the observer's dynamics under typical conditions. Placement of the observer eigenvalues should also consider the eigenvalues of the feedback controller. Experience has shown that a good choice is to place the observer eigenvalues slightly to the left of the controller eigenvalues. If the time constants of the observer are made too small, the resulting high gains can cause the observer's estimates not to converge to the true state values because the high gains do not allow noise effects to be filtered out.

*Example 10.3*

The state feedback controller of Example 10.1 required measurements of the liquid heights $h_1$ and $h_2$. The control law for the controlled flow rate $q$ into the first tank was found to be

$$q = -22h_1 - 55h_2 \tag{10.2-14}$$

This gives closed-loop eigenvalues of $s = -5$ and $-20$. Now suppose that the measurement of $h_1$ is not available. Design an observer for the system.

The output equation is $y = h_2$. Thus $C = [0 \quad 1]$. The plant's equations are given by (10.1-19) and the system's matrices are

$$\mathbf{A} = \begin{bmatrix} -1 & 1 \\ 1 & -2 \end{bmatrix}, \qquad \mathbf{B} = \begin{bmatrix} 1 \\ 0 \end{bmatrix}$$

First check the system's observability. Since $n = 2$ and $\mathbf{CA} = [1 \quad -2]$, (10.2-12) gives

$$\text{rank } \mathbf{Q}^T = \text{rank} \begin{bmatrix} 0 & 1 \\ 1 & -2 \end{bmatrix} = 2$$

The system is observable with the output $y = h_2$.

Since the controller poles are at $s = -5, -20$, we select the observer's poles to be to the left of $s = -20$ – say, at $s = -25$ and $-30$. The matrix $\mathbf{K}_o$ must now be computed to give these poles. Let $\mathbf{K}_o$ be written as

$$\mathbf{K}_o = \begin{bmatrix} k_{o1} \\ k_{o2} \end{bmatrix}$$

The product $\mathbf{K}_o\mathbf{C}$ is $(2 \times 2)$ since $\mathbf{K}_o$ is $(2 \times 1)$ and $\mathbf{C}$ is $(1 \times 2)$. Thus

$$\mathbf{A} - \mathbf{K}_o\mathbf{C} = \begin{bmatrix} -1 & 1 \\ 1 & -2 \end{bmatrix} - \begin{bmatrix} k_{o1} \\ k_{o2} \end{bmatrix} [0 \quad 1]$$

$$= \begin{bmatrix} -1 & (1 - k_{o1}) \\ 1 & -(2 + k_{o2}) \end{bmatrix}$$

From (10.2-9)

$$|s\mathbf{I} - \mathbf{A} + \mathbf{K}_o\mathbf{C}| = \begin{vmatrix} (s+1) & (k_{o1}-1) \\ -1 & (s+2+k_{o2}) \end{vmatrix} = 0$$

or

$$(s+1)(s+2+k_{o2}) + k_{o1} - 1 = 0$$

The desired poles give

$$(s+25)(s+30) = s^2 + 55s + 750 = 0$$

Comparison of the last two equations shows that $k_{o1} = 697$ and $k_{o2} = 52$, and thus

$$\mathbf{A} - \mathbf{K}_o\mathbf{C} = \begin{bmatrix} -1 & -696 \\ 1 & -54 \end{bmatrix} \tag{10.2-15}$$

The performance of the observer can be simulated with the error equation (10.2-10). The errors are $e_1 = h_1 - \hat{h}_1$ and $e_2 = h_2 - \hat{h}_2$. The results are shown in Figure 10.5. The initial conditions $e_1(0) = e_2(0) = 1$ correspond to initial heights of $h_1(0) = h_2(0) = 1$ and initial estimates of $\hat{h}_1(0) = \hat{h}_2(0) = 0$. The choice of initial values for the observer estimates is somewhat arbitrary, with zero being a neutral and therefore common choice. The initial values $\hat{\mathbf{h}}(0)$ represent the initial voltages placed on the observer circuit for analog implementation, or the starting values of $\hat{\mathbf{h}}$ in the observer's computer program, for digital implementation.

Case 1 in Figure 10.5 indicates that the error in the estimate of $h_1$ is much larger than for $h_2$. This is to be expected since $h_2$ appears in the measurement. The peak error can be reduced by increasing the observer's time constants. For example, let the observer's eigenvalues be $s = -8, -10$ instead of $s = -25, -30$. Then $k_{o1} = 64$ and $k_{o2} = 15$. This response is shown in case 2. The peak error for $h_1$ is much less than before but now takes longer to decay. Now, however, both observer eigenvalues do not lie to the left of the closed-loop controller eigenvalues, so a simulation of the combined observer-controller system is advised.

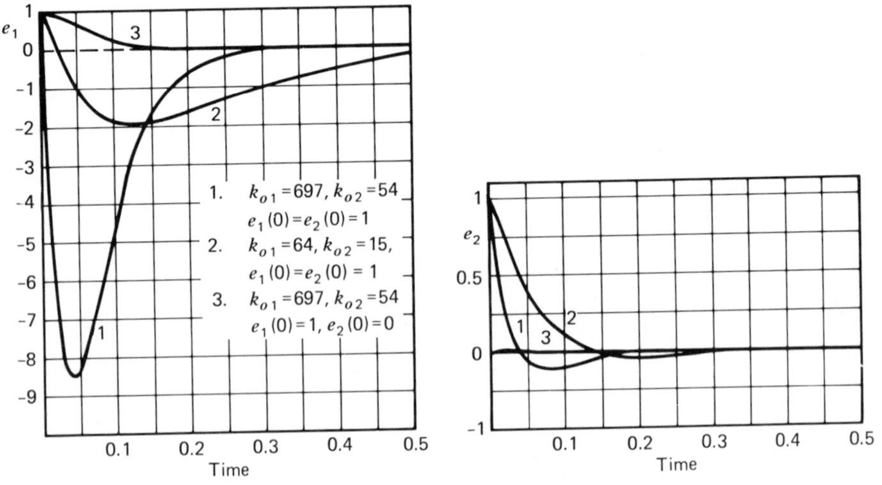


**Figure 10.5** Error dynamics for the observer of Example 10.3.

Since $h_2$ constitutes the measurement we can construct a reduced-order observer by using $h_2$ instead of $\hat{h}_2$ in the feedback loop of the controller.* We also know $h_2(0)$. Case 3 shows a simulation with $h_2(0) = \hat{h}_2(0) = 1\,[e_2(0) = 0]$. The improved knowledge leads to a reduced error in both estimates.

While the solution for the gains $k_{o1}$ and $k_{o2}$ was easy in this case, design of higher-order observers requires more effort. However, note that we can transform the observer state vector $\hat{\mathbf{x}}$ to the phase-variable form, as was done for the eigenvalue placement method of Section 10.1. With the observer eigenvalues specified, $\mathbf{K}_o$ can be found more easily with this approach.

## Feedback Control with an Observer

The use of an observer with a feedback controller of the form given in Section 10.1 is shown in Figure 10.6. The linear feedback law $\mathbf{u} = -\mathbf{K}\mathbf{x}$ is replaced with $\mathbf{u} = -\mathbf{K}\hat{\mathbf{x}}$, since $\mathbf{x}$ is now presumed to be unavailable. We now show that the controller matrix $\mathbf{K}$ for both feedback laws can be computed in exactly the same way. That is, the use of $\hat{\mathbf{x}}$ instead of $\mathbf{x}$ does not affect the choice of $\mathbf{K}$.

To see this, write the equations for the plant-controller subsystem shown in Figure 10.6.

$$\dot{\mathbf{x}} = \mathbf{A}\mathbf{x} + \mathbf{B}\mathbf{u} \tag{10.2-16}$$

$$\mathbf{y} = \mathbf{C}\mathbf{x} \tag{10.2-17}$$

$$\mathbf{u} = -\mathbf{K}\hat{\mathbf{x}} + \mathbf{G}\mathbf{r} \tag{10.2-18}$$

where $\mathbf{r}$ is the vector of reference inputs, and $\mathbf{G}$ is a gain matrix that can be selected to give a zero steady-state error for step changes in $\mathbf{r}$. The observer is described by

$$\dot{\hat{\mathbf{x}}} = \mathbf{A}\hat{\mathbf{x}} + \mathbf{B}\mathbf{u} + \mathbf{K}_o(\mathbf{y} - \mathbf{C}\hat{\mathbf{x}}) \tag{10.2-19}$$

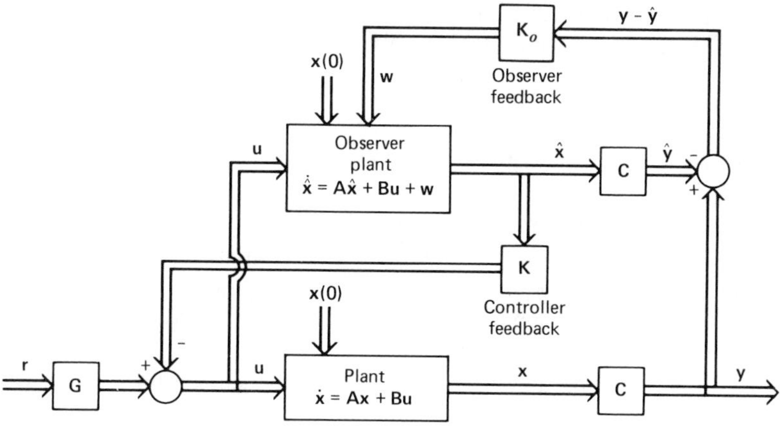


**Figure 10.6 Structure of a feedback controller with an observer.**

* However, when noise is present in the measurements, the identity observer will often give a performance superior to that of the reduced-order observer.

Substitution of (10.2-17) and (10.2-18) into (10.2-16) and (10.2-19) gives

$$\dot{\mathbf{x}} = \mathbf{A}\mathbf{x} - \mathbf{B}\mathbf{K}\hat{\mathbf{x}} + \mathbf{B}\mathbf{G}\mathbf{r} \tag{10.2-20}$$

$$\dot{\hat{\mathbf{x}}} = (\mathbf{A} - \mathbf{B}\mathbf{K} - \mathbf{K}_o\mathbf{C})\hat{\mathbf{x}} + \mathbf{K}_o\mathbf{C}\mathbf{x} + \mathbf{B}\mathbf{G}\mathbf{r} \tag{10.2-21}$$

or

$$\begin{bmatrix} \dot{\mathbf{x}} \\ \dot{\hat{\mathbf{x}}} \end{bmatrix} = \begin{bmatrix} \mathbf{A} & -\mathbf{B}\mathbf{K} \\ \mathbf{K}_o\mathbf{C} & (\mathbf{A} - \mathbf{K}_o\mathbf{C} - \mathbf{B}\mathbf{K}) \end{bmatrix} \begin{bmatrix} \mathbf{x} \\ \hat{\mathbf{x}} \end{bmatrix} + \begin{bmatrix} \mathbf{B}\mathbf{G} \\ \mathbf{B}\mathbf{G} \end{bmatrix} \mathbf{r} \tag{10.2-22}$$

The eigenvalues of this system of order $2n$ are given by

$$\begin{vmatrix} s\mathbf{I} - \mathbf{A} & \mathbf{B}\mathbf{K} \\ -\mathbf{K}_o\mathbf{C} & s\mathbf{I} - \mathbf{A} + \mathbf{K}_o\mathbf{C} + \mathbf{B}\mathbf{K} \end{vmatrix} \triangleq \begin{vmatrix} \mathbf{M}_1 & \mathbf{M}_2 \\ \mathbf{M}_3 & \mathbf{M}_4 \end{vmatrix} = 0$$

where we have partitioned the determinant as shown. Subtract $\mathbf{M}_1$ from $\mathbf{M}_3$ and $\mathbf{M}_2$ from $\mathbf{M}_4$ to obtain

$$\begin{vmatrix} s\mathbf{I} - \mathbf{A} & \mathbf{B}\mathbf{K} \\ -s\mathbf{I} + \mathbf{A} - \mathbf{K}_o\mathbf{C} & s\mathbf{I} - \mathbf{A} + \mathbf{K}_o\mathbf{C} \end{vmatrix} \triangleq \begin{vmatrix} \mathbf{M}_5 & \mathbf{M}_6 \\ \mathbf{M}_7 & \mathbf{M}_8 \end{vmatrix} = 0$$

Add $\mathbf{M}_5$ to $\mathbf{M}_6$ and $\mathbf{M}_7$ to $\mathbf{M}_8$. This gives

$$\begin{vmatrix} s\mathbf{I} - \mathbf{A} + \mathbf{B}\mathbf{K} & \mathbf{B}\mathbf{K} \\ \mathbf{0} & s\mathbf{I} - \mathbf{A} + \mathbf{K}_o\mathbf{C} \end{vmatrix} = |s\mathbf{I} - \mathbf{A} + \mathbf{B}\mathbf{K}|\,|s\mathbf{I} - \mathbf{A} + \mathbf{K}_o\mathbf{C}| = 0$$

The determinant $|s\mathbf{I} - \mathbf{A} + \mathbf{B}\mathbf{K}|$ is that of the controller, while $|s\mathbf{I} - \mathbf{A} + \mathbf{K}_o\mathbf{C}|$ is due to the observer. The gain matrices $\mathbf{K}$ and $\mathbf{K}_o$ appear in different determinants. Thus we can compute $\mathbf{K}$ to achieve any desired eigenvalues for the controller and find $\mathbf{K}_o$ to give any set of eigenvalues for the observer. The controller design does not depend on how the feedback information on the state $\mathbf{x}$ is generated.

## 10.3 REGULATOR DESIGN WITH A QUADRATIC PERFORMANCE INDEX

As we have seen, the design of a controller for systems with more than one manipulated variable is challenging. In complex systems with high performance specifications, we cannot treat each controller as being uncoupled from the others, and controller design must be approached by considering the total system response. A technique that has found practical applications in recent years is based on an extension of the ISE performance index to multivariable systems where it is referred to as the *quadratic index.* It provides a systematic way of computing the controller's feedback gains. Here we present the method as applied to regulator design – that is, the problem of keeping the output variables near constant, prescribed values.

When a performance index is used, the resulting design is sometimes called *optimal.* Thus the following regulator algorithm is called the optimal linear regulator. Other advanced control algorithms are called "optimal control." We must keep in mind that the design is optimal only with respect to the particular performance index used and to the assumptions made. Other optimal control formulations include the problem of driving a system to attain a desired state in minimum time (such as in an interception problem) or by using minimum energy. Treatment of such formulations is beyond our scope here; however, a minimum-time design is hardly ever a minimum-energy design. So the two are "optimal" designs only with respect to their own performance index. Also, the assumed form of the system model must be considered. The following technique produces an optimal design relative to the quadratic index only if the state equations are linear. Any nonlinear effects should therefore be modeled with a linearized description. This usually poses no difficulties because the regulator is designed to preserve the accuracy of the linearization.

The standard description of the plant and output is

$$\dot{\mathbf{x}} = \mathbf{A}\mathbf{x} + \mathbf{B}\mathbf{u} \tag{10.3-1}$$

$$\mathbf{y} = \mathbf{C}\mathbf{x} \tag{10.3-2}$$

where $\mathbf{x}$ is $(n \times 1)$, $\mathbf{u}$ is $(m \times 1)$, and $\mathbf{y}$ is $(q \times 1)$, as before. We assume that the system model has been linearized, or if it is already linear that the origins of the state, control, and output vectors have been translated to the values corresponding to the desired values of these variables. After this translation has been made, the desired values are represented by $\mathbf{y} = \mathbf{0} = \mathbf{x}$. The value of the control signal required to keep $\mathbf{x}$ and $\mathbf{y}$ at $\mathbf{0}$ is $\mathbf{u} = \mathbf{0}$.

## A Scalar Illustration

Suppose we have a first-order plant with a state variable $w$ and a control input $v$, such that

$$\dot{w} = -3w + v \tag{10.3-3}$$

Assume that we wish to use $v$ to drive $w$ to $w_d = 2$ and keep it there. With $w = 2$, the input required to keep it there is $v_d = 6$, from (10.3-3). Therefore let $x = w - 2$ and $u = v - 6$. Then (10.3-3) becomes

$$\dot{x} = -3x + u \tag{10.3-4}$$

and the desired condition is $x = 0 = u$.

Temporarily ignore the fact that the methods of previous chapters can deal with this first-order problem very easily, and consider how we might approach it in a general way. Assume that we do not want to apply any more control effort than is necessary. For example, we might wish to avoid saturation of the control elements or to use as little power as possible. Thus we want to keep $u$ as well as $x$ near zero. The following extension of the ISE index expresses this mathematically. We want to find the feedback control law $u = f(x)$ that minimizes

$$J = \frac{1}{2}\int_0^\infty (qx^2 + ru^2)\,dt \tag{10.3-5}$$

The weighting factors $q$ and $r$ express the relative importance of keeping $x$ and $u$ near zero. If we place more importance on $x$, then we select $q$ to be larger than $r$, and so forth.

Although we are interested in minimizing $J$, the actual value of $J$ that results is usually not of interest. Therefore, we can introduce the factor $\frac{1}{2}$ for numerical convenience later. This also means that we can set either $q$ or $r$ to unity for convenience because it is their relative weight that is important. This step also reduces the number of weighting factors to be selected.

We indicate later that the feedback-control law that minimizes $J$ is a linear law. In the current example, this means that $u$ is related to $x$ by

$$u = -Kx \tag{10.3-6}$$

where we assume that the output is $x$. Later we will see how the feedback-gain matrix **K** can be computed for a general problem, but for now let us substitute (10.3-6) into (10.3-5). This results in

$$J = \frac{1}{2}\int_0^\infty (qx^2 + ru^2)\,dt = \tfrac{1}{2}(q + K^2)\int_0^\infty x^2\,dt \tag{10.3-7}$$

where from (10.3-4)

$$\dot{x} = -3x - Kx = -(K+3)x \tag{10.3-8}$$

The solution of (10.3-8) for constant $K$ is

$$x(t) = x(0)e^{-(K+3)t} \tag{10.3-9}$$

Substituting $x(t)$ into (10.3-7) gives

$$J = \tfrac{1}{2}(q + K^2)x^2(0)\int_0^\infty e^{-2(K+3)t}\,dt$$

$$= \frac{(q + K^2)}{4(K+3)}x^2(0) \tag{10.3-10}$$

To minimize $J$ for fixed $q$ and $x(0)$, we compute $\partial J/\partial K$ and set it to zero. This gives

$$K^2 + 6K - q = 0 \tag{10.3-11}$$

For a minimum, $\partial^2 J/\partial K^2 \geqslant 0$. For $q > 0$, this implies that

$$K + \frac{54 + 6q}{18 + 2q} \geqslant 0$$

or

$$K + 3 \geqslant 0 \tag{10.3-12}$$

This last condition is of interest. It says that the value of $K$ that gives a minimum for $J$ must be such that the closed-loop system (10.3-8) will be stable. One root of (10.3-11) will satisfy this condition, while the other will not. In the former case, $x(t) \to 0$ and $J$ has a minimum value at $J = 0$. In the latter case $[(K+3) < 0]$, $x(t) \to \infty$. Therefore, $J \to \infty$, and no minimum exists.

The requirement (10.3-12) now seems obvious from the stability condition for (10.3-8) with the advantage of hindsight, but the lessons of this example will be useful later. Finally, we note that the design problem has been transformed into one of selecting a value for $q$. The larger $q$ is, the larger will be the gain $K$, and the faster will $x(t)$ approach zero. However, the peak magnitude of $u$ will be larger. The parameter $q$ is selected to achieve a compromise between these effects. We indicate later a general procedure for doing this.

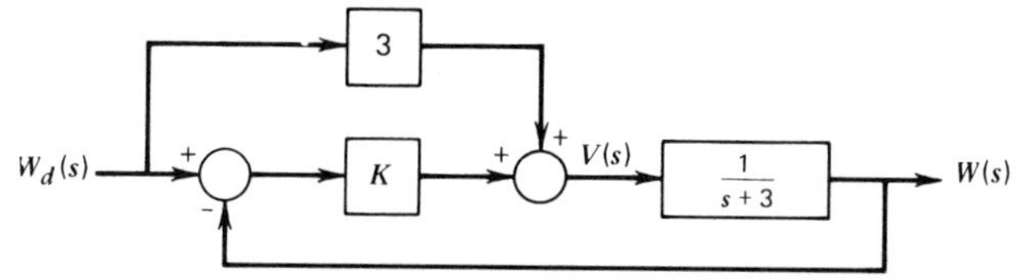


**Figure 10.7 Feedback controller with feed-forward compensation of the input to achieve zero steady-state error.**

With $K$ computed and $w$ the measured variable, the controller design is

$$v = 6 + u = 6 - Kx = 6 + K(2 - w)$$

or, in general, for the system (10.3-3)

$$v = 3w_d + K(w_d - w) \tag{10.3-13}$$

where $K$ is found from (10.3-11) and (10.3-12). The result is independent of the value of $w_d$.

This controller is illustrated in Figure 10.7. The multiplier block of gain 3 constitutes feed-forward compensation of the command input. This assures a zero steady-state error in response to a step command.

## The Role of Controllability

Some more insight into the regulator problem is given by the following problem. Let the state model be

$$\dot{x}_1 = x_1 \tag{10.3-14a}$$

$$\dot{x}_2 = x_2 + u \tag{10.3-14b}$$

Let the performance index be

$$J = \frac{1}{2}\int_0^\infty (q_1 x_1^2 + q_2 x_2^2 + u^2)\,dt \tag{10.3-15}$$

Suppose that we are primarily interested in driving $x_1$ to zero. Therefore, we select $q_2 = 0$. It is easy to see from (10.3-14*a*) that no minimum for $J$ will exist because the state $x_1$ is uncontrollable and also represents an unstable mode. Thus the response of $x_1$ will be $x_1(t) = x_1(0)e^t$ regardless of what $u$ and $x_2$ do. Therefore $x_1(t) \to \infty$, and no matter what $u(t)$ does,

$$J = \frac{1}{2}\int_0^\infty [q_1 x_1^2(0)e^{2t} + u^2]\,dt \to \infty$$

if $q_1 \neq 0$.

We can summarize our observations. No minimum exists for the index $J$ in this problem because of the following.

1. The state $x_1$ is uncontrollable.
2. The uncontrollable state also contains an unstable mode.
3. The unstable part of the state vector appears in the index $J$.

A minimum for $J$ would exist if any one of these three situations did not occur. The algorithm to be developed is capable of stabilizing any mode that is controllable and also appears in the performance index in some way.

Our previous definition of controllability simply required that the state can be transferred from one point to another in some finite time. This is called *complete controllability*. A state is *totally controllable* if it can be transferred from one point to another as quickly as desired. The controllability tests given in Section 9.4 also constitute tests for total controllability, which is a stronger condition. In light of these observations, we make the following assumption.

*Assumption 1*

All the states of the system (10.3-1) are totally controllable.

## Formulation of the Performance Index

It is convenient at this point to introduce a vector-matrix formulation for the quadratic index. We note that the index given by (10.3-15) can be written as

$$J = \frac{1}{2}\int_0^\infty (\mathbf{x}^T\mathbf{Q}\mathbf{x} + ru^2)\,dt$$

where

$$\mathbf{Q} = \begin{bmatrix} q_1 & 0 \\ 0 & q_2 \end{bmatrix}$$

We allow for the possibility of more than one input, so that the general form of the quadratic index is

$$J = \frac{1}{2}\int_0^\infty (\mathbf{x}^T\mathbf{Q}\mathbf{x} + \mathbf{u}^T\mathbf{R}\mathbf{u})\,dt \tag{10.3-16}$$

Since $J$ is a scalar, it is easy to show that the weighting matrices $\mathbf{Q}$ and $\mathbf{R}$ must be symmetric; that is, $\mathbf{Q}^T = \mathbf{Q}$ and $\mathbf{R}^T = \mathbf{R}$. The matrices $\mathbf{Q}$ and $\mathbf{R}$ appear most often in diagonal form, but there is no inherent restriction to such a form.

If the output $\mathbf{y}$, and not the state $\mathbf{x}$, is to be controlled to approach $\mathbf{0}$, then

$$J = \frac{1}{2}\int_0^\infty (\mathbf{y}^T\mathbf{Q}_o\mathbf{y} + \mathbf{u}^T\mathbf{R}\mathbf{u})\,dt \tag{10.3-17}$$

where $\mathbf{Q}_o$ is the output weighting matrix. Substituting $\mathbf{y} = \mathbf{C}\mathbf{x}$, we see that $\mathbf{Q} = \mathbf{C}^T\mathbf{Q}_o\mathbf{C}$. Thus, given $\mathbf{Q}_o$ and $\mathbf{C}$, we can use the results based on the form (10.3-16).

The control law that minimizes (10.3-16) will be found to be of the form

$$\mathbf{u} = -\mathbf{K}\mathbf{x} \tag{10.3-18}$$

where $\mathbf{K}$ is a constant matrix if $\mathbf{A}, \mathbf{B}, \mathbf{C}, \mathbf{Q}$, and $\mathbf{R}$ are constant. We restrict ourselves to this time-invariant case. Before we can develop a procedure for computing $\mathbf{K}$, we need some additional assumptions.

*Assumption 2*

The weighting matrices $\mathbf{Q}$ and $\mathbf{R}$ are symmetric. The matrix $\mathbf{R}$ is positive definite, while the matrix $\mathbf{Q}$ is positive semidefinite.

A matrix **M** is *positive definite* if $\mathbf{p}^T\mathbf{M}\mathbf{p} > 0$ for all nonzero $(n \times 1)$ vectors **p**. If $\mathbf{p}^T\mathbf{M}\mathbf{p} \geqslant 0$, **M** is *positive semidefinite.* Assumption 2 is required for a minimum of $J$ to exist. For example, it prevents problem formulations such as

$$J = \frac{1}{2}\int_0^\infty (x^2 - u^2)\,dt$$

$$\dot{x} = -x + u$$

or

$$J = \frac{1}{2}\int_0^\infty (-x^2 + u^2)\,dt$$

$$\dot{x} = -x + u$$

In both cases, the formulation leads to $|u| \to \infty$. The former case causes $u \to \infty$, while in the latter case $u \to -\infty$ so as to drive $x$ to infinity.

The differing requirements on **Q** and **R** can be explained as follows. The formulation

$$J = \frac{1}{2}\int_0^\infty u^2\,dt$$

$$\dot{x} = -x + u$$

is acceptable. Its solution, though uninteresting, is $u = 0$. On the other hand, the formulation

$$J = \frac{1}{2}\int_0^\infty x^2\,dt$$

$$\dot{x} = -x + u$$

is unacceptable since $u$ can go to infinity to drive $x$ to 0 in an infinitesimal amount of time.

*Sylvester's criterion* can be used to determine quickly whether or not a matrix is positive definite. For the matrix

$$\mathbf{M} = \begin{bmatrix} m_{11} & m_{12} \\ m_{21} & m_{22} \end{bmatrix}$$

the criterion states that **M** is positive definite if and only if all the principal minors of **M** are positive. The principal minors are the determinants formed by starting at the upper left corner of **M** and expanding outward; that is, the principal minors of the given matrix **M** are

$$m_{11}$$

and

$$\begin{vmatrix} m_{11} & m_{12} \\ m_{21} & m_{22} \end{vmatrix} = m_{11}m_{22} - m_{12}m_{21}$$

These must both be positive for **M** to be positive definite. If one is positive and the other is zero, **M** is positive semidefinite. The extension of this procedure to matrices of higher dimension is straightforward. It can be used to show that any matrix with positive elements along the main diagonal and with zeros elsewhere is positive definite. If at least one of the diagonal elements is also zero, the matrix is positive semidefinite.

We have previously seen that the manner in which the state variables appear (or do not appear) in the performance index is important. Consider again another simple example. Let the state equation and index be

$$\dot{x} = x + u$$

$$J = \frac{1}{2} \int_0^\infty u^2\, dt$$

The system is totally controllable but unstable. It is obvious that the choice of $u$ that minimizes $J$ is $u(t) = 0$ for all $t$. Putting $u = 0$ in the state equation gives the solution $x(t) = x(0)e^t$. This simple example serves to point out that the control signal $u$ cannot stabilize an unstable state that does not appear in the performance index.

This situation is avoided if the following assumption is made.

*Assumption 3*

The pair $(\mathbf{A}, \mathbf{F})$ is completely observable, where $\mathbf{F}$ is any matrix such that $\mathbf{F}\mathbf{F}^T = \mathbf{Q}$.

The partitioning of the matrix $\mathbf{F}$ does not matter. For example, if

$$\mathbf{Q} = \begin{bmatrix} 1 & 0 \\ 0 & q \end{bmatrix}$$

then

$$\mathbf{F} = \begin{bmatrix} 1 & 0 \\ 0 & \sqrt{q} \end{bmatrix}$$

is a possible choice for $\mathbf{F}$.

Assumption 3 says that *all* the states will appear in some form in the performance index; that is, all the states will be "observed" by the performance index. It happens that Assumption 3 is stronger than necessary because stable states that do not appear in the index cause no difficulty. Kalman has shown that Assumption 3 guarantees that the closed-loop system $\dot{\mathbf{x}} = (\mathbf{A} - \mathbf{B}\mathbf{K})\mathbf{x}$ is stable if $\mathbf{K}$ is computed as shown in the following (Reference 5).

## Calculation of the Gain Matrix

We develop an algorithm for computing $\mathbf{K}$ by following the procedure by Stear, as outlined by Gelb (References 6 and 7). Substitution of (10.3-18) into (10.3-1) gives the closed-loop system

$$\dot{\mathbf{x}} = (\mathbf{A} - \mathbf{B}\mathbf{K})\mathbf{x} \tag{10.3-19}$$

We assume that $\mathbf{K}$ is such that this system is stable. Assume the existence of a symmetric differentiable matrix $\mathbf{P}(t)$ such that $\mathbf{x}^T\mathbf{P}\mathbf{x}$ is a scalar. Therefore $\mathbf{P}$ is $(n \times n)$. Differentiate $\mathbf{x}^T\mathbf{P}\mathbf{x}$ with respect to time using (10.3-1), add and subtract the integrand of (10.3-16), and collect terms to obtain

$$\begin{aligned}\frac{d}{dt}(\mathbf{x}^T\mathbf{P}\mathbf{x}) &= \dot{\mathbf{x}}^T\mathbf{P}\mathbf{x} + \mathbf{x}^T\dot{\mathbf{P}}\mathbf{x} + \mathbf{x}^T\mathbf{P}\dot{\mathbf{x}} \\ &= \mathbf{x}^T(\mathbf{A}^T\mathbf{P} + \mathbf{P}\mathbf{A} + \dot{\mathbf{P}})\mathbf{x} + \mathbf{u}^T\mathbf{B}^T\mathbf{P}\mathbf{x} + \mathbf{x}^T\mathbf{P}\mathbf{B}\mathbf{u} \\ &= (\mathbf{x}^T\mathbf{P}\mathbf{B} + \mathbf{u}^T\mathbf{R})\mathbf{R}^{-1}(\mathbf{B}^T\mathbf{P}\mathbf{x} + \mathbf{R}\mathbf{u}) - \mathbf{x}^T\mathbf{Q}\mathbf{x} - \mathbf{u}^T\mathbf{R}\mathbf{u} \qquad (10.3\text{-}20)\end{aligned}$$

The latter equality results if we stipulate that **P** must satisfy the following equation.

$$\mathbf{A}^T\mathbf{P} + \mathbf{P}\mathbf{A} + \dot{\mathbf{P}} + \mathbf{Q} = \mathbf{P}\mathbf{B}\mathbf{R}^{-1}\mathbf{B}^T\mathbf{P} \qquad (10.3\text{-}21)$$

This equation is called the *matrix Riccati differential equation*, in analogy with the scalar equation of the same name. Carrying out the multiplication in (10.3-20) is the best way to convince yourself of the validity of it and (10.3-21).

Note that if $\mathbf{P}(t)$ does not approach infinity as $t \to \infty$, then

$$\begin{aligned}\int_0^\infty \frac{d}{dt}(\mathbf{x}^T\mathbf{P}\mathbf{x})\,dt &= \mathbf{x}^T(\infty)\mathbf{P}(\infty)\mathbf{x}(\infty) - \mathbf{x}^T(0)\mathbf{P}(0)\mathbf{x}(0) \\ &= -\mathbf{x}^T(0)\mathbf{P}(0)\mathbf{x}(0)\end{aligned}$$

since $\mathbf{x}(\infty) \to \mathbf{0}$ from the stability assumption. From this and (10.3-20),

$$\begin{aligned}\frac{1}{2}\int_0^\infty \frac{d}{dt}(\mathbf{x}^T\mathbf{P}\mathbf{x})\,dt &= \frac{1}{2}\int_0^\infty (\mathbf{x}^T\mathbf{P}\mathbf{B} + \mathbf{u}^T\mathbf{R})\mathbf{R}^{-1}(\mathbf{B}^T\mathbf{P}\mathbf{x} + \mathbf{R}\mathbf{u})\,dt \\ &\quad - \frac{1}{2}\int_0^\infty (\mathbf{x}^T\mathbf{Q}\mathbf{x} + \mathbf{u}^T\mathbf{R}\mathbf{u})\,dt \\ &= -\mathbf{x}^T(0)\mathbf{P}(0)\mathbf{x}(0)\end{aligned}$$

Thus

$$J = \mathbf{x}^T(0)\mathbf{P}(0)\mathbf{x}(0) + \frac{1}{2}\int_0^\infty (\mathbf{x}^T\mathbf{P}\mathbf{B} + \mathbf{u}^T\mathbf{R})\mathbf{R}^{-1}(\mathbf{B}^T\mathbf{P}\mathbf{x} + \mathbf{R}\mathbf{u})\,dt \qquad (10.3\text{-}22)$$

But $\mathbf{x}(0)$ and $\mathbf{P}(0)$ are independent of the control input **u**, which is to be chosen. Thus $J$ can be minimized by considering only the integrand in (10.3-22). This is nonnegative and therefore has a minimum of zero. This minimum occurs at

$$\mathbf{B}^T\mathbf{P}\mathbf{x} + \mathbf{R}\mathbf{u} = \mathbf{0}$$

and this implies that

$$\mathbf{u} = -\mathbf{R}^{-1}\mathbf{B}^T\mathbf{P}\mathbf{x} \qquad (10.3\text{-}23)$$

Thus the optimal control law consists of a linear feedback of the state vector **x**.

The specification of $\mathbf{P}(t)$ by (10.3-21) is incomplete, however, until we state a boundary condition for $\mathbf{P}(t)$. We note that the time-invariant nature of the problem implies that **P** should be constant (this insight is borne out by a rigorous derivation). Therefore, we set $\dot{\mathbf{P}} = 0$ in (10.3-21), and the control law that minimizes (10.3-16) is given by

$$\mathbf{u} = -\mathbf{K}\mathbf{x} \qquad (10.3\text{-}24)$$

where

$$\mathbf{K} = \mathbf{R}^{-1}\mathbf{B}^T\mathbf{P} \qquad (10.3\text{-}25)$$

$$\mathbf{A}^T\mathbf{P} + \mathbf{P}\mathbf{A} + \mathbf{Q} - \mathbf{P}\mathbf{B}\mathbf{R}^{-1}\mathbf{B}^T\mathbf{P} = \mathbf{0} \qquad (10.3\text{-}26)$$

## Solution of the Matrix Riccati Equation

Given **A**, **B**, **Q**, and **R**, the solution of the algebraic matrix Riccati equation (10.3-26) can be accomplished in several ways (References 8–11, plus Reference 2 of Chapter Nine). It is not difficult to see that as the order of the system increases, the nonlinear algebra that results from the term $\mathbf{PBR}^{-1}\mathbf{B}^T\mathbf{P}$ can be formidable. At the level of this presentation the best solution method is to obtain **P** as the steady-state solution of the Riccati differential equation (10.3-21). To this end we quote some pertinent results without proof (Reference 2 of Chapter Nine).

The matrix **P** is the symmetric positive-definite steady-state solution of the differential equation (10.3-21) with the boundary condition $\mathbf{P}(\infty) = \mathbf{0}$. This implies that the equation is solved backward in time from $t = \infty$. To make this computationally possible, we change the independent variable to run backward in time. Let $\tau = T - t$, where $T$ is the final time: $T \to \infty$. Then $d\mathbf{P}/d\tau = -d\mathbf{P}/dt$ and (10.3-21) becomes

$$\frac{d\mathbf{P}}{d\tau} = \mathbf{A}^T\mathbf{P} + \mathbf{PA} + \mathbf{Q} - \mathbf{PBR}^{-1}\mathbf{B}^T\mathbf{P}, \quad \tau \geqslant 0 \tag{10.3-27}$$

with the trial initial value

$$\mathbf{P}(\tau = 0) = \mathbf{P}_0 \tag{10.3-28}$$

The true boundary condition is matched by selecting $\mathbf{P}_0 = \mathbf{0}$. But if we have a good guess for the steady-state solution, the convergence can be speeded up by using this guess as the initial condition $\mathbf{P}_0$.

The following theorems aid in computing **P**.

*Theorem 1*

If the states of the system $\dot{\mathbf{x}} = \mathbf{Ax} + \mathbf{Bu}$ that are not stable are controllable, then a steady-state solution **P** of (10.3-27) exists and is constant and positive definite. Furthermore, there is only one positive definite solution for **P**.

*Theorem 2*

If the solution for **P** exists for a time-invariant system, then the closed-loop system is stable.

Thus if our assumptions are satisfied, a stable closed-loop system always results when the quadratic performance index is used. This is a good engineering justification for its use. One should contrast this feature with other design methods that do not guarantee stability.

*Theorem 3*

Under the same conditions guaranteeing stability of the closed-loop system, the Riccati differential equation (10.3-27) is globally stable relative to the steady-state positive definite solution **P** if the initial guess $\mathbf{P}_0$ is positive definite.

This means that the solution of the differential equation (10.3-27) with the initial condition (10.3-28) will always approach the solution of (10.3-26) as $\tau \to \infty$. We add a cautionary note here, however. When (10.3-27) is solved numerically, the errors due to truncation and round-off can sometimes lead to a numerically unstable solution. This generally occurs only for very high-order systems. If so, another solution method based on eigenvector methods is available (Reference 9).

## Selection of the Weighting Matrices

The selection of the weighting matrices **Q** and **R** is usually made on the basis of experience together with simulations of the results for different trial values. The following guidelines have emerged. Usually **Q** and **R** are selected to be diagonal so that specific components of the state and control vectors can be penalized individually if their response is undesirable. Therefore the state and control vectors should represent sets of variables that are easily identifiable physically, rather than a set of transformed variables such as the modal variables. This enables the designer to visualize the effects of the trial values for **Q** and **R**.

The larger the elements of **Q** are, the larger the gain matrix **K** elements, and the faster the state variables approach zero. On the other hand, the larger the elements of **R**, the smaller the elements of **K** and the slower the response.

Often some state variables are simply derivatives of others, such as the displacement and velocity of a mass. If only the displacement variable is penalized by **Q**, the response tends to be oscillatory. To reduce overshoots, therefore, include in **Q** a weighting term for the velocity variable.

Often the linear state equations are the result of a linearization. In this case it is sometimes possible to obtain a clue as to the proper weights by examining the second-order term in the Taylor expansion used to obtain the linearization. These terms are quadratic forms [see (A.1-20) and (A.1-22)], and the relative size of their components can indicate which variables are most likely to jeopardize the accuracy of the linearization. These variables would then be weighted more heavily in **Q** or **R**.

### *Example 10.4*

Consider the problem

$$J = \frac{1}{2}\int_0^\infty (x^2 + u^2)\,dt$$

$$\dot{x} = u$$

Find the control law that minimizes $J$.

The Riccati equation (10.3-27) reduces to

$$\frac{dP}{d\tau} = 1 - P^2$$

with $P(0) = P_o$. Direct integration gives

$$\int_0^{P(t_1)} \frac{dP}{1-P^2} = \int_0^{t_1} d\tau$$

or

$$P(t_1) = \tanh(\tanh^{-1} P_0 + t_1)$$

As $t_1 \to \infty$, $P(t_1) \to 1$ regardless of the initial value $P_0$. Therefore, the steady-state solution for $P$ is 1. This could have been obtained immediately from the algebraic equation, but the example illustrates the convergence property of the Riccati differential equation. Note that the algebraic equation gives two roots $P = \pm 1$, with the positive definite root as the correct one. The optimal control law from (10.3-24) is $u = -x$.

*Example 10.5*

The following system is to be controlled with the input $u$ so that $x_1$ remains near zero. Devise a feedback-control law.

$$\dot{x}_1 = x_2$$

$$\dot{x}_2 = u$$

Although the controller is to keep $x_1$ near zero, we include $x_2$ in the index to maintain control over the overshoots in the response. Thus choose

$$J = \frac{1}{2}\int_0^\infty (x_1^2 + qx_2^2 + ru^2)\,dt$$

Assumption 3 requires that the pair (**A**, **F**) be observable, where

$$\mathbf{Q} = \begin{bmatrix} 1 & 0 \\ 0 & q \end{bmatrix}, \qquad \mathbf{F} = \begin{bmatrix} 1 & 0 \\ 0 & \sqrt{q} \end{bmatrix}, \qquad \mathbf{A} = \begin{bmatrix} 0 & 1 \\ 0 & 0 \end{bmatrix}$$

Thus the following matrix must be of rank 2.

$$\begin{bmatrix} F \\ \hline FA \end{bmatrix} = \begin{bmatrix} 1 & 0 \\ 0 & \sqrt{q} \\ 0 & 1 \\ 0 & 0 \end{bmatrix}$$

Since this contains a nonzero $(2 \times 2)$ determinant, it is of rank 2. From Kalman's result, we know that the closed-loop system will be stable.

The Riccati matrix **P** will be $(2 \times 2)$ and symmetric. Denote it by

$$\mathbf{P} = \begin{bmatrix} p_1 & p_2 \\ p_2 & p_3 \end{bmatrix}$$

The terms in the algebraic Riccati equation are

$$\mathbf{A}^T\mathbf{P} = \begin{bmatrix} 0 & 0 \\ p_1 & p_2 \end{bmatrix}, \qquad \mathbf{PA} = \begin{bmatrix} 0 & p_1 \\ 0 & p_2 \end{bmatrix}, \qquad \mathbf{PBR}^{-1}\mathbf{B}^T\mathbf{P} = \begin{bmatrix} p_2^2 & p_2p_3 \\ p_2p_3 & p_3^2 \end{bmatrix} r^{-1}$$

Equation (10.3-26) results in three equations to be solved.

$$1 - p_2^2 r^{-1} = 0$$

$$p_1 - p_2 p_3 r^{-1} = 0$$

$$2p_2 + q - p_3^2 r^{-1} = 0$$

The first equation gives $p_2 = \pm\sqrt{r}$. Using each value and solving for $p_1$ and $p_3$, we can employ Sylvester's criterion to show that only the solution $p_2 = +\sqrt{r}$ gives a positive definite matrix $\mathbf{P}$. The remaining elements, in terms of $b = \sqrt{r}$, are

$$p_1 = \sqrt{2b+q}, \qquad p_3 = b\sqrt{2b+q}$$

Thus

$$\mathbf{K} = \mathbf{R}^{-1}\mathbf{B}^T\mathbf{P} = r^{-1}[0 \quad 1]\begin{bmatrix} p_1 & p_2 \\ p_2 & p_3 \end{bmatrix} = r^{-1}[p_2 \quad p_3]$$

The feedback-control law is

$$u = -\mathbf{K}\mathbf{x} = -r^{-1}(p_2 x_1 + p_3 x_2)$$

If the control effort is weighted equally with the position error $x_1$, then $r = 1$ and

$$u = -x_1 - \sqrt{2+q}\, x_2$$

The term $qx_2^2$ was included in the performance index to allow the system's damping to be adjusted. This can be seen from the damping ratio of the resulting closed-loop system, which is $\zeta = \sqrt{2+q}/2$. Increasing $q$ will increase $\zeta$.

## Design With Multiple Control Variables

As with any of the design methods, when the system order or the number of control variables increases, it is impossible to obtain a closed-form solution. In the case of the linear-quadratic method, the Riccati equation must be solved numerically. However, this method has advantages over the pole-placement and modal-control methods in that it provides a more systematic way of computing the control matrix $\mathbf{K}$. To see why, consider an $n$th-order plant with $m$ control variables. Then the gain matrix $\mathbf{K}$ will have $nm$ elements. If all the states and control variables appear in the quadratic performance index, then $(n + m - 1)$ weighting factors must be selected (one less than $n + m$ because the weights can all be made *relative* to one of the variables). Once these are selected, all the elements of $\mathbf{K}$ are computed automatically by the linear-quadratic algorithm. With the pole-placement method, the characteristic equation will have $n$ coefficients, which give $n$ equations that can be used to find $n$ of the elements of $\mathbf{K}$. This leaves $nm - n$, or $n(m - 1)$ elements of $\mathbf{K}$ yet to be found, *without* any general algorithm available to guide the selection.

For a scalar input, $m = 1$, and no gains are left undetermined with the pole-placement method, regardless of the system order. However, for $m > 1$, $\mathbf{K}$ is not completely determined by pole placement. For example, a fifth-order plant with three control variables has 15 gains, 10 of which are not determined by the eigenvalue specifications. For such a problem the linear-quadratic method requires that seven weighting factors be selected. However, the designer is guided in this selection by the performance index and the use of simulation to see what variables must have their weighting changed to improve the system's performance.

The designer should always attempt to decouple a multivariable control problem into a set of single-input problems if possible. Even if not possible, this procedure often shows what feedback variables are important. For example, a two-mass system with a controlled force on each mass has $n = 4$, $m = 2$, and $\mathbf{K}$ is $(2 \times 4)$ with eight elements. With pole placement, four gains will be unspecified. Suppose the force $f_i$

acts on the mass $m_i$, and the state vector is

$$\mathbf{x} = (y_1, v_1, y_2, v_2)^T$$

where $y_i$ is the displacement and $v_i$ the velocity of $m_i$. Then we might try a decoupled control law of the form

$$\begin{bmatrix} f_1 \\ f_2 \end{bmatrix} = - \begin{bmatrix} k_{11} & k_{12} & 0 & 0 \\ 0 & 0 & k_{23} & k_{24} \end{bmatrix} \begin{bmatrix} y_1 \\ v_1 \\ y_2 \\ v_2 \end{bmatrix} \tag{10.3-29}$$

since $f_1$ affects $y_2$ and $v_2$ only indirectly, with a similar relation between $f_2, y_1$, and $v_1$. Now no gains will be left unspecified, assuming enough flexibility remains to achieve the desired pole placement.

If further simplification is required, the physical coupling of the subsystems can be temporarily ignored. For the two-mass example, this approach leads to the plant model.

$$\begin{bmatrix} \dot{y}_1 \\ \dot{v}_1 \\ \dot{y}_2 \\ \dot{v}_2 \end{bmatrix} = \left[\begin{array}{c|c} \mathbf{A}_1 & \mathbf{0} \\ \hline \mathbf{0} & \mathbf{A}_2 \end{array}\right] \begin{bmatrix} y_1 \\ v_1 \\ y_2 \\ v_2 \end{bmatrix} + \left[\begin{array}{c|c} \mathbf{B}_1 & \mathbf{0} \\ \hline \mathbf{0} & \mathbf{B}_2 \end{array}\right] \begin{bmatrix} f_1 \\ f_2 \end{bmatrix} \tag{10.3-30}$$

which consists of two uncoupled subsystems. Of course, the model ignores the physics of the situation, but the gains computed with (10.3-29) and (10.3-30) often provide a good starting point for determining the gains to be used with the original multivariable coupled model. The linear-quadratic method can be used for the latter calculations.

Many variations of the linear-quadratic problem have been developed (Reference 8). We have treated the problem only in its most basic form. When the state measurement is incomplete, the separation theorem of Section 10.2 can be invoked, and the observer algorithm can be used to provide an estimate of $\mathbf{x}$. Because of the widespread availability of computer packages for the numerical solution of differential equations, we have presented an algorithm for the linear-quadratic problem that utilizes the solution of a differential equation, the Riccati equation, for computing the optimal gain matrix $\mathbf{K}$. The calculation of $\mathbf{K}$ can also be formulated as a matrix eigenvalue problem (Reference 9). This approach has been found to be useful when the round-off errors in the numerical differential-equation solution cause convergence problems. Good results have been reported with the use of the eigenvector method to provide an accurate initial guess for the solution of the Riccati equation (Reference 2 of Chapter Nine).

## 10.4 MATRIX METHODS FOR DIGITAL CONTROL

Since the matrix analysis of linear discrete-time systems parallels that of continuous-time systems, one would expect that the design methods for state-vector feedback control would also be similar. This is indeed the case. For a linear feedback law of the form

$$\mathbf{u}(k) = -\mathbf{K}\mathbf{x}(k) \tag{10.4-1}$$

the state equation (9.5-4) becomes

$$\mathbf{x}(k+1) = \mathbf{P}\mathbf{x}(k) + \mathbf{Q}\mathbf{u}(k) = (\mathbf{P} - \mathbf{Q}\mathbf{K})\mathbf{x}(k) \tag{10.4-2}$$

Since the eigenvalue and modal transformation methods are the same for discrete and continuous-time systems, the methods of eigenvalue placement and modal control are developed in a similar way (see Section 10.1).

The design of a state observer in Section 10.2 utilized a state feedback-control principle and the observability criterion. Therefore, we can approach the design of state observers for discrete systems in the same manner.

Because of the similarities we will not pursue these topics further. Instead, we now introduce some control techniques with emphasis on their digital implementation.

## The Linear-Quadratic Regulator Problem

The quadratic performance index used in Section 10.3 involved an integral over time. Its counterpart for digital control is a summation

$$J = \frac{1}{2}\sum_{i=0}^{\infty}[\mathbf{x}^T(i+1)\mathbf{S}\mathbf{x}(i+1) + \mathbf{u}^T(i)\mathbf{R}\mathbf{u}(i)] \tag{10.4-3}$$

where we denote the state weighting matrix by $\mathbf{S}$ to avoid confusion with the matrix $\mathbf{Q}$ of (9.5-4), repeated here

$$\mathbf{x}(k+1) = \mathbf{P}\mathbf{x}(k) + \mathbf{Q}\mathbf{u}(k) \tag{10.4-4}$$

Note that the indices in $J$ are different for $\mathbf{x}$ and $\mathbf{u}$. This is because $\mathbf{x}(0)$ is a given quantity and therefore cannot be selected, whereas $\mathbf{u}(0)$ is to be chosen.

The assumptions required to obtain a feedback-control law of the form (10.4-1) are similar to those given in Section 10.3 and the development of the optimal regulator algorithm for discrete time is given in Reference 8. The control law that minimizes $J$ is given by (10.4-1) where the gain matrix is

$$\mathbf{K} = (\mathbf{R} + \mathbf{Q}^T\mathbf{H}\mathbf{Q})^{-1}\mathbf{Q}^T\mathbf{H}\mathbf{P} \tag{10.4-5}$$

The Riccati matrix is now denoted by $\mathbf{H}$. It is the steady-state solution of the following difference equation

$$\mathbf{H}(k) = \mathbf{P}^T\mathbf{H}(k+1)\mathbf{P} - \mathbf{P}^T\mathbf{H}(k+1)\mathbf{Q}[\mathbf{R} + \mathbf{Q}^T\mathbf{H}(k+1)\mathbf{Q}]^{-1}\mathbf{Q}^T\mathbf{H}(k+1)\mathbf{P} + \mathbf{S}$$

$$k = 0, 1, 2, \ldots, \infty \tag{10.4-6}$$

with the following boundary condition at the final time ($k = \infty$),

$$\mathbf{H}(\infty) = \mathbf{0} \tag{10.4-7}$$

As with the continuous-time case, (10.4-6) can be converted to a difference equation whose time index $j$ runs forward, $j = 0, 1, 2, \ldots$, by the relation $j = T - k$, where $T \to \infty$. The initial condition (10.4-7) becomes

$$\mathbf{H}(j = 0) = \mathbf{H}_0$$

where $\mathbf{H}_0$ is chosen to speed up the convergence.

## Finite Settling Time Control

The finite settling time algorithm treated in Section 8.10 was obtained from the $z$-transform of the desired response. The controller transfer function required to achieve this response was computed with this transform and the transfer function of the plant. However, the controller algorithm so obtained might not be realizable.

Here we present another approach to the problem in which the control law is specified to be a linear state feedback law of the form (10.4-1) where $u$ is a scalar. This law is easy to implement, of course. For a plant of the form (9.5-4), it is possible to find a matrix $\mathbf{K}$ that will bring the state vector $\mathbf{x}$ from an arbitrary starting value $\mathbf{x}(0)$ to the origin $\mathbf{x} = 0$ in at most $n$ time steps, if the plant is controllable and all the states are available for feedback. The method is by Mullin and Barbeyrac (Reference 12) and is outlined in Chapter 11 of Reference 1.

Since $u$ is assumed to be a scalar, the input matrix $\mathbf{Q}$ is an $(n \times 1)$ column vector and the gain matrix $\mathbf{K}$ is a $(1 \times n)$ row vector. The closed-loop system is described by (10.4-2). Let $\mathbf{G}$ denote the resulting closed-loop state matrix, so $\mathbf{x}(k+1) = \mathbf{G}\mathbf{x}(k)$, and

$$\mathbf{G} = \mathbf{P} - \mathbf{Q}\mathbf{K} \tag{10.4-8}$$

On the first time step, $\mathbf{x}(1) = \mathbf{G}\mathbf{x}(0)$. On the second and third steps,

$$\begin{aligned} \mathbf{x}(2) &= \mathbf{G}\mathbf{x}(1) = \mathbf{G}^2\mathbf{x}(0) \\ \mathbf{x}(3) &= \mathbf{G}^3\mathbf{x}(0) \end{aligned}$$

But $\mathbf{G}^2$ and $\mathbf{G}^3$ can be expanded as

$$\begin{aligned} \mathbf{G}^2 &= \mathbf{P}^2 - \mathbf{PQK} - \mathbf{QK}(\mathbf{P} - \mathbf{QK}) \\ \mathbf{G}^3 &= \mathbf{P}^3 - \mathbf{P}^2\mathbf{QK} - \mathbf{PQK}(\mathbf{P} - \mathbf{QK}) - \mathbf{QK}(\mathbf{P} - \mathbf{QK})^2 \\ &= \mathbf{P}^3 - [\mathbf{P}^2\mathbf{Q}, \mathbf{PQ}, \mathbf{Q}] \begin{bmatrix} \mathbf{K} \\ \mathbf{K}(\mathbf{P} - \mathbf{QK}) \\ \mathbf{K}(\mathbf{P} - \mathbf{QK})^2 \end{bmatrix} \end{aligned}$$

Continuing in this manner, the result is

$$\begin{aligned} \mathbf{x}(n) &= \mathbf{G}^n\mathbf{x}(0) \\ \mathbf{G}^n &= \mathbf{P}^n - [\mathbf{P}^{n-1}\mathbf{Q}, \ldots, \mathbf{PQ}, \mathbf{Q}] \begin{bmatrix} \mathbf{K} \\ \mathbf{K}(\mathbf{P} - \mathbf{QK}) \\ \vdots \\ \mathbf{K}(\mathbf{P} - \mathbf{QK})^{n-1} \end{bmatrix} \end{aligned}$$

Therefore, if $\mathbf{x}(n)$ is to be $\mathbf{0}$, $\mathbf{G}^n$ must be $\mathbf{0}$. We can use this equation to find $\mathbf{K}$ as follows. Use the preceding expression for $\mathbf{G}^n$ and solve for the matrix containing $\mathbf{K}$. This gives

$$\begin{bmatrix} \mathbf{K} \\ \mathbf{K}(\mathbf{P} - \mathbf{QK}) \\ \vdots \\ \mathbf{K}(\mathbf{P} - \mathbf{QK})^{n-1} \end{bmatrix} = [\mathbf{P}^{n-1}\mathbf{Q}, \ldots, \mathbf{PQ}, \mathbf{Q}]^{-1}\mathbf{P}^n \tag{10.4-9}$$

This is possible only if the indicated inverse matrix exists. The condition for the existence of this inverse reduces to the controllability condition (9.5-22). To see this, we note that the rank of a matrix equals the number of linearly independent column vectors in the matrix. Therefore, if we change the ordering of the columns, the rank remains unchanged.

The gain matrix **K** is found from the first row on the right-hand side of (10.4-9).

### *Example 10.6*

The pulse transfer function for a dc motor-amplifier system shown in Figure 5.39 is given by (5.11-5), repeated here.

$$G_p(z) = \frac{K}{b^2}\frac{(bT-1+a)z+1-a-bTa}{(z-1)(z-a)}$$

where $a$, $b$, and $K$ depend on the inertia, damping, and the amplifier gain as well as the sampling period $T$. This transfer function is of the form

$$G_p(z) = \frac{cz+d}{(z-1)(z-a)} = \frac{\Theta(z)}{E(z)} \tag{10.4-10}$$

where $\theta$ is the load displacement and $e$ is the input voltage to the amplifier. Assume that the gain has been selected to eliminate the numerator dynamics, so that $c = 0$. Design a feedback controller to take the load displacement and velocity to zero in no more than two time steps.

First convert the model to state-variable form. With $c = 0$, the difference equation for $\theta$ is

$$\theta(k+2)-(1+a)\theta(k+1)+a\theta(k) = de(k)$$

Define $x_1(k) = \theta(k)$, $x_2(k) = \theta(k+1)$, and $u(k) = e(k)$. Thus,

$$x_1(k+1) = x_2(k)$$

$$x_2(k+1) = -ax_1(k)+(1+a)x_2(k)+du(k)$$

and

$$\mathbf{P} = \begin{bmatrix} 0 & 1 \\ -a & (1+a) \end{bmatrix} \qquad \mathbf{Q} = \begin{bmatrix} 0 \\ d \end{bmatrix}$$

It is desired to reach $\mathbf{x} = \mathbf{0}$ in two steps, so $n = 2$. Thus (10.4-9) becomes

$$\begin{bmatrix} \mathbf{K} \\ \mathbf{K}(\mathbf{P}-\mathbf{QK}) \end{bmatrix} = [\mathbf{PQ}, \mathbf{Q}]^{-1}\mathbf{P}^2$$

where

$$\mathbf{P}^2 = \begin{bmatrix} -a & (1+a) \\ -a(1+a) & -a+(1+a)^2 \end{bmatrix} \qquad \mathbf{PQ} = \begin{bmatrix} d \\ d(1+a) \end{bmatrix}$$

$$[\mathbf{PQ}, \mathbf{Q}] = \begin{bmatrix} d & 0 \\ d(1+a) & d \end{bmatrix}$$

The rank of $[\mathbf{PQ}, \mathbf{Q}] = 2$ if $d \neq 0$. If so, then $[\mathbf{PQ}, \mathbf{Q}]^{-1}$ exists and we can solve for **K** as follows.

$$\begin{bmatrix} \mathbf{K} \\ \mathbf{K}(\mathbf{P}-\mathbf{Q}\mathbf{K}) \end{bmatrix} = \begin{bmatrix} \frac{1}{d} & 0 \\ \frac{-(1+a)}{d} & \frac{1}{d} \end{bmatrix} \begin{bmatrix} -a & (1+a) \\ -a(1+a) & 1+a+a^2 \end{bmatrix}$$

$$= \begin{bmatrix} -\frac{a}{d} & \frac{1+a}{d} \\ 0 & -\frac{a}{d} \end{bmatrix}$$

The matrix **K** is the first row on the right-hand side.

$$\mathbf{K} = \begin{bmatrix} -\frac{a}{d} & \frac{1+a}{d} \end{bmatrix}$$

Thus the desired control law is

$$u(k) = \frac{a}{d} x_1(k) - \frac{(1+a)}{d} x_2(k)$$

In practice, the next step would be to investigate the peak value of the manipulated variable to see if it is feasible.

The algorithm is useful in starting operations in which the initial state is not close to the desired state. Under such conditions the PID-type control laws will often not perform well. We note that the finite settling time algorithm depends heavily on an accurate model of the system. Also, if disturbances or a nonzero command input occur, the algorithm often will produce a steady-state error. In this case it can be combined with another form of compensation such as feed-forward compensation of the command input, as in Figure 10.7, or combined with integral action.

The algorithm described here has been extended to handle vector inputs as well as cases in which only incomplete state-vector feedback is available (Reference 1).

## 10.5 SUMMARY

The design methods of classical control theory are very useful for relatively low-order systems with a single control variable. However, the increasing demands of high-performance systems require that multiinput control systems be designed for such applications. In addition, the design models must often be of higher order if they are to provide the necessary predictive capability. In this chapter we have attempted to demonstrate the advantages of matrix methods based on the state-variable formulation for such problems. These advantages can be categorized as providing (1) improved insight into the system response, (2) more general design procedures for control systems, and (3) increased computational power and efficiency.

The matrix methods allow general control-system structures and algorithms to be formulated without regard to system order. For example, the algorithms for feedback control by pole placement, modal control, and the linear-quadratic regulator can be concisely expressed in matrix form. The conciseness of the matrix formulations means that they are ideally suited for computer implementation, and this is their primary advantage. In addition, modal analysis leads to a better understanding of the controller's influence on the system. The associated concepts of controllability and observability are very useful in the design of state feedback controllers and observers.

For these reasons, the state-variable matrix methods of modern control theory have an important role to play. Anyone engaged in the modeling, analysis, and control of dynamic systems should be familiar with these techniques, as well as those of classical control theory, and should know which is most advantageous to use in a given application.

## REFERENCES

1. Y. Takahashi, M. J. Rabins, and D. M. Auslander, *Control*, Addison-Wesley, Reading, Mass., 1970.
2. J. L. Melsa and S. K. Jones, *Computer Programs for Computational Assistance in the Study of Linear Control Theory*, McGraw-Hill, New York, 1973.
3. D. G. Luenberger, "Observers for Multivariable Systems," *IEEE Trans. Automatic Control*, Vol. AC-11, No. 2, April 1966, pp. 190–197.
4. D. G. Luenberger, "An Introduction to Observers," *IEEE Trans. on Automatic Control*, Vol. AC-16, No. 6, December 1971, pp. 596–602.
5. R. E. Kalman, "When is a Linear System Optimal?" *Trans. ASME Ser. D: J. Basic Engineering*, Vol. 86, March 1964, pp. 1–10.
6. E. B. Stear, Notes for a short course on Guidance and Control of Tactical Missiles, presented at the University of California at Los Angeles, July 21–25, 1969.
7. A. Gelb, ed., *Applied Optimal Estimation*, MIT Press, Cambridge, Mass., 1974.
8. H. Kwakernaak and R. Sivan, *Linear Optimal Control Systems*, John Wiley, New York, 1972.
9. G. F. Franklin and J. D. Powell, *Digital Control of Dynamic Systems*, Addison-Wesley, Reading, Mass., 1980.
10. B. D. O. Anderson and J. B. Moore, *Linear Optimal Control*, Prentice-Hall, Englewood Cliffs, N.J., 1971.
11. A. E. Bryson and Y. C. Ho, *Applied Optimal Control*, Blaisdell, Waltham, Mass., 1969.
12. F. J. Mullin and J. DeBarbeyrac, "Linear Digital Control," *Trans. ASME Ser. D: J. Basic Engineering*, Vol. 86, March 1964, pp. 61–66.

## PROBLEMS

**10.1** Consider the tank system shown in Figure 10.2. Let $C_1 = C_2 = 1$, $R_1 = \frac{1}{2}$, $R_2 = 1$. Then the system matrices are

$$\mathbf{A} = \begin{bmatrix} -2 & 2 \\ 2 & -6 \end{bmatrix} \qquad \mathbf{B} = \begin{bmatrix} 1 \\ 0 \end{bmatrix} \qquad \mathbf{C} = \begin{bmatrix} 1 & 0 \\ 0 & 1 \end{bmatrix}$$

Find the state-vector feedback gains to place the closed-loop poles at $s = -5$, $-10$.

**10.2** Consider the system of Problem 10.1, except that now there are two inflow rates under control, $q_1$ and $q_2$. Then

$$\mathbf{B} = \begin{bmatrix} 1 & 0 \\ 0 & 1 \end{bmatrix}$$

Find the feedback gains for a modal controller such that the closed-loop modal state matrix is

$$\mathbf{\Lambda}_d = \begin{bmatrix} -5 & 0 \\ 0 & -10 \end{bmatrix}$$

**10.3** Compare the responses of the controllers designed in Problems 10.1 and 10.2. Use the initial conditions $h_1(0) = -2$, $h_2(0) = -1$. The simulation should show the flow rate $q_1$ for 10.1, and $q_1$, $q_2$ for 10.2, as functions of time. Which design requires the least peak flow rate?

**10.4** Pole placement control can be achieved by substituting $\mathbf{u} = -\mathbf{Kx}$ into the state equation $\dot{\mathbf{x}} = \mathbf{Ax} + \mathbf{Bu}$ and solving for the elements of $\mathbf{K}$, without putting $\mathbf{A}$ and $\mathbf{B}$ into phase-variable form. However, the resulting nonlinear algebra that must be solved will be formidable for many problems. An algorithm by Kalman is available for transforming a controllable plant to phase-variable form using a nonsingular linear transformation matrix $\mathbf{T}$ such that $\mathbf{x} = \mathbf{Tx}_p$, where $\mathbf{x}_p$ denotes the phase variables. Then the matrices for the phase variable model are [see (10.1-9)]

$$\mathbf{A}_p = \mathbf{T}^{-1}\mathbf{AT} \qquad \mathbf{B}_p = \mathbf{T}^{-1}\mathbf{B} \qquad \mathbf{C}_p = \mathbf{T}^T\mathbf{C}$$

where $\mathbf{B}$ and $\mathbf{C}$ are column and row vectors, respectively. Let $\mathbf{v}_i$ be the $i$th column of $\mathbf{T}$. Kalman's algorithm consists of the following recursion relations.

$$\mathbf{v}_n = \mathbf{B}$$

$$\mathbf{v}_{n-i} = \mathbf{A}\mathbf{v}_{n-i+1} + a_i\mathbf{B}, \quad i = 1, 2, \ldots, n-1$$

where the plant's characteristic polynomial is

$$s^n + a_1 s^{n-1} + \ldots + a_{n-1}s + a_n$$

Apply this algorithm to determine the phase-variable model for the matrices $\mathbf{A}$, $\mathbf{B}$, and $\mathbf{C}$ given in Problem 10.1.

10.5 Consider a single-input, single-output control system with state feedback and proportional control such that the control variable $u$ is

$$u = g(r - \mathbf{K}\mathbf{x})$$

where $g$ is a scalar constant and $r$ is the reference input.

(a) Use a procedure similar to that in Section 10.2 to obtain the equations to be solved for the elements of **K**.

(b) Find the transfer function between the output $y$ and the input $r$. What value must $g$ have in order to achieve zero steady-state position error?

10.6 For the system shown in Figure P9.1, let $k_1 = k_2 = m_1 = m_2 = 1$, $k_3 = f_1 = c_1 = c_2 = 0$, and $u = f_2$. Is it possible to stabilize the system by using one state variable as feedback; that is, by using $u = -k_i x_i$? (*Hint.* Try vector feedback $u = -\mathbf{K}\mathbf{x}$ to see what is required for stability.) What are the restrictions on the pole placement with this control law?

10.7 Consider the system

$$\dot{\mathbf{x}} = \begin{bmatrix} -1 & 0 \\ 2 & -2 \end{bmatrix}\mathbf{x} + \begin{bmatrix} 1 & 0 \\ 0 & 2 \end{bmatrix}\mathbf{u}, \qquad \mathbf{y} = \begin{bmatrix} 0 & 1 \\ 1 & -1 \end{bmatrix}\mathbf{x}$$

Design a modal control system so that the eigenvalues of the closed-loop system are $s_1 = -5, s_2 = -6$.

10.8 Suppose that the measurement equation in Problem 10.7 were instead

$$y = [1 \quad -1]\mathbf{x}$$

(a) Without actually designing an observer, how do you know whether or not it is possible to do so?

(b) Design an observer for this system. The observer eigenvalues are to be at $s = -10$ and $-15$.

10.9 The linearized equations for the translational and rotational motion of the inverted pendulum-cart system shown in Figure P10.9 are

$$(M + m)\ddot{x} + mL\ddot{\theta} = f$$

$$\ddot{x} + L\ddot{\theta} - g\theta = 0$$

Assume that $M = 50$ slugs, $m = 10$ slugs, $L = 30$ ft, and $g = 32.2$ ft/sec$^2$.

(a) Choose the state variables to be $\theta, \dot{\theta}, x, \dot{x}$ and put the preceding model into state-variable form.

(b) Can the pendulum be kept near the vertical if the controlled force $f$ is a linear-feedback function of the variables $\theta$ and $\dot{\theta}$? What happens to the translational motion $y$ with such a control scheme?

(c) Use feedback of all four state variables to keep $\theta$ and $y$ near 0. Find the gains required to place the closed-loop poles at $s = -0.5, -0.5, -0.5 \pm i0.5$.

(d) Consider the full state-feedback scheme of part c. If possible, construct a full state observer with poles at $s = -2, -2, -2 \pm i2$, if the measurement $y$ is one of the following.

(1) $y = x$

(2) $y = \theta$

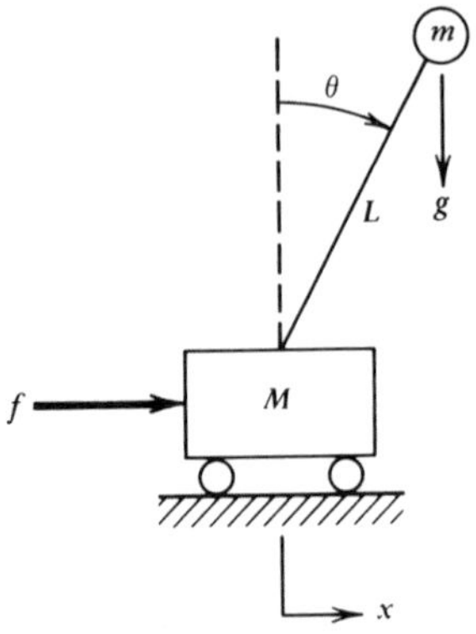

Figure P10.9

**10.10** Given the system model and the performance index

$$\dot{x}_1 = x_2 + u_1$$

$$\dot{x}_2 = u_2$$

$$J = \frac{1}{2}\int_0^\infty (x_1^2 + qx_2^2 + u_1^2 + u_2^2)\,dt$$

(a) Use Kalman's result to show that the closed-loop system will be stable if $\mathbf{u} = -\mathbf{Kx}$ and $\mathbf{K}$ is selected to minimize $J$.

(b) Obtain the differential equations for the elements of the Riccati matrix $\mathbf{P}$ and find $\mathbf{K}$ in terms of these elements.

(c) Let $q = 3$ and solve for $\mathbf{K}$.

**10.11** Given the system model and performance index

$$\dot{x}_1 = x_2$$

$$\dot{x}_2 = -x_2 + u$$

$$J = \frac{1}{2}\int_0^\infty (x_1^2 + qx_2^2 + u^2)\,dt$$

(a) Find the gain matrix $\mathbf{K}$ where $\mathbf{u} = -\mathbf{Kx}$ to minimize $J$.

(b) Compute the dominant time constant and oscillation frequency of the closed-loop system for the following.

(1) $q = 0$

(2) $q = 1$

(3) $q = 97$

**10.12** The roll dynamics of a missile can be characterized by the following equations of motion (Reference 11).

$$\dot{\delta} = u$$

$$\dot{\omega} = -\frac{1}{\tau}\omega + \frac{a}{\tau}\delta$$

$$\dot{\phi} = \omega$$

*Where*

$\phi$ = roll angle

$\omega$ = roll rate

$\delta$ = aileron deflection

$u$ = command signal to the aileron actuators

$\tau$ = roll time constant

$a$ = aileron effectiveness constant

It is desired to maintain $\phi$ near 0. Also physical restrictions on the actuators require that the aileron deflection $\delta$ and the deflection rate $\dot{\delta}$ be kept small. Thus the following index is chosen.

$$J = \frac{1}{2}\int_0^{\infty}\left[\left(\frac{\phi}{\phi_m}\right)^2 + \left(\frac{\delta}{\delta_m}\right)^2 + \left(\frac{u}{u_m}\right)^2\right] dt$$

where $\phi_m$, $\delta_m$, and $u_m$ are the maximum desired values of $\phi$, $\delta$, and $u$.

Let $\tau = 1$ sec, $a = 10\ \text{sec}^{-1}$, $\phi_m = \pi/180$ rad, $\delta_m = \pi/12$ rad, and $u_m = \pi$ rad/sec. Solve for the feedback gains required to minimize $J$.

**10.13** Find the feedback gains necessary for state-variable feedback to bring both states of the following system to zero in two time steps.

$$\mathbf{x}(k+1) = \begin{bmatrix} 0 & 0 \\ 1 & 0.9 \end{bmatrix}\mathbf{x}(k) + \begin{bmatrix} 0.1 \\ 0 \end{bmatrix} u(k)$$

# APPENDIX A
# Some Useful Analytical Techniques

## A.1 THE TAYLOR SERIES EXPANSION

Taylor's theorem states that a function $f(y)$ can be represented in the vicinity of $y = a$ by the expansion

$$f(y) = f(a) + f'(a)(y-a) + \frac{1}{2!}f''(a)(y-a)^2 + \ldots + \frac{1}{k!}f^{(k)}(a)(y-a)^k + \ldots + R_n \tag{A.1-1}$$

where the $k$th derivative evaluated at $y = a$ is denoted by

$$f^{(k)}(a) = \left.\frac{d^k f}{dy^k}\right|_{y=a} \tag{A.1-2}$$

The term $R_n$ is the remainder and is given by

$$R_n = \frac{f^{(n)}(b)(y-a)^n}{n!} \tag{A.1-3}$$

where $b$ lies between $a$ and $y$. These results hold if $f(y)$ has continuous derivatives through order $n$. If $R_n$ approaches zero for large $n$, (A.1-1) is called the Taylor series for $f(y)$ about $y = a$. If $a = 0$, the series is sometimes called the Maclaurin series.

Some common examples of the Taylor series are

$$\sin y = y - \frac{y^3}{3!} + \frac{y^5}{5!} - \frac{y^7}{7!} + \ldots \tag{A.1-4}$$

$$e^y = 1 + y + \frac{y^2}{2!} + \frac{y^3}{3!} + \ldots \tag{A.1-5}$$

where $a = 0$ in both examples.

Generally the series converges for all values of $y$ in some interval called the interval of convergence and diverges for all values of $y$ outside this interval. The accuracy of the expansion depends on how close $y$ is to $a$ and on how many terms are used in the expansion, in addition to the nature of the function in question. If only the first-order term is retained, the function is approximated by a straight line.

$$f(y) \cong f(a) + f'(a)(y-a) \tag{A.1-6}$$

The error is given by the remainder term

$$R_2 = \frac{f''(b)(y-a)^2}{2} \tag{A.1-7}$$

This result is the basis of the technique used to linearize a nonlinear function or nonlinear differential equation. If we define the deviations

$$g = f(y) - f(a) \tag{A.1-8}$$

$$x = y - a \tag{A.1-9}$$

then (A.1-6) can be expressed more simply as

$$g(x) \cong mx, \qquad m = f'(a) \tag{A.1-10}$$

where $m$ is the slope of the straight-line approximation at the reference point $y = a$.

## Multivariable Case

The Taylor series can be generalized to any number of variables. For two variables, it is

$$\begin{aligned} f(y_1, y_2) &= f(a_1, a_2) + f_{y_1}(a_1, a_2)(y_1 - a_1) + f_{y_2}(a_1, a_2)(y_2 - a_2) \\ &\quad + \frac{1}{2!} f_{y_1 y_1}(a_1, a_2)(y_1 - a_1)^2 + \frac{1}{2!} f_{y_1 y_2}(a_1, a_2)(y_1 - a_1)(y_2 - a_2) \\ &\quad + \frac{1}{2!} f_{y_2 y_2}(a_1, a_2)(y_2 - a_2)^2 + \ldots \end{aligned} \tag{A.1-11}$$

where

$$f_{y_i y_j}(a_1, a_2) = \left. \frac{\partial^2 f}{\partial y_i \partial y_j} \right|_{\substack{y_1 = a_1 \\ y_2 = a_2}} \tag{A.1-12}$$

The linearizing approximation is

$$f(y_1, y_2) \cong f(a_1, a_2) + f_{y_1}(a_1, a_2)(y_1 - a_1) + f_{y_2}(a_1, a_2)(y_2 - a_2) \tag{A.1-13}$$

or

$$g(x_1, x_2) \cong m_1 x_1 + m_2 x_2 \tag{A.1-14}$$

*Where*

$$g = f(y_1, y_2) - f(a_1, a_2) \tag{A.1-15}$$

$$x_1 = y_1 - a_1 \tag{A.1-16}$$

$$x_2 = y_2 - a_2 \tag{A.1-17}$$

$$m_1 = f_{y_1}(a_1, a_2) \tag{A.1-18}$$

$$m_2 = f_{y_2}(a_1, a_2) \tag{A.1-19}$$

The linearizing approximation for two variables is thus seen to be a plane passing through the point $(a_1, a_2)$ and having the slopes $m_1$ and $m_2$ in the $y_1$ and $y_2$ directions, respectively.

The two-variable case can easily be generalized to the case with any number of variables. This is represented most easily in matrix notation (see Section A.4 for a review of matrix methods).

## The Matrix Form

In terms of matrix notation, the Taylor series can be expressed as

$$f(\mathbf{y}) = f(\mathbf{a}) + (\mathbf{y} - \mathbf{a})^T \nabla_{\mathbf{y}} f(\mathbf{a}) + \frac{1}{2!} (\mathbf{y} - \mathbf{a})^T \mathbf{H}(\mathbf{a})(\mathbf{y} - \mathbf{a}) + \ldots \tag{A.1-20}$$

where the *gradient* of $f(\mathbf{y})$ with respect to $\mathbf{y}$ evaluated at $\mathbf{y} = \mathbf{a}$ is

$$\nabla_{\mathbf{y}} f(\mathbf{a}) = \frac{\partial f}{\partial \mathbf{y}}(\mathbf{a}) = \begin{bmatrix} \dfrac{\partial f}{\partial y_1} \\ \dfrac{\partial f}{\partial y_2} \\ \vdots \end{bmatrix}_{\mathbf{y}=\mathbf{a}} \doteq \begin{bmatrix} m_1 \\ m_2 \\ \vdots \end{bmatrix} \tag{A.1-21}$$

The vectors $\mathbf{y}$ and $\mathbf{a}$ are column vectors, as is the gradient. (Sometimes the gradient is defined as a row vector; in this case the form of (A.1-20) will be different.) The *Hessian* matrix is

$$\mathbf{H}(\mathbf{a}) = \left[\frac{\partial^2 f}{\partial y_i \partial y_j}\right]_{\mathbf{y}=\mathbf{a}} = \begin{bmatrix} f_{y_1 y_1} & f_{y_1 y_2} & \cdots \\ f_{y_2 y_1} & f_{y_2 y_2} & \cdots \\ \cdot & \cdot & \cdots \end{bmatrix}_{\mathbf{y}=\mathbf{a}} \tag{A.1-22}$$

where $i = 1, 2, \ldots, n$ and $j = 1, 2, \ldots, n$.

The linearization of a set of differential equations requires that a set of functions $f_1, f_2, \ldots,$ be linearized. Let

$$\mathbf{f} = \mathbf{f}(\mathbf{y}) = \begin{bmatrix} f_1(y_1, y_2, \ldots) \\ f_2(y_1, y_2, \ldots) \\ \vdots \end{bmatrix} \tag{A.1-23}$$

With (A.1-20) applied to each component of $\mathbf{f}$, the result is

$$\mathbf{f}(\mathbf{y}) = \mathbf{f}(\mathbf{a}) + \mathbf{J}(\mathbf{a})(\mathbf{y} - \mathbf{a}) + \ldots \tag{A.1-24}$$

where the *Jacobian* matrix is

$$\mathbf{J}(\mathbf{a}) = [J_{ij}] \tag{A.1-25}$$

$$J_{ij} = \left.\frac{\partial f_i}{\partial y_j}\right|_{\mathbf{y}=\mathbf{a}} \tag{A.1-26}$$

In terms of deviation variables,

$$\mathbf{g}(\mathbf{x}) \cong \mathbf{J}(\mathbf{a})\mathbf{x} \tag{A.1-27}$$

where

$$\mathbf{g} = \mathbf{f}(\mathbf{y}) - \mathbf{f}(\mathbf{a}) \tag{A.1-28}$$

$$\mathbf{x} = \mathbf{y} - \mathbf{a} \tag{A.1-29}$$

## A.2 NEWTON ITERATION

A useful application of the Taylor series occurs in the numerical solution of nonlinear equations. The problem is as follows. Find the value $y$ such that

$$f(y) = 0 \tag{A.2-1}$$

where $f(y)$ is some given function. Let $y^*$ be an initial guess. If it is assumed that $y$ is close enough to $y^*$ so that the linearized form (A.1-6) is valid, then

$$f(y) \cong f(y^*) + f'(y^*)(y - y^*) \tag{A.2-2}$$

Using (A.2-1) we can solve for the estimate of the solution

$$y \cong y^* - \frac{f(y^*)}{f'(y^*)} \tag{A.2-3}$$

The solution $y$ from (A.2-3) gives a new $y^*$, and the process is repeated until (A.2-1) is satisfied to the desired degree of accuracy. In difference-equation form, the algorithm is

$$y_k \cong y_{k-1} - \frac{f(y_{k-1})}{f'(y_{k-1})} \tag{A.2-4}$$

The initial guess occurs at $k = 0$ and is $y_0$. The usual criterion for termination is

$$|f(y_k)| < \epsilon \tag{A.2-5}$$

where $\epsilon$ is a small number chosen by the analyst. Often in practice, however, the iteration is stopped when the solution stops changing in the number of significant figures required of the answer.

When the initial guess is reasonably close to the correct value, Newton's method converges very rapidly. It is possible that convergence will not occur, however. One cannot overestimate the importance of sketching the function $f(y)$ in order to obtain a reasonable idea of its shape and the number and locations of its roots. For example, consider the function

$$f(y) = 10y\, e^{-y} - 1 \tag{A.2-6}$$

Assuming that only positive values of $y$ are of interest, we see that $f(0) = -1, f(1) = 2.7$, and $f(y) \to 0$ as $y \to \infty$. From this we can tell that a root lies in the range $0 < y < 1$ and that no other roots occur. Thus a reasonable starting guess would be in this range. For a starting value of $y_0 = 0$, the solution sequence is

$$\begin{aligned} y_0 &= 0 \\ y_1 &= 0.1 \\ y_2 &= 0.1117 \\ y_3 &= 0.1118 \end{aligned}$$

where more significant figures have been retained in the calculation than are shown. The last value gives $f(0.1118) = -0.000258$. If we are satisfied with this accuracy, we can stop. If the guess is farther away, say $y_0 = 0.5$, the convergence will take longer. These results are

$$\begin{aligned} y_0 &= 0.5 \\ y_1 &= -0.1703 \\ y_2 &= 0.0473 \\ y_3 &= 0.1077 \\ y_4 &= 0.1118 \end{aligned}$$

However, there is a peak in the function at $y = 1$. If we guess $y_0 > 1$, the iteration will not converge but will approach $y = \infty$. The reason for this is that the method approximates the function with a straight line whose slope is the slope of $f(y)$ at the current value of $y_k$. Here, for $y > 1$, the slope is always negative, and the method is deceived into thinking that the curve is approaching an axis-crossing (a root), when in fact the function is approaching zero asymptotically. Near the peak at $y = 1$, the slope is very small. Thus for $y_0$ close to but less than 1, the method will at first head for large negative values of $y$ before converging, and more iterations will be required.

Of course, we will not always be able to determine the shape of the function as well as in the previous example but the insights generated from it are important to keep in mind. The two most important applications of Newton's method of interest to us are (1) the solution for polynomial roots required to determine the location of breakaway and break-in points on the root-locus plot (Chapter Eight), and (2) the solution of nonlinear equations that specify the equilibrium points of a nonlinear differential or difference equation. In the root-locus application, we know where the locus can exist on the real axis, and in general where the breakaway and break-in points will occur. Thus we can usually make a good initial guess, and there is no difficulty due to multiple or complex roots. (Newton iteration cannot find complex roots and might converge to the wrong root if more than one exists. It must then be restarted with a new guess.)

### The Matrix Formulation

The equilibrium points of the dynamic model

$$\dot{\mathbf{y}} = \mathbf{f}(\mathbf{y}) \tag{A.2-7}$$

are found from

$$\mathbf{f}(\mathbf{y}) = \mathbf{0} \tag{A.2-8}$$

Newton iteration can be applied to this problem with the use of the truncated matrix series (A.1-24). The extension of (A.2-2) is

$$\mathbf{f}(\mathbf{y}) \cong \mathbf{f}(\mathbf{y}^*) + \mathbf{J}(\mathbf{y}^*)(\mathbf{y} - \mathbf{y}^*) \tag{A.2-9}$$

In light of (A.2-8), the estimate of $\mathbf{y}$ can be obtained from (A.2-9) as

$$\mathbf{y} \cong \mathbf{y}^* - [\mathbf{J}(\mathbf{y}^*)]^{-1}\mathbf{f}(\mathbf{y}^*) \tag{A.2-10}$$

where the inverse of the Jacobian matrix is assumed to exist. This algorithm can be converted to difference-equation form in the same manner as (A.2-4).

## A.3 FOURIER SERIES

The Taylor series is useful for representing a function with an approximate straight line, while the Fourier series is a technique for representing a periodic function as a sum of sines, cosines, and a constant. With such a description, the sinusoidal response of a given linear system can be used with the principle of superposition to obtain the system's response to the periodic function. The concept of system bandwidth is helpful in reducing the number of terms to be considered.

Consider a periodic function $f(y)$ of period $2L$ such that $f(y + 2L) = f(y)$. The

Fourier series representation of this function defined on the interval $c \leqslant y \leqslant c + 2L$, where $c$ and $L > 0$ are constants, is defined as

$$f(y) = \frac{a_0}{2} + \sum_{n=1}^{\infty} \left( a_n \cos \frac{n\pi y}{L} + b_n \sin \frac{n\pi y}{L} \right) \tag{A.3-1}$$

where

$$a_n = \frac{1}{L} \int_c^{c+2L} f(y) \cos \frac{n\pi y}{L} \, dy \tag{A.3-2}$$

$$b_n = \frac{1}{L} \int_c^{c+2L} f(y) \sin \frac{n\pi y}{L} \, dy \tag{A.3-3}$$

If $f(y)$ is defined outside the specified interval by a periodic extension of period $2L$, and if $f(y)$ and $f'(y)$ are piecewise continuous, then the series converges to $f(y)$ if $y$ is a point of continuity, and to the average value $\frac{1}{2}[f(y_+) + f(y_-)]$ otherwise.

For example, a train of unit pulses of width $\pi$ and alternating in sign is described by

$$f(y) = \begin{cases} 1 & 0 < y < \pi \\ -1 & -\pi < y < 0 \end{cases} \tag{A.3-4}$$

Here the period is $2\pi$, so that $L = \pi$. We can take the constant $c$ to be zero. From (A.3-2) and (A.3-3), with the aid of a table of integrals, the coefficients are found to be

$$a_n = 0 \quad \text{for all } n$$

$$b_n = \frac{4}{n\pi} \quad \text{for } n \text{ odd}$$

$$b_n = 0 \quad \text{for } n \text{ even}$$

The resulting Fourier series is

$$f(y) = \frac{4}{\pi} \left( \frac{\sin y}{1} + \frac{\sin 3y}{3} + \frac{\sin 5y}{5} + \ldots \right) \tag{A.3-5}$$

The cosine terms and the constant $a_0$ will not appear in the series if the function is odd; that is, if $f(-y) = -f(y)$. If the function is even, $f(-y) = f(y)$, and no sine terms will appear. If the function is neither even nor odd, both sines and cosines, and perhaps the constant $a_0$, will appear in the series. For some applications it is convenient to define the function outside its basic interval to be either odd or even in order to obtain a special series form, but this technique is of little interest here.

## A.4 VECTORS AND MATRICES

The use of vector-matrix notation allows simplified representation of complex mathematical expressions. For example, a set of many equations can be represented as one vector-matrix equation. This has advantages in presenting analytical concepts and in developing techniques that are applicable to models with an arbitrary number of variables.

The set of linear algebraic equations

$$\left.\begin{aligned} a_{11}x_1 + a_{12}x_2 + \ldots + a_{1n}x_n &= b_1 \\ a_{21}x_1 + a_{22}x_2 + \ldots + a_{2n}x_n &= b_2 \\ \cdots\cdots\cdots\cdots\cdots\cdots\cdots\cdots\cdots& \\ a_{n1}x_1 + a_{n2}x_2 + \ldots + a_{nn}x_n &= b_n \end{aligned}\right\} \quad \text{(A.4-1)}$$

can be represented in compact form as

$$\mathbf{Ax} = \mathbf{b} \quad \text{(A.4-2)}$$

The *matrix* **A** is an array of numbers that correspond in an ordered fashion to the coefficients $a_{ij}$. It is ordered as follows.

$$\mathbf{A} = \begin{bmatrix} a_{11} & a_{12} & \ldots & a_{1n} \\ a_{21} & a_{22} & \ldots & a_{2n} \\ \cdot & \cdot & \cdots & \cdot \\ a_{n1} & a_{n2} & \ldots & a_{nn} \end{bmatrix} \quad \text{(A.4-3)}$$

This matrix has $n$ rows and $n$ columns, so it is a *square* matrix. Its dimension is expressed as $(n \times n)$.

The matrix should not be confused with the *determinant*, which is a single number derived from the rows and columns of (A.4-3) by a prescribed set of reduction rules. A determinant is denoted by a pair of parallel lines; square brackets denote matrices and vectors. The determinant of the matrix **A** and also of the set of equations (A.4-1) is

$$\det(\mathbf{A}) = |\mathbf{A}| = \begin{vmatrix} a_{11} & a_{12} & \ldots & a_{1n} \\ a_{21} & a_{22} & \ldots & a_{2n} \\ \cdot & \cdot & \cdots & \cdot \\ a_{n1} & a_{n2} & \ldots & a_{nn} \end{vmatrix} \quad \text{(A.4-4)}$$

Determinant dimensions are expressed in the same way as matrices.

The $(2 \times 2)$ determinant is evaluated as follows.

$$|\mathbf{A}| = \begin{vmatrix} a_{11} & a_{12} \\ a_{21} & a_{22} \end{vmatrix} = a_{11}a_{22} - a_{12}a_{21} \quad \text{(A.4-5)}$$

Higher-order determinants are defined in a similar manner. Let $M_{ij}$ denote the *minor* of the element $a_{ij}$ and $C_{ij}$ the *cofactor*. The minor is the determinant obtained by crossing out the row and column of the element $a_{ij}$. The cofactor is defined as

$$C_{ij} = (-1)^{i+j}M_{ij} \quad \text{(A.4-6)}$$

A determinant is evaluated by selecting a row or column (preferably the one with the most zero elements), computing the product $a_{ij}C_{ij}$ for each element, and summing the results. In equation form, this may be expressed as follows. For expansion about row $i$,

$$\det(\mathbf{A}) = \sum_{k=1}^{n} a_{ik}C_{ik} \quad \text{(A.4-7)}$$

For expansion about column $j$,

$$\det(\mathbf{A}) = \sum_{k=1}^{n} a_{kj}C_{kj} \quad \text{(A.4-8)}$$

These results are independent of which row or column is selected. A determinant is always square and is said to be *singular* if its value is zero. If it is the determinant of a matrix, the matrix is also said to be singular.

For example, consider the determinant

$$|\mathbf{A}| = \begin{vmatrix} 1 & 0 & -1 \\ 0 & 2 & 5 \\ 1 & 2 & 4 \end{vmatrix} \tag{A.4-9}$$

For expansion about the second row, the minors are

$$M_{21} = \begin{vmatrix} 0 & -1 \\ 2 & 4 \end{vmatrix}, \quad M_{22} = \begin{vmatrix} 1 & -1 \\ 1 & 4 \end{vmatrix}, \quad M_{23} = \begin{vmatrix} 1 & 0 \\ 1 & 2 \end{vmatrix}$$

or $M_{21} = 2$, $M_{22} = 5$, $M_{23} = 2$. The cofactors are $C_{21} = -2$, $C_{22} = 5$, $C_{23} = -2$. The value of the determinant is

$$\det(\mathbf{A}) = 0C_{21} + 2C_{22} + 5C_{23} = 0$$

The determinant is singular.

## Types of Matrices

A *vector* is a special case of a matrix; it has only one row or one column. Thus a *row vector* has the dimension $(1 \times n)$, since it has one row and $n$ columns. A *column vector* has one column and is of dimension $(n \times 1)$. In this text a vector is taken to be a column vector unless otherwise specified. Examples of column vectors are the vectors **x** and **b** in (A.4-1), where

$$\mathbf{x} = \begin{bmatrix} x_1 \\ x_2 \\ \cdot \\ \cdot \\ \cdot \\ x_n \end{bmatrix} \quad \mathbf{b} = \begin{bmatrix} b_1 \\ b_2 \\ \cdot \\ \cdot \\ \cdot \\ b_n \end{bmatrix} \tag{A.4-10}$$

A matrix with only one element is a *scalar*.

A column vector can be converted into a row vector by the *transpose* operation, in which the rows and columns are interchanged. Here we denote this operation by the superscript $T$. For an $(n \times m)$ matrix **A** with $n$ rows and $m$ columns, $\mathbf{A}^T$ (read "*A* transpose") is an $(m \times n)$ matrix. If **A** is given by

$$\mathbf{A} = \begin{bmatrix} 2 & 4 & 1 \\ 5 & 3 & 0 \end{bmatrix}$$

its transpose is

$$\mathbf{A}^T = \begin{bmatrix} 2 & 5 \\ 4 & 3 \\ 1 & 0 \end{bmatrix}$$

If $\mathbf{A}^T = \mathbf{A}$, the matrix is *symmetric*. Only a square matrix can be symmetric.

It is sometimes convenient to consider a matrix to consist of submatrices or vectors. The matrix is said to be *partitioned*. For example, if the matrix given by (A.4-4) is partitioned into column vectors $\mathbf{a}_i$,

$$\mathbf{A} = [\mathbf{a}_1 \quad \mathbf{a}_2 \quad \ldots \quad \mathbf{a}_n] \tag{A.4-11}$$

where the column vectors are

$$\mathbf{a}_1 = \begin{bmatrix} a_{11} \\ a_{21} \\ \cdot \\ \cdot \\ \cdot \\ a_{n1} \end{bmatrix}, \quad \mathbf{a}_2 = \begin{bmatrix} a_{12} \\ a_{22} \\ \cdot \\ \cdot \\ \cdot \\ a_{n2} \end{bmatrix}, \quad \ldots, \quad \mathbf{a}_n = \begin{bmatrix} a_{1n} \\ a_{2n} \\ \cdot \\ \cdot \\ \cdot \\ a_{nn} \end{bmatrix} \tag{A.4-12}$$

A similar procedure is used to partition by rows.

A *diagonal* matrix is one in which all the elements $a_{ij} = 0$ for $i \neq j$. The only non-zero elements can be the elements $a_{ii}$. A diagonal matrix must be square and is automatically symmetric. An example is

$$\mathbf{A} = \begin{bmatrix} 1 & 0 & 0 \\ 0 & -2 & 0 \\ 0 & 0 & 5 \end{bmatrix}$$

The *identity* or *unity* matrix is a square matrix of any order, which is diagonal with all the diagonal elements equal to 1. The symbol **I** is reserved for the identity matrix.

$$\mathbf{I} = \begin{bmatrix} 1 & 0 & 0 & \ldots & 0 \\ 0 & 1 & 0 & \ldots & 0 \\ \cdot & \cdot & \cdot & \ldots & \cdot \\ 0 & \cdot & \cdot & 0 & 1 \end{bmatrix} \tag{A.4-13}$$

The *null* matrix is a matrix of any dimension (not necessarily square) whose elements are all zero. Its symbol is **0**.

## Matrix Addition

Two matrices **A** and **B** are equal to one another if they have the same dimensions and the corresponding elements are equal; that is, if $a_{ij} = b_{ij}$ for every $i$ and $j$.

Two or more matrices can be added if they all have the same dimensions. The resulting matrix is obtained by adding all the corresponding elements; that is,

$$\mathbf{A} + \mathbf{B} = \mathbf{C} \tag{A.4-14}$$

implies that

$$c_{ij} = a_{ij} + b_{ij} \tag{A.4-15}$$

The matrix **C** has the same dimensions as **A** and **B**.

For example,

$$\begin{bmatrix} 2 & 5 \\ 4 & 1 \end{bmatrix} + \begin{bmatrix} 3 & 8 \\ 6 & 7 \end{bmatrix} = \begin{bmatrix} 5 & 13 \\ 10 & 8 \end{bmatrix}$$

Matrix subtraction is defined in a similar way, so that

$$\begin{bmatrix} 2 & 5 \\ 4 & 1 \end{bmatrix} - \begin{bmatrix} 3 & 8 \\ 6 & 7 \end{bmatrix} = \begin{bmatrix} -1 & -3 \\ -2 & -6 \end{bmatrix}$$

Matrix addition and subtraction are both associative and commutative. This means that for addition

$$(\mathbf{A} + \mathbf{B}) + \mathbf{C} = \mathbf{A} + (\mathbf{B} + \mathbf{C}) \tag{A.4-16}$$

$$\mathbf{A} + \mathbf{B} + \mathbf{C} = \mathbf{B} + \mathbf{C} + \mathbf{A} = \mathbf{A} + \mathbf{C} + \mathbf{B} \tag{A.4-17}$$

## Matrix Multiplication

Equation (A.4-2), the matrix form of (A.4-1), implies a multiplication rule for matrices. Consider an $(n \times p)$ matrix **A** and a $(p \times q)$ matrix **B**. The $(n \times q)$ matrix **C** is the product **AB** where

$$c_{ij} = \sum_{k=1}^{p} a_{ik} b_{kj} \tag{A.4-18}$$

for all $i = 1, \ldots, n$ and $j = 1, \ldots, q$. In order for the product to be defined, the matrices **A** and **B** must be *conformable*; that is, the number of rows in **B** must equal the number of columns in **A**.

The algorithm defined by (A.4-18) is easily remembered. Each element in the $i$th row of **A** is multiplied by the corresponding element in the $j$th column of **B**. The sum of the products is the element $c_{ij}$. For example,

$$\begin{bmatrix} 2 & 5 \\ 3 & 6 \\ 8 & 9 \end{bmatrix} \begin{bmatrix} 1 & 4 \\ 0 & 7 \end{bmatrix} = \begin{bmatrix} 2 & 43 \\ 3 & 54 \\ 8 & 95 \end{bmatrix}$$

The element $c_{11}$ is obtained from $2(1) + 5(0) = 2$. The element $c_{32}$ is $8(4) + 9(7) = 95$.

Matrix multiplication does not commute; that is, in general $\mathbf{AB} \neq \mathbf{BA}$. A simple counterexample is all that is needed to be convinced of this fact. Overlooking this property is a common mistake for beginning students.

Since vectors are also matrices, they are multiplied in the fashion shown. You should convince yourself of the equivalence of the two forms (A.4-1) and (A.4-2) at this point.

The only time two matrices need not be conformable for multiplication is when one of them is a scalar. In this case the multiplication is defined as

$$w\mathbf{A} = [wa_{ij}] \tag{A.4-19}$$

where $w$ is a scalar. Multiplying **A** by a scalar $w$ produces a matrix whose elements are found by multiplying each element of **A** by $w$.

The general exceptions to the noncommutative property are the null matrix and the identity matrix. For the latter,

$$\mathbf{IA} = \mathbf{AI} = \mathbf{A} \tag{A.4-20}$$

Verify this fact.

The associative and distributive laws hold for matrix multiplication. The former states that

$$\mathbf{A}(\mathbf{B}+\mathbf{C}) = \mathbf{AB}+\mathbf{AC} \tag{A.4-21}$$

while for the latter

$$(\mathbf{AB})\mathbf{C} = \mathbf{A}(\mathbf{BC}) \tag{A.4-22}$$

The transpose operation affects multiplication and addition in the following ways.

1. $(w\mathbf{A})^T = w\mathbf{A}^T$, where $w$ is a scalar **(A.4-23)**
2. $(\mathbf{A}+\mathbf{B})^T = \mathbf{A}^T + \mathbf{B}^T$ **(A.4-24)**
3. $(\mathbf{AB})^T = \mathbf{B}^T\mathbf{A}^T$ **(A.4-25)**

The last property is particularly important to remember.

## Inverse of a Matrix

The division operation of scalar algebra has an analogous operation in matrix algebra. For example, to solve (A.4-2) for $\mathbf{x}$ in terms of $\mathbf{b}$ and $\mathbf{A}$, we must somehow "divide" $\mathbf{b}$ by $\mathbf{A}$. The procedure for doing this is developed from the concept of a *matrix inverse*. The inverse of a matrix $\mathbf{A}$ is defined only if $\mathbf{A}$ is square and nonsingular; it is denoted by $\mathbf{A}^{-1}$ and has the property that

$$\mathbf{A}^{-1}\mathbf{A} = \mathbf{A}\mathbf{A}^{-1} = \mathbf{I} \tag{A.4-26}$$

Using this property, we multiply both sides of (A.4-2) from the left by $\mathbf{A}^{-1}$ to obtain

$$\mathbf{A}^{-1}\mathbf{A}\mathbf{x} = \mathbf{A}^{-1}\mathbf{b}$$

or

$$\mathbf{x} = \mathbf{A}^{-1}\mathbf{b} \tag{A.4-27}$$

Evidently the solution for $\mathbf{x}$ is given by (A.4-27). We now need to know how to compute $\mathbf{A}^{-1}$.

We state without proof that $\mathbf{A}^{-1}$ is found as follows. The *cofactor* matrix cof($\mathbf{A}$) is the matrix consisting of the cofactors of the matrix $\mathbf{A}$ (see (A.4-6).) The *adjoint* matrix adj($\mathbf{A}$) is the transpose of cof($\mathbf{A}$).

$$\text{adj}(\mathbf{A}) = [\text{cof}(\mathbf{A})]^T = [C_{ij}]^T \tag{A.4-28}$$

The inverse is computed from

$$\mathbf{A}^{-1} = \frac{\text{adj}(\mathbf{A})}{|\mathbf{A}|} \tag{A.4-29}$$

For example, the inverse of

$$\mathbf{A} = \begin{bmatrix} a_{11} & a_{12} \\ a_{21} & a_{22} \end{bmatrix}$$

is

$$\mathbf{A}^{-1} = \frac{1}{a_{11}a_{22} - a_{12}a_{21}} \begin{bmatrix} a_{22} & -a_{12} \\ -a_{21} & a_{11} \end{bmatrix} \tag{A.4-30}$$

Consider another example.

$$\mathbf{A} = \begin{bmatrix} 1 & 1 & 0 \\ -1 & 0 & 3 \\ 2 & 0 & 1 \end{bmatrix}, \qquad |\mathbf{A}| = 7$$

The inverse is

$$\mathbf{A}^{-1} = \tfrac{1}{7}\begin{bmatrix} 0 & 7 & 0 \\ -1 & 1 & 2 \\ 3 & -3 & 1 \end{bmatrix}^T = \begin{bmatrix} 0 & -1/7 & 3/7 \\ 1 & 1/7 & -3/7 \\ 0 & 2/7 & 1/7 \end{bmatrix}$$

These calculations are tedious, and you should check the result to see if $\mathbf{A}^{-1}\mathbf{A} = \mathbf{I}$.

Some important properties of the inverse are

$$(\mathbf{A}^{-1})^{-1} = \mathbf{A} \tag{A.4-31}$$

$$(\mathbf{AB})^{-1} = \mathbf{B}^{-1}\mathbf{A}^{-1} \tag{A.4-32}$$

if the inverses exist. Also, if $\mathbf{A}^{-1}$ exists, then

$$\mathbf{AB} = \mathbf{AC} \tag{A.4-33}$$

implies that $\mathbf{B} = \mathbf{C}$. Otherwise, this is not true in general.

## Solution of Linear Algebraic Equations

With the matrix inverse we can, in theory at least, solve the equation set (A.4-2). For example, suppose the equations to be solved for $x_1, x_2$, and $x_3$ are

$$x_1 + x_2 + 0x_3 = 2$$

$$-x_1 + 0x_2 + 3x_3 = 0$$

$$2x_1 + 0x_2 + x_3 = 1$$

The inverse of the coefficient matrix was computed in the previous example. From (A.4-27) we obtain

$$\mathbf{x} = \begin{bmatrix} x_1 \\ x_2 \\ x_3 \end{bmatrix} = \begin{bmatrix} 0 & -1/7 & 3/7 \\ 1 & 1/7 & -3/7 \\ 0 & 2/7 & 1/7 \end{bmatrix}\begin{bmatrix} 2 \\ 0 \\ 1 \end{bmatrix} = \begin{bmatrix} 3/7 \\ 11/7 \\ 1/7 \end{bmatrix}$$

This procedure can fail in three ways. If the number of elements in the solution vector $\mathbf{x}$ exceeds the number of elements in the vector of the right-hand side $\mathbf{b}$, the matrix $\mathbf{A}$ will not be square and will not possess an inverse. The system of equations is said to be *underdetermined*. Not enough equations have been given to determine a solution. The second case occurs when the dimension of $\mathbf{x}$ is less than that of $\mathbf{b}$, and the system is *overdetermined*. More equations have been given than is necessary to specify a solution. If they are contradictory, no solution will exist. The last case occurs when the number of equations equals the number of unknowns ($\mathbf{A}$ is square), but $|\mathbf{A}| = 0$. This indicates that at least two of the equations are identical; thus not enough information is given to obtain a solution. This case can be recognized when one or more rows of $\mathbf{A}$ can be obtained by multiplying another row by a constant.

It is instructive to consider the case when $\mathbf{b} = \mathbf{0}$. If $\mathbf{A}$ is nonsingular, the system $\mathbf{Ax} = \mathbf{0}$ has only the solution $\mathbf{x} = \mathbf{0}$, and the system $\mathbf{Ax} = \mathbf{b}$ has a unique solution. When $\mathbf{A}$ is singular, the system $\mathbf{Ax} = \mathbf{0}$ has no unique solution, and the system $\mathbf{Ax} = \mathbf{b}$ has either no solution or an infinite number of solutions.

The preceding observations are of importance to the eigenvalue problem discussed in Chapter Nine, Section 9.2. The problem is to find a scalar $\lambda$ and a vector $\mathbf{x}$, given $\mathbf{A}$, such that

$$\mathbf{Ax} = \lambda\mathbf{x} \tag{A.4-34}$$

This can be rewritten as

$$\lambda\mathbf{Ix} - \mathbf{Ax} = \mathbf{0}$$

or

$$(\lambda\mathbf{I} - \mathbf{A})\mathbf{x} = \mathbf{0} \tag{A.4-35}$$

A nonzero solution for $\mathbf{x}$ exists only if

$$\det(\lambda\mathbf{I} - \mathbf{A}) = 0 \tag{A.4-36}$$

This equation is used to find $\lambda$. The solution for $\mathbf{x}$ is nonunique.

## Rank and Linear Independence

A set of $(n \times 1)$ vectors $\mathbf{e}_1, \mathbf{e}_2, \ldots, \mathbf{e}_n$ is said to be *linearly independent* if no constants $c_1, c_2, \ldots, c_n$ exist other than $c_1 = c_2 = \ldots = c_n = 0$, such that

$$c_1\mathbf{e}_1 + c_2\mathbf{e}_2 + \ldots + c_n\mathbf{e}_n = \mathbf{0} \tag{A.4-37}$$

If such a set of constants can be found, the vectors are linearly dependent.

Equation (A.4-37) is equivalent to

$$\mathbf{Ac} = \mathbf{0} \tag{A.4-38}$$

where

$$\mathbf{A} = [\mathbf{e}_1 \quad \mathbf{e}_2 \quad \ldots \quad \mathbf{e}_n] \tag{A.4-39}$$

$$\mathbf{c} = \begin{bmatrix} c_1 \\ c_2 \\ \cdot \\ \cdot \\ \cdot \\ c_n \end{bmatrix} \tag{A.4-40}$$

From (A.4-38) the vector $\mathbf{c}$ will be $\mathbf{0}$ unless $\mathbf{A}$ is singular. Thus the condition for linear independence is that the matrix $\mathbf{A}$ given by (A.4-39) must be nonsingular.

If the vectors are linearly independent, any $(n \times 1)$ vector $\mathbf{x}$ can be represented in terms of a linear combination of the vectors, as

$$\mathbf{x} = \sum_{i=1}^{n} a_i\mathbf{e}_i \tag{A.4-41}$$

where the $a_i$ are the scalar components of $\mathbf{x}$ with respect to the $\mathbf{e}_i$ vectors. The most familiar example is the set of mutually perpendicular three-dimensional vectors.

$$\mathbf{e}_1 = \begin{bmatrix} 1 \\ 0 \\ 0 \end{bmatrix} \quad \mathbf{e}_2 = \begin{bmatrix} 0 \\ 1 \\ 0 \end{bmatrix} \quad \mathbf{e}_3 = \begin{bmatrix} 0 \\ 0 \\ 1 \end{bmatrix} \tag{A.4-42}$$

Vectors need not be perpendicular to be linearly independent. The extension of the perpendicularity concept to $n$-dimensional space is orthogonality. Two vectors $\mathbf{x}$ and $\mathbf{y}$ are *orthogonal* if $\mathbf{x}^T\mathbf{y} = 0$. A vector is a *unit vector* if its length is unity. The length of a vector $\mathbf{x}$ can be found from $\sqrt{\mathbf{x}^T\mathbf{x}}$ if the elements of $\mathbf{x}$ are measured relative to an orthogonal coordinate system. The length formula is a generalization of the Pythagorean theorem to $n$-dimensional space.

The *rank* of a matrix $\mathbf{A}$ is the maximum number of linearly independent columns of $\mathbf{A}$. Equivalently, it is the dimension of the largest nonsingular matrix contained in $\mathbf{A}$. Let $R(\mathbf{A})$ denote the rank of $\mathbf{A}$. We note that

$$R(\mathbf{A}^T) = R(\mathbf{A}) \tag{A.4-43}$$

$$R(\mathbf{A}) = R(\mathbf{A}^T\mathbf{A}) = R(\mathbf{A}\mathbf{A}^T) \tag{A.4-44}$$

The rank of $\mathbf{A}$ in (A.4-2) indicates the number of equations, if any, that are redundant. If $R(\mathbf{A}) = n - 1$, one equation is redundant, if $R(\mathbf{A}) = n - 2$, two are redundant, etc. The rank is also useful in determining the controllability and observability of a system (see Chapter Nine, Section 9.4).

### Vector-Matrix Calculus

The operations of differentiation and integration of a matrix with respect to an independent variable are applied to each element of the matrix. Thus

$$\dot{\mathbf{x}} = \begin{bmatrix} \dot{x}_1 \\ \dot{x}_2 \\ \cdot \\ \cdot \\ \cdot \\ \dot{x}_n \end{bmatrix} \qquad \int \mathbf{x}\,dt = \begin{bmatrix} \int x_1\,dt \\ \int x_2\,dt \\ \cdot \\ \cdot \\ \cdot \\ \int x_n\,dt \end{bmatrix}$$

Differentiation of a product is performed as follows.

$$\frac{d}{dt}(\mathbf{AB}) = \mathbf{A}\frac{d\mathbf{B}}{dt} + \frac{d\mathbf{A}}{dt}\mathbf{B} \tag{A.4-45}$$

$$\frac{d}{dt}(w\mathbf{A}) = w\frac{d\mathbf{A}}{dt} + \frac{dw}{dt}\mathbf{A} \tag{A.4-46}$$

where $w$ is a scalar.

The derivatives of a scalar with respect to a vector are the gradient vector and the Hessian matrix for the first- and second-order derivatives. The first derivative of a vector with respect to another vector is the Jacobian matrix. These were discussed in Section A.1.

## A.5 COMPLEX NUMBER ALGEBRA

Complex numbers are ones that have a real part and an imaginary part. The primary way of representing a complex number is in the form

$$s = a + ib \tag{A.5-1}$$

where $s$ is the complex number, $a$ is its real part, $b$ is the imaginary part ($b$ is a real number), and $i = \sqrt{-1}$. The abbreviations *Re* and *Im* are used to denote real and imaginary parts, respectively. Thus, $a = Re(s)$ and $b = Im(s)$. The number $\bar{s} = a - ib$ is the *complex conjugate* of $a + ib$.

## Operations on Complex Numbers

One advantage of the representation (A.5-1) is that it allows the rules of ordinary algebra to be applied to complex numbers. To do this, consider $a + ib$ to be like any algebraic quantity. Let $s_1 = a_1 + ib_1$, and $s_2 = a_2 + ib_2$. Addition of $s_1$ and $s_2$ is performed as follows.

$$\begin{aligned} s_1 + s_2 &= a_1 + ib_1 + a_2 + ib_2 \\ &= (a_1 + a_2) + i(b_1 + b_2) \end{aligned} \tag{A.5-2}$$

where it is the usual practice to group the resulting real and imaginary parts. Subtraction is done in a similar way.

The product of $s_1$ and $s_2$ is

$$\begin{aligned} s_1 s_2 &= (a_1 + ib_1)(a_2 + ib_2) \\ &= (a_1 a_2 - b_1 b_2) + i(a_1 b_2 + a_2 b_1) \end{aligned} \tag{A.5-3}$$

where we have used the fact that $i^2 = -1$.

The ratio obtained by division can be simplified by multiplying top and bottom by the complex conjugate of the denominator.

$$\frac{s_1}{s_2} = \frac{a_1 + ib_1}{a_2 + ib_2}\,\frac{a_2 - ib_2}{a_2 - ib_2} = \frac{(a_1 a_2 + b_1 b_2) + i(a_2 b_1 - a_1 b_2)}{a_2^2 + b_2^2} \tag{A.5-4}$$

From the denominator result, note that the product of a number with its conjugate is the sum of the squares of the real and imaginary parts.

Two complex numbers are equal if and only if their real parts are equal and their imaginary parts are equal; that is, $s_1 = s_2$ if and only if $a_1 = a_2$ and $b_1 = b_2$.

## Alternate Representations

The representation given in (A.5-1) is the *rectangular* form. Two other forms are commonly used. Each has its own advantages.

The *polar* form treats the complex number as a two-dimensional vector whose components are the real and imaginary parts. This is shown in Figure A.1. The *modulus* $r$ is the length of the vector and is the magnitude of the complex number. Its angle is $\phi$. From the figure we see that

$$r = \sqrt{a^2 + b^2} \tag{A.5-5}$$

$$\phi = \tan^{-1} \frac{b}{a} \tag{A.5-6}$$

The polar representation is

$$s = r(\cos \phi + i \sin \phi) \tag{A.5-7}$$

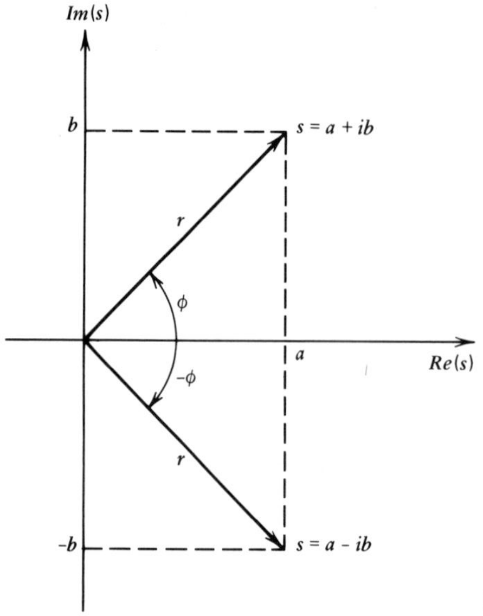


**Figure A.1 Polar representation of complex numbers.**

Note that the complex conjugate is the mirror image of $s$ about the real axis. It has the same modulus, but its angle is opposite in sign. Thus

$$\bar{s} = r(\cos\phi - i\sin\phi) \tag{A.5-8}$$

The reciprocal of $s$ is

$$\frac{1}{s} = \frac{1}{a+ib} = \frac{\bar{s}}{r^2} = \frac{1}{r^2}(a - ib)$$

$$= \frac{1}{r}(\cos\phi - i\sin\phi) \tag{A.5-9}$$

Thus the reciprocal lies along the complex conjugate but has a magnitude of $1/r$.

The Taylor series expansion can be applied to functions of complex numbers. For example,

$$e^s = e^{(a+ib)} = e^a e^{ib}$$

Since $a$ is real we can evaluate $e^a$, but as yet we do not know how to evaluate $e^{ib}$. To see how this is done, use (A.1-5).

$$e^{ib} = 1 + ib + \frac{1}{2}(ib)^2 + \frac{1}{3!}(ib)^3 + \frac{1}{4!}(ib)^4 + \ldots$$

$$= 1 - \frac{1}{2}b^2 + \frac{1}{4!}b^4 - \ldots + i\left(b - \frac{1}{3!}b^3 + \ldots\right)$$

From (A.1-4) the imaginary part can be recognized as $\sin b$. The real part is the Taylor series for $\cos b$. Thus

$$e^{ib} = \cos b + i\sin b \tag{A.5-10}$$

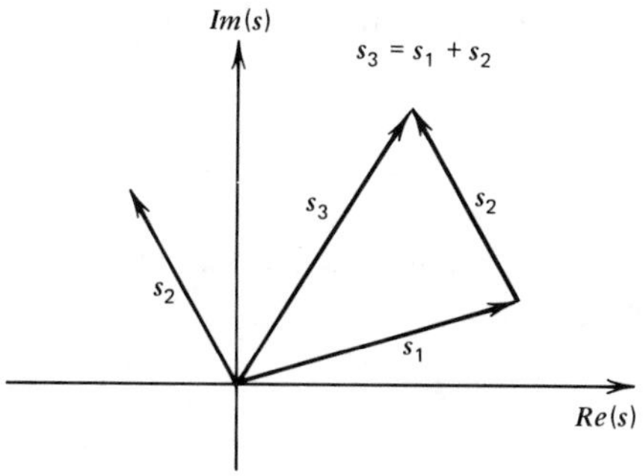


**Figure A.2 Vector addition of two numbers.**

This is *Euler's identity* and is frequently used in the text. Letting $(-b)$ replace $b$, we obtain

$$e^{-ib} = \cos b - i \sin b \tag{A.5-11}$$

With Euler's identity we can derive the *complex exponential* representation of a complex number. From (A.5-7)

$$s = re^{i\phi} \tag{A.5-12}$$

From (A.5-8) and (A.5-9)

$$\bar{s} = re^{-i\phi} \tag{A.5-13}$$

$$\frac{1}{s} = \frac{1}{r}e^{-i\phi} \tag{A.5-14}$$

The exponential form greatly simplifies multiplication and division.

$$s_1 s_2 = r_1 e^{i\phi_1} r_2 e^{i\phi_2} = r_1 r_2 e^{i(\phi_1 + \phi_2)} \tag{A.5-15}$$

$$\frac{s_1}{s_2} = \frac{r_1 e^{i\phi_1}}{r_2 e^{i\phi_2}} = \frac{r_1}{r_2} e^{i(\phi_1 - \phi_2)} \tag{A.5-16}$$

## Vector Interpretation

A concept that is useful in the root-locus method is the vector addition and subtraction of complex numbers. Equation (A.5-2) shows that two complex numbers can be added in component form. The vector interpretation is shown in Figure A.2, where $s_3 = s_1 + s_2$. This is the parallelogram law of vector addition. The law for subtraction is

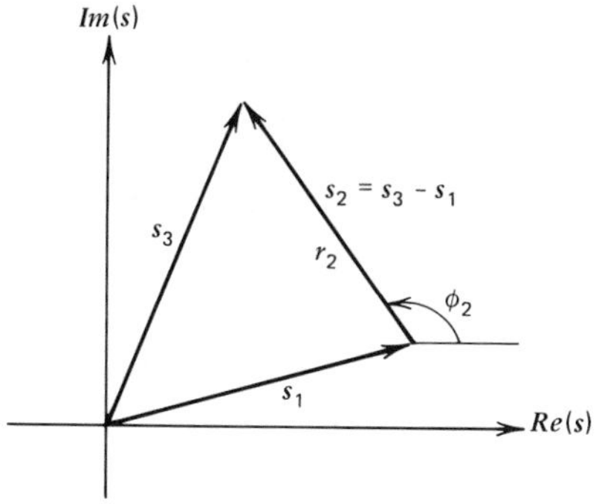


**Figure A.3 Vector subtraction of two numbers.**

shown in Figure A.3 and can be obtained from Figure A.2. From this figure note that

$$s_2 = s - s_1 = r_2 e^{i\phi_2} \tag{A.5-17}$$

## DeMoivre's Theorem

The analysis of discrete-time systems requires that complex numbers be raised to a power. DeMoivre's theorem provides the means. It states that

$$s^p = r^p e^{ip\phi} = r^p(\cos p\phi + i \sin p\phi) \tag{A.5-18}$$

and is a consequence of Euler's identity.

# APPENDIX B
# The Laplace Transform

The Laplace transform is a technique for converting a functional dependence on one variable, say time, into a functional dependence on another variable, denoted by $s$. Just as the logarithmic transformation makes the operations of multiplication and division easier to perform, so the Laplace transform makes operations such as differentiation and integration easier to perform in the $s$-domain. Here we give a brief introduction to the topic sufficient for the purposes of this text. The application of Laplace transforms to the solution of differential equations is treated in Chapters Three, Five, and Nine for first-, second-, and higher-order equations, respectively.

The definition of the Laplace transform of the function $f(t)$ is

$$\mathscr{L}[f(t)] = F(s) = \int_0^\infty f(t)e^{-st}\,dt \tag{B-1}$$

where $s$ is in general a complex number. The definition presumes that the integral will exist for some range of values of $s$. The question of existence is usually no problem for our purposes; the functions of interest to us have well-defined transforms. Equation (B-1) is the definition of the *one-sided* transform. It assumes that the function $f(t)$ is zero for $t < 0$. This assumption is satisfied for our applications, and we will not need the two-sided transform, which has a lower limit of $t = -\infty$.

## Transforms of Common Functions

A table of Laplace transforms for common functions can easily be generated by application of the definition (B-1). Consider the transform of a constant $f(t) = A$ for $t \geqslant 0$. From (B-1)

$$\mathscr{L}(A) = F(s) = \int_0^\infty Ae^{-st}\,dt = A\int_0^\infty e^{-st}\,dt = \left.\frac{-Ae^{-st}}{s}\right|_{t=0}^{t=\infty}$$

The integration gives

$$\mathscr{L}(A) = F(s) = \frac{A}{s} \tag{B-2}$$

if we assume that the real part of $s$ is greater than zero so that the limit of $e^{-st}$ exists as $t \to \infty$.

For the *pulse function*, defined as

$$f(t) = \begin{cases} A & \text{for } 0 \leqslant t \leqslant T \\ 0 & \text{elsewhere} \end{cases} \tag{B-3}$$

the transform is

$$F(s) = \int_0^\infty Ae^{-st}\,dt = -\frac{Ae^{-st}}{s}\bigg|_{t=0}^{t=T}$$

$$= A\frac{(1-e^{-sT})}{s} \tag{B-4}$$

The area under the pulse function is $AT$. If we let this area remain constant at the value $B$ and let the pulse width $T$ approach zero, we obtain the *impulse function.* Its transform is thus

$$F(s) = \lim_{T\to 0}\left[\frac{B(1-e^{-sT})}{sT}\right]$$

From L'Hôpital's limit rule,

$$F(s) = \lim_{T\to 0}\left[\frac{Bse^{-sT}}{s}\right] = B \tag{B-5}$$

If $B = 1$, the function is a *unit impulse.*

The transforms for other common functions are obtained similarly, with the help of integration theorems and a table of integrals. Table B.1 lists the results.

## Properties of the Transform

The Laplace transform has several properties that are useful. These are listed in Table B.2. The transformation is a linear operation because

$$\mathcal{L}[af_1(t) + bf_2(t)] = a\mathcal{L}[f_1(t)] + b\mathcal{L}[f_2(t)] \tag{B-6}$$

if $a$ and $b$ are constants. This property follows from the linearity property of the integral in (B-1). Note that the property has two elements; namely, that the transform of a sum equals the sum of the transforms, and that the transform of a constant times a function equals the constant times the function's transform.

The initial and final value theorems are useful for evaluating the initial and steady-state values of a function. If the limit of $f(t)$ as $t \to \infty$ exists, the final value theorem states that

$$f(\infty) = \lim_{s\to 0} sF(s) \tag{B-7}$$

For a function whose limit exists as $t$ approaches zero from positive values of time, the initial value theorem gives

$$f(0+) = \lim_{s\to\infty} sF(s) \tag{B-8}$$

The proofs are straightforward applications of integration theorems and are given in most texts dealing with advanced engineering mathematics.

Solution of differential equations with Laplace transforms requires the following differentiation and integration properties. Let $f(t) = dx/dt$, where $x = x(t)$. From (B-1)

$$\mathcal{L}\left(\frac{dx}{dt}\right) = \int_0^\infty \frac{dx}{dt}e^{-st}\,dt = x(t)e^{-st}\bigg|_{t=0}^{t=\infty} + s\int_0^\infty x(t)e^{-st}\,dt$$

where we have used integration by parts. If $x(t)$ is transformable, this gives

$$\mathcal{L}\left(\frac{dx}{dt}\right) = sX(s) - x(0) \tag{B-9}$$

**TABLE B.1 Laplace Transform Pairs**

| $F(s)$ | $f(t), t \geqslant 0$ |
|---|---|
| 1. $1$ | $\delta(t)$, the unit impulse at $t = 0$ |
| 2. $\dfrac{1}{s}$ | 1, the unit step |
| 3. $\dfrac{n!}{s^{n+1}}$ | $t^n$ |
| 4. $\dfrac{1}{s+a}$ | $e^{-at}$ |
| 5. $\dfrac{1}{(s+a)^n}$ | $\dfrac{1}{(n-1)!} t^{n-1} e^{-at}$ |
| 6. $\dfrac{a}{s(s+a)}$ | $1 - e^{-at}$ |
| 7. $\dfrac{1}{(s+a)(s+b)}$ | $\dfrac{1}{(b-a)}(e^{-at} - e^{-bt})$ |
| 8. $\dfrac{s+p}{(s+a)(s+b)}$ | $\dfrac{1}{(b-a)}[(p-a)e^{-at} - (p-b)e^{-bt}]$ |
| 9. $\dfrac{1}{(s+a)(s+b)(s+c)}$ | $\dfrac{e^{-at}}{(b-a)(c-a)} + \dfrac{e^{-bt}}{(c-b)(a-b)} + \dfrac{e^{-ct}}{(a-c)(b-c)}$ |
| 10. $\dfrac{s+p}{(s+a)(s+b)(s+c)}$ | $\dfrac{(p-a)e^{-at}}{(b-a)(c-a)} + \dfrac{(p-b)e^{-bt}}{(c-b)(a-b)} + \dfrac{(p-c)e^{-ct}}{(a-c)(b-c)}$ |
| 11. $\dfrac{b}{s^2+b^2}$ | $\sin bt$ |
| 12. $\dfrac{s}{s^2+b^2}$ | $\cos bt$ |
| 13. $\dfrac{b}{(s+a)^2+b^2}$ | $e^{-at} \sin bt$ |
| 14. $\dfrac{s+a}{(s+a)^2+b^2}$ | $e^{-at} \cos bt$ |
| 15. $\dfrac{\omega_n^2}{s^2 + 2\zeta\omega_n s + \omega_n^2}$ | $\dfrac{\omega_n}{\sqrt{1-\zeta^2}} e^{-\zeta\omega_n t} \sin \omega_n\sqrt{1-\zeta^2}\, t, \quad \zeta < 1$ |
| 16. $\dfrac{\omega_n^2}{s(s^2 + 2\zeta\omega_n s + \omega_n^2)}$ | $1 + \dfrac{1}{\sqrt{1-\zeta^2}} e^{-\zeta\omega_n t} \sin(\omega_n\sqrt{1-\zeta^2}\, t + \phi)$<br>$\phi = \tan^{-1} \dfrac{\sqrt{1-\zeta^2}}{\zeta} + \pi$<br>(third quadrant) |

TABLE B.2 Properties of the Laplace Transform

| $f(t)$ | $F(s) = \int_0^\infty f(t)e^{-st}\,dt$ |
|---|---|
| 1. $af_1(t) + bf_2(t)$ | $aF_1(s) + bF_2(s)$ |
| 2. $\dfrac{df}{dt}$ | $sF(s) - f(0)$ |
| 3. $\dfrac{d^2f}{dt^2}$ | $s^2F(s) - sf(0) - \left.\dfrac{df}{dt}\right\vert_{t=0}$ |
| 4. $\dfrac{d^nf}{dt^n}$ | $s^nF(s) - \sum_{k=1}^{n} s^{n-k}g_{k-1}$ <br> $g_{k-1} = \left.\dfrac{d^{k-1}f}{dt^{k-1}}\right\vert_{t=0}$ |
| 5. $\int_0^t f(t)\,dt$ | $\dfrac{F(s)}{s} + \dfrac{h(0)}{s}$ <br> $h(0) = \left.\int f(t)\,dt\right\vert_{t=0}$ |
| 6. $\begin{Bmatrix} 0, & t < D \\ f(t-D), & t \geqslant D \end{Bmatrix}$ | $e^{-sD}F(s)$ |
| 7. $e^{-at}f(t)$ | $F(s+a)$ |
| 8. $f\left(\dfrac{t}{a}\right)$ | $aF(as)$ |
| 9. $f(t) = \int_0^t x(t-\tau)y(\tau)\,d\tau = \int_0^t y(t-\tau)x(\tau)\,d\tau$ | $F(s) = X(s)Y(s)$ |
| 10. $f(\infty) = \lim_{s\to 0} sF(s)$ | |
| 11. $f(0+) = \lim_{s\to\infty} sF(s)$ | |

Extending this procedure to the second derivative, we obtain

$$\mathscr{L}\left(\frac{d^2x}{dt^2}\right) = s^2X(s) - sx(0) - \left.\frac{dx}{dt}\right|_{t=0} \tag{B-10}$$

The initial value of the derivative is commonly abbreviated as $\dot{x}(0)$. In general,

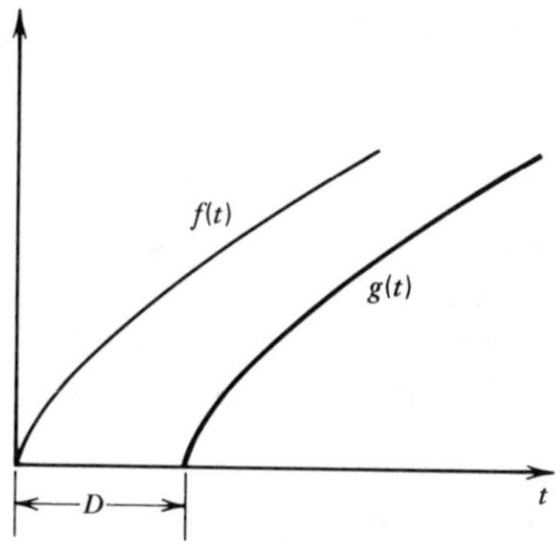


**Figure B.1 Translated function.**

$$\mathscr{L}\left(\frac{d^n x}{dt^n}\right) = s^n X(s) - \sum_{k=1}^{n} s^{n-k} g_{k-1} \quad \text{(B-11)}$$

where

$$g_{k-1} = \left.\frac{d^{k-1}x}{dt^{k-1}}\right|_{t=0}$$

The following property can be derived by applying integration by parts to the transform of $\int f(t)\,dt$. The result is

$$\mathscr{L}\left[\int f(t)\,dt\right] = \frac{F(s)}{s} + \frac{h(0)}{s} \quad \text{(B-12)}$$

where $h(0) = \int f(t)\,dt$ evaluated at $t = 0+$.

The transform of a time-shifted function is useful for discrete-time systems analysis and for systems with dead-time elements. The shifted, or translated, function is defined as

$$g(t) = \begin{cases} 0 & \text{for} \quad t < D \\ f(t-D) & \text{for} \quad t \geqslant D \end{cases} \quad \text{(B-13)}$$

where $f(t)$ is some arbitrary but transformable function. This is shown in Figure B.1. The transform is

$$\mathscr{L}[g(t)] = \int_0^\infty f(t-D)e^{-st}\,dt = \int_D^\infty f(t-D)e^{-st}\,dt$$

$$= \int_0^\infty f(q)e^{-s(q+D)}\,dq = e^{-sD}\int_0^\infty f(q)e^{-sq}\,dq$$

where the integration variable was changed to $q = t - D$. Thus

$$\mathscr{L}[g(t)] = e^{-sD}F(s) \quad \text{(B-14)}$$

If $u(t)$ is the unit step function,

$$u(t) = \begin{cases} 0 & \text{for} \quad t < 0 \\ 1 & \text{for} \quad t \geqslant 0 \end{cases} \quad \text{(B-15)}$$

we note that the translated function $g(t)$ can be expressed as $f(t-D)u(t-D)$.

The final property of use to us is the convolution theorem. If $F(s) = X(s)Y(s)$ and the time function $f(t)$ corresponding to $F(s)$ is written as

$$f(t) = \mathscr{L}^{-1}[F(s)] \quad \text{(B-16)}$$

the theorem states that

$$f(t) = \int_0^t x(t-\tau)y(\tau)\,d\tau$$

$$= \int_0^t y(t-\tau)x(\tau)\,d\tau \quad \text{(B-17)}$$

The application of the convolution theorem to the analysis of linear dynamic systems is described in Chapter Three, along with its derivation.

## The Inverse Transformation and Partial Fraction Expansion

When the transform of the solution to a differential equation has been obtained, it is necessary to invert the transformation to obtain the solution as a function of time. Unless the transform is a simple one appearing in the transform table, it will have to be represented as a combination of simple transforms. Most transforms occur in the form of a ratio of two polynomials, such as

$$F(s) = \frac{N(s)}{D(s)} = \frac{b_m s^m + b_{m-1}s^{m-1} + \ldots + b_1 s + b_0}{s^n + a_{n-1}s^{n-1} + \ldots + a_1 s + a_0} \tag{B-18}$$

If this transform results from a dynamic system, integral causality (Chapter One) requires that $m \leqslant n$. If $F(s)$ is of the form in (B-18), the method of partial fraction expansion can be used. We assume that the coefficient $a_n$ has been absorbed into the $b_i$ coefficients. The first step is to solve for the $n$ roots of the denominator (several methods for doing this are discussed in the text). If the $a_i$ coefficients are real, the roots will fall into one of the following categories.

1. Real and distinct.
2. Real and repeated.
3. Complex conjugate pairs.

If all the roots are real and distinct, we can express $F(s)$ in factored form as follows.

$$F(s) = \frac{N(s)}{(s+r_1)(s+r_2)\ldots(s+r_n)} \tag{B-19}$$

where the roots are $s = -r_1, -r_2, \ldots, -r_n$. This form can be expanded as

$$F(s) = \frac{C_1}{(s+r_1)} + \frac{C_2}{(s+r_2)} + \ldots + \frac{C_n}{(s+r_n)} \tag{B-20}$$

where

$$C_i = \lim_{s \to -r_i} [F(s)(s+r_i)] \tag{B-21}$$

Multiplication by the factor $(s + r_i)$ cancels that term in the denominator before the limit is taken. This is a good way to remember (B-21). Each factor corresponds to an exponential function of time, and the solution is

$$f(t) = C_1 e^{-r_1 t} + C_2 e^{-r_2 t} + \ldots + C_n e^{-r_n t} \tag{B-22}$$

Now suppose that $p$ of the roots have the same value, $s = -r_1$, and that the remaining $(n-p)$ roots are real and distinct. Then $F(s)$ is of the form

$$F(s) = \frac{N(s)}{(s+r_1)^p (s+r_{p+1})(s+r_{p+2})\ldots(s+r_n)} \tag{B-23}$$

The expansion is

$$F(s) = \frac{C_1}{(s+r_1)^p} + \frac{C_2}{(s+r_1)^{p-1}} + \ldots + \frac{C_p}{s+r_1} + \ldots + \frac{C_{p+1}}{s+r_{p+1}} + \ldots + \frac{C_n}{s+r_n} \tag{B-24}$$

The coefficients for the repeated roots are found from

$$C_1 = \lim_{s \to -r_1} [F(s)(s + r_1)^p] \tag{B-25}$$

$$C_2 = \lim_{s \to -r_1} \left\{ \frac{d}{ds} [F(s)(s + r_1)^p] \right\} \tag{B-26}$$

$$\vdots$$

$$C_i = \lim_{s \to -r_1} \left\{ \frac{1}{(i-1)!} \frac{d^{i-1}}{ds^{i-1}} [F(s)(s + r_1)^p] \right\} \qquad i = 1, 2, \ldots, p \tag{B-27}$$

The coefficients for the distinct roots are found from (B-21). The solution for the time function is

$$f(t) = C_1 \frac{t^p}{p!} e^{-r_1 t} + C_2 \frac{t^{p-1}}{(p-1)!} e^{-r_1 t} + \ldots + C_p e^{-r_1 t} + \ldots$$
$$+ C_{p+1} e^{-r_{p+1} t} + \ldots + C_n e^{-r_n t} \tag{B-28}$$

When the roots are complex conjugates the expansion has the same form as (B-20), since the roots are in fact distinct. The coefficients $C_i$ can be found from (B-21). However, these will be complex numbers, and the solution form given by (B-22) will not be convenient to use. Euler's identity

$$e^{i\theta} = \cos\theta + i \sin\theta \tag{B-29}$$

can be used to convert the solution into a usable form consisting of exponentially damped sines and cosines. This procedure is discussed in more detail in Section 5.7.

When the roots contain a mixture of the three root types the partial fraction expansion is a combination of the preceding forms, as was done in (B-24). Examples are given in Chapter Three, Sections 3.3 and 3.4, and in Chapter Five, Sections 5.7 and 5.8.

# APPENDIX C
# A Runge-Kutta Subroutine

The fourth-order Runge-Kutta method is widely used in engineering applications because of its accuracy and ease with which it can be applied. It does not require a starting solution, unlike some other powerful methods. Runge-Kutta algorithms are currently available in magnetic card and in solid-state memory modules for programmable hand-held calculators. These are generally limited to a system of third order or less. For higher-order systems, or for applications in which the output is to be plotted, the following FORTRAN subroutine can be used. Many computer installations have a graphic plotting routine. If not, one such program is given in Reference C.1.

The following subroutine was adapted by F. M. White of the University of Rhode Island from a similar routine in use with the MAD language at the University of Michigan.* It is an implementation of Gill's method and has been found to be very accurate for a wide variety of problems.

The user is required only to write the expressions for the derivatives, with his model expressed in state variable form. His must select the step size $H$, and this value can be varied by the programmer during the run. The subroutine is called by the statement

CALL RUNGE(N, Y, F, T, H, M, K)

The arguments N to K may be described in order of appearance.

N = Number of differential equations to be solved (set by the programmer).

Y = Array of N dependent variables (with initial values set by the programmer).

F = Array of the N derivatives of the variables Y [the programmer must provide an expression in the main program for the calculation of each F(I)].

T = Independent variable (initialized by the programmer).

H = Step size $\Delta T$ (set by the programmer).

M = Index used in the subroutine that must be set equal to zero by the programmer before the first CALL.

K = Integer from the subroutine that is used as the argument of a computed GO TO statement in the main program, such as GO TO (10,20), K. Statement 10 calculates the derivatives F(I) and statement 20 prints the answers, T and Y(I).

For example, a schematic main program that makes use of subroutine RUNGE might look like this.

---

* F. M. White, *Viscous Fluid Flow*, McGraw-Hill, New York, 1974. Used with permission.

```
      DIMENSION Y(10), F(10)
    4 FORMAT(3F10.0, 2I5/(8F10.0))
    5 FORMAT(1X, 6E15.8)
      READ (5, 4) T, TLIM, H, M, N, (Y(I), I = 1, N)
    8 IF(T-TLIM) 6, 6, 7
    6 CALL RUNGE (N, Y, F, T, H, M, K)
      GO TO (10, 20), K
   10 F(1) = the derivative of Y(1)
      F(2) = the derivative of Y(2)
      .......................
      F(N) = the derivative of Y(N)
      GO TO 6
   20 WRITE (6, 5)T, (Y(I), I = 1, N)
      GO TO 8
    7 STOP
      END
```

It is only necessary that one place the proper expressions for the derivatives in the statements for F(I) = . . . .

Note that it is not necessary to increment T in the main program. This is automatically done correctly in the subroutine. One could change H at any time by inserting a logic expression for changing H underneath the WRITE statement (statement 20).

In this example, Y and F are dimensioned only to a size of 10. There is no inherent reason why more equations could not be handled. The only limitations would be accuracy (round-off error) and computer memory space. It would only be necessary to redimension Y and F in both the main program and subroutine and to redimension Q in the subroutine. It is not necessary to have the same dimensions in both the main program and subroutine as long as dimensions in the subroutine equal or exceed those in the main program.

A listing of the subroutine follows.

```
      SUBROUTINE RUNGE (N, Y, F, T, H, M, K)
C     THIS ROUTINE PERFORMS RUNGE-KUTTA CALCULATION
C     BY GILLS METHOD
      DIMENSION Y(10), F(10), Q(10)
      M = M + 1
      GO TO (1, 4, 5, 3, 7), M
    1 DO 2 I = 1, N
    2 Q(I) = 0
      A = .5
      GO TO 9
    3 A = 1.707107
C     IF YOU NEED MORE ACCURACY, USE
C     A = 1.7071067811865475244
    4 T = T + .5*H
    5 DO 6 I = 1, N
```

```
         Y(I) = Y(I) + A*(F(I)*H - Q(I))
       6 Q(I) = 2.*A*H*F(I) + (1. - 3.*A)*Q(I)
         A = .2928932
C        IF YOU NEED MORE ACCURACY SET
C        A = .29289321881345247 56
         GO TO 9
       7 DO 8 I = 1, N
       8 Y(I) = Y(I) + H*F(I)/6. - Q(I)/3.
         M = 0
         K = 2
         GO TO 10
       9 K = 1
      10 RETURN
         END
```

Some examples will illustrate the use of the program.

*Example C.1*

Consider the $n$th-order equation

$$\frac{d^n y}{dt^n} = f\left(t, y, \frac{dy}{dt}, \frac{d^2 y}{dt^2}, \ldots, \frac{d^{n-1} y}{dt^{n-1}}\right) \tag{C-1}$$

One way of converting into state variable form is to define

$$\begin{aligned} y_1 &= \frac{d^{n-1} y}{dt^{n-1}} \\ y_2 &= \frac{d^{n-2} y}{dt^{n-2}} \\ &\vdots \\ y_{n-1} &= \frac{dy}{dt} \\ y_n &= y \end{aligned}$$

Then with the dot notation for derivatives, (C-1) becomes

$$\begin{aligned} \dot{y}_1 &= f(t, y_n, y_{n-1}, \ldots, y_1) \\ \dot{y}_2 &= y_1 \\ \dot{y}_3 &= y_2 \\ &\vdots \\ \dot{y}_n &= y_{n-1} \end{aligned}$$

The functions to be inserted into the main program are

```
F(1) = f[T, Y(N), Y(N-1), Y(N-2), . . . , Y(2), Y(1)]
F(2) = Y(1)
F(3) = Y(2)
  .
  .
  .
F(N-1) = Y(N-2)
F(N) = Y(N-1)
```

Of course, any other choice of state variables can be used.

*Example C.2*

The subroutine can also be used to compute integrals. For example,

$$z = \int\int f(t)\,dt$$

is equivalent to

$$\ddot{z} = f(t)$$

The programming looks like

```
F(1) = f(T)
F(2) = Y(1)
```

*Example C.3*

Example C.2 shows that the subroutine can be used to evaluate a performance index (IAE, ISE, etc., from Chapter Seven) while solving the system's differential equations. For example, the system model

$$\dot{y} = -y + u(t)$$

and the index

$$J = \int_0^t y^2\,dt$$

can be reduced to the system

$$\dot{y}_1 = -y_1 + u(t)$$

$$\dot{y}_2 = y_1^2$$

by defining $y_1 = y$, and $y_2 = J$. For most indices, the upper limit is $t = \infty$. Here we can handle this by using a value for TLIM that is large enough to allow $J$ to become approximately constant. Several trial values might have to be used before a suitable one is found.

## Reference

C.1. J. L. Melsa and S. K. Jones, *Computer Programs for Computational Assistance in the Study of Linear Control Theory*, McGraw-Hill, New York, 1973.

# APPENDIX D

# Mason's Rule for Diagram Reduction

When the number of loops or forward paths in a diagram is large, repeated application of the loop reduction and cascaded element formulas becomes tedious. In these instances, the use of Mason's rule simplifies the task of finding the transfer function. The rule applies to both block diagrams and signal flow graphs, but we present it here in terms of the former.

To establish the terminology, consider the single-loop path shown in Figure D.1. The transfer function of the forward path is $G(s)$. The loop transfer function is $G(s)H(s)$. The closed-loop transfer function is that of the forward path divided by one minus the loop transfer function; that is,

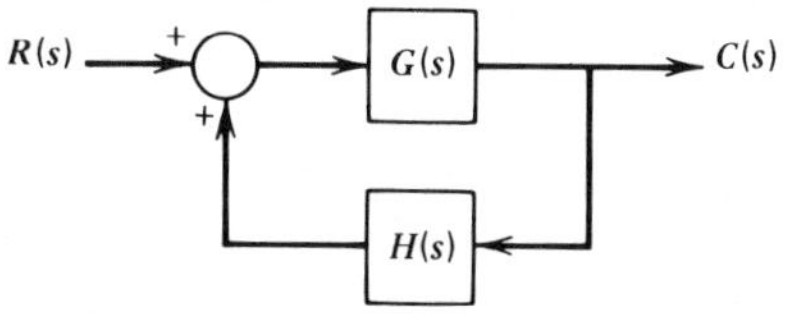


**Figure D.1 Single-loop system.**

$$\frac{C(s)}{R(s)} = \frac{G(s)}{1 - G(s)H(s)} \tag{D-1}$$

Note that the sign of the feedback loop is positive. It is wise when using Mason's rule to absorb any minus signs into the feedback element $H(s)$.

A block diagram is *tightly connected* if each forward path touches every loop path, as is shown in Figure D.2. Mason's rule may be stated as follows (Reference D.1). *The transfer function of a tightly connected block diagram is the sum of the forward-path transfer functions divided by one minus the sum of all the loop transfer functions.* Applying this rule to Figure D.2, we obtain

$$\frac{C(s)}{R(s)} = \frac{b_0 + b_1 s^{-1} + b_2 s^{-2}}{1 - (a_1 s^{-1} + a_2 s^{-2})} \tag{D-2}$$

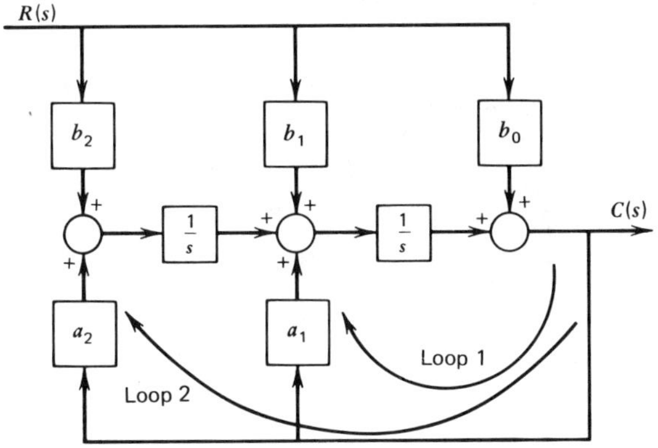


**Figure D.2 Second-order system in observer canonical form.**

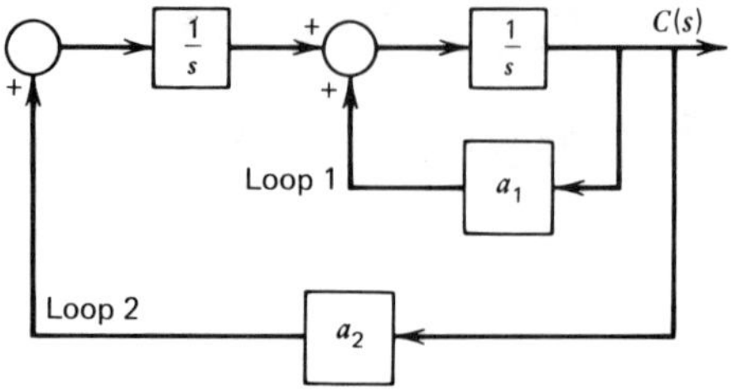


**Figure D.3 Zero-input version of Figure D.2 with loops separated.**

or in the usual form

$$\frac{C(s)}{R(s)} = \frac{b_0 s^2 + b_1 s + b_2}{s^2 - a_1 s - a_2} \tag{D-3}$$

Note that the input signal can reach the output in three ways while traveling only in the forward direction. The product of the transfer functions along such a path is the forward-path transfer function. For example, the path from $R(s)$ through $b_2$ also passes through two integrators to give a forward-path transfer function of $b_2(s^{-1})(s^{-1}) = b_2 s^{-2}$.

A loop transfer function is found by ignoring the other loops and the input, and following the signal around the loop. For example, ignore the input in Figure D.2 and separate loop 1 from loop 2 to give Figure D.3. From this figure, it is clear that the loop transfer function for loop 2 is $a_2(s^{-1})(s^{-1}) = a_2 s^{-2}$. Similarly, that for loop 1 is $a_1 s^{-1}$.

Now consider the system shown in Figure D.4. Mason's rule is easily applied to show that the system transfer function is that given by (D-2).

The diagram in Figure D.2 is the *observer canonical* form because all the feedback loops come from the output or "observed" signal. Figure D.4 shows the *control canonical* form because all the feedback loops return to the comparator where the actuating or error signal is formed for use by the controller. Both forms are called *direct canonical* forms because the gains in the diagram are coefficients in the numerator and denominator polynomials of the system transfer function. In *cascade*

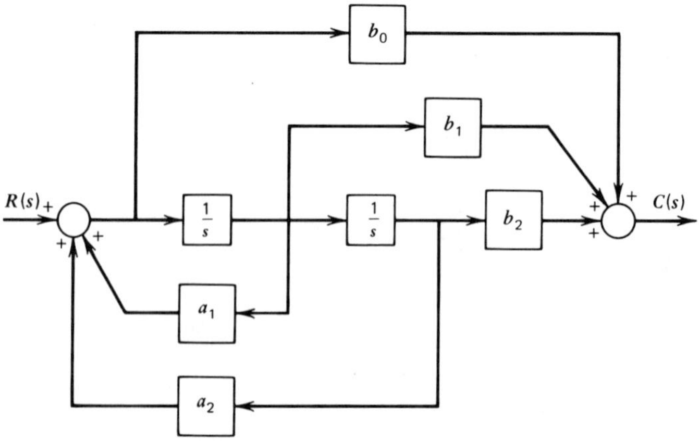


**Figure D.4 Second-order system in control canonical form.**

*canonical* form the poles and zeros of the transfer function appear in the diagram, and the transfer function appears as a ratio of the products of these factors.

A diagram that is not tightly connected can be treated with Mason's rule by reducing the diagram until subdiagrams appear that are tightly connected.

## Reference

D.1. S. J. Mason, "Feedback Theory: Further Properties of Signal Flow Graphs," *Proceedings of the IRE*, **44**, 1956, pp. 970–976.

# APPENDIX E
# Simulation with an Analog Computer

The operational amplifier can be used to solve the differential equation model of a dynamic system. The essence of the technique is to construct an op-amp circuit whose governing equations are the same as those of the system under study, except for magnitude and time-scale factors. The resulting circuit is said to be an *analog computer*.

Analog computation has lost some of its importance because of the increasing availability of digital computers. Nevertheless, one can identify three advantageous features of analog simulation.

1. Selection of a time step is not required, as opposed to digital methods. This has an advantage for fast systems, which require small time steps and many iterations in order to achieve accurate results with digital methods.
2. Oscilloscopes and $xy$-recorders, which are widely available, can be used to present the analog simulation results in graphic form, whereas graphics terminals for digital computers are not as commonly available.
3. Often the actual components to be used in the real system can be employed in the analog simulation. This provides an excellent way to ensure that the simulation includes the important characteristics of the system.

## Analog Computer Elements

The basic element of most analog computers is the op amp. In Section 6.8 the characteristics of this device were explored, and there it was shown how to implement such functions as multiplication by a constant, summation, and integration. These constitute all the operations necessary to develop an analog simulation of a linear system. In fact, the electronic implementation of the PID control laws are analog computer circuits. The only additional considerations required to develop a general procedure for analog simulation are those relating to magnitude and time scale factors.

Table E.1 lists the relevant op-amp circuits and their functions. The potentiometer is useful for multiplication by a fraction without the use of an amplifier. The computing diagram symbols are helpful for setting up the analog circuit. The block symbols represent active elements (those containing an amplifier). Another commonly used set of symbols is shown in Table E.2. They do not display the sign inversion associated with each amplifier, and we must be careful not to forget its existence when using these symbols.

Some commercially available analog computers use standard resistance and capacitance values so that only certain multiplication factors are available. In such cases a potentiometer can be used in series with the multiplier to obtain any other desired factor.

**TABLE E.1**

ANALOG COMPUTING ELEMENTS

| Wiring diagram | Function | Computing diagram |
|---|---|---|
| a. General op-amp relation | $\dfrac{E_o(s)}{E_i(s)} = -\dfrac{T_f(s)}{T_i(s)}$ | None |
| b. Multiplier | $e_o = -\dfrac{R_2}{R_1}\, e_i$ | |
| c. Summer | $e_o = -\dfrac{R_3}{R_1}\, e_1 - \dfrac{R_3}{R_2}\, e_2$ | |
| d. Inverter | $e_o = -e_i$ | |

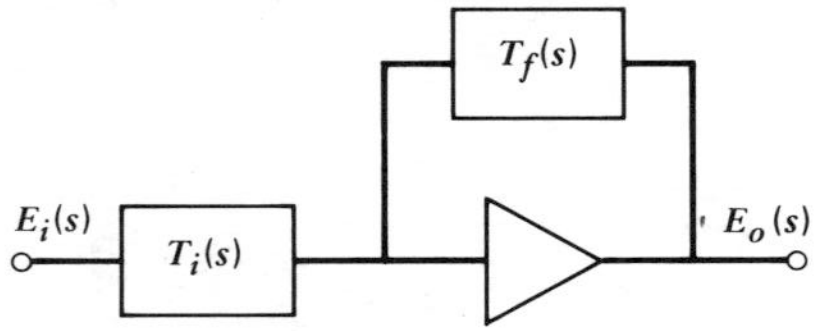

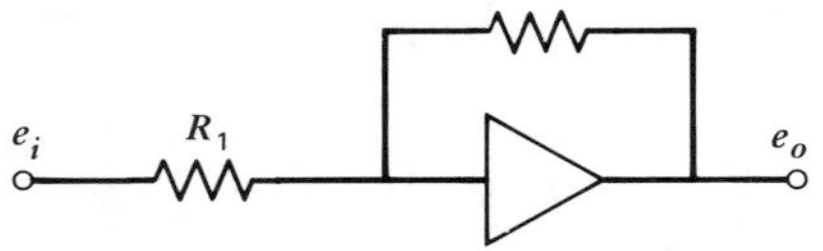

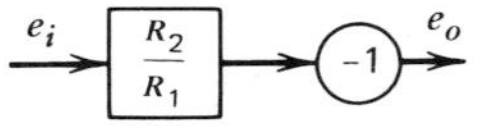

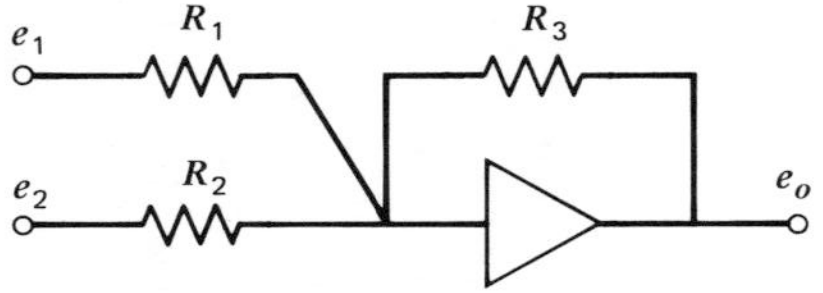

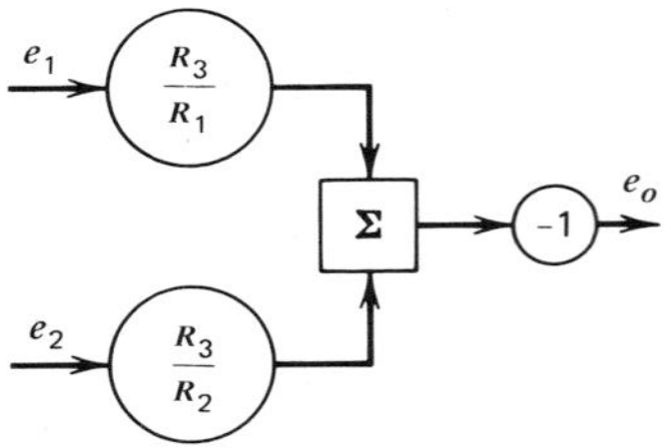

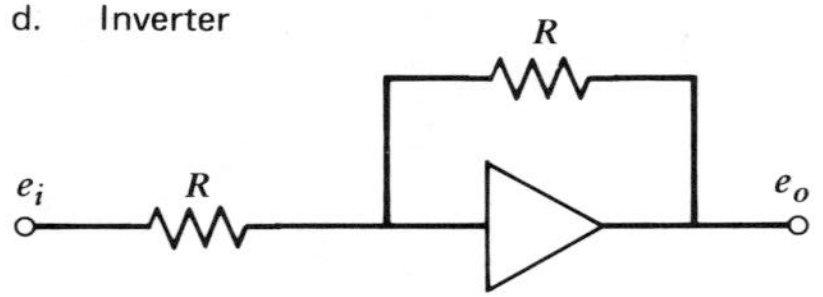

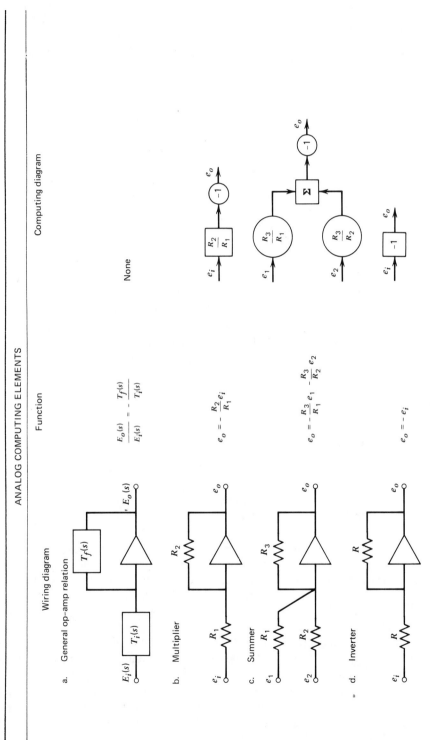

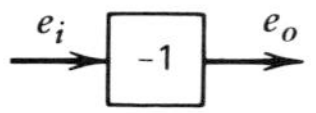

e. Integrator

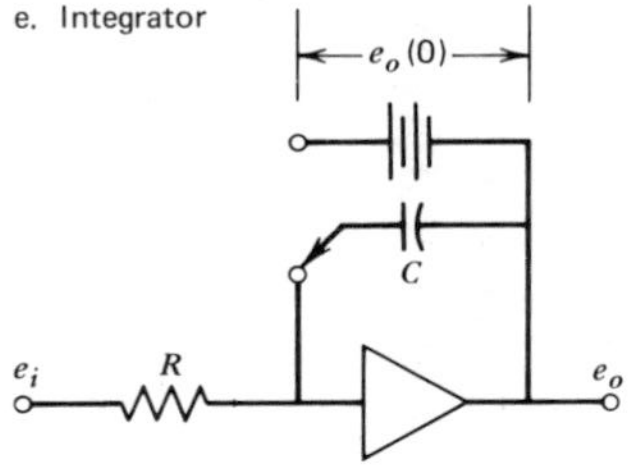


$$e_o = e_o(0) - \frac{1}{RC}\int_0^t e_i \, dt$$

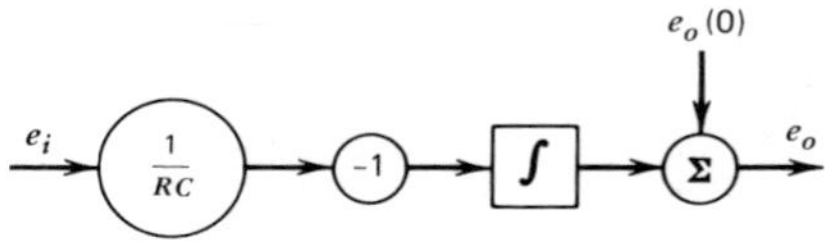


f. Summing integrator

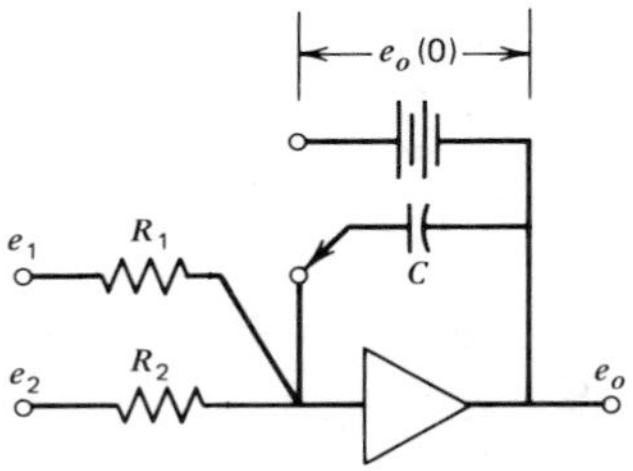


$$e_o = e_o(0) - \int_0^t \left( \frac{1}{R_1C}e_1 + \frac{1}{R_2C}e_2 \right) dt$$

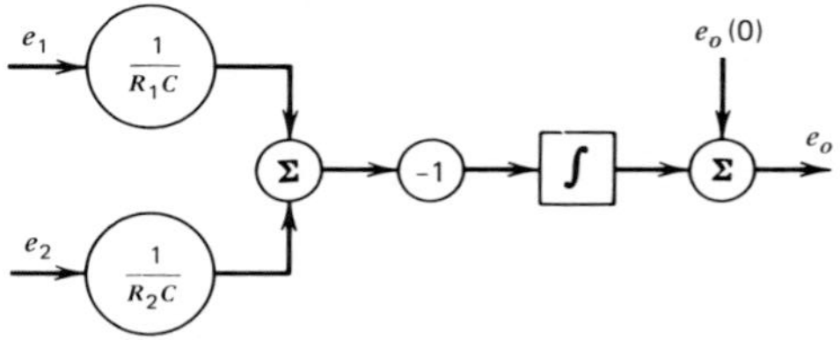


g. Fractional multiplier (pot)

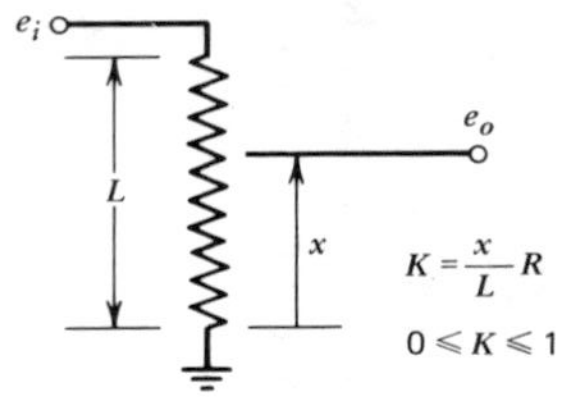


$$e_o = Ke_i$$

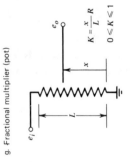

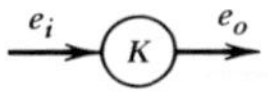

**TABLE E.2 Simplified Analog Computing Symbols**

| Symbol | Function |
|---|---|
| 1. Multiplier | |
| 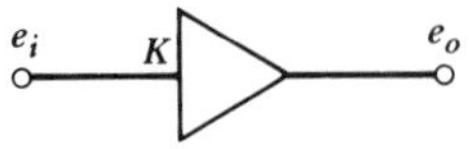 | $e_o = -Ke_i$ |
| 2. Integrator | |
| 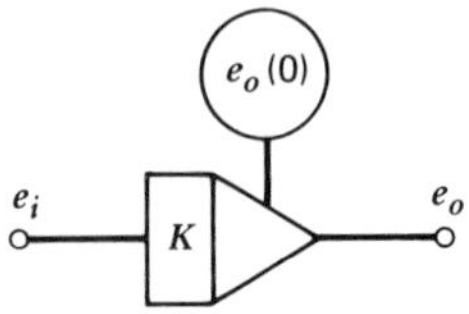 | $e_o = e_o(0) - K\int_0^t e_i\,dt$ |
| 3. Summer | |
| 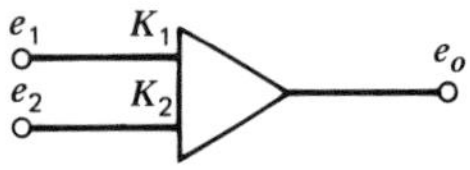 | $e_o = -K_1e_1 - K_2e_2$ |
| 4. Summing integrator | |
| 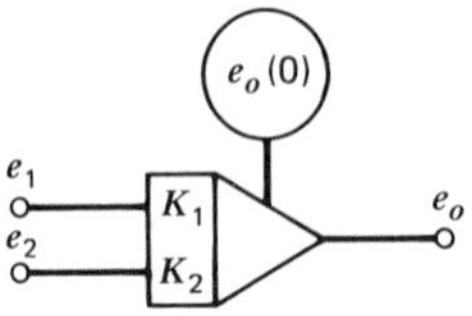 | $e_o = -\int_0^t (K_1e_1 + K_2e_2)\,dt$ |

## Designing The Computer Circuit

The first step in obtaining an analog computer circuit is to express the differential equation in block diagram form. This is then converted to a computing diagram. As an example, consider the equation

$$\begin{aligned} \ddot{y} + 2\dot{y} + 4y &= f(t) \\ y(0) &= 1 \\ \dot{y}(0) &= 3 \end{aligned} \tag{E-1}$$

The time-domain block diagram is shown in Figure E.1*a*. Using the symbols given in Table E.1, we construct the computing diagram shown in Figure E.1*b*. An alternative diagram is shown in Figure E.1*c*. Note that the sign of $\dot{y}(0)$ must be reversed in Figures E.1*b* and E.1*c* because of where it enters the circuit.

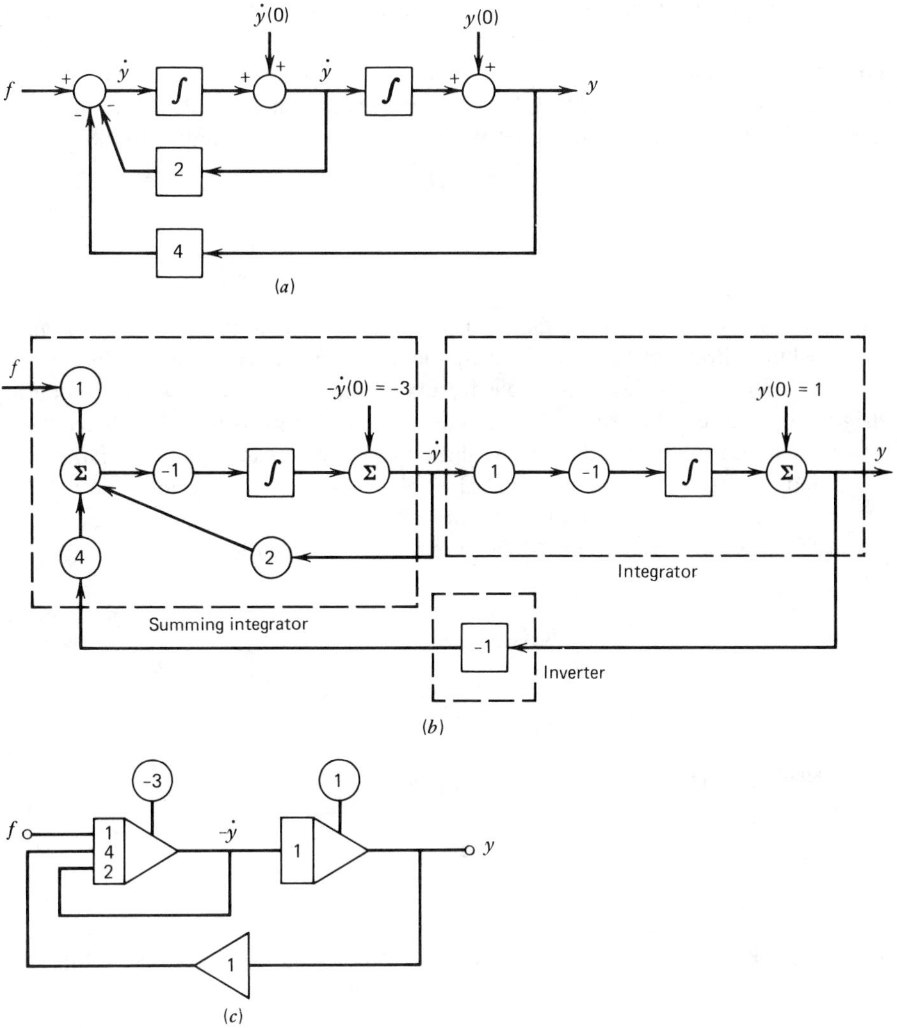


**Figure E.1** **Analog computing diagrams for the equation $\ddot{y} + 2\dot{y} + 4y = f$, $y(0) = 1$, $\dot{y}(0) = 3$. (*a*) Block diagram. (*b*) Computation diagram. (*c*) Simplified computation diagram.**

The main differences between the block diagrams and the computing diagrams are (1) the computing diagram accounts for the sign inversions due to the op amps, and (2) the computing diagram does not allow subtraction. Subtraction must be developed with a summer and an inverter.

## Magnitude Scale Factors

The circuit shown in Figure E.1 will implement the solution of (E-1) provided the voltage saturation limits of the amplifiers are not exceeded. These limits are typically either ± 10 v or ± 100 v for most commercial units. In addition, the voltages should be large enough to prevent noise from introducing errors. For these reasons magnitude scale factors are usually required. These relate the voltage levels in the circuit to the corresponding values of the variables $f$, $y$, and $\dot{y}$. For example, a large enough value of $f$ will produce values of $|y|$ and $|\dot{y}|$ that exceed the saturation limits. Therefore we let these variables be represented by the voltages $v_1$, $v_2$, and $v_f$ according to the relations

$$
\begin{aligned}
y &= k_1 v_1 \\
\dot{y} &= k_2 v_2 \\
f &= k_f v_f
\end{aligned}
\tag{E-2}
$$

where $k$ is the magnitude scale factor. These factors are determined by the maximum expected magnitudes of $y, \dot{y}$, and $f$ and by the saturation limits of the machine.

In order to see how the scale factors are chosen, we redraw the computing diagram of Figure E.1*b* except that we now leave the gains in terms of the resistance and capacitance values. We denote the diagram variables as $v_1$, $v_2$, and $v_f$, rather than $y$, $\dot{y}$, and $f$, as shown in Figure E.2. The voltages $v_1$ and $v_2$ are the result of integrations, which suggests that they can be taken as the state variables of the circuit. From the diagram we can immediately write the voltage equations as

$$\dot{v}_1 = -\frac{1}{R_4 C_2} v_2 \tag{E-3}$$

$$\dot{v}_2 = -\frac{1}{R_1 C_1} v_f + \frac{1}{R_2 C_1} v_1 - \frac{1}{R_3 C_1} v_2 \tag{E-4}$$

Next write (E-1) in state-variable form as

$$\dot{x}_1 = x_2 \tag{E-5}$$

$$\dot{x}_2 = f - 4x_1 - 2x_2 \tag{E-6}$$

where $x_1 = y$ and $x_2 = \dot{y}$. From (E-2), $x_1 = k_1 v_1$ and $x_2 = k_2 v_2$. Substituting these

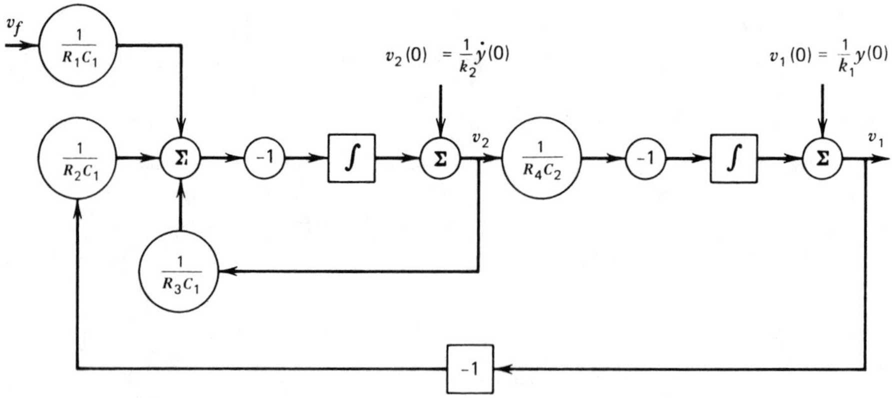


**Figure E.2 Computing diagram with magnitude scale factors.**

into (E-5) and (E-6) we obtain

$$\dot{v}_1 = \frac{k_2}{k_1} v_2 \tag{E-7}$$

$$\dot{v}_2 = \frac{1}{k_2}(k_f v_f - 4k_1 v_1 - 2k_2 v_2) \tag{E-8}$$

Comparison of (E-3), (E-4), (E-7), and (E-8) shows that

$$\left.\begin{array}{ll} \dfrac{k_2}{k_1} = -\dfrac{1}{R_4C_2}, & \dfrac{k_f}{k_2} = -\dfrac{1}{R_1C_1} \\ \dfrac{-4k_1}{k_2} = \dfrac{1}{R_2C_1}, & -2 = -\dfrac{1}{R_3C_1} \end{array}\right\} \tag{E-9}$$

Examination of Figures E.1*b* and E.2 shows that if $\dot{y} > 0$ then $v_2 < 0$. Thus $k_2$ must be negative. In a similar way we can see that $k_1 > 0$ and $k_f > 0$.

Let us suppose that our machine has voltage saturation limits of ± 10 v. In order to have a safety margin, we will thus attempt to keep the voltage magnitudes less than, say, 9 v, but well above zero to maintain an adequate signal-to-noise ratio. Therefore from (E-2) we have (noting that $k_2 < 0$),

$$\left.\begin{array}{lll} |y|_{max} &= k_1|v_1|_{max} &= 9k_1 \\ |\dot{y}|_{max} &= k_2|v_2|_{max} &= -9k_2 \\ |f|_{max} &= k_f|v_f|_{max} &= 9k_f \end{array}\right\} \tag{E-10}$$

Suppose that we can estimate the maximum values as follows.

$$|y|_{max} = 6, \quad |\dot{y}|_{max} = 3, \quad |f|_{max} = 5 \tag{E-11}$$

This information might come from insight concerning the physical process being simulated. Estimation of the maximum values of the model's variables is the trickiest aspect of analog simulation, since it requires knowledge of the very solution being sought. But with physical insight and some trial and error, a useful set of scale factors can usually be found.

Substitution of (E-11) into (E-10) gives

$$k_1 = \tfrac{2}{3}, \quad k_2 = -\tfrac{1}{3}, \quad k_f = \tfrac{5}{9} \tag{E-12}$$

These values are then substituted into (E-9) to obtain

$$\left.\begin{array}{ll} \dfrac{1}{R_4C_2} = \dfrac{1/3}{2/3} = \dfrac{1}{2}, & \dfrac{1}{R_1C_1} = \dfrac{5/9}{1/3} = \dfrac{5}{3} \\ \dfrac{1}{R_2C_1} = \dfrac{4(2/3)}{1/3} = 8, & \dfrac{1}{R_3C_1} = 2 \end{array}\right\} \tag{E-13}$$

These are the gains in the computing diagram of Figure E.2. We are fortunate here because the gains given by (E-13) are reasonably close in value. Large differences in gain values should be avoided because they require disproportionate sizes for the resistances and capacitances. Such wide ranges are not available on commercial analog computers. Gains on such machines are generally obtainable in the range from 0.1 to 10.

Equations (E-13) contain six unknowns, so there is some flexibility in the resulting circuit. Choosing $C_1 = C_2 = 10^{-6} f$ as a commonly available value, we obtain

$$\left.\begin{aligned} R_1 &= \tfrac{3}{5} \times 10^6\,\Omega, & R_2 &= \tfrac{1}{8} \times 10^6\,\Omega \\ R_3 &= \tfrac{1}{2} \times 10^6\,\Omega, & R_4 &= 2 \times 10^6\,\Omega \end{aligned}\right\} \tag{E-14}$$

These values are roughly the same size, so the design would seem to be acceptable. However, before drawing the final wiring diagram, we will investigate some additional considerations.

## Choice of Time Scale

Often it is desired to speed up the dynamics of the process under study so that many simulations can be made within a reasonable period. On the other hand, it is sometimes necessary to slow down the simulation to be compatible with the response of an $xy$-plotter. Denote the analog computer's time scale by $T$ and that of the process by $t$. The two scales are related by the scale factor $k_t$ as follows.

$$t = k_t T \tag{E-15}$$

Then the rate of change of voltage in the two scales are related as

$$\frac{dV}{dT} = \frac{dV}{dt}\frac{dt}{dT} = k_t \frac{dV}{dT} = k_t \dot{V} \tag{E-16}$$

In terms of the new scale $T$, the voltage equations (E-3) and (E-4) become

$$\left.\begin{aligned} \frac{dv_1}{dT} &= -\frac{1}{R_4C_2} v_2 \\ \frac{dv_2}{dT} &= -\frac{1}{R_1C_1} v_f + \frac{1}{R_2C_1} v_1 - \frac{1}{R_3C_1} v_2 \end{aligned}\right\} \tag{E-17}$$

With (E-16) we obtain

$$\left.\begin{aligned} \frac{dv_1}{dt} &= -\frac{1}{k_t}\frac{1}{R_4C_2} v_2 \\ \frac{dv_2}{dt} &= -\frac{1}{k_t}\left(\frac{1}{R_1C_1} v_f - \frac{1}{R_2C_1} v_1 + \frac{1}{R_3C_1} v_2\right) \end{aligned}\right\} \tag{E-18}$$

Comparison of (E-18) with (E-7) and (E-8) gives the following requirements.

$$\left.\begin{aligned} \frac{k_2}{k_1} &= -\frac{1}{k_tR_4C_2}, & \frac{k_f}{k_2} &= -\frac{1}{k_tR_1C_1} \\ \frac{-4k_1}{k_2} &= \frac{1}{k_tR_2C_1}, & -2 &= -\frac{1}{k_tR_3C_1} \end{aligned}\right\} \tag{E-19}$$

Note the similarity to (E-9).

Suppose that we wish to slow down the simulation by a factor of two. Then $k_t = 1/2$. Using the previous values for $k_1$, $k_2$, and $k_f$, we obtain the required resistances as before. From (E-19),

$$\left.\begin{aligned} \frac{1}{R_4C_2} &= \frac{1}{4}, \qquad \frac{1}{R_1C_1} = \frac{5}{6} \\ \frac{1}{R_2C_1} &= 4, \qquad \frac{1}{R_3C_1} = 1 \end{aligned}\right\} \tag{E-20}$$

With $C_1 = C_2 = 10^{-6} f$, we have $R_1 = 6/5\,M\Omega$, $R_2 = 1/4\,M\Omega$, $R_3 = 1\,M\Omega$, and $R_4 = 4\,M\Omega$.

## Use of Potentiometers

Thus far we have made the implicit assumption that a continuous range of values are available for the resistances and capacitances of the multipliers, summers, and integrators of the computer. Usually this is not the case, and only discrete values are available. If so, the required gain values can still be obtained by choosing an available resistance or capacitance value lower than the computed value and using a potentio-

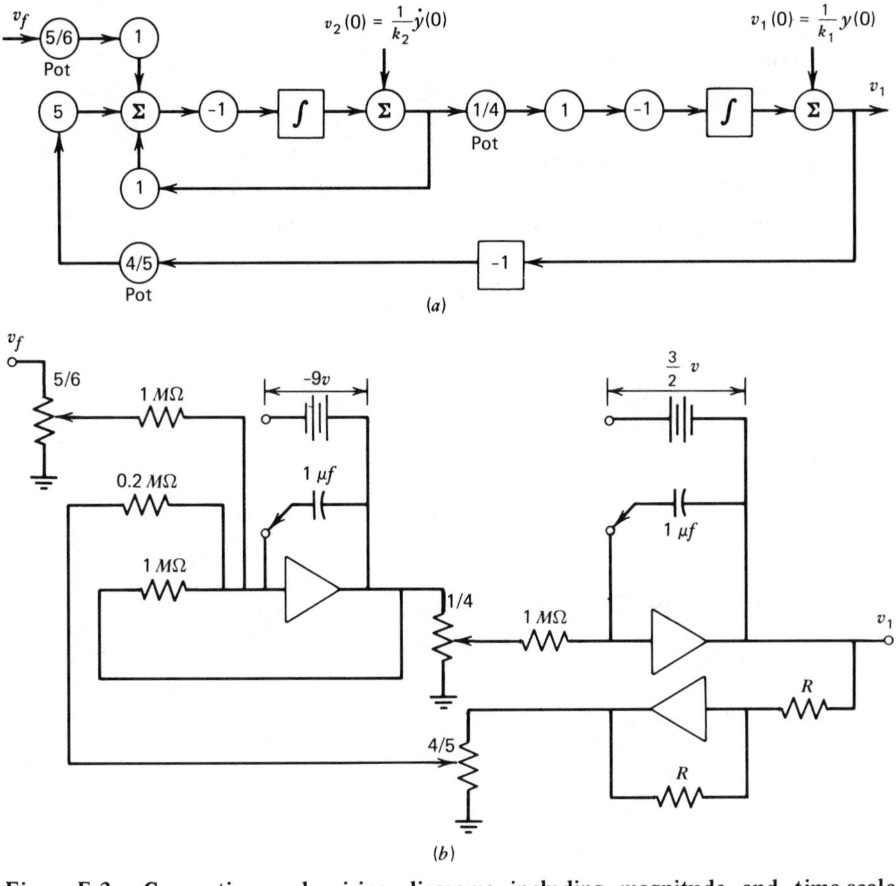


Figure E.3 Computing and wiring diagrams including magnitude and time-scale factors.

meter in series to reduce the gain to the required value. The technique also has the advantage of producing gains that are adjustable. This is convenient if we wish to modify the scale factors somewhat.

For example, suppose that $10^{-6} f$ is an available capacitance and that the available resistances are integer multiples of 100 $k\Omega$. Then the gain $1/R_4C_2 = 1/4$ can be realized by choosing $R_4 = 1\,M\Omega$ (instead of $1/4\,M\Omega$) and by using a series potentiometer with a fractional gain of 1/4. The other resistances are chosen similarly. The resulting computing and wiring diagrams are shown in Figure E.3. This circuit simulates (E.1) with the scale factors $k_1 = 2/3$, $k_2 = -1/3$, $k_f = 5/9$, and $k_t = 1/2$, and with the given initial conditions.

## A General Procedure

The preceding approach to the design of a simulation circuit can be concisely stated in terms of state-variable notation, as follows.

1. Write the model of the process to be simulated in state variable form as

$$\frac{d\mathbf{x}}{dt} = \mathbf{A}\mathbf{x} + \mathbf{B}\mathbf{u} \tag{E-21}$$

2. Use these equations to draw the state-variable block diagram. The computing diagram is drawn from this. The output of each integrator in the block diagram corresponds to a voltage output of an integrator in the computing diagram. The only differences between the two diagrams are that the computing diagram must use a summer and inverter to implement subtraction, and must show the sign inversions produced by the amplifiers.

3. From the computing diagram, find the equations for the electrical circuit. These equations will be of the form

$$\frac{d\mathbf{V}}{dT} = \mathbf{A}_E\mathbf{V} + \mathbf{B}_E\mathbf{V}_u \tag{E-22}$$

where $T$ is the machine time scale. The voltages $\mathbf{V}$ and $\mathbf{V}_u$ represent $\mathbf{x}$ and $\mathbf{u}$ through the scale factors as follows.

$$\mathbf{x} = \mathbf{K}\mathbf{V} \tag{E-23}$$

$$\mathbf{u} = \mathbf{K}_u\mathbf{V}_u \tag{E-24}$$

$$t = k_tT \tag{E-25}$$

The diagonal matrices $\mathbf{K}$ and $\mathbf{K}_u$ contain the scale factors.

$$\mathbf{K} = \begin{bmatrix} k_1 & 0 & 0 & \cdots & . & 0 \\ 0 & k_2 & 0 & \cdots & . & 0 \\ . & . & . & \cdots & . & . \\ 0 & . & . & \cdots & 0 & k_n \end{bmatrix} \tag{E-26}$$

$$\mathbf{K}_u = \begin{bmatrix} k_{1u} & 0 & 0 & \cdots & . & 0 \\ 0 & k_{2u} & 0 & \cdots & . & 0 \\ . & . & . & \cdots & . & . \\ 0 & . & . & \cdots & 0 & k_{mu} \end{bmatrix} \tag{E-27}$$

The elements of the matrices $\mathbf{A}_E$ and $\mathbf{B}_E$ are functions of the circuit's resistances and capacitances.

4. Substitution of (E-23) to (E-25) into (E-21), and comparison of the result with (E-22) shows that

$$\mathbf{A}_E = k_t \mathbf{K}^{-1} \mathbf{A} \mathbf{K} \tag{E-28}$$

$$\mathbf{B}_E = k_t \mathbf{K}^{-1} \mathbf{B} \mathbf{K}_u \tag{E-29}$$

These are the scaling relations that relate the resistances and capacitances to the scale factors. Since $\mathbf{K}$ and $\mathbf{K}_u$ are diagonal, their inverses are easily computed.

5. Pick the scale factors so that the given specifications are satisfied. These include the given saturation limits, the desired time scale, and the limits on the available coefficient ranges on the machine. The latter usually requires that the magnitudes of the elements of $\mathbf{A}_E$ and $\mathbf{B}_E$ lie within the range 0.1 to 10.

6. Once $\mathbf{K}$, $\mathbf{K}_u$, and $k_t$ have been selected, use the scaling relations (E-28) and (E-29) to obtain the resistances and capacitances. Some flexibility occurs here since there are usually more unknowns than constraints.

7. If only discrete resistance and capacitance values are available, use potentiometers to obtain the desired gains.

For the example problem given by (E-1), the state equations (E-5) and (E-6) with $u = f$, give the following matrices.

$$\mathbf{A} = \begin{bmatrix} 0 & 1 \\ -4 & -2 \end{bmatrix} \qquad \mathbf{B} = \begin{bmatrix} 0 \\ 1 \end{bmatrix} \tag{E-30}$$

From the computing diagram shown in Figure E.2 we obtained the circuit equations (E-17). These give the matrices $\mathbf{A}_E$ and $\mathbf{B}_E$.

$$A_E = \begin{bmatrix} 0 & -\dfrac{1}{R_4C_2} \\ \dfrac{1}{R_2C_1} & -\dfrac{1}{R_3C_1} \end{bmatrix} \qquad B_E = \begin{bmatrix} 0 \\ -\dfrac{1}{R_1C_1} \end{bmatrix} \tag{E-31}$$

where

$$\mathbf{V} = \begin{bmatrix} v_1 \\ v_2 \end{bmatrix} \qquad V_u = v_f$$

$$\mathbf{x} = \begin{bmatrix} k_1 & 0 \\ 0 & k_2 \end{bmatrix} \mathbf{V} \qquad u = k_u v_f \qquad k_u = k_f \tag{E-32}$$

The scaling equations (E-28) and (E-29) become

$$\mathbf{A}_E = k_t \begin{bmatrix} 1/k_1 & 0 \\ 0 & 1/k_2 \end{bmatrix} \begin{bmatrix} 0 & 1 \\ -4 & -2 \end{bmatrix} \begin{bmatrix} k_1 & 0 \\ 0 & k_2 \end{bmatrix}$$

$$= k_t \begin{bmatrix} 0 & k_2/k_1 \\ -4k_1/k_2 & -2 \end{bmatrix} \tag{E.33}$$

$$\mathbf{B}_E = k_t \begin{bmatrix} 1/k_1 & 0 \\ 0 & 1/k_2 \end{bmatrix} \begin{bmatrix} 0 \\ 1 \end{bmatrix} k_f = k_t k_f \begin{bmatrix} 0 \\ 1/k_2 \end{bmatrix} \tag{E.34}$$

Comparison of the right-hand sides with the expressions for $\mathbf{A}_E$ and $\mathbf{B}_E$ in (E-31) gives the scaling equations obtained in (E-19). In steps 5, 6, and 7, the scale factors, resistances, capacitances, and potentiometer settings are found as before.

Even with the assumption that one can roughly estimate the maximum values of each variable, the preceding procedure might still require some trial-and-error adjustment of the scale factors in order to obtain the proper magnitudes for the elements of $\mathbf{A}_E$ and $\mathbf{B}_E$. If this fails to work, the formulation of the problem should be reviewed. Such a failure can occur if the model to be simulated has very fast and very slow response modes (some small and some large time constants). A solution is to separate the system model into two parts, one for the fast response and one for the slow response. Each model is then simulated separately.

## Analog Simulation of Nonlinear Systems

Simulation of nonlinear systems is possible with analogs, but it requires a special circuit for each type of nonlinearity encountered. For example, a saturation nonlinearity can be simulated by using diodes to limit the op-amp multiplier's output, as in Figure 6.45. Many nonlinear functions can be represented well enough by a series of straight-line segments, each generated by a multiplier-diode combination. References 7 and 10 of Chapter Six contain examples of such circuits.

# APPENDIX F
# The Routh-Hurwitz Criterion: The General Case

Consider the following characteristic equation.

$$b_n s^n + b_{n-1} s^{n-1} + b_{n-2} s^{n-2} + \ldots + b_2 s^2 + b_1 s + b_0 = 0 \qquad \text{(F-1)}$$

We assume that this equation has been normalized so that $b_n > 0$. The Routh-Hurwitz criterion consists of several rules applied to the coefficients of (F-1). The first rule is as follows.

*Rule 1*

If any of the coefficients $b_i$, $i = 0, 1, 2, \ldots, n-1$ are zero or negative, the system is not stable. It can be either unstable or neutrally stable.

For example, the equation $s^2 + 1 = 0$ has two imaginary roots, and represents a neutrally stable system. The equation $s^2 - 1 = 0$ represents an unstable system, as does the equation $s^2 - 3s + 1 = 0$. The reader is cautioned that Rule 1 cannot be used to prove stability, because its converse is not necessarily true. A polynomial with all positive coefficients can represent an unstable or neutrally stable system. For example, the polynomial $s^3 + 2s^2 + s + 2$ has the roots: $s = -2, \pm i$, and so it gives neutral stability.

## The Routh Array

The remaining rules can be expressed in terms of the following array, suggested by Routh. Arrange the coefficients in rows and columns as follows.

$$\begin{array}{l|lllll}
s^n & b_n & b_{n-2} & b_{n-4} & b_{n-6} & \cdots \\
s^{n-1} & b_{n-1} & b_{n-3} & b_{n-5} & b_{n-7} & \cdots \\
s^{n-2} & c_1 & c_2 & c_3 & c_4 & \cdots \\
s^{n-3} & d_1 & d_2 & d_3 & d_4 & \cdots \\
\cdots & \cdots & \cdots & \cdots & \cdots & \cdots \\
s^3 & e_1 & e_2 & 0 & & \\
s^2 & f_1 & f_2 & 0 & & \\
s^1 & g_1 & 0 & & & \\
s^0 & h_1 & 0 & & &
\end{array} \qquad \text{(F-2)}$$

The powers of $s$ ($s^n$, $s^{n-1}$, . . . , $s^0$) are used for indexing purposes. The array has $(n + 1)$ rows. The entries in the third and succeeding rows are obtained from the following procedure.

$$c_1 = \frac{-\begin{vmatrix} b_n & b_{n-2} \\ b_{n-1} & b_{n-3} \end{vmatrix}}{b_{n-1}} = \frac{b_{n-1}b_{n-2} - b_n b_{n-3}}{b_{n-1}}$$

$$c_2 = \frac{-\begin{vmatrix} b_n & b_{n-4} \\ b_{n-1} & b_{n-5} \end{vmatrix}}{b_{n-1}} = \frac{b_{n-1}b_{n-4} - b_n b_{n-5}}{b_{n-1}} \tag{F-3}$$

The entry $c_3$ is found by forming a determinant whose first column consists of $b_n$ and $b_{n-1}$ and whose second column consists of the two $b_i$ entries above and immediately to the right of $c_3$. The negative of this determinant is then divided by the pivotal element $b_{n-1}$, the element above the first entry in the $c_i$ row. The result is

$$c_3 = \frac{-\begin{vmatrix} b_n & b_{n-6} \\ b_{n-1} & b_{n-7} \end{vmatrix}}{b_{n-1}} = \frac{b_{n-1}b_{n-6} - b_n b_{n-7}}{b_{n-1}}$$

This process is easily generalized to produce the remaining entries. Moving down a row, the $d_i$ entries are obtained as

$$d_1 = \frac{-\begin{vmatrix} b_{n-1} & b_{n-3} \\ c_1 & c_2 \end{vmatrix}}{c_1} = \frac{c_1 b_{n-3} - c_2 b_{n-1}}{c_1}$$

$$d_2 = \frac{-\begin{vmatrix} b_{n-1} & b_{n-5} \\ c_1 & c_3 \end{vmatrix}}{c_1} = \frac{c_1 b_{n-5} - c_3 b_{n-1}}{c_1} \tag{F-4}$$

and so forth until the $(n + 1)$ row is obtained.

Consider the polynomial.

$$s^4 + 2s^3 + s^2 + 4s + 2$$

Rule 1 gives no information, so we form the array.

| | | | |
|---|---|---|---|
| $s^4$ | 1 | 1 | 2 |
| $s^3$ | 2 | 4 | 0 |
| $s^2$ | −1 | 4 | |
| $s^1$ | 12 | 0 | |
| $s^0$ | 4 | | |

Zeros may be included to fill in missing elements in any determinant used to find the entries in the array.

*Rule 2*

The number of roots of (F-1) with positive real parts is equal to the number of sign changes in the entries in the first column of the array.

*Rule 3*

The necessary and sufficient condition for all roots of (F-1) to lie in the left-half plane (a stable system) is that Rule 1 is satisfied and that the first column of the array contain no sign changes.

*Example F.1*

Determine the stability criteria for the second-order system whose characteristic equation is

$$b_2 s^2 + b_1 s + b_0 = 0$$

Assume that $b_2 > 0$. (If not, multiply through the equation by $-1$.)

Rule 1 requires that $b_1 > 0$ and $b_0 > 0$. To apply Rule 2, form the array.

$$\begin{array}{c|cc} s^2 & b_2 & b_0 \\ s^1 & b_1 & 0 \\ s^0 & b_0 & \end{array}$$

Since $b_2 > 0$, no sign changes occur if Rule 1 is satisfied. The necessary and sufficient condition for a stable second-order system is that all three coefficients must be positive.

*Example F.2*

Derive the stability conditions for the third-order polynomial

$$b_3 s^3 + b_2 s^2 + b_1 s + b_0 = 0 \tag{F-5}$$

where $b_3 > 0$.

Rule 1 requires that $b_2 > 0$, $b_1 > 0$ and $b_0 > 0$. The Routh array is

$$\begin{array}{c|cc} s^3 & b_3 & b_1 \\ s^2 & b_2 & b_0 \\ s^1 & \dfrac{b_1 b_2 - b_0 b_3}{b_2} & \\ s^0 & b_0 & \end{array}$$

Given the conditions already imposed on the signs of $b_i$, no sign changes will occur in the first column if $(b_1 b_2 - b_0 b_3) > 0$. Thus the necessary and sufficient conditions for stability are

$$\left.\begin{array}{llll} b_3 > 0, & b_2 > 0, & b_1 > 0, & b_0 > 0 \\ b_1 b_2 - b_0 b_3 > 0 & & & \end{array}\right\} \tag{F-6}$$

## Special Cases

Several situations can occur that require additional interpretation.

### *Rule 4*

If a zero occurs in the first column, and the remaining elements in that row are not all zero, replace the zero element with a small positive number $\epsilon$, and continue forming the array. When completed, let $\epsilon \to 0$.

### *Rule 5*

If the entries above and below a zero in the first column have the same sign, a pair of purely imaginary roots exists. If the signs are opposite, there is one sign change (and therefore at least one root in the right half plane).

To illustrate Rules 4 and 5, consider the polynomial

$$s^3 + 2s^2 + s + 2$$

The array has a zero in the third row. This is replaced with $\epsilon$ and the rest of the array is constructed. This gives

$$\begin{array}{c|cc} s^3 & 1 & 1 \\ s^2 & 2 & 2 \\ s^1 & \epsilon & \\ s^0 & 2\epsilon/\epsilon = 2 & \end{array}$$

As $\epsilon \to 0$, we see $+2$ occurs above and below the zero. Thus the polynomial has a pair of imaginary roots, while the third root must lie in the left-half plane, since no sign changes occur in the first column. We can use this information to determine all three roots easily. Let the imaginary root factors be written as $s^2 + b$, where $b > 0$. Let the real root factor be $s + a$, $a > 0$. Then

$$\begin{aligned} (s+a)(s^2+b) &= s^3 + as^2 + bs + ab \\ &= s^3 + 2s^2 + s + 2 \end{aligned}$$

Comparison of the two forms shows that $a = 2$, $b = 1$. The roots are $s = -2, \pm i$.

The polynomial

$$s^3 + 2s^2 - s - 2$$

has the array

$$\begin{array}{c|cc} s^3 & 1 & -1 \\ s^2 & 2 & -2 \\ s^1 & \epsilon & \\ s^0 & -2 & \end{array}$$

The polynomial has one root in the right-half plane. It must therefore be a real root. However, the array gives no other information in this case.

The existence of one or more roots at the origin is obvious by the absence of the term $b_0$, or both terms $b_0$ and $b_1$, etc. This is also indicated by the array if the last row is zero, or if the last two rows are zero, etc. However, the occurrence of a zero row followed by a nonzero row is another matter. We now consider this case. (Note that Rules 4 and 5 do not apply to identically zero rows of more than one element. "Filler" zeros are not counted as zero elements.)

## *Rule 6*

If any derived row is identically zero, the system is not stable. It may be unstable or neutrally stable. The zero row corresponds to roots located symmetrically about the origin. These can be pairs of real roots with the same magnitude but opposite sign ($\pm r$), purely imaginary complex pairs ($\pm i\omega$), or two pairs of complex conjugate roots ($a \pm ib$, $-a \pm ib$). The auxiliary polynomial containing these roots is obtained from the row above the zero row. The coefficients of the derivative of the auxiliary polynomial are used to replace the zero row, and the rest of the array is then constructed.

If the auxiliary polynomial is even and of order $2m$, there will be $m$ pairs of equal and opposite roots.* Knowledge of these auxiliary roots, and the use of synthetic division often enables us to determine all the characteristic roots. For example, consider the polynomial

$$s^6 + 8s^5 + 18s^4 + 24s^3 + 41s^2 - 32s - 60$$

From Rule 1, this represents an unstable system, but sometimes we wish to know the degree of instability; that is, how many roots lie in the right-half plane, and how close are all of the roots to the imaginary axis. The Routh array is, for the first four rows,

| | | | | |
|---|---|---|---|---|
| $s^6$ | 1 | 18 | 41 | −60 |
| $s^5$ | 8 | 24 | −32 | |
| $s^4$ | 15 | 45 | −60 | |
| $s^3$ | 0 | 0 | | |

The appearance of the zero row requires the use of Rule 6. The auxiliary equation from the $s^4$ row is

$$A(s) = 15s^4 + 45s^2 - 60 = 0$$

This indicates the usefulness of the $s^i$ terms for indexing each row. The derivative gives

$$\frac{dA(s)}{ds} = 60s^3 + 90s$$

---

* The auxiliary polynomial is not always even. For example, the characteristic equation $s(s^2 - 1)(s^2 + 2s + 4) = 0$ produces the auxiliary polynomial $A(s) = 4s^3 - 4s$. The author is grateful to a reviewer for pointing this out.

The zero row is now replaced with the preceding coefficients, and the array becomes

$$\begin{array}{c|cccc}
s^6 & 1 & 18 & 41 & -60 \\
s^5 & 8 & 24 & -32 & \\
s^4 & 15 & 45 & -60 & \\
s^3 & 60 & 90 & & \\
s^2 & 22.5 & -60 & & \\
s^1 & 250 & & & \\
s^0 & -60 & & &
\end{array}$$

In this example we need not complete the array in order to determine the roots. Four of them are given by the roots of the auxiliary equation (after dividing by 15)

$$s^4 + 3s^2 - 4 = 0$$

or

$$s^2 = -4, +1$$

The auxiliary roots are $s = \pm 2i, \pm 1$. The remaining two characteristic roots can be found by synthetic division. Dividing the auxiliary polynomial into the characteristic polynomial gives

$$\begin{array}{r}
s^2 + 8s + 15 \\
s^4 + 3s^2 - 4 \,\big/\, s^6 + 8s^5 + 18s^4 + 24s^3 + 41s^2 - 32s - 60 \\
\underline{s^6 \qquad + 3s^4 \qquad\qquad - 4s^2 \qquad\qquad} \\
8s^5 + 15s^4 + 24s^3 + 45s^2 - 32s - 60 \\
\underline{8s^5 \qquad\quad + 24s^3 \qquad\quad - 32s \qquad} \\
15s^4 \qquad\quad + 45s^2 \qquad\quad - 60 \\
\underline{15s^4 \qquad\quad + 45s^2 \qquad\quad - 60}
\end{array}$$

The other factor in the characteristic polynomial is seen to be $s^2 + 8s + 15$. This gives the roots $s = -3, -5$. We have thus found the roots of a sixth-order polynomial, without recourse to numerical methods.

Note in the preceding example that if we divide $A(s)$ by 15, and then compute the derivative, we obtain $4s^3 + 6s$. The entries to replace the zero row are now 4 and 6, not 60 and 90. Both are correct, because of Rule 7.

*Rule 7*

Any derived row in the Routh array can be multiplied or divided by any nonzero positive number without changing the results.

This can be used to simplify the calculations by reducing the size of the numbers to be manipulated.

## Relative Stability

The Routh-Hurwitz criterion provides a yes-or-no answer to the stability question. Thus it gives information on the *absolute stability* of the system. However, for design purposes we would also like to know "how close" is the system to being unstable, if the criterion predicts a stable system. That is, since the numerical values of the coefficients of the characteristic equation are not known exactly, then what appears to be a stable system mathematically might in reality be an unstable one. Thus we are also interested in the *relative stability* of the system, which is indicated by the proximity of the roots to the imaginary axis.

We have seen that all roots lying on a vertical line in the complex plane have the same time constant, which is the reciprocal of the distance from the line to the imaginary axis (Figure 5.15*d*). Any root lying to the right of this line has a larger time constant and is closer to being an unstable root than those roots to its left. Thus a measure of relative stability is given by the system's largest time constant; that is, the time constant of the dominant root. The Routh-Hurwitz criterion can be used to estimate the relative stability and sometimes to find the location of the dominant root. If we create a new polynomial by shifting the imaginary axis to the left, the Routh-Hurwitz criterion can be applied to this new polynomial to determine how many roots lie to the right of the new imaginary axis. This technique is stated in Rule 8.

*Rule 8*

The number of roots lying to the right of the vertical line $s = -\sigma$ can be determined by substituting $s = p - \sigma$ into (F-1) and applying the Routh-Hurwitz criteria to the resulting polynomial in $p$.

*Example F.3*

The characteristic polynomial of a third-order system is given to be

$$s^3 + 9s^2 + 26s + K$$

Find the value of $K$ required so that the dominant time constant is no larger than 1/2. It is desired to design the system to be slightly underdamped. Determine whether or not this is possible, and if so, find the value of $K$ needed and the resulting damping ratio.

The largest time constant can be 1/2. This means that no root can lie to the right of the line $s = -2$. From Rule 8 we have $\sigma = 2$, and we substitute $s = p - 2$. This gives

$$(p-2)^3 + 9(p-2)^2 + 26(p-2) + K$$

or

$$p^3 + 3p^2 + 2p + K - 24$$

We see immediately from Rule 1 that $K > 24$ is required but does not guarantee that no root lies to the right of $s = -2$. To obtain the complete set of requirements, construct the array from the polynomial in $p$.

| | | |
|---|---|---|
| $p^3$ | 1 | 2 |
| $p^2$ | 3 | $(K - 24)$ |
| $p^1$ | $(10 - K/3)$ | |
| $p^0$ | $(K - 24)$ | |

No sign changes will occur in the first column if $24 \leqslant K \leqslant 30$. This condition guarantees that the dominant time constant will be no larger than 1/2.

If $K = 24$, the last row is zero and there exists a single root at the origin of the $p$ coordinate system (at $p = 0$) or, equivalently, a root at $s = -2$. If $K = 30$, the third row is zero, with positive entries above and below. From Rule 5, a pair of purely imaginary roots exists (in the $p$ coordinates). Therefore, if $K = 30$, two roots are $s = -2 \pm ib$, where $b$ is unknown at this point. Note that $K = 30$ is the only value that will give underdamped behavior, given the restriction on the dominant time constant. To compute the damping ratio for $K = 30$, write the $p$ polynomial in factored form (we now know the third root must be real).

$$\begin{aligned} p^3 + 3p^2 + 2p + 6 &= (p + a)(p^2 + b^2) \\ &= p^3 + ap^2 + b^2p + ab^2 \end{aligned}$$

Comparing the two forms we see that

$$\begin{aligned} a &= 3 \\ b^2 &= 2 \\ ab^2 &= 6 \end{aligned}$$

or

$$\begin{aligned} a &= 3 \\ b &= \sqrt{2} \end{aligned}$$

In terms of $s$, this gives $s = p - 2 = -a - 2, \pm ib - 2$, or $s = -5, -2 \pm i\sqrt{2}$. The damping ratio of the dominant root is found from Figure 5.14 to be $\zeta = 0.82$.

In practical problems the coefficients of the characteristic equation rarely produce an exactly zero entry or row in the Routh array. However, as Example F.3 shows, it is important to be able to interpret the significance of zeros, when considered as limiting cases.

The Routh-Hurwitz criterion is intimately related to the geometry of the complex plane, as shown by Rule 8. Some other transformations are sometimes useful, and we briefly mention these before concluding our discussion. If we let $s = -q$ in the original polynomial, the Routh-Hurwitz criteria applied to the resulting polynomial in $q$ will tell how many roots lie in the *left*-half plane. This is useful for determining the number of purely imaginary roots. The transformation $s = re^{-i\theta}$ rotates the axes through an angle $\theta$. The resulting polynomial in $r$ will reveal how many roots of the original polynomial lie to the right of the lines making angles of $\pm(90° + \theta)$ relative to the positive real axis in the $s$-plane. This can be used to determine the minimum value of the damping ratio $\zeta = \sin\theta$, but the required algebra can sometimes be prohibitive.

# Index

# Index